Lecture Notes in Computer Science 16752

The series Lecture Notes in Computer Science (LNCS), including its subseries Lecture Notes in Artificial Intelligence (LNAI) and Lecture Notes in Bioinformatics (LNBI), has established itself as a medium for the publication of new developments in computer science and information technology research, teaching, and education.

LNCS enjoys close cooperation with the computer science R & D community, the series counts many renowned academics among its volume editors and paper authors, and collaborates with prestigious societies. Its mission is to serve this international community by providing an invaluable service, mainly focused on the publication of conference and workshop proceedings and postproceedings. LNCS commenced publication in 1973.

Jie Lu · Javier Montero · Yi Zhang · Junyu Xuan ·
Tianrui Li · Luis Martínez · Etienne Kerre
Editors

Machine Learning and Knowledge Engineering for Decision Making

The 17th International FLINS Conference on Fuzzy Logic for Intelligent Systems and The 21st International Conference on Intelligent Systems and Knowledge Engineering, FLINS-ISKE 2026
Sydney, NSW, Australia, July 15–19, 2026, Proceedings, Part I

Springer

Editors
Jie Lu
University of Technology Sydney
Sydney, NSW, Australia

Yi Zhang
University of Technology Sydney
Sydney, NSW, Australia

Tianrui Li
Southwest Jiaotong University
Chengdu, China

Etienne Kerre
Ghent University
Ghent, Belgium

Javier Montero
Universidad Complutense de Madrid
Madrid, Spain

Junyu Xuan
University of Technology Sydney
Sydney, NSW, Australia

Luis Martínez
University of Jaén
Jaén, Spain

ISSN 0302-9743 ISSN 1611-3349 (electronic)
Lecture Notes in Computer Science
ISBN 978-981-92-2368-8 ISBN 978-981-92-2369-5 (eBook)
https://doi.org/10.1007/978-981-92-2369-5

This Springer imprint is published by the registered company Springer Nature Singapore Pte Ltd.
The registered company address is: 152 Beach Road, #21-01/04 Gateway East, Singapore 189721, Singapore

If disposing of this product, please recycle the paper.

Preface

Originally an acronym for Fuzzy Logic and Intelligent Technologies in Nuclear Science, FLINS was launched in 1994 by Da Ruan of the Belgian Nuclear Research Center (SCK-CEN). His vision was to provide early-career researchers with the opportunity to conduct future-oriented research. He then founded the FLINS conference series and established a global, multidisciplinary consortium dedicated to computational intelligence. Since 2011, his successors have honored his spirit of openness, collaboration, and innovation, continuing to consolidate and expand this research consortium by developing new theories and applications. Over more than 30 years, the scope of FLINS has grown to encompass the full spectrum of theoretical and practical developments in computational intelligent systems, decision support systems, and more recently artificial intelligence.

FLINS 2026 followed a long line of successful biennial conferences held across the globe: FLINS 1994 and FLINS 1996 in Mol (Belgium), FLINS 1998 in Antwerp (Belgium), FLINS 2000 in Bruges (Belgium), FLINS 2002 in Ghent (Belgium), FLINS 2004 in Blankenberge (Belgium), FLINS 2006 in Genoa (Italy), FLINS 2008 in Madrid (Spain), FLINS 2010 in Chengdu (China), FLINS 2012 in Istanbul (Turkey), FLINS 2014 in João Pessoa (Brazil), FLINS 2016 in Roubaix (France), FLINS 2018 in Belfast (UK), FLINS 2020 in Cologne (Germany), FLINS 2022 in Tianjin (China), and FLINS 2024 in Madrid (Spain).

ISKE (Intelligent Systems and Knowledge Engineering), launched in 2006, serves as a premier annual forum for researchers and practitioners to explore the synergy between computational systems and knowledge engineering. ISKE focuses on the integration of intelligent systems, machine learning, and soft computing to solve complex, uncertain, dynamic real-world problems. Over 20 years, the conference has evolved alongside the AI revolution, expanding its scope to include big data analytics, cognitive computing, and autonomous systems. By bridging the gap between theoretical knowledge representation and practical intelligent applications, ISKE has built a global reputation for fostering multidisciplinary innovation and academic excellence.

Because FLINS and ISKE share the common goal of linking the theoretical foundations of computational intelligence with applications in complex engineering and multidisciplinary decision-making, the two conferences are now organized jointly every two years. This successful collaboration includes the 2020, 2022, and 2024 editions, and continues with FLINS-ISKE 2026, held in Sydney, Australia, from July 15–19, 2026.

FLINS-ISKE 2026 served as an international platform bringing together researchers in mathematics, AI, computer science, information technology, and engineering. This year, the joint conference received 275 submissions, of which 160 papers were selected for inclusion in these proceedings following a rigorous single-blind peer-review process. This reflects an acceptance rate of approximately 58%. The contributing authors represent over 15 countries and regions, including Australia, China, Spain, New Zealand, Germany, France, India, and the UK. To ensure the highest academic standards, each submission underwent a comprehensive evaluation by at least two independent reviewers, totaling over 753 individual reviews with an average of 2.7 reviews per submission. The technical program was further enriched by three keynote speeches from world-renowned experts: Lei Chen from the Hong Kong University of Science and Technology, Dongrui Wu from the Huazhong University of Science and Technology, and Anton van den Hengel from the University of Adelaide.

These proceedings are organized into five chapters: (1) Fuzzy Machine Learning and Decision Making, (2) Knowledge Engineering, LLMs, and Reasoning, (3) Computer Vision and Medical Imaging, (4) Graph Neural Networks and Relational Intelligence, and (5) Trustworthy, Federated, and Robust Learning Systems. Reflecting the rapid evolution of Artificial Intelligence, 14 special sessions were organized to address cutting-edge topics, such as AI for Agriculture, Health, and Optimization, Emerging Cross-Modal Data Science for Generative AI, Evolving Deep and Transfer Learning Models for Computer Vision and Medical Imaging, Human-Centric Intelligent Control: Behavioral Dynamics and Social Sensing for Resilience, Information Aggregation Tools for AI, Integrating Data and Knowledge for Trustworthy AI, and Towards Responsible Intelligent Decision Support Systems.

We are confident that the research presented in the FLINS-ISKE 2026 proceedings will provide readers with the essential insights needed to navigate and excel in the current AI era. Beyond presenting technical excellence, this volume serves as a testament to the power of international research collaboration in tackling the complex challenges of modern intelligent systems. We offer our sincere gratitude to the authors, referees, special session organizers, and session chairs, as well as the members of the International Scientific Committee. It is through their visionary contributions, rigorous peer reviews, and shared dedication to academic merit that FLINS-ISKE 2026 successfully bridged the gap between foundational theory and the future of artificial intelligence.

May 2026

Jie Lu
Javier Montero
Yi Zhang
Junyu Xuan
Tianrui Li
Luis Martínez
Etienne Kerre

Organization

Founding Chair

Da Ruan (1960–2011)	University of Technology Sydney, Australia

Steering Committee

Etienne Kerre (Chair)	Ghent University, Belgium
Guoqing Chen	Tsinghua University, China
Cengiz Kahraman	Istanbul Technical University, Turkey
Tianrui Li	Southwest Jiaotong University, China
Zhong Li	FernUniversität in Hagen, Germany
Jun Liu	University of Ulster, UK
Jie Lu	University of Technology Sydney, Australia
Luis Martinez López	University of Jaén, Spain
Javier Montero	Universidad Complutense de Madrid, Spain
Ronei Macros de Moraes	Federal University of Paraíba, Brazil
Yang Xu	Southwest Jiaotong University, China
Xianyi Zeng	ENSAIT Textile Institute, France

General Chairs

Jie Lu	University of Technology Sydney, Australia
Javier Montero	Universidad Complutense de Madrid, Spain
Yang Xu	Southwest Jiaotong University, China
Guoqing Chen	Tsinghua University, China

Program Chairs

Junyu Xuan	University of Technology Sydney, Australia
Tianrui Li	Southwest Jiaotong University, China
Luis Martinez López	University of Jaén, Spain

Organization Chairs

Yi Zhang	University of Technology Sydney, Australia
Zhong Li	FernUniversität in Hagen, Germany
Farookh Hussain	University of Technology Sydney, Australia
Bo Liu	University of Technology Sydney, Australia

Special Session Chairs

Cengiz Kahraman	Istanbul Technical University, Turkey
Guangquan Zhang	University of Technology Sydney, Australia
Qian Liu	University of Auckland, New Zealand
Zhen Fang	University of Technology Sydney, Australia

Workshop Chairs

Philipp Mayr	GESIS - Leibniz Institute for the Social Sciences, Germany
Chengzhi Zhang	Nanjing University of Science and Technology, China
Sang Nguyen	International University, VNU-HCMC, Vietnam
Hang Yu	Shanghai University, China

Tutorial Chairs

Alan Porter	Georgia Institute of Technology, USA
Scott Cunningham	University of Strathclyde, UK
María Victoria López	CUNEF Universidad, Spain
Feng Liu	University of Melbourne, Australia

Publicity Chairs

Arash Hajikhani	VTT Technical Research Centre of Finland, Finland
Guanjin Wang	Murdoch University, Australia
Ronei Macros de Moraes	Federal University of Paraíba, Brazil

Li Zou	Shandong Jianzhu University, China
Arho Suominen	VTT Technical Research Centre of Finland, Finland

PhD Symposium Chairs

Xianyi Zeng	ENSAIT Textile Institute, France
Jun Liu	University of Ulster, UK
Yiliao Song	University of Adelaide, Australia
Chris Nugent	Ulster University, UK

Poster Chairs

Hua Zuo	University of Technology Sydney, Australia
Maoying Qiao	University of Technology Sydney, Australia
Hui Wang	Queen's University of Belfast, UK
Guidong Zhang	Guangdong University of Technology, China

Web Chair

Mengjia Wu	University of Technology Sydney, Australia

Contents

Fuzzy Machine Learning and Decision Making (16752)

Drive with Causal Insight: A Neuro-Symbolic Agent for Interpretable Real-Time Decision Making

Shan Yang[1], Wenjun Hou[1,2,3](✉), and Runnan Li[1,2,3]

[1] School of Digital Media and Design Arts, Beijing University of Posts and Telecommunications, Beijing, China
{yangshan_stella,hwj1505,runnan.li}@bupt.edu.cn
[2] Beijing Key Laboratory of Network System and Network Culture, Beijing, China
[3] Key Laboratory of Interactive Technology and Experience System, Ministry of Culture and Tourism, Beijing, China

Abstract. Vision-Language Models (VLMs) show great potential in autonomous driving. However, they often operate as black boxes and fail in safety-critical scenarios. Physical hallucinations and causal confusion cause these failures. To address these Explainable AI challenges, we propose PG-Agent (Physics-Grounded Agent). This framework enhances robust decision making under uncertainty. It uses a neuro-symbolic approach to align neural perception with symbolic causality. Traditional methods rely on implicit feature matching. Our approach explicitly grounds the VLM through a dual-stream mechanism. First, we employ WideGazeNet. This lightweight network generates semantic-aware visual prompts to simulate human visual attention and filter environmental noise. Second, we align this visual stream with a Structural Causal Model derived from simulator ground truth. This step translates expert physical laws into Chain-of-Thought supervision. We also introduce a visual counterfactual training strategy. It physically removes risk objects from the scene. This intervention forces the agent to learn the true causal link between object existence and actions. Experiments on the DATAD benchmark show that PG-Agent achieves top performance in lateral control. It demonstrates highly interpretable and real-time decision making in complex scenarios.

Keywords: Autonomous driving · Explainable AI · Causal Decision

1 Introduction

Autonomous driving relies increasingly on end-to-end (E2E) machine learning. Models like UniAD [9] map sensor data directly to actions. These systems enable real-time decision making in dynamic environments. However, they operate as uninterpretable black boxes. They optimize trajectory accuracy but cannot explain their decisions. Researchers have introduced Large Vision-Language

J. Lu et al. (Eds.): FLINS-ISKE 2026, LNCS 16752, pp. 3–16, 2027.
https://doi.org/10.1007/978-981-92-2369-5_1

Models (VLMs) like DriveVLM [19] to address this. VLMs offer reasoning capabilities. They generate textual explanations for driving actions.

Despite these advances, VLM agents face two critical challenges in safety-critical scenarios. The first challenge is physical hallucination. Pure neural networks lack explicit physical constraints. They often misjudge spatial metrics like Time-to-Collision (TTC) [18]. The second challenge is Causal Confusion [7]. Imitation learning models often learn spurious correlations instead of true causal mechanisms. For instance, a model might associate braking with a distant red light instead of a cutting-in vehicle. Existing VLMs cannot perform counterfactual reasoning to verify these causal links.

We introduce PG-Agent (Physics-Grounded Agent) to solve these problems. This neuro-symbolic framework combines neural perception with symbolic knowledge engineering. We model the driving process through a clear causal graph: Environment → Risk → Attention → Action. Specifically, we construct a Causal Decision Tree based on simulator ground-truth physics. This tree distills expert causal logic into the VLM. Our system explicitly models human-like peripheral awareness. This addresses "Gaze-Action Misalignment", where drivers react to risks without direct foveal vision. We also apply Visual Counterfactual Alignment. We physically remove risk objects from the scene and flip the control labels. This intervention forces the agent to identify the true cause of an action. It effectively eliminates spurious correlations.

Our main contributions are summarized as follows:

1. We design WideGazeNet. This lightweight spatiotemporal network simulates human visual attention. It filters texture noise and provides visual prompts. It supports real-time decision making with a 1.88ms inference speed.
2. We develop a data generation engine governed by a Causal Decision Tree. It generates high-quality neuro-symbolic data. It resolves gaze-action misalignment through modeling implicit peripheral awareness.
3. We introduce a visual counterfactual training strategy. Experiments on the DATAD [10] benchmark show that PG-Agent achieves better performance in lateral control. It provides robust and interpretable decisions in long-tail scenarios.

2 Related Work

2.1 Machine Learning-Based End-to-End Decision Making

Researchers are moving from modular pipelines to end-to-end (E2E) learning. These systems optimize driving decisions directly from sensor inputs [4]. Recent Transformer-based architectures have significantly advanced real-time decision making. UniAD [9] integrates perception and planning to handle complex interactions. VAD [11] uses vectorized representations. This structure improves inference speed. TransFuser [16] fuses LiDAR and camera inputs via attention mechanisms to improve safety.

These machine learning models achieve strong closed-loop performance. However, they remain opaque black boxes. They optimize trajectory accuracy but cannot explain their decisions. Furthermore, they struggle with long-tail corner cases. These uncertain scenarios require semantic understanding beyond simple geometric patterns.

2.2 Enhancing Interpretability with Vision-Language Models

Large Vision-Language Models (VLMs) treat driving as a reasoning task. Foundational models like GPT-4V [24] accelerated this trend. Recent works integrate these models into the driving decision loop. DriveGPT4 [22] answers questions about the environment using video frames. DriveLM [17] introduces Graph Visual Question Answering. This models logical dependencies between perception and planning. DriveVLM [19] proposes a dual-system architecture. It uses a VLM for slow reasoning and a traditional planner for fast reactions. This trend extends beyond driving-specific agents. Researchers explore Vision-Language-Action (VLA) models for general embodied control. For instance, the π_0 model [3] integrates vision and language to provide explicit decision guidance. Similarly, VGAS [21] explores value-guided action selection. This approach improves robustness under uncertainty. These embodied control works share our motivation. They aim to move beyond purely implicit end-to-end action prediction. Current VLM agents possess reasoning capabilities. Yet, they suffer from two critical flaws. The first flaw is physical hallucination. VLMs often generate physically inaccurate estimations without explicit constraints [18]. The second flaw is weak causal links. Most methods rely on text-based Chain-of-Thought. They do not verify the causal necessity of objects. A model might correctly predict a braking action but attribute it to distant traffic lights instead of a cutting-in vehicle.

2.3 Modeling Human-Like Visual Cognition

Understanding human visual attention is essential for intent prediction. Datasets like DR(eye)VE [15] and BDD-A [1] provide large-scale data for gaze prediction. Recently, PILOT [10] proposed a programmatic imitation learning approach. It derives explicit rules to predict discrete gaze shifts based on hand-crafted physical features.

Existing works predominantly treat gaze as a final prediction target. In contrast, our work treats gaze as an intermediate visual prompt. We argue that knowing where to look is a prerequisite for knowing how to drive. The discrete action space of PILOT is too coarse for visual prompting. We employ a continuous heatmap estimator instead. This provides fine-grained spatial guidance to the VLM. Furthermore, our WideGazeNet is a pure-vision module. It does not rely on upstream object detection or hand-crafted features.

2.4 Neuro-Symbolic Systems and Causal Knowledge Engineering

Standard imitation learning suffers from Causal Confusion. De Haan et al. [7] formally defined this phenomenon. Policies learn spurious correlations instead

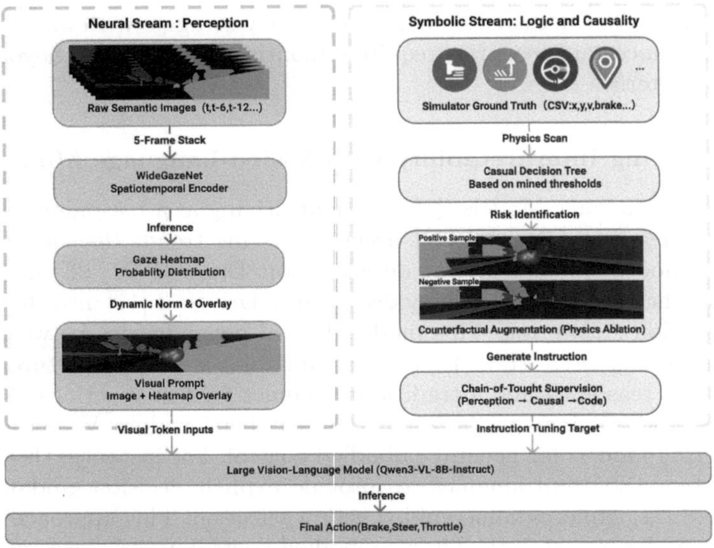

Fig. 1. The Framework Overview of PG-Agent. The system operates via a dual-stream architecture: (Left) The **Neural Stream** employs WideGazeNet to process spatiotemporal semantic frames (t to $t-12$) and generate a probability-based gaze heatmap as a visual prompt. (Right) The **Symbolic Stream** utilizes a Causal Decision Tree derived from simulator ground truth to generate rigorous CoT supervision and counterfactual augmentation samples. Both streams are aligned via instruction tuning on a Large Vision-Language Model to output the final control actions.

of true causal mechanisms. To address this, CausalCity [14] uses intervention-based methods to disentangle causal factors. In the VLM domain, neuro-symbolic approaches combine neural perception with symbolic logic. Recent work on Counterfactual VLM [25] demonstrates that generating hypothetical samples significantly reduces bias.

Our PG-Agent applies Counterfactual Data Augmentation to VLM-based driving. We construct a Causal Decision Tree. This structure acts as a knowledge engineering tool. It explicitly models both explicit and implicit causal pathways. This method resolves the misalignment between human gaze and control actions, providing interpretable decision making.

3 Methodology

3.1 Framework Overview

We propose PG-Agent (Physics-Grounded Agent). This neuro-symbolic framework connects visual perception with logical control in autonomous driving. Figure 1 illustrates our approach. The framework operates via a dual-stream architecture during the training phase. This structure establishes causal alignment:

1. Neural Stream (Input): A lightweight visual encoder processes semantic segmentation sequences. It generates Visual Prompts. This stream simulates human-like visual cognition and attention mechanisms.
2. Symbolic Stream (Supervision): A Structural Causal Model extracts knowledge from simulator ground-truth states. It constructs a rigorous Causal Decision Tree. This tree generates Chain-of-Thought (CoT) annotations and control codes.
3. Alignment: We fine-tune a Large Vision-Language Model (VLM). The VLM maps the visual inputs to the logical outputs. This step distills physical laws into the reasoning process.

3.2 Modeling Human-Like Visual Cognition

Spurious correlations in RGB images often cause causal confusion [7]. We utilize Semantic Segmentation Maps as the primary visual modality to mitigate this issue. We introduce WideGazeNet. It serves as a spatiotemporal attention estimator.

Spatiotemporal Encoding. We define a sequence of T semantic frames $I = \{I_t, I_{t-k}, ..., I_{t-(N-1)k}\}$. Here, k represents the stride. We stack these frames channel-wise. This creates a tensor $X \in \mathbf{R}^{B\times(C\cdot N)\times H\times W}$. This design allows 2D convolutional layers to capture optical flow and object motion trends implicitly. It avoids computationally expensive 3D modules.

Gaze Heatmap Generation. The network predicts a gaze heatmap $H \in [0,1]^{H\times W}$. Prior works often model gaze as discrete classification [10]. In contrast, we model gaze as a continuous probability distribution. We apply dynamic normalization and adaptive thresholding to the heatmap. Then, we overlay it onto the semantic map. These steps ensure the VLM attends to the prompted regions effectively. We define the final visual input as $V_{prompt} = I_t \oplus H$.

3.3 Knowledge Engineering via Causal Decision Trees

Existing VLM-based driving agents [17,19] often rely on implicit imitation. This reliance causes hallucinations in physical estimation. We construct a Data Engine governed by a Causal Decision Tree to address this problem. This engine acts as a knowledge engineering tool. It converts ground-truth physical states into structured reasoning chains.

Physical Risk Assessment We define an objective risk state S_{risk}. We base this state on two data-driven thresholds. We mine these thresholds from the dataset distribution:

$$S_{risk} = \begin{cases} \text{CRITICAL, if TTC} < \tau_{crit} \vee \text{Acc}_{obj} < \alpha_{hard} \\ \text{WARNING, if } \tau_{crit} \leq \text{TTC} < \tau_{warn} \\ \text{SAFE,} \quad \text{otherwise} \end{cases} \tag{1}$$

Here, $\tau_{crit} = 1.7s$ represents the emergency braking limit. The variable $\tau_{warn} = 7.2s$ represents the perceptual awareness horizon. This formulation handles boundary uncertainty systematically.

Threshold Sensitivity and Adaptation. The thresholds τ_{crit} and τ_{warn} originate from the specific driving behaviors in the DATAD training set. Minor variations in these exact values do not disrupt the overall causal logic. The symbolic component remains robust under uncertainty. Furthermore, researchers can easily re-derive these thresholds for different datasets. Users apply K-Means clustering to separate braking intensities. Then, users fit a shallow decision tree to the Time-to-Collision data. This data-driven extraction process ensures the framework adapts to various driving environments flexibly.

Causal Pathway Modeling. We explicitly model the causal graph $E \rightarrow R \rightarrow A \rightarrow Y$. In this graph, E is the environment, and R is the risk. A represents attention, and Y represents the action. The decision tree handles two distinct causal pathways:

Explicit Causal Path ($R \rightarrow A \rightarrow Y$): The driver's gaze A might align with the risk object O_{risk} (i.e., $\text{IoU}(A, O_{\text{risk}}) > 0$). In this case, visual attention mediates the action Y. The generated CoT explicitly states: Visual confirmation of [Object ID] triggers braking.

Implicit Causal Path ($R \rightarrow Y$): The gaze A might not align with O_{risk}, and also a significant braking action Y is observed. For example, a driver performs defensive braking for a side cut-in. We attribute this uncertainty to Peripheral Awareness. The engine generates a specific reasoning text. It states: Foveal gaze is on the road, and also peripheral awareness detected risk from the [Left/Right]. This requires a defensive action. This mechanism handles long-tail samples. In these cases, human gaze behavior separates from immediate motor control.

3.4 Robust Decision Making via Counterfactual Alignment

The agent must learn the true causal mechanism (*Object* $\rightarrow$ *Action*). It should not memorize background correlations (*Background* $\rightarrow$ *Action*). We introduce Visual Counterfactual Training to ensure this.

We define a high-risk sample x with a label y_{brake}. We apply an intervention operator $do(\cdot)$ on the visual input:

$$I_{cf} = do(I \setminus O_{risk}) \tag{2}$$

The object O_{risk} is physically ablated. We inpaint this area with road semantics based on the heatmap mask. We then flip the logical label to y_{safe}. We train the VLM on both factual pairs (I, y_{brake}) and counterfactual pairs (I_{cf}, y_{safe}). This contrastive objective forces the model to identify the presence of the risk object. The model recognizes this presence as the necessary and sufficient cause for the braking decision.

3.5 Explainable AI via Neuro-Symbolic Tuning

We fine-tune the Qwen3-VL-8B-Instruct model [23]. We use Low-Rank Adaptation (LoRA) for this process. The instruction tuning objective minimizes the negative log-likelihood of the generated tokens. These tokens include natural language reasoning and executable code:

$$\mathcal{L} = -\sum_{t} \log P(\text{token}_t | I_{\text{prompt}}, \text{Instruction}) \tag{3}$$

We structure the target output as a Python function. This format ensures deterministic generated control signals (brake, steer, throttle). The vehicle control interface can execute these signals directly. This provides a transparent and explainable decision process.

4 Experiments

4.1 Experimental Setup

Dataset and Protocol. We evaluate our framework on the DATAD dataset [10]. This high-fidelity benchmark provides human-machine takeover scenarios. We adopt a Leave-Subjects-Out protocol to ensure rigorous evaluation. This prevents data leakage. We use data from Testers 1–20 (approximately 74,000 frames) for training. We reserve Testers 21–25 exclusively for testing. This setting challenges the model to generalize its decision making to unseen driving behaviors.

Evaluation Metrics. We assess the system across three dimensions of decision making:

1. **Robust Decision Making:** We report the Mean Absolute Error (MAE) for steering angle, brake pressure, and throttle. We also introduce Critical Event Recall. This metric defines the percentage of correctly predicted braking actions in safety-critical scenarios (Time-to-Collision $< 3.0s$).
2. **Human-like Visual Cognition:** We evaluate the spatial accuracy of WideGazeNet using Mean Squared Error (MSE) and Bounding-Box Hit Rate. These metrics follow the PILOT [10] standard. We also measure Average Inference Time per Frame (AIPF). This assesses the capacity for real-time decision making.
3. **Interpretable Causal Validation:** We propose the Flip Rate. This metric quantifies the effectiveness of our counterfactual training. It measures the probability of the model correctly altering its decision when the causal risk object is physically absent from the input.

Baselines. We compare PG-Agent against three representative baselines:

1. **ResNet-50 (BC)** [13]: A standard behavior cloning baseline. It represents traditional black-box end-to-end models.
2. **DriveVLM-Style** [19]: A reproduced baseline using Qwen3-VL. It is fine-tuned with standard Chain-of-Thought (CoT) prompts. It lacks our knowledge-engineered causal tree guidance and visual prompts.
3. **PILOT** [10]: A rule-based gaze prediction method. It relies on hand-crafted physical features. We use it for visual cognition comparison.

Implementation Details. Our WideGazeNet processes a 5-frame stacked semantic tensor (512 × 128). The PG-Agent utilizes Qwen3-VL-8B-Instruct as the backbone. We fine-tune it via LoRA with 4-bit quantization on a single NVIDIA RTX 3080 GPU. We derive the causal decision tree thresholds from data: $\tau_{crit} = 1.7s$ and $\tau_{warn} = 7.2s$.

4.2 Main Results

Robust End-to-End Decision Making. Table 1 presents the comparison of control decisions. PG-Agent achieves superior performance in lateral control. It records a Steer MAE of 0.0148. This significantly outperforms the ResNet baseline and DriveVLM-Style. This precision stems from the semantic-aware visual prompting. The prompt highlights lane topology and road geometry effectively.

Table 1. Control Performance Comparison on DATAD Test Set. (Note: Critical Recall refers to the detection rate of braking actions in safety-critical scenarios where TTC < 3.0 s.)

Method	Modality	Steer MAE ↓	Brake MAE ↓	Critical Recall ↑
ResNet-50 (BC) [3]	Vision Only	0.0207	0.2990	81.63%
DriveVLM-Style [4]	Vision-Language	0.0382	0.1105	85.36%
PG-Agent (Ours)	**Neuro-Symbolic**	**0.0148**	**0.0805**	**96.50%**

PG-Agent successfully identifies 96.5% of hazardous situations. This is shown in the Critical Event Recall metric. In contrast, ResNet often fails to brake in long-tail corner cases. The MAE metrics reflect the agent's tendency for defensive deceleration in uncertain scenarios. This aligns with safety-critical requirements. It does not merely fit the average human behavior. The specific results are presented in Table 1.

Efficiency of Human-Like Visual Cognition. We compare WideGazeNet with PILOT and other models [10] in Table 2. Our MSE (716k) is slightly higher than PILOT (634k). However, our method is purely vision-based. It eliminates the dependency on upstream object detection. It avoids the hand-crafted features required by PILOT. Crucially, WideGazeNet achieves an inference speed of 1.88 ms. This is $\sim 30\times$ faster than PILOT (68.5 ms). This speed ensures negligible latency for the downstream VLM. Furthermore, our At-Least-Once Hit Rate reaches 100%. This demonstrates robust coverage across all testing scenarios.

Table 2. Visual Cognition Performance Comparison. (Note: ALO Hit refers to At-Least-Once Hit Rate. Features indicate hand-crafted physical features. Our Hit Rate uses a strict bounding box approximation threshold of 400px.)

Model & Frame Rate	Input	MSE per Frame ↓	Average Hit Rate ↑	ALO Hit Rate ↑	Speed
SalM2-60fps [25]	RGB	1,033,065	81.82%	90.07%	919.8 ms
PGNet-60fps [20]	RGB	1,055,746	74.73%	96.08%	116.7 ms
DRIVE-30fps [2]	RGB	5,294,155	33.66%	67.33%	934.8 ms
DR(eye)VE-60fps [15]	RGB	1,265,572	63.02%	63.17%	> 1 min
DR(eye)VE-25fps [15]	RGB	1,146,465	67.79%	68.09%	> 1 min
PILOT-60fps [10]	Features	634,401	88.7%	95.50%	68.5 ms
OURS(WideGazeNet)-60fps	**Sem. Map**	**716,819**	**68.4%**	**100%**	**1.88ms**

The predicted gaze heatmaps generated by the model are shown in Fig. 2.

Fig. 2. The predicted gaze heatmaps demonstrating simulated visual cognition.

4.3 Explainable AI: Causal Reasoning and Physics Grounding

We conduct two specific analyses. These tests validate whether our model truly understands the scene rather than hallucinating text.

Physics Grounding Analysis. As visualized in Fig. 3, we compare the visual results with the physical ground-truth values. This comparison determines which branch of the decision tree they belong to.

Causal Consistency Test. We validate the model's causal reasoning. We apply our Visual Counterfactual Intervention for this test. We select risk scenarios. Then, we physically ablate the target vehicle from the input image.

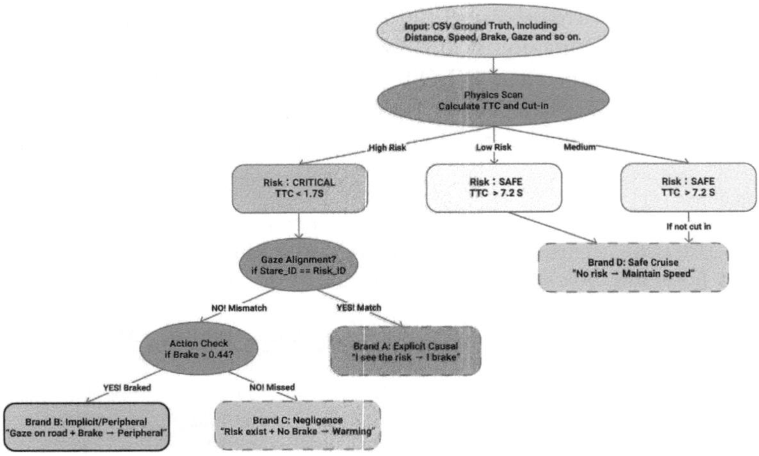

Fig. 3. Comparison of visual results and physical ground-truth.

1. **Result:** PG-Agent achieves a Flip Rate of 92.7%. In 92.7% of cases, the model correctly switches its decision from 'Brake' to 'Cruise' when the risk object is removed. It also updates its reasoning to 'Object removed, path clear'.
2. **Comparison:** The ResNet baseline achieves an 81.63% Critical Recall. However, it often continues to brake due to background correlations. This exhibits the Causal Confusion phenomenon described by de Haan et al. [7]. This test proves that PG-Agent learns the true causal mechanism: Object Existence → Action.

4.4 Ablation Study

We perform ablation studies to isolate the contribution of each component. The results are in Table 3.

Table 3. Ablation Study on Key Components. (Note: Flip Rate (Causal) measures the percentage of correct decision reversals when the risk object is physically ablated.)

Configuration	Steer MAE ↓	Critical Recall ↑	Flip Rate (Causal) ↑
Full PG-Agent	**0.0148**	**96.5%**	**92.7%**
w/o Visual Prompt	0.0184	90.2%	88.0%
w/o Causal Tree	0.0176	89.1%	65.0%
w/o Counterfactual	0.0221	94.0%	45.0%

1. **w/o Visual Prompt:** Removing WideGazeNet leads to a drop in Critical Recall (-6.3%). The VLM struggles to locate small risk objects in the wide-view image without attention guidance.
2. **w/o Causal Tree:** Replacing our structured CoT with generic GPT-4 descriptions significantly increases Physics Hallucination. This validates the necessity of symbolic knowledge engineering.
3. **w/o Counterfactual Training:** Excluding counterfactual samples results in a major drop in Flip Rate. This indicates the model becomes highly susceptible to spurious correlations.

The final comparison results are shown in Fig. 4 on the next page.

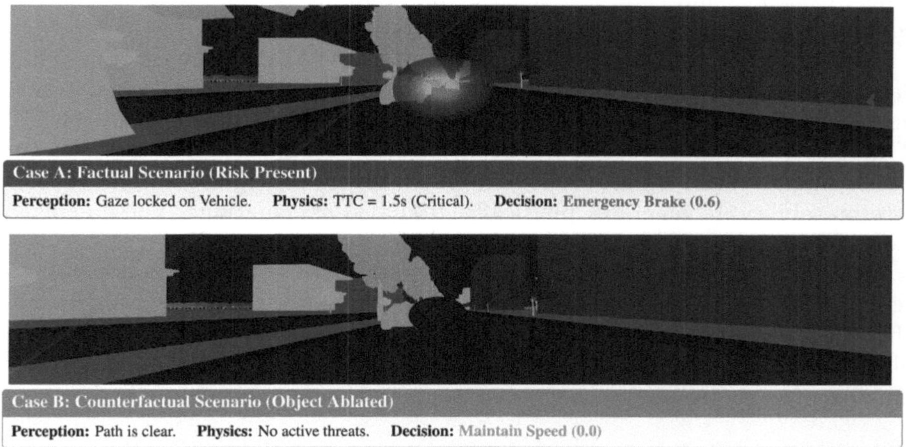

Fig. 4. Visual Counterfactual Consistency Test. (a) In the factual scenario, the agent correctly identifies the risk and initiates braking. (b) When the risk object is physically ablated (inpainted), the agent immediately updates its reasoning to 'Path clear' and cancels braking. This confirms the decision is causally linked to the object's presence, not background correlations.

5 Conclusion

Experimental results show an important fact. Eye movement data and causal reasoning support robust decision making in lateral control. Vision-Language Models (VLMs) face Explainable AI challenges in autonomous driving. Physical hallucinations and causal confusion are the main problems. We address these problems with PG-Agent. This neuro-symbolic framework aligns neural visual perception with symbolic knowledge engineering. Traditional end-to-end models rely on implicit feature matching. PG-Agent is different. It explicitly grounds its decisions in physical reality. We use a Causal-Tree Guided Data Engine for this process.

Our work has three main contributions. First, we demonstrate Semantic-Aware Visual Prompting. Our lightweight WideGazeNet enables this feature. It filters environmental noise. It provides precise attentional guidance. The latency is negligible (1.88 ms). This fast perception supports real-time decision making. Second, we construct a Structural Causal Model (SCM). We derive this model from simulator ground truth. We successfully model complex human-like driving behaviors. This includes the implicit peripheral awareness pathway. Third, we propose Visual Counterfactual Training. This strategy mitigates spurious correlations effectively. We physically remove risk objects from the images during training. This intervention forces the agent to learn the true causal link between object existence and control actions. It prevents overfitting to the background context [7].

We conducted extensive experiments on the DATAD benchmark. The results validate our approach. PG-Agent achieves top performance in lateral control accuracy (Steer MAE: 0.0148). It reaches a high safety-critical recall (96.5%). These metrics surpass traditional behavior cloning baselines [13], and also surpass standard VLM agents [19]. More importantly, the agent exhibits superior interpretability. It generates physically grounded explanations. These explanations are consistent with its control execution.

PG-Agent represents a significant step toward trustworthy autonomous systems. It connects the reasoning capabilities of Large Language Models with the safety requirements of kinetic control. This connection offers a generalizable paradigm for future research in embodied AI and neuro-symbolic decision making.

6 Limitation and Future Work

PG-Agent shows promising performance. However, several limitations remain. First, our system relies on upstream semantic segmentation. Extreme weather can cause perception failures. The system is vulnerable to these failures. End-to-end joint training can mitigate this error propagation. We plan to explore this in the future. Second, WideGazeNet operates in real-time. However, the VLM's inference latency aligns with slow 'System 2' thinking [12]. This latency is insufficient for high-frequency reflex control. A parallel 'System 1' module is necessary for immediate reactions. Third, we use an open-loop evaluation. This follows standard practices [17,19]. Open-loop tests cannot capture compounding errors in long-horizon interactions.

Future research will address these gaps. We will focus on closed-loop deployment and Sim-to-Real transfer. Semantic representations have a domain-invariant nature. We will use this feature to adapt our model to real-world datasets like nuScenes [6]. Technically, we aim to replace static risk thresholds. We will use differentiable causal discovery [5] instead. This will allow dynamic context adaptation for robust decision making under uncertainty. Finally, we plan to apply model distillation [8]. This technique will transfer the VLM's causal reasoning capabilities into lightweight student models. It will enable real-time edge execution for autonomous vehicles.

Acknowledgments. This work is supported by the Beijing Key Laboratory of Network System and Network Culture and the Key Laboratory of Interactive Technology and Experience System.The authors also thank Gemini for its assistance in language polishing and grammatical refinement.

Disclosure of Interests. The authors have no competing interests to declare that are relevant to the content of this article.

References

1. Angelakis, I., Gooding, P.A., Panagioti, M.: Suicidality in body dysmorphic disorder (BDD): a systematic review with meta-analysis. Clin. Psychol. Rev. **49**, 55–66 (2016)
2. Bao, W., Yu, Q., Kong, Y.: Drive: deep reinforced accident anticipation with visual explanation. In: Proceedings of the IEEE/CVF International Conference on Computer Vision, pp. 7619–7628 (2021)
3. Black, K., et al.: π_0: A vision-language-action flow model for general robot control (2026)
4. Bojarski, M., et al.: End to end learning for self-driving cars. arXiv preprint arXiv:1604.07316 (2016)
5. Brouillard, P., Lachapelle, S., Lacoste, A., Lacoste-Julien, S., Drouin, A.: Differentiable causal discovery from interventional data. Adv. Neural. Inf. Process. Syst. **33**, 21865–21877 (2020)
6. Caesar, H., et al.: nuscenes: a multimodal dataset for autonomous driving. In: Proceedings of the IEEE/CVF Conference on Computer Vision and Pattern Recognition, pp. 11621–11631 (2020)
7. De Haan, P., Jayaraman, D., Levine, S.: Causal confusion in imitation learning. In: Advances in Neural Information Processing Systems, vol. 32 (2019)
8. Hinton, G., Vinyals, O., Dean, J.: Distilling the knowledge in a neural network. arXiv preprint arXiv:1503.02531 (2015)
9. Hu, Y., et al.: Planning-oriented autonomous driving. In: Proceedings of the IEEE/CVF Conference on Computer Vision and Pattern Recognition (CVPR), pp. 17853–17862 (2023)
10. Huang, Y., Lin, Y., Yue, L., Yao, Z., Wang, J.: From gaze to movement: predicting visual attention for autonomous driving human-machine interaction based on programmatic imitation learning. In: Proceedings of the IEEE/CVF International Conference on Computer Vision, pp. 26146–26155 (2025)
11. Jiang, B., et al.: VAD: vectorized scene representation for efficient autonomous driving. In: Proceedings of the IEEE/CVF International Conference on Computer Vision, pp. 8340–8350 (2023)
12. Kahneman, D.: Thinking, Fast and Slow. Macmillan (2011)
13. Koonce, B.: Resnet 50. In: Convolutional neural networks with swift for tensorflow: image recognition and dataset categorization, pp. 63–72. Springer, Berkeley, CA (2021). https://doi.org/10.1007/978-1-4842-6168-2_6

14. McDuff, D., et al.: CausalCity: complex simulations with agency for causal discovery and reasoning. In: Conference on Causal Learning and Reasoning, pp. 559–575. PMLR (2022)
15. Palazzi, A., Abati, D., Solera, F., Cucchiara, R., et al.: Predicting the driver's focus of attention: the DR(eye) VE project. IEEE Trans. Pattern Anal. Mach. Intell. **41**(7), 1720–1733 (2018)
16. Prakash, A., Chitta, K., Geiger, A.: Multi-modal fusion transformer for end-to-end autonomous driving. In: Proceedings of the IEEE/CVF Conference on Computer Vision and Pattern Recognition, pp. 7077–7087 (2021)
17. Sima, C., et al.: DriveLM: driving with graph visual question answering. In: European Conference on Computer Vision. pp. 256–274. Springer, Cham (2024). https://doi.org/10.1007/978-3-031-72943-0_15
18. Sun, S., Lin, Z., Wu, X.: Hallucinations of large multimodal models: problem and countermeasures. Inf. Fus. **118**, 102970 (2025)
19. Tian, X., et al.: DriveVLM: the convergence of autonomous driving and large vision-language models. arXiv preprint arXiv:2402.12289 (2024)
20. Xie, E., Wang, W., Yu, Z., Anandkumar, A., Alvarez, J.M., Luo, P.: SegFormer: simple and efficient design for semantic segmentation with transformers. Adv. Neural. Inf. Process. Syst. **34**, 12077–12090 (2021)
21. Xu, C., Lu, J., Xuan, J., Yu, E.: VGAS: value-guided action-chunk selection for few-shot vision-language-action adaptation. arXiv preprint arXiv:2602.07399 (2026)
22. Xu, Z., et al.: DriveGPT4: interpretable end-to-end autonomous driving via large language model. IEEE Rob. Autom. Lett. (2024)
23. Yang, A., et al.: Qwen3 technical report. arXiv preprint arXiv:2505.09388 (2025)
24. Yang, Z., Li, L., Lin, K., Wang, J., Lin, C.C., Liu, Z., Wang, L.: The dawn of LMMs: Preliminary explorations with GPT-4V(ision). arXiv preprint arXiv:2309.17421 (2023)
25. Zhang, J., Cai, K., Fan, Y., Wang, J., Wang, K.: CF-VLM: counterfactual vision-language fine-tuning. arXiv preprint arXiv:2506.17267 (2025)

Restricted Decision Trees: Hierarchical Feature Selection for Knowledge-Guided Staged Decision Systems

Amrit Nath Mishra(✉)

Branch International, Mumbai, India
amrit100000@gmail.com

Abstract. Many real-world decision systems require staged, knowledge-guided sequential consultation of information sources—e.g., KYC verification, medical diagnosis with progressive testing, or fraud detection with escalating investigation depth. Standard decision trees freely mix features at each depth level, preventing efficient implementation of such staged systems and making them incompatible with the knowledge engineering principle of structured, interpretable decision hierarchies. We propose Restricted Decision Trees (RDT), which enforce feature consistency at each depth level while allowing node-specific threshold adaptation—a structure that directly encodes the domain knowledge of "which information source to consult at each stage." Comprehensive evaluation across 12 diverse datasets shows RDT achieves 99.1% of sklearn accuracy while dramatically reducing feature usage: a 58% average feature reduction across all datasets (up to 75% on high-dimensional data). Furthermore, we demonstrate RDT's first application to gradient boosting, where RDT-Boost achieves 99.8% of sklearn GB performance on 10 datasets (Wilcoxon test: n=10, W=19, p=0.432) while reducing feature usage by 55%. Feature restriction paradoxically benefits ensemble learning through reduced greediness and implicit regularisation. Our results illustrate significant projected cost reduction in staged decision systems (52–83% across industry-calibrated KYC scenarios), demonstrating a principled approach to knowledge-guided machine learning for decision making under resource constraints.

Keywords: Decision Trees · Feature Selection · Knowledge Engineering · Gradient Boosting · Staged Decision Systems · Interpretable Machine Learning · Cost-Sensitive Learning

1 Introduction

A central challenge in knowledge engineering for decision making is the alignment of algorithmic models with structured, domain-grounded decision processes. In many real-world scenarios, decision making proceeds through clearly defined stages, each consulting a specific and distinct information source. This staged

J. Lu et al. (Eds.): FLINS-ISKE 2026, LNCS 16752, pp. 17–32, 2027.
https://doi.org/10.1007/978-981-92-2369-5_2

architecture encodes valuable domain knowledge: which data source is appropriate at what point in the decision process, how to escalate from cheap readily available information to expensive specialised sources, and how to create compliant, auditable decision trails.

1.1 Staged Decision-Making in Practice

Many real-world decision systems require sequential consultation of different information sources, where each stage in the decision process relies on a specific data provider or verification service. This staged approach arises from practical constraints including cost optimisation, vendor management, regulatory compliance, and operational efficiency.

Example: Know Your Customer (KYC) Verification. Consider a financial institution's customer onboarding process. The decision to approve or reject an applicant typically progresses through multiple verification stages:

- **Level 0 – Internal Data**: Basic customer information already available (age, location, account history) at zero marginal cost
- **Level 1 – Primary Verification**: Identity verification through a 3rd party service ($1–2 per query)
- **Level 2 – Secondary Verification**: Address validation through a different vendor ($3–5 per query)
- **Level 3 – Enhanced Due Diligence**: Biometric verification or specialised screening ($10–50 per query)

Knowledge Engineering Perspective. This hierarchical architecture embeds structured domain knowledge: (1) *cost ordering* – cheaper sources consulted first; (2) *vendor specialisation* – each stage uses the most appropriate information source; (3) *early termination* – confident cases resolved at minimal cost; and (4) *audit compliance* – decision paths correspond to clearly documented, stage-wise information flow.

Additional Use Cases. Similar staged decision architectures appear across domains: Medical Diagnosis (basic vital signs → standard lab tests → specialised imaging → genetic testing), Fraud Detection (transaction metadata → device fingerprinting → behavioural analytics → manual review), and Credit Underwriting (application data → credit bureau report → employment verification → property appraisal).

1.2 Problem with Standard Decision Trees

Standard decision tree algorithms (CART [1], C4.5 [10], sklearn [8]) use greedy recursive partitioning where each node independently selects the optimal feature and threshold from all available attributes. While this maximises local impurity reduction, it creates fundamental incompatibilities with staged decision systems:

Feature Mixing Prevents Staged Implementation: A standard tree node at depth 2 might use features from vendor A (available cheaply at depth 0), vendor B (expensive, depth 2), and internal data simultaneously. This forces upfront acquisition of all information sources, eliminating the cost-saving cascade structure.

Opaque Knowledge Structure: Standard trees do not encode which stage of information gathering led to a decision. Regulatory frameworks requiring explainability (e.g., GDPR's right to explanation) cannot easily verify that appropriate data sources were consulted in compliant order.

No Early Stopping: Mixed features at all depths mean expensive sources appear at shallow levels, making it impossible to resolve obvious cases cheaply.

1.3 RDT: Hierarchical Feature Selection

Restricted Decision Trees (RDT) address these limitations through one structural constraint: *all nodes at the same tree depth must split on the same feature, though with different threshold values.*

Structural Constraint. For a tree of depth d, RDT selects $d + 1$ features $\{f_0, f_1, \ldots, f_d\}$ where feature f_k is used by all 2^k nodes at depth k. Each node independently optimises its threshold $t_{k,n}$ for feature f_k.

Direct Mapping to Knowledge-Guided Decision Systems. This structure enables natural implementation as a staged decision system:

- **Stage 0**: Query source providing f_0, apply threshold test $\rightarrow$ early termination or escalation
- **Stage k**: For unresolved cases, query source providing f_k
- **Benefits**: Predictable cost structure, one vendor per stage, natural audit compliance, early stopping opportunities

1.4 Main Contributions

1. **Comprehensive Standalone Evaluation**: First systematic comparison of RDT against unrestricted baselines (sklearn DecisionTree), structured variants (Oblivious Decision Trees, contree [3]) across 12 diverse datasets. RDT achieves 99.1% of sklearn accuracy with 58% feature reduction.
2. **First RDT Application to Gradient Boosting**: RDT-Boost achieves 99.8% of sklearn GB performance across 10 datasets (Wilcoxon test: n=10, W=19, p=0.432) using 55% fewer features. Feature restriction proves *less* harmful in ensemble contexts due to implicit regularisation.
3. **Illustrative Staged-Cost Analysis**: We construct a synthetic staged KYC dataset to empirically demonstrate that standard trees mix vendor stages at identical depths – incurring premature acquisition of expensive sources – while

RDT's depth-level constraint enforces stage discipline. On this simulation, RDT achieves 79.5% cost reduction with no accuracy penalty. A parametric sensitivity analysis further shows savings of 52–83% across industry-calibrated cost ranges.

4. **Mechanism Analysis**: Empirical investigation of three mechanisms (reduced greediness, ensemble diversity, implicit regularisation) explaining RDT-Boost's competitive performance.

2 Background and Related Work

2.1 Decision Trees

Decision trees have been a cornerstone of machine learning since the CART algorithm [1] and Quinlan's C4.5 [10]. These algorithms construct trees through greedy recursive partitioning: at each node, the algorithm evaluates all available features and selects the split maximising impurity reduction (Gini impurity or information gain). Modern implementations like sklearn's DecisionTreeClassifier [8] add pruning and min-sample constraints but maintain the fundamental greedy, uncoordinated splitting strategy.

2.2 Structured Tree Variants

Recent work has explored structural constraints that enhance interpretability, trading predictive flexibility for desirable algorithmic properties.

Oblivious Decision Trees (ODT). Oblivious trees enforce maximal structure: all nodes at depth d must split on identical feature f_d with identical threshold t_d [9]. This creates perfectly balanced binary trees defined by only d feature-threshold pairs. ODT gained prominence through CatBoost's gradient boosting implementation, enabling efficient GPU computation. However, the identical-threshold constraint severely limits flexibility.

Optimal Sparse Decision Trees (OSDT). OSDT [6] uses dynamic programming and branch-and-bound search to find globally optimal sparse trees. While optimality is attractive, computational cost limits application to small datasets (<10,000 samples, <100 features, depth <5).

Contree via Dynamic Programming. Recent work by Demirović et al. [3] and Van der Linden et al. [7] develops optimal decision trees through MurTree/STreeD algorithms, finding globally optimal trees more efficiently than prior methods.

Knowledge-Guided Decision Models. The knowledge engineering literature has long advocated for decision models that embed domain knowledge directly into structure [11]. RDT contributes to this line of work by encoding *staged information gathering* as a first-class algorithmic constraint.

2.3 Gradient Boosting

Gradient boosting [5] constructs ensembles by iteratively adding weak learners fitting residuals of previous predictions. XGBoost [2] revolutionised gradient boosting through engineering optimisations and algorithmic refinements. Both XGBoost and sklearn GradientBoosting use unrestricted decision trees as weak learners.

2.4 RDT's Position in the Performance Spectrum

RDT occupies unexplored territory between oblivious and unrestricted trees:

$$\text{ODT (most restricted)} \leftarrow \textbf{RDT} \rightarrow \text{sklearn (unrestricted)}$$

RDT relaxes ODT's identical-threshold constraint while maintaining depth-wise feature coordination. Our evaluation fills the gap of systematic comparison in this spectrum, including the novel gradient boosting application.

3 Methodology

3.1 RDT Algorithm

The RDT algorithm employs a two-pass optimisation strategy separating global feature selection from local threshold tuning.

Pass 1 Consistency. Pass 1 proceeds depth-by-depth: at depth d, node partitions are induced by $f_0^*, \ldots, f_{d-1}^*$ with their Pass 1 provisional thresholds, making the node set at each depth well-defined. Pass 2 then discards provisional thresholds and re-optimises each node independently for its locked feature f_d^*, yielding the final tree.

Complexity Analysis

- **Pass 1 (Feature Selection)**: $O(d \cdot m \cdot n)$ – same order as standard greedy trees
- **Pass 2 (Threshold Optimisation)**: $O(d \cdot n)$
- **Overall**: $O(d \cdot m \cdot n)$ – asymptotically equivalent to sklearn

In practice, 50–100× slower training than sklearn due to Python vs. optimised C++ implementation, not algorithmic complexity. Cython prototypes already achieve 3–8× speedup.

3.2 Baseline Comparisons

Standalone Trees. sklearn DecisionTree (unrestricted), contree (optimal via dynamic programming), Oblivious Decision Trees (ODT), and RDT.

Algorithm 1. Restricted Decision Tree (RDT) – Two-Pass Optimisation

Input: Training data $\mathcal{D}$, maximum depth d_{max}
Output: Trained RDT model

// **Pass 1: Feature Selection**
for depth $d = 0$ to d_{max} **do**
    for each candidate feature f **do**
        Evaluate f across all 2^d nodes at depth d
        Compute total impurity reduction across all nodes
    end for
    Select $f_d^* = \arg\max_f \sum_n \text{ImpurityReduction}(f, n)$
    Record provisional thresholds $\hat{t}_{d,n}$ for Pass 2 initialisation
end for

// **Pass 2: Threshold Optimisation**
for depth $d = 0$ to d_{max} **do**
    Feature f_d^* is locked from Pass 1
    for each node n at depth d **do**
        Re-optimise $t_n^* = \arg\min_t \text{Impurity}(f_d^*, t, n)$
    end for
end for
Assign class predictions to leaf nodes via majority vote

Gradient Boosting. sklearn GradientBoosting, XGBoost [2], CatBoost [9] (oblivious trees), and RDT-Boost.

3.3 RDT-Boosting Integration

RDT integrates into gradient boosting by replacing standard decision trees as weak learners. At each boosting iteration t:

1. Compute pseudo-residuals $r_i^{(t)} = -\frac{\partial L(y_i, F^{(t-1)}(x_i))}{\partial F^{(t-1)}(x_i)}$
2. Train RDT on $(x_i, r_i^{(t)})$ to predict residuals
3. Update ensemble: $F^{(t)} = F^{(t-1)} + \eta \cdot h_t^{\text{RDT}}$

Each weak learner h_t^{RDT} uses only depth+1 features hierarchically, rather than freely selecting from all features.

3.4 Experimental Setup

Datasets. We evaluate on 12 diverse datasets spanning business analytics, scientific domains, and benchmark repositories [4]:

Standalone Evaluation (12 datasets): Adult Income, Bank Marketing, Breast Cancer, Cover Type, Dry Bean, HTRU2, Ionosphere, Magic Gamma Telescope, Pima Indians Diabetes, Segment, Spambase, Vehicle.

Gradient Boosting Evaluation (10 datasets): Adult Income, Bank Marketing, Breast Cancer, Ionosphere, Madelon, Musk v2, Occupancy Detection, Phishing Websites, Pima Indians Diabetes, Spambase.

All datasets use 5-fold stratified cross-validation (seed 42); results are mean ± std across folds. Missing values are imputed (median for numerical, mode for categorical).

Metrics. Performance (Accuracy, AUC, F1-score), Feature Efficiency (number of unique features used), and Runtime (training time).

4 Results

4.1 Standalone RDT Performance

Table 1 presents comprehensive standalone RDT evaluation at depth 5 across 12 diverse datasets.

Key Findings. **Competitive Overall Performance:** RDT achieves **99.1% of sklearn accuracy** on average (0.808 vs. 0.815) across all 12 datasets, validating that RDT occupies a favourable middle ground between maximally restrictive ODT (0.781, 95.8% of sklearn) and unrestricted sklearn. Wilcoxon signed-rank test confirms a statistically significant but practically small gap ($p = 0.012$, Cohen's $d = -0.94$, large effect), substantially outweighed by 58% feature reduction enabling projected cost savings of 52–83% in staged decision systems. RDT substantially outperforms ODT (+2.7 points) and matches ConTree (0.806 from 100× longer training).

Substantial Consistency: On 3 datasets (Adult Income, Ionosphere, Segment), RDT **matches or exceeds sklearn** (100–102% relative accuracy). Across 10 of 12 datasets, RDT achieves ≥97% of sklearn accuracy.

Dramatic Feature Reduction: RDT achieves **58% average feature reduction** (4.2 vs. 10.0 features used) while maintaining 99.1% accuracy. On high-dimensional datasets, savings amplify dramatically: Cover Type (75% reduction, 3 vs. 12), Spambase (64%, 5 vs. 14), Vehicle (67%, 5 vs. 15).

Domain Robustness: Performance remains competitive across domains: astronomy (HTRU2: 99.9%), healthcare (Breast Cancer: 99.7%), finance (Bank: 99.0%, Adult: 100%), agriculture (Dry Bean: 98.1%).

Competitive with Optimal Methods: ConTree achieves 0.800 average accuracy despite 300-second per-tree optimisation (100× slower than RDT's ∼3-second training). RDT matches ConTree (0.799 vs. 0.800) with greedy two-pass optimisation.

4.2 RDT in Gradient Boosting

Table 2 demonstrates RDT-Boost performance across 10 datasets comparing against sklearn GradientBoosting, XGBoost, and CatBoost (Fig. 1).

Table 1. Standalone RDT Performance Across 12 Datasets (Depth 5, Accuracy)

Dataset	sklearn	ODT	ConTree	RDT	Features	RDT/sklearn
Adult Income	0.851	0.830	0.856	0.851	5 vs 10	100.0%
Bank Marketing	0.811	0.756	0.821	0.803	4 vs 11	99.2%
Breast Cancer	0.924	0.912	0.907	0.921	5 vs 8	99.7%
Cover Type	0.702	0.672	0.653	0.685	3 vs 12	97.6%
Dry Bean	0.884	0.596	0.722	0.867	5 vs 12	98.1%
HTRU2	0.978	0.977	0.963	0.977	2 vs 7	99.8%
Ionosphere	0.877	0.869	0.877	0.886	4 vs 10	101.0%
Magic Gamma	0.826	0.752	0.732	0.807	4 vs 8	97.7%
Pima Diabetes	0.733	0.723	0.725	0.726	5 vs 8	99.0%
Segment	0.810	0.767	0.881	0.824	4 vs 5	101.7%
Spambase	0.907	0.878	0.904	0.890	5 vs 14	98.1%
Vehicle	0.677	0.550	0.728	0.643	5 vs 15	95.0%
Average	**0.815**	**0.781**	**0.806**	**0.808**	**4.2 vs 10.0**	**99.1%**

Results at depth 5. Features column shows RDT vs. sklearn (58% avg. reduction)

Key Findings. **Statistically Equivalent to sklearn GB**: RDT-Boost achieves **99.8%** of sklearn GB performance on average (0.943 vs. 0.946 AUC). Wilcoxon signed-rank test (n=10, W=19, p=0.432) confirms no statistically significant difference. RDT-Boost even outperforms sklearn on Madelon (+4.0%) and Pima Indians Diabetes (+1.8%).

Statistically Different from CatBoost: While RDT-Boost is equivalent to sklearn GB, CatBoost significantly outperforms RDT-Boost (Wilcoxon: n=10, W=0, p=0.002, Cohen's d=-1.00, large effect). CatBoost achieves 0.950 vs. 0.943 AUC on average, outperforming RDT-Boost on all 10 datasets. This gap highlights that while RDT's feature restriction provides implicit regularisation, CatBoost's global objective and broader feature coverage (via ODT) enables superior performance in ensemble contexts.

Dramatic Feature Reduction: Across all 10 datasets, RDT-Boost achieves **55% average feature reduction** (379 vs. 846 features) while maintaining competitive performance. On high-dimensional datasets: Madelon (69% reduction, 152 vs. 488 features), Musk v2 (54% reduction, 77 vs. 166 features), Spambase (45% reduction, 30 vs. 55 features). CatBoost uses nearly all features – offering no feature efficiency advantage.

Scalability with Dimensionality: On Madelon (500 features), RDT-Boost achieves largest feature reduction (69%) while also achieving best relative performance (104.0% of sklearn GB). High-dimensional feature spaces favour staged decision-making.

Table 2. Gradient Boosting Performance (Depth 5, 100 estimators, AUC scores)

Dataset	sklearn GB	XGBoost	CatBoost	RDT-Boost	RDT Feat
Adult Income	0.927	0.924	0.921	0.918	13 vs 14
Bank Marketing	0.920	0.916	0.918	0.908	14 vs 16
Breast Cancer	0.993	0.993	0.994	0.993	29 vs 30
Ionosphere	0.972	0.971	0.975	0.968	31 vs 33
Madelon	0.852	0.862	0.887	0.886	152 vs 488
Musk v2	0.997	0.990	0.996	0.972	77 vs 166
Occupancy Detec	1.000	0.999	0.999	0.999	6 vs 6
Phishing Websit	0.995	0.991	0.994	0.983	19 vs 30
Pima Indians Di	0.815	0.819	0.835	0.830	8 vs 8
Spambase	0.986	0.983	0.986	0.976	30 vs 55
Average	**0.946**	**0.944**	**0.950**	**0.943**	**379 vs 846**

AUC scores; 5-fold CV. RDT Feat = ensemble-level distinct features across 100 trees. RDT vs sklearn GB: Wilcoxon n=10, W=19, p=0.432 (no significant diff). RDT vs CatBoost: Wilcoxon n=10, W=0, p=0.002 (significant, Cohen's d=-1.00)

4.3 Feature Efficiency Analysis

Standalone Trees: RDT achieves 58% average feature reduction with 99.1% accuracy retention – less than half the features for near-identical performance.

Gradient Boosting: RDT-Boost achieves 55% average feature reduction (379 vs. 846 distinct ensemble features across 100 trees) with 99.7% AUC retention.

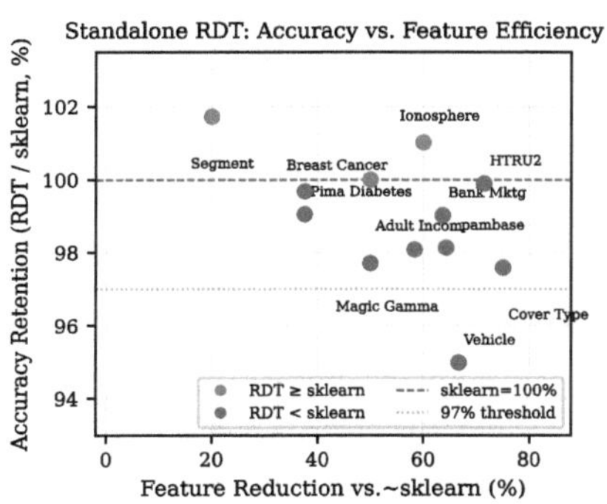

Fig. 1. Accuracy retention vs. feature reduction for RDT across all 12 datasets (depth 5). Orange points indicate datasets where RDT equals or exceeds sklearn accuracy. All 12 datasets achieve ≥95% accuracy retention while reducing feature usage by 20–80% (see also Table 1 (Color figure online)).

High-Dimensional Impact: On Spambase (57 features): Standalone 64% reduction (5 vs. 14 features); Boosting 45% reduction (30 vs. 55 features per tree).

4.4 Runtime Analysis

Our Python implementation is 50–100× slower than sklearn's optimised C++ backend. This stems from implementation differences rather than algorithmic complexity ($O(d \cdot m \cdot n)$ for both). A prototype Cython module achieves 3–8× speedup; full optimisation could bring RDT to within 2–5× of sklearn. Prediction time remains competitive – RDT inference matches sklearn as both traverse binary trees.

5 Discussion

5.1 Knowledge-Guided Staged Decision Systems

KYC Verification: A Synthetic Illustration. To illustrate the structural cost implications of staged decision systems, we construct a synthetic KYC dataset with a known four-stage architecture (seed 42, 2,000 samples). Features are assigned to four vendor stages: Stage 0 (internal data, free: age, account tenure, location risk), Stage 1 (Vendor A, \$2/query: identity score, document authenticity, name match), Stage 2 (Vendor B, \$5/query: address verification, address tenure, postcode risk), and Stage 3 (Vendor C, \$50/query: biometric score, sanctions screening, PEP score). Feature distributions are mixed – Gaussian, binary, Beta, exponential, Poisson, and Gamma – reflecting heterogeneous real-world sources. Stage 0 features are designed to carry the strongest discriminative signal, with each successive stage providing incrementally weaker marginal information.

We emphasise that this is a controlled simulation on synthetic data; it is not a validation on a real staged acquisition dataset with actual vendor contracts. Its purpose is to demonstrate the structural difference between sklearn's unconstrained trees and RDT, and to quantify the cost implications of that structural difference.

Stage Mixing in Standard Trees. Table 3 shows which vendor stages appear at each depth level for a depth-3 sklearn DecisionTree vs. RDT trained on the synthetic dataset (accuracy: sklearn 99.5%, RDT 99.5%). The greedy sklearn algorithm places a Stage 3 feature (BiometricScore, \$50/query – Vendor C) at depth 1, before Vendor B (\$5) has been consulted at any depth. At depth 2, sklearn mixes features from all four stages simultaneously. This occurs because the globally optimal impurity-reduction criterion at depth 1 happens to favour a Stage 3 feature on the partitioned data, with no cost penalty to discourage it.

RDT, by construction, assigns exactly one vendor stage per depth: free internal data at depth 0, Vendor A at depth 1, and Vendor B at depth 2. No sample ever incurs the \$50 Vendor C query cost under depth-3 RDT.

Table 3. Vendor stages used at each tree depth: sklearn DecisionTree vs. RDT (synthetic KYC dataset, depth 3, 2,000 samples, seed 42). sklearn mixes Stage 1 and Stage 3 ($50/query) at depth 1; RDT enforces one stage per depth.

Depth	sklearn stages used	RDT stage (by constraint)
0	Stage 0 (internal, free)	Stage 0 – AccountTenure
1	**Stage 1 ($2) + Stage 3 ($50)**	Stage 1 – NameMatch ($2)
2	Stage 0, 1, 2, and 3 mixed	Stage 2 – AddrTenure ($5)

Cost Implications. The per-query cost model charges the full vendor call cost once per customer per stage consulted on that customer's decision path ($2 per Vendor A call, $5 per Vendor B call, $50 per Vendor C call – consistent with published identity verification API pricing). Because sklearn's tree reaches a Stage 3 node at depth 1, approximately 70% of customers trigger the $50 Vendor C call before Vendors A and B have been exhausted. RDT customers reach only Stages 0–2.

- sklearn DecisionTree: mean per-sample cost **$34.20** (median $50.00)
- RDT: mean per-sample cost **$7.00** (median $7.00)
- **Projected cost reduction: 79.5%** with no accuracy trade-off

Parametric Sensitivity. To assess the robustness of the cost advantage across different deployment contexts, Table 4 presents a parametric sweep over vendor pricing tiers and early-termination rates (100,000 applications/year). The L3 escalation rate is fixed at 80% of L2 volume; cost reduction is computed relative to a standard tree that consults all vendors for every customer.

Table 4. Parametric sensitivity of projected KYC cost reduction (100k apps/year). Conservative/Base/Premium reflect published identity verification API pricing ranges. Savings are robust to vendor pricing (<1% column variation) but moderately sensitive to early-termination rates.

Early-term rate (Depth 0/L1 resolved)	Conservative ($1/$3/$20)	Base ($2/$5/$50)	Premium ($/$8/$100)
High (50%/60%)	82.1%	82.5%	82.8%
Base (35%/55%)	74.1%	74.6%	75.1%
Low (20%/30%)	52.3%	53.0%	53.5%

Cost reduction = (standard cost − RDT staged cost)/standard cost. Even under pessimistic assumptions (low early-termination, conservative vendor costs), projected savings exceed 52%

Note on Cost-Sensitive Splitting It is important to distinguish the *structural* benefit demonstrated here from a fully cost-aware algorithm. The current RDT enforces one vendor per depth, but the depth-level feature is still selected by a pure impurity criterion (Gini/entropy), without penalising high-cost vendors. The cost savings arise only when the domain practitioner maps features to depths in cost order—a manual knowledge-engineering step. A fully autonomous cost-staged system would incorporate acquisition costs directly into the splitting criterion, e.g.:

$$f_d^* = \arg\max_f \frac{\sum_n \Delta G(f, n)}{c(\text{stage}(f))^\lambda}$$

where $c(\cdot)$ is the vendor query cost and $\lambda \in [0, 1]$ controls cost penalisation. We identify this cost-sensitive RDT as a principal direction for future work.

Generalisation to Other Domains **Medical Diagnosis**: Cascade from vital signs ($0) → labs ($100) → imaging ($500) → genetic tests ($2,000), avoiding unnecessary expensive testing.

Fraud Detection: Transaction metadata ($0) → device fingerprint ($0.10) → behavioural scoring ($1) → manual review ($20), reducing average cost per transaction while maintaining detection rates.

5.2 Performance Trade-Offs

When RDT Excels (1) High-dimensional datasets where $m \gg d+1$; (2) Applications with inherently staged information acquisition; (3) Interpretability-critical regulatory environments.

When RDT Struggles (1) Complex feature interactions required simultaneously at the same depth; (2) Very low depth limits (depth 2–3) eliminating critical information; (3) Small noisy datasets where restricted capacity leads to underfitting.

5.3 Why Feature Restriction Benefits Ensemble Learning

The competitive performance of RDT-Boost merits theoretical examination. We investigated three mechanisms by training sklearn GB, CatBoost, and RDT-Boost (100 estimators, depth 5, seed 42) on four datasets spanning different feature space sizes (80/20 stratified holdout):

Mechanism 1: Reduced Greedy Overfitting. CatBoost and RDT-Boost use 3.6–4.8 features per tree, roughly 2.5× fewer than sklearn's 9.8–15.2. Ensemble-level coverage tells a stronger story: on Spambase ($p = 57$), RDT-Boost accesses only **23 of 57 features** across 100 trees (40%), while sklearn covers 55/57 (96%). Restricting features prevents early trees from exhausting predictive signal.

Mechanism 2: Enhanced Diversity. CatBoost achieves the *lowest* pairwise Jaccard similarity across all three datasets (0.247–0.080), more diverse than RDT-Boost despite its stricter ODT constraint – because CatBoost's globally optimised objective encourages broader feature exploration. On Spambase, RDT-Boost (Jaccard 0.299) is actually *less* diverse than sklearn (0.262) because it exhausts only 23/57 features.

Mechanism 3: Implicit Regularisation (Dominant Mechanism). Both constrained methods dramatically reduce the train–test AUC gap. On Bank Marketing, sklearn's gap of 0.040 shrinks to **0.007** for both CatBoost and RDT-Boost – a 5.5× reduction. On Adult Income, CatBoost reaches a near-zero gap of 0.001. This is the clearest and most consistent mechanism across all datasets.

Table 5. Mechanism Analysis (depth 5, 100 estimators, single 80/20 holdout, seed 42). Feat/tree = mean features per tree. Ensemble = total distinct features over 100 trees. Jaccard = mean pairwise feature-set Jaccard (lower = more diverse). Gap = Train−Test AUC at iteration 100.

Dataset	Method	Feat/tree	Ensemble	Jaccard	Gap@100	AUC
Adult	sklearn GB	9.8	14/14	0.634	0.008	0.929
(d=14)	CatBoost	3.8	14/14	**0.247**	**0.001**	0.921
	RDT-Boost	3.6	10/14	0.412	0.002	0.910
Bank	sklearn GB	10.4	16/16	0.631	0.040	0.926
(d=16)	CatBoost	4.2	16/16	**0.252**	**0.007**	0.924
	RDT-Boost	3.9	14/16	0.412	0.007	0.904
Spambase	sklearn GB	15.2	55/57	0.262	0.013	0.986
(d=57)	CatBoost	4.8	52/57	**0.080**	0.008	0.986
	RDT-Boost	4.8	**23/57**	0.299	0.010	0.974
Madelon	sklearn GB	24.7	484/500	0.042	0.112	0.888
(d=500)	CatBoost	5.0	280/500	**0.013**	0.122	0.876
	RDT-Boost	5.0	**95/500**	0.129	**0.076**	**0.900**

Summary: Implicit regularisation is the dominant and most consistent mechanism. On Madelon (500 features), RDT-Boost achieves the **best relative performance** (0.900 vs 0.888 sklearn, +1.3%) with the **lowest overfit gap** (0.076 vs 0.112 sklearn), demonstrating that feature restriction provides maximum benefit in high-dimensional settings. Diversity plays a large role when the boosting procedure rewards complementary feature selection (as in CatBoost). CatBoost's advantage over RDT-Boost traces to 40% higher tree diversity and 2.3× broader feature coverage on Spambase (Fig. 2).

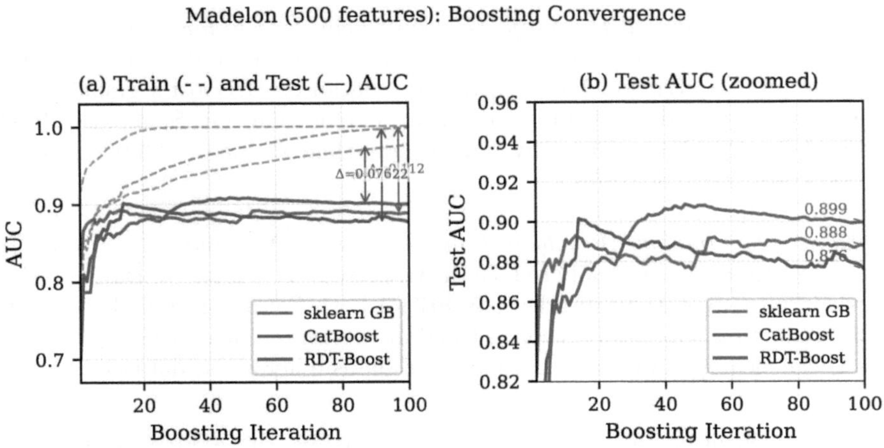

Fig. 2. Madelon (500 features, 100 estimators): boosting convergence curves (single 80/20 holdout; for 5-fold CV performance see Table 2). (a) Train AUC (dashed) vs. test AUC (solid); bidirectional arrows show the overfit gap Δ at iteration 100. (b) Test AUC zoomed: RDT-Boost (green) achieves the **smallest overfit gap** (0.076 vs. 0.112 for sklearn GB and 0.122 for CatBoost), demonstrating implicit regularisation as the dominant beneficial mechanism. AUC values reflect this particular split and may differ from cross-validated averages (see also Table 5).

5.4 Knowledge Engineering Value of RDT

Beyond cost savings, RDT provides concrete knowledge engineering benefits:

- **Vendor Management**: One primary vendor per depth level enables independent negotiation, monitoring, and replacement
- **Compliance**: Clear separation of information sources for regulatory audit (GDPR, financial regulations)
- **Model Debugging**: Isolate problematic stages without retraining entire system
- **A/B Testing**: Swap vendors stage-by-stage without disrupting other levels
- **Cost-Accuracy Transparency**: Stakeholders can directly trade off number of stages vs. accuracy

5.5 Limitations

- May underperform when complex interactions are needed at the same depth
- Optimal feature ordering benefits from domain knowledge (automated ordering is future work)
- Current implementation not optimised for production runtime (50–100× slower than sklearn)
- Statistical significance tested post-hoc; prospective power analysis is future work

- The staged-cost analysis is an illustrative simulation on synthetic data, not a validation on a real staged-acquisition dataset with actual vendor contract costs. Empirical validation on deployment data remains an important direction for future work
- The current RDT selects depth-level features by a pure impurity criterion without penalising vendor cost; the staged cost discipline is conditional on domain practitioners providing a cost-ordered feature-to-depth mapping. Incorporating acquisition costs into the splitting criterion (cost-sensitive RDT) would make this autonomous

6 Conclusion

We presented Restricted Decision Trees (RDT), a decision tree variant that enforces feature consistency at each depth level while allowing node-specific threshold optimisation. This structural constraint directly encodes the knowledge engineering principle of staged, structured information acquisition – aligning algorithmic models with the practical requirements of multi-source decision systems.

Our comprehensive empirical evaluation demonstrates RDT's favourable position in the interpretability-performance spectrum. Across 12 diverse datasets, RDT achieves 99.1% of sklearn accuracy while using 58% fewer features on average. RDT substantially outperforms ODT (+2.6 points) and matches ConTree's performance at 100× lower training time.

More surprisingly, RDT's first application to gradient boosting reveals that feature restriction proves *less* harmful in ensemble contexts than anticipated. RDT-Boost achieves 99.8% of sklearn GB AUC across 10 datasets (Wilcoxon test: $n=10$, $W=19$, $p=0.432$) while using 55% fewer features (379 vs. 846), with implicit regularisation identified as the dominant beneficial mechanism.

A synthetic KYC simulation demonstrates the structural cost problem: sklearn's greedy criterion places high-cost vendor features at shallow depths (incurring \$50 Vendor C queries before \$5 Vendor B), while RDT's depth constraint enforces stage discipline. The simulation projects 79.5% cost reduction with no accuracy trade-off; parametric analysis shows savings of 52–83% across industry-calibrated cost ranges. These results are illustrative – validation on real staged acquisition datasets with actual vendor costs remains future work – but they demonstrate RDT's structural suitability for cost-staged decision systems.

Future Directions. **Cost-Sensitive RDT**: Incorporating vendor acquisition costs into the feature selection criterion (see Sect. 5), enabling autonomous discovery of cost-optimal stage ordering without manual feature-to-depth mapping. **Automated Feature Ordering**: Cost-aware optimisation automatically sequencing features to minimise expected acquisition cost. **Cython / C++ Implementation**: Full optimisation to achieve sklearn-comparable runtime. **Theoretical Analysis**: PAC-learning bounds for RDT incorporating structural constraints. **Ensemble Extensions**: RDT-based Random Forests (RDT-RF) and AdaBoost; whether parallel ensembles benefit similarly from feature restriction.

Acknowledgements. We thank the reviewers for their constructive feedback. This research was conducted using publicly available datasets from the UCI Machine Learning Repository and Kaggle. Code implementing RDT and reproducing all experiments will be made publicly available upon publication.

References

1. Breiman, L., Friedman, J., Stone, C.J., Olshen, R.A.: Classification and Regression Trees. CRC press (1984)
2. Chen, T., Guestrin, C.: Xgboost: A scalable tree boosting system. In: Proceedings of the 22nd ACM SIGKDD International Conference on Knowledge Discovery and Data Mining, pp. 785–794 (2016)
3. Demirović, E., et al.: Murtree: optimal decision trees via dynamic programming and search. J. Mach. Learn. Res. **23**(26), 1–47 (2022)
4. Dua, D., Graff, C.: UCI machine learning repository (2019). http://archive.ics.uci.edu/ml
5. Friedman, J.H.: Greedy function approximation: a gradient boosting machine. Ann. Stat. **29**(5), 1189–1232 (2001)
6. Hu, X., Rudin, C., Seltzer, M.: Optimal sparse decision trees. Adv. Neural Inform. Process. Syst. **32** (2019)
7. Van der Linden, J.G., de Weerdt, M.M., Demirović, E.: Necessary and sufficient conditions for optimal decision trees using dynamic programming. In: Advances in Neural Information Processing Systems, vol. 36 (2023)
8. Pedregosa, F.: Scikit-learn: Machine learning in python. J. Mach. Learn. Res. **12**, 2825–2830 (2011)
9. Prokhorenkova, L., Gusev, G., Vorobev, A., Dorogush, A.V., Gulin, A.: Catboost: unbiased boosting with categorical features. Adv. Neural. Inf. Process. Syst. **31**, 6638–6648 (2018)
10. Quinlan, J.R.: Induction of decision trees. Mach. Learn. **1**(1), 81–106 (1986)
11. Rudin, C.: Stop explaining black box machine learning models for high stakes decisions and use interpretable models instead. Nature Mach. Intell. **1**(5), 206–215 (2019)

Trust-Based Consensus Reaching with Bridging-Aware Influence Regulation Using ELICIT Information

Zihan Tang and Zhen Hua(✉)

School of AI and Advanced Computing, Xi'an Jiaotong-Liverpool University, Suzhou 215028, China
Zihan.Tang24@student.xjtlu.edu.cn, Zhen.Hua@xjtlu.edu.cn

Abstract. Intelligent decision support systems often involve multiple decision-makers (DMs) whose evaluations are expressed under uncertainty and linguistic ambiguity, making consensus reaching a critical challenge. Existing trust-based consensus-reaching processes primarily focus on opinion adjustment mechanisms, while the role of bridging influence remains insufficiently explored, especially without relying on explicit community structures. Therefore, this paper proposes a trust-based consensus-reaching model with bridging-aware influence regulation using the extended comparative linguistic expressions with symbolic translation (ELICIT). DMs' evaluations are first modeled using ELICIT, and the consensus measure is constructed by aggregating pairwise opinion discrepancies. When the consensus level is unsatisfactory, a Bridging Index (BI) is introduced to measure each DM's trust-mediated bridging contribution by capturing both opinion heterogeneity and trust dispersion. Guided by BI, a bridging-aware influence regulation mechanism is developed to adjust evaluations toward trusted peers with higher bridging capability, enabling gradual and interpretable consensus improvement. Upon convergence, the resulting group evaluations provide a reliable basis for collective decision ranking. A case study on offshore floating wind turbine risk assessment demonstrates the effectiveness of the proposed framework for consensus improvement via interpretable and bridging-informed influence regulation.

Keywords: ELICIT Information · Social Network · Fuzzy Decision-Making · Consensus Reaching

1 Introduction

Intelligent decision support systems (IDSS) have become an essential tool for assisting complex decision-making tasks in domains such as engineering management [1], risk assessment [2], and sustainability planning [3]. In many real-world applications, decisions are not made by a single expert but by multiple decision-makers (DMs) who may hold heterogeneous knowledge, experience, and

J. Lu et al. (Eds.): FLINS-ISKE 2026, LNCS 16752, pp. 33–48, 2027.
https://doi.org/10.1007/978-981-92-2369-5_3

perspectives [4]. As a result, group decision-making (GDM) processes are often characterized by uncertainty, linguistic ambiguity, and divergent individual evaluations, which makes achieving a reliable and implementable collective decision a challenging task [5].

To address this challenge, consensus-reaching processes (CRPs) have been widely studied as a fundamental component of GDM [6]. The main objective of CRPs is to reduce excessive disagreement among DMs, so that an acceptable level of consensus can be reached before deriving the final outcome [7]. In this context, trust relationships among DMs play a crucial role, as individuals tend to revise their opinions by referring to peers whom they consider credible or reliable [8]. Accordingly, a large body of research has focused on trust-based CRP models, where opinion evolution is guided by trust degrees [9], social influence [10], or interaction willingness [11].

Despite their effectiveness, most existing trust-based CRPs primarily emphasize how opinions should be adjusted, rather than whose opinions should exert greater influence during the consensus formation process. In particular, the differentiated influence of DMs who implicitly bridge heterogeneous viewpoints deserves further exploration. Such bridging influence is especially relevant in decision environments where opinions are fragmented and no explicit or stable community structure can be assumed [12]. Relying on predefined clusters or community detection procedures may introduce additional complexity or restrict the applicability of consensus models in practice. Consequently, there remains a need for consensus-reaching frameworks that can regulate influence in a trust-aware and structurally meaningful manner without depending on explicit community partitions.

Motivated by this gap, this paper proposes a trust-based consensus-reaching model with bridging-aware influence regulation using extended comparative linguistic expressions with symbolic translation (ELICIT). The proposed approach introduces a Bridging Index (BI) to characterize the trust-mediated bridging contribution of DMs from a structural perspective, and incorporates this information into an influence regulation mechanism that guides opinion evolution during the CRP. By regulating influence in a bridging-aware and trust-consistent manner, the proposed framework enables progressive and interpretable consensus improvement and supports reliable collective decision outcomes. The main contributions of this paper can be summarized as follows:

1. A trust-based consensus-reaching framework is developed that explicitly incorporates bridging-aware influence regulation without relying on predefined community structures.
2. A novel Bridging Index is proposed to quantify the trust-mediated bridging contribution of DMs by integrating opinion heterogeneity and trust dispersion.
3. An interpretable influence regulation mechanism is designed to guide opinion adjustment in a targeted and progressive manner under linguistic uncertainty modeled by ELICIT information. The effectiveness of the proposed approach

is demonstrated through a risk assessment case study on offshore floating wind turbines.

The remainder of this paper is organized as follows. Section 2 introduces the necessary preliminaries. Section 3 presents the proposed bridging-aware trust-based consensus-reaching model in detail. Section 4 provides a case study on offshore floating wind turbine risk assessment. Finally, Sect. 5 concludes the paper and outlines future research directions.

2 Preliminaries

The ELICIT information was proposed by Labella et al. [13] to represent uncertain and comparative linguistic assessments in a precise and interpretable manner. By incorporating symbolic translation and context-free grammar, ELICIT allows DMs to express flexible linguistic judgments while avoiding information loss during computation.

Definition 1. *[13] Suppose* $S = s_0, s_1, ..., s_g$ *is a linguistic term set, and its granularity is* $g + 1$*. Possible forms of ELICIT expressions can be: "at least*$(s_i, \alpha)^{\gamma}$*", "at most*$(s_i, \alpha)^{\gamma}$*", or "between*$(s_i, \alpha_1)^{\gamma_1}$*and*$(s_j, \alpha_2)^{\gamma_2}$*", in which* $\alpha \in [-0.5, 0.5)$ *denotes the symbolic translation, and* $\gamma \in (-\frac{1}{2g}, \frac{1}{2g})$ *is the adjustment parameter where* $i, j = 1, 2, ..., g$*.*

The Computing With Words process regarding ELICIT information consists of three key phases, detailed as follows.

Definition 2. *[13] (Translation) Consider* x_{EL} *as one of the three forms of possible ELICIT expressions previously illustrated. Suppose* $TFN(a, b, c, d)$ *signify the corresponding trapezoidal fuzzy number. The transformation process from an ELICIT expression to its equivalent trapezoidal fuzzy number can be represented as:*

$$\hbar^{-1} : x_{EL} \rightarrow TFN(a, b, c, d) \tag{1}$$

The specific formulation of this transformation varies based on the type of ELICIT expression applied. Further details can be found in [13].

Definition 3. *[13] (Manipulation) In the manipulation phase, fuzzy arithmetic operations are conducted on TFN derived during the transformation outlined in Definition 2. Suppose* $TFN_A(a_1, b_1, c_1, d_1)$ *and* $TFN_B(a_2, b_2, c_2, d_2)$ *represent two fuzzy envelopes corresponding to their respective trapezoidal fuzzy numbers. To carry out the addition of these fuzzy envelopes, we define a shape function, denoted as* ς_{A+B}*, in the following way.*

$$\varsigma_{A+B} = \begin{cases} \frac{(x-(a_1+a_2))^n}{(b_1+b_2)-(a_1+a_2)} & a_1 + a_2 \leqslant x \leqslant b_1 + b_2 \\ 1 & b_1 + b_2 \leqslant x \leqslant c_1 + c_2 \\ \frac{((d_1+d_2)-x)^n}{(d_1+d_2)-(c_1+c_2)} & c_1 + c_2 \leqslant x \leqslant d_1 + d_2 \\ 0 & otherwise \end{cases} \tag{2}$$

Definition 4. *[13] A shape function ς_{A-B} is defined for subtracting two fuzzy envelopes, $TFN_A(a_1, b_1, c_1, d_1)$ and $TFN_B(a_2, b_2, c_2, d_2)$, as follows.*

$$\varsigma_{A-B} = \begin{cases} \frac{(x-(a_1-d_2))^n}{(b_1-a_1)+(d_2-c_2)} & a_1 - d_2 \leqslant x \leqslant b_1 - c_2 \\ 1 & b_1 - c_2 \leqslant x \leqslant c_1 - b_2 \\ \frac{((d_1-a_2)-x)^n}{(d_1-c_1)-(b_2-a_2)} & c_1 - b_2 \leqslant x \leqslant d_1 - a_2 \\ 0 & otherwise \end{cases} \tag{3}$$

Definition 5. *[13] (Re-translation) In this step, the trapezoidal fuzzy number $\tilde{\delta}$ obtained during manipulation must be converted back into its corresponding ELICIT expression using the inverse function $\hbar : \tilde{\delta} \to x_{EL}$, as described below.*

(1) If $\tilde{\delta} = TFN(0, 0, c, d)$, then $\hbar(\tilde{\delta}) = atmost(s_i, \alpha)^\gamma$
(2) If $\tilde{\delta} = TFN(a, b, 1, 1)$, therefore $\hbar(\tilde{\delta}) = atleast(s_i, \alpha)^\gamma$
(3) If $\tilde{\delta} = TFN(a, b, c, d)$, then $\hbar(\tilde{\delta}) = between(s_i, \alpha_1)^{\gamma_1} and (s_j, \alpha_2)^{\gamma_2}$

Definition 6. *[13] Let x_{el1} and x_{el2} be two ELICIT expressions, and let $\widetilde{\beta_1} = Tr_1(a_1, b_1, c_1, d_1)$ and $\widetilde{\beta_2} = Tr_2(a_2, b_2, c_2, d_2)$ be their equivalent fuzzy numbers obtained from $\hbar^{-1}(x_{el1})$ and $\hbar^{-1}(x_{el2})$, respectively. Then, the distance between x_{el1} and x_{el2} can be determined as:*

$$dev\left(x_{el1}, x_{el2}\right) = d\left(\widetilde{\beta_1}, \widetilde{\beta_2}\right) = \sqrt{\frac{(a_1-a_2)^2 + (b_1-b_2)^2 + (c_1-c_2)^2 + (d_1-d_2)^2}{4}} \tag{4}$$

3 Bridging-Aware Trust-Based Consensus-Reaching Model

This section presents the proposed bridging-aware trust-based consensus model. First, the decision problem and consensus measurements are introduced to quantify disagreement at different levels. Then, a bridging index is developed to characterize the trust-mediated bridging contribution of DMs without relying on explicit community structures. On this basis, a bridging-aware influence regulation mechanism is designed to guide opinion adjustment during CRP.

3.1 Problem Description

A standard GDM problem considering social relationships can be formulated as follows. $A = \{a_i | i = 1, 2, ..., m\}$ denotes a set of m potential alternatives. $C = \{c_j | j = 1, 2, ..., n\}$ represents a set of n criteria. $E = \{e_k | k = 1, 2, ..., r\}$ indicates a group of DMs that provide evaluations. The weight of criteria is denoted as $w_j (j = 1, 2, ..., n)$ with $w_j \in (0, 1)$ and $\sum_{j=1}^{n} w_j = 1$. The ELICIT information is employed to describe DMs' evaluations based on a 7-level linguistic term set $S = \{s_0 =$ very bad, $s_1 =$ bad, $s_2 =$ slightly bad, $s_3 =$ medium, $s_4 =$ slightly good, $s_5 =$ good, $s_6 =$ very good$\}$. The assessments are organized into a matrix $X = [x_{ij}^k]_{m \times n}$, where x_{ij}^k denotes an ELICIT expression reflecting the evaluation made by e_k for alternative a_i regarding criterion c_j.

$$X^k = [x_{ij}^k]_{m\times n} = \begin{array}{c} \\ a_1 \\ a_2 \\ \vdots \\ a_m \end{array} \begin{array}{c} \begin{array}{cccc} c_1 & c_2 & \cdots & c_n \end{array} \\ \begin{pmatrix} x_{11}^k & x_{12}^k & \cdots & x_{1n}^k \\ x_{21}^k & x_{22}^k & \cdots & x_{2n}^k \\ \vdots & \vdots & \ddots & \vdots \\ x_{m1}^k & x_{m2}^k & \cdots & x_{mn}^k \end{pmatrix} \end{array} (k = 1, 2, \cdots, r) \tag{5}$$

A directed graph $G = (E, L)$ is utilized to model the social network among DMs, where E represents the set of DMs and L denotes the trust relationships among them. $T = [t_{kl}]_{r\times r}$ is the adjacency matrix of the network, where $t_{kl} \in [0, 1]$ represents the degree of trust from DM e_k to DM e_l, with larger values indicating stronger trust relationships.

3.2 Consensus Measurement and Bridging-Aware Regulation Framework

Definition 7. *Let x_{ij}^k and x_{ij}^l represent the individual assessments provided by e_k $(k = 1, 2, ..., r)$ and e_l $(l = 1, 2, ..., r)$ regarding a_i $(i = 1, 2, ..., m)$ on c_j $(j = 1, 2, ..., n)$, respectively. The consensus on evaluation level is calculated as:*

$$CL_{ij}^k = \frac{1}{r-1} \sum_{l=1, l\neq k}^{r} (1 - dev(x_{ij}^k, x_{ij}^l)), \tag{6}$$

where $dev(x_{ij}^k, x_{ij}^l)$ indicates the variation between the individual assessments x_{ij}^k and x_{ij}^l.

Definition 8. *The DM level consensus refers to the agreement of $e_k (k = 1, 2, ..., r)$ on the set of alternatives across all criteria, and it can be calculated as:*

$$CL^k = \frac{1}{mn} \sum_{i=1}^{m} \sum_{j=1}^{n} CL_{ij}^k, \tag{7}$$

where CL_{ij}^k represents the individual assessment consensus level.

Definition 9. *Building on Definitions 7 and 8, the group consensus level (GCL) can be derived as:*

$$GCL = \frac{1}{r} \sum_{k=1}^{r} CL^k, \tag{8}$$

where $GCL \in (0, 1)$, with a larger GCL representing stronger alignment within the group.

Subsequently, the GCL is compared with a predefined threshold μ. If $GCL \geq \mu$, alternative ranking is performed; otherwise, a trust-driven consensus mechanism is activated.

To support the trust-driven CRP, the bridging-aware influence regulation is introduced. Consensus measures at the DM and evaluation levels are then used to identify DMs and assessments contributing least to the GCL, defined as follows:

$$\Psi = \{k \,|CL^k < \mu\} \tag{9}$$

The evaluations from the DMs in Ψ that do not meet the threshold of consensus are identified as:

$$\Phi = \{(i,j,k) \,|k \in \Psi \wedge CL_{ij}^k < \mu\} \tag{10}$$

In trust-based CRP, DMs exert differentiated influence on consensus formation, and some play a bridging role by connecting heterogeneous opinions through trust relationships. This role is particularly important in fragmented decision environments where explicit or stable community structures cannot be assumed.

To characterize this trust-mediated bridging capability, a Bridging Index (BI) is introduced, capturing two key aspects: (i) the degree to which a DM is exposed to heterogeneous opinions through trust relationships, and (ii) the dispersion of its trust structure, which prevents influence from being confined to a narrow echo chamber.

Trust-Conditioned Opinion Heterogeneity. For DM e_k, the opinion heterogeneity perceived through its trust relationships is defined as

$$D_k^{(op)} = \sum_{l=1,l\neq k}^{r} t_{kl} \cdot \frac{1}{mn} \sum_{i=1}^{m} \sum_{j=1}^{n} dev(x_{ij}^k, x_{ij}^l), \tag{11}$$

where x_{ij}^k and x_{ij}^l denote the ELICIT evaluations provided by e_k and e_l for a_i under c_j, respectively, and $dev(\cdot,\cdot)$ represents the variation between two assessments. $D_k^{(op)}$ measures the extent to which DM e_k is exposed to heterogeneous viewpoints among the peers it trusts. A larger value indicates that the trusted peers of e_k hold more diverse opinions, which constitutes a necessary condition for bridging heterogeneous perspectives.

Trust Dispersion. To avoid overestimating bridging capability in cases where trust is highly concentrated, the dispersion of trust associated with DM e_k is determined as

$$D_k^{(tr)} = \sum_{l=1,l\neq k}^{r} \left|t_{kl} - \bar{t}_k\right|, \qquad \bar{t}_k = \frac{1}{r-1} \sum_{l=1,l\neq k}^{r} t_{kl}. \tag{12}$$

This term reflects how evenly the trust of e_k is distributed across its peers. A higher value of $D_k^{(tr)}$ indicates a more diversified trust structure, whereas a lower value suggests that trust is concentrated on a limited subset of DMs.

Bridging Index (BI). By jointly considering trust-conditioned opinion heterogeneity and trust dispersion, the BI of e_k can be calculated as

$$BI_k = D_k^{(op)} \cdot D_k^{(tr)}. \tag{13}$$

It is worth noting that the proposed BI differs fundamentally from purely topological centrality measures commonly used in social network analysis. In unweighted or binary networks, a node's bridging role is typically characterized by its position in the graph topology, such as lying on shortest paths between clusters. However, in trust-based GDM, influence propagation is inherently conditioned on both the strength and the distribution of trust relationships, as well as the actual opinion landscape perceived through those relationships. A DM who occupies a topologically central position but trusts only a narrow, like-minded subset of peers does not genuinely bridge heterogeneous viewpoints. By coupling opinion heterogeneity with trust dispersion, the proposed BI captures a trust-conditioned and content-aware notion of bridging that is specifically tailored to weighted directed trust networks arising in CRP. A DM exhibits strong bridging capability only when it is simultaneously exposed to heterogeneous opinions and maintains a diversified trust structure, ensuring that high-BI DMs are functionally positioned to integrate and transmit diverse viewpoints rather than merely reflecting topological prominence or local agreement.

Bridging-Aware Opinion Update Rule. Based on the proposed BI, a bridging-aware influence regulation mechanism is developed to guide opinion adjustment. Unlike conventional trust-based update rules relying solely on trust degrees or uniform averaging, the proposed mechanism regulates influence by prioritizing trusted DMs with stronger bridging capability.

Let Ψ denote the set of DMs whose consensus levels do not satisfy the predefined threshold μ, and let Φ represent the set of specific evaluations requiring adjustment. For each DM $e_k \in \Psi$ and each identified evaluation $(i, j, k) \in \Phi$, the opinion update at iteration $t+1$ is defined as

$$x_{ij}^{k,(t+1)} = (1 - \lambda_k)\, x_{ij}^{k,(t)} + \lambda_k \sum_{l=1}^{r} \omega_{kl}\, x_{ij}^{l,(t)}, \tag{14}$$

where $\lambda_k \in (0, 1)$ is an adjustment parameter controlling the degree to which DM e_k revises its opinion.

The influence weight ω_{kl} can be determined by both trust relationships and bridging capability as

$$\omega_{kl} = \frac{t_{kl}\, BI_l}{\sum_{q=1}^{r} t_{kq}\, BI_q}, \tag{15}$$

which satisfies $\sum_{l=1}^{r} \omega_{kl} = 1$.

This formulation ensures that opinion adjustment is guided toward peers who are not only trusted by DM e_k, but also exhibit stronger bridging capability in the group. Therefore, influence is regulated in a trust-consistent and structurally informed manner, rather than being uniformly or arbitrarily distributed.

The proposed CRP proceeds iteratively. After each round of opinion adjustment, the GCL is recalculated according to Definition 9, and the process terminates once the condition $GCL \geq \mu$ is satisfied. By regulating influence based on bridging capability, the iterative process progressively reduces excessive disagreement while preserving trust consistency and structural interpretability, thereby avoiding indiscriminate averaging and promoting alignment through structurally meaningful opinion references.

Alternative Ranking. Once consensus is achieved, the final group evaluation of alternative a_i under criterion c_j is obtained by aggregating individual assessments with differentiated influence weights, reflecting the distinct structural roles of DMs in the trust network.

$$\bar{x}_{ij} = \sum_{k=1}^{r} \alpha_k \, x_{ij}^k, \qquad i = 1, \ldots, m, \; j = 1, \ldots, n, \tag{16}$$

where $\alpha_k = \frac{BI_k}{\sum_{q=1}^{r} BI_q}$ denotes the aggregation weight of e_k and satisfies $\sum_{k=1}^{r} \alpha_k = 1$.

The overall performance of a_i is calculated as

$$V(a_i) = \sum_{j=1}^{n} w_j \, \bar{x}_{ij}, \tag{17}$$

where w_j denotes the importance of c_j and satisfies $\sum_{j=1}^{n} w_j = 1$.

Finally, alternatives are ranked in descending order of $V(a_i)$, and the overall framework of the proposed method is illustrated in Fig. 1.

4 Case Study: Offshore Floating Wind Turbine Risk Assessment

4.1 Problem Background

The floating offshore wind turbine is made up of interconnected subsystems, which face negative effects under harsh operating conditions [14]. Therefore, conducting risk analysis to mitigate system failures and facilitate maintenance is crucial. During the analysis, expert opinions are indispensable. Due to the inherent ambiguity in human cognition, employing linguistic models to assess qualitative risk factors more effectively captures the uncertainty associated with the evaluation information. After preliminary screening, six of the most critical failure modes $a_i (i = 1, 2, ..., 6)$ were selected for further analysis, as shown in Table 1. Seven risk factors $c_j (j = 1, 2, ..., 7)$ reflecting occurrence, detectability, and severity are collected in Table 2. A group of risk analysis experts, $e_k (k = 1, 2, ..., 20)$, were invited from the Goldwind Technology Floating Offshore Wind Turbine Innovation Base to assess the risk levels of failure modes concerning the seven risk factors using ELICIT information.

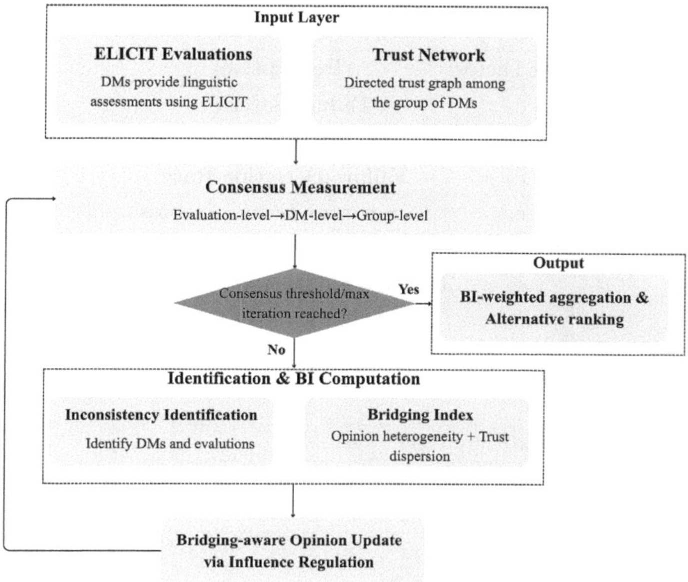

Fig. 1. The overall framework of the proposed method.

Table 1. Typical Failure Modes, Causes, and Impacts

Item	Failure Mode	Causes	Impacts
a_1	Crack in connector	Fatigue, corrosion	Potential collapse
a_2	Falling object impact	High winds, waves, biological collisions	System failure, shutdown
a_3	Seal failure	Pipe fatigue, welding failure	System failure
a_4	Pitch angle error	Improper calibration	Reduced power production efficiency
a_5	Yaw system leakage	Seal wear	Rotor failure
a_6	Anchor chain rupture	Winch failure, buoy friction chain wear	System shutdown

4.2 Implementation of the Proposed CRP

Given the inherent subjectivity of human judgment and the qualitative nature of evaluation criteria, assessments are represented using the 7-scale linguistic

Table 2. Risk Factors and Corresponding Evaluation Criteria

Risk Factor	Description
c_1	Failure Severity
c_2	Detection Difficulty
c_3	Failure Detection Rate
c_4	Component Economic Loss
c_5	Energy Production Loss
c_6	Damage to Other Components
c_7	Maintenance Effort

Table 3. Risk assessments given by e_1

	a_1	a_2	a_3	a_4	a_5	a_6
c_1	s_1	$bets_1\&s_2$	s_3	s_2	$bets_2\&s_3$	s_3
c_2	s_2	s_3	s_2	$bets_2\&s_3$	s_4	s_3
c_3	$atleasts_1$	s_3	s_2	s_3	$bets_1\&s_2$	s_4
c_4	s_3	$bets_1\&s_2$	s_4	s_0	s_4	s_2
c_5	s_4	s_3	s_1	$bets_1\&s_3$	s_2	s_5
c_6	$bets_3\&s_4$	s_5	s_0	s_2	$atleasts_4$	s_0
c_7	s_4	$atmosts_2$	s_2	s_3	s_3	s_1

term set $S = \{s_0 = \text{very bad}, s_1 = \text{bad}, s_2 = \text{slightly bad}, s_3 = \text{medium}, s_4 = \text{slightly good}, s_5 = \text{good}, s_6 = \text{very good}\}$. Table 3 presents the evaluations provided by e_1 for the failure modes across risk factors for brevity. Figure 2 illustrates the social network formed by DMs.

Consensus measures across assessment, individual, and group levels are determined using Eqs (6)-(8). Due to space limitations, the detailed calculation process is omitted. The initial GCL is calculated as 0.783, falling short of the predefined threshold of 0.85, which ensures alignment with the majority of DMs. To address this gap, the CRP model needs to be activated. According to the identification criteria, the following risk assessments are identified.

$$\Phi = \left\{ \begin{array}{l} (1,4,4), (3,6,4), (3,7,4), (1,7,5), (3,4,5), \\ (4,6,5), (5,2,5), (1,3,6), (2,5,6), (3,4,6), \\ (4,5,6), (2,4,9), (3,6,9), (5,2,9), (2,5,18), \\ (4,4,18), (3,7,18), (5,4,18), (6,7,18) \end{array} \right\} \tag{18}$$

Based on the proposed bridging-aware influence regulation mechanism, the identified DMs are guided to adjust their evaluations by selectively referring to trusted peers with stronger bridging capability. Specifically, the influence exerted by each trusted DM is jointly determined by the trust relationship and BI, which captures the DM's trust-mediated ability to bridge heterogeneous opinions in the

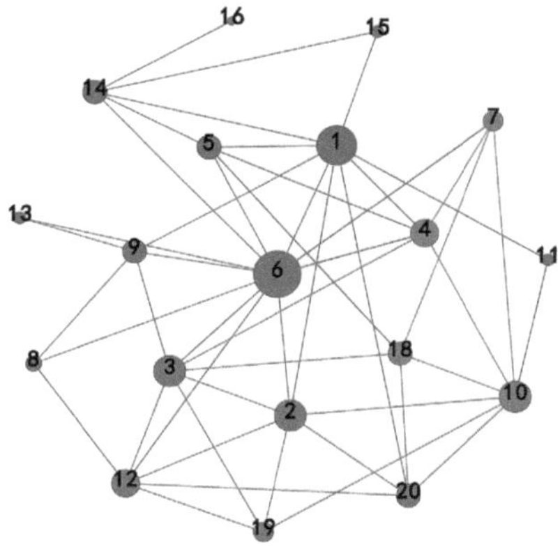

Fig. 2. Trust-based social network of the DMs.

network. Take the assessment (1, 4, 4) as an example, the modified ones can be computed as $bet(s_3, 0.314)^{-0.003} \& (s_4, 0.028)^{0.042}$ using Eq. (14).

Using the revised risk assessments, the three consensus levels are re-evaluated, yielding a GCL of 0.796 after the first iteration. Although improved, this value remains below the predefined threshold. By iteratively applying the proposed bridging-aware influence regulation mechanism, the GCL increases to 0.852 after another seven iterations, thereby satisfying the consensus requirement. The evolution of GCL during this process is shown in Fig. 3. The final group evaluations are reported in Table 4. Assuming equal importance for all risk factors, the aggregated results indicate that a_1 (Crack in connector) is the most critical failure mode.

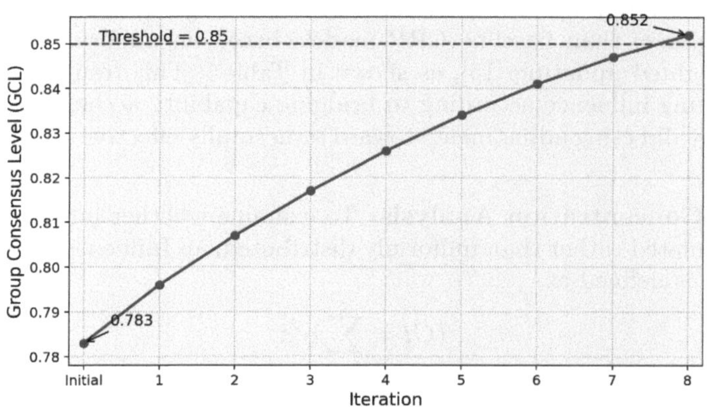

Fig. 3. The evolution of GCL during the CRP.

Table 4. The final group risk assessments and ranking

	Group risk assessments	Ranking
a_1	$bet(s_5, 0.067)^{-0.173}\&(s_6, 0.150)^{0.045}$	1
a_2	$bet(s_3, 0.055)^{0.003}\&(s_4, 0.108)^{0.017}$	3
a_3	$bet(s_1, 0.156)^{0.133}\&(s_2, 0.482)^{0.047}$	5
a_4	$bet(s_0, 0.002)^{0.012}\&(s_1, 0.045)^{0.237}$	6
a_5	$bet(s_2, -0.019)^{0.078}\&(s_3, 0.234)^{0.072}$	4
a_6	$bet(s_4, 0.223)^{0.342}\&(s_5, 0.047)^{0.014}$	2

4.3 Discussion and Comparative Analysis

To further evaluate the effectiveness of the proposed method, the discussion and comparative analysis focus on three aspects: consensus improvement efficiency, influence regulation characteristics, and opinion adjustment stability.

Consensus Improvement Efficiency. Consensus improvement efficiency reflects how rapidly a CRP achieves the predefined consensus threshold. It is defined as

$$\eta = \frac{GCL^{(T)} - GCL^{(0)}}{T}, \tag{19}$$

where $GCL^{(0)}$ and $GCL^{(T)}$ denote the initial and final GCLs, respectively, and T is the number of iterations required to satisfy $GCL \geq \mu$.

A larger value of η indicates faster convergence toward consensus. In this case study, the proposed method increases the GCL from 0.783 to 0.852 within eight iterations. The evolution of GCL across iterations is depicted in Fig. 3, which shows a rapid initial improvement followed by gradual convergence. This yields a higher η than baseline CRP models based on uniform averaging [14] or trust-weighted updating [15], as shown in Table 5. This result demonstrates that regulating influence according to bridging capability accelerates consensus formation by directing adjustments toward structurally effective opinion sources.

Influence Concentration Analysis. To examine whether influence is meaningfully regulated rather than uniformly distributed, an Influence Concentration Index (ICI) is defined as

$$ICI = \sum_{k=1}^{r} \alpha_k^2, \tag{20}$$

where α_k denotes the normalized aggregation weight of e_k, derived from the BI.

A higher ICI indicates that influence is concentrated on a smaller subset of DMs, while a lower value corresponds to near-uniform influence distribution. The proposed method yields a moderate ICI (shown in Table 5), indicating that influence is neither excessively centralized nor uniformly diffused. Instead, influence is selectively allocated to DMs with stronger trust-mediated bridging capability, which aligns with their structural role during the CRP. In contrast, the uniform averaging-based CRP produces very low ICI values, while the community-based approach [16] leads to excessive influence concentration within predefined clusters.

Opinion Adjustment Magnitude. To assess the stability and interpretability of CRP, the average opinion adjustment magnitude is defined as

$$\Delta = \frac{1}{|\Phi|} \sum_{(i,j,k)\in\Phi} dev\left(x_{ij}^{k,(T)}, x_{ij}^{k,(0)}\right), \tag{21}$$

where $x_{ij}^{k,(0)}$ and $x_{ij}^{k,(T)}$ denote the initial and final evaluations of e_k on a_i under c_j, respectively, and Φ is the set of identified evaluations.

Smaller values of Δ indicate that consensus is achieved through limited and targeted opinion modifications. The proposed method yields lower Δ than most of the baseline CRPs, suggesting that consensus improvement is realized through structurally informed influence regulation rather than drastic opinion shifts. This property enhances the interpretability and acceptability of the CRP in practical decision-making contexts.

Table 5. Comparison of Consensus Performance

Method	η	ICI	Δ
Uniform averaging CRP [14]	0.0048	0.050	0.087
Trust-weighted CRP [15]	0.0065	0.082	0.061
Community-based CRP [16]	0.0069	0.176	0.045
Proposed method	**0.00863**	**0.123**	0.048

The comparative results in Table 5 indicate that the proposed bridging-aware influence regulation mechanism achieves faster consensus improvement while maintaining a more balanced trade-off between influence concentration and opinion adjustment magnitude. Unlike community-based approaches, it does not rely on explicit cluster structures, which enhances its applicability in fragmented and dynamically evolving decision environments. Overall, the results suggest that explicitly regulating influence according to trust-mediated bridging capability enables efficient yet interpretable consensus formation in CRP.

Ablation Analysis. To further examine the contribution of each component in the proposed Bridging Index, an ablation analysis is conducted by comparing the full model with two degraded variants:

- **Variant 1 (Trust-only):** The BI is entirely removed from the influence weight. The weight ω_{kl} in Eq. (15) is replaced by $\omega_{kl} = t_{kl}/\sum_{q=1}^{r} t_{kq}$, so that opinion adjustment is guided solely by trust degrees without considering bridging capability.
- **Variant 2 (Heterogeneity-only):** The trust dispersion component $D_k^{(tr)}$ is removed, and the BI is replaced by $BI_k = D_k^{(op)}$. In this variant, bridging capability is measured only by trust-conditioned opinion heterogeneity, without accounting for how evenly a DM's trust is distributed.

Table 6. Ablation Analysis of the Bridging Index

Variant	η	ICI	Δ
Trust-only (no BI)	0.00655	0.081	0.066
Heterogeneity-only (no $D_k^{(tr)}$)	0.00762	0.152	0.054
Proposed method (full BI)	**0.00863**	**0.123**	**0.048**

Table 6 reports the results of the ablation analysis using the same case study setting. The results reveal distinct performance degradation patterns when individual components of the BI are removed. Variant 1 (Trust-only) achieves the lowest η (0.00655) among the three configurations, indicating that trust degrees alone are insufficient to guide opinion adjustments toward structurally effective sources, which substantially slows down convergence. Its ICI (0.081) is also the lowest, suggesting that without bridging-aware differentiation, influence distribution becomes nearly uniform and fails to prioritize DMs with meaningful structural roles. In addition, the highest Δ (0.066) implies that the lack of targeted influence regulation leads to broader and less efficient opinion modifications.

Variant 2 (Heterogeneity-only) improves η (0.00762) relative to Variant 1, confirming that incorporating opinion heterogeneity contributes to consensus efficiency. However, its ICI rises to 0.152, which is notably higher than the full model and approaches the level of community-based methods in Table 5. This over-concentration occurs because, without the trust dispersion filter, DMs who are exposed to diverse opinions but concentrate their trust on a narrow subset of peers receive inflated bridging scores, leading to disproportionate influence allocation. The elevated Δ (0.054) further indicates that such concentrated influence does not translate into more targeted adjustments.

The full BI consistently yields the best trade-off: the highest η (0.00863), the lowest Δ (0.048), and a moderate ICI (0.123) that avoids both excessive uniformity and over-concentration. This confirms that both opinion heterogeneity and trust dispersion are necessary components for characterizing the trust-mediated bridging capability of DMs in the proposed framework.

5 Conclusion

This paper proposes a bridging-aware trust-based consensus-reaching framework under linguistic uncertainty, in which DMs' influence is regulated according to their trust-mediated bridging capability rather than uniform averaging or predefined community structures. By integrating trust-conditioned opinion heterogeneity and trust dispersion into a unified BI, the proposed approach enables structurally informed and interpretable influence regulation throughout the CRP. Case study results demonstrate that the proposed method improves consensus efficiency while preserving stable and interpretable opinion adjustments.

Despite its effectiveness, the framework does not explicitly distinguish between conformity-driven adjustments and genuine belief-updating behaviors among DMs. Incorporating such behavioral considerations into influence regulation mechanisms represents a promising direction for future research to further enhance the realism and applicability of CRP models.

Acknowledgments. The authors thank the organizers of the FLINS-ISKE 2026, especially Professor Luis Martínez for organizing the special session on "Towards Responsible Intelligent Decision Support Systems." The authors also appreciate the constructive comments provided by the anonymous reviewers. This work is supported by the Research Development Fund (No. RDF-23-02-028), and Teaching Development Fund at Xi'an Jiaotong-Liverpool University (No. TDF2425-R29-266).

Disclosure of Interests. The authors have no competing interests to declare that are relevant to the content of this article.

References

1. Zhu, J., Xiao, X., Rao, C.: Quantum probabilistic group consensus decision-making model based on matrix fluctuation grey correlation for engineering bid evaluation. Expert Syst. Appl. **298**, 129600 (2026)
2. Hua, Z., Jing, X., Martínez, L.: An ELICIT information-based ORESTE method for failure mode and effect analysis considering risk correlation with GRA-DEMATEL. Inform. Fusion **93**, 396–411 (2023)
3. Liu, Z., Weng, Z., Letchmunan, S., Deveci, M., Huang, Y., Pamucar, D., Siarry, P.: Assessing sustainable energy systems using interval-valued Fermatean fuzzy sets-based decision-making model. Appl. Soft Comput. **186**, 114261 (2026)
4. Yan, S., Xu, Y., Gong, Z., Herrera-Viedma, E.: Fuzzy quantum group decision making and its application in meteorological disaster emergency. IEEE Trans. Fuzzy Syst. **33**, 1441–1454 (2025)
5. Hua, Z., Jing, X., Martínez, L.: Consensus reaching for social network group decision making with ELICIT information: A perspective from the complex network. Inf. Sci. **627**, 71–96 (2023)
6. García-Zamora, D., Labella, Á., Rodríguez, R.M., Martínez, L.: A linguistic metric for consensus reaching processes based on ELICIT comprehensive minimum cost consensus models. IEEE Trans. Fuzzy Syst. **31**, 1676–1688 (2023)

7. Zhou, Y., Zhou, L.: Dynamic adaptive consensus reaching process with risk attitude for multi-criteria group decision making. Inform. Fusion **127**, 103760 (2026)
8. Wang, Y., Zhan, J.: Group consensus formation under fuzzy preferences: Addressing hypocritical trust and irrational confidence. IEEE Trans. Fuzzy Syst. **33**, 3808–3822 (2025)
9. Zhang, H., Liu, S., Li, C.-C., Dong, Y., Chiclana, F., Herrera-Viedma, E.: Addressing the influence of limited tolerance and compromise behaviors on the social trust network consensus-reaching process. Inf. Sci. **686**, 121292 (2025)
10. Hua, Z., Gou, X., Martínez, L.: Dynamic clustering-based consensus model for large-scale group decision-making considering overlapping communities. Inform. Fusion **115**, 102743 (2025)
11. Teng, F., Liu, X., Liu, P., Wu, X.: Fuzzy opinion dynamics-driven consensus model within hybrid mesoscale structure in social trust network. IEEE Trans. Fuzzy Syst. **33**, 2295–2307 (2025)
12. Acemoglu, D., Ozdaglar, A.: Opinion dynamics and learning in social networks. Dynamic Games Appl. **1**, 3–49 (2011)
13. Romero, A.L., Rodríguez, R.M., Martínez, L.: Computing with comparative linguistic expressions and symbolic translation for decision making: ELICIT information. IEEE Trans. Fuzzy Syst. **28**, 2510–2522 (2020)
14. Yang, W., Zhang, L.: A multi-period intuitionistic fuzzy consensus reaching model for group decision making problem in social network. Compl. Intell. Syst. **10**, 7213–7234 (2024)
15. Mao, J., Gou, X., Hua, Z.: An improved Laplacian gravity centrality-based consensus method for social network group decision-making with incomplete ELICIT information. Mathematics **12**, 2013 (2024)
16. Ding, R.-X., Yang, B., Huang, Y., Zhang, Y., Chiclana, F.: Social network-based overlapping community clustering and feedback mechanism for large-scale group decision making. Eur. J. Oper. Res. **329**, 518–535 (2026)

The Ordered Weighted Average Budget

Ihab Alkasrawi(✉), José M. Merigó Lindahl, and Li Guan

School of Computer Science, Faculty of Engineering and Information Technology, University of Technology Sydney, 81 Broadway, Ultimo, NSW 2007, Australia
ihab.alkasrawi@student.uts.edu.au, {jose.merigo,li.guan}@uts.edu.au

Abstract. This study introduces the ordered weighted average budget (OWAB) operator, a novel aggregation framework that integrates objective financial data with hierarchical attitudinal preferences in budgeting. Building on classical ordered weighted averaging (OWA) operators and advances in fuzzy logic and generalized aggregation, OWAB provides a flexible mechanism for synthesizing budget components across multiple organizational levels and stakeholder perspectives. The framework encompasses traditional budget aggregation methods as special cases and extends them through generalized and quasi-arithmetic variants, enabling nuanced modeling of optimistic, pessimistic, and neutral attitudes in budget evaluation. Applications to multi-person and multi-country budget aggregation demonstrate the operator's ability to handle complex, multitiered decision environments. An empirical case study of Australian regional and international research and development (R&D) budgets illustrates the practical usefulness of OWAB in supporting transparent, consensus-oriented resource allocation. Taken together, the proposed framework advances both the theoretical foundations and practical methodologies of budget analysis in diverse economic and policy settings.

Keywords: Budget · averages · OWA operator · multi-person aggregation · multi-country analysis

1 Introduction

Budgeting is a foundational socio-technical process that aligns resource allocation with strategic organizational goals. However, traditional deterministic models often struggle to manage the uncertainty and diverse stakeholder preferences inherent in modern, multi-tiered institutions [2,18]. While traditional arithmetic methods offer transparency, they lack the flexibility required for complex decision-making contexts [8].

To address these limitations, researchers have leveraged ordered weighted averaging (OWA) and fuzzy set theory to better model subjective attitudes and data ambiguity [7,22]. Despite these advancements, a gap remains for a framework that merges objective financial data with hierarchical attitudinal preferences [9]. This paper introduces the ordered weighted average budget (OWAB)

J. Lu et al. (Eds.): FLINS-ISKE 2026, LNCS 16752, pp. 49–62, 2027.
https://doi.org/10.1007/978-981-92-2369-5_4

operator, a novel model for synthesizing fiscal data and stakeholder perspectives across organizational levels. The following sections detail the OWAB formulation, its generalized variants, and its practical application in government R&D budgeting.

2 Literature Review

Modern organizational strategy is fundamentally underpinned by budgeting, which functions as a multidimensional framework for prioritizing resources and managing institutional constraints. Beyond its fiscal role, budgeting operates as a socio-technical artifact that integrates governance mechanisms with stakeholder expectations and embedded cultural values [15]. In specialized domains such as healthcare, budget impact analyses are particularly critical for evaluating the economic feasibility of emerging technologies and ensuring that resource allocation remains aligned with evolving policy objectives [21].

Historically, resource allocation has relied on deterministic aggregation models, most notably arithmetic means and weighted averages, which have been favored for their computational simplicity and transparency [19,20]. Despite these advantages, such linear approaches often prove inadequate in dynamic, multi-agent environments characterized by data imprecision, uncertainty, and heterogeneous stakeholder preferences [8]. To address these challenges, recent scholarship has increasingly incorporated fuzzy set theory and linguistic decision making frameworks, enabling more nuanced representations of qualitative judgment and subjective uncertainty [6,14]. In parallel, artificial intelligence and machine-learning techniques have enhanced forecast accuracy and optimization capabilities, further improving decision support in complex financial ecosystems [3,17].

Within this evolving methodological landscape, aggregation theory has increasingly emphasized the Ordered Weighted Averaging (OWA) operator and its advanced extensions, including the Generalized OWA (GOWA) and OWA Weighted Average (OWAWA) operators. The original OWA formulation introduced by Yager (1988) [22] established the foundational framework, while subsequent developments expanded its attitudinal and generalized modeling capabilities [23,24]. These extensions provide a flexible mechanism for integrating objective fiscal data with subjective attitudinal considerations, thereby supporting consensus formation in multi-criteria and hierarchical decision environments [14,26]. Building on this line of research, the Ordered Weighted Average Budget (OWAB) operator represents a context-specific advancement that adapts these OWA principles to budgeting. By explicitly incorporating attitudinal preferences into fiscal aggregation, OWAB provides a structured, interpretable approach to synthesizing complex budgetary information across organizational levels.

3 Preliminaries

This section reviews core aggregation operators that form the theoretical foundation of the proposed Ordered Weighted Average Budget (OWAB) framework.

3.1 Average Budgets

Baseline financial synthesis typically relies on two primary operators. The Average Budget (AB) assigns uniform importance to all budget components B_i, providing a neutral benchmark commonly used in macro fiscal analysis and policy evaluation [4]. Formally, it is defined as

$$\mathrm{AB}(B_1, \ldots, B_n) = \frac{1}{n} \sum_{i=1}^{n} B_i. \tag{1}$$

To reflect heterogeneous levels of reliability, relevance, or strategic priority among budget items, the Weighted Average Budget (WAB) introduces a weight vector $\mathbf{w} = (w_1, \ldots, w_n)$, where $w_j \in [0, 1]$ and $\sum_{j=1}^{n} w_j = 1$. The operator is defined as

$$\mathrm{WAB}(\mathbf{B}) = \sum_{j=1}^{n} w_j \, B_j. \tag{2}$$

These linear aggregation models are widely employed in consensus-based control systems and resource allocation problems due to their transparency and ease of interpretation [20,27].

3.2 OWA Operators

Unlike linear aggregation methods, the Ordered Weighted Averaging (OWA) operator [22] assigns weights according to the ordered magnitudes of the input data rather than to fixed budget components. Given a vector of inputs $A = (a_1, \ldots, a_n)$, the values are reordered in descending order to obtain $\mathbf{B} = (b_1, \ldots, b_n)$, where b_j denotes the j-th largest element of A.

The OWA aggregation is defined as

$$\mathrm{OWA}(a_1, \ldots, a_n) = \sum_{j=1}^{n} w_j \, b_j, \tag{3}$$

where the weights satisfy $w_j \in [0, 1]$ and $\sum_{j=1}^{n} w_j = 1$. By adjusting the weight vector, the OWA operator enables decision-makers to model a wide range of attitudinal behaviors, from extreme optimism (maximum-oriented aggregation) to extreme pessimism (minimum-oriented aggregation).

3.3 Generalized and Quasi-OWA Variants

To enhance modeling flexibility, Yager (2004) [24] introduced the Generalized OWA (GOWA) operator by incorporating a power parameter $\lambda \neq 0$:

$$\mathrm{GOWA}(a_1, \ldots, a_n) = \left(\sum_{j=1}^{n} w_j \, b_j^{\lambda} \right)^{1/\lambda}. \tag{4}$$

This formulation allows different degrees of compensation among inputs, with larger values of λ emphasizing higher-ranked elements and smaller values increasing sensitivity to lower-ranked values.

Further generalization is achieved through the Quasi-OWA operator, which applies aggregation in a transformed domain using a strictly monotonic and invertible function g [5]:

$$\text{Quasi-OWA}(a_1, \ldots, a_n) = g^{-1}\left(\sum_{j=1}^{n} w_j\, g(b_j)\right). \tag{5}$$

Quasi-OWA formulations permit nonlinear compensation effects and are particularly valuable in decision environments where extreme values or contextual sensitivity must be emphasized [13].

3.4 Multi-person OWA (MP-OWA)

In collaborative decision-making contexts, aggregation must account for both individual attitudes and the relative importance of multiple experts. The Multi-Person OWA (MP-OWA) operator addresses this requirement by combining subjective evaluations across decision-makers [10].

The approach follows a two-stage process:

(i) an OWA operator aggregates each expert's evaluation vector to capture individual attitudinal preferences; and
(ii) the resulting values are aggregated using expert-specific importance weights ν_j, with $\nu_j \in [0, 1]$ and $\sum_{j=1}^{m} \nu_j = 1$:

$$\text{MP-OWA} = \sum_{j=1}^{m} \nu_j \left(\sum_{i=1}^{n} w_i\, b_i^{(j)}\right). \tag{6}$$

Merigó introduces a more sophisticated operator called the Multi-Person OWAIMAM (Index of Maximum and Minimum Level). Let X_j and Y_j be two sets of evaluations from decision-maker j. The MP-OWAIMAM operator is:

$$\text{MP-OWAIMAM}(X, Y) = \sum_{j=1}^{m} \nu_j \, \text{OWAIMAM}(X_j, Y_j), \tag{7}$$

where $\text{OWAIMAM}(X_j, Y_j)$ applies the OWA operator to the IMAM distance between the elements of X_j and Y_j.

4 The Ordered Weighted Average Budget (OWAB)

4.1 Theoretical Foundations

The Ordered Weighted Average Budget (OWAB) operator extends the classical Ordered Weighted Averaging (OWA) framework introduced by Yager (1988)

[22] to the aggregation of budgetary data. Unlike traditional linear aggregation methods, which assign weights directly to predefined budget categories regardless of their realized magnitudes, OWAB reorders budget components by their values prior to aggregation. This value-based ordering enables the explicit incorporation of decision-maker attitudes, such as optimism or pessimism, into the budgeting process, allowing aggregated outcomes to reflect institutional priorities and risk preferences more transparently.

Formally, let $\mathbf{B} = (B_1, B_2, \ldots, B_n)$ denote a vector of budget components, and let $b_1 \geq b_2 \geq \cdots \geq b_n$ represent the same components reordered in non-increasing order. The OWAB operator is defined as

$$\mathrm{OWAB}(B_1, \ldots, B_n) = \sum_{j=1}^{n} w_j \, b_j, \tag{8}$$

where the weight vector $\mathbf{w} = (w_1, \ldots, w_n)$ satisfies $w_j \in [0, 1]$ and $\sum_{j=1}^{n} w_j = 1$.

A defining feature of the OWAB operator is its attitudinal characteristic, measured by the *orness* index, originally formalized by Yager (1993) [23]:

$$\mathrm{Orness}(\mathbf{w}) = \frac{1}{n-1} \sum_{j=1}^{n} (n-j) \, w_j. \tag{9}$$

The orness measure quantifies the degree of optimism embedded in the aggregation process. A fully optimistic aggregation corresponds to Orness $= 1$, achieved when $w_1 = 1$ and all other weights are zero, yielding the maximum budget value. Conversely, a fully pessimistic aggregation corresponds to Orness $= 0$, attained when $w_n = 1$, yielding the minimum budget value. A neutral attitude is represented by uniform weights $w_j = 1/n$, resulting in Orness $= 0.5$ and reducing OWAB to the arithmetic mean.

Consistent with the theoretical properties of OWA operators, OWAB satisfies several desirable mathematical characteristics. It is *idempotent*, meaning that if all budget components are equal, the aggregated result equals that common value. It is *commutative*, ensuring invariance under permutations of the input vector, and *monotonic*, such that an increase in any budget component cannot decrease the aggregated result. Finally, OWAB is *bounded*, guaranteeing that the aggregate lies between the minimum and maximum input values [22, 23].

4.2 Specialized Families of OWAB Operators

To accommodate the complexity of real-world budgeting environments, a number of specialized OWAB-based operators can be defined.

The Probabilistic Budget OWA (PBOWA) integrates probability information into the weighting structure, allowing aggregation to reflect the likelihood of different fiscal outcomes. By associating ordered budget components with probability weights, PBOWA provides an effective mechanism for modeling budgetary uncertainty in stochastic or forecast-driven contexts [10].

In situations where the numerical magnitude of budget components is not the primary ordering criterion, the Induced Budget OWA (IBOWA) offers greater flexibility. In this variant, the ordering of budget items is determined by an external inducing variable, such as political urgency, perceived risk, or social impact, rather than by fiscal size alone. This induced ordering framework originates from induced OWA operators introduced by Yager and Filev (1999) [25] and enables context-dependent prioritization in policy-driven budgeting processes.

The Generalized Budget OWA (GBOWA) extends OWAB by introducing a compensation parameter $\lambda \neq 0$, yielding

$$\text{GBOWA} = \left(\sum_{j=1}^{n} w_j \, b_j^{\lambda} \right)^{1/\lambda}. \tag{10}$$

By adjusting λ, decision-makers can modulate sensitivity to extreme budget values: larger values of λ emphasize higher components, while smaller values give greater influence to lower components. This formulation is grounded in the generalized OWA framework proposed by Yager (2004) [24].

Finally, the Hybrid Budget OWA (HBOWA) addresses scenarios in which both objective importance and subjective attitude must be represented simultaneously. HBOWA combines classical weighted averaging, capturing the intrinsic importance of budget categories such as mandatory versus discretionary expenditures, with ordered weighting that reflects attitudinal preferences. This hybridization yields a comprehensive aggregation tool capable of balancing structural constraints and decision-maker risk tolerance, consistent with generalized hybrid aggregation models [11,13]. Its general formulation is as follows:

$$\text{HBOWA}(A) = \sum_{i=1}^{n} v_i \, w_i \, b_i, \tag{11}$$

where v_i indicates the relative significance of each budget attribute (such as mandatory versus discretionary expenditures), and w_i represents the decision-maker's risk preferences or attitude toward risk.

5 Average Budgets with Generalized and Quasi-Arithmetic Means

Traditional methods for combining budgets, such as the arithmetic mean or weighted average, are often used because they are simple to understand. However, these methods may not fully capture the complexities of real-world budgeting, especially when decision-makers seek to prioritize certain values or reflect specific attitudes toward risk. To address these limitations, researchers have proposed using generalized and quasi-arithmetic means in aggregation processes, offering greater flexibility and adaptability in modeling [11,14].

5.1 Generalized Ordered Weighted Average Budget (GOWAB)

The Generalized Ordered Weighted Average Budget (GOWAB) is an extension of the classical OWAB that incorporates generalized means. Let $a = (a_1, a_2, \ldots, a_n)$ be a set of budget components, and $\mathbf{w} = (w_1, \ldots, w_n)$ a weighting vector such that $\sum_{j=1}^{n} w_j = 1$. The GOWAB is defined as

$$\text{GOWAB}_p(a) = \left(\sum_{j=1}^{n} w_j \, b_j^p\right)^{1/p}, \tag{12}$$

where b_j is the j-th most significant element of a, and $p \in R$ is a parameter that determines the type of mean:

- $p = 1$: arithmetic mean (standard OWAB).
- $p = 2$: quadratic mean.
- $p \to 0$: geometric mean.
- $p = -1$: harmonic mean.

By using this formulation, the aggregation process can emphasize different aspects of the distribution. For example, higher values of p place more weight on larger budget components, whereas lower values place more weight on smaller budget components. This flexibility is helpful in situations where risk sensitivity or strategic emphasis is needed [11].

5.2 Quasi-Ordered Weighted Average Budget (Quasi-OWAB)

The Quasi-Ordered Weighted Average Budget (Quasi-OWAB) operator builds on the GOWAB by replacing the power function with a strictly monotonic and continuous function g. It is defined as

$$\text{Quasi-OWAB}_g(a) = g^{-1}\left(\sum_{j=1}^{n} w_j \, g(b_j)\right). \tag{13}$$

This formulation includes the GOWAB as a special case when $g(x) = x^p$ and allows for even greater modeling flexibility.

5.3 Particular Cases

- **Quadratic OWAB** ($p = 2$):

$$\text{Quadratic OWAB}(a) = \left(\sum_{j=1}^{n} w_j \, b_j^2\right)^{1/2}. \tag{14}$$

- **Harmonic OWAB** ($p = -1$):

$$\text{Harmonic OWAB}(a) = \left(\sum_{j=1}^{n} w_j \, b_j^{-1} \right)^{-1}. \tag{15}$$

- **Geometric OWAB** ($p \to 0$ or $g(x) = \log(x)$):

$$\text{Geometric OWAB}(a) = \prod_{j=1}^{n} b_j^{w_j}. \tag{16}$$

These operators enable analysts to tailor the aggregation process to the specific characteristics of the budget data and the decision-making context, thereby creating a more nuanced and robust framework for budget analysis.

6 Multi-person and Multi-level Aggregation with OWAB Operators

6.1 Multi-person Aggregation

Budgeting decisions in complex organizational settings often require aggregating information from multiple experts or stakeholders, each with distinct perspectives, expertise levels, and risk attitudes. Classical aggregation mechanisms, such as arithmetic or weighted averages, provide transparent summaries but often fail to capture the interaction between individual influence and collective attitudinal behavior. To address this shortcoming, ordered aggregation frameworks have been extended to multi-person environments, allowing attitudes such as optimism or conservatism to be explicitly modeled [16,22,23].

Multi-person aggregation operators combine individual budget evaluations into a single representative outcome by accounting for both stakeholder importance and aggregation attitude [10]. The simplest formulation is the Multi-Person Average Budget (MP-AB), which treats all participants symmetrically and computes the arithmetic mean of their inputs. While intuitive, this approach implicitly assumes homogeneous expertise and influence, an assumption rarely tenable in practice [12].

To accommodate heterogeneity among decision-makers, the Multi-Person Weighted Average Budget (MP-WAB) assigns importance weights to individuals, thereby reflecting differences in authority, credibility, or relevance. This structure is widely applied in collaborative planning and forecasting contexts where expert hierarchy plays a central role [12].

Moving beyond fixed individual weights, the Multi-Person Ordered Weighted Average Budget (MP-OWAB) operator incorporates attitudinal aggregation by ordering participants' inputs before weighting. Rooted in the foundational OWA framework [22,23], MP-OWAB enables group-level attitudes, such as consensus-seeking, optimism, or pessimism, to be encoded via the ordered weighting vector.

This flexibility is particularly valuable under uncertainty, where collective risk preferences influence final budget outcomes [11].

Further generalization is achieved through the Multi-Person Generalized Ordered Weighted Average Budget (MP-GOWAB), which integrates generalized means into the ordered aggregation process. By introducing a parameter that controls the degree of compensation among ordered inputs, MP-GOWAB supports a broad range of collective behaviors, encompassing arithmetic, geometric, and harmonic aggregation regimes. Such flexibility allows the modeling of diverse group attitudes toward dominance, compromise, and risk in multi-stakeholder budgeting contexts [11,13,24].

6.2 Multi-country and Multi-regional Budget Aggregation

The extension of OWAB-based aggregation to hierarchical environments enables coherent budget synthesis across multiple organizational levels, including countries, regions, enterprises, and individual actors. A comprehensive framework for such multi-level aggregation systems has been formalized by Merigo (2015) [12], integrating arithmetic, weighted, and ordered operators within a unified structure.

Within this framework, the Average Budget (AB) and Weighted Average Budget (WAB) operators provide baseline aggregation mechanisms, while OWAB introduces attitudinal flexibility by weighting ordered budget positions. The Generalized OWAB (GOWAB) further enriches this structure through parametric control of compensation effects, enabling arithmetic, geometric, or quadratic aggregation within the same formal setting. These operators can be systematically adapted to multi-country (MC), multi-region (MR), and combined MC–MR contexts.

At the highest level of complexity, aggregation systems such as the Multi-Country, Multi-Enterprise, Multi-Person OWA integrate structural weights across hierarchical units with attitudinal weights over ordered outcomes. This architecture supports transparent, adaptable, and context-sensitive budget analysis for large-scale organizations and international policy environments [12].

7 Case Study: Regional R&D Budget Aggregation in Australia

7.1 Theoretical Framework

Effective allocation of research and development (R&D) budgets across higher education, business, and government sectors requires aggregation methods capable of integrating heterogeneous financial data and stakeholder preferences. This need is particularly pronounced in federated systems such as Australia, where budget decisions are taken at both national and regional levels, including New South Wales (NSW), Victoria (VIC), Queensland (QLD), Western Australia (WA), South Australia (SA), Tasmania (TAS), the Australian Capital Territory (ACT), and the Northern Territory (NT).

Conventional aggregation approaches, such as arithmetic and weighted averages, provide transparent benchmarks but often fail to capture uncertainty, sectoral dominance, and attitudinal considerations. To address these limitations, this study employs advanced aggregation operators Minimum Budget (Min-B), Maximum Budget (Max-B), Average Budget (AB), Weighted Average Budget (WAB), and Ordered Weighted Average Budget (OWAB), which enable flexible modeling of pessimistic, neutral, and optimistic perspectives [12,22,23].

7.2 Data and Methodology

Regional R&D expenditure data for higher education, business, and government sectors were obtained from the Australian Bureau of Statistics [1]. For each region, the sectoral budget vector

$$\mathbf{B} = (b_1,\, b_2,\, b_3) \tag{17}$$

is aggregated using the following operators:

- **Min-B:** $\min\{b_1, b_2, b_3\}$, representing a pessimistic assessment.
- **Max-B:** $\max\{b_1, b_2, b_3\}$, representing an optimistic assessment.
- **AB:** $\frac{1}{3}(b_1 + b_2 + b_3)$, providing a neutral benchmark.
- **WAB:** $\sum_{i=1}^{3} w_i\, b_i$, incorporating sectoral importance.
- **OWAB:** $\sum_{i=1}^{3} w_i\, b_{(i)}$, where ordered values and attitudinal weights capture decision-maker preferences [23].

A moderately optimistic attitude is modeled using the weight vector $\mathbf{w} = (0.5,\, 0.3,\, 0.2)$.

7.3 Illustrative Example: New South Wales

Let the sectoral budgets be: higher education b_1 = 4,398, business b_2 = 7,501, and government b_3 = 958 (all values in USD millions). The maximum and minimum budgets have been obtained from the Australian Bureau of Statistics (ABS), while AB, WAB, and OWAB have been calculated by the author in Table 1.

- The minimum budget is:

$$\text{Min-B} = \min\{4{,}398;\ 7{,}501;\ 958\} = 958. \tag{18}$$

- The maximum budget is:

$$\text{Max-B} = \max\{4{,}398;\ 7{,}501;\ 958\} = 7{,}501. \tag{19}$$

- The average budget is:

$$AB = \frac{4{,}398 + 7{,}501 + 958}{3} = 4{,}285.67. \tag{20}$$

- The weighted average budget, using sector weights $\mathbf{w} = (0.3758, 0.5080, 0.1162)$, is:

$$WAB = 0.3758 \times 4{,}398 + 0.5080 \times 7{,}501 + 0.1162 \times 958 = 1{,}652.20 + 3{,}813.09 + 109.40 = 5{,}574.69. \tag{21}$$

- The ordered weighted average budget, with ordered values $7{,}501 \geq 4{,}398 \geq 958$, is:

$$OWAB = 0.5 \times 7{,}501 + 0.3 \times 4{,}398 + 0.2 \times 958 = 3{,}750.50 + 1{,}319.40 + 191.60 = 5{,}261.50. \tag{22}$$

7.4 Aggregated Regional Results

Table 1. Aggregated R&D Budget Results for Australia's Eight Regions (USD millions).

Region	Max-B	Min-B	AB	WAB	OWAB
NSW	7,501.00	958.00	4,285.67	5,574.69	5,261.50
VIC	5,948.00	1,099.00	3,713.00	4,687.15	4,421.40
QLD	2,560.00	763.00	1,865.00	2,243.01	2,114.20
WA	1,690.00	308.00	1,035.00	1,310.34	1,238.70
SA	946.00	578.00	729.67	760.51	788.10
TAS	285.00	155.00	219.00	244.34	238.60
ACT	860.00	192.00	479.33	465.60	584.20
NT	98.00	68.00	81.00	80.32	85.70

7.5 Interpretation and Policy Implications

The results demonstrate substantial regional heterogeneity in Australian R&D investment. NSW and VIC consistently dominate across all operators, driven primarily by strong business sector expenditures. TAS and NT exhibit considerably lower values, reflecting smaller and more evenly distributed R&D systems.

Comparisons among AB, WAB, and OWAB illustrate how aggregation attitude influences interpretation. While AB provides a neutral reference, WAB increases values in regions aligned with sectoral priority weights. OWAB yields intermediate outcomes that favor dominant sectors without disregarding secondary contributions, offering a balanced, policy-relevant synthesis. These findings indicate that OWAB is particularly suitable for regional R&D evaluation under uncertainty, supporting transparent, consensus-driven resource allocation.

8 Conclusions

8.1 Theoretical Contributions

This paper introduces the Ordered Weighted Average Budget (OWAB) operator as a unified framework for budget aggregation. A significant theoretical contribution of this work is the demonstration that the OWAB operator subsumes classical indicators—specifically, the minimum, maximum, arithmetic mean, and traditional weighted average—as special cases corresponding to distinct attitudinal settings. This consolidation enables a coherent interpretation of traditional budget measures within a single, consistent aggregation paradigm.

The framework explicitly integrates objective budget data with subjective decision-maker attitudes through fixed attitudinal weights determined by the Orness measure. This architectural choice clearly distinguishes OWAB from argument-dependent OWA (DOWA) operators: while DOWA weights vary as a function of input values, OWAB maintains interpretability and attitudinal control through a stable weighting vector. Furthermore, the introduction of GOWAB and Quasi-arithmetic OWAB families allows analysts to modulate compensation effects and sensitivity to extreme fiscal values. The hierarchical variants developed (MC-OWAB and MR-OWAB) confirm that the framework preserves essential mathematical properties—monotonicity, boundedness, and idempotency—across multi-actor and multi-level decision environments.

8.2 Practical Implications

From an applied perspective, the OWAB framework serves as a transparent and flexible decision-support tool for budgeting in public policy, research management, and organizational planning. By allowing explicit representation of optimistic, neutral, and pessimistic attitudes, OWAB supports robust evaluation of resource-allocation strategies under fiscal uncertainty.

The empirical analysis of Australian regional R&D data (e.g., the New South Wales case study) illustrates the utility of the operator for benchmarking regional performance and identifying structural imbalances across the higher education, business, and government sectors. Operationally, OWAB can be embedded into institutional decision-support systems to enhance the traceability and justification of budget decisions, particularly in settings where multiple stakeholders hold competing interests.

8.3 Future Research Directions

Future work should focus on extending the OWAB framework through induced-aggregation mechanisms, where external factors such as policy priorities or risk drive the ordering process. Additionally, the development of time-dependent operators for multi-period planning and the integration of the framework with machine learning forecasting models represent significant opportunities for growth. Finally, the creation of comprehensive Budgeting Aggregation Systems that combine prediction, stakeholder negotiation, and OWAB aggregation will further advance evidence-based, transparent budgeting practices in complex economic landscapes.

Acknowledgments. The authors gratefully acknowledge the support of the School of Computer Science, Faculty of Engineering and Information Technology at the University of Technology Sydney, which provided an excellent academic environment for conducting this research. The authors also express their appreciation to colleagues and reviewers for their valuable comments and constructive suggestions, which have helped improve the quality and clarity of this paper. Any errors or omissions remain the sole responsibility of the author.

Disclosure of Interests. The authors declare that they have no competing interests. The authors have no financial or non-financial conflicts of interest to disclose that could have influenced the work reported in this paper.

Declaration of Generative AI and AI-Assisted Technologies in the Manuscript Preparation Process. An advanced large language model was employed in the editing phase to improve grammar and clarity. The author independently verified and approved all scientific content to guarantee its accuracy and integrity, taking full responsibility for the manuscript.

References

1. Australian Bureau of Statistics: Research and experimental development, government and private non-profit organisations, Australia. ABS (2022–2023), https://www.abs.gov.au/statistics/industry/technology-and-innovation/research-and-experimental-development-government-and-private-non-profit-organisations-australia/2022-23
2. Brahim Belhaouari, S., Bensmail, H., Mehrdoust, F.: Topics in mathematics and applications. In: Discrete Mathematics, Probability Theory and Stochastic Processes: For Applications in Engineering and Computer Science, pp. 1–97. Springer (2025)
3. Capone, C., Talgat, S., Hazir, O., Abdrasheva, K., Kozhakhmetova, A.: Artificial intelligence models for predicting budget expenditures. Eurasian J. Econ. Bus. Stud. **68**(1), 32–43 (2024)
4. Cebula, R.J., Boylan, R.: An empirical analysis for the US of the impact of federal budget deficits and the average effective personal income tax rate on the ex post real interest rate yield on ten-year treasuries. PSL Q. Rev. **72**(288), 41–52 (2019)
5. Fodor, J., Marichal, J.L., Roubens, M.: Characterization of the ordered weighted averaging operators. IEEE Trans. Fuzzy Syst. **3**(2), 236–240 (1995). https://doi.org/10.1109/91.388176
6. Hussain, W., Merigó, J.M.: Onsite/offsite social commerce adoption for SMEs using fuzzy linguistic decision making in complex framework. J. Ambient. Intell. Humaniz. Comput. **14**(9), 12875–12894 (2023)
7. Kacprzyk, J., Yager, R., Merigo, J.: Towards human-centric aggregation via ordered weighted aggregation operators and linguistic data summaries: a new perspective on Zadeh's inspirations. IEEE Comput. Intell. Mag. **14**(1), 16–30 (2019)

8. Kar, R., Le, D.N., Mukherjee, G., Mallik, B.B., Shaw, A.K.: Fuzzy Logic Applications in Computer Science and Mathematics. John Wiley & Sons (2023)
9. Libório, M.P., Ekel, P.I., Laudares, S., Martins, C.: A goal-based weighting for composite indicators constructed through ordered weighted averaging (OWA) operator. MethodsX **12**, 102575 (2024)
10. Merigó, J.M.: Probabilities in the OWA operator. Expert Syst. Appl. **39**(13), 11456–11467 (2012)
11. Merigó, J.M., Gil-Lafuente, A.M.: The induced generalized OWA operator. Inf. Sci. **179**(6), 729–741 (2009)
12. Merigó, J.M., Palacios-Marques, D., Ribeiro-Navarrete, B.: Aggregation systems for sales forecasting. J. Bus. Res. **68**(11), 2299–2304 (2015)
13. Merigó, J.M., Yager, R.R.: Generalized moving averages, distance measures and OWA operators. Internat. J. Uncertain. Fuzziness Know. Based Syst. **21**(4), 533–559 (2013)
14. Merigó, J.M., Zhou, L.G., Yu, D.J., Alrajeh, N., Alnowibet, K.: Probabilistic OWA distances applied to asset management. Soft. Comput. **22**(15), 4855–4878 (2018)
15. Neumair, J.: The quantitative research on government budgeting: a literature review on local culture budgets. Urban Aff. Rev. (2025). https://doi.org/10.1177/10780874251357499
16. O'Hagan, M.: Aggregating template or rule antecedents in real-time expert systems with fuzzy set logic. In: Proceedings of the Twenty-Second Asilomar Conference on Signals, Systems and Computers. vol. 2, pp. 681–689. IEEE (1988)
17. Radhika, E.G., Sadasivam, G.S.: Budget optimized dynamic virtual machine provisioning in hybrid cloud using fuzzy analytic hierarchy process. Expert Syst. Appl. **183**, 115398 (2021)
18. Sergi, D., Sari, I.U., Senapati, T.: Extension of capital budgeting techniques using interval-valued Fermatean fuzzy sets. J. Intell. Fuzzy Syst. **42**(1), 365–376 (2022)
19. Tavassoli, N., Moorman, C.: The end of averages for marketing budgets. MIT Sloan Manag. Rev. **65**(1), 1–3 (2023)
20. Wang, L., Xi, J.X., Yu, Z.Y., Liu, X., Liu, G.B.: Limited-budget finite-time average consensus design for multi-agent systems. IET Control Theory Appl. **14**(15), 2197–2204 (2020)
21. Wenzel, A., Roeder, J., Poos, T., Dusse, F., Kron, F.: Economic effects of next-generation sequencing diagnostics in unspecific sepsis patients: a budget impact analysis from the healthcare providers' perspective in Germany. Eur. J. Clin. Microbiol. Infect. Dis. **43**(12), 2311–2321 (2024)
22. Yager, R.R.: On ordered weighted averaging aggregation operators in multicriteria decision-making. IEEE Trans. Syst. Man Cybern. **18**(1), 183–190 (1988)
23. Yager, R.R.: Families of OWA operators. Fuzzy Sets Syst. **59**(2), 125–148 (1993)
24. Yager, R.R.: Generalized OWA aggregation operators. Fuzzy Optim. Decis. Making **3**(1), 93–107 (2004)
25. Yager, R.R., Filev, D.P.: Induced ordered weighted averaging operators. IEEE Trans. Syst. Man Cybern. Part B (Cybern.) **29**(2), 141–150 (1999)
26. Yan, H., Ma, T., Huynh, V.: On qualitative multi-attribute group decision making and its consensus measure: a probability-based perspective. Omega Int. J. Manag. Sci. **70**, 94–117 (2017)
27. Zhang, H., Zheng, X.Y., Xia, Y., Li, M.Q.: Workflow scheduling in the cloud with weighted upward-rank priority scheme using random walk and uniform spare budget splitting. IEEE Access **7**, 60359–60375 (2019)

The Ordered Weighted Average Gross Domestic Product

Hadeel S. Baraheem[1,2(✉)] and José M. Merigó[1]

[1] School of Computer Science, Faculty of Engineering and Information Technology, University of Technology Sydney, 81 Broadway, Ultimo, NSW 2007, Australia
Jose.Merigo@uts.edu.au

[2] Department of Information System, Faculty of Computing and Information Technology, King Abdulaziz University, Jeddah, Saudi Arabia
hbaraheem@kau.edu.sa

Abstract. Aggregation methods for Gross Domestic Product (GDP) often rely on risk-neutral assumptions and therefore may fail to capture asymmetric policy concerns, where decision-makers may weight downside risks and upside potential differently. This paper applies the Ordered Weighted Averaging (OWA) operator originally introduced by Yager (1988) to the analysis of GDP and introduces the Ordered Weighted Average Gross Domestic Product (OWAGDP) as a preference-sensitive aggregation framework. The approach generates a family of GDP aggregates ranging from worst-case (extreme pessimism) to best-case (extreme optimism), with the arithmetic mean emerging as a special case corresponding to attitudinal neutrality. The paper further extends the framework to generalized and quasi-arithmetic forms, as well as to the multi-person OWAGDP operator, in order to aggregate heterogeneous GDP estimates from multiple experts in non-linear environments. An illustrative example using the European Union GDP per inhabitant data shows that different attitudinal weights produce distinct GDP distributions. This highlights how the OWAGDP framework provides a more flexible, preference-sensitive aggregation approach for decision-making under uncertainty.

Keywords: Gross domestic product · averages · OWA operator · multi-person aggregation · multi-country analysis

1 Introduction

The Gross Domestic Product (GDP) is widely recognised as the primary indicator for measuring the economic performance of nations, regions, and industrial sectors [6,13]. Since its inception, it has served as a fundamental tool for policymakers, governments, and financial institutions to assess economic health, allocate resources, and formulate fiscal policies. The need for GDP aggregation arises frequently in practice: when combining GDP across regions or sectors,

J. Lu et al. (Eds.): FLINS-ISKE 2026, LNCS 16752, pp. 63–77, 2027.
https://doi.org/10.1007/978-981-92-2369-5_5

summarising projections across scenarios, or consolidating forecasts from multiple institutions. Traditionally, analysts rely on classical statistical techniques such as the arithmetic mean or the weighted average for this purpose [9]. While these methods are computationally simple and intuitive, they share a major limitation: they are attitudinally neutral. They assume a risk-neutral position that does not account for the subjective attitudes of decision-makers, nor do they distinguish between favourable and adverse economic scenarios [1].

Yet in modern global economics, policymaking is rarely neutral. Governments and financial analysts routinely assess economic scenarios under varying degrees of risk, and economic uncertainty itself has measurable consequences for output and growth [3,11]. For instance, a central bank planning for financial stability may wish to analyse a "pessimistic" scenario, focusing on underperforming sectors or regions to establish a safety baseline. Conversely, an investment board seeking to attract foreign capital might favour an "optimistic" aggregation that highlights potential capacity and high-growth opportunities. Standard aggregation operators cannot encode this attitudinal character, motivating the use of more flexible aggregation methods [12].

To address this limitation, the field of computational intelligence offers advanced aggregation operators. One of the most successful techniques is the Ordered Weighted Averaging (OWA) operator, introduced by Yager (1988) [23]. Unlike the weighted average, which weights values based on their source (e.g., the size of a country), the OWA operator weights values based on their position in the ordered set of data. This reordering process allows the decision-maker to emphasise either the highest values (optimism) or the lowest values (pessimism), thereby integrating the decision-maker's attitudinal character directly into the computation. While the OWA operator has been successfully applied to health [4], sales forecasting [15], and supply chain [20], its application to macroeconomic indicators such as GDP remains largely unexplored.

The objective of this paper is to formulate the Ordered Weighted Average Gross Domestic Product (OWAGDP) for GDP aggregation and analysis. The paper adapts OWA-family operators to the GDP setting in order to summarise GDP values under different attitudinal character. Through this framework, policymakers can generate a family of GDP values ranging from the minimum (extreme pessimism) to the maximum (extreme optimism), with the traditional average appearing as merely one special case. In this way, the analysis of GDP is interpreted not only as a static reporting tool but also as a preference-sensitive decision support tool.

This paper makes three contributions. First, it formulates the OWAGDP in the context of GDP and identifies its basic properties, including boundedness, monotonicity, and idempotency. Second, it considers generalized and quasi-arithmetic extensions, as well as a multi-person formulation for combining GDP estimates or forecasting from multiple experts. Third, it provides an illustrative example using the European Union (EU) GDP data to demonstrate how different attitudinal weightings affect the resulting aggregated values. This paper should therefore be viewed as an application of established aggregation operators to the analysis of GDP rather than as a demonstration of new theoretical results.

The remainder of the paper is organised as follows. Section 2 reviews the related literature. Section 3 presents the preliminaries on average GDP and OWA operators. Section 4 describes the OWAGDP framework and discusses its properties. In Sect. 5, we extend the framework to generalized and quasi-arithmetic forms, followed by the multi-person aggregation model in Sect. 6. Section 7 provides an illustrative example using EU GDP data, with a sensitivity analysis, and Sect. 8 concludes the paper by outlining implications, limitations, and directions for future research.

2 Literature Review

Gross Domestic Product (GDP) has long been the dominant indicator for measuring economic performance at the national, regional, and sectoral levels [6,13]. One underexplored issue concerns how GDP values should be aggregated when multiple observations are available. This situation arises frequently in practice: aggregating GDP across regions or sectors, summarising GDP projections across scenarios, or consolidating forecasts from multiple institutions. In such cases, analysts typically rely on neutral aggregation rules such as the arithmetic mean, or the weighted average [9]. These methods implicitly assume risk neutrality and symmetry with respect to high and low values, even though economic decision-making often involves asymmetric concerns, such as emphasising downside risks or highlighting upside potential. This gap motivates the need for aggregation tools that retain GDP's interpretability while allowing for preference-sensitive summaries.

Meanwhile, the literature on GDP estimation and forecasting has expanded rapidly in recent years. Traditional econometric approaches have been complemented, and in some cases replaced, by machine learning and data-driven models that can capture nonlinear relationships and exploit large, heterogeneous datasets. For example, Agu et al. (2022) [2] and Wu et al. (2021) [21] employ neural networks and hybrid learning approaches to predict GDP growth, whereas Ye et al. (2022) [25] leverage patent and innovation data to improve forecasting performance. Recent contributions also emphasise spatial and scenario-based GDP estimation. Xu et al. (2023) [22] develop gridded GDP estimates using ensemble learning techniques, allowing GDP to be mapped at fine geographic resolution. Other studies construct alternative GDP scenarios under different macroeconomic assumptions, policy interventions, or shock conditions [19]. These approaches generate sets of GDP values rather than a single point estimate, reflecting uncertainty, alternative futures, or expert disagreement and therefore create a natural aggregation problem that standard neutral averaging methods are not equipped to handle with attitudinal sensitivity.

Ordered Weighted Averaging (OWA) operators, introduced by Yager (1988) [23], offer a flexible aggregation framework that explicitly incorporates attitudinal character such as optimism and pessimism. By assigning weights to ordered positions rather than to data sources, OWA operators generate a family of aggregation measures ranging from the minimum to the maximum, with the arithmetic mean as a particular case. Subsequent research extended the OWA framework in several directions. Generalized OWA (GOWA) operators incorporate power means, enabling control over sensitivity with respect to large and small values [14,24]. Quasi-OWA operators replace power functions with general monotonic transformations, allowing greater flexibility [8]. These operators have been widely applied in multi-criteria decision-making, information fusion, and evaluation problems.

In the socio-economic domain, OWA-based methods have been used to aggregate indicators such as inflation [7], sustainable development goals [18], taxation [10], energy supply [5], and human development index [16], among others. These studies demonstrate that ordered weighted aggregation can meaningfully incorporate policy attitudes while remaining interpretable. However, despite this extensive literature, GDP aggregation itself has received little attention from the OWA perspective. None of the existing studies explicitly integrates OWA aggregation with GDP; they either focus on GDP estimation or apply OWA to other socio-economic domains. In particular, there is a lack of an explicit formulation and application of OWA operators for GDP aggregation across scenarios, experts, or regions. This gap motivates the OWAGDP framework proposed in this paper.

3 Preliminaries

This section outlines the fundamental concepts used throughout the paper.

3.1 Average GDPs

The Gross Domestic Product (GDP) represents the monetary value of all finished goods and services made within a country during a specific period. When analysing a group of countries or regions, the most common aggregation methods are the arithmetic mean and the weighted average. Let $G = \{g_1, g_2, ...g_n\}$ denote a set of GDP values for n distinct regions or countries. The Average GDP (AGDP) is defined as:

$$AGDP(g_1, g_2, ..., g_n) = \frac{1}{n}\sum_{i=1}^{n} g_i. \tag{1}$$

Meanwhile, the Weighted Average GDP (WAGDP) allows for the assignment of importance to each region (e.g., based on population size or geographic area). It is defined as:

$$WAGDP(g_1, g_2, ..., g_n) = \sum_{i=1}^{n} v_i g_i, \tag{2}$$

where v_i denotes the importance of the i-th region, such that $v_i \geq 0$ for all i and $\sum_{i=1}^{n} v_i = 1$.

3.2 OWA Operators

The Ordered Weighted Averaging (OWA) operator was introduced by Yager (1988) [23] to provide a parameterised family of aggregation operators that include the maximum, the minimum, and the average as special cases.

An OWA operator of dimension n aggregates a set of input value $\{a_1, a_2, ...a_n\}$. It has an associated weighting vector $W = (w_1, w_2, ...w_n)$ of dimension n, where $w_j \in [0, 1]$ and $\sum_{j=1}^{n} w_j = 1$, defined as follows:

$$OWA(a_1, a_2, ...a_n) = \sum_{j=1}^{n} w_j b_j, \tag{3}$$

where b_j denotes the j-th largest value in the input set.

A fundamental characteristic of the OWA operator is the reordering step. In the context of GDP, this implies that the weights are not associated with a specific country (as in WAGDP) but rather with the magnitude of the GDP value (e.g., the highest GDP, the second-highest, etc.).

3.3 Generalized OWA Operators

The Generalized OWA (GOWA) operator, introduced by Yager (2004) [24], adds a parameter λ to the aggregation, generalizing the arithmetic mean to include geometric, harmonic, and quadratic means. A GOWA operator of dimension n is defined as:

$$GOWA(a_1, a_2, ..., a_n) = \left(\sum_{j=1}^{n} w_j {b_j}^{\lambda} \right)^{1/\lambda}, \tag{4}$$

where $\lambda \in (-\infty, \infty)$ is a parameter such that $\lambda \neq 0$.

Furthermore, Fodor et al. (1995) [8] introduced the Quasi-OWA operator, which replaces the power function with a continuous and monotonic function f:

$$QuasiOWA(a_1, a_2, ..., a_n) = f^{-1} \left(\sum_{j=1}^{n} w_j f(b_j) \right). \tag{5}$$

4 Ordered Weighted Average GDP

This section introduces the Ordered Weighted Average Gross Domestic Product (OWAGDP) operator and its basic properties. This section adapts the OWA aggregation mechanism to the analysis of GDP, allowing the GDP values to be aggregated across different scenarios, regions, or expert assessments while explicitly incorporating the decision-maker's attitudinal character. By doing so, the OWAGDP generalizes traditional average GDP measures and provides a flexible framework ranging from pessimistic to optimistic economic evaluations.

4.1 Definition

Let g_i denote GDP values corresponding to n economic units, scenarios, or regions, and let $W = (w_1, w_2, ...w_n)$ be an ordered weighting vector such that $w_i \geq 0$ and $\sum_{i=1}^{n} w_i = 1$. The Ordered Weighted Average GDP (OWAGDP) operator is defined as:

$$OWAGDP(g_1, g_2, ...g_n) = \sum_{i=1}^{n} w_i g_{(i)}, \tag{6}$$

where $g_{(i)}$ is the i-th largest GDP value in the ordered set $\{g_1, g_2, ...g_n\}$.

The choice of the weighting vector W reflects the aggregation attitude. An optimistic aggregation assigns higher weights to larger GDP values, emphasising economic potential and growth leaders. A pessimistic aggregation assigns higher weights to smaller GDP values, emphasising economic vulnerability and downside risk. Neutral attitudes are represented by approximately uniform weights.

A widely used measure to quantify this attitudinal character is the orness measure [23], defined as:

$$Orness(w) = \frac{1}{n-1} \sum_{i=1}^{n} (n-i) w_i. \tag{7}$$

An orness value close to 1 indicates an optimistic aggregation, whereas a value close to 0 indicates a pessimistic one. An orness of 0.5 corresponds to a neutral, average-type aggregation. The OWAGDP operator includes several classical GDP aggregation measures depending on the weighting vector W. It reduces to the Arithmetic Mean GDP when weights $w_i = \frac{1}{n}$ for all i, the Maximum GDP when $w_1 = 1$, and the Minimum GDP when $w_n = 1$.

4.2 Properties

The following properties are inherited from the classical OWA operator and are stated here in the GDP setting.

1. **Boundedness:** The OWAGDP always produces a value between the minimum and maximum GDP observations.

$$\min(g_1, g_2, ..., g_n) \leq OWAGDP(g_1, g_2, ..., g_n) \leq \max(g_1, g_2, ..., g_n)$$

2. **Monotonicity:** If $g_i \leq g_i'$ for all i, then

$$OWAGDP(g_1, g_2, ..., g_n) \leq OWAGDP(g_1', g_2', ..., g_n')$$

3. **Idempotency:** If all GDP values are equal, $g_i = g$, then

$$OWAGDP(g_1, g_2, ..., g_n) = g$$

5 Average GDPs with Generalized and Quasi-arithmetic Means

5.1 Generalized Ordered Weighted Average GDP (GOWAGDP)

The Generalized Ordered Weighted Average (GOWA) operator introduces the parameter λ which controls the type of mean used in the aggregation [24]. By integrating the GOWA operator into GDP analysis, we obtain the Generalized Ordered Weighted Average GDP (GOWAGDP). Let g_i be a set of GDP values and let $W = (w_1, w_2, ...w_n)$ be an ordered weighting vector such that $w_i \geq 0$ and $\sum_{i=1}^{n} w_i = 1$.

$$GOWAGDP(g_1, g_2, ..., g_n) = \left(\sum_{i=1}^{n} w_i g_{(i)}^{\lambda} \right)^{1/\lambda}, \quad \lambda \neq 0. \tag{8}$$

The parameter λ controls the aggregation behaviour. Larger values of λ increase the influence of higher GDP values, while smaller values emphasise lower GDP values. This allows decision-makers to jointly model attitude (through the OWA weights) and non-linearity (via λ). The following are notable special cases:

- The Quadratic OWAGDP ($\lambda = 2$), which emphasises large values
- The standard OWAGDP ($\lambda = 1$)
- The Geometric ($\lambda \to 0$)
- The Harmonic ($\lambda = -1$) averages.

5.2 Quasi-ordered Weighted Average GDP (Quasi-OWAGDP)

While the GOWAGDP relies on power means, the Quasi-OWA framework allows for even greater flexibility by replacing the power function with a strictly monotonic and continuous function [8,14]. This leads to the Quasi-Ordered Weighted Average GDP (Quasi-OWAGDP) operator. Formally, let f be a continuous and strictly monotonic function.

$$Quasi - OWAGDP(g_1, g_2, ..., g_n) = f^{-1} \left(\sum_{i=1}^{n} w_i f(g_{(i)}) \right). \tag{9}$$

This formulation generalizes both the OWAGDP and the GOWAGDP operators. By selecting different functions $f(\cdot)$, a wide range of aggregation behaviours can be obtained, allowing the analyst to capture non-linear economic relationships, risk attitudes, or diminishing returns. The Quasi-OWAGDP encompasses several particular cases depending on the choice of the generating function f. Notable cases include the standard OWAGDP ($f(x) = x$), the Geometric average ($f(x) = \ln(x)$), the Harmonic average ($f(x) = x^{-1}$), and the Quadratic average ($f(x) = x^2$).

6 Aggregation Systems with Multiple Experts

In real-world economic analysis, GDP estimates are often produced by multiple experts, institutions, or forecasting models. These sources may rely on different assumptions, methodologies, or scenario constructions, which can lead to heterogeneous GDP estimates. Aggregating such information requires methods that not only combine multiple inputs but also account for differences in expert importance and decision-maker preferences. To address this issue, this section introduces the multi-person ordered weighted average GDP (MP-OWAGDP) operator and its extensions. These operators are designed to aggregate GDP estimates provided by multiple experts across different scenarios, while incorporating attitudinal character such as optimism or pessimism.

6.1 The Multi-person Ordered Weighted Average GDP (MP-OWAGDP)

The MP-OWAGDP operator extends the OWAGDP operator to accommodate GDP inputs provided by multiple experts. When only one expert is considered, the MP-OWAGDP reduces to the standard OWAGDP operator.

Let us consider m experts providing GDP estimates across n scenarios for a given country, region, or economic unit. Let g_{ij} represent the GDP estimate provided by expert j for scenario i, where $i = 1, 2, ..., n$ and $j = 1, 2, ..., m$. Each expert is assigned a degree of importance α_j, satisfying $\alpha_j \geq 0$ and $\sum_{j=1}^{m} \alpha_j = 1$. The MP-OWAGDP operator is defined as:

$$MP - OWAGDP(G) = \sum_{i=1}^{n} w_i \tilde{g}_{(i)}, \tag{10}$$

where $G = \{\tilde{g}_1, \tilde{g}_2, ..., \tilde{g}_n\}$ is the set of aggregated GDP estimates across experts for each scenario, w_i are the OWA weights satisfying $w_i \geq 0$ and $\sum_{i=1}^{n} w_i = 1$, and $\tilde{g}_{(i)}$ denotes the i-th largest value of $\tilde{g}_i$ after ordering. The aggregated GDP estimate for each scenario is obtained as:

$$\tilde{g}_i = \sum_{j=1}^{m} \alpha_j g_{ij}. \tag{11}$$

This formulation allows expert opinions to first be combined using their degrees of importance, after which the OWAGDP operator is applied across scenarios to reflect the decision-maker's attitudinal character.

Similarly, the multi-person generalized OWAGDP (MP-GOWAGDP) operator incorporates a parameter λ to allow for generalized means:

$$MP - GOWAGDP(G) = \left(\sum_{i=1}^{n} w_i \tilde{g}_{(i)}^{\lambda} \right)^{1/\lambda}. \tag{12}$$

This operator jointly models expert importance, scenario ordering, and sensitivity to large versus small GDP values.

7 Illustrative Example

This section demonstrates how the proposed OWAGDP framework can be applied in practice, using the EU as an example. This example combines multiple expert opinions and multiple scenario assumptions into a single preference-sensitive aggregated GDP value.

Step 1: Collect Expert Estimates. We consider 27 EU countries and three experts (optimistic, neutral, and pessimistic) providing GDP per inhabitant estimates across seven scenarios. Scenario 4 serves as the baseline anchored in Eurostat 2024 data, while the remaining scenarios represent alternative macroeconomic conditions ranging from Scenario 1 (extremely good), characterised by favourable productivity and demand, to Scenario 7 (extremely bad), reflecting sustained adverse shocks. Scenarios 2–3 represent "Very good" and "Good" conditions, while Scenarios 5–6 cover "Bad" and "Very bad" outlooks, respectively. To construct the numerical scenario values transparently, let $g^{(0)}$ be the Eurostat 2024 baseline for a country. We set Scenario 4 for each expert as: neutral $= g^{(0)}$, optimistic $= 1.02g^{(0)}$, pessimistic $= 0.98g^{(0)}$. Scenarios 1 to 3 and 5 to 7 are then generated as steps around each expert's Scenario 4 (i.e., +6%, +4%, +2%, 0%, -2%, -4%, -6%).

Step 2: Assign Expert Importance. We assign degrees of importance to experts 1 to 3 using $(\alpha_1, \alpha_2, \alpha_3) = (0.35, 0.45, 0.20)$, where $\sum \alpha_j = 1$. These expert importances, as well as the OWA weights applied in Step 4, were chosen for illustrative purposes to demonstrate clear optimistic and pessimistic aggregation bounds. In a real-world policy application, these weights would be elicited directly from the risk profiles of decision-makers, central banks, or specific institutional mandates.

Step 3: Compute Collective Scenario Values. For each country and scenario, we compute the collective value $\tilde{g}_i = \sum_{j=1}^{3} \alpha_j g_{ij}$.

Step 4: Compute Aggregated GDP Measures. Using the collective scenario values for each country, we compute the following summary measures. The aggregated results reported in Table 1 illustrate how the GDP changes across all member states, demonstrating how the aggregated GDP values respond to shifts in the decision-maker's attitudinal character weighting (from strictly pessimistic to strictly optimistic):

- Max: maximum value (best-case across scenarios).
- Min: minimum value (worst-case across scenarios).
- AGDP: arithmetic mean across the seven scenarios.
- WAGDP: weighted average of the seven collective scenario values using fixed, scenario-specific probability weights $p = (0.04, 0.08, 0.16, 0.28, 0.22, 0.14,$

0.08), with $\sum_{i=1}^{7} p_i = 1$. Unlike OWA weights, these are assigned to named scenarios without any ordering, reflecting the empirical likelihood of each macroeconomic state rather than the decision-maker's attitudinal stance. The mild downside skew is consistent with probability-weighted scenario frameworks used in central bank fan-chart practice [1].

- OWAGDP: ordered weighted aggregation reflecting an attitudinal character. We use the OWA weight vector $w = (0.20, 0.18, 0.20, 0.16, 0.12, 0.08, 0.06)$, $\sum_{i=1}^{7} w_i = 1$. The associated orness is 0.6167, which is greater than 0.5, indicating an optimistic weighting.
- QOWAGDP: quadratic extension (a particular case of the generalized framework) computed using the same weights w.
- OWAGDP*: OWAGDP computed using the reversed OWA weights $w^* =$ (0.06, 0.08, 0.12, 0.16, 0.20, 0.18, 0.20) with orness = 0.3833, which is less than 0.5, indicating a pessimistic weighting attitudinal character.

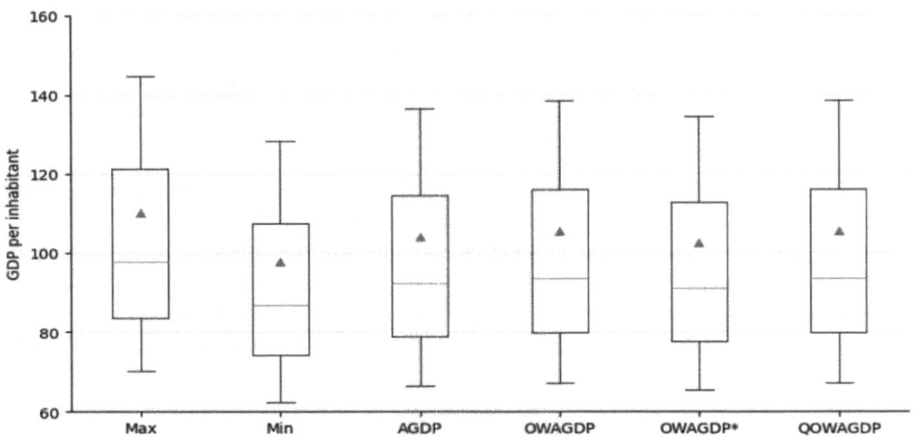

Fig. 1. Distribution of aggregated GDP values across EU countries. Note: Outlier points are not displayed in the boxplots for readability, but all observations are included in the calculations.

The boxplot in Fig. 1 summarises how the distribution of GDP per inhabitant across EU countries changes when different aggregation attitudes are applied to the same set of collective scenario values. As expected, Max produces the highest central tendency and upper spread because it selects the best-case scenario for each country, while Min shifts the distribution downward by focusing on worst-case outcomes. The AGDP (arithmetic mean) lies between these extremes and provides a neutral benchmark. Comparing OWA-based measures highlights the role of attitudinal weighting. OWAGDP (with optimistic weights) lies above AGDP, indicating that placing more weight on higher-scenario outcomes increases the aggregated GDP per inhabitant for many countries. In

Table 1. GDP Aggregation Results under Different Methods.

Country	Max	Min	AGDP	WAGDP	OWAGDP	QOWAGDP	OWAGDP*
Belgium	123.33	109.37	116.35	115.65	117.98	118.05	114.72
Bulgaria	70.17	62.23	66.20	65.80	67.12	67.17	65.27
Czechia	96.75	85.80	91.27	90.73	92.55	92.61	90.00
Denmark	136.09	120.68	128.38	127.61	130.18	130.26	126.59
Germany	122.27	108.42	115.35	114.65	116.96	117.03	113.73
Estonia	83.99	74.48	79.24	78.76	80.35	80.40	78.13
Ireland	224.33	198.94	211.63	210.36	214.60	214.73	208.67
Greece	74.42	66.00	70.21	69.79	71.19	71.24	69.23
Spain	97.81	86.74	92.28	91.72	93.57	93.63	90.98
France	105.25	93.34	99.30	98.70	100.69	100.75	97.91
Croatia	81.87	72.60	77.23	76.77	78.31	78.36	76.15
Italy	104.19	92.40	98.29	97.70	99.67	99.73	96.92
Cyprus	101.00	89.57	95.29	94.71	96.62	96.68	93.95
Latvia	75.49	66.94	71.21	70.79	72.21	72.26	70.22
Lithuania	93.56	82.97	88.26	87.73	89.50	89.56	87.03
Luxembourg	257.29	228.16	242.73	241.27	246.12	246.28	239.33
Hungary	81.87	72.60	77.23	76.77	78.31	78.36	76.15
Malta	115.89	102.77	109.33	108.67	110.86	110.93	107.80
Netherlands	144.59	128.22	136.41	135.59	138.32	138.40	134.50
Austria	123.33	109.37	116.35	115.65	117.98	118.05	114.72
Poland	83.99	74.48	79.24	78.76	80.35	80.40	78.13
Portugal	87.18	77.31	82.25	81.75	83.40	83.45	81.09
Romania	82.93	73.54	78.23	77.76	79.33	79.38	77.14
Slovenia	96.75	85.80	91.27	90.73	92.55	92.61	90.00
Slovakia	79.74	70.71	75.23	74.77	76.28	76.33	74.17
Finland	109.51	97.11	103.31	102.69	104.76	104.82	101.86
Sweden	120.14	106.54	113.34	112.66	114.93	115.00	111.75

contrast, OWAGDP* (computed with reversed weights) shifts the distribution downward relative to OWAGDP, reflecting a more cautious emphasis on lower scenario outcomes.

7.1 Sensitivity Analysis to Attitudinal Parameters

A brief sensitivity analysis is conducted to illustrate how the OWAGDP framework responds to the chosen attitudinal parameter settings. We isolate a single country (Germany) and analyse how its aggregated GDP measures respond to changes in the attitudinal character. Figure 2 plots the main aggregated measures reported in Table 1 together with their corresponding orness values.

As shown in Fig. 2, the aggregated value ranges from the worst-case lower bound (Min = 108.42) up to the best-case upper bound (Max = 122.27). The figure also shows how different attitudinal settings produce distinct aggregated outcomes relative to the neutral baseline (AGDP = 115.35 at an orness of 0.50). The pessimistic weighting (OWAGDP*) pulls the evaluation below the neutral mean, yielding 113.73 at an orness of 0.38. Conversely, the optimistic weighting (OWAGDP) pushes the evaluation above the neutral mean, yielding 116.96 at an orness of 0.62. This figure indicates that OWAGDP changes consistently with the decision-maker's risk attitude. This analysis is illustrative and focuses on a representative country; a broader cross-country sensitivity analysis is left for future research.

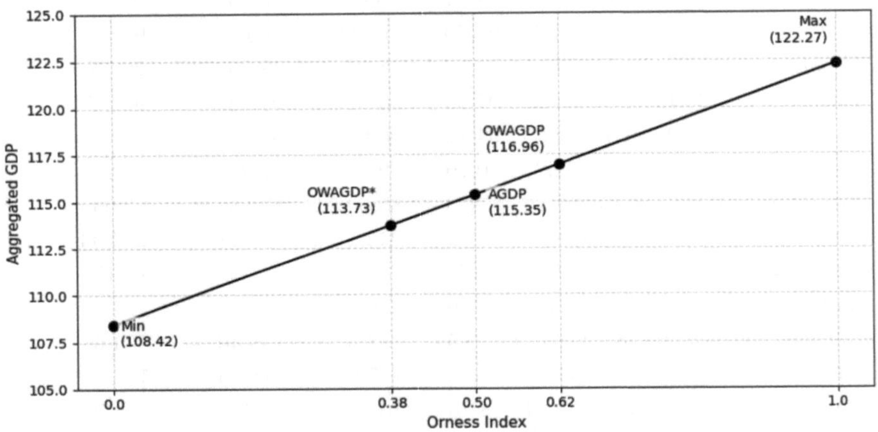

Fig. 2. Sensitivity analysis: The impact of varying the orness measure on the aggregated GDP measures for Germany.

8 Conclusions

8.1 Methodological Contribution

This paper presents the Ordered Weighted Average Gross Domestic Product (OWAGDP) as a framework for analysing GDP based on OWA-family aggregation operators. In particular, when the weights are uniform (i.e., $w_i = 1/n$) OWAGDP reduces to the arithmetic mean, which corresponds to a neutral attitude (orness = 0.5). The framework enables the aggregation to reflect the decision-maker's attitudinal character through the orness measure. The paper also discusses generalized, quasi-arithmetic, and multi-person extensions in the same setting.

We showed that the GOWAGDP allows for the parameterisation of non-linear averaging behaviours, encompassing geometric, harmonic, and quadratic aggregations, controlled by the parameter λ. Additionally, the Quasi-OWAGDP provides a unified structure that utilises continuous monotonic functions to capture complex economic behaviours, such as diminishing returns or specific utility preferences. The contribution lies primarily in domain-specific adaptation and interpretation, rather than in a new operator with new theorems.

8.2 Practical Implications

OWAGDP provides a transparent tool for policy settings in which GDP information is routinely communicated, but decisions are made under uncertainty. Rather than reporting only a neutral average, institutions can report a small set of complementary aggregation operators and their special cases (e.g., Min, AGDP, OWAGDP, OWAGDP*) that summarise the same scenario information under different attitudes.

In scenario planning and dynamic reporting, the OWAGDP framework allows institutions to move beyond a single GDP figure by providing a range of aggregated GDP values under different attitudinal character (optimistic, neutral, pessimistic) tailored to different strategic needs. Central banks can use pessimistic weights to simulate worst-case scenarios for stress testing, investment boards can apply optimistic weights to identify growth ceilings, and statistical offices can serve users with varying risk profiles, all from the same underlying scenario data.

The MP-OWAGDP proposed in Sect. 6 further supports consensus building by providing a structured method for aggregating heterogeneous forecasts from diverse experts. This is particularly valuable for supranational organisations (such as the EU or the OECD) that must reconcile conflicting economic projections from different member states or agencies.

8.3 Open Questions for Future Research

While this paper presents the OWAGDP framework, several avenues for future research remain open. First, the illustrative example relies on simulated expert elicitation to demonstrate mathematical mechanics rather than reflecting actual policy consensus. Future applications should therefore incorporate structured elicitation from real economic stakeholders to fully explore the practical utility of this approach. Second, the current study uses point estimates of GDP, but future research should extend the framework to handle uncertain environments by using interval numbers, fuzzy logic, or probabilistic information, thereby allowing the aggregation of ranges rather than precise values.

There are several methodological extensions that should be considered. A promising extension would be Induced OWAGDP (IOWAGDP), where the reordering is induced by a secondary variable, such as inflation rates, unemployment figures, or debt-to-GDP ratios, rather than the GDP value itself. This would enable complex policy analyses, such as aggregating GDPs conditional on high unemployment, rather than solely on the magnitude of GDP. Future work could address complex interdependencies between economic units. Since the standard OWAGDP assumes independence among input values, integrating Bonferroni means, for example, would allow researchers to model synergies and correlations between different economic regions. Finally, although this paper focuses on GDP, the methodology is equally applicable to other macroeconomic indicators such as trade balances and human development indices, offering a broader toolkit for economic policy analysis.

Acknowledgments. The study used EU 2024 GDP data, available on Eurostat.

Disclosure of Interests. The authors declare that they have no known competing interests that could have influenced the work reported in this paper.

Declaration of Generative AI and AI-Assisted Technologies in the Manuscript Preparation Process. During the preparation of this work, the author used ChatGPT to improve the readability and language of the work. After using this tool, the author reviewed and edited the content as needed and takes full responsibility for the content of the published article.

References

1. Adrian, T., Boyarchenko, N., Giannone, D.: Vulnerable growth. Am. Econ. Rev. **109**(4), 1263–1289 (2019). https://doi.org/10.1257/aer.20161923
2. Agu, S.C., Onu, F.U., Ezemagu, U.K., Oden, D.: Predicting gross domestic product to macroeconomic indicators. Intell. Syst. Appl. **14**, 200082 (2022). https://doi.org/10.1016/j.iswa.2022.200082
3. Ahir, H., Bloom, N., Furceri, D.: The world uncertainty index. NBER Working Paper 29763, National Bureau of Economic Research (2022). https://ideas.repec.org/p/nbr/nberwo/29763.html
4. Ahmad, S., Alnowibet, K., Alqasem, L., Merigó, J.M., Zaindin, M.: Generalized OWA operators for uncertain queuing modeling with application in healthcare. Soft. Comput. **25**(6), 4951–4962 (2021). https://doi.org/10.1007/s00500-020-05507-1
5. Badea, A.C., Rocco, C.M., Tarantola, S., Bolado, R.: Composite indicators for security of energy supply using ordered weighted averaging. Reliab. Eng. Syst. Saf. **96**(6), 651–662 (2011). https://doi.org/10.1016/j.ress.2010.12.025
6. Cohen Kaminitz, S.: The significance of GDP: a new take on a century-old question. J. Econ. Methodol. **30**(1), 1–14 (2023). https://doi.org/10.1080/1350178X.2023.2167228
7. Espinoza-Audelo, L.F., León-Castro, E., Olazabal-Lugo, M., Merigó, J.M., Gil-Lafuente, A.M.: Using ordered weighted average for weighted averages inflation. Int. J. Inf. Technol. Decis. Making **19**(2), 601–628 (2020). https://doi.org/10.1142/S0219622020500066
8. Fodor, J., Marichal, J.L., Roubens, M.: Characterization of the ordered weighted averaging operators. IEEE Trans. Fuzzy Syst. **3**(2), 236–240 (1995). https://doi.org/10.1109/91.388176
9. Genre, V., Kenny, G., Meyler, A., Timmermann, A.: Combining expert forecasts: can anything beat the simple average? Int. J. Forecast. **29**(1), 108–121 (2013). https://doi.org/10.1016/j.ijforecast.2012.06.004

10. León-Castro, E., Blanco-Mesa, F., Hussain, W., Flores-Sosa, M., Pérez-Arellano, L.A.: Tax revenue measurement using OWA operators. Cybern. Syst. **55**(1), 230–244 (2024). https://doi.org/10.1080/01969722.2022.2110687
11. Liu, N., Gao, F.: The world uncertainty index and GDP growth rate. Financ. Res. Lett. **49**, 103137 (2022). https://doi.org/10.1016/j.frl.2022.103137
12. Luukka, P., Stoklasa, J.: Similarity based fuzzy TOPSIS with OWA: reflecting risk attitudes in multiple-criteria and multi-expert evaluation under uncertainty. Comput. Indus. Eng. **204**, 111081 (2025)
13. MacFeely, S., van de Ven, P., Peltola, A.: To GDP and beyond: the past and future history of the world's most powerful statistical indicator. Stat. J. IAOS **40**(3), 685–711 (2024). https://doi.org/10.3233/SJI-240003
14. Merigó, J.M., Gil-Lafuente, A.M.: The induced generalized OWA operator. Inf. Sci. **179**(6), 729–741 (2009). https://doi.org/10.1016/j.ins.2008.11.013
15. Merigó, J.M., Palacios-Marqués, D., Ribeiro-Navarrete, B.: Aggregation systems for sales forecasting. J. Bus. Res. **68**(11), 2299–2304 (2015). https://doi.org/10.1016/j.jbusres.2015.06.015
16. Pereira Libório, M., Ekel, P., D'Angelo, M.F.S.V., Martínez, L., Rabiei, H.: Constructing composite indicators through extreme values reductions–ordered weighted averaging: Human development index. IEEE Access **13**, 48306–48317 (2025). https://doi.org/10.1109/ACCESS.2025.3546529
17. Rossi-Hansberg, E., Zhang, J.: Local GDP estimates around the world. NBER Working Paper 33458, National Bureau Econ. Res. (2025). https://www.nber.org/papers/w33458
18. Ruiz-Morales, B., Espitia-Moreno, I.C., Alfaro-García, V.G., León-Castro, E.: Sustainable development goals analysis with ordered weighted average operators. Sustainability **13**(9), 5240 (2021). https://doi.org/10.3390/su13095240
19. Shams, M.Y., Tarek, Z., El-Kenawy, E.S.M., Eid, M.M., Elshewey, A.M.: Predicting gross domestic product (GDP) using a PC-LSTM-RNN model in urban profiling areas. Comput. Urban Sci. **4**(1), 3 (2024). https://doi.org/10.1007/s43762-024-00116-2
20. Shen, Y., Qian, Y.: Expert system and entropy-ordered weighted average method in cold supply chain risk evaluation. Expert. Syst. **40**(5), e13163 (2023). https://doi.org/10.1111/exsy.13163
21. Wu, X., Zhang, Z., Chang, H., Huang, Q.: A data-driven gross domestic product forecasting model based on multi-indicator assessment. IEEE Access **9**, 99495–99503 (2021). https://doi.org/10.1109/ACCESS.2021.3062671
22. Xu, Z., Wang, Y., Sun, G., Chen, Y., Ma, Q., Zhang, X.: Generating gridded gross domestic product data for China using geographically weighted ensemble learning. ISPRS Int. J. Geo Inf. **12**(3), 123 (2023). https://doi.org/10.3390/ijgi12030123
23. Yager, R.R.: On ordered weighted averaging aggregation operators in multicriteria decision making. IEEE Trans. Syst. Man Cybern. **18**(1), 183–190 (1988). https://doi.org/10.1109/21.87068
24. Yager, R.R.: Generalized OWA aggregation operators. Fuzzy Optim. Decis. Making **3**(1), 93–107 (2004). https://doi.org/10.1023/B:FODM.0000013074.68765.97
25. Ye, Y., Xu, S., Mariani, M.S., Lü, L.: Forecasting countries' gross domestic product from patent data. Chaos, Solitons Fractals **160**, 112234 (2022). https://doi.org/10.1016/j.chaos.2022.112234

A Simplified Multilevel SVPWM Strategy for Cascaded H-Bridge Converters

Jiawei Wang[1], Yunhe Wang[1], Wenjie Ma[1(✉)], Guidong Zhang[1], Lihao Wu[1], and Zhong Li[2]

[1] School of Automation, Guangdong University of Technology, Guangzhou, China
3089143130@qq.com, 512886496@qq.com,
{epwjma,guidong.zhang,wu_lihao}@gdut.edu.cn

[2] Faculty of Mathematics and Computer Science, Fernuniversität in Hagen, Hagen, Germany
zhong.li@fernuni-hagen.de

Abstract. With the increasing number of levels in cascaded H-bridge multilevel converter (CHBMC), traditional space vector pulse width modulation (SVPWM) faces severe challenges, such as the enormous computational burden of reference vector location and the accompanying high-frequency common-mode voltage (CMV). This paper proposes a simplified SVPWM and CMV suppression strategy based on a J-K coordinate system for three-phase CHBMCs. First, a fast location algorithm based on J and K parameters is proposed in the Cartesian coordinate system. By mapping the space vectors onto integer coordinates, the underlying computational logic is significantly simplified. Second, through the algebraic analysis of redundant switching states, the optimal switching states with the minimum absolute CMV value are calculated, which effectively reduces the CMV. Simulations based on a 6-cell 13-level CHBMC fully verify the correctness and effectiveness of the proposed strategy, demonstrating that this method achieves excellent CMV suppression performance while remarkably reducing the consumption of computational resources.

Keywords: Space vector pulse width modulation · Common-mode voltage suppression · Cascaded H-bridge converter

1 Introduction

Multilevel converters have evolved into the core fundamental architecture in the field of medium- and high-voltage, high-power AC/DC power conversion [1]. Compared with conventional two-level inverters, multilevel topologies synthesize output voltage waveforms with multiple steps through the coordinated operation of multiple low-voltage switching devices [2], thereby significantly reducing the voltage stress of devices and the total harmonic distortion (THD) of output currents [3]. Among them, the cascaded H-bridge converter has become a topology with in-depth theoretical research in academia and mature applications in industry [4].

J. Lu et al. (Eds.): FLINS-ISKE 2026, LNCS 16752, pp. 78–89, 2027.
https://doi.org/10.1007/978-981-92-2369-5_6

In cascaded H-bridge converters, the modulation strategy, which accurately discretizes the continuous reference voltage and maps it into the switching drive pulses of the underlying power devices, is the core control part of the system [5]. The SVPWM strategy features high DC voltage utilization, favorable harmonic performance, and ease of digital control [6], but is difficult to extend to high-level topologies. The method in [7] proposes a simplified multilevel space vector modulation (SVM) algorithm based on orthogonal unit vectors. This method decouples the reference voltage into an orthogonal coordinate system, and can directly locate the nearest vectors using only simple rounding and extremum operations, and then directly generate optimized switching sequences through a simplified mapping table. The method in [8] presents a multilevel SVPWM scheme based on non-orthogonal reference frames. Abandoning the traditional orthogonal coordinate system, this method defines three non-orthogonal static reference frames, and directly calculates the target switching sequences and duty cycles only through low-dimensional and simple algebraic matrix operations.

Aiming at the common-mode voltage (CMV) issue, Ref. [9] proposes a novel SVPWM-based CMV suppression scheme for CHB motor drive systems. This method introduces an optimal factor related to the zero-sequence component, and directly determines the redundant switching states with minimum CMV online via algebraic operations, minimizing the inverter switching loss on the premise of ensuring low CMV. Ref. [10] presents a new pulse width modulation scheme for three-level inverters to collaboratively suppress optimal current ripple and CMV. Based on an accurate quantitative analysis of the generation mechanism of current ripple, this method employs four specific voltage vectors (large, medium, small, and zero vectors) to jointly synthesize the reference voltage.

To address the above problems, this paper proposes a J-K coordinate system. Through coordinate transformation, space vectors are mapped to integer coordinate points, thereby reducing the difficulty of reference vector positioning. Furthermore, by selecting the switching states corresponding to the minimum common-mode voltage for each space vector, the common-mode voltage is suppressed to its theoretical minimum. Finally, simulations verify the effectiveness of the proposed method.

2 Principles of Cascaded H-Bridge Converters and SVPWM

2.1 CHBMC Topology

The cascaded H-bridge converter achieves high-voltage and large-capacity power conversion by series-connecting multiple independent low-voltage power cells in the AC circuit. Its physical essence lies in the discrete on-off control of the power devices within each cell, enabling the algebraic superposition of multiple low-voltage DC levels to approximate the desired low-frequency AC sinusoidal wave in a staircase manner. The overall topology of the three-phase cascaded H-bridge converter is shown in Fig. 1. This converter adopts a Y-connected configuration, where its three phases, A, B, and C, are each composed of n identical H-bridge cells cascaded in series on the AC side. To ensure that each cell can independently output the staircase voltage without causing phase-to-phase or cell-to-cell short circuits, the DC side of each H-bridge cell must be equipped with an electrically isolated independent DC power source E.

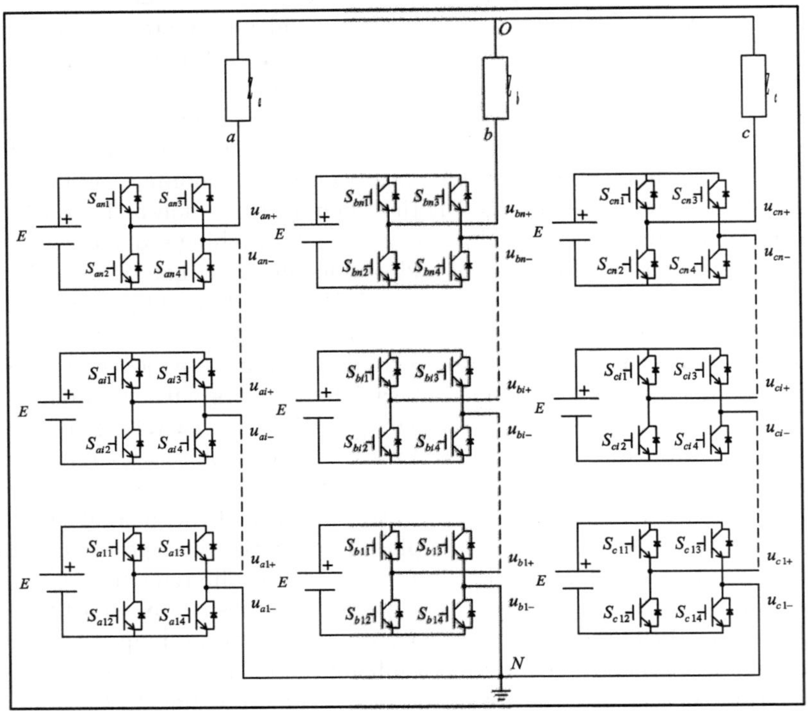

Fig. 1. Topology of Three-phase Cascaded H-bridge Converter

2.2 Principle of SVPWM

Let the three-phase reference phase voltages, u_{ra}, u_{rb}, and u_{rc}, be the desired continuous three-phase sinusoidal output voltages of the CHB. Their mathematical expressions are given as follows:

$$\begin{cases} u_{ra} = U_{rm} \cos \omega t \\ u_{rb} = U_{rm} \cos \left(\omega t - \frac{2}{3}\pi \right) \\ u_{rc} = U_{rm} \cos \left(\omega t + \frac{2}{3}\pi \right) \end{cases} \tag{1}$$

where U_{rm} represents the amplitude of the sinusoidal voltage, and ω is its angular frequency.

The matrix expression of the reference space vector corresponding to the reference voltage signals in the $\alpha-\beta$ coordinate system is given by:

$$\begin{bmatrix}\alpha_r\\ \beta_r\end{bmatrix}=\frac{1}{3E}\begin{bmatrix}2u_{ra}-u_{rb}-u_{rc}\\ \sqrt{3}(u_{rb}-u_{rc})\end{bmatrix}=\frac{1}{3E}\begin{bmatrix}2 & -1 & -1\\ 0 & \sqrt{3} & -\sqrt{3}\end{bmatrix}\begin{bmatrix}u_{ra}\\ u_{rb}\\ u_{rc}\end{bmatrix}\tag{2}$$

Assuming the number of levels of the actual output phase voltages for phases A, B, and C are a, b, and c, respectively, let $S(a,b,c)$ denote the switching state of the CHB. The corresponding matrix equation for the space vector in the $\alpha-\beta$ coordinate system is:

$$\begin{bmatrix}\alpha\\ \beta\end{bmatrix}=\frac{1}{3E}\begin{bmatrix}2u_a-u_b-u_c\\ \sqrt{3}(u_b-u_c)\end{bmatrix}=\frac{1}{3}\begin{bmatrix}2 & -1 & -1\\ 0 & \sqrt{3} & -\sqrt{3}\end{bmatrix}\begin{bmatrix}a\\ b\\ c\end{bmatrix}\tag{3}$$

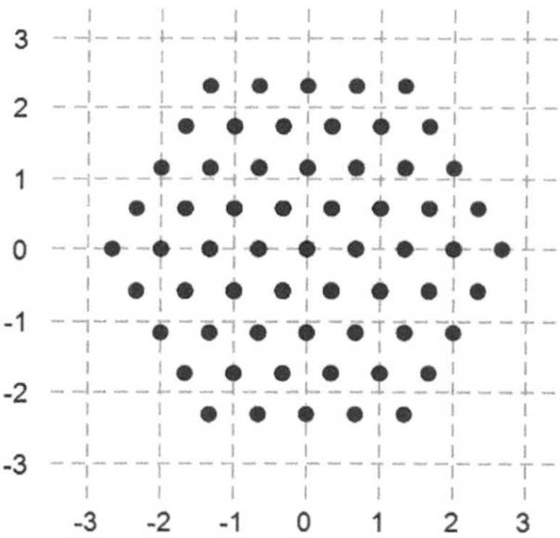

Fig. 2. $\alpha-\beta$ coordinate system

In a CHB with n cells per phase, the maximum number of output phase voltage levels is $2n+1$, and all space vectors constitute $2n$ nested regular hexagons. If the zero vector corresponding to the origin is denoted as layer 0, the subsequent layers from the inside out are sequentially numbered as 1, 2, ..., $2n$. The q-th hexagonal layer contains $6q$ space vectors, resulting in a total of $1+6\sum_{q=1}^{2n}q$ space vectors, which correspond to $(2n+1)^3$ switching states. When $n=2$, the space vector diagram is shown in Fig. 2.

To simplify the location of space vectors and reduce the implementation complexity, this paper simplifies the SVPWM strategy based on the J-K coordinate system. The complex nested hexagonal grid in the complex plane is mapped into an orthogonal discrete integer lattice. The transformation process from the traditional $\alpha - \beta$ coordinate system to the proposed equivalent-intercept J-K coordinate system can be abstracted as a linear transformation matrix T_{eq}. The matrix equation for this process can be expressed as:

$$\begin{bmatrix} J \\ K \end{bmatrix} = T_{eq} \begin{bmatrix} \alpha \\ \beta \end{bmatrix} = \begin{bmatrix} \frac{3}{2} & -\frac{\sqrt{3}}{2} \\ \frac{3}{2} & \frac{\sqrt{3}}{2} \end{bmatrix} \begin{bmatrix} \alpha \\ \beta \end{bmatrix} \tag{4}$$

where the column vector $[\alpha, \beta]^T$ represents the coordinate components in the $\alpha - \beta$ coordinate system prior to the transformation, and the column vector $[J, K]^T$ represents the coordinate components of the space vector in the $J - K$ coordinate system.

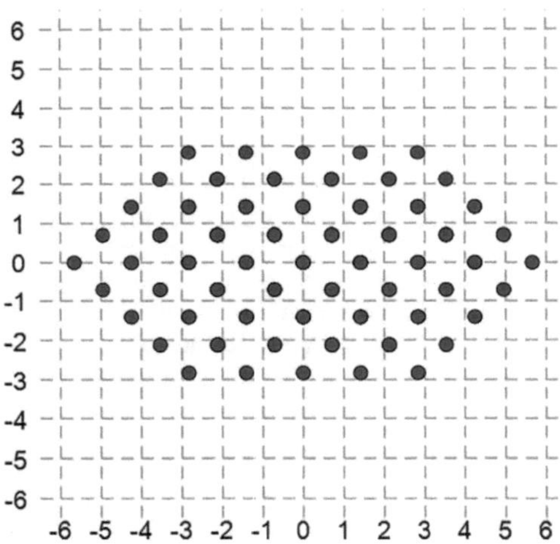

Fig. 3. Scaling by T_c

To reveal the underlying essence of this transformation matrix, the affine transformation theory from linear algebra is introduced to decouple the matrix structure. Assuming that the aforementioned linear transformation matrix

T_{eq} is composed of a scaling matrix T_c and a rotation matrix T_r, and strictly specifying the transformation sequence as coordinate axis scaling followed by global rotation, the following relationship holds:

$$T_{eq} = T_r T_c \tag{5}$$

where the scaling matrix T_c and the rotation matrix T_r are defined as follows:

$$T_c = \begin{bmatrix} \frac{3\sqrt{2}}{2} & 0 \\ 0 & \frac{\sqrt{6}}{2} \end{bmatrix}, \quad T_r = \begin{bmatrix} \frac{\sqrt{2}}{2} & -\frac{\sqrt{2}}{2} \\ \frac{\sqrt{2}}{2} & \frac{\sqrt{2}}{2} \end{bmatrix} \tag{6}$$

Figure 3 illustrates the effect of the scaling matrix T_c on the $\alpha - \beta$ coordinate system when $n = 2$, and Fig. 4 shows the result after further rotation.

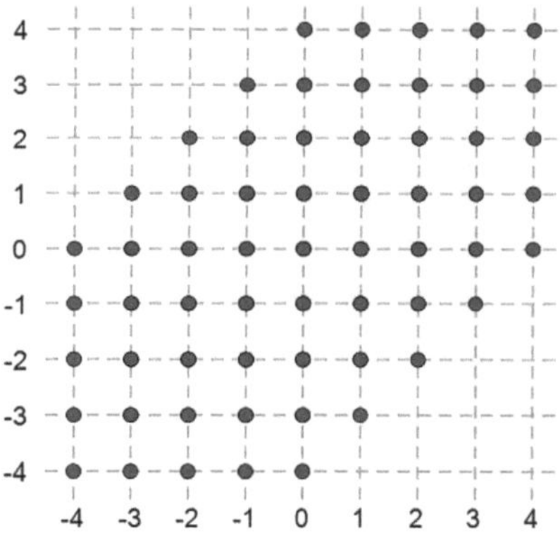

Fig. 4. Result after further rotation

Furthermore, by combining with the amplitude-invariant Clarke transformation matrix T_{Clark}, the transformation matrix mapping the level numbers a, b, and c to the $J - K$ coordinate system can be derived as follows:

$$\begin{bmatrix} J \\ K \end{bmatrix} = T_{eq} T_{clark} \begin{bmatrix} a \\ b \\ c \end{bmatrix} = \begin{bmatrix} \frac{3}{2} & -\frac{\sqrt{3}}{2} \\ \frac{3}{2} & \frac{\sqrt{3}}{2} \end{bmatrix} \begin{bmatrix} \frac{2}{3} & -\frac{1}{3} & -\frac{1}{3} \\ 0 & \frac{\sqrt{3}}{3} & -\frac{\sqrt{3}}{3} \end{bmatrix} \begin{bmatrix} a \\ b \\ c \end{bmatrix} = \begin{bmatrix} 1 & -1 & 0 \\ 1 & 0 & -1 \end{bmatrix} \begin{bmatrix} a \\ b \\ c \end{bmatrix} \tag{7}$$

Similarly, the reference space vector can be expressed as follows:

$$\begin{bmatrix} J_r \\ K_r \end{bmatrix} = \frac{1}{E} \begin{bmatrix} \frac{3}{2} & -\frac{\sqrt{3}}{2} \\ \frac{3}{2} & \frac{\sqrt{3}}{2} \end{bmatrix} \begin{bmatrix} \frac{2}{3} & -\frac{1}{3} & -\frac{1}{3} \\ 0 & \frac{\sqrt{3}}{3} & -\frac{\sqrt{3}}{3} \end{bmatrix} \begin{bmatrix} u_{ra} \\ u_{rb} \\ u_{rc} \end{bmatrix} = \begin{bmatrix} 1 & -1 & 0 \\ 1 & 0 & -1 \end{bmatrix} \begin{bmatrix} a_r \\ b_r \\ c_r \end{bmatrix} \tag{8}$$

In the $J - K$ coordinate system, the reference vector $V_r^*(J_r, K_r)$ at any given instant falls within a unit square grid formed by four adjacent integer coordinate points, as shown in Fig. 5. The base point coordinates $V_0^*(J_0, K_0)$ at the bottom-left corner of this grid uniquely determine the unit square region where the reference vector is located. These coordinates can be directly obtained by performing a floor operation on the coordinates of the reference vector:

$$\begin{cases} J_0 = \text{floor}(J_r) \\ K_0 = \text{floor}(K_r) \end{cases} \tag{9}$$

where floor(·) is the floor function.

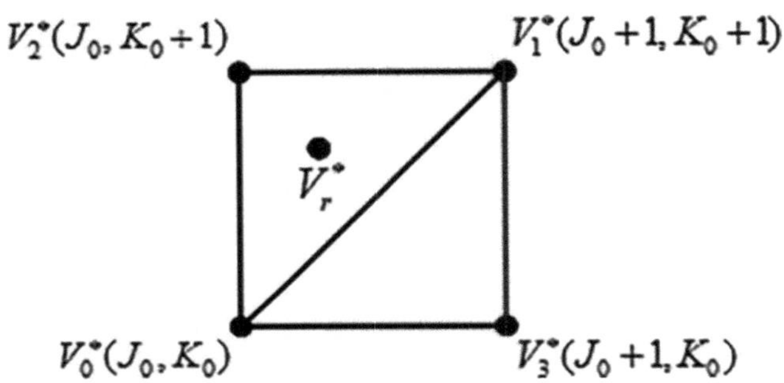

Fig. 5. Reference vector localization in $J - K$ grid

Since the converter typically employs three adjacent basic vectors to equivalently synthesize the reference vector within one switching period, the aforementioned unit square needs to be further divided along its main diagonal into two triangular regions: the top-left and the bottom-right.

When $J_r - J_0 \leq K_r - K_0$, V_r^* is located in the top-left triangular region shown in Fig. 5. The three basic space vectors participating in the synthesis are $V_0^*(J_0, K_0)$, $V_1^*(J_0 + 1, K_0 + 1)$, and $V_2^*(J_0, K_0 + 1)$, with their corresponding action times being t_0, t_1, and t_2, respectively. According to the volt-second balance principle within a sampling period T_s, t_0, t_1, and t_2 are given by:

$$\begin{cases} t_0 = [1 - (K_r - K_0)]T_s \\ t_1 = (J_r - J_0)T_s \\ t_2 = [(K_r - K_0) - (J_r - J_0)]T_s \end{cases} \tag{10}$$

Similarly, when $J_r - J_0 > K_r - K_0$, the reference vector V_r^* falls within the bottom-right triangular region shown in Fig. 5. The basic space vectors participating in the synthesis become $V_0^*(J_0, K_0)$, $V_1^*(J_0+1, K_0+1)$, and $V_3^*(J_0+1, K_0)$, with their corresponding action times being t_0, t_1, and t_3, respectively:

$$\begin{cases} t_0 = [1 - (J_r - J_0)]T_s \\ t_1 = (K_r - K_0)T_s \\ t_3 = [(J_r - J_0) - (K_r - K_0)]T_s \end{cases} \tag{11}$$

To suppress the common-mode voltage, after calculating the three space vectors $V_0(J_0, K_0)$, $V_1(J_1, K_1)$, and $V_z(J_z, K_z)$ (where $z = 2, 3$) for synthesizing the reference vector $V_r^*(J_r, K_r)$, the optimal switching states $S_0(a_0, b_0, c_0)$, $S_1(a_1, b_1, c_1)$, and $S_z(a_z, b_z, c_z)$ with the minimum common-mode voltage are determined under the constraint of $\min(|a+b+c|)$. Finally, a five-segment algorithm is employed to switch among the three space vectors.

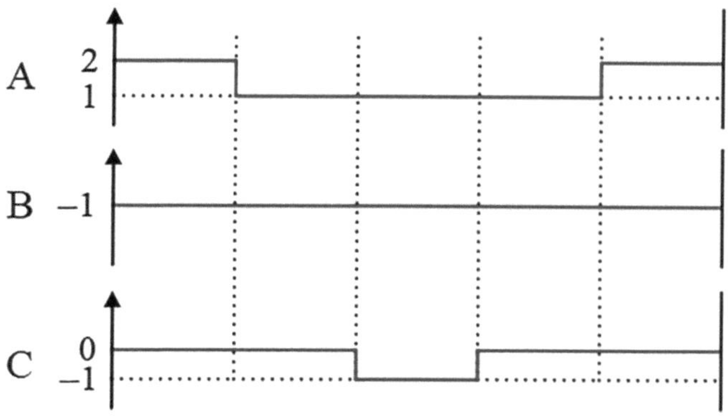

Fig. 6. Switching sequence and time allocation

Taking $S_0(1, -1, 0)$, $S_1(2, -1, 0)$, and $S_z(1, -1, -1)$ as an example, to reduce switching losses, the switching sequence is arranged as $S_1(2, -1, 0) \rightarrow S_0(1, -1, 0) \rightarrow S_z(1, -1, -1) \rightarrow S_0(1, -1, 0) \rightarrow S_1(2, -1, 0)$. The switching sequence and time allocation are shown in Fig. 6.

3 Simulation Verification

To verify the correctness of the proposed SVPWM strategy and the common-mode voltage (CMV) suppression method, a simulation platform for the cascaded H-bridge (CHB) converter is built in MATLAB/Simulink. The number of sub-modules per phase is 6, the DC-side voltage of each sub-module is 30 V, and the reference signal frequency is 50 Hz. Subsequently, the $J-K$ coordinate system-based SVPWM strategy with CMV suppression is simulated and verified.

At a modulation index of 0.8, the waveforms and FFT analysis of the phase and line voltages are shown in Fig. 7. As can be observed, the phase voltage waveform consists of 11 levels with a total harmonic distortion (THD) of 12.73%, and the fundamental root-mean-square (RMS) value of the phase voltage is 146.3 V. The line voltage waveform exhibits 19 levels with a THD of 9.50%, and its fundamental RMS value is 254.3 V.

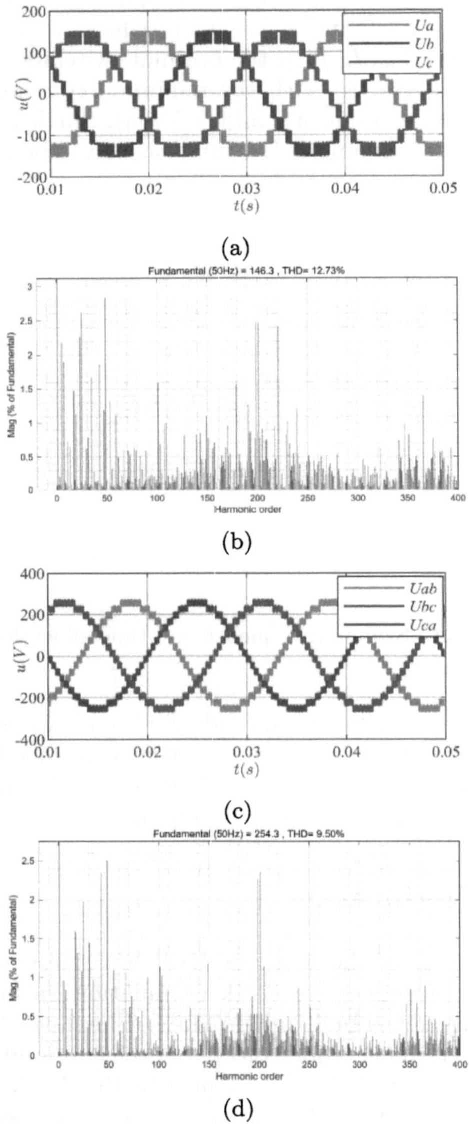

Fig. 7. Simulation results ($m = 0.8$): (a) Three-phase phase voltage; (b) FFT of phase voltage; (c) Line-to-line voltage; (d) FFT of line-to-line voltage.

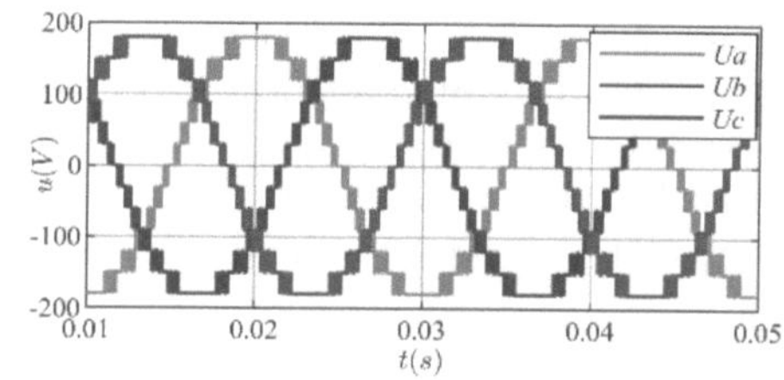

(a)

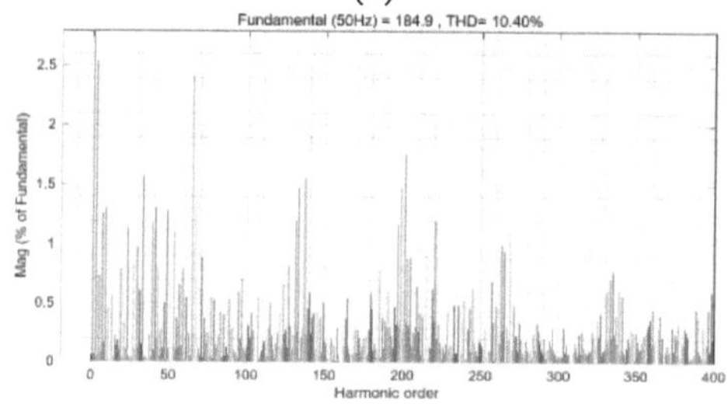

(b)

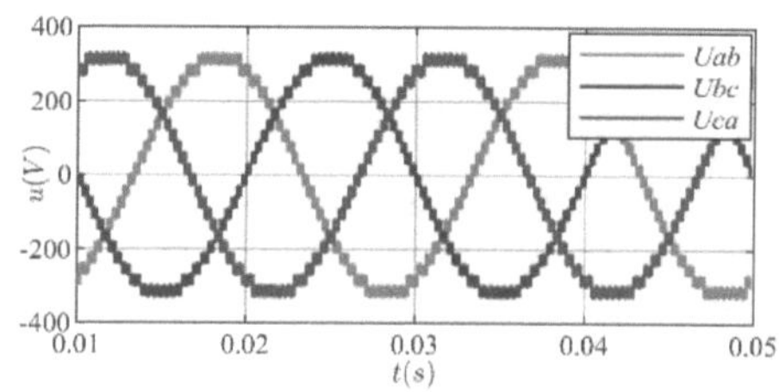

(c)

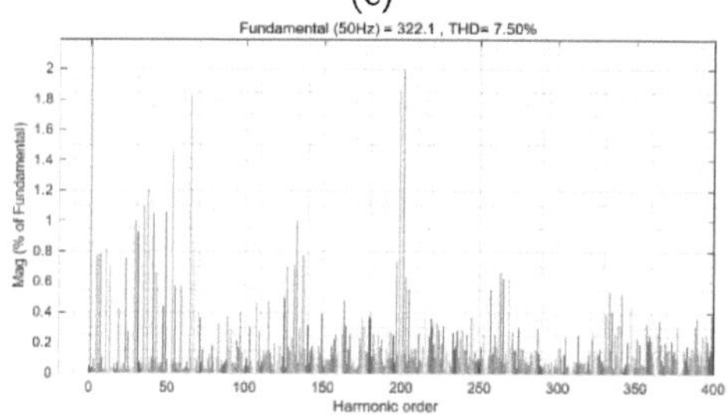

(d)

Fig. 8. Simulation results ($m = 0.9$): (a) Three-phase phase voltage; (b) FFT of phase voltage; (c) Line-to-line voltage; (d) FFT of line-to-line voltage.

At a modulation index of 0.9, the waveforms and FFT analysis of the phase and line voltages are shown in Fig. 8. As can be observed, the phase voltage waveform consists of 6 levels with a THD of 10.40%, and the fundamental RMS value of the phase voltage is 146.3 V. The line voltage waveform exhibits 23 levels with a THD of 7.50%, and its fundamental RMS value is 322.1 V.

Figure 9 shows the waveforms of the CMV. At a modulation index of 0.8, the CMV amplitude is 10 V, corresponding to $E/3$. At a modulation index of 0.9, the CMV amplitude is 20 V, corresponding to $2E/3$.

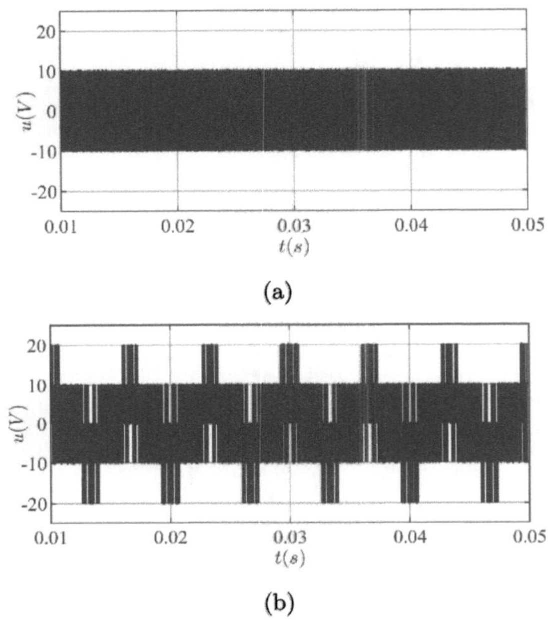

Fig. 9. Simulation Results of Common-mode Voltage: (a) m=0.8; (b) m=0.9.

4 Conclusion

This paper proposes a simplified SVPWM and CMV suppression strategy based on a $J-K$ coordinate system for CHB multilevel converters. By mapping the space vectors onto integer coordinate points, the computational complexity of reference vector location and modulation implementation is significantly reduced. Furthermore, by leveraging redundant switching states to select the optimal vectors, the CMV is effectively suppressed. Simulation results indicate that the proposed method achieves excellent CMV suppression performance while ensuring high-quality output voltage waveforms and low harmonic distortion, thereby demonstrating certain value for engineering applications.

References

1. Bieber, L., Wang, L., Jatskevich, J.: Dual-arm modular multilevel converter with a compact footprint for VSC-HVDC applications. IEEE Trans. Power Deliv. **39**(1), 504–517 (2024)
2. Yuan, X.: Ultimate generalized multilevel converter topology. IEEE Trans. Power Electron. **36**(8), 8634–8639 (2021)
3. Deng, F., Lü, Y., Liu, C., Heng, Q., Yu, Q., Zhao, J.: Overview on submodule topologies, modeling, modulation, control schemes, fault diagnosis, and tolerant control strategies of modular multilevel converters. Chin. J. Electr. Eng. **6**(1), 1–21 (2020)
4. Gray, P.A., Lehn, P.W., Yakop, N.: A modular multilevel DC–DC converter with flying capacitor converter like properties. IEEE Trans. Industr. Electron. **69**(7), 6774–6783 (2022)
5. Kim, I.-D., Nho, E.-C., Kim, H.-G., Ko, J.S.: A generalized undeland snubber for flying capacitor multilevel inverter and converter. IEEE Trans. Industr. Electron. **51**(6), 1290–1296 (2004)
6. Huang, M., Li, W., Zou, J., Ma, X.: Analysis and design of a novel hybrid modular multilevel converter with time-sharing alternative arm converter. IEEE Trans. Industr. Electron. **71**(1), 14–26 (2024)
7. Deng, Y., Wang, Y., Teo, K.H., Harley, R.G.: A simplified space vector modulation scheme for multilevel converters. IEEE Trans. Power Electron. **31**(3), 1873–1886 (2016)
8. Ebrahimi, J., Karshenas, H., Eren, S., Bakhshai, A.: A fast-decoupled space vector modulation scheme for flying capacitor-based multilevel converters. IEEE Trans. Power Electron. **36**(12), 14539–14549 (2021)
9. Yao, H., Yan, Y., Shi, T., Zhang, G., Wang, Z., Xia, C.: A novel SVPWM scheme for field-oriented vector-controlled PMSM drive system fed by cascaded H-bridge inverter. IEEE Trans. Power Electron. **36**(8), 8988–9000 (2021)
10. Liu, X., et al.: Optimal current ripple PWM for three-level inverter with common mode voltage reduction. IEEE Trans. Industr. Electron. **69**(5), 4890–4900 (2022)

A Switching Control for LLC Converter Under Pulse-Load Conditions

Bingran Liang[1], Yuzhe Bie[1], Guidong Zhang[1(✉)], Wenjie Ma[1], Lihao Wu[1], and Zhong Li[2]

[1] School of Automation, Guangdong University of Technology, Guangzhou, China
bieyuzhe1@mails.gdut.edu.cn, {guidong.zhang,epwjma,wu_lihao}@gdut.edu.cn
[2] Faculty of Mathematics and Computer Science, Fernuniversität in Hagen, Hagen, Germany
zhong.li@fernuni-hagen.de

Abstract. This paper addresses the pulse-load disturbances introduced by the downstream inverter-side H-bridge. The front-end LLC converter typically suffers from large output-voltage drop, long recovery time, and significant steady-state ripple under such operating conditions. Conventional linear control cannot achieve a satisfactory trade-off between dynamic response and steady-state accuracy. To solve this problem, a switching control strategy based on H-bridge mode identification is proposed. First, a state-plane model of the LLC converter under pulse-load conditions is established to analyze the optimal trajectory radius and switching instants during the transition from no-load to full-load operation. Then, the switching states of the H-bridge are used to construct a load-mode criterion, according to which the controller switches among three operating modes: no-load, full-load, and load-transition modes. Fast regulation is achieved through state-trajectory transition at the instant of load variation. The results show that the proposed strategy effectively suppresses voltage overshoot and drop without requiring additional hardware. The dynamic settling time is reduced by 50%, and the output-voltage ripple is maintained below 0.6 V. Therefore, the robustness and power quality of the system under severe pulse-load conditions are significantly improved.

Keywords: LLC Converter · Switching Control · Pulse Load · Simplified Optimal Trajectory Control

1 Introduction

With the increasing severity of the global energy crisis and environmental pollution, renewable-energy generation technologies represented by solar energy have attracted extensive attention [1]. In recent years, the Common DC-Bus Cascaded H-Bridge (CDB-CHB) photovoltaic grid-connected structure has received increasing interest [2,3]. In the CDB-CHB topology reported in [4], the input

J. Lu et al. (Eds.): FLINS-ISKE 2026, LNCS 16752, pp. 90–104, 2027.
https://doi.org/10.1007/978-981-92-2369-5_7

sides of the isolated DC–DC converters in all modules are connected in parallel to form a common DC bus. The front-end converters provide isolated power to the downstream H-bridges and stabilize their DC-side voltages. Under this architecture, the intermediate isolated converter mainly performs electrical isolation and output-voltage regulation. Owing to its soft-switching capability over the full load range, the LLC converter has become an attractive choice for this application [5–8].

However, the operating mode of the downstream H-bridge inverter introduces a special pulse-load challenge to the front-end LLC converter. When the H-bridge switches between zero and non-zero voltage levels according to the modulation command, the DC-side power demand changes abruptly. As a result, the equivalent load seen by the front-end LLC converter undergoes a step transition between no-load and full-load conditions. The corresponding output waveform is shown in Fig. 1. In most conventional designs, a proportional-integral (PI) controller is used to regulate the DC voltage. However, its loop bandwidth is limited by the switching frequency and is therefore insufficient to track severe pulse-load disturbances [9]. Although other methods, such as average current control [10], Bang-Bang charge control [11], and sliding-mode control [12], have also been proposed, they still suffer from strict sampling requirements or steady-state errors under such extreme pulse-load conditions.

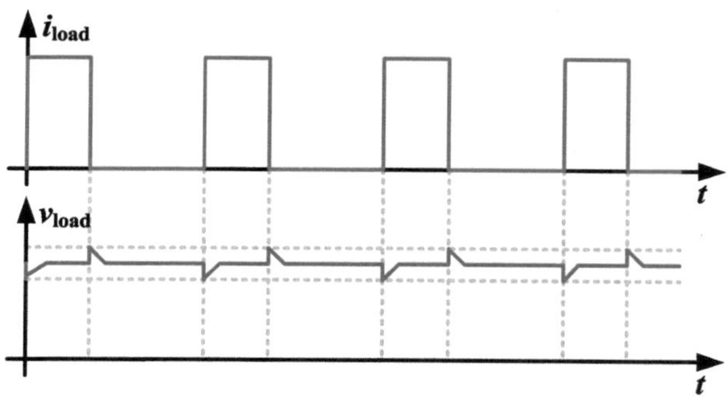

Fig. 1. Output waveforms under pulse load.

To improve the dynamic response of the LLC converter during load transients, the simplified optimal trajectory control (SOTC) [13] and its improved trajectory calculation methods [14,15] have been proposed, enabling the state variables to rapidly transition to the steady-state trajectory. However, these methods rely on voltage threshold judgments, resulting in a certain delay in the dynamic response.

Regarding the aforementioned issues, this paper proposes a switching control strategy for LLC converters under pulse-load conditions. The proposed strategy determines the load condition according to the switching states of the H-bridge. PFM control is adopted under steady-state full-load conditions. Burst mode is introduced under no-load conditions to reduce switching loss. At the instant of load step change, SOTC control is employed to calculate the optimal pulse width, so that the system state trajectory can move to the new steady-state circle in the shortest possible time. In this way, output-voltage fluctuation is reduced and the recovery time is shortened.

The remainder of this paper is organized as follows: Sect. 2 introduces the topology of the LLC and H-bridge converter, establishes the LLC model based on the state-plane method, and analyzes the generation mechanism of the pulse load. Section 3 details the proposed switching control strategy based on H-bridge mode identification. Section 4 conducts simulation verification and analysis of the proposed control strategy. Finally, Sect. 5 concludes the paper.

2 System Structure and Pulse-Load Mechanism

2.1 System Modeling

The topology of the LLC converter and the downstream H-bridge investigated in this paper is shown in Fig. 2. The front-end employs a full-bridge LLC converter. Its primary side consists of an input capacitor C_{in} and a full-bridge inverter network formed by switches S_1–S_4. The resonant tank consists of the resonant inductor L_{r}, resonant capacitor C_{r}, and magnetizing inductor L_{m}. The transformer turns ratio is n. The secondary side consists of the rectifier diodes D_1–D_4 and the output filter capacitor C_{o}. The downstream H-bridge is formed by switches Q_1–Q_4 and is connected to an output resistance R_{L}.

To highlight the step disturbance in the DC-side power demand caused by H-bridge switching and to avoid additional coupling introduced by the grid impedance, the output of the downstream H-bridge is modeled as an equivalent resistive load R_{eq} in this section. It serves as a controllable equivalent model of the pulse load. To accurately describe the dynamic behavior of the system under pulse-load conditions, the LLC converter is modeled by the state-plane method.

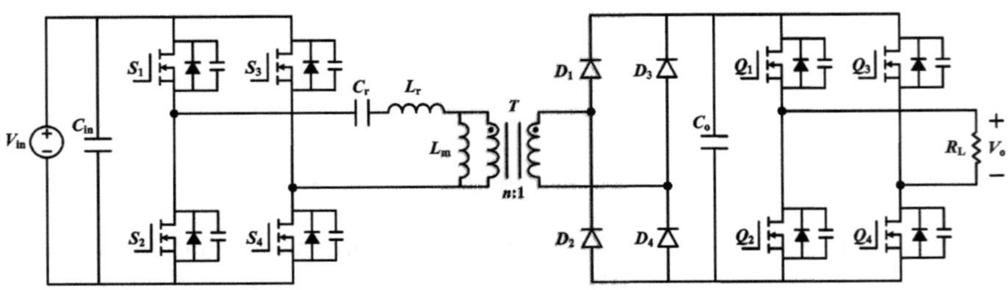

Fig. 2. Topology of the LLC converter and H-bridge inverter.

When the downstream H-bridge inverter outputs a non-zero voltage level, the LLC converter operates in the full-load condition. According to KVL and

KCL, it can be obtained that:

$$\begin{cases} C_{\rm r}\dfrac{dv_{\rm Cr}}{dt} = i_{\rm Lr} \\ L_{\rm r}\dfrac{di_{\rm Lr}}{dt} = V_{\rm in} - nV_{\rm o} - v_{\rm Cr} \end{cases} \tag{1}$$

Solving the above second-order differential equations yields:

$$\begin{cases} v_{\rm Cr} = I_{\rm Lr0}Z_{\rm r}\sin[\omega_{\rm r}(t-t_0)] + V_{\rm Cr0}\cos[\omega_{\rm r}(t-t_0)] \\ i_{\rm Lr} = I_{\rm Lr0}\cos[\omega_{\rm r}(t-t_0)] - \dfrac{V_{\rm Cr0}}{Z_{\rm r}}\sin[\omega_{\rm r}(t-t_0)] \end{cases} \tag{2}$$

where $V_{\rm Cr0}$ and $I_{\rm Lr0}$ are the initial values of the resonant capacitor voltage and the resonant inductor current at the initial moment, respectively, and $\omega_{\rm r}$ is the resonant angular frequency of $L_{\rm r}$ and $C_{\rm r}$, $\omega_{\rm r} = 2\pi f_{\rm r}$.

By normalizing (2), the voltage variables are normalized by $V_{\rm in}$, and the current variables are normalized by $V_{\rm in}/Z_{\rm r}$:

$$\begin{cases} v_{\rm CrN} = I_{\rm Lr0N}\sin[\omega_{\rm r}(t-t_0)] + V_{\rm Cr0N}\cos[\omega_{\rm r}(t-t_0)]\,, \\ i_{\rm LrN} = I_{\rm Lr0N}\cos[\omega_{\rm r}(t-t_0)] - V_{\rm Cr0N}\sin[\omega_{\rm r}(t-t_0)]\,. \end{cases} \tag{3}$$

Rearranging the above equations yields the trajectory equation:

$$v_{\rm CrN}^2 + i_{\rm LrN}^2 = V_{\rm Cr0N}^2 + I_{\rm Lr0N}^2 \tag{4}$$

During the full-load steady-state operation of the LLC converter, the radius ρ of the circular state trajectory is equal to the per-unit peak value of the resonant current $i_{\rm Lr}$. The root-mean-square (RMS) value of the resonant inductor current at full load is given by:

$$I_{\rm Lr,rms} = \frac{1}{4\sqrt{2}}\frac{V_{\rm o}}{nR_{\rm L}}\sqrt{\frac{n^4R_{\rm L}^2T^2}{L_{\rm m}^2} + 4\pi^2} = \frac{1}{4\sqrt{2}}\sqrt{\frac{n^4V_{\rm o}^2T^2}{L_{\rm m}^2} + 4\pi^2\left(\frac{i_{\rm o}}{n}\right)^2} \tag{5}$$

The radius of the trajectory circle is given by:

$$\rho = \frac{\sqrt{2}\,I_{\rm Lr,rms}}{V_{\rm in}/Z_{\rm r}} \tag{6}$$

The state-plane trajectory equation at full load is expressed as:

$$v_{\rm CrN}^2 + i_{\rm LrN}^2 = \left(\frac{\sqrt{2}\,I_{\rm Lr,rms}}{V_{\rm in}/Z_{\rm r}}\right)^2 \tag{7}$$

When the H-bridge inverter outputs a zero voltage level, the LLC converter is equivalent to an open-circuit no-load condition:

$$\begin{cases} \dfrac{dv_{\rm Cr}}{dt} = \dfrac{i_{\rm Lr}}{C_{\rm r}} \\ \dfrac{di_{\rm Lr}}{dt} = \dfrac{V_{\rm in} - v_{\rm Cr}}{L_{\rm r} + L_{\rm m}} \end{cases} \tag{8}$$

Solving these second-order equations yields:

$$\begin{cases} v_{\mathrm{Cr}} = V_{\mathrm{in}} + I_{\mathrm{Lr0}} Z_{\mathrm{m}} \sin[\omega_{\mathrm{m}}(t - t_0)] + (V_{\mathrm{Cr0}} - V_{\mathrm{in}}) \cos[\omega_{\mathrm{m}}(t - t_0)] \\ i_{\mathrm{Lr}} = I_{\mathrm{Lr0}} \cos[\omega_{\mathrm{m}}(t - t_0)] - \dfrac{V_{\mathrm{Cr0}} - V_{\mathrm{in}}}{Z_{\mathrm{m}}} \sin[\omega_{\mathrm{m}}(t - t_0)] \end{cases} \quad (9)$$

where Z_{m} and ω_{m} are the characteristic impedance and the resonant angular frequency, respectively, of the resonant tank formed by L_{r}, C_{r}, and L_{m}.

$$\begin{cases} Z_{\mathrm{m}} = \sqrt{\dfrac{L_{\mathrm{r}} + L_{\mathrm{m}}}{C_{\mathrm{r}}}} \\ \omega_{\mathrm{m}} = \dfrac{1}{\sqrt{(L_{\mathrm{r}} + L_{\mathrm{m}}) C_{\mathrm{r}}}} \end{cases} \quad (10)$$

After normalization, it can be obtained that:

$$\begin{cases} v_{\mathrm{CrN}} - 1 = \dfrac{I_{\mathrm{Lr0N}}}{Z_{\mathrm{r}}/Z_{\mathrm{m}}} \sin[\omega_{\mathrm{m}}(t - t_0)] + (V_{\mathrm{Cr0N}} - 1) \cos[\omega_{\mathrm{m}}(t - t_0)] \\ \dfrac{i_{\mathrm{LrN}}}{Z_{\mathrm{r}}/Z_{\mathrm{m}}} = \dfrac{I_{\mathrm{Lr0N}}}{Z_{\mathrm{r}}/Z_{\mathrm{m}}} \cos[\omega_{\mathrm{m}}(t - t_0)] - (V_{\mathrm{Cr0N}} - 1) \sin[\omega_{\mathrm{m}}(t - t_0)] \end{cases} \quad (11)$$

Rearranging the equations yields the no-load state trajectory equation:

$$(v_{\mathrm{CrN}} - 1)^2 + \left(\frac{i_{\mathrm{LrN}}}{Z_{\mathrm{r}}/Z_{\mathrm{m}}}\right)^2 = (V_{\mathrm{Cr0N}} - 1)^2 + \left(\frac{I_{\mathrm{Lr0N}}}{Z_{\mathrm{r}}/Z_{\mathrm{m}}}\right)^2 \quad (12)$$

2.2 Pulse-Load Characteristic Analysis

In DC-DC converter applications where an H-bridge serves as the downstream inverter unit, the front-end LLC converter is responsible not only for providing a stable DC bus voltage but also for enduring the dynamic load variations induced by the downstream switching operations.

When the H-bridge outputs a non-zero voltage level and transfers power to the AC side, the front-end LLC converter is effectively connected to the nominal load, operating in the full-load mode with an equivalent load impedance of $R_{\mathrm{eq}} = R_{\mathrm{L}}$. Conversely, when the H-bridge outputs a zero voltage level, the energy transmission is interrupted. In this case, the output port of the LLC converter is equivalent to an open circuit ($R_{\mathrm{eq}} \to \infty$), and the system operates in the no-load mode. The specific correspondence between the switching modes of the H-bridge and the load states of the LLC converter is summarized in Table 1.

Therefore, as the H-bridge frequently switches between zero and non-zero voltage levels, the equivalent load of the front-end LLC converter experiences rapid and repeated transitions between the no-load and full-load conditions. This dynamic toggling constitutes a severe pulse-load disturbance for the LLC converter.

Table 1. Relationship between H-bridge switching states and equivalent load states of the LLC converter

Q_1	Q_2	Q_3	Q_4	Output State	Load State
1	0	0	1	Positive level	Full load
0	1	1	0	Negative level	Full load
1	0	1	0	Zero level	No load
0	1	0	1	Zero level	No load

3 Proposed Switching Control Strategy

To address the pulse-load disturbances introduced by the H-bridge, this paper proposes a switching control strategy based on H-bridge mode identification. According to the switching states of the H-bridge, the controller adaptively switches the control mode under no-load and full-load conditions, thereby reducing the output-voltage ripple and shortening the recovery time.

3.1 Overall Control Strategy

When the output voltage of the H-bridge periodically switches among positive, negative, and zero levels, from the perspective of the front-end LLC converter, the different operating states of the H-bridge directly map to the equivalent load it bears. To quantitatively analyze the generation mechanism of the pulse load, the four switches of the H-bridge inverter are defined as Q_1, Q_2, Q_3, and Q_4. The switching function of the H-bridge, $m(t)$, is defined as:

$$m(t) = \begin{cases} 0, & Q_1 = Q_3 = 1 \text{ or } Q_2 = Q_4 = 1 \\ 1, & Q_1 = Q_4 = 1 \text{ or } Q_2 = Q_3 = 1 \\ 2, & m = 0 \rightarrow m = 1 \end{cases} \tag{13}$$

Under the constraint of this switching function, the equivalent load impedance R_{eq} viewed from the front-end LLC converter towards the downstream stage exhibits discrete characteristics highly correlated with the H-bridge operating modes.

The control architecture of this strategy is illustrated in Fig. 3. After the system is powered on, the LLC converter starts and establishes the basic operating voltage and resonant state, while the H-bridge switches begin to conduct according to the predefined timing sequence. The core of the control system lies in the introduction of the H-bridge state identification module. This module monitors and identifies the current operating state of the H-bridge in real time and quantizes it into a state indicator variable $m(t)$. Consequently, the operating mode of the LLC converter is divided into three distinct intervals, seamlessly switching among Burst mode, PFM control, and SOTC control to achieve optimal performance across the entire operating range. Figure 4 shows the proposed switching control under pulse-load conditions.

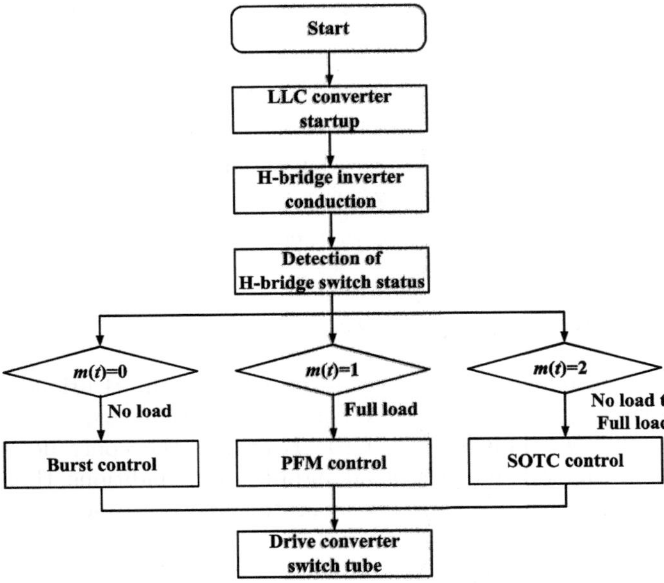

Fig. 3. Control flowchart of the proposed control.

No-load mode ($m(t) = 0$): The H-bridge outputs a zero voltage level, and the converter is equivalent to an open-circuit no-load condition. The system enters the Burst mode.

Full-load mode ($m(t) = 1$): The H-bridge outputs an effective non-zero voltage level, and the converter is equivalent to the rated load. The system enters the closed-loop PFM control.

Load-switching transient mode ($m(t) = 2$): When the H-bridge operating mode switches, indicating a sudden load step from no load to full load, the system is triggered to enter the SOTC control.

After the respective loop calculations of the aforementioned three control strategies, specific switching signals are finally generated to directly drive the primary-side switches of the LLC resonant converter, ensuring the stable operation of the closed-loop system.

3.2 PFM and Burst Control Under Steady-State Conditions

When $m(t) = 0$, the H-bridge operates at the zero voltage level, and the LLC converter enters the no-load state. Under no-load conditions, the quality factor Q approaches zero. According to the DC gain characteristics, the gain curve rises near the resonant frequency, leading to an increase in the output voltage. The controller will inevitably increase the switching frequency to step down the output voltage. However, the high-frequency gain inflection point caused by

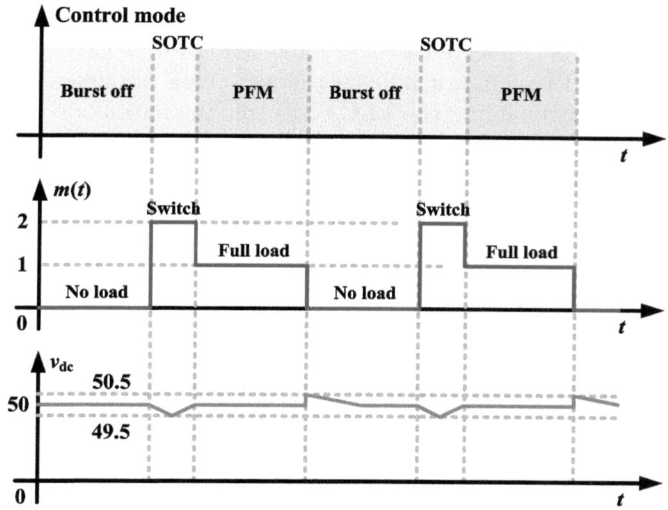

Fig. 4. Proposed switching control under pulse load.

the parasitic capacitance renders the voltage control of the converter ineffective under extreme light loads.

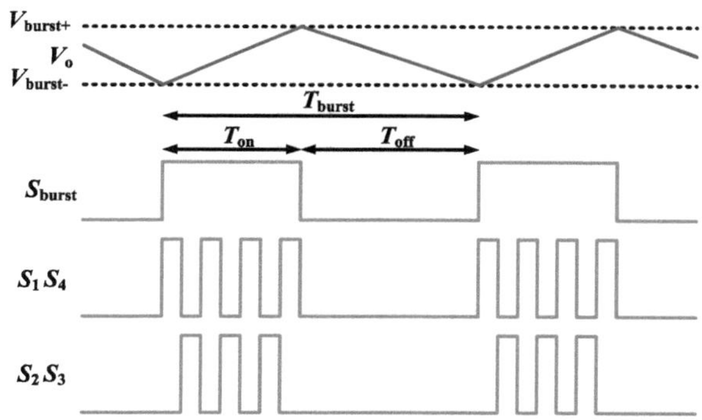

Fig. 5. Burst mode of the LLC converter.

Therefore, Burst control is introduced during no-load operation. The LLC converter switches between the burst-on and burst-off states based on the preset upper and lower voltage thresholds, V_{burst+} and V_{burst-}, as shown in Fig. 5. In the burst-on state, driving signals are provided to the switches to maintain the energy output, causing the output voltage to rise. In the burst-off state, the driving signals are blocked to cut off energy transfer, causing the output voltage to naturally fall. This method significantly reduces switching losses, improves light-load efficiency, and strictly restricts the voltage within a safe range.

When $m(t) = 1$, the H-bridge operates at the non-zero voltage level, and the LLC converter is connected to the rated load. The controller switches to PFM control. The PI regulator adjusts the switching frequency based on the output voltage error, enabling the LLC converter to continuously operate near the resonant frequency. This ensures excellent soft-switching characteristics while maintaining the strict regulation accuracy of the output voltage.

3.3 Switching Control Under Pulse-Load Conditions

When the H-bridge switches from the zero voltage level to a non-zero voltage level, $m(t) = 2$, the load experiences a sudden step change from no load to full load. Due to the bandwidth limitations of the PI controller, the state variables cannot respond promptly, resulting in a severe voltage drop. The linear controller requires multiple switching cycles to adjust the switching frequency. Figure 6 shows the conventional linear frequency control during the load transient from no load to full load. To solve this problem, SOTC is introduced at the load-switching instant. By applying two optimal switching pulse widths, the state trajectory is forced to transition from the no-load orbit to the full-load steady-state circle within a single switching cycle. Figure 7 illustrates the state trajectory transition from no load to full load under SOTC.

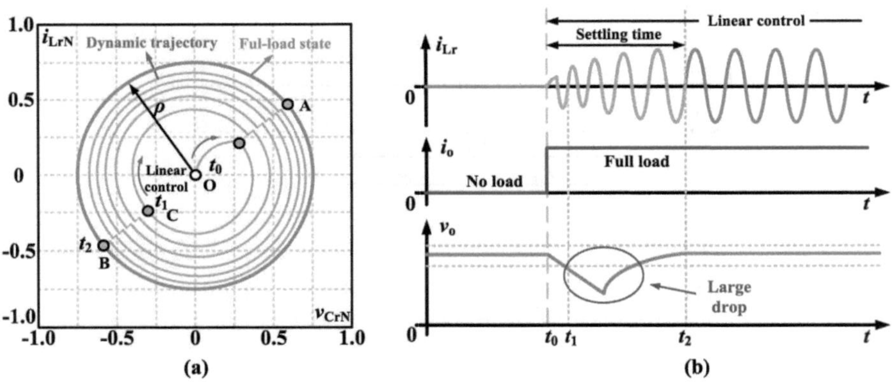

Fig. 6. Conventional linear frequency control from no load to full load. (a) State trajectory. (b) Typical waveforms.

Trajectory OA can be written in the form of time-parameterized equations as follows:

$$\begin{cases} v_{\mathrm{CrN}}(t) = 1 + \cos(-\omega_{\mathrm{m}} t + \pi) \\ i_{\mathrm{LrN}}(t) = \dfrac{Z_{\mathrm{r}}}{Z_{\mathrm{m}}} \sin(-\omega_{\mathrm{m}} t + \pi) \end{cases} \tag{14}$$

The initial phase conditions of the time-parameterized equations correspond to the starting point O of trajectory OC. The abscissa of point A is defined as

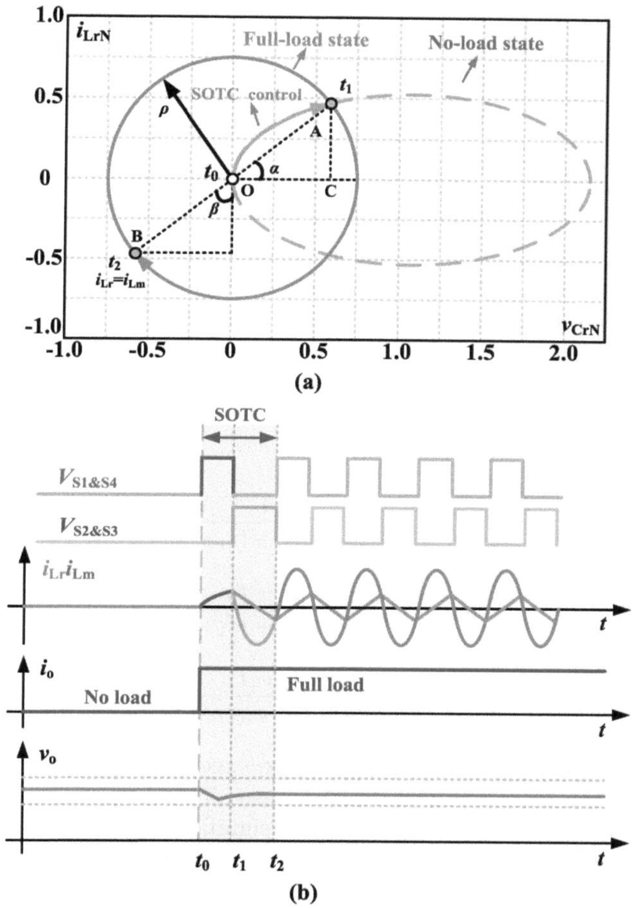

Fig. 7. SOTC control from no load to full load. (a) State trajectory. (b) Typical waveforms.

v_{CrNA}:

$$v_{CrNA} = \frac{\sqrt{L_r^2 + \rho^2 L_m (L_r + L_m)} - L_r}{L_m} \tag{15}$$

Consequently, the first switching pulse width can be calculated as:

$$t_1 - t_0 = \sqrt{(L_r + L_m) C_r} \left[\pi - \cos^{-1}(v_{CrNA} - 1)\right] \tag{16}$$

At point B, $i_{Lr} = i_{Lm}$, and at this moment, i_{Lm} reaches its maximum value. The ordinate of point B is defined as i_{LrNB}:

$$i_{LrNB} = -i_{LmN,max} = -\frac{nV_o}{4f_r L_m} \frac{1}{V_{in}/Z_r} = -\frac{Z_r}{4f_r L_m} \tag{17}$$

The trajectory AB from t_1 to t_2 can be divided into three angular segments, namely α, β, and $\pi/2$. They are expressed respectively as:

$$\begin{cases} \alpha = \cos^{-1}\left(\dfrac{v_{\mathrm{CrNA}}}{\rho}\right) \\ \beta = \cos^{-1}\left(\dfrac{|i_{\mathrm{LrNB}}|}{\rho}\right) \end{cases} \tag{18}$$

Since the angular frequency corresponding to trajectory OB is ω_{r}, the second switching pulse width is obtained as:

$$t_2 - t_1 = \frac{1}{\omega_{\mathrm{r}}}\left(\alpha + \beta + \frac{\pi}{2}\right) \tag{19}$$

Rearranging the above equation yields:

$$\begin{cases} f_{\mathrm{sotc}} = \dfrac{1}{(t_2 - t_1) + (t_1 - t_0)} \\ d_{\mathrm{sotc1}} = \dfrac{t_1 - t_0}{(t_2 - t_1) + (t_1 - t_0)} \\ d_{\mathrm{sotc2}} = \dfrac{t_2 - t_1}{(t_2 - t_1) + (t_1 - t_0)} \end{cases} \tag{20}$$

Once these two optimal switching pulses are executed, the state variables will precisely reach the new steady-state trajectory. The controller is subsequently handed back to the PFM control, thereby significantly reducing the dynamic error caused by the sudden load change and achieving seamless mode switching.

4 Simulation Results

To verify the effectiveness of the proposed switching control strategy based on H-bridge mode identification, a simulation model is established in this section. The dynamic response characteristics of the cascaded LLC converter and H-bridge system are investigated under no-load operation, full-load operation, and load-switching transients. Furthermore, comparative analysis with the traditional linear PI control are conducted. The main circuit parameters of the system are listed in Table 2.

Figure 8 illustrates the response waveforms of the system under the conventional linear control during pulse-load transients. When the equivalent pulse load R_{eq} of the downstream H-bridge switches at a frequency of 100 Hz, the switching frequency cannot be effectively adjusted within a short time due to the limited loop bandwidth of the PI controller. This limitation leads to a significant drop in the output voltage, with a maximum drop of 3.5 V, and the system requires 4 ms to recover to the steady-state value.

Table 2. Parameters of the LLC Resonant Converter

Parameter	Value
Input voltage V_{dc}	100 V
Output voltage V_o	50 V
Rated power P_o	200 W
Transformer turns ratio n	2
Resonant inductor L_r	36 μH
Magnetizing inductor L_m	133 μH
Resonant capacitor C_r	82 nF
Resonant frequency f_r	92 kHz

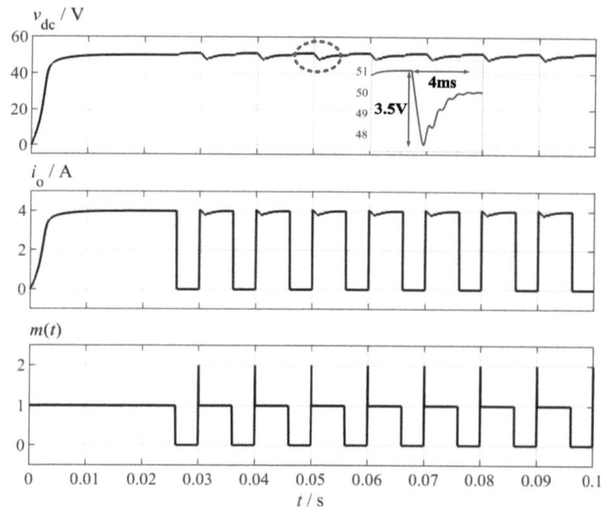

Fig. 8. Dynamic response under conventional linear control.

In contrast, the proposed strategy can identify the operating modes of the H-bridge and trigger three control modes via the switching function $m(t)$. As shown in Fig. 9, the mode identification module can instantaneously capture the load transition. Simulation results demonstrate that the proposed method can suppress the output voltage fluctuation within 0.6 V and reduce the settling time to 2 ms. Compared with the conventional control, the recovery speed is improved by 50%.

Figure 10 presents the zoomed-in waveforms of the SOTC process. When the load transitions from no load to full load, based on the previous calculation, the switching period of the LLC converter is 10 μs. At this point, the switching function $m(t)$ outputs 2, and the controller outputs two optimal driving pulses with duty cycles of 0.457 and 0.543, respectively. Through these two pulses, the energy

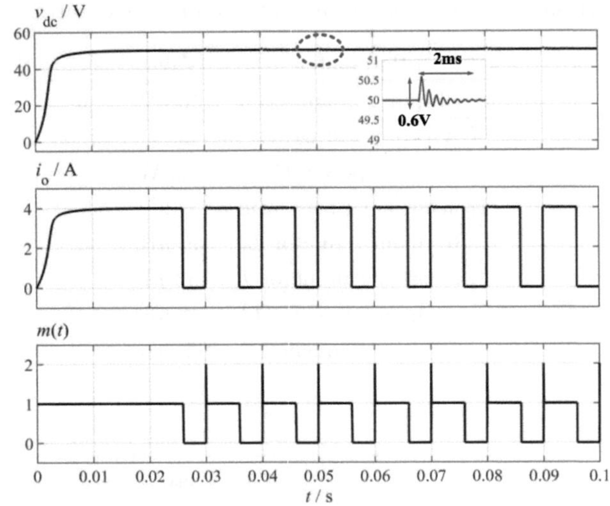

Fig. 9. Dynamic response under the proposed switching control strategy.

in the resonant tank is rapidly built up, forcing the state variables to quickly transition from the no-load trajectory to the full-load steady-state trajectory within a single switching cycle. Once the SOTC is completed, the system switches back to the PFM control. When the load suddenly drops from full load to no load, the controller enters the Burst mode to reduce switching losses.

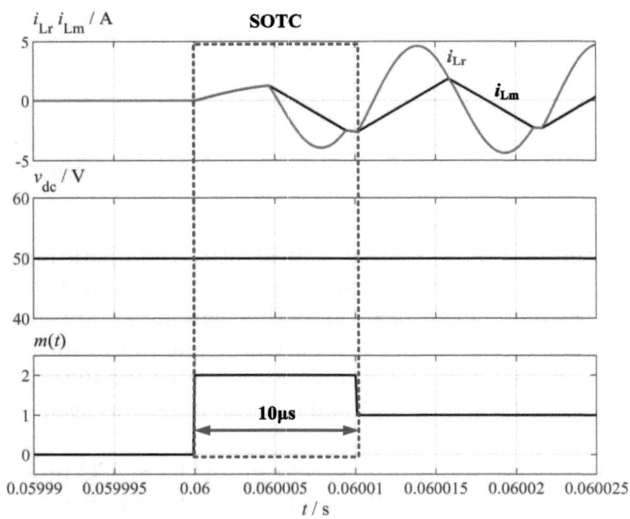

Fig. 10. Operating waveforms under SOTC control.

5 Conclusion

This paper addresses the pulse-load disturbances encountered by the front-end LLC resonant converter under the operation of the downstream H-bridge and proposes a switching control strategy based on H-bridge mode identification. This strategy overcomes the inherent delay limitations of traditional voltage-threshold-triggered methods. By directly monitoring the switching states of the H-bridge as the criterion for load transitions, the system achieves seamless transitions among three operating modes: Burst mode for no-load conditions, PFM control for full-load steady state, and SOTC control during the load-switching transients. Simulation results demonstrate that the proposed control strategy effectively mitigates voltage overshoots and drops induced by sudden pulse variations, without requiring any additional hardware. Compared to conventional control methods, the proposed strategy significantly reduces the dynamic settling time by 50% and maintains the steady-state output voltage ripple within 0.6 V, thereby substantially enhancing the dynamic performance and output power quality of the cascaded LLC and H-bridge system.

References

1. Sechilariu, M., Wang, B., Locment, F.: Building integrated photovoltaic system with energy storage and smart grid communication. IEEE Trans. Industr. Electron. **60**(4), 1607–1618 (2013)
2. Zhao, T., Sun, Z., Feng, Z., Wang, M., Wu, M., Zhang, X.: An optimized active power backflow suppression strategy for cascaded H-bridge PV grid-connected inverter during inter-phase short-circuit fault. IEEE Trans. Power Electron. **38**(7), 9127–9142 (2023)
3. Zhang, X., Zhao, T., Mao, W., Tan, D., Chang, L.: Multilevel inverters for grid-connected photovoltaic applications: examining emerging trends. IEEE Power Electron. Mag. **5**(4), 32–41 (2018)
4. Kouro, S., Fuentes, C., Perez, M., Rodriguez, J.: Single DC-link cascaded H-bridge multilevel multistring photovoltaic energy conversion system with inherent balanced operation. In: IECON 2012 – 38th Annual Conference on IEEE Industrial Electronics Society, Montreal, QC, Canada, pp. 4998–5005 (2012)
5. Beiranvand, R., Rashidian, B., Zolghadri, M.R., Hossein Alavi, S.M.: A design procedure for optimizing the LLC resonant converter as a wide output range voltage source. IEEE Trans. Power Electron. **27**(8), 3749–3763 (2012)
6. Fei, C., Lee, F.C., Li, Q.: High-efficiency high-power-density LLC converter with an integrated planar matrix transformer for high-output current applications. IEEE Trans. Industr. Electron. **64**(11), 9072–9082 (2017)
7. Chen, Y., Liu, Y.-F.: Latest advances of LLC converters in high current, fast dynamic response, and wide voltage range applications. CPSS Trans. Power Electron. Appl. **2**(1), 59–67 (2017)
8. Shang, M., Wang, H., Cao, Q.: Reconfigurable LLC topology with squeezed frequency span for high-voltage bus-based photovoltaic systems. IEEE Trans. Power Electron. **33**(5), 3688–3692 (2018)

9. Chatterjee, D., Kapat, S., Mallik, R., Kar, I.N., Jain, A.: PID controller tuning in voltage mode controlled LLC converters for fast transient under PFM. In: 2024 IEEE Transportation Electrification Conference and Expo (ITEC), Chicago, IL, USA, pp. 1–6 (2024)
10. Jang, J., Joung, M., Choi, S., Choi, Y., Choi, B.: Current mode control for LLC series resonant dc-to-dc converters. In: 2011 Twenty-Sixth Annual IEEE Applied Power Electronics Conference and Exposition (APEC), Fort Worth, TX, USA, pp. 21–27 (2011)
11. Hu, Z., Liu, Y.F., Sen, P.C.: Bang-Bang charge control for LLC resonant converters. IEEE Trans. Power Electron. **30**(2), 1093–1108 (2015)
12. Ma, H., Liu, Q., Guo, J.: A sliding-mode control scheme for LLC resonant DC/DC converter with fast transient response. In: IECON 2012 – 38th Annual Conference on IEEE Industrial Electronics Society, pp. 162–167 (2012)
13. Feng, W., Lee, F.C., Mattavelli, P.: Simplified optimal trajectory control (SOTC) for LLC resonant converters. IEEE Trans. Power Electron. **28**(5), 2415–2426 (2013)
14. Yao, K., Zhang, Z., Tang, J., Gao, Z., Zhu, W., Gu, Z., Ren, X., Chen, Q.: Linear-nonlinear optimal step control for 1-kV SiC LLC converters with pulse loads. IEEE Trans. Power Electron. **36**(10), 12008–12018 (2021)
15. Liu, W., Duan, S., Luo, S., Peng, T., Wang, Y.: Nonlinear optimal control for 2.5-kW LLC converters with pulse loads. In: 2022 IEEE 3rd China International Youth Conference on Electrical Engineering (CIYCEE), Wuhan, China, pp. 1–6 (2022)

Main-Auxiliary Cooperative Minimum-Time Control for Asymmetric Multiphase Buck Converters

Yaopei Liang[1], Chen Zhang[1], Lihao Wu[1](✉), Guidong Zhang[1], Wenjie Ma[1], and Zhong Li[2]

[1] School of Automation, Guangdong University of Technology, Guangzhou, China
2112404536@mail2.gdut.edu.cn, {wu_lihao,guidong.zhang,epwjma}@gdut.edu.cn
[2] Faculty of Mathematics and Computer Science, Fernuniversität in Hagen, Hagen, Germany
zhong.li@fernuni-hagen.de

Abstract. This paper investigates the transient control problem of asymmetric multiphase buck converters used in high-performance voltage regulator modules (VRMs). Conventional control strategies suffer from limited transient response due to the inherent trade-off between steady-state efficiency and dynamic performance. To address this issue, a mainauxiliary cooperative minimum-time control (MACMTC) strategy is proposed for a three-phase asymmetric parallel buck converter consisting of two large-inductance main phases and one small-inductance auxiliary phase. Based on the output capacitor charge balance principle, analytical models are established for two typical transient conditions, namely load current step-up and step-down, and the optimal turn-on and turn-off timing of the main and auxiliary phases are derived. Simulation results obtained from MATLAB/Simulink demonstrate that the proposed MACMTC scheme significantly improves the transient response and reduces output voltage deviation compared with the conventional average current mode control (ACMC) method.

Keywords: Asymmetric parameters · Multiphase parallel buck converter · Minimum-time control · Transient response

1 Introduction

With the explosive growth of big data and artificial intelligence technologies, high-performance processors have imposed extremely stringent requirements on the voltage accuracy, power density, and dynamic response performance of power supply systems [1]. The load current of processors frequently exhibits step changes with large amplitudes and high slew rates, posing severe challenges to the transient response capability of dedicated VRMs: how to achieve rapid recovery while minimizing output voltage overshoot or undershoot has become a core challenge in VRM design [2].

J. Lu et al. (Eds.): FLINS-ISKE 2026, LNCS 16752, pp. 105–117, 2027.
https://doi.org/10.1007/978-981-92-2369-5_8

Multiphase parallel buck converters have become the mainstream topology choice for VRMs due to their current-sharing characteristics and ripple-cancellation advantages [3,4]. However, traditional symmetric-parameter multiphase topologies face inherent performance trade-offs in their design: while a larger inductance can effectively suppress steady-state current ripple and improve efficiency, it severely limits the slew rate of the inductor current, resulting in a lagging transient response; conversely, a smaller inductance can enhance dynamic response speed but introduces significant output ripple [5]. This contradiction between steady-state efficiency and transient speed makes it difficult for a single inductance parameter to achieve global performance optimization.

To overcome the aforementioned bottlenecks, both academia and industry have proposed asymmetric-parameter multiphase parallel buck converter architectures [6,7], as shown in Fig. 1. This topology achieves functional decoupling by configuring a large-inductance main phase and a small-inductance auxiliary phase: during steady-state operation, the large-inductance main phase handles the power output to ensure system efficiency; when a transient occurs, the small-inductance auxiliary phase is rapidly activated, utilizing its high current slew rate to achieve rapid energy injection or absorption [8]. The parallel scheme of a "high-efficiency channel + high-speed channel" proposed in [9], as well as the collaborative control of Normal and Turbo phases designed for different CPU operating modes in [10], have both verified the potential of the asymmetric-parameter architecture in balancing steady-state efficiency and transient performance.

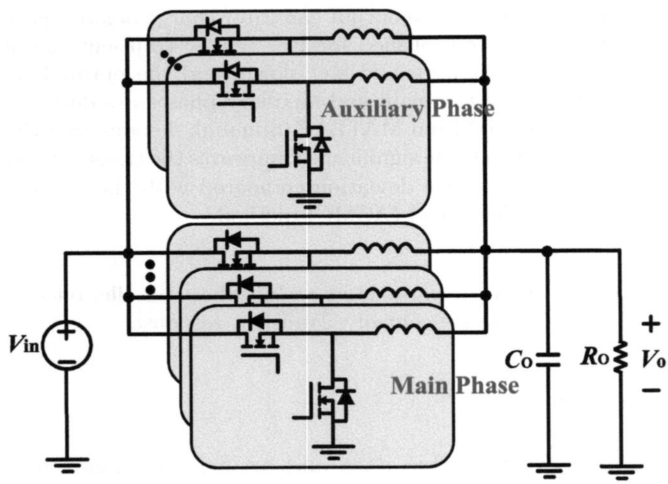

Fig. 1. Multiphase Interleaved Buck Converter with Asymmetric Parameters.

Fast transient response is a critical requirement for VRMs in modern high-performance processors. Asymmetric multiphase buck converters have been proposed to improve dynamic performance by combining large-inductance main phases with small-inductance auxiliary phases. However, existing control strategies cannot fully exploit the transient capability of such topologies. Addressing this issue, this paper investigates a three-phase asymmetric parallel buck converter composed of a two-phase large-inductance main phase and a one-phase small-inductance auxiliary phase. To optimize its dynamic performance, a MACMTC strategy is proposed. Based on the output capacitor charge balance principle, mathematical models are established for two typical transient conditions: load current step-up and step-down. Subsequently, analytical expressions for the precise turn-on and turn-off times of the main and auxiliary phases are derived. Finally, simulation results obtained from the MATLAB/Simulink platform demonstrate that the proposed MACMTC scheme achieves faster transient recovery and smaller output voltage deviation compared with the conventional ACMC method.

The remainder of this paper is organized as follows. Section 2 introduces the operating principles of the asymmetric multiphase buck converter and the transient charge balance analysis. Section 3 presents the proposed MACMTC control strategy and derives the analytical expressions for the switching timing. Section 4 provides simulation results and performance comparisons with the conventional ACMC method. Finally, Sect. 5 concludes the paper.

2 Minimum-Time Control

When the system operates in a steady state, the load current corresponds to the DC component of the inductor current, and its value equals the average inductor current; meanwhile, the capacitor current primarily reflects the AC component of the inductor current, and its average value over a switching cycle is zero. According to the principle of capacitor charge balance, under steady-state conditions, the average value of the output capacitor current within each switching cycle must remain zero. Only when this condition is satisfied can the output voltage at the beginning and end of a switching cycle remain identical, thereby maintaining the steady-state operation of the system. When a sudden change in load current occurs, the output current changes. However, since the inductor current cannot change abruptly, the variation in the output current is borne by the output capacitor, which consequently manifests as an overshoot or undershoot in the load voltage. To mitigate the output voltage deviation and shorten the regulation time of the buck converter during load disturbances, relevant studies have proposed a class of optimal control strategies based on Pontryagin's Minimum Principle.

This strategy aims to minimize the state transition time and is therefore commonly referred to as the MTC strategy. In terms of specific implementation, the system continues to employ traditional linear control strategies during the steady-state operation phase to ensure the steady-state accuracy and stability of the output voltage; when a transient load change is detected, a transient control mechanism based on the charge balance concept is introduced, enabling the system to transition from its initial operating point to the new target state in the shortest possible time while minimizing the output voltage fluctuation range. The transient regulation process is typically divided into two control intervals, namely the switch turn-on time and turn-off time. By rationally planning these two periods, the transient regulation is completed within a single switching cycle, thereby achieving a time-optimal load step trajectory at the control level. Figure 2 illustrates the basic principle of the MTC strategy for a single-phase buck converter.

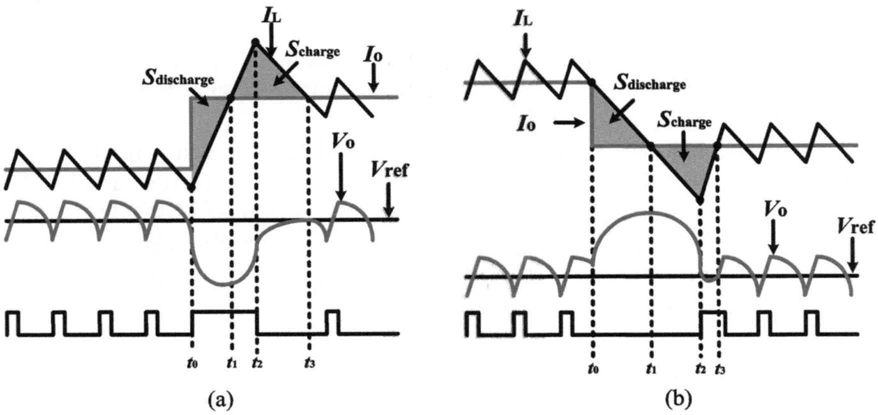

Fig. 2. The basic principle of minimum time control. (a) Load current step-up. (b) Load current step-down.

Taking the scenario shown in Fig. 2(a) as an example, the implementation process of the minimum-time control strategy under a load current step-up condition is analyzed. At time t_0, the load current experiences a step increase. However, restricted by the energy storage characteristics of the inductor, the inductor current cannot complete the corresponding change instantaneously. Consequently, the newly added load current demand must be temporarily supplied by the output capacitor, leading to a drop in the output capacitor voltage to maintain the load power supply. To reduce the amplitude of the output voltage undershoot, it is necessary to decrease the discharge amount of the output capacitor, which is achieved by increasing the rising slew rate of the inductor current. Therefore, during the time interval t_0-t_1, by turning on the high-side switch of the buck converter, the maximum forward voltage is applied across the inductor, thereby causing the inductor current to rise at its maximum slope. The moment when the inductor current rises to equal the load current is denoted as t_1. At this point, the output capacitor ceases discharging, and the output voltage undershoot reaches its maximum value. The magnitude of the output voltage drop is directly proportional to the amount of charge released by the capacitor; thus, suppressing the voltage undershoot can be achieved simply by minimizing the discharge area $S_{\text{discharge}}$ shown in Fig. 2(a).

Since the capacitor bore the energy difference between the inductor current and the load current in the previous stage, the inductor current must subsequently continue to rise to compensate for the charge released by the output capacitor. Equating the entire transient regulation process to a single cycle, and according to the principle of capacitor charge balance, the inductor current continues to rise during the interval t_1-t_2 and decreases during the interval t_2-t_3, ultimately returning to equal the load current. This ensures that the discharge area and charge area of the capacitor in Fig. 2(a) satisfy $S_{\mathrm{discharge}} = S_{\mathrm{charge}}$. Consequently, accurately determining the instant t_2 becomes the key to realizing minimum-time control. When a load current step-down occurs, its dynamic process is the exact opposite of the aforementioned scenario. At this time, the inductor current is greater than the load current, and the excess energy will be absorbed by the output capacitor, resulting in an output voltage overshoot. Similarly, by treating the transient response process as a single regulation cycle, the variation process of the inductor current is controlled by calculating the turn-off and turn-on durations of the buck converter. This control principle is fundamentally consistent with the analysis method for a load current step-up, and its schematic is illustrated in Fig. 2(b).

Therefore, introducing the minimum-time control concept into the circuit topology and control strategy of asymmetric-parameter multiphase parallel buck converters, based on charge balance theory, holds significant research value. Under the premise of ensuring the steady-state voltage regulation accuracy and excellent current-sharing characteristics of the system, conducting in-depth research on the load transient response characteristics of such converters is of great significance for further enhancing their dynamic performance.

3 Main-Auxiliary Cooperative Minimum-Time Control

To ensure the steady-state performance of the asymmetric-parameter multiphase parallel buck converter and further improve its load dynamic performance, the concept of minimum-time control is introduced at the control strategy level. The topology of the proposed asymmetric-parameter three-phase parallel buck converter is shown in Fig. 3. The proposed Main-Auxiliary Cooperative Minimum-Time Control scheme is illustrated in Fig. 4. This control scheme is essentially a steady-state/transient/steady-state switching control strategy: during steady-state operation, average current control takes over, whereas during transients, the main-auxiliary cooperative minimum-time control assumes control. This hybrid control approach provides flexibility to the control of asymmetric-parameter multiphase buck converters, enabling the controller to better achieve excellent system performance under varying operating conditions. Furthermore, this scheme is implemented using a single digital controller, which effectively reduces the cost of the control circuit.

To determine the turn-on and turn-off instants of the switching devices under the minimum-time control strategy, this paper treats the asymmetric-parameter three-phase interleaved buck converter as an integrated system and establishes a

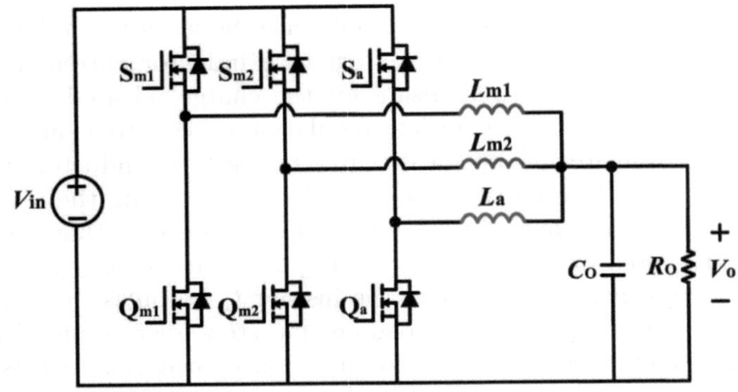

Fig. 3. Asymmetric-Parameter Three-Phase Interleaved Buck Converter.

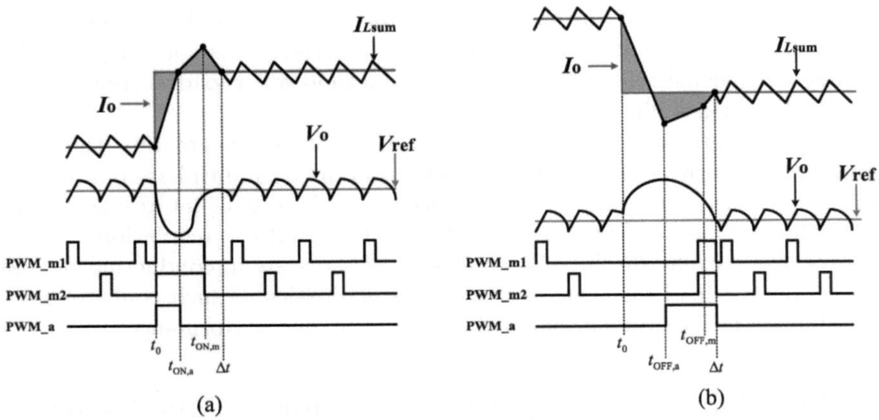

Fig. 4. Principle of the main-auxiliary cooperative minimum-time control. (a) Load current step-up. (b) Load current step-down.

calculation method for the switching sequence based on charge balance theory. To facilitate theoretical derivation and analysis, the influence of the parasitic parameters of the output filter network on the system's dynamic process is neglected during the transient response of the asymmetric-parameter three-phase parallel buck converter.

When the load experiences a step increase of ΔI_o, according to Kirchhoff's Current Law, the output node satisfies:

$$\sum_{i=1}^{2} i_{L\mathrm{m},i}(t) + i_{L\mathrm{a}}(t) = i_{\mathrm{c}}(t) + i_{\mathrm{o}}(t) \tag{1}$$

Consequently, the output capacitor charge and the load-side charge are given by:

$$Q_{\mathrm{c}} = C\Delta V = \int_{t_0}^{t_0+\Delta t} i_c dt \tag{2}$$

$$\Delta Q_{\mathrm{o}} = \int_{t_0}^{t_0+\Delta t} i_{\mathrm{o}} dt \tag{3}$$

Integrating both sides of equation (1) over the transient regulation time Δt yields the charge balance equation:

$$\sum_{i=1}^{2} Q_{L\mathrm{m},i} + Q_{L\mathrm{a}} = Q_c + \Delta Q_{\mathrm{o}} \tag{4}$$

To achieve single-cycle, oscillation-free recovery under minimum-time control, it is required that the capacitor voltage deviation ΔV at the end of the transient equals zero, meaning the net absorbed charge Q_c of the capacitor is zero. Simultaneously, the total charge consumed by the load during Δt can be approximated by a rectangular area:

$$\Delta Q_{\mathrm{o}} = \int_{t_0}^{t_0+\Delta t} i_{\mathrm{o}}(t) dt \approx (2I_{\mathrm{m0}} + \Delta I_{\mathrm{o}})\Delta t \tag{5}$$

Combining the above equations yields the core charge conservation constraint equation for a load current step-up:

$$\sum_{i=1}^{2} Q_{L\mathrm{m},i} + Q_{L\mathrm{a}} = (2I_{\mathrm{m0}} + \Delta I_{\mathrm{o}})\Delta t \tag{6}$$

where $Q_{L\mathrm{m},i}$ denotes the charge of the i-th main-phase inductor, $Q_{L\mathrm{a}}$ is the charge of the auxiliary-phase inductor, and I_{m0} represents the initial current of the main-phase inductor.

Assume the inductances of the two main phases are $L_1 = L_2 = L_{\mathrm{m}}$, and the inductance of the auxiliary phase is L_{a}. During steady-state operation, the main phases equally share the load current, meaning the initial steady-state currents of the main-phase inductors are $I_{\mathrm{m1,0}} = I_{\mathrm{m2,0}} = I_{\mathrm{m}}$. Meanwhile, the auxiliary phase remains in a dormant state, meaning its initial steady-state inductor current is $I_{\mathrm{a,0}} = 0$. During the transient period, the high-side switches of the main phases are forcibly turned on for a duration of $t_{\mathrm{ON,m}}$, after which the inductors enter a discharging state. The corresponding time-domain piecewise expression is given by:

$$i_{L\mathrm{m},i}(t) = \begin{cases} I_{\mathrm{m0}} + \frac{V_{\mathrm{in}}-V_{\mathrm{o}}}{L_{\mathrm{m}}} t & (0 \le t < t_{\mathrm{ON,m}}) \\ I_{\mathrm{m0}} + \frac{V_{\mathrm{in}}-V_{\mathrm{o}}}{L_{\mathrm{m}}} t_{\mathrm{ON,m}} - \frac{V_{\mathrm{o}}}{L_{\mathrm{m}}}(t - t_{\mathrm{ON,m}}) & (t_{\mathrm{ON,m}} \le t \le \Delta t) \end{cases} \tag{7}$$

Integrating this piecewise function over the transient regulation time Δt, expanding the polynomial, and performing algebraic simplification yield the analytical expression for the charge of the main-phase and auxiliary-phase.

$$Q_{Lm,i} = -\frac{V_{in}}{2L_m}t_{ON,m}^2 + \frac{V_{in}}{L_m}t_{ON,m}\Delta t - \frac{V_o}{2L_m}\Delta t^2 + I_{m0}\Delta t \tag{8}$$

$$Q_{La} = -\frac{V_{in}}{2L_a}t_{ON,a}^2 + \frac{V_{in}}{L_a}t_{ON,a}\Delta t - \frac{V_o}{2L_a}\Delta t^2 \tag{9}$$

Upon the completion of the transient, the main phases must fully take over the load step increment, with the final current increment of each phase being $\Delta I_o/2$; the auxiliary phase does not bear the load current, and its final current must return to zero. Based on the geometric relationships of the inductor current slopes, the physical constraints for the time variables are established:

$$L_m\frac{\Delta I_o}{2} = (V_{in} - V_o)t_{ON,m} - V_o(\Delta t - t_{ON,m}) \tag{10}$$

$$L_a \cdot 0 = (V_{in} - V_o)t_{ON,a} - V_o(\Delta t - t_{ON,a}) \tag{11}$$

Introducing the steady-state duty cycle $D = V_o/V_{in}$ yields the turn-on time expressions for the main and auxiliary phases:

$$t_{ON,m} = D\Delta t + \frac{L_m\Delta I_o}{2V_{in}} \tag{12}$$

$$t_{ON,a} = D\Delta t \tag{13}$$

Substituting (12) into (8) yields the single-phase charge equation of the main phase, which, after rearrangement, gives the total charge equation for the two main phases:

$$\sum_{i=1}^{2} Q_{Lm,i} = \frac{V_{in}D(1-D)}{L_m}\Delta t^2 + (1-D)\Delta I_o\Delta t - \frac{L_m\Delta I_o^2}{4V_{in}} + 2I_{m0}\Delta t \tag{14}$$

Similarly, substituting (13) into (9) yields the charge equation for the auxiliary phase:

$$Q_{La} = \frac{V_{in}D(1-D)}{2L_a}\Delta t^2 \tag{15}$$

Substituting (14) and (15) into (6) yields the solution for the optimal regulation time Δt under a load current step-up:

$$\Delta t = \frac{\Delta I_o\left(D + \sqrt{D\left[1 + (1-D)\frac{L_m}{2L_a}\right]}\right)}{V_{in}D(1-D)\left(\frac{2}{L_m} + \frac{1}{L_a}\right)} \tag{16}$$

Following a similar derivation to the load current step-up, the turn-off and turn-on times of the main and auxiliary phases during a load current step-down are obtained as follows:

$$t_{OFF,m} = (1-D)\Delta t + \frac{L_m\Delta I_o}{2V_{in}} \tag{17}$$

$$t_{\mathrm{OFF,a}} = (1 - D)\Delta t \tag{18}$$

Similarly, the optimal regulation time Δt under a load current step-down can be expressed by (19):

$$-\Delta t^2 \frac{V_{\mathrm{in}} D(1-D)}{2}\left(\frac{2}{L_{\mathrm{m}}} + \frac{1}{L_{\mathrm{a}}}\right) + \Delta t(1-D)\Delta I_{\mathrm{o}} + \frac{L_{\mathrm{m}}\Delta I_{\mathrm{o}}^2}{4V_{\mathrm{in}}} = 0 \tag{19}$$

By applying the quadratic formula and discarding the negative root, the solution for the optimal regulation time Δt under a load current step-down is obtained:

$$\Delta t = \frac{\Delta I_{\mathrm{o}}\left((1-D) + \sqrt{(1-D)\left[1 + D\frac{L_{\mathrm{m}}}{2L_{\mathrm{a}}}\right]}\right)}{V_{\mathrm{in}} D(1-D)\left(\frac{2}{L_{\mathrm{m}}} + \frac{1}{L_{\mathrm{a}}}\right)} \tag{20}$$

Based on the preceding charge balance derivation, the turn-off and turn-on times for the main and auxiliary phases during the transient response under a load current step-down are obtained.

4 Simulation Results

To verify that the asymmetric-parameter multiphase parallel buck converter employing the proposed MACMTC scheme meets the design requirements of VRMs, simulations are conducted using MATLAB/Simulink software. These simulations aim to validate the applicability, feasibility, and effectiveness of the proposed MACMTC scheme in enhancing load transient performance. The circuit simulation parameters of the proposed asymmetric-parameter three-phase parallel buck converter are listed in Table 1.

Table 1. Simulation parameters

Simulation Parameter	Symbol	Value
Input voltage	V_{in}	12 V
Output voltage	V_{o}	3 V
Main phase inductance	$L_1 = L_2 = L_{\mathrm{m}}$	20 μH
Auxiliary phase inductance	L_{a}	10 μH
Switching frequency	f_{s}	100 kHz
Output capacitance	C_{o}	220 μF

To demonstrate the applicability and effectiveness of the proposed MACMTC scheme in handling load current transients, this section conducts simulations of load current steps under two different operating conditions. Figure 5 illustrates the scenario where the load current steps from 10 A to 20 A, while Fig. 6 shows the case where the load current steps from 20 A to 10 A. Figure 5a and Fig. 6a display the operating waveforms of the main-phase and auxiliary-phase inductor currents under the proposed MACMTC scheme during the transients. Furthermore, Fig. 5b and Fig. 6b present the corresponding waveforms of the output voltage and the total inductor current.

As can be observed from Figs. 5 and 6, the performance of the proposed MACMTC scheme is fundamentally consistent with the theoretical calculations. Once the transient control takes over, the overall response concludes within a single regulation cycle, and no secondary regulation occurs after the new steady state is reached.

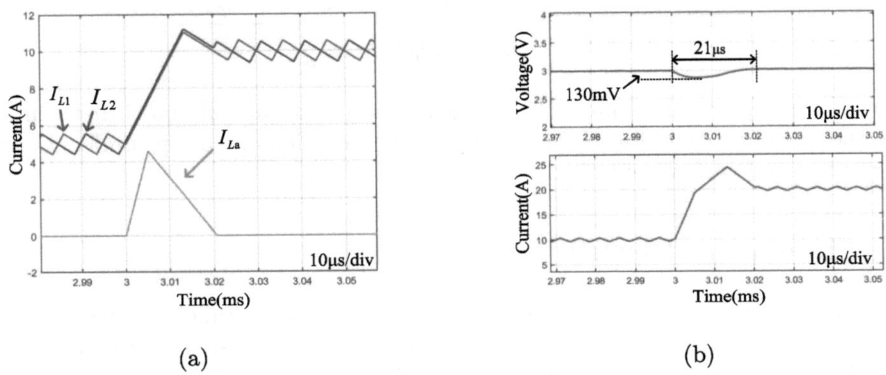

Fig. 5. Simulation results of the proposed MACMTC scheme under a load current step from 10 A to 20 A: (a) Main phase and auxiliary phase inductor current. (b) Output voltage and total inductor current.

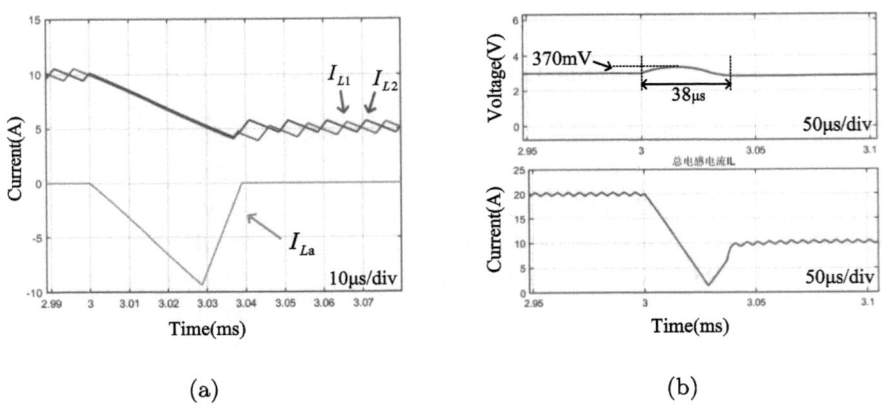

Fig. 6. Simulation results of the proposed MACMTC scheme under a load current step from 20 A to 10 A: (a) Main phase and auxiliary phase inductor current. (b) Output voltage and total inductor current.

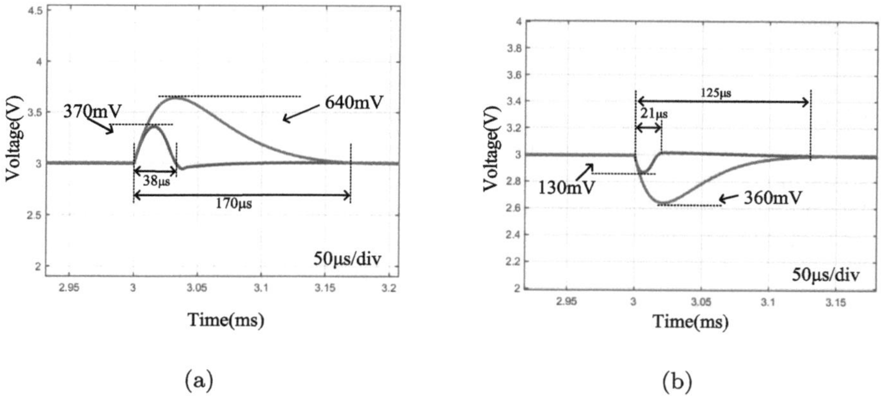

Fig. 7. Comparison of the proposed MACMTC scheme and ACMC: (a) Load current 20 A-10 A step-down. (b) Load current 10 A-20 A step-up.

Subsequently, the simulation compares the performance between the proposed MACMTC and the traditional ACMC during load current steps. As illustrated in Fig. 7, the blue line represents the output voltage waveform of the proposed MACMTC, while the red line depicts the output voltage waveform of the traditional ACMC. It can be observed that under both load current step-up and step-down conditions, the proposed MACMTC exhibits a smaller voltage overshoot and a shorter transient response time. This clearly demonstrates the applicability, feasibility, and effectiveness of the proposed control strategy in enhancing the load transient performance of the asymmetric-parameter three-phase parallel buck converter.

5 Conclusion

This paper focuses on optimizing the transient performance of asymmetric-parameter multiphase parallel buck converters and deeply investigates a transient current control strategy based on MTC. To address the sluggish response of traditional control methods during load transients, a MACMTC scheme is proposed. By leveraging the small-inductance advantage of the auxiliary phase in the asymmetric topology, a hybrid control architecture with smooth transitions between steady-state and transient operations is designed. In this scheme, the steady-state control unit dominates during normal operation, whereas the transient control unit takes over upon encountering a load step, utilizing the auxiliary phase to achieve ultra-fast current injection or absorption. Furthermore, a transient charge balance mathematical model is constructed. Based on the principle of output capacitor charge balance, detailed mathematical modeling is conducted for two typical transient conditions: load current step-up and step-down. Through rigorous algebraic derivations, the analytical expressions for the precise turn-on and turn-off times of the main and auxiliary phases during the transient response are solved, ensuring that the net charge absorbed by the output capacitor is zero at the end of the transient, thereby achieving a rapid, single-cycle, oscillation-free recovery. Finally, to verify the effectiveness of the transient control strategy, a system model of the asymmetric three-phase parallel buck converter was built in the Simulink platform to comparatively analyze the dynamic performance of the proposed MACMTC scheme and the traditional ACMC under various load step conditions. Simulation results demonstrate that the MACMTC scheme significantly reduces the overshoot and undershoot amplitudes of the output voltage, and the entire current regulation process reaches a new steady state after a single switching action without secondary voltage oscillations, thereby validating the effectiveness and superiority of the proposed control strategy in enhancing load transient performance.

References

1. Wang, L., Li, C., Liang, J., Ng, W.T., Wang, H.: Analysis and design of ultrafast series capacitor trans-inductor voltage regulator with constant on-time control. IEEE Trans. Power Electron. **40**(6), 8315–8327 (2025)
2. Yao, K., Ren, Y., Lee, F.C.: Critical bandwidth for the load transient response of voltage regulator modules. IEEE Trans. Power Electron. **19**(6), 1454–1461 (2004)
3. Abu-Qahouq, J., Mao, H., Batarseh, I.: Multiphase voltage-mode hysteretic controlled DC–DC converter with novel current sharing. IEEE Trans. Power Electron. **19**(6), 1397–1407 (2004)
4. Panov, Y., Jovanovic, M.M.: Design considerations for 12-V/1.5-V, 50-A voltage regulator modules. IEEE Trans. Power Electron. **16**(6), 776–783 (2001)
5. Liu, H., Zhang, D., Wang, D.: Design considerations for output capacitance under inductance mismatches in multiphase buck converters. IEEE Trans. Power Electron. **32**(7), 5004–5015 (2017)
6. Kim, D., Hong, M., Baek, J., Lee, J., Shin, J., Shin, J.-W.: Soft-switching auxiliary current control for improving load transient response of buck converter. IEEE Trans. Power Electron. **36**(3), 2488–2494 (2021)
7. Barrado, A., Lazaro, A., Vazquez, R., Salas, V., Olias, E.: The fast response double buck DC–DC converter (FRDB): operation and output filter influence. IEEE Trans. Power Electron. **20**(6), 1261–1270 (2005)
8. Qian, Y., Xu, S., Wang, X., Li, L., Sun, W.: A reconfigurable auxiliary circuit for transient enhancement in processor power systems with multiple voltage rails. IEEE Trans. Industr. Electron. **72**(6), 6383–6392 (2025)
9. Kim, D., Baek, J., Lee, J., Shin, J., Shin, J.-W.: Implementation of soft-switching auxiliary current control for faster load transient response. IEEE Access **9**, 7092–7106 (2021)
10. Su, Y., Cheng, K.-Y.B., Wu, W.: High-efficiency multiphase DC–DC converters for powering processors with turbo mode based on configurable current sharing ratios and intelligent phase management. In: 2017 IEEE Applied Power Electronics Conference and Exposition (APEC), pp. 191–196. IEEE (2017)
11. Meyer, E., Liu, Y.-F.: A practical minimum time control method for buck converters based on capacitor charge balance. In: 2008 Twenty-Third Annual IEEE Applied Power Electronics Conference and Exposition (APEC), pp. 10–16. IEEE (2008)

12. Yousefzadeh, V., Babazadeh, A., Ramachandran, B., Alarcon, E., Pao, L., Maksimovic, D.: Proximate time-optimal digital control for synchronous buck DC–DC converters. IEEE Trans. Power Electron. **23**(4), 2018–2026 (2008)
13. Cheng, P.M., Vasić, M., Garcia, O., Oliver, J.A., Alou, P., Cobos, J.A.: Multiphase buck converter with minimum time control strategy for RF envelope modulation. In: 2011 Twenty-Sixth Annual IEEE Applied Power Electronics Conference and Exposition (APEC), pp. 904–909. IEEE (2011)
14. Halivni, B., Peretz, M.M.: Digital controller for high-performance multiphase VRM with current balancing and near-ideal transient response. In: 2020 IEEE Applied Power Electronics Conference and Exposition (APEC), pp. 2206–2213. IEEE (2020)

Research on the Impact of Green Innovation on the Performance of Agricultural Enterprises ESG

Yuewen Wang[1], Sixian Cheng[1], Siqi Hou[2], and Limin Du[2(✉)]

[1] Advanced Institute of Finance, Henan University, Zhengzhou 45003, Henan Province, China

[2] International Institute of Science and Technology Institute, Henan University, Zhengzhou 45003, Henan Province, China

dulimin@vip.henu.edu.cn

Abstract. From a green innovation perspective, this study explores its impact on the ESG performance of agricultural enterprises. Based on data from listed agricultural companies in China, the empirical findings reveal that green innovation significantly improves corporate ESG performance by enhancing information transparency. This research provides new evidence to help us understand how green innovation drives the sustainable development of agricultural firms, offering both theoretical insights and practical guidance to support their green transformation and high-quality development.

Keywords: Green innovation · Agribusiness · ESG performance · Corporate information transparency · Corporate size

1 Introduction

Agriculture is the foundation of China's economic and social development. However, it has long been plagued by challenges such as overproduction, significant environmental strain and limited sustainability. As the sector undergoes a crucial shift from traditional to modern practices, a central question is how to build a production system that aligns with ecological security [1]. In response, China has adopted the ESG framework as a key metric for evaluating corporate sustainability, supported by national strategies including the 'dual carbon' goals and rural revitalization [2]. Policies such as the 'Accelerating the Construction of a Powerful Agricultural Country (2024–2025)' further promote the green transition. However, Leite de Almeida et al. note that ESG research in agriculture is still in its infancy, particularly with regard to the social (S) and governance (G) dimensions, such as oversight of the supply chain, policy incentives and transparency of information, while also overlooking marginalized groups such as smallholder farmers in developing countries [3].

Strong ESG performance can enhance the efficiency and sustainability of agricultural enterprises, enabling them to fulfill their responsibilities, align with

J. Lu et al. (Eds.): FLINS-ISKE 2026, LNCS 16752, pp. 118–131, 2027.
https://doi.org/10.1007/978-981-92-2369-5_9

national policy and strike a balance between economic and social benefits [4]. Green innovation serves as a key driver to achieve these goals. By investing in green research and development (R&D), improving processes, and innovating products, firms can reduce their use of resources and pollution while boosting their productivity and competitiveness. For agricultural enterprises in particular, green innovation is critical for mitigating non-point source pollution, ensuring product quality and safety, and restoring ecosystems. However, given the long agricultural cycle, its heavy reliance on natural conditions and its pronounced ecological externalities, the mechanisms linking green innovation to ESG performance are more complex than in manufacturing. Thus, a tailored analysis based on the developmental logic of agricultural enterprises is necessary and timely.

Agricultural enterprises play a key role in resource utilization, environmental protection, and social responsibility. Their ESG performance is influenced by multiple factors: internally, by green technology application [5], industrial chain coordination [6], and scale operations [7]; externally, by digital development [8], green consumption, and supply chain collaboration, as well as market-based instruments like green credit [9], carbon trading [10], and green certification [11], alongside administrative regulations. In this context, green innovation embeds ecological adaptability into production, advancing emission reduction and food security. It motivates firms to pursue economic benefits while prioritizing ecological protection and rural livelihoods, enhancing ESG impact through farmer participation. Unlike rigid regulations, it provides flexible, context-sensitive solutions. Green innovation optimizes resource allocation, boosts competitiveness, and drives comprehensive ESG improvement, fostering sustainable development for enterprises and rural industries.

Using unbalanced panel data from 2015 to 2024, this study examines the impact of green innovation on the environmental, social and governance (ESG) performance of agricultural listed companies in China's A-share market. The research aims to deepen understanding of agricultural green transition pathways and provide empirical evidence for policy refinement. Compared to existing studies, this paper's marginal contributions are reflected primarily in three areas. Firstly, by integrating green innovation with agricultural ESG, it expands the scope of research on green innovation's economic effects. Secondly, focusing on agricultural enterprises, it addresses the lack of attention to manufacturing sectors in prior literature. Thirdly, it reveals the underlying mechanisms through which green innovation influences ESG performance, offering policy insights for advancing agricultural green transformation.

2 Theoretical Analysis and Research Hypotheses

2.1 Direct Impact

The role of green innovation in promoting the Environmental, Social, and Governance (ESG) performance of agricultural enterprises can be demonstrated from multiple theoretical perspectives, including Technological Innovation Theory, Dynamic Capability Theory, Stakeholder Theory, Resource-Based Theory, and Pigou Tax Theory. Technological innovation theory points out that green innovation is the core of agricultural technological progress. It reduces resource consumption and non-point source pollution by improving agronomy, upgrading equipment, and optimizing industrial chain management. It also strengthens the mechanism of linking interests and supply chain transparency, synergistically improving environmental performance, social responsibility, and governance. This promotes sustainable agricultural development based on the concept of "innovation-driven green agricultural development" [?]. Based on this, dynamic capability theory posits that enterprises must adapt to rapidly changing external environments through continuous learning and resource reorganization. Based on this, the dynamic capability theory posits that enterprises must adapt to rapidly changing external environments through continuous learning and reorganization of resources [13]. Green innovation is key to enterprises adapting to changes in the external environment. It allows agricultural enterprises to convert environmental regulations, market demand, and rural revitalization development requirements into internal driving forces and respond promptly to external pressure regarding agricultural green development. It maintains advantages in supply chain cooperation.The resource view theory highlights that scarce and non-replicable resources and capabilities constitute the foundation of sustainable competitive advantages for enterprises. Through green innovation, agricultural enterprises accumulate unique green technologies and management expertise, such as stable partnerships with farmers and local ecological governance experience. These resources enhance ESG performance and secure sustainable competitive advantages [14].

Similarly, stakeholder theory emphasizes that enterprises must consider the expectations and needs of all stakeholders. Green innovation is a strategic measure through which enterprises can address the concerns of consumers, rural communities, farmers, and investors regarding agricultural ecological protection and rural development [15]. Improving environmental (E) and social (S) performance enhances corporate brand reputation and stakeholder satisfaction, promoting long-term corporate value creation and improving agricultural industry chain governance (G). From the perspective of Pigou tax theory, the external cost of environmental pollution in agricultural production can be internalized through policy or market mechanisms, such as carbon sink trading, emission trading, and subsidies. This market-based environmental regulation provides economic incentives and imposes environmental cost constraints on agricultural enterprises, forcing them to adopt green innovation models to reduce emissions and costs. In summary, green innovation reduces the environmental external costs brought about by agricultural production and promotes the synergistic improvement of enterprises in the dimensions of environment (E), society (S), and governance (G) under multiple internal and external drivers. Based on this, this paper proposes the following hypotheses:

H1: Green innovation can substantially enhance the ESG performance of agricultural enterprises.

2.2 Mediation Effect

Corporate transparency refers to the adequacy and authenticity of information disclosed by a company to external stakeholders regarding its operations, management, financial status and fulfillment of social responsibilities. Higher corporate transparency can reduce information asymmetry, enhance the capital market, and increase public awareness of, and trust in, corporate operating conditions. This can have an important impact on corporate resource acquisition capabilities and governance levels [17]. In the case of agricultural enterprises, the production process is subject to obvious seasonality and uncertainty, and is greatly affected by fluctuations in the natural environment and market prices. Consequently, business information tends to be more complex and unstable. Therefore, insufficient information disclosure and low transparency can easily exacerbate information asymmetry between enterprises and external investors, thereby limiting enterprises' access to resources and affecting their ability to develop sustainably [16].

Against this backdrop, green innovation is a vital means of advancing the industrial upgrading of agriculture, significantly enhancing information transparency within agribusinesses. Green innovation often involves activities such as research and development of green technology, environmental protection investments and improvements to green production processes. Such activities usually necessitate enterprises engaging in more standardized and comprehensive information disclosure, thereby increasing the transparency of their business operations. Secondly, as enterprises pursue green innovation, they place greater emphasis on publicly disclosing their environmental performance, social responsibility and governance structures. This fosters greater recognition among external investors, government agencies and the public of the enterprise's capacity for sustainable development. As the level of green innovation increases, the quality of information disclosure by agricultural enterprises improves, consequently strengthening corporate transparency.

Enhanced corporate transparency can further improve the environmental, social and governance (ESG) performance of agricultural enterprises. This mechanism can theoretically be explained from multiple perspectives, including information asymmetry theory, the resource-based view, dynamic capability theory and stakeholder theory. Information asymmetry theory [17] posits that green innovation encourages agricultural enterprises to disclose more environmental and social responsibility information. This reduces the information gap between enterprises and external stakeholders, thereby enhancing their ESG ratings. The resource-based view theory emphasizes that increased transparency helps agricultural enterprises gain greater access to policy, market, and social resources, providing crucial support for their sustained green development. Dynamic Capabilities Theory [13]suggests that green innovation demonstrates the capacity of agricultural enterprises to optimize resource allocation and operational models continuously in order to adapt to green agricultural development. This enhanced capability also motivates enterprises to establish more standardized and transparent governance systems. Stakeholder theory[18] posits that increased corporate transparency strengthens trust among stakeholders, including governments, investors, consumers and farmers, prompting enterprises to fulfill their environmental protection and social responsibilities more actively and driving sustained improvements in ESG performance.

In summary, green innovation enhances agricultural enterprises' ESG performance by improving information disclosure and corporate transparency, thereby reducing information asymmetry and strengthening stakeholder trust. Based on the above analysis, this paper proposes the following research hypothesis:

H2: Green innovation can indirectly promote the improvement of agricultural enterprises' ESG performance by enhancing corporate transparency.

3 Model Design and Variable Selection

3.1 Empirical Model Construction

Two-Way Fixed Effects Model. The model is specified as follows:

$$Y_{it} = \beta_0 + \beta_1 X_{it} + \lambda \sum Controls_{it} + \nu_i + \tau_t + \varepsilon_{it} \tag{1}$$

In Model (1), Y_{it} represents the environmentals, social and governance (ESG) performance of agricultural firm i in year t, while X_{it} is the core explanatory variable of this paper: agricultural green innovation. Controls denotes the set of control variables, including all firm-level control variables selected in this study. Following the specification requirements of the fixed-effects model, this paper controls for the fixed effects of individual agricultural firms and time. ε_{it} is the random error term.

Mediating Effect Model. The Mediated Effect Model is a classic analytical method used to explore the intrinsic transmission pathways and operational mechanisms through which independent variables influence dependent variables via mediating variables. First proposed by Baron and Kenny, the model's primary value lies in its ability to clearly dissect indirect relationships between variables and elucidate the role of mediating variables in transmitting effects between independent and dependent variables. This paper uses the classic three-step mediation model to examine the specific mechanism through which agricultural green innovation influences the environmental, social and governance(ESG) performance of agricultural enterprises. The specific model specifications as follows:

$$\begin{aligned} Y_{it} &= \beta_0 + \beta_1 X_{it} + \lambda \sum Controls_{it} + \nu_i + \tau_t + \varepsilon_{it} \\ M_{it} &= \beta_0 + \beta_1 X_{it} + \lambda \sum Controls_{it} + \nu_i + \tau_t + \varepsilon_{it} \\ Y_{it} &= \beta_0 + \beta_1 X_{it} + \beta_2 X_{it} + \lambda \sum Controls_{it} + \nu_i + \tau_t + \varepsilon_{it} \end{aligned} \tag{2}$$

The core mediating variable in this study is the transparency of agricultural enterprise information.

3.2 Variable Selection and Data Resource

Dependent Variables Agricultural enterprise ESG performance (AESG): As specialized ESG rating systems for agricultural enterprises in China are still under development, this study uses ESG rating scores from a publicly listed company database tailored to China's capital markets and covering agricultural enterprise-specific dimensions as a proxy variable. This rating system considers agricultural-specific ESG dimensions such as ecological conservation, farmer benefit linkage and food security safeguards, making it more aligned with the research context.

Explanatory Variables. Agricultural green innovation (GI): Reflecting the nature of agricultural innovation, GI focuses on the research and development (R&D) and application of green agricultural technologies. The mainstream academic metric for GI is the number of green agricultural patents granted. This study uses the sum of green agricultural invention and utility model patents granted to the enterprise in the current year as the core indicator. These green patents focus on agricultural fields such as green technologies for crop cultivation and animal husbandry, controlling non-point source pollution in agriculture,circular ecological agriculture, and agricultural carbon sinks. To mitigate data skewness, the sum of agricultural green invention and utility model patents is increased by one and then log-transformed.

Mediating Variable. Information transparency (trans) There are multiple metrics for measuring information transparency, each suited to different research contexts and corporate characteristics. According to Xin Qingquan and Kong Dongmin (2014), information transparency is defined as the extent to which external information users can access specific information about publicly traded companies (e.g. annual reports, disclosure announcements, analyst reports and voluntary corporate disclosures) [19]. Taking into account the characteristics of agricultural enterprises, such as production seasonality and weak asset specificity, the information transparency index used in this study is based on multiple dimensions, including the adequacy and authenticity of corporate disclosures. This index provides a more comprehensive and accurate reflection of the level of information transparency of agricultural enterprises. Therefore, it is selected as the proxy variable for information transparency in agricultural enterprises, with higher values indicating greater transparency

The Control Variable. To enhance the robustness of the model and the interpretability of the empirical results, this study incorporates a series of firm-level characteristics as control variables, based on the operational features of agricultural enterprises and existing research. These include debt-to-asset ratio, return on assets, cash flow ratio, revenue growth rate, top five shareholder ownership ratio, dual-role positions, fixed asset ratio, firm age and financing constraints. Table 1 provides detailed definitions of each control variable.

Table 1. Key Variable Definitions

Variable Type	Variable Name	Symbol	Definition
Dependent Variable	ESG Performance of Agricultural-related Firms	AESG	Huazheng ESG Rating Index.
Independent Variable	agricultural green innovation	GI	The natural logarithm of the sum of the number of granted patents for agricultural green inventions and utility models plus one
control variable	asset-liability ratio	lev	Total corporate debt / total corporate assets
	return on assets	roa	Net profit of enterprise / total assets of enterprise
	cash flow ratio	cashflow	Net cash flow from operating activities / total assets of the enterprise.
	revenue growth rate	growth	(Current year's operating revenue-Previous year's operating revenue) / previous year's operating revenue
	top five shareholders' shareholding ratio	Topbig	Number of shares held by top five shareholders / total shares of the company
	dual-role integration	dual	If the chairman and general manager are the same person, assign 1; otherwise, assign 0.
	fixed assets ratio	fixed	Net fixed assets of the enterprise / total assets of the enterprise
	firm age	firmage	Observation year-enterprise Establishment Year
	financing constraints	WW	The WW index was calculated using the method proposed by Whited and Wu [20].
Mediating Variable	Information Transparency	Trans	Measured following the method of Xin and Kong [19]

Sample Selection and Data Resource. As this study focuses on agricultural enterprises, all others were excluded. To avoid distortion from extreme values, a two-tailed trimming at the 5% and 95% levels was applied. Sample validity was ensured through three steps: First, Chinese A-share listed agricultural enterprises across supply chain segments were identified using the "CSRC 2012 Industry Classification." Second, firms labelled ST, *ST, or delisted were excluded to avoid financial distress bias. Third, observations with missing or discontinuous key variables were removed; partial gaps were imputed using linear interpolation and moving averages to maintain data integrity and continuity. Ultimately, unbalanced panel data for 633 agriculture-related enterprises from 2015 to 2024 were obtained. Data sources include the Listed Companies ESG Rating Database, Chinese government websites, and the CSMAR database. Table 2 reports descriptive statistics and variable sources.

Table 2. Descriptive Statistics

Variable	Sample Size	Mean	Standard Deviation	Minimum	Maximum
AESG	6330	72.507	6.039	47.8	96.64
GI	6330	1.165	1.41	0	6.526
lev	6330	0.436	0.242	0.002	6.348
roa	6330	0.043	0.086	−1.387	2.637
cashflow	6330	0.16	7.552	−591.618	34.757
growth	6330	1.461	29.309	−2.78	1207.915
topbig	6330	−0.026	0.069	−1.622	0.642
dual	6330	0.046	0.21	0	1
fixed	6330	0.222	0.144	−0.005	0.863
firmage	6330	27.635	5.045	15	51
ww	6330	−1.043	0.838	−66.934	−0.761
Trans	6270	0.324	0.194	0.005	0.97

4 Empirical Results

4.1 Benchmark Regression Analysis

This paper constructs a panel regression model by gradually adding control variables to test the baseline impact effect of agricultural green innovation (GI) on the ESG performance of agricultural enterprises (AESG). The regression results are presented in four models, as shown in Table 3. Model (1) is the baseline model that only includes the core explanatory variables, where the coefficient of GI is 0.177 and is significant at the 1% level. (2) to (4) are extended models that gradually add control variables, and all models control for the fixed effects of year (Year FE), industry (Industry FE), and enterprise individual (Stkcd FE). We found that the coefficient of GI and its significance did not change significantly. Column (4) indicates the regression results including all control variables, and the coefficient of GI remains significantly positive. In conclusion, green innovation can effectively enhance the ESG performance of agricultural enterprises, thereby verifying the validity of H1.

Table 3. Benchmark Regression Analysis

	(1) AESG	(2) AESG	(3) AESG	(4) AESG
GI	0.177**	0.173**	0.181**	0.183**
	(0.0808)	(0.0808)	(0.0804)	(0.0802)
lev		0.246	0.295	0.469
		(0.423)	(0.466)	(0.466)
roa		-2.475***	−2.584***	−2.891***
		(0.820)	(0.824)	(0.825)
cashflow			0.0875	0.0849
			(0.0990)	(0.0988)
growth			0.00493**	0.00442**
			(0.00208)	(0.00208)
topbig			0.0661	−0.160
			(1.169)	(1.168)
dual				0.704**
				(0.347)
fixed				−3.726***
				(0.810)
firmage				−0.130***
				(0.0285)
ww				0.0549
				(0.0700)
Constant	72.18***	72.13***	72.09***	75.13***
	(0.199)	(0.275)	(0.282)	(0.699)
Year FE	YES	YES	YES	YES
Industry FE	YES	YES	YES	YES
Stkcd FE	YES	YES	YES	YES
Observations	6,330	6,330	6,277	6,277
R-squared	0.026	0.028	0.028	0.033

4.2 Robustness Analysis

Replacing Green Innovation Measurement Indicators. The benchmark regression of this paper takes the logarithm of green patent applications as the proxy variable of green innovation. In order to avoid the measurement deviation caused by a single index, this paper uses the logarithm of the total number of green invention patents and utility model patents plus 1 as the substitution variable of green innovation (GI_replace), and re-estimates the regression. As shown in column (1) of Table 4, the coefficient of GI_replace is 0.290, and it is significantly positive at the 1% level. It can be seen that after replacing the green innovation measurement method, the positive promotion effect of green innovation on the ESG performance of agriculture-related enterprises is still significant, and the core conclusions of this paper have good robustness.

Introducing Explanatory Variables Lags One Period. To mitigate the potential lag effects of economic cycles, industry trends, and policy adjustments, this paper re-evaluates lagged green innovation (L.GI) as the core explanatory variable. As shown in column (3) of Table 4, the coefficient of lagged green innovation is 0.283, significantly positive at the 1% level. This indicates that, after excluding the potential lag effects, the positive impact of green innovation on the

Table 4. Robustness Tests

	AESG	
	(1) Green invention patent authorization	(2) Lagging by one period
GI_replace	0.290***	
	(3.35)	
L. GI		0.283***
		(3.30)
lev	0.470	−0.072
	(1.01)	(−0.14)
roa	-2.903***	−1.388
	(−3.52)	(−1.51)
cashflow	0.083	0.109
	(0.84)	(1.04)
growth	0.004**	0.004**
	(2.06)	(1.96)
topbig	−0.162	−0.758
	(−0.14)	(−0.57)
dual	0.722**	0.765**
	(2.08)	(2.08)
fixed	−3.780***	−4.281***
	(−4.36)	(−4.95)
firmage	−0.136***	−0.125***
	(−4.81)	(−3.91)
ww	0.055	−0.532
	(0.78)	(−0.92)
_cons	75.154***	74.442***
	(108.70)	(76.35)
Year FE	YES	YES
Industry FE	YES	YES
Stkcd FE	YES	YES
Observations	6,276	5,644
R-squared	0.034	0.036

ESG performance of agricultural enterprises remains robust, further confirming the reliability and stability of the conclusions presented in this paper.

4.3 Mechanism Analysis

Information Transparency Columns (1) and (2) in the transparency of information table present the mechanism effect test with information transparency (Trans) as the mediator. In column (1), the coefficient of agricultural green innovation (GI) is 0.00442, positive at the 5% level, indicating that GI significantly

Table 5. Mechanism Test

	(2) AFSG	(3) AFSG
GI	0.00442**	0.149*
	(0.00198)	(0.0787)
Trans		7.924***
		(0.532)
lev	0.00576	0.300
	(0.0115)	(0.457)
roa	-0.0795***	−2.267***
	(0.0204)	(0.812)
cashflow	−0.00163	0.103
	(0.00243)	(0.0968)
growth	0.0000601**	0.00393**
	(0.000211)	(0.00203)
topbig	0.0202	−0.307
	(0.0287)	(0.341)
dual	0.00599	0.567*
	(0.00858)	(0.265)
fixed	-0.0086	−3.811***
	(0.0203)	(0.806)
firmage	-0.00921***	−0.125***
	(0.00172)	(0.0284)
ww	−0.00204	0.0719
	(0.000703)	(0.0684)
Constant	0.510***	70.99***
	(0.0172)	(0.737)
Year FE	YES	YES
Industry FE	YES	YES
Stkcd FE	YES	YES
Observations	6,219	6,219
R-squared	0.063	0.069

enhances enterprise information transparency, improves disclosure quality, and reduces asymmetry with external stakeholders. In column (2), the coefficient of Trans is 7.924, significantly positive at the 1% level, while GI remains significant. Together, these results verify hypothesis H2: agricultural green innovation indirectly promotes ESG performance by improving information transparency (Table 5).

5 Conclusions and Recommendations

Improving the ESG performance of agricultural enterprises not only helps them fulfill their social responsibilities such as ecological protection and rural revitalization, but also enhances their long-term competitiveness. Green innovation, as a key force driving sustainable agricultural development, is an important path to improving ESG performance. This paper, based on data from Chinese A-share listed agricultural enterprises from 2011 to 2022, systematically examines the impact and underlying mechanisms of green innovation on the ESG performance of agricultural enterprises. The main conclusions are as follows: First, green innovation ignificantly promotes the improvement of the ESG performance of agricultural enterprises. This conclusion remains valid after controlling for multidimensional fixed effects and undergoing a series of robustness tests. Second, mechanism testing shows that green innovation mainly promotes the optimization of the ESG performance of agricultural enterprises indirectly through enhancing corporate information transparency. Based on the above conclusions, this paper proposes the following policy recommendations:

Firstly, the government should continue to strengthen policy support for green innovation in agriculture, focusing on key areas such as ecological farming, agricultural non-point source pollution control, circular agriculture, and low-carbon production. It should increase financial subsidies, tax incentives, and project support to encourage agricultural enterprises to conduct green technology research and development and promote its application, thereby driving the overall improvement of the ESG level of agricultural enterprises through green innovation.

Secondly, relying on the intermediary mechanism of enterprise information transparency, agricultural enterprises build a virtuous cycle mechanism of'innovation-driven disclosure and disclosure-enabled development' : taking green technology innovation achievements as the core content of information disclosure to reduce information asymmetry ; at the same time, a standardized and quantifiable non-financial indicator system is established to transform implicit green efforts into explicit transparent signals, so as to attract long-term capital, ease financing constraints, and realize the two-way promotion of ESG performance and market value.

Thirdly, the market and society should cooperate to cultivate green value identity ecology. stakeholders such as investors, consumers and rural communities should strengthen the preference expression and supervision participation of green agricultural products, and force enterprises to improve environmental and social performance ; the media and third-party rating agencies also need to strengthen the dissemination and evaluation of green innovation and ESG practices of agricultural enterprises, and jointly create a market atmosphere of " emphasizing green, rewarding green, and using green. "

References

1. Kexin, Z., Yongxiu, B.: New quality productivity empowering synergistic enhancement of pollution reduction and carbon emission reduction in agriculture: theoretical mechanisms and empirical research [J/OL]. In: Journal of Southwest University (Social Sciences Edition), pp. 1–12 [2026]. https://link.cnki.net/urlid/50.1188.C.20260121.1123.002
2. Jiajing, W., Jinli, L.: The impact of green innovation on ESG performance of manufacturing enterprises. Bus. Res. (05), 121–130+152 (2025). https://doi.org/10.13902/j.cnki.syyj.20250901.001
3. Leite de Almeida, A. C., Dale, A., Hay, R., Everingham, Y., Lockie, S.: Environmental, social and governance (ESG) in agriculture: trends and gaps on research. Australasian J. Environ. Manag. **32**(3): 257–286 (2025). https://doi.org/10.1080/14486563.2024.2430313
4. Chunxiao, Y., Yawen, L., Dixing, Z.: Can ESG ratings promote corporate green transformation?. Contemp. Finan. Res. **6**(06), 1–13 (2023). https://doi.org/10.20092/j.cnki.ddjryj.2023.06.001
5. Zankui, D., Feiyan, D., Siyi, L.: The impact of digital literacy on farmers' adoption of green production technologies and its mechanism: an empirical analysis based on CLES mixed cross-sectional data . J. Ecology Rural Environ. Sci.,1–14 (2026). https://doi.org/10.19741/j.issn.1673-4831.2025.0326
6. Zaili, L., Xue, W.: An analysis of the impact of agricultural industrial chain integration on rural economic growth. Times Econom. Trade **22**(05), 167–169 (2025). https://doi.org/10.19463/j.cnki.sdjm.2025.05.014
7. Jing, L., Qian, L.: Research on development strategies for agricultural scale operations to enhance labor productivity. Agricult. Technol. **36**(21), 154–156 (2016)
8. Rongjun, L., Hui, W.: Agricultural digitalization, spatial spillovers, and farmer income growth. Chinese Agricult. Res. Region. Plann, 1– 16 (2026). https://link.cnki.net/urlid/11.3513.S.20250328.1508.016
9. Yiheng, Z.: Green credit, agricultural mechanization, and agricultural output value: an empirical analysis based on China's provincial PVAR model. Technol. Market **32**(02), 123–130 (2025)
10. Feng, H.: Normative logic and institutional optimization for realizing agricultural carbon sink value under the "dual carbon" goals. Southern Fin. **03**, 87–98 (2024)
11. Chengqing, L.: Empirical study on the development mechanism of green brands. Shandong University (2010)
12. Chengqing, L.: Empirical study on the development mechanism of green brands. Shandong University (2010)
13. Yuechao, D., Honghao, H., Zhongming, W.: How digital leadership stimulates breakthrough innovation: a dynamic capabilities theory perspective. In: Science and Technology Progress and Policy, pp. 1–10 (2026). https://link.cnki.net/urlid/42.1224.G3.20251204.1629.002

14. Jianzhu, W., Wenkang, Y.: The impact of intelligent transformation on green technology innovation in manufacturing enterprises: a resource-based view approach. Eng. Manag. Sci. Front, 1–14 (2026). https://link.cnki.net/urlid/34.1336.N.20260202.1610.003
15. Daoyuan, L., Yajie, L.: Research on corporate interest protection from the perspective of conflict of interest theory. In: Journal of Zhejiang Gongshang University, pp. 1–11 (2026). https://link.cnki.net/urlid/33.1337.C.20251225.1243.002
16. Xiaojuan, H., Guanxia, X.: ESG information transparency and green finance governance: evidence from Chinese listed companies. Int. Finance Res. (01), 108–120 (2026). https://doi.org/10.16475/j.cnki.1006-1029.2026.01.008
17. Yong, W., Xinge, F., Chenyu, Z.: Digital finance, information asymmetry, and SME financing: evidence from loan facilitators. J. Quantit. Econom. Techn. Econom. **43**(01), 242–264 (2026). https://doi.org/10.13653/j.cnki.jqte.2026.01.006
18. Yubin, H.: Review and future outlook of ESG research from a stakeholder perspective. China Bus. Rev. **34**(23), 160–164 (2025). https://doi.org/10.19699/j.cnki.issn2096-0298.2025.23.160
19. Qingquan, X., Dongmin, K., Ying, H.: Corporate transparency and stock price volatility. J. Finan. Res. **3**, 193–206 (2014)
20. Whited, T.M., Wu, G.: Financial constraints risk. Rev. Finan. Stud. **19**(2), 531–559 (2006)

A Crest Factor-Based Chaotic Synchronization Error Energy Method for Early Bearing Degradation Detection

Junying Niu[1(✉)] and Zhong Li[2]

[1] Shunde Polytechnic University, Foshan, Guangdong 528300, China
niujunying@foxmail.com
[2] Faculty of Mathematics and Computer Science, FernUniversität in Hagen, Hagen, Germany
Zhong.Li@fernuni-hagen.de

Abstract. Addressing the critical challenge of detecting incipient bearing faults in low signal-to-noise ratio (SNR) environments, a Health Indicator (HI) framework is formulated utilizing Crest Factor-driven chaotic synchronization. By exploiting the inherent sensitivity of chaotic systems to infinitesimal perturbations, the methodology nonlinearly amplifies weak, fault-related impulses typically submerged in heavy background noise. The Crest Factor (CF) is integrated as a dynamic driver for a Lorenz chaotic system, where the resulting synchronization error energy serves as a high-sensitivity HI for degradation assessment. Empirical validation on the XJTU-SY dataset yields an Early Detection Percentage (EDP) of 18.6%, identifying degradation 25.7%–55.7% earlier than eleven conventional baselines. This approach offers a robust diagnostic solution for capturing the earliest stages of rolling element bearing failure.

Keywords: Chaotic synchronization · Crest factor · Error energy · Early degradation

1 Introduction

Rolling element bearings are critical yet vulnerable components, with failures accounting for nearly half of industrial mechanical breakdowns [1]. While early detection is essential for proactive maintenance [2], incipient fault signatures are frequently submerged 20–40 dB below the noise floor, rendering traditional diagnostics ineffective. Conventional time-domain methods often fail to reflect degradation until 60%–80% of component life has elapsed, while frequency and time-frequency methods [3,4] are constrained by noise sensitivity and the requirement for prior kinematic knowledge. Furthermore, complexity-based estimators [5] suffer from prohibitive computational costs and parameter dependency, leading to the degraded reliability of direct detection paradigms under severe background interference.

J. Lu et al. (Eds.): FLINS-ISKE 2026, LNCS 16752, pp. 132–146, 2027.
https://doi.org/10.1007/978-981-92-2369-5_10

To address these limitations, nonlinear methodologies such as the Duffing oscillator [6] and stochastic resonance [7] have been explored for weak signal enhancement. However, these approaches are inherently limited by their narrowband nature and reliance on prior fault frequency information. In contrast, chaotic systems offer a superior alternative by exploiting their extreme sensitivity to infinitesimal perturbations, where minute deviations trigger exponential divergence in dynamic trajectories. This characteristic facilitates the detection of primordial, weak signatures that remain indiscernible within linear or narrowband frameworks.

Leveraging this mechanism, a Chaotic Synchronization Error-Driven Health Indicator (CSED-HI) is proposed utilizing a master-slave Lorenz synchronization system. By functionally mapping Crest Factor (CF) deviations into the synchronization error space, CSED-HI exploits the broadband nature of chaos to respond to systematic degradation without requiring frequency priors. The synchronization error energy (E) is implemented as the primary indicator, providing a robust, time-invariant measure of the system's deviation from synchronization. Its quadratic formulation ensures that subtle changes in the driving signal produce a measurable response in E, while its inherent time-averaging confers resilience against transient noise and ensures physical interpretability.

2 Chaotic Synchronization

2.1 Definition

To establish the master-slave synchronization system, the Lorenz dynamical model is employed. This canonical three-dimensional autonomous dissipative system [8] is defined as:

$$\begin{cases} \dot{x}_1 &= \sigma(x_2 - x_1) \\ \dot{x}_2 &= \rho x_1 - x_2 - x_1 x_3 \\ \dot{x}_3 &= x_1 x_2 - \beta x_3 \end{cases} \tag{1}$$

Two Lorenz systems are coupled through the drive-response architecture [9]:

$$\text{Master:} \quad \dot{\mathbf{x}} = \mathbf{f}(\mathbf{x}) \tag{2}$$

$$\text{Slave:} \quad \dot{\mathbf{y}} = \mathbf{f}(\mathbf{y}) + \mathbf{u}(\mathbf{e}) \tag{3}$$

where $\mathbf{x}, \mathbf{y} \in \mathbb{R}^3$ are state vectors and $\mathbf{u}(\mathbf{e})$ is the synchronization controller. The synchronization error $\mathbf{e} = \mathbf{y} - \mathbf{x}$ evolves as $\dot{\mathbf{e}} = \mathbf{f}(\mathbf{y}) - \mathbf{f}(\mathbf{x}) + \mathbf{u}$.

Applying active control $\mathbf{u}(\mathbf{e}) = -\mathbf{K}\mathbf{e} + \mathbf{v}(\mathbf{x}, \mathbf{e})$, where $\mathbf{K} = \text{diag}(k_1, k_2, k_3)$ provides linear feedback and $\mathbf{v}$ cancels nonlinear coupling terms, the linearized error dynamics reduce to:

$$\dot{\mathbf{e}} = \mathbf{A}\mathbf{e}, \quad \mathbf{A} = \begin{bmatrix} -\sigma & \sigma - k_1 & 0 \\ \rho & -(1 + k_2) & 0 \\ 0 & 0 & -(\beta + k_3) \end{bmatrix} \tag{4}$$

Asymptotic synchronization $\|\mathbf{e}(t)\| \to 0$ is guaranteed when all eigenvalues of $\mathbf{A}$ are strictly negative, achievable by appropriate selection of $\mathbf{K}$.

2.2 Signal-Driven Synchronization Dynamics

When external driving signals are injected into the coupled Lorenz system, the governing equations evolve as:

$$\dot{\mathbf{x}} = \mathbf{f}(\mathbf{x}) + \alpha \mathbf{d}_{\text{ref}} \tag{5}$$

$$\dot{\mathbf{y}} = \mathbf{f}(\mathbf{y}) + \alpha \mathbf{d}_{\text{test}} + \mathbf{u}(\mathbf{e}) \tag{6}$$

where $\alpha > 0$ denotes the coupling strength, and $\mathbf{d}_{\text{ref}}, \mathbf{d}_{\text{test}} \in \mathbb{R}^3$ encode the reference and test feature vectors, respectively. Consequently, the error dynamics are dictated by:

$$\dot{\mathbf{e}} = \mathbf{A}\mathbf{e} + \alpha \Delta \mathbf{d}, \quad \Delta \mathbf{d} = \mathbf{d}_{\text{test}} - \mathbf{d}_{\text{ref}} \tag{7}$$

From a dynamical systems perspective, two operational regimes are identified according to the synchronization behavior. In the **synchronized regime** ($\Delta\mathbf{d} \to \mathbf{0}$), the controller asymptotically drives the error trajectory toward the synchronization manifold ($\|\mathbf{e}\| \to 0$), with the residual error reflecting only ambient measurement noise. In the **desynchronized regime** ($\Delta\mathbf{d} \neq \mathbf{0}$), the persistent perturbation $\Delta\mathbf{d}$ forces the error system into a non-trivial attractor. The steady-state error energy is theoretically bounded by

$$E_{\text{energy}} = \frac{1}{N}\sum_{t=1}^{N} \|\mathbf{e}(t)\|^2 \propto \frac{\alpha^2 \|\Delta \mathbf{d}\|^2}{|\lambda_{\text{min}}^{\perp}|^2} \tag{8}$$

where $|\lambda_{\text{min}}^{\perp}|$ represents the magnitude of the dominant conditional Lyapunov exponent. The quadratic scaling with respect to the input perturbation ($\|\Delta\mathbf{d}\|^2$) ensures that E_{energy} exhibits a substantially stronger response to incipient degradation than linear metrics ($E_{\text{rms}} \propto \|\Delta\mathbf{d}\|$), providing the system with high sensitivity to early faults.

3 Chaotic Synchronization Error-Driven Health Indicator

The overall architecture of the proposed CSED-HI is illustrated in Fig. 1. The methodology bridges statistical signal processing and nonlinear dynamical modeling.

3.1 Kinematic Feature Encoding via Crest Factor

Direct application of raw vibration waveforms to drive the chaotic system is precluded by inherent random noise, which significantly obscures degradation signatures (preliminary tests yielded an error detection probability $\approx 89\%$). To address this, the crest factor (CF) is adopted as the fundamental driving dimension:

$$\text{CF}(x) = \frac{\max|x|}{\sqrt{\frac{1}{N}\sum_{n=1}^{N} x_n^2}} \tag{9}$$

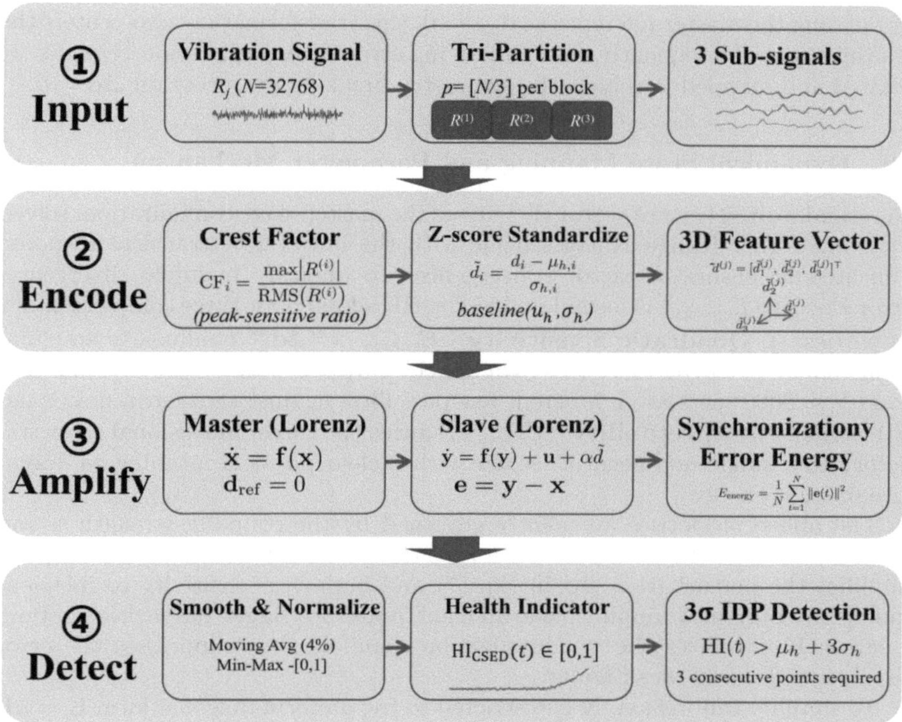

Fig. 1. The proposed CSED-HI theoretical framework

Physically, CF is advantageous because its numerator is highly sensitive to the impulsive transients of early degradation, while its ratio-based structure normalizes baseline energy fluctuations, thereby inherently suppressing broadband noise.

To achieve continuous mapping into the three-dimensional Lorenz state space, each raw recording R_j (N = 32,768 samples) is evenly partitioned into three independent observation windows of length $p = \lfloor N/3 \rfloor$. The localized CF formulations construct a 3D driving vector $\mathbf{d}^{(j)} = [d_1^{(j)}, d_2^{(j)}, d_3^{(j)}]^\top$:

$$d_i^{(j)} = \frac{\max |R_j^{(i)}|}{\mathrm{RMS}(R_j^{(i)})}, \quad i = 1, 2, 3 \tag{10}$$

Subsequently, Z-score standardization is implemented using the statistical baseline ($\mu_{h,i}$, $\sigma_{h,i}$) established exclusively from the pristine operation stage:

$$\tilde{d}_i^{(j)} = \frac{d_i^{(j)} - \mu_{h,i}}{\sigma_{h,i}}, \quad i = 1, 2, 3 \tag{11}$$

By defining the master reference as $\mathbf{d}_{\text{ref}} = \mathbf{0}$, this transformation zero-centers the healthy state. Consequently, any underlying structural degradation is explicitly isolated and mapped purely as the symmetry-breaking perturbation $\Delta\mathbf{d} = \tilde{\mathbf{d}}^{(j)}$.

3.2 Dynamical State Mapping and Parameter Mechanism

The standardized feature vector $\tilde{\mathbf{d}}^{(j)}$ drives the master-slave configuration, solved via a high-order Runge–Kutta scheme with the initial 20% transient response truncated to ensure attractor convergence. To quantify manifold divergence, Error Energy (E_{energy}) is formulated as the HI, adopted for three complementary properties: (i) **Quadratic Sensitivity**—$E_{\text{energy}} \propto \|\Delta\mathbf{d}\|^2$ nonlinearly amplifies infinitesimal perturbations; (ii) **Stochastic Suppression**—temporal integration ($N \approx 8000$) acts as an inherent low-pass filter against zero-mean noise; and (iii) **Scalar Interpretability**—E_{energy} reduces the three-dimensional error trajectory to a single non-negative scalar with a clear physical meaning as desynchronization energy.

The above properties are jointly governed by the coupling strength α and the feedback gain $\mathbf{K}$, which define a sensitivity–stability trade-off. A larger α amplifies the perturbation $\Delta\mathbf{d}$ in Eq. (7) and improves sensitivity to incipient faults, but may also amplify measurement noise or trigger desynchronization. A larger $\mathbf{K}$ stabilizes the synchronization manifold, but suppresses the error response and masks weak faults.

To simplify calibration, $\mathbf{K}$ is restricted to the uniform-diagonal form $\mathbf{K} = k\mathbf{I}$, reducing the gain space to a single scalar k. This is justified by the symmetry of the three driving segments (Sect. 3.1), the natural dissipativeness of the z-channel in Eq. (4) [9], and an exploratory 3-D search showing the optimum lies on the diagonal. The reduction also lowers the search complexity from $O(n^3)$ to $O(n)$.

Calibration then proceeds in two steps: k is first chosen so that all eigenvalues of $\mathbf{A}$ have strictly negative real parts (ensuring asymptotic stability), and α is subsequently tuned by a one-dimensional grid search on the healthy-stage baseline to maximize $E_{\text{fault}}/E_{\text{healthy}}$ while keeping E_{healthy} below a prescribed noise floor.

3.3 HI Synthesis and Incipient Fault Nucleation

The holistic HI is defined by the extracted energy metric $\text{HI}_{\text{CSED}}(t) = E_{\text{energy}}(t)$. To isolate the macro-degradation trend from microscopic fluctuations, a moving average filter (4% temporal window) is applied, followed by min-max normalization into the $[0,1]$ bounding generalized range.

The initial degradation point (IDP) identification is approached as a strict statistical hypothesis test against the null hypothesis (system health). It is rigorously defined by the 3σ threshold criteria:

$$\text{IDP} = \min\{t : \text{HI}(t+i) > \mu_h + 3\sigma_h,\ i = 0, 1, 2\} \tag{12}$$

where μ_h and σ_h characterize the Gaussian distribution of the HI during early steady-state operation. The auxiliary 3-consecutive-point constraint acts as an essential low-pass logic gate, suppressing false alarms triggered by isolated anomalous impulses.

4 Experimental Validation

The XJTU-SY accelerated life test dataset [10] (under the 37.5 Hz / 11 kN operating condition) provides five bearings with diverse failure modes (Table 1), sampled at 25.6 kHz (32,768 samples/recording, 1-min intervals).

Table 1. XJTU-SY dataset, under the 37.5 Hz / 11 kN condition

Bearing	Failure Mode	Life (min)	# Rec.
B2_1	Inner race	491	~491
B2_2	Outer race	161	~161
B2_3	Cage	533	~533
B2_4	Outer race	42	~42
B2_5	Outer race	339	~339

4.1 Implementation Setup

All experiments share the configuration summarized in Table 2 to ensure reproducibility. The Lorenz system parameters are set to the canonical values $(\sigma, \rho, \beta) = (10, 28, 8/3)$. The synchronization controller adopts the uniform-diagonal feedback gain $\mathbf{K} = \mathrm{diag}(12, 12, 12)$ and the coupling strength $\alpha = 0.1$, determined as described in Sect. 3.2; the corresponding error matrix $\mathbf{A}$ has eigenvalues $\{-11.5 \pm j\,7.33,\ -14.67\}$, guaranteeing global asymptotic stability of the synchronization manifold. The drive–response system is integrated by a fourth-order Runge–Kutta scheme with step size $\Delta t = 0.005$ and total integration length $T = 40$; the initial 20% of the trajectory is discarded as the transient. The HI smoothing window is set to 4% of the recording sequence length, and the 3σ baseline (μ_h, σ_h) is estimated from the first 10% of recordings of each bearing.

Table 2. CSED-HI hyperparameter configuration

Module	Parameter	Value
Lorenz system	(σ, ρ, β)	(10, 28, 8/3)
Coupling	α	0.1
Feedback gain	$\mathbf{K} = \text{diag}(k_1, k_2, k_3)$	diag(12, 12, 12)
Stability check	Eigenvalues of $\mathbf{A}$	$\{-11.5 \pm j7.33, -14.67\}$
RK4 integrator	step size Δt	0.005
	total time T	40
	transient truncation	20%
HI processing	moving-average window	4%
	healthy baseline ratio	10%
IDP detection	threshold	$\mu_h + 3\sigma_h$
	confirmation	3 consecutive points

4.2 Dynamical Evolution of Synchronization Error

To illustrate the dynamical mechanism of the proposed method, the synchronization error vector is extracted across six operational stages, namely Healthy (early), Healthy (late), Near IDP, Mild degradation, Moderate degradation, and Severe degradation.

Figure 2 presents the temporal evolution of the error norm $\|e(t)\|$. During the healthy stages, the master and slave systems remain in the synchronized regime ($\|e(t)\| \approx 0$). Once the bearing crosses the IDP, weak impulsive features introduced by early micro-cracks act as persistent symmetry-breaking perturbations. These perturbations induce a measurable departure from synchronization, and the peak amplitude of $\|e(t)\|$ increases progressively through the moderate and severe stages.

The corresponding 3D phase portrait evolution, plotted on a uniform scale in Fig. 3, provides a complementary topological perspective that reveals distinct physical mechanisms underlying different failure modes.

Topological Expansion (Impact-Driven Faults). For bearings exhibiting impact-driven failure modes, such as inner race (B2_1) and outer race (B2_2, B2_4, B2_5) faults, the phase portraits during the healthy operational regime reveal that the error trajectories (e_1, e_2, e_3) remain confined to a topologically trivial neighborhood of the origin, consistent with the asymptotic synchronization condition $\|\mathbf{e}(t)\| \to 0$ established in Regime I (Sect. 2.2). Upon the onset of localized surface exfoliation, repetitive high-frequency impact transients introduce persistent symmetry-breaking perturbations $\Delta\mathbf{d} \neq \mathbf{0}$ into the error dynamics governed by Eq. (7). In accordance with the quadratic amplification mechanism of Eq. (8), these perturbations drive the error system into Regime II. The trajectories undergo a pronounced topological expansion, departing from

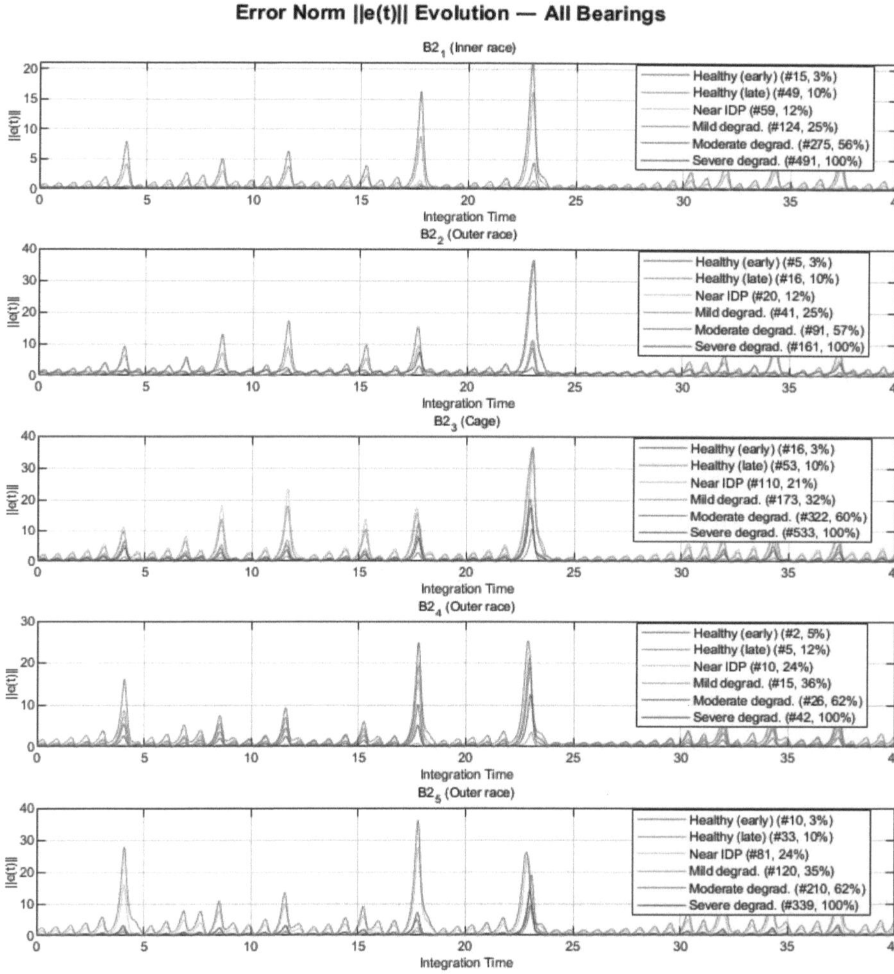

Fig. 2. Evolution of the synchronization error norm $\|e(t)\|$ across six degradation stages

the trivial limit set and evolving into spatially extensive, geometrically complex chaotic attractors whose occupied phase volume scales monotonically with fault severity. This expansion constitutes a direct geometric manifestation of the nonlinear amplification predicted by the theoretical framework and stands in stark contrast to the collapse phenomenon observed under friction-driven degradation.

Topological Collapse (Friction-Driven Faults). An extraordinary divergent phenomenon occurs for bearing B2_3, which suffered a cage failure. Instead of expanding outward, its trajectory shrinks toward the origin as the degradation becomes severe. This dynamic behavior intrinsically aligns with the phys-

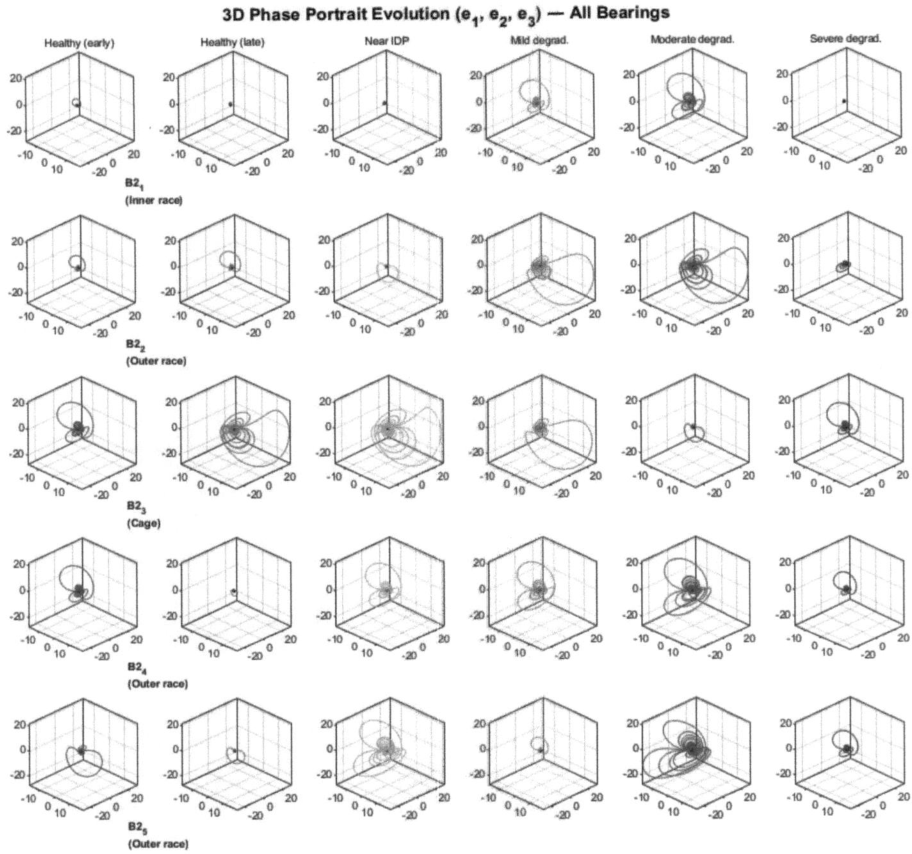

Fig. 3. 3D phase portrait evolution (e_1, e_2, e_3) mapped to uniform coordinate scales

ical reality of cage failures. Unlike race faults that produce repetitive sharp impacts (increasing CF), severe cage fracture causes rolling elements to crowd and slide. This sliding friction smears out initial high-frequency structural resonances, returning the input signal closer to a Gaussian distribution. Consequently, the loss of these symmetry-breaking impulsive features dynamically removes the external disturbance, forcing the chaotic error systems to mutually collapse back into perfect mathematical synchronization.

4.3 Full-Lifecycle Degradation Assessment

Figure 4 presents the full-lifecycle CSED-HI degradation curves for all five bearings under the 37.5 Hz / 11 kN condition, where the dashed vertical lines mark the identified initial degradation points (IDPs) determined by the 3σ criterion defined in Eq. (12).

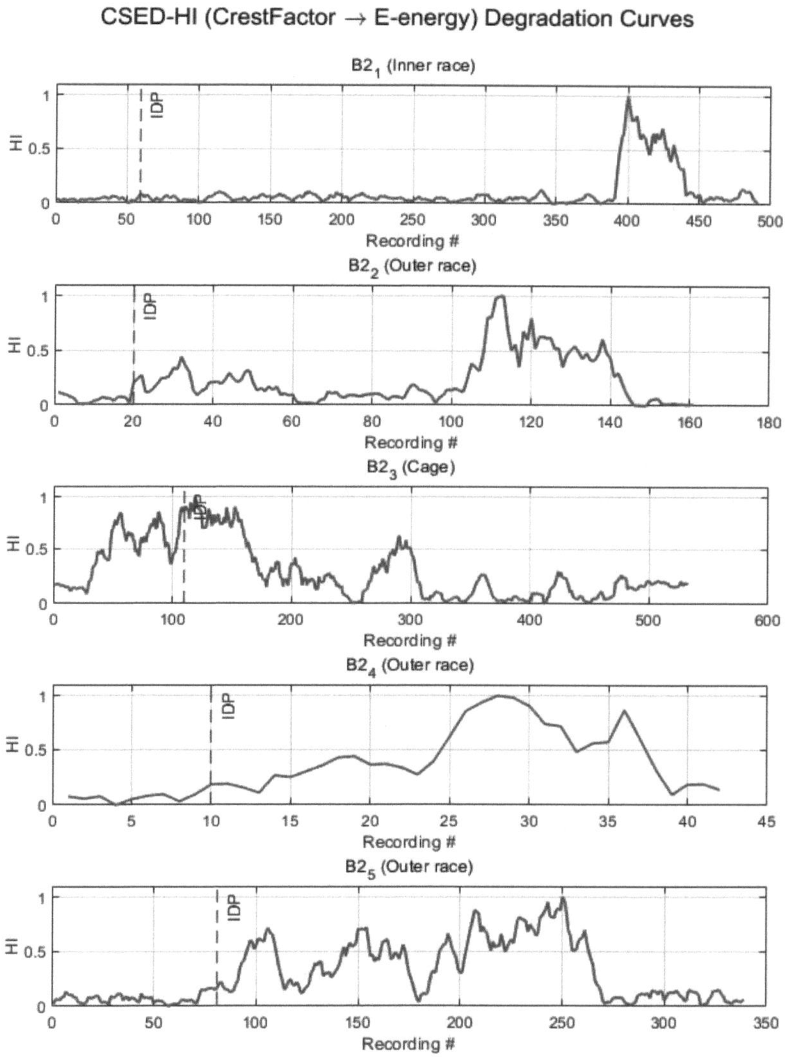

Fig. 4. CSED-HI (CrestFactor → E_{energy}) degradation curves for all five bearings

Several key observations emerge from the degradation trajectories. For bearings with impact-driven failure modes (B2_1, B2_2, B2_4, B2_5), the HI maintains a near-zero baseline during the healthy operational phase, with the 3σ threshold effectively suppressing sporadic noise-induced fluctuations. Once the IDP is triggered, the HI exhibits sharp transient spikes corresponding to localized impulsive events, followed by sustained elevated energy levels as the fault propagates. Notably, the IDPs are consistently identified within the first 12%–24%

of total bearing life, confirming the high sensitivity of the chaotic amplification mechanism to incipient fault nucleation.

For bearing B2_3 (cage fault), the degradation curve reveals a distinctive non-monotonic profile: an initial spike near the IDP is followed by a gradual decline in HI amplitude during the later stages. This trajectory is consistent with the topological collapse phenomenon discussed in Sect. 4.2, where the transition from impulsive to friction-dominated dynamics reduces the symmetry-breaking perturbation and drives partial re-synchronization of the master-slave system. Despite this non-monotonic behavior, the IDP is still reliably captured at approximately 20.6% of total life, demonstrating the robustness of the 3σ detection criterion across fundamentally different failure mechanisms.

4.4 Comparative Study

Eleven signal-based indicators spanning five categories are implemented under identical conditions: time-domain (RMS, kurtosis, peak, crest factor), frequency-domain (spectral RMS, spectral centroid), time-frequency (wavelet packet energy, WPE, 3-level Daubechies-4), entropy, permutation entropy, and nonlinear dynamics (maximum Lyapunov exponent, correlation dimension).

Analysis on EDP. Table 3 confirms that CSED-HI achieves the earliest detection across all five bearings and all failure modes, as measured by early detection percentage $\mathrm{EDP} = t_{\mathrm{IDP}}/t_{\mathrm{total}} \times 100\%$ (lower is better). With a mean EDP of 18.6%, CSED-HI uniformly outperforms all eleven baseline indicators, yielding an average advancement of 25.7–55.7% over existing methods regardless of the underlying failure mechanism.

Table 3. EDP (%) per bearing and detection time advantage

Method	B2_1	B2_2	B2_3	B2_4	B2_5	Mean	Adv.
CSED-HI	**12.0**	**12.4**	**20.6**	**23.8**	**23.9**	**18.6**	—
CorrDim	79.6	39.1	43.0	42.9	16.8	44.3	+25.7
WPE	79.6	37.3	42.6	57.1	15.6	46.5	+27.9
RMS	79.6	39.1	43.0	57.1	16.8	47.1	+28.5
FreqRMS	79.6	39.1	43.0	57.1	16.8	47.1	+28.5
PermEn	79.4	37.3	46.9	57.1	18.0	47.7	+29.1
Peak	79.6	37.9	46.9	59.5	16.5	48.1	+29.5
Lyap-Dir	79.4	39.1	47.1	57.1	18.0	48.2	+29.6
Kurtosis	44.2	23.0	47.8	100.0	52.8	53.6	+35.0
Freq-Ctr	21.6	68.3	100.0	100.0	18.0	61.6	+43.0
Crest F.	79.8	15.5	22.1	100.0	100.0	63.5	+44.9
SampEn	100.0	100.0	100.0	40.5	31.3	74.3	+55.7

Table 4. Comparison on HI of different methods

Method	Category	Mon	Tre	Rob	Score	EDP%
RMS	Time	0.219	0.266	0.965	0.483	47.1
Kurtosis	Time	0.019	0.229	0.969	0.406	53.6
Peak	Time	0.092	0.270	0.960	0.441	48.1
CrestFactor	Time	0.043	0.361	0.964	0.456	63.5
FreqRMS	Freq	0.216	0.266	0.965	0.482	47.1
FreqCentroid	Freq	0.027	0.400	0.964	0.464	61.6
WPE	T-F	0.219	0.271	0.965	0.485	46.5
SampEn	Entropy	0.033	0.338	0.950	0.440	74.3
PermEn	Entropy	0.186	0.352	0.963	0.500	47.7
LyapDirect	NL	0.162	0.341	0.963	0.489	48.2
CorrDim	NL	0.232	0.272	0.966	0.490	44.3
CSED		0.035	0.373	0.960	0.456	**18.6**

(Score = $\frac{\text{Mon}+\text{Tre}+\text{Rob}}{3}$)

Analysis on HI Quality. Table 4 presents the HI quality metrics averaged across all five bearings. The Monotonicity of CSED-HI is comparatively low (0.035), attributable to non-monotonic excursions during intermediate degradation stages. As the fault topography evolves from localized impulsive pitting to distributed surface sliding, the CF driving signal undergoes transient reductions that trigger partial re-synchronization of the master and slave system, producing temporary reversals in the HI trajectory. The Trendability (0.373) and Robustness (0.960) remain at moderate and high levels, respectively.

These results indicate that CSED-HI is a sensitivity-oriented indicator optimized for incipient fault detection rather than for full-lifecycle trend tracking. These complementary characteristics motivate a *two-stage monitoring strategy* for practical deployment, in which CSED-HI and trend-based indicators play distinct, sequential roles. In Stage I (*detection*), CSED-HI continuously monitors the bearing and triggers an alarm at the IDP through the 3σ rule, exploiting its quadratic sensitivity to incipient perturbations. Once the IDP is confirmed, Stage II (*prognosis*) is activated: a trend-based indicator such as RMS or wavelet-packet energy is re-baselined at the IDP and used to track the monotonic propagation of damage toward end-of-life, supporting remaining useful life estimation. The two stages operate on the same vibration stream and can be implemented in parallel; CSED-HI thus serves as a sensitive trigger, while the trend indicator provides the long-horizon degradation curve required by prognostic algorithms.

4.5 Validation under an Alternative Operating Condition

To assess the generalization capability of the proposed CSED-HI beyond the single operating point used in the previous subsections, the entire pipeline is

Table 5. Comparison on HI of different methods under the 35 Hz / 12 kN condition

Method	Category	Mon	Tre	Rob	Score	EDP%
RMS	Time	0.594	0.925	0.993	0.837	35.3
Kurtosis	Time	0.501	0.943	0.987	0.810	32.8
Peak	Time	0.205	0.934	0.984	0.708	39.5
CrestFactor	Time	0.090	0.921	0.972	0.661	39.5
FreqRMS	Freq	0.594	0.926	0.993	0.838	35.3
FreqCentroid	Freq	0.074	0.914	0.970	0.653	39.4
WPE	T-F	0.594	0.925	0.993	0.837	35.3
SampEn	Entropy	0.026	0.555	0.888	0.490	71.4
PermEn	Entropy	0.641	0.934	0.991	0.855	34.5
LyapDirect	NL	0.646	0.935	0.991	0.857	39.5
CorrDim	NL	0.566	0.926	0.993	0.828	39.5
CSED		0.098	0.908	0.954	0.653	**23.6**

(Score = $\frac{\text{Mon+Tre+Rob}}{3}$)

Table 6. Per-bearing EDP (%) under the 35 Hz / 12 kN condition and detection time advantage

Method	B1_1	B1_2	B1_3	B1_4	B1_5	Mean	Adv.
CSED-HI	**11.4**	**13.7**	40.5	**13.9**	38.5	**23.6**	—
Kurtosis	39.8	39.8	17.1	28.7	38.5	32.8	+9.2
PermEn	39.8	39.8	39.9	39.3	13.5	34.5	+10.9
RMS	39.8	39.8	19.0	39.3	38.5	35.3	+11.7
FreqRMS	39.8	39.8	19.0	39.3	38.5	35.3	+11.7
WPE	39.8	39.8	19.0	39.3	38.5	35.3	+11.7
Freq-Ctr	40.7	40.4	39.9	39.3	36.5	39.4	+15.8
Peak	39.8	39.8	39.9	39.3	38.5	39.5	+15.9
Crest F.	39.8	39.8	39.9	39.3	38.5	39.5	+15.9
Lyap-Dir	39.8	39.8	39.9	39.3	38.5	39.5	+15.9
CorrDim	39.8	39.8	39.9	39.3	38.5	39.5	+15.9
SampEn	100.0	19.3	100.0	37.7	100.0	71.4	+47.8

re-run on the XJTU-SY dataset under the 35 Hz / 12 kN condition. Table 5 reports the HI quality metrics averaged over the five bearings of this condition, and Table 6 provides the corresponding per-bearing EDP breakdown.

The cross-condition results confirm that CSED-HI retains its early-detection advantage under a markedly different load–speed regime. From the mean-EDP perspective (Table 5), CSED-HI achieves 23.6%, outperforming the best baseline (Kurtosis, 32.8%) by roughly nine percentage points and the worst one (SampEn, 71.4%) by nearly 50%. The per-bearing breakdown in Table 6 further reveals that

CSED-HI ranks first on three of the five bearings (B1_1, B1_2, B1_4) with EDPs as low as 11.4%–13.9%, and remains competitive on the remaining two (B1_3, B1_5) where Kurtosis and PermEn occasionally win on specific failure modes. Notably, the eleven baselines exhibit a strong clustering pattern around 39%–40% on this condition, suggesting that their EDPs are largely dominated by the same gross degradation event rather than reflecting genuine sensitivity to incipient signatures; CSED-HI breaks out of this cluster on the bearings where early micro-impacts are present, which is consistent with its quadratic-amplification mechanism.

Consistent with the trend observed under the 37.5 Hz / 11 kN condition, CSED-HI exhibits comparatively low Monotonicity (0.098), reflecting transient re-synchronization during intermediate degradation stages rather than a deficiency of the indicator itself. Its Trendability and Robustness remain at competitive levels (0.908 and 0.954). This pattern reaffirms that CSED-HI is a sensitivity-oriented indicator well suited to incipient fault detection.

5 Conclusion

This paper proposes CSED-HI, a Health Indicator framework that integrates the Crest Factor of vibration signatures into a master–slave Lorenz synchronization architecture. By mapping CF deviations from the healthy baseline as symmetry-breaking perturbations into the chaotic error space, the framework achieves a quadratic sensitivity to incipient fault impulses that remain undetectable by linear indicators (e.g., RMS, Peak) and narrow-band nonlinear estimators (e.g., Lyapunov exponent, correlation dimension).

Validation on the XJTU-SY dataset under the 37.5 Hz / 11 kN condition yields a mean EDP of 18.6%, advancing detection by 25.7–55.7 percentage points over eleven baselines across diverse failure modes. Cross-condition validation under the 35 Hz / 12 kN condition produces a mean EDP of 23.6% and again outperforms all eleven baselines, demonstrating that the proposed framework remains effective across distinct load–speed regimes. The HI quality analysis further shows that CSED-HI maintains moderate Trendability and high Robustness, while its low Monotonicity reflects transient re-synchronization caused by evolving fault morphology rather than a deficiency of the indicator itself.

The above characteristics position CSED-HI as a sensitivity-oriented early-warning trigger rather than a standalone prognostic tool, naturally motivating a two-stage monitoring strategy in which CSED-HI raises the IDP alarm via the 3σ rule and a trend-based indicator (e.g., RMS or wavelet-packet energy) is subsequently re-baselined at the IDP to support remaining useful life estimation. Several directions remain open for future investigation. First, the present implementation relies on an offline calibration procedure on the healthy-stage baseline; an online, data-driven self-tuning mechanism would allow the framework to adapt continuously to heterogeneous and time-varying operating conditions. Second, integrating CSED-HI with downstream prognostic models, such as Wiener-process or deep learning–based RUL estimators, may further consolidate the two-stage strategy into a unified detection-and-prediction pipeline deployable in real industrial settings.

References

1. Sahu, D., Dewangan, R.K., Matharu, S.P.S.: An investigation of fault detection techniques in rolling element bearing. J. Vibrat. Eng. Technol. **12**(4), 5585–5608 (2024)
2. Zifei, X., et al.: A novel health indicator for intelligent prediction of rolling bearing remaining useful life based on unsupervised learning model. Comput. Indust. Eng. **176**, 108999 (2023)
3. Randall, R.B.: Vibration-Based Condition Monitoring: Industrial, Aerospace and Automotive Applications. John Wiley & Sons (2011)
4. Dragomiretskiy, K., Zosso, D.: Variational mode decomposition. IEEE Trans. Signal Process. **62**(3), 531–544 (2014)
5. Bandt, C., Pompe, B.: Permutation entropy: a natural complexity measure for time series. Phys. Rev. Lett. **88**(17), 174102 (2002)
6. Niu, J., Li, Z.: Chaos-based fault diagnosis: methodologies and practices. Int. J. Bifurcation Chaos **35**(09), 2550112 (2025)
7. Wang, Z., Jianhua Yang, Yu., Guo, T.G., Shan, Z.: Nonstationary feature extraction based on stochastic resonance and its application in rolling bearing fault diagnosis under strong noise background. Rev. Sci. Instrum. **94**(1), 015110 (2023)
8. Lorenz, E.N.:. Deterministic nonperiodic flow. J. Atmosp. Sci. **20**(2), 130–141 (1963)
9. Pecora, L.M., Carroll, T.L.: Synchronization in chaotic systems. Phys. Rev. Lett. **64**(8), 821–824 (1990)
10. Lei, Y., Han, T., Wang, B., Li, N., Yan, T., Yang, J.: XJTU-SY rolling element bearing accelerated life test datasets: a tutorial. J. Mech. Eng. **55**(16), 1–6 (2019)

Fuzzy Partition Kernels in Machine Learning and Neural Networks

Irina Perfilieva[1,2(✉)] and Jin Hee Joon[3]

[1] University of Ostrava, IRAFM, Ostrava, Czech Republic
irina.perfilieva@osu.cz
[2] Cracow University of Technology, Faculty CS and M, Krakow, Poland
[3] Sejong University, Seoul, South Korea
jin9135@sejong.ac.kr
https://ifm.osu.cz/irina-perfiljeva/13263/

Abstract. We contribute to the general theory of supervised machine learning by describing a feature extraction methodology using fuzzy partitions and their generative kernels. We explain how the Representer theorem applies to RKHS, whose kernel is related to a uniform fuzzy partition. We show that the connection with fuzzy modeling helps impart semantic meaning to the parameters defining a kernel-based regression model. We developed a neural network architecture that validates our theoretical results and solves the regression problem, including prediction. We describe all the steps of our method and demonstrate how to initialize the parameters of the corresponding neural network. This contributes to the explainability of the neural network design and execution.

Keywords: Regression problem · Representer theorem · Fuzzy partition · p-d kernel · Machine learning

1 Introduction

In recent years, machine learning has become one of the most analyzed topics from various theoretical, algorithmic, and technological perspectives. However, in our view, a significant gap remains between the various theories surrounding machine learning and the practical methods used in this field. This is due to the very general formulation of the primary focus of machine learning, which is the analysis of experimental data and all possible *generalizations* based on it. In this context, *generalization* is the ability to process and reason about new, previously unencountered objects or properties of data after gaining experience with a training dataset.

Since data objects can belong to datasets of varying nature (numerical measurements, signals, images, etc.), their processing also varies and is governed by the specific physical laws of the respective domain. Consequently, the data-driven generalization should be less dependent on the physical nature of the data and

J. Lu et al. (Eds.): FLINS-ISKE 2026, LNCS 16752, pp. 147–159, 2027.
https://doi.org/10.1007/978-981-92-2369-5_11

more focused on the problems to be solved with it. These are problems of classification, recognition, prediction, etc. The methods for solving these problems are based on collecting objects into groups according to their common properties (or "features" in machine learning terminology).

Thus, an important implication of the general approach to machine learning as a theory focused on data-driven generalization is that feature extraction/selection is an essential step that involves transforming the original data when solving the above-mentioned problems [1].

In particular, in this paper, we propose a detailed explanation of the theoretical solution to the regression problem and how the theory can be applied to a given data sample. We also explain a methodology for feature extraction using *fuzzy partition kernels*. The proposed material will be applied for solving the problem of prediction where predictors or explanatory variables are features determined by a fuzzy partition of a given dataset. The novelty of the proposed approach lies in the following:

- We answered the question: does a uniform fuzzy partition define a p-d kernel?
- We characterize the features extracted from the fuzzy partition kernel;
- We develop a represented kernel Hilbert space whose elements are represented as linear combinations of fuzzy sets in a fuzzy partition;
- We explain all the steps of our method and demonstrate how to initialize the parameters of the corresponding neural network.

2 Preliminaries

First, we will characterize the so-called *feature space* as an abstract mathematical concept [1]. A *feature vector* as an element of a feature space is an n-dimensional numerical vector representing some object. For example, if the original data are images, the features might correspond to the types of edges (linear, curved, etc.) in the image. In linear regression, feature vectors are equivalent to variable vectors. Good features are those that have *discriminative* properties: they help a machine learning algorithm identify patterns and distinguish between instances of the data. The vector space associated with feature vectors is often called a *feature space*.

In theoretical machine learning, a feature space is a functional or vector space, chosen depending on the task at hand, taking into account the complexity of processing the extracted features. Thus, to connect the general concept of a feature space with solutions represented in terms of this space, we will narrow our general approach to solving specific problems. In this paper, we will focus on prediction and/or classification problems. Since both of these problems are closely related to regression analysis, we will analyze machine learning methods in the context of this topic. We will explain how features extracted from a fuzzy partition of a given domain can be useful for solving these problems.

Remark 1. Our contribution will be based on *supervised machine learning* as one of three learning strategies. As a general characteristic, supervised learning develops algorithms that learn to map inputs to specific outputs based on given examples of input-output pairs.

2.1 Regression Analysis

The goal of regression analysis is to estimate a function $f(X_i, \beta)$ that best fits the data (X_i, Y_i), $i = 1, \ldots, n$, i.e.

$$Y_i = f(X_i, \beta) + e_i, \tag{1}$$

where the additive error term e_i represents random statistical noise. To do this, it is necessary to determine the form of f. However, the choice of f is an *aleatoric source of uncertainty* that cannot be reduced by increasing the amount of data. Sometimes, the form of f is based on additional knowledge about the relationship between the dependent and independent variables. Otherwise (if no additional knowledge is available), any convenient form of f can be chosen. For example, simple univariate regression yields a function f that minimizes the sum of squared errors.

In many regression models, the given independent variables X_i are first transformed into a feature vector representation using a user-specified feature map. The latter is any map $\phi : \Re^d \to H$ such that

$$\phi : \mathcal{X} \subset \Re^d \to H,\ x \mapsto \phi(x),$$

where $\mathcal{X}$ contains values of independent variables X_i and H is usually a Hilbert space of functions on $\mathcal{X}$.

If a feature map ϕ is applied to the independent variables, then the dependent variables in (1) can be expressed using simpler linear forms with respect to the feature vectors $\phi(X_i)$, i.e.,

$$Y_i = (\alpha_i \phi(X_i) + \beta_i) + e_i. \tag{2}$$

Thus, the correct choice of feature maps helps decompose the regression model (1) into a simplified form (2), which is less computationally expensive.

2.2 Kernels in Machine Learning Focused on Regression

In machine learning, kernels[1] help transform data into a feature space to better distinguish between classes or simplify data representation (Hofmann et al., 2006; 2008). In other words, a kernel serves as a source of feature maps used in the regression model (2).

Based on Mercer's theorem, a continuous symmetric positive-semidefinite kernel defines feature maps whose values are in the Hilbert space ℓ^2. Furthermore,

[1] In the context of machine learning, a kernel (kernel function) is a symmetric function defined on the Cartesian product of the input space $\mathcal{X}$.

these feature maps can be used to reconstruct the original kernel. The latter has the reproducibility property, allowing any function from the corresponding reproducing kernel Hilbert space (RKHS) to be derived from its reproducing kernel.

In the context of regression analysis and the goal of estimating a function f that best fits the data (X_i, Y_i), a possible solution is to select f from a RKHS, whose reproducing kernel distinguishes between the independent variables X_i. Below, we discuss how, according to the *Representer Theorem* [9], one can construct a formal solution to the optimization problem related to regression model selection.

Theorem 1. *(Schölkopf, Herbrich, and Smola) Consider a positive-definite real-valued kernel* $k : \mathcal{X} \times \mathcal{X} \to \Re$ *on a non-empty set* $\mathcal{X} \subset \Re^d$ *with a corresponding reproducing kernel Hilbert space* H_k. *Let there be given*

- *a training sample* $(\mathbf{x}_1, y_1), \dots, (\mathbf{x}_n, y_n) \in \mathcal{X} \times \Re$;
- *a strictly increasing real-valued function* g *on* $[0, +\infty)$,
- *an arbitrary error function* $E : (\mathcal{X} \times \Re^2)^n \to \Re \cup \{+\infty\}$,

which together define the following regularized empirical risk functional on H_k

$$f \mapsto E((\mathbf{x}_1, y_1, f(\mathbf{x}_1)), \dots, (\mathbf{x}_n, y_n, f(\mathbf{x}_n))) + g(\|f\|_{H_k}). \tag{3}$$

Then, any minimizer of the empirical risk

$$f^* = \operatorname*{argmin}_{f \in H_k} E((\mathbf{x}_1, y_1, f(\mathbf{x}_1)), \dots, (\mathbf{x}_n, y_n, f(\mathbf{x}_n))) + g(\|f\|_{H_k}), \tag{4}$$

admits a representation of the form:

$$f^*(\cdot) = \sum_{i=1}^{n} w_i \, k(\cdot, \mathbf{x}_i), \tag{5}$$

where $w_i \in \Re$, $i = 1, \dots, n$.

The role of the kernel k in the above formulation is primarily that it significantly simplifies the problem of minimizing the regularized empirical risk (3). By this, we mean that instead of searching in an infinite-dimensional space H_k, the problem is reduced to finding the optimal n-dimensional vector of coefficients $\alpha = (\alpha_1, \dots, \alpha_n) \in \Re^n$. Since α can be obtained using any standard minimization algorithm, the representative theorem provides a theoretical basis for reducing the general problem of machine learning to algorithms that can actually be implemented on computers.

In this contribution, we propose to construct a kernel k based on fuzzy modeling methods. Specifically, we demonstrate how a positive-definite kernel can be related to a fuzzy partition of a certain set. Our proposal is based, first, on a semantic analysis of the positive-definite kernel as a measure of *similarity* between objects in a given set. Second, the expression of the minimizing function (5) is similar to the expression of the inverse F-transform, introduced by the author in [6]. Below, we provide the necessary details of the proposed construction.

3 Fuzzy Partition and Its Representations

We will consider two approaches to the notion of a fuzzy partition: one as a special collection of membership functions and the second one as a kernel-generated functions with a compact support. Let us start with the original definition [6] of the concept of *fuzzy partition*, which refers to the partition of a closed interval on the real line.

Definition 1. *(Fuzzy partition) [6] Let* $[a, b]$ *be an interval of real numbers and* $t_1 < \cdots < t_n$ *be fixed* nodes *within* $[a, b]$ *such that* $t_1 = a,\ t_n = b$ *and* $n \geq 2$*. We say that the collection of fuzzy sets* $\{A_1, \cdots, A_n\}$*, identified with their membership functions* $A_1, \cdots, A_n : [a, b] \to [0, 1]$*, form a* fuzzy partition *of* $[a, b]$*, if they fulfill the following conditions for* $i = 1, \cdots, n$*:*

(1) $A_i : [a, b] \to [0, 1],\quad A_i(t_i) = 1 \quad i = 1, \cdots, n;$
(2) $A_i(t) = 0$ *if* $t \notin (t_{i-1}, t_{i+1}),\quad i = 2, \cdots, n-1;$
(3) $A_i(t)$ *is continuous function on* $[a, b]$*;*
(4) $A_i(t)$ *monotonically increases on* $[t_{i-1}, t_i]$ *for* $i = 2, \cdots, n$ *and monotonically decreases on* $[t_i, t_{i+1}]$ *for* $i = 1, \cdots, n-1$*;*
(5) For all $t \in [a, b]$*,* $\sum_{i=1}^{n} A_i(t) > 0$.

We say that a fuzzy partition $\{A_1, \cdots, A_n\}$ *of* $[a, b]$ *is of Ruspini type if property* (5) *has the form* $\sum_{i=1}^{n} A_i(t) = 1,\ t \in [a, b]$.

The membership functions $A_1(t), \cdots, A_n(t)$ *are called* basic functions. *The* support *of the basic function* A_i *is the closure of the set where* A_i *has positive values. A fuzzy partition* $A_1(t), \cdots, A_n(t),\ n > 2$*, is* h-uniform *if nodes* $t_1, \cdots, t_n$ *are* h*-uniformly distant, i.e.* $t_i = a + h(i-1),\ i = 1, 2, \cdots, n$*, where* $h = (b-a)/(n-1)$*, and two more properties are fulfilled; Optionally (Ruspini condition)*

(6) $A_i(t_i - t) = A_i(t_i + t)$*, for all* $t \in [0, h]$ *and* $i = 2, \cdots, n-1;$
(7) $A_{i+1}(t) = A_i(t - h)$*, for all* $t \in [t_i, t_{i+1}]$ *and* $i = 1, \cdots, n-1$.

The basic functions $A_i,\ i = 1, \ldots, n$*, of an* h*-uniform fuzzy partition are rescaled translations of a* generating function *[6]* $A_0 : [-1, 1] \to [0, 1]$*, i.e.* $A_i(t) = A_0(\frac{t-t_i}{h})$*. Moreover, these basic functions are linearly independent.*

Example 1. In what follows, the following h-uniform *triangular* fuzzy partition $A_1, \cdots, A_n,\ n \geq 2$, of $[a, b]$ with nodes $t_1 = a, t_2, \ldots, t_n = b$ such that $t_i = t_{i-1} + h$, and $h = \frac{b-a}{n-1}$ will be used:

$$A_1(t) = \begin{cases} 1 - \frac{t-t_1}{h}, & t \in [t_1, t_2], \\ 0 & otherwise, \end{cases}$$
$$A_i(t) = \begin{cases} 1 - \frac{|t-t_i|}{h}, & t \in [t_{i-1}, t_{i+1}], \\ 0 & otherwise, \end{cases} \qquad i = 2, 3, ..., n-1 \tag{6}$$
$$A_n(t) = \begin{cases} \frac{t-t_{n-1}}{h}, & t \in [t_{n-1}, t_n], \\ 0 & otherwise, \end{cases}$$

Figure 1 shows the graph of a 1-uniform triangular fuzzy partition of the interval $[0, 10]$.

It is easy to see that the *generating function* of the h-uniform triangular fuzzy partition (6) is $A_0(t) = \max(1 - |t|, 0)$.

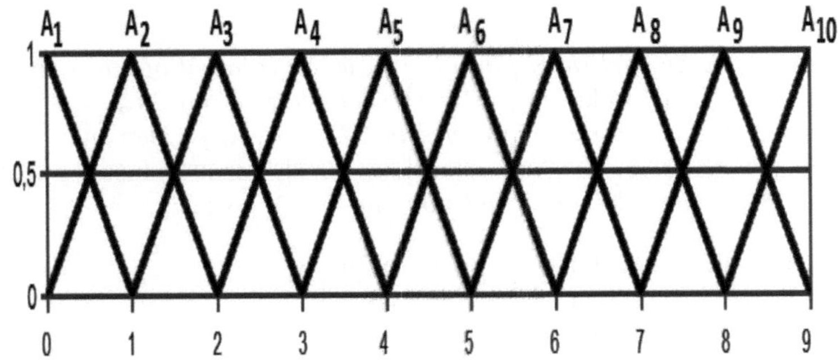

Fig. 1. Plot of the 1-uniform triangular fuzzy partition of $[0, 10]$.

3.1 Kernel-Based Fuzzy Partition

Suppose that the functions $A_1, \cdots, A_n : [a, b] \to [0, 1]$ defined on $[a, b]$ form a uniform fuzzy partition of $[a, b]$ with generating continuous function $A_0 : \Re \to [0, 1]$. We recall that the notion of a *generating function* was axiomatically characterised in [2,7] as a continuous function $A_0 : \Re \to [0, 1]$, which is even, bell-shaped, and vanishing outside of $(-1, 1)$. It is moreover demanded that A_0 is *normalized.* A generating function A_0 can be taken as a *membersip function* of a certain fuzzy set on $\Re$.

If A_0 is a generating function of a uniform fuzzy partition $A_1, \cdots, A_n : [a, b] \to [0, 1]$ of $[a, b]$ with equidistant nodes $t_1, \cdots, t_n$, then the general element A_k of this fuzzy partition is represented as $A_0(t - t_k)$, $k = 1, \ldots, n$.

Based on the above arguments, we define the kernel function $k_{A_0}(t, s) : \Re^2 \to [0, 1]$ as $k_{A_0}(t, s) = A_0(t - s)$ and use this function to create a fuzzy partition $k_{A_0}(t, t_1), \ldots k_{A_0}(t, t_n)$ of the interval $[a, b]$ with nodes $t_1, \cdots, t_n$.

Thus, we came to the following definition.

Definition 2. *Suppose that* $A_0 : \Re \to [0, 1]$ *is a generating function and* $k_{A_0}(t, s) = A_0(t - s)$ *is the corresponding kernel. Let* $t_1, \ldots, t_n$ *be nodes in* $[a, b]$. *Then the functions* $k_{A_0}(t, t_1), \ldots, k_{A_0}(t, t_n)$ *form a kernel-based fuzzy partition of* $[a, b]$ *if its basic functions are restricted on this interval. A kernel-based fuzzy partition is uniform (h-uniform) if the nodes* $t_1, \ldots, t_n$ *are equidistant (h-uniformly distant) and, moreover,* $a = t_1$, $b = t_n$.

Note that a kernel-based fuzzy partition is not necessarily of the Ruspini type.

In Example 1, the h-uniform triangular fuzzy partition of $[a, b]$ is a kernel-based where the kernel is given by the h-rescaled generating function $A_{0,h} = \frac{1}{h}\max(1 - \frac{|t|}{h}, 0))$. In a general d-dimensional space $\Re^d$, $d \geq 1$, the notion of a kernel-based fuzzy partition was introduced in [11].

4 From a Kernel-Based Fuzzy Partition to RKHS

Recall that the use of the Representer Theorem 1 imposes certain restrictions on the kernel and the space in which the solution must reside. The kernel must be positive definite, and the function space must be a Hilbert space with a reproducing kernel. Definitions and explanations are given below.

Definition 3. *Let $\mathcal{X}$ be a nonempty set. A symmetric function k is called a positive-definite (p-d) kernel on $\mathcal{X}$, if*

$$\sum_{i=1}^{n}\sum_{j=1}^{n} c_i c_j k(x_i, x_j) \geq 0, \tag{7}$$

holds for any $x_1, \ldots, x_n \in \mathcal{X}$, given $n \in \mathbb{N}$, $c_1, \ldots, c_n \in \Re$.

We are interested in the answer to the question: does the concept of a uniform fuzzy partition with generating function A_0 define a p-d kernel expressed as $k_{A_0}(t, s) = A_0(t - s)$. We found a positive answer to this question in the article [10], where the author analyzed the positive definiteness of radial functions with compact support. In particular, it has been shown that if $A_0 = \max(1 - |t|, 0)$ generates a uniform triangular fuzzy partition then the corresponding kernel $k_{A_0}(t, s) = A_0(t-s)$ is positive-definite in $\Re^1$. The latter means that the inequality in (7) is valid for any pairwise distinct points $x_1, \ldots, x_n \in \Re^1$.

Our next step is to develop a model of a Hilbert space whose elements are candidates for the role of "solvers" in the regression problem described in Theorem 1. According to the assumptions of this theorem and the optimization problem in (4), the Hilbert space must be associated with the chosen kernel by the property of reproducibility. Below, we give necessary explanations.

4.1 Reproducing Kernel Hilbert Space

We will describe a closed relationship between p-d kernels and *reproducing kernel Hilbert spaces* and *feature maps*.

Let $\mathcal{X}$ be a set, H the Hilbert space of functions $f : \mathcal{X} \to \mathbb{R}$, and $(\cdot, \cdot)_H : H \times H \to \mathbb{R}$ the corresponding inner product in H. For any $x \in X$, define the *evaluation functional* $e_x : H \to \mathbb{R}$ using the formula $f \mapsto e_x(f) = f(x)$. We now say that H is a reproducing kernel Hilbert space (abbreviated RKHS) if all evaluation functionals are continuous.

Every RKHS has a special function associated to it, namely the *reproducing kernel*:

Definition 4. *Reproducing kernel is a function $k : \mathcal{X} \times \mathcal{X} \to \mathbb{R}$ such that*

1. *$\forall x \in \mathcal{X}$, $k_x(\cdot) \in H$, and*
2. *for all $f \in H$ and $x \in \mathcal{X}$, $(f, k_x)_H = f(x)$.*

The latter property is called the "reproducing property".

The following result shows equivalence between RKHS and their reproducing kernels.

Theorem 2. *Every reproducing kernel k induces a unique RKHS, and every RKHS has a unique reproducing kernel.*

The following theorem gives a connection between a p-d kernels and the corresponding RKHS.

Theorem 3. *Every reproducing kernel is positive-definite, and every positive definite kernel defines a unique RKHS, of which it is the unique reproducing kernel.*

Thus, given a positive-definite kernel k, it is possible to build an associated RKHS with k as a reproducing kernel. This fact can be used to connect p-d kernels with another interesting object that arises in machine learning applications, namely the *feature map.*

Let H be a Hilbert space, and $(\cdot, \cdot)_H$ the corresponding inner product. Any map $\Phi : \mathcal{X} \to H$ is called a feature map. In this case we call H the feature space. It is not difficult to show that every feature map defines a unique p-d kernel by

$$k(x, y) = (\Phi(x), \Phi(y))_H. \tag{8}$$

This suggests a new way of conceptualizing p-d kernels as scalar products in the corresponding Hilbert spaces, or, in other words, p-d kernels can be viewed as similarity maps that effectively quantify the degree of similarity between two points x and y using the value $k(x, y)$. Moreover, due to the equivalence of p-d kernels and their corresponding RKHSs, any feature map can be used to construct an RKHS.

4.2 Moore-Aronszajn Theorem

The Moore-Aronszajn theorem (Theorem 4) is a kind of inverse statement to the above relation (8) between kernels and the corresponding reproducing kernel Hilbert spaces. It formulates conditions for $k : \mathcal{X} \times \mathcal{X} \to \Re$ that are sufficient to determine a Hilbert space H_k of functions on $\mathcal{X}$ for which k is a reproducing kernel.

Theorem 4. *Suppose k is a symmetric, positive definite kernel on a set $\mathcal{X}$. Then there exists a unique Hilbert space H_k of functions on $\mathcal{X}$ for which k is a reproducing kernel.*

The proof of this theorem is constructive in nature, so we will highlight only the part where the elements and the inner product in H_k are characterized.

Denote $k_x = k(x, \cdot)$, $x \in \mathcal{X}$. Let H_0 be the linear span of $\{k_x : x \in \mathcal{X}\}$. Define an inner product on H_0 as follows:

$$\left\langle \sum_{j=1}^{n} b_j k_{y_j}, \sum_{i=1}^{m} a_i k_{x_i} \right\rangle_{H_0} = \sum_{i=1}^{m} \sum_{j=1}^{n} a_i b_j k(y_j, x_i). \tag{9}$$

Let H_k be the completion of H_0 with respect to this inner product. Then H_k consists of functions of the form

$$f(x) = \sum_{i=1}^{\infty} a_i k_{x_i}(x) \quad \text{where} \quad \lim_{n\to\infty} \sup_{p\geq 0} \left\| \sum_{i=n}^{n+p} a_i k_{x_i} \right\|_{H_0} = 0.$$

4.3 RKHS Induced by a Kernel-Based Fuzzy Partition

In this section, we apply all the above theoretical results to the special case where $\mathcal{X} = [0,1] \subseteq \Re$ and an h-uniform fuzzy partition $A_1, \cdots, A_n : [0,1] \to [0,1]$ of the domain $[0,1]$ has generating function $A_{0,h}(t) = \frac{1}{h}\max(1 - \frac{|t|}{h}, 0)$. At the beginning of the Sect. 4, we showed that this partition is kernel-based, the corresponding kernel is expressed as $k(t,s) = A_{0,h}(t-s)$, and it is a p-d kernel in $\Re$. Since the nodes $t_1, \ldots, t_n$ of fuzzy partition $A_1, \cdots, A_n$ are at distance h from each other, then $A_i(t) = k(t, t_i)$.

Note that the kernel k we extracted from this fuzzy partition can generate an infinite set of other fuzzy partitions of the same interval $[0,1]$. To obtain a new fuzzy partition generated by kernel k, it suffices to choose nodes $z_1, \ldots, z_m \in [0,1]$, create the set of basic functions $\{k(t, z_i) \mid i = 1, \ldots, m\}$, and check condition (5) from Definition 1.

By the Moore-Aronszajn theorem (Theorem 4), the kernel $k(t,s) = A_{0,h}(t - s)$, defines unique functional space RKHS H_k, consisting of all functions $f : [0,1] \to \Re$ represented as follows:

$$f(t) = \sum_{i=1}^{m} b_i k(t, z_i), \tag{10}$$

where $z_1, \ldots, z_m \in [0,1]$ are nodes of the corresponding fuzzy partition $\{k(t, z_i) \mid i = 1, \ldots, m\}$.

Since for each $z_i \in [0,1]$ the function $k(\cdot, z_i)$ belongs to H_k, it is an instance of a feature map. Therefore, the representation (10) confirms that every function from RKHS H_k can be expressed as a linear combination of features. On the other hand, every feature is a basic function of some fuzzy partition of $[0,1]$. Therefore, we can say that the representation (10) shows that every function from RKHS H_k can be expressed as a linear combination of basic functions from some fuzzy partition of $[0,1]$.

To complete the description of the RKHS induced by the kernel-based fuzzy partition, it is necessary to define the inner product in this space. According to

(9), the inner product of two functions $f, h \in H_k$ is completely determined by the coefficients in their corresponding representations (10) and the kernel values at the nodes.

5 Addressing the Regression Problem in RKHS Induced by a Kernel-Based Fuzzy Partition

Recall that the formal solution to the regression problem (1) was given in the Representer Theorem (Theorem 1). According to this theorem, the solution is an element of the RKHS whose kernel is positive definite. Further, using the Moore-Aronszajn theorem (Theorem 4), this solution can be expressed using (10), where the kernel k can be constructed from a uniform fuzzy partition. Finally, according to (4), the search space is an m-dimensional vector space, so the solution expressed using (10) is completely determined by the coefficient vector $b^* = (b_1, \ldots, b_m) \in \Re$.

Below we give an example [9] of solving the problem of finding a minimizer whose existence is guaranteed by the representer theorem. Let us assume that in Theorem 1 we choose: $\mathcal{X} = [0,1]$, the error function $E : (\mathcal{X} \times \Re^2)^n \to \Re \cup \{+\infty\}$ is the least squares error function, i.e.

$$E((\mathbf{x}_1, y_1, f(\mathbf{x}_1)), \ldots, (\mathbf{x}_n, y_n, f(\mathbf{x}_n))) = \sum_{i=1}^{n} (y_i - f(\mathbf{x}_i))^2, \tag{11}$$

and the regularization function $g(x) = \lambda x^2$ for some $\lambda > 0$. The error function E in (11) is an estimate of the standard error used in regression analysis.

Let the kernel $k(t,s) = A_{0,h}(t-s)$ correspond to a uniform fuzzy partition of the domain $[0,1]$ with the generating function $A_{0,h}(t) = \frac{1}{h}\max(1 - \frac{|t|}{h}, 0)$. Denote $K_{ij} = k(\mathbf{x}_i, \mathbf{x}_j)$, $i,j = 1, \ldots, n$, and $\mathbf{y} = (y_1, \ldots, y_n)$. Then, the solution f is expressed using (10) where the coefficient vector $b^* = (b_1, \ldots, b_m)$ is equal to

$$b^* = (K^T K + \lambda K)^{-1} K^T \mathbf{y}.$$

It can be shown that b^* is the *global minimum* to the empirical risk (3) which aggregates the error function E in (11) and the regularization function $g(x) = \lambda x^2$. If $\lambda = 0$ then the global minimum of the empirical risk (3) is achieved by the function f expressed using (10) where the coefficient vector $b^* = (K^T K)^{-1} K^T \mathbf{y}$, and in the case where K is a square matrix $b^* = K^{-1}\mathbf{y}$.

6 Regression in Machine Learning and Computation Using ANNs

As we characterised in Remark 1, machine learning is typically divided into three main learning paradigms: supervised learning, unsupervised learning, and reinforcement learning. Each is tailored to a specific learning task. Regression and classification tasks are supervised machine learning tasks.

In neural networks, machine learning is used as a method to help adapt the network to solve a problem more efficiently by taking into account sample observations. Adaptation is accomplished through *learning*. Learning involves adjusting the network weights (and, if necessary, thresholds) to improve the accuracy of the result. This is achieved by minimizing observed errors.

Based on this general description, solving a regression problem represents a specific way of implementing neural network training. Thus, neural network computations focus on solving a variety of problems, where regression and approximation issues (either independently or as part of a more general problem) are always at the forefront [8].

This article demonstrates how to design a neural network architecture to effectively solve a regression problem. In particular, the proposed analysis helps select neural network hyperparameters, namely the number of neurons in each layer, the type of activation function, and the initialization of neuron weights and biases. To confirm this assertion, consider the following example [5].

Example 2. Assume that the training sample $(x_1, y_1), \ldots, (x_n, y_n) \in \mathcal{X} \times \Re$ consists of argument-value pairs of the function

$$f(x) = e^{\frac{\pi x}{100}} \sin(\frac{\pi x}{10}),\ x \in [0, \pi], \tag{12}$$

so that $n = 100$ and the nodes $x_1, \ldots, x_n$ are equidistant in $x \in [0, \pi]$. Precisely, the training sample is given as

$$S = \{(\frac{k\pi}{100}, e^{\frac{k\pi}{100}} \sin(\frac{k\pi}{10})) \mid k = 1, \ldots, 100\}.$$

Let us develop the corresponding regression model based on the uniform triangular-shaped fuzzy partition (6) whose generating function is $A_{0,h}(t) = \frac{1}{h}\max(1 - \frac{|t|}{h}, 0)$, $h = \frac{\pi}{100}$. The corresponding kernel is expressed as $k(t, s) = A_{0,h}(t - s)$, and it is a p-d kernel in $\Re$ as we showed in Sect. 4.3. On the basis of the proposed in Sect. 4 theory, the regression model is expressed in (10) as

$$g(t) = \sum_{i=1}^{100} b_i k(t, x_i),$$

where $x_1, \ldots, x_{100} \in [0, 1]$ are the sample nodes of the corresponding fuzzy partition $\{k(t, x_i) \mid i = 1, \ldots, 100\}$. This solution helps to design the fully connected neural network with the following hyperparameters: 3 layers, 100 neurons in the hidden layer, triangular-shaped activation function. We use the ELM learning method [3] to calculate the output weights b_i, $i = 1, \ldots, 100$ of the hidden layer neurons.

In Fig. 2 we show the visualization of the quality assessment of the results of two predictions computed at intermediate points of S by different ANNs trained on the same input-output data set S, but with two different initializations associated with the Cybenko [4] and the proposed regression model based on the kernel-based fuzzy partition. We see that the ANN initialized based on

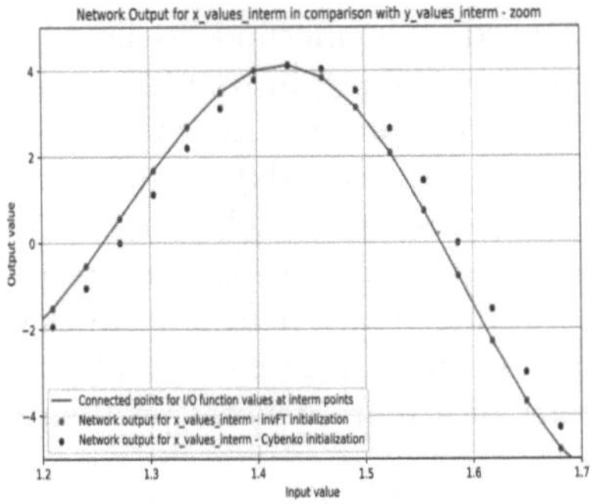

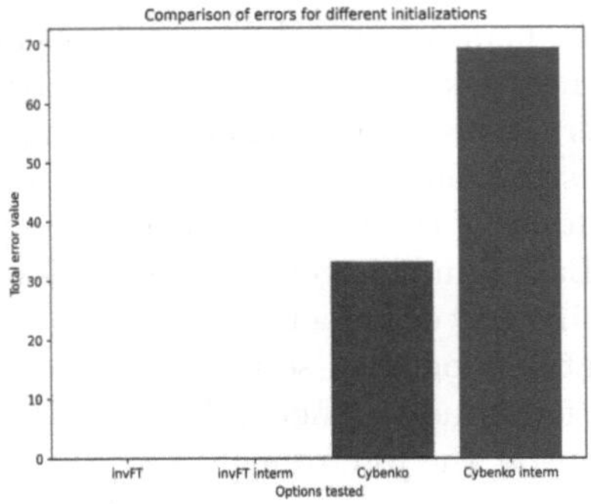

Fig. 2. *Left.* Plot of two predictions computed at intermediate points of S by two different ANNs trained on S but using different initializations associated with the Cybenko (blue dots) and the proposed regression model (red dots). *Right.* Histogram of errors.

the proposed regression model outperforms the ANN initialized based on the Cybenko-type approximation.

In Table 1, we show the cumulative absolute differences of the two predictions computed at intermediate points of S by two different ANNs trained on S.

Table 1. Accumulating Errors over training and testing sets

Approximation model	Abs Error Train	Abs Error Prediction
Proposed	1.9806E-12	2.045E-12
Cybenko	33.081	69.228

7 Conclusion

This paper demonstrates how fuzzy modeling based on a uniform fuzzy partition of the universe facilitates theoretical machine learning focused on the regression problem. The latter is formulated as an optimization problem whose solution is characterized by the Representer theorem. Our key finding is that the uniform triangular fuzzy partition has a kernel representation, and, moreover, this kernel is positive definite. This confirms the applicability of the general theory based on the Representer theorem to RKHS, whose kernel is generated by the fuzzy partition. Finally, we developed a neural network architecture that validates our theoretical results and solves the regression problem, including prediction. The proposed analysis helps select the neural network hyperparameters, namely, the number of neurons in each layer, the type of activation function, and the initialization of neuron weights and biases.

Acknowledgments. This study was partially funded by the project of the University of Ostrava: Social Dimension of New Technologies in the Energy Sector in the Ostrava Metropolitan Area (reg. number No. CZ.02.01.01/00/23_021/000859 with financial support from the European Union through the Jan Amos Komensky Operational Program.

Disclosure of Interests. The author has received travel grant from the MDPI.

References

1. Bishop, C.: Pattern Recognition and Machine Learning. Springer, Berlin (2006)
2. Holčapek, M., Perfilieva, I., Novák, V., Kreinovich, V.: Necessary and sufficient conditions for generalized uniform fuzzy partitions. Fuzzy Sets Syst. **277**, 97–121 (2015)
3. Huang, G.B., Zhu, Q.Y., Siew, C.K.: Extreme learning machine: theory and applications. Neurocomputing **70**(1), 489–501 (2006)
4. Lorke, A., Schneider, F., Heck, J., Nitter, P.: Cybenko's theorem and the capability of a neural network as function approximator (2019). https://www.mathematik.uni-wuerzburg.de/fileadmin/10040900/2019/Seminar-Artificial-Neural-Network-24-9-.pdf
5. Niemczynowicz, A., Perfilieva, I., Garcia-Raffi, L., Kycia, R.: Complex, Hypercomplex and Fuzzy-Valued Neural Networks: New Perspectives and Applications. Routledge, New York (2025)
6. Perfilieva, I.: Fuzzy transforms: theory and applications. Fuzzy Sets Syst. **157**, 993–1023 (2006)
7. Perfilieva, I., Holčapek, M., Kreinovich, V.: A new reconstruction from the F-transform components. Fuzzy Sets Syst. **288**, 3–25 (2016)
8. Russell, S., Norvig, P.: Artificial Intelligence: A Modern Approach. 2nd edn. Prentice Hall (2003)
9. Schölkopf, B., Herbrich, R., Smola, A.J.: A generalized representer theorem. In: Helmbold, D., Williamson, B. (eds.) Computational Learning Theory, pp. 416–426. Springer, Berlin Heidelberg, Berlin, Heidelberg (2001)
10. Wu, Z.: Compactly supported positive-definite radial functions. Adv. Comput. Math. **4**, 283–292 (1995)
11. Zamecnikova, H., Cammarasana, S., Perfilieva, I., Patane, G.: Generalized fuzzy transform and non-local Laplace operator. IEEE Trans. Fuzzy Syst. 1–13 (2024). https://doi.org/10.1109/TFUZZ.2024.3489227

Consensus-Trust-Driven Choquet Aggregation for Probabilistic Opinion Fusion

Han Wang[1], Zhen-Song Chen[2](✉), Rosa M. Rodríguez[3], and Luis Martínez[3]

[1] School of Transportation and Logistics, Southwest Jiaotong University, Chengdu 611756, China
wanghan@my.swjtu.edu.cn

[2] School of Civil Engineering, Wuhan University, Wuhan 430072, China
zschen@whu.edu.cn

[3] Department of Computer Science, University of Jaen, Jaen, Spain
{rmrodrig,martin}@ujaen.es

Abstract. In group decision-making (GDM), expert opinions are increasingly represented as probability distributions to capture inherent uncertainty. Existing dominant aggregation approaches, including linear combination and quantile averaging, are predominantly additive and rely on exogenously specified weights, limiting their ability to reflect interaction patterns among experts. Moreover, although trust has been widely studied in social networkbased GDM, its integration into distribution-based aggregation remains underexplored. To address these issues, this paper proposes a consensustrust-driven probabilistic aggregation framework. First, distributional conflict between experts is quantified via the 1-Wasserstein distance in the quantile domain, and a conflict-sensitive trust function is constructed to capture relational credibility decay. Second, consensus and trust are fused into cooperation intensities, which are embedded into a 2-additive capacity through normalized Möbius coefficients. Finally, a Choquet-based quantile averaging operator is developed to obtain the aggregated collective distribution while preserving distributional validity and modeling interaction effects. An automotive complaint severity assessment example involving experts and large language models illustrates its effectiveness. The proposed model bridges probabilistic uncertainty modeling and interaction-aware aggregation theory, offering a structurally interpretable and flexible framework for complex decision scenarios.

Keywords: Group decision-making · Probabilistic opinion aggregation · Trust modeling · Choquet integral

1 Introduction

In many complex decision environments, group decision-making (GDM) plays a central role in synthesizing heterogeneous expertise and reaching collective

J. Lu et al. (Eds.): FLINS-ISKE 2026, LNCS 16752, pp. 160–177, 2027.
https://doi.org/10.1007/978-981-92-2369-5_12

judgments [1]. In practical decision processes, expert assessments are often subject to substantial uncertainty, which has motivated the use of probabilistic representations to capture subjective beliefs and variability in expert opinions [2]. Consequently, instead of relying on single-point estimates, expert evaluations can be expressed as probability distributions that reflect both central tendency and inherent uncertainty. Aggregating such probabilistic opinions has therefore become an important research topic in decision analysis and risk assessment [3,4]. Distribution-based aggregation thus provides a more expressive and realistic modeling paradigm for uncertainty-aware collective decision-making [5]. Under this framework, the key task is to combine individual probability density functions (PDFs) into a collective distribution that preserves distributional validity. However, a fundamental challenge is to ensure that the resulting aggregation mechanism remains statistically coherent and structurally meaningful.

Existing probabilistic aggregation approaches commonly adopt additive weighting schemes, including linear combination (LC) [6], logarithmic opinion pooling (LOP) [7], and quantile averaging (QA) [8]. Although these methods are computationally efficient and theoretically well-established, they typically rely on exogenously specified weights and implicitly assume independence among experts. As a result, the aggregation process remains structurally additive, regardless of the underlying distributional relationships. Moreover, consensus is often evaluated using overlap-based similarity measures [4], which quantify distributional alignment but do not account for how discrepancies among experts may influence the overall aggregation structure. Consequently, these additive schemes are limited in their ability to reflect interaction patterns or structural dependencies within the group.

Trust-based models have attracted increasing attention in GDM, particularly within social network environments. Li et al. [9] modeled trust risk behaviors in large-scale social networks to derive dynamic aggregation weights, Liu et al. [10] incorporated social trust network completeness into consensus optimization models, and Gai et al. [11] proposed trust propagation mechanisms to regulate noncooperative behavior. Trust has been widely recognized as a critical factor influencing collective decision outcomes. However, these approaches are primarily developed under social network or behavioral regulation settings, where opinions are typically expressed through preference relations or linguistic variables. In probabilistic GDM settings where expert opinions are represented as probability distributions, aggregation focuses on combining PDFs or quantile functions while preserving distributional validity. Within this distribution-based paradigm, trust has not been systematically embedded into the aggregation structure, nor has its interaction with distributional discrepancies been explicitly modeled. Consequently, the role of trust in reshaping influence patterns under probabilistic opinion representations remains an open research problem.

To address these issues, we propose a consensustrust-driven probabilistic aggregation framework. The main contributions of this paper are summarized as follows:

(1) We introduce a distributional conflict-sensitive trust modeling mechanism grounded in 1-Wasserstein geometry.

(2) We propose a consensustrust-driven 2-additive capacity construction that embeds relational structure into non-additive aggregation.

(3) We develop a Choquet-based quantile aggregation framework for probabilistic severity assessment, enabling interaction-aware and distribution-preserving collective evaluation.

The rest of the paper is organized as follows. Section 2 presents preliminaries, Sect. 3 presents the proposed consensus–trust-driven aggregation framework, including conflict modeling, capacity construction, and Choquet-based quantile aggregation. Section 4 provides illustrative examples and analysis. Finally, Sect. 5 concludes the paper.

2 Preliminaries

2.1 Probabilistic Foundations for Consensus-Based Aggregation

In probabilistic GDM, each expert's assessment of a given event is represented as a probability distribution to capture both its central tendency and inherent uncertainty. Let N experts participate in the decision process. The opinion of the i-th expert is modeled as a PDF $f_i(x)$ defined on the evaluation domain $X \subseteq \mathbb{R}$, with corresponding cumulative distribution function (CDF) $F_i(x)$. Each individual distribution satisfies the normalization condition $\int_X f_i(x)\,dx = 1$. The aggregation of the individual PDFs $f_i(x)$ into a collective distribution $f(x)$ can be formally expressed as

$$f(x) = \Phi\big(f_1(x), \ldots, f_N(x)\big), \tag{1}$$

where $\Phi : \mathbb{R}^N \to \mathbb{R}^+$ denotes an aggregation mapping, and the resulting $f(x)$ also satisfies the normalization condition over X.

A fundamental concept in distribution-based aggregation is consensus, which reflects the degree of agreement between probabilistic opinions. A commonly adopted measure of pairwise distributional consensus was proposed by Liu et al. [4], which evaluates the overlapping area between two PDFs. For the i-th and j-th experts, the pairwise consensus level is defined as

$$\mathcal{C}_{ij} = \int_X \min\{f_i(x),\ f_j(x)\}\,dx. \tag{2}$$

where larger values of $\mathcal{C}_{ij}$ indicate a stronger degree of agreement. Owing to its intuitive geometric interpretation with distributional representations, this overlapping-area measure has been widely adopted in probabilistic aggregation models, including the works of Chen et al. [12–14], Cai et al. [5], and Ji et al. [8]. Although the overlapping-area-based consensus effectively captures distributional similarity, it does not explicitly characterize how discrepancies among distributions reshape relational influence structures.

2.2 Quantile Averaging Aggregation

Building upon the abstract aggregation mapping Φ, several concrete realizations have been widely adopted in the literature, among which LC, LOP, and QA are the most representative. Under the LC scheme, Φ is implemented as a weighted LC of individual CDFs, i.e., $F(x) = \sum_{i=1}^{N} w_i F_i(x)$ with $w_i \geq 0$ and $\sum_{i=1}^{N} w_i = 1$. Despite its simplicity, LC aggregation may fail to preserve unimodality, potentially resulting in multimodal collective distributions. The LOP scheme defines Φ via a weighted geometric mean of individual PDFs, but it suffers from the absorbing-zero problem [15].

To address these limitations, QA aggregation has been proposed as a more robust alternative. In this paradigm, the aggregation operator Φ operates in the quantile domain by linearly combining individual quantile functions. Specifically, the aggregated quantile function is defined as

$$F_{QA}^{-1}(p) = \sum_{i=1}^{N} w_i F_i^{-1}(p), \qquad p \in [0, 1], \tag{3}$$

with the aggregated CDF obtained via inversion and the corresponding PDF given by differentiation. QA aggregation enjoys several desirable theoretical properties: it preserves unimodality when individual PDFs are unimodal, and it maintains distributional family consistency under common parametric settings [8].

2.3 The 2-Additive Choquet Integral Model

The Choquet integral is defined with respect to a capacity (also called a fuzzy measure). Let $\mathcal{N} = \{1, \ldots, N\}$ denote the index set of experts. A capacity is a set function $v : 2^{\mathcal{N}} \to [0, 1]$ satisfying $v(\varnothing) = 0$, $v(\mathcal{N}) = 1$, and the monotonicity condition $A \subseteq B \Rightarrow v(A) \leq v(B)$ for all $A, B \subseteq \mathcal{N}$.

The Möbius transform of a capacity v is denoted by m. A capacity is called *2-additive* if its Möbius transform satisfies $m(T) = 0$ for all $T \subseteq \mathcal{N}$ with $|T| > 2$. In this case, only singleton and pairwise terms are nonzero. For a vector $\mathbf{x} = (x_1, \ldots, x_N) \in \mathbb{R}^N$, where $\mathbf{x}$ represents the vector of evaluations from all experts, the Choquet integral with respect to a 2-additive capacity admits the representation [16]:

$$C_v(\mathbf{x}) = \sum_{i \in \mathcal{N}} m_i x_i + \sum_{i<j} m_{ij} \min(x_i, x_j), \tag{4}$$

where $m_i = m(\{i\})$ and $m_{ij} = m(\{i, j\})$ denote the singleton and pairwise Möbius coefficients, respectively. The 2-additive Choquet integral therefore generalizes linear aggregation by incorporating pairwise interactions while avoiding higher-order terms, providing a parsimonious non-additive aggregation structure.

It is worth noting that while the traditional Choquet integral requires defining capacities for 2^n subsets, leading to exponential complexity in parameter identification, our framework utilizes a 2-additive capacity. This approach restricts interactions to pairwise levels, effectively reducing the number of required parameters to n^2. Consequently, the computational burden remains manageable even as the number of experts n scales up, ensuring the framework's efficiency in decision-making scenarios.

3 The Proposed Consensus–Trust-Driven Aggregation Framework

3.1 Problem Description and Framework Overview

A central challenge is that distributional disagreement among experts is not merely a loss of consensus, but may induce different levels of credibility among experts and thus should affect how their opinions are aggregated. Conventional PDF aggregation schemes typically adopt exogenously specified weights and rely primarily on overlap-based consensus measures, which makes it difficult to reflect how conflict patterns reshape the effective influence structure in the group. Moreover, when expert opinions exhibit interactions, purely additive weighting is insufficient to capture such dependencies.

To address these issues, we propose a consensus–trust-driven aggregation framework that links distributional conflict to trust, and further embeds consensus and trust into a non-additive aggregation mechanism. The overall architecture of the proposed framework is illustrated in Fig. 1. The framework consists of three stages: **(1) First,** we quantify distributional conflict between experts based on their probabilistic opinions and construct a conflict-driven trust structure that reflects how trust decays with increasing disagreement. **(2) Second,** we integrate consensus and trust into a 2-additive capacity, yielding both singleton and pairwise interaction coefficients that encode the interaction influence among experts. **(3) Third,** we perform aggregation via a Choquet-based QA operator under the constructed capacity, producing the final collective distribution.

3.2 Conflict-Driven Trust Modeling

In GDM, disagreements among experts are reflected by discrepancies between their opinion distributions. To quantify such discrepancies and construct an endogenous trust structure, we first measure pairwise distributional conflict using the 1-Wasserstein distance and then map conflict intensity to trust via a conflict-sensitive decay function. This will be used in the subsequent capacity construction and aggregation steps.

Distributional Conflict Measure. Let $f_i(x)$ and $f_j(x)$ denote the probabilistic opinions provided by the i-th and j-th experts, respectively, with corresponding CDFs F_i and F_j defined on the evaluation domain $[a, b]$. In GDM, discrepancies are commonly used to quantify opinion conflicts. The 1-Wasserstein distance

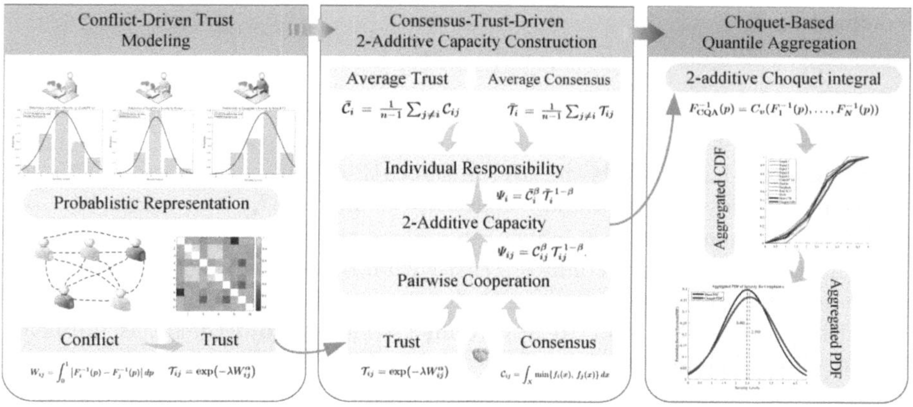

Fig. 1. The Proposed Consensus–Trust Driven 2-Additive Choquet Aggregation Framework

measures the minimal transportation cost between two distributions, capturing differences in both location and shape. It has been applied by Valaei et al. [17] to characterize deviations within uncertainty sets. Motivated by this perspective, we adopt the 1-Wasserstein distance to quantify distributional conflicts among experts.

In the quantile setting, the pairwise distributional conflict between the i-th and j-th experts is defined as

$$W_{ij} = \int_0^1 \left|F_i^{-1}(p) - F_j^{-1}(p)\right| dp, \tag{5}$$

where $p \in [0,1]$ denotes the quantile level. The quantity $W_{ij} \geq 0$ increases as the two distributions become more separated and directly serves as the endogenous conflict measure driving the subsequent trust construction.

Since the evaluation scale is bounded in a compact domain $[a,b]$, the Wasserstein distance satisfies $0 \leq W_{ij} \leq b-a$. To obtain a scale-invariant conflict index and ensure comparability across different evaluation domains, we define the normalized conflict level

$$\tilde{W}_{ij} = \frac{1}{b-a} \int_0^1 \left|F_i^{-1}(p) - F_j^{-1}(p)\right| dp, \tag{6}$$

which ensures $\tilde{W}_{ij} \in [0,1]$. Such a scale-free formulation allows the conflict intensity to be interpreted purely as a relative discrepancy between distributions, which facilitates consistent comparison across different evaluation domains.

Conflict-Sensitive Trust Function. Based on the normalized 1-Wasserstein conflict measure $\tilde{W}_{ij}$, the pairwise trust from the i-th expert to the j-th expert is defined as

$$\mathcal{T}_{ij} = \left(1 - \tilde{W}_{ij}^{\alpha}\right)^{\lambda}, \tag{7}$$

where $\mathcal{T}_{ij} \in [0,1]$, $\alpha > 0$ is the perception curvature parameter that regulates how disagreement is nonlinearly interpreted, and $\lambda > 0$ is the penalization intensity parameter that controls the overall strength of trust attenuation. This formulation ensures strict boundary consistency, namely $\mathcal{T}_{ij} = 1$ when $\tilde{W}_{ij} = 0$ and $\mathcal{T}_{ij} = 0$ when $\tilde{W}_{ij} = 1$, which guarantees that trust completely vanishes under maximal normalized conflict. Trust within this model functions as a purely endogenous variable originating from distributional discrepancies. Specifically, the initial trust state, prior to any conflict-sensitive adjustment, is implicitly defined as a unit baseline where $\mathcal{T}_{ij} = 1$ when $\tilde{W}_{ij} = 0$. This setup represents a neutral prior where all experts are granted full credibility at the outset, with any subsequent erosion of trust being strictly driven by the statistical misalignments in their assessments.

Proposition 1. *Let $\tilde{W}_{ij} \in [0,1]$, $\alpha > 0$, and $\lambda > 0$. The trust function $\mathcal{T}_{ij} = \left(1 - \tilde{W}_{ij}^{\alpha}\right)^{\lambda}$ satisfies: (i) $\mathcal{T}_{ij} \in [0,1]$ for all $\tilde{W}_{ij} \in [0,1]$; (ii) $\mathcal{T}_{ij} = 1$ if and only if $\tilde{W}_{ij} = 0$; (iii) $\mathcal{T}_{ij} = 0$ if and only if $\tilde{W}_{ij} = 1$; (iv) $\mathcal{T}_{ij}$ is strictly decreasing on $(0,1)$ and monotonically decreasing on $[0,1]$.*

Proof. *(i)* Since $\tilde{W}_{ij} \in [0,1]$ and $\alpha > 0$, we have $\tilde{W}_{ij}^{\alpha} \in [0,1]$. Hence $1 - \tilde{W}_{ij}^{\alpha} \in [0,1]$. Because $\lambda > 0$, raising a number in $[0,1]$ to a positive power preserves the interval, which implies $\mathcal{T}_{ij} \in [0,1]$.

(ii) If $\tilde{W}_{ij} = 0$, then $\mathcal{T}_{ij} = (1-0)^{\lambda} = 1$. Conversely, if $\mathcal{T}_{ij} = 1$, then $(1 - \tilde{W}_{ij}^{\alpha})^{\lambda} = 1$, which implies $1 - \tilde{W}_{ij}^{\alpha} = 1$, and therefore $\tilde{W}_{ij} = 0$.

(iii) If $\tilde{W}_{ij} = 1$, then $\mathcal{T}_{ij} = (1-1)^{\lambda} = 0$. Conversely, if $\mathcal{T}_{ij} = 0$, then $1 - \tilde{W}_{ij}^{\alpha} = 0$, which implies $\tilde{W}_{ij} = 1$.

(iv) For $\tilde{W}_{ij} \in (0,1)$, the derivative is $\frac{d\mathcal{T}_{ij}}{d\tilde{W}_{ij}} = -\lambda\alpha\tilde{W}_{ij}^{\alpha-1}\left(1 - \tilde{W}_{ij}^{\alpha}\right)^{\lambda-1}$. Since $\lambda > 0$, $\alpha > 0$, $\tilde{W}_{ij}^{\alpha-1} > 0$ on $(0,1)$, and $\left(1 - \tilde{W}_{ij}^{\alpha}\right)^{\lambda-1} > 0$, the derivative is strictly negative on $(0,1)$. Hence $\mathcal{T}_{ij}$ is strictly decreasing on $(0,1)$ and therefore monotonically decreasing on $[0,1]$.

α and λ capture two complementary behavioral mechanisms: conflict perception curvature and penalization intensity. Their joint specification enables independent control over how disagreement is interpreted and how severely it influences relational trust, thereby enhancing modeling flexibility for diverse GDM scenarios, as illustrated in Fig. 3. The left panel of Fig. 3 depicts the influence of α under a fixed $\lambda = 1$. It can be observed that different values of α reshape the curvature of the trust decay pattern while preserving the boundary conditions. For example, at a moderate conflict level around $\tilde{W}_{ij} = 0.6$, the trust level varies substantially across α values: when $\alpha = 0.5$, trust drops rapidly to a relatively low level, whereas for $\alpha = 3$, trust remains significantly higher before sharply decreasing near the high-conflict region. This indicates that α regulates how

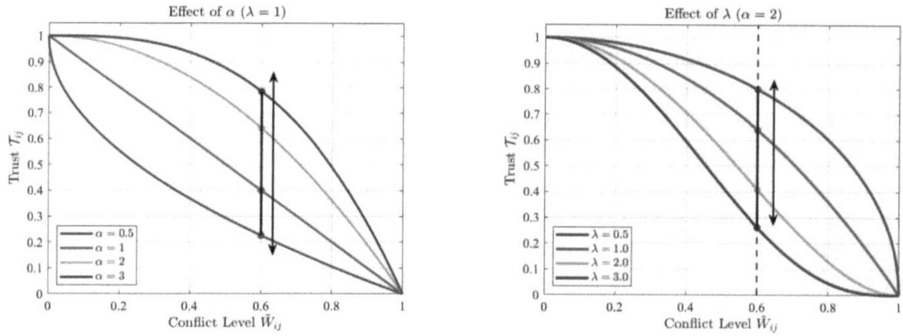

Fig. 2. Sensitivity of the Conflict-Sensitive Trust Function

early and how strongly disagreement begins to erode trust. In GDM, a larger α corresponds to a tolerant attitude toward mild discrepancies but stricter penalization of severe divergence, whereas a smaller α reflects heightened sensitivity even to minor inconsistencies among experts.

The right panel of Fig. 2 illustrates the effect of λ under a fixed $\alpha = 2$. Unlike α, which modifies curvature, λ controls the overall attenuation intensity across the entire conflict domain. For instance, at $\tilde{W}_{ij} = 0.6$, increasing λ from 0.5 to 3 leads to a pronounced downward shift in trust levels, while preserving the same qualitative decay shape. This demonstrates that λ acts as a global penalization parameter, uniformly compressing trust as conflict increases. In GDM applications, larger λ values represent stricter trust discounting policies, whereas smaller values correspond to more tolerant aggregation environments.

3.3 ConsensusTrust-Driven 2-Additive Capacity Construction

To capture the intertwined effects of consensus alignment and relational trust within a coherent non-additive framework, we integrate these two dimensions into a unified capacity generation mechanism. Such a 2-additive construction allows individual influence and pairwise interactions to be modeled simultaneously, providing a structurally consistent foundation for Choquet-based aggregation.

Consensus–Trust Fusion. Within the 2-additive Choquet capacity framework, the pairwise interaction term m_{ij} is introduced to capture the cooperative strength between expert pairs. In the present study, we focus exclusively on modeling the intensity of cooperation rather than redundancy effects, thereby constructing a purely synergistic interaction structure. It is important to note that such interaction cannot be determined solely by distributional agreement. If the capacity were constructed only from the consensus measure $\mathcal{C}_{ij}$ defined by distributional overlap, the resulting interaction structure would reduce to a similarity, failing to reflect relational reliability under conflict conditions. On the other hand, the trust function $\mathcal{T}_{ij}$ is derived from the 1-Wasserstein conflict measure and fundamentally represents a conflict-driven relational attenuation mechanism. Although trust effectively captures relational stability under disagreement, conflict intensity alone cannot fully substitute for structural consistency between probability distributions. Two experts may exhibit substantial global overlap while differing locally in certain regions, leading to discrepancies between consensus and conflict-based trust.

Therefore, in the capacity construction process, relying solely on consensus would overlook conflict-sensitive relational information, whereas relying solely on trust could weaken structural similarity information embedded in the distributions. To ensure that the interaction term simultaneously reflects distributional alignment and relational credibility under conflict, we integrate consensus and trust into a unified cooperative intensity. To this end, we introduce a cooperative intensity Ψ_{ij} through a fusion mechanism:

$$\Psi_{ij} = \mathcal{C}_{ij}^{\beta}\, \mathcal{T}_{ij}^{1-\beta}, \tag{8}$$

where C_{ij} denotes the pairwise consensus level and $\mathcal{T}_{ij}$ represents the conflict-sensitive trust degree. The quantity Ψ_{ij} is interpreted as the cooperative intensity between expert i and expert j, measuring the strength of their synergistic interaction. The parameter $\beta \in [0, 1]$ regulates the relative contribution of consensus and trust in shaping cooperative intensity. When β approaches 1, the cooperative mechanism becomes consensus-dominant, emphasizing structural alignment of distributions. When β approaches 0, trust becomes dominant, highlighting relational credibility under conflict conditions. Thus, β serves as a structural balance parameter that controls the relative influence of agreement consistency and relational reliability in determining cooperative strength.

From a GDM perspective, this fusion mechanism reflects two essential behavioral principles. First, cooperative influence between experts should depend not only on the proximity of their probabilistic assessments, but also on the stability of their relational credibility under conflict. Second, effective cooperation in collective aggregation emerges from the joint reinforcement of agreement and trust, rather than from either factor alone. Therefore, the proposed consensus–trust fusion provides a structurally coherent and behaviorally interpretable measure of cooperative intensity, which serves as the foundational building block for constructing a conflict-aware 2-additive capacity in Choquet-based aggregation.

We then construct the individual-level cooperative intensity by aggregating the pairwise cooperative strengths. Specifically, since Ψ_{ij} represents the cooperative intensity between ith and jth experts, the overall cooperative strength of expert i is defined as the average of its pairwise cooperative intensities: $\Psi_i = \frac{1}{N-1}\sum_{j\neq i}\Psi_{ij}$. The quantities Ψ_{ij} and Ψ_i serve as interaction- and individual-level cooperative signals, respectively, for determining the Möbius coefficients in the 2-additive capacity.

2-Additive Capacity Generation. To embed the consensus–trust cooperative intensities into the 2-additive Choquet framework introduced in Sect. 2.3, we construct the Möbius coefficients directly from Ψ_i and Ψ_{ij}. Since the 2-additive Choquet representation in Eq. (4) requires normalized and nonnegative Möbius coefficients, the cooperative intensities must be globally rescaled while preserving their relative magnitudes.

To embed the cooperative intensities into a valid 2-additive capacity, the Möbius coefficients must satisfy the normalization condition $\sum_{i\in N} m_i + \sum_{i<j} m_{ij} = 1$, which guarantees that the total mass of singleton and pairwise interactions equals one. We therefore define the singleton and pairwise Möbius coefficients through a unified normalization scheme:

$$m_i = \frac{\Psi_i}{Z}, \qquad m_{ij} = \frac{\Psi_{ij}}{Z}, \tag{9}$$

where $Z = \sum_{i\in N} \Psi_i + \sum_{i<j} \Psi_{ij}$. Substituting this definition yields $\sum_{i\in N} m_i + \sum_{i<j} m_{ij} = \frac{1}{Z}\left(\sum_{i\in N} \Psi_i + \sum_{i<j} \Psi_{ij}\right) = 1$, so the normalization requirement is satisfied by construction. To verify monotonicity under the standard 2-additive framework, it is sufficient to check the condition $m_i + \sum_{j\neq i} m_{ij} \geq 0, \quad \forall i \in N$. Substituting the normalization scheme gives $m_i + \sum_{j\neq i} m_{ij} = \frac{1}{Z}\left(\Psi_i + \sum_{j\neq i} \Psi_{ij}\right)$. Since $\Psi_i \geq 0$ and $\Psi_{ij} \geq 0$ by construction and $Z > 0$, the above quantity is nonnegative. Hence, the monotonicity requirement of a valid 2-additive capacity is satisfied.

Therefore, the proposed normalization scheme preserves both normalization and monotonicity while maintaining the relative magnitudes of cooperative intensities, yielding a coherent and purely synergistic non-additive representation of collective influence.

3.4 Choquet-Based Quantile Averaging Aggregation

Building upon the constructed 2-additive Möbius coefficients, we develop a Choquet-based quantile averaging (CQA) scheme to obtain the final collective distribution. For each quantile level $p \in [0, 1]$, the aggregated quantile is defined via the 2-additive Choquet integral as

$$F_{\text{CQA}}^{-1}(p) = C_v(F_1^{-1}(p), \ldots, F_N^{-1}(p)) \tag{10}$$

which admits the explicit form

$$F_{\text{CQA}}^{-1}(p) = \sum_{i\in N} m_i F_i^{-1}(p) + \sum_{i<j} m_{ij} \min\bigl(F_i^{-1}(p), F_j^{-1}(p)\bigr). \tag{11}$$

The aggregated CDF is obtained by numerical inversion of the quantile function. Specifically, the CDF is defined as $F_{CQA}(x) = \max\left\{p \in [0, 1] \ : \ F_{CQA}^{-1}(p) \leq x\right\}$. In practice, this inversion is performed numerically since $F_{CQA}^{-1}(p)$ is obtained in closed form while its inverse may not admit an analytic expression, and the PDF is obtained by $f_{CQA}(x) = \frac{d}{dx} F_{CQA}(x)$.

4 Illustrative Examples

4.1 Scenario Overview

In recent years, collaborative human–AI evaluation has emerged as a growing trend in safety-critical decision-making domains. In the automotive industry, customer complaints constitute an important source of risk signals for manufacturers and regulatory agencies. Accurate severity assessment of such complaints is essential for identifying potential safety hazards and prioritizing corrective actions. In this study, we consider a human–LLM collaborative assessment scenario in which domain experts and large language models(LLMs) jointly evaluate the severity distributions of automotive complaint texts. The data are collected from the National Highway Traffic Safety Administration (NHTSA) and CheZhiNet, covering issues such as engine malfunction, safety defects, and performance risks. The proposed distribution-based aggregation framework in Sect. 3 is then applied to integrate human and machine evaluations for collaborative severity assessment of automotive complaints.

4.2 Method Application

In the following content, we will describe the data collection, assessment process, and assessment results in detail.

First, data collection and distribution fitting. We collected three representative automotive complaint texts from the NHTSA and CheZhiNet (see Table 1 in the Supplementary Materials for details). The severity evaluation criteria follow the Chinese national standard "Guidelines for Risk Assessment and Risk Control of Motor Vehicle Products" (GB/T), where severity levels are classified into five intervals: $(0, 1]$, $(1, 2]$, $(2, 3]$, $(3, 4]$, and $(4, 5]$, each corresponding to increasing levels of impact (see Table 1 in the Supplementary Materials).

To construct a human expert–LLM collaborative assessment setting, five LLMs (ChatGPT 5.2, Doubao, DeepSeek, Kimi K2.5, and Qwen) and five human experts independently evaluated each complaint according to the same severity criteria. The LLMs generated severity probability distributions using the prompts provided in Table 3 of the Supplementary Materials, while the human experts provided their assessments based on domain knowledge and practical experience. We then performed probability density fitting using MATLAB, as illustrated in Fig. 3 for **Complaint a**, and in Supplementary Materials Figs. 1 and 2 for **Complaint b** and **Complaint c**. For notation consistency, the five human experts are denoted as e_1, e_2, e_3, e_4, e_5, and the five LLMs as $e_6, e_7, e_8, e_9, e_{10}$. In the subsequent interaction modeling, we focus exclusively on cross-group human expert–LLM interactions to characterize cooperative behavior between human experts and intelligent models.

Second, conflict and trust calculation. For **Complaints a, b, and c**, we first computed the pairwise distributional conflict W_{ij} between each human expert and LLM based on their probabilistic evaluation discrepancies. The raw conflict measure W_{ij} was calculated using the 1-Wasserstein distance defined in Eq. (5),

Table 1. Severity Assessment of **Complaint a**

LLM and Expert	[0,1]	(1,2]	(2,3]	(3,4]	(4,5]	Mean	Deviation
Expert 1 (e_1)	10%	20%	40%	20%	10%	2	1.101
Expert 2 (e_2)	5%	22%	50%	19%	5%	1.9703	0.8995
Expert 3 (e_3)	6%	25%	45%	18%	6%	1.93	0.95616
Expert 4 (e_4)	5%	30%	35%	20%	10%	2	1.0541
Expert 5 (e_5)	7%	27%	42%	18%	6%	1.89	0.98365
ChatGPT 5.2 (e_6)	5%	25%	40%	20%	10%	2.05	1.0286
Doubao (e_7)	0%	20%	60%	15%	5%	2.05	0.74366
DeepSeek (e_8)	10%	20%	40%	30%	0%	1.9	0.94815
Kimi K2.5 (e_9)	0%	15%	25%	45%	15%	2.6	0.92113
Qwen (e_{10})	5%	15%	50%	20%	10%	2.15	0.96792

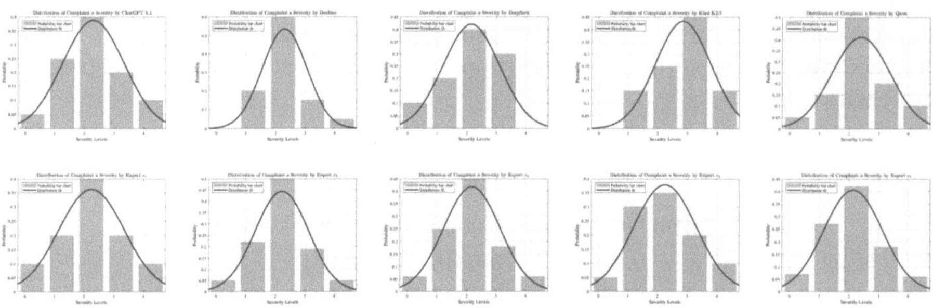

Fig. 3. Individual PDF Fitting for **Complaint a**

which quantifies distributional divergence in the quantile domain. To ensure scale invariance and comparability across evaluation domains, the conflict values were further normalized according to Eq. (6), yielding the normalized conflict levels $\tilde{W}_{ij} \in [0,1]$. The complete pairwise normalized conflict matrices $\tilde{W}_{ij}$, jointly presenting the results for all three complaints, are provided in Supplementary Materials Table 5.

Based on the normalized conflict values, the corresponding conflict-sensitive trust degrees $\mathcal{T}_{ij}$ were then derived through the trust function defined in Eq. (7), with parameters $\alpha = 1$ and $\lambda = 1$. The resulting pairwise trust matrices $\mathcal{T}_{ij}$ for **Complaints a, b, and c** are reported in Table 2.

Third, construction of the 2-additive capacity. The 2-additive capacity is constructed by mapping the consensustrust cooperative intensities into Mbius coefficients. The pairwise consensus ($\mathcal{C}_{ij}$) is computed via Eq. (2) and reported in Supplementary Materials Table 6. It is then combined with the trust degrees ($\mathcal{T}_{ij}$) (Table 2) to obtain the pairwise cooperative intensities (Ψ_{ij}) using Eq. (8) with ($\beta = 0.5$), as reported in Supplementary Materials Table 7. By averaging the corresponding pairwise cooperative intensities, the individual cooperative

Table 2. Pairwise Conflict-Sensitive Trust $\mathcal{T}_{ij}$ for Complaints a–c

	Complaint a					**Complaint b**					**Complaint c**				
	e_6	e_7	e_8	e_9	e_{10}	e_6	e_7	e_8	e_9	e_{10}	e_6	e_7	e_8	e_9	e_{10}
e_1	0.95	0.65	0.90	0.40	0.85	0.91	0.78	0.78	0.53	0.81	0.88	0.78	0.94	0.89	0.83
e_2	0.83	0.83	0.80	0.34	0.79	0.81	0.80	0.80	0.61	0.73	0.88	0.78	0.88	0.75	0.73
e_3	0.88	0.78	0.83	0.33	0.78	0.44	0.19	0.59	0.74	0.54	0.91	0.83	0.91	0.76	0.76
e_4	0.95	0.65	0.80	0.40	0.85	0.86	0.81	0.71	0.50	0.76	0.88	0.78	0.90	0.75	0.75
e_5	0.84	0.74	0.81	0.29	0.74	0.85	0.86	0.70	0.55	0.75	0.80	0.80	0.80	0.91	0.89

intensities (Ψ_i) are obtained and reported in Supplementary Materials Table 8. Finally, the singleton and pairwise Möbius coefficients (m_i) and (m_{ij}) are determined via Eq. (9), with the final results reported in Tables 3–4, respectively, completing the 2-additive capacity construction for subsequent Choquet aggregation. Tables 18 collectively summarize the experimental setup and the intermediate results supporting the proposed consensustrust-driven aggregation framework.

Table 3. m_i values for **Complaints a, b, and c**

Agent	Complaint a	Complaint b	Complaint c
e_1	0.0302	0.0348	0.0318
e_2	0.0295	0.0344	0.0305
e_3	0.0292	0.0268	0.0314
e_4	0.0290	0.0342	0.0309
e_5	0.0283	0.0345	0.0312
e_6	0.0346	0.0355	0.0322
e_7	0.0294	0.0322	0.0297
e_8	0.0318	0.0346	0.0331
e_9	0.0183	0.0297	0.0306
e_{10}	0.0320	0.0327	0.0301

Fourth, aggregated severity assessment. The final severity distributions are obtained via the 2-additive CQA scheme. The aggregated quantile at each $p \in [0, 1]$ is computed according to Eq. (11), which integrates individual and interaction effects through the learned Möbius coefficients. The operator is implemented in MATLAB by pointwise evaluation over dense quantile grids, followed by inversion to recover the CDF and numerical differentiation to derive the PDF. The aggregated CDFs and PDFs for **Complaints a, b, and c** are reported in Figs. 4 and 5, respectively. In these figures, the black curves represent the aggregation results obtained by the conventional QA operator, while the blue curves correspond to the proposed 2-additive CQA results.

Table 4. m_{ij} for **Complaints a, b, and c**

	Complaint a					Complaint b					Complaint c				
	e6	e7	e8	e9	e10	e6	e7	e8	e9	e10	e6	e7	e8	e9	e10
e1	0.0360	0.0274	0.0341	0.0201	0.0332	0.0395	0.0356	0.0364	0.0274	0.0352	0.0321	0.0288	0.0344	0.0330	0.0306
e2	0.0330	0.0328	0.0311	0.0177	0.0327	0.0360	0.0354	0.0370	0.0311	0.0323	0.0324	0.0297	0.0329	0.0289	0.0287
e3	0.0345	0.0309	0.0317	0.0174	0.0316	0.0253	0.0155	0.0308	0.0344	0.0280	0.0335	0.0313	0.0335	0.0289	0.0298
e4	0.0360	0.0265	0.0304	0.0201	0.0323	0.0386	0.0369	0.0345	0.0268	0.0343	0.0326	0.0295	0.0337	0.0294	0.0291
e5	0.0337	0.0296	0.0315	0.0163	0.0303	0.0380	0.0378	0.0345	0.0288	0.0336	0.0302	0.0293	0.0312	0.0330	0.0322

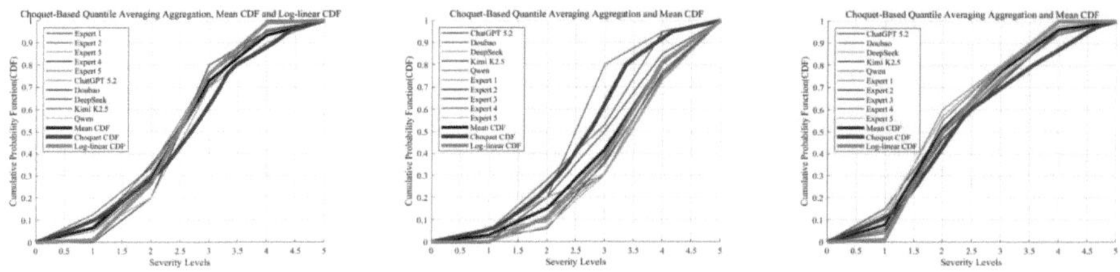

Fig. 4. Aggregated CDF Results for **Complaints a, b and c**

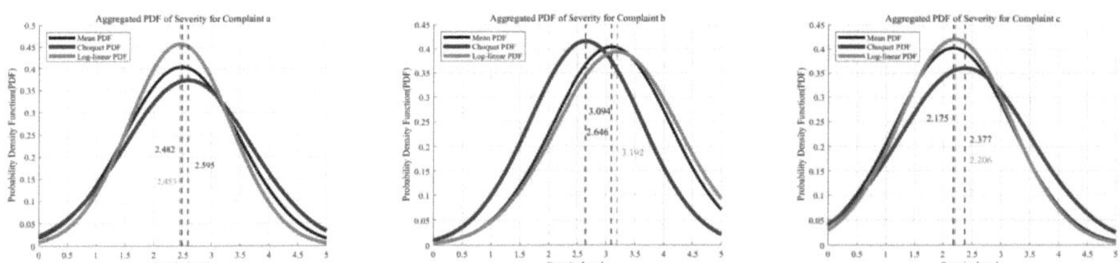

Fig. 5. Aggregated PDF Results for **Complaints a, b and c**

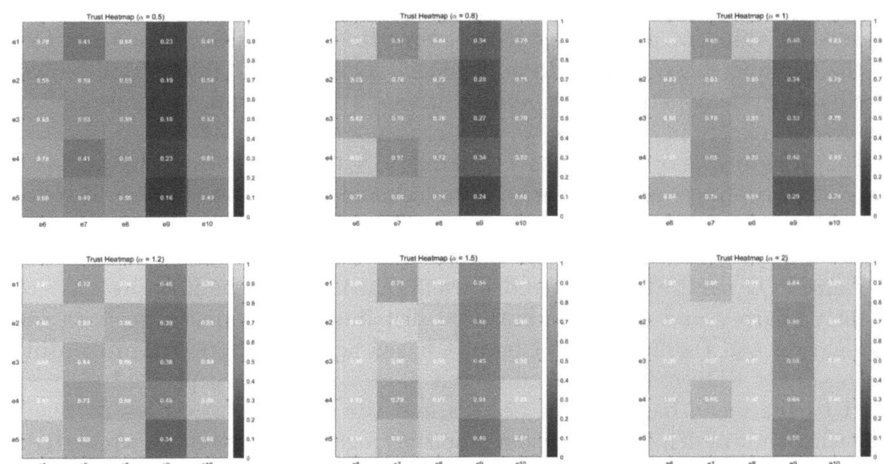

Fig. 6. Sensitivity analysis under different α values

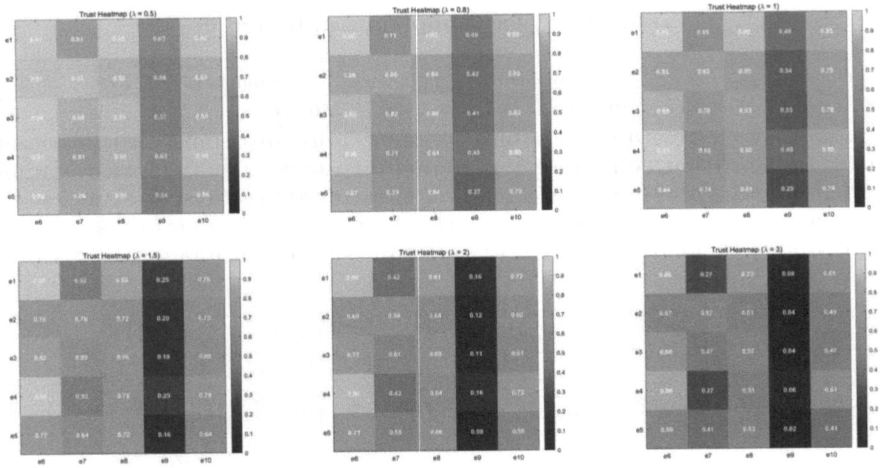

Fig. 7. Sensitivity analysis under different λ values

4.3 Analysis and Discussion

The aggregated CDF and PDF results in Figs. 4 and 5 reveal consistent yet non-trivial differences between the arithmetic mean aggregation and the 2-additive Choquet-based aggregation. For **Complaint a**, the Choquet PDF exhibits a slightly right-shifted peak (2.595) compared to the mean-based result (2.482), indicating a moderately higher collective severity assessment under interaction-aware aggregation. In contrast, for **Complaint b**, the Choquet peak (2.646) is notably lower than the mean peak (3.094), suggesting that accounting for trust-weighted interactions attenuates the influence of relatively severe evaluations. For **Complaint c**, a similar but milder leftward adjustment is observed, reflecting redistribution of probability mass under the non-additive structure. Overall, the Choquet-based aggregation produces smoother and interaction-adjusted distributions, demonstrating its ability to balance individual influence and relational trust effects in collective severity assessment.

To provide a more concrete comparison with probabilistic fusion strategies, we incorporate a log-linear pooling scheme as a representative Bayesian-inspired aggregation method, where the aggregated distribution is constructed as $f_{\log}(x) \propto \prod_{i=1}^{n} f_i(x)^{w_i}$ with equal normalized weights $w_i = \frac{1}{n}$. The aggregated CDF and PDF results in Figs. 4 and 5 reveal that log-linear pooling produces consistently sharper and more concentrated distributions compared to both mean aggregation and the 2-additive Choquet-based aggregation. For **Complaint a**, the log-linear PDF exhibits a higher and more concentrated peak (2.453) compared to the mean-based result (2.482) and the Choquet-based result (2.595), indicating that multiplicative fusion amplifies regions of agreement while slightly pulling the distribution toward the consensus center. In contrast, for **Complaint b**, the log-linear peak (3.192) is further right-shifted relative to both the mean peak (3.094) and the Choquet peak (2.646), suggesting that multiplicative aggregation reinforces relatively severe evaluations when they are consistently supported across experts. For **Complaint c**, the log-linear peak (2.206) shows a mild rightward shift compared to the mean result (2.175) while remaining lower than the Choquet peak (2.377), reflecting a moderate concentration effect without substantial redistribution toward higher severity levels. Overall, log-linear pooling produces more peaked and less dispersed distributions, demonstrating its tendency to reinforce shared probability mass through multiplicative aggregation, whereas the Choquet-based approach yields smoother and interaction-adjusted distributions by explicitly accounting for pairwise relationships, thereby providing a more balanced integration of individual influence and relational trust effects.

Figures 6 and 7 provide a visual sensitivity analysis of the conflict-sensitive trust function $\mathcal{T}_{ij} = \left(1 - \tilde{W}_{ij}^{\alpha}\right)^{\lambda}$,illustrating the respective roles of the curvature parameter α and the decay parameter λ. Figure 6 presents the sensitivity with respect to α (with $\lambda = 1$). When $\alpha = 0.5$, the heatmap exhibits pronounced color contrast, and several moderate-conflict entries already shift toward cooler tones, indicating heightened sensitivity to intermediate discrepancies. As α increases (e.g., $\alpha = 0.8, 1, 1.2, 1.5, 2$), the trust values in low-to-moderate conflict regions remain comparatively high and visually more uniform, while only high-conflict entries undergo noticeable attenuation. This behavior confirms that α governs the curvature of trust decay: smaller α amplifies mild conflicts, whereas larger α attenuates minor deviations and concentrates penalization on substantial discrepancies. Importantly, the relative spatial pattern of high- and low-trust regions remains stable across different α values, indicating structural robustness of the trust ordering. Figure 7 depicts the sensitivity with respect to λ (with $\alpha = 1$). As λ increases (e.g., $\lambda = 0.5, 0.8, 1, 1.5, 2, 3$), the overall trust matrix undergoes a progressive global compression, with warm colors gradually transitioning toward cooler tones. This confirms that λ functions as a global attenuation parameter, scaling the overall intensity of trust without altering its underlying distributional structure. Despite the change in magnitude, the relative configuration of high- and low-trust interactions remains largely unchanged, which is consistent with the monotonicity and strict decay properties established in Proposition 1.

From a broader perspective, the proposed consensus–trust mechanism can be interpreted as a form of relational inference among decision makers. Specifically, the pairwise trust values T_{ij} can be viewed as encoding the strength of interactions among experts, thereby forming an implicit network that governs the aggregation process. This view is closely aligned with Bayesian ego-graph inference in networked multi-agent systems, where inter-agent relationships are inferred from observed data [18]. While such approaches typically rely on explicit probabilistic modeling of latent connections, our method derives interaction strengths directly from distributional consistency, enabling the aggregation process to dynamically emphasize reliable sources and suppress inconsistent ones. This connection highlights the broader applicability of the proposed mechanism beyond traditional group decision-making settings.

5 Conclusion

This paper introduced a novel consensustrust-driven aggregation framework, which combines distributional conflict and relational trust to provide a robust mechanism for aggregating expert and LLM evaluations. By leveraging a 2-additive Choquet integral, our approach effectively incorporates both individual influence and pairwise interactions. Experimental results on **Complaints a, b, and c** demonstrate the ability of our framework to produce consistent and interpretable aggregated severity distributions. While the method shows promising results, future work will focus on addressing the scalability challenges associated with large numbers of experts and datasets, as well as refining the aggregation process for dynamic contexts where trust and conflict parameters may evolve over time.

Acknowledgments. This work was supported by the National Natural Science Foundation of China under grant 72571203.

Disclosure of Interests. The authors have no competing interests to declare that are relevant to the content of this article.

References

1. Li, G., Kou, G., Peng, Y.: A group decision making model for integrating heterogeneous information. IEEE Trans. Syst. Man Cybern. Syst. **48**(6), 982–992 (2016)
2. Roger Cooke. Experts in Uncertainty: Opinion and Subjective Probability in Science. Oxford university Press (1991)
3. Clemen, R.T., Winkler, R.L.: Combining probability distributions from experts in risk analysis. Risk Analy. **19**(2), 187–203 (1999)
4. Liu, X., Ghorpade, A., Tu, Y., Zhang, W.: A novel approach to probability distribution aggregation. Info. Sci. **188**, 269–275 (2012)
5. Cai, M., Lin, Y., Han, B., Liu, C., Zhang, W.: On a simple and efficient approach to probability distribution function aggregation. IEEE Trans. Syst. Man Cybern. Syst. **47**(9), 2444–2453 (2016)
6. Stratigakos, A., Pineda, S., Morales, J.M.:. Decision-focused linear pooling for probabilistic forecast combination. Int. J. Forecasting **41**(3), 1112–1125 (2025)
7. Yu, K., Taylor, J.W., Meng, X.: Quantile-adaptive probabilistic forecast combining. European J. Operat. Res. (2025)
8. Ji, C., Xiwen, L., Zhang, W.: A biobjective optimization model for expert opinions aggregation and its application in group decision making. IEEE Syst. J. **15**(2), 2834–2844 (2020)
9. Li, M., Qiu, L., Yejun, X., Liu, X., Kou, G., Herrera-Viedma, E.: A consensus model based on social network analysis in large-scale group decision making: mining and managing trust risk behaviors. IEEE Trans. Syst. Man Cybern.: Syst. **53**(10), 6204–6219 (2023)

10. Liu, P., Li, Y., Wang, P.: Social trust-driven consensus reaching model for multiattribute group decision making: exploring social trust network completeness. IEEE Trans. Fuzzy Syst. **31**(9), 3040–3054 (2023)
11. Gai, T., Chiclana, F., Jin, W., Zhou, M., Jian, W.: A transformation method of non-cooperative to cooperative behavior by trust propagation in social network group decision making. IEEE Trans. Fuzzy Syst. **33**(7), 2238–2250 (2025)
12. Chen, Z.-S., Zhou, M.-D., Chin, K.-S., Darko, A., Wang, X.-J., Pedrycz, W.: Optimized decision support for BIM maturity assessment. Autom. Constr. **149**, 104808 (2023)
13. Chen, Z., Chen, K., Xu, Y., Pedrycz, W., Skibniewski, M.J.: Multiobjective optimization-based decision support for building digital twin maturity measurement. Adv. Eng. Info. **59**, 102245 (2024)
14. Chen, S., Wang, Z., Deveci, M., Ding, W., Pedrycz, W., Skibniewski, M.J.: Optimization-based probabilistic decision support for assessing building information modelling (BIM) maturity considering multiple objectives. Info. Fus. **102**, 102026 (2024)
15. Koliander, G., El-Laham, Y., Djurić, P.M., Hlawatsch, F.: Fusion of probability density functions. Proc. IEEE **110**(4), 404–453 (2022)
16. Mayag, B., Bouyssou, D.: Necessary and possible interaction between criteria in a 2-additive Choquet integral model. Eur. J. Oper. Res. **283**(1), 308–320 (2020)
17. Valaei, M., Saif, A., Sarhadi, H., Afshari, H.: Robust prepositioning and allocation of maritime search and rescue vessels with incident location uncertainty. Transport. Res. Part E: Logist. Transport. Rev. **209**, 104693 (2026)
18. Duan, W., Lu, J., Xuan, J.: Bayesian ego-graph inference for networked multi-agent reinforcement learning. arXiv preprint. arXiv:2509.16606, 2025. to appear in NeurIPS (2025)

Calibrated Fuzzy-Measure Reward Shaping for Safe UAV Obstacle Avoidance

Xiang Luo[1], Hongbing Luo[1], Jian Di[2(✉)], Tao Wang[3], Yu Kang[1,3], and Haibo Ji[1]

[1] Department of Automation, University of Science and Technology of China, Hefei 230026, China

[2] Institute of Advanced Technology, University of Science and Technology of China, Hefei 230088, China
dijian@mail.ustc.edu.cn

[3] Department of Automation, Hefei University of Technology, Hefei 230009, China

Abstract. In this paper, a learning-based obstacle avoidance problem for unmanned aerial vehicles (UAVs) under perceptual uncertainty is studied, particularly when range-like observations and velocity-related cues are jointly corrupted. In such regimes, nominal safety potentials can become over-optimistic near obstacles, producing dense feedback that underestimates risk and weakens corrective signals from sparse failures. To address this, a plug-in potential-based reward shaping module is proposed. It constructs a bounded safety potential from aligned distance and closing-speed cues, evaluates it on a fixed perturbation set, and aggregates the resulting scores via an ordered Choquet integral induced by a λ-fuzzy measure. This design captures non-additive interactions between coupled perturbations and yields interpretable Choquet weights. To avoid using λ as an ad-hoc conservatism knob, conservatism is defined operationally and the fuzzy-measure parameters are calibrated offline with a near-obstacle objective that penalizes overly high potentials. In pilot-scale simulations on `SimpleAvoid` across multiple uncertainty levels, the proposed shaping was associated with higher task completion under non-zero noise while maintaining comparable collision rates.

Keywords: Fuzzy measure · Choquet integral · Reward shaping · Safe reinforcement learning · UAV obstacle avoidance

1 Introduction

Autonomous UAV navigation in cluttered environments must reconcile task efficiency with strict safety requirements. Model-free reinforcement learning offers a pragmatic route to end-to-end obstacle avoidance, but safety concerns and

This work was supported in part by the National Natural Science Foundation of China under Grant 62503450 and 62503147, and the Anhui Provincial Natural Science Foundation under Grant 2508085QF230.

J. Lu et al. (Eds.): FLINS-ISKE 2026, LNCS 16752, pp. 178–194, 2027.
https://doi.org/10.1007/978-981-92-2369-5_13

deployment gaps remain central challenges in real systems [3,6]. These challenges are amplified by perceptual uncertainty: range-like cues and velocity-related estimates are noisy, and their errors can interact in ways that are not well captured by simple additive assumptions. As a result, a controller that appears tolerant to each noise source in isolation can still behave unsafely when the two sources co-occur, especially in near-obstacle situations where decisions must be made quickly.

Existing safe and robust RL approaches often enforce safety through constrained objectives or robust optimization, for example via constraint-based policy updates [1] or robust MDP formulations [11,15]. While effective, these methods can require non-trivial changes to the learning objective, optimization procedure, or system assumptions. In this work, we focus on a complementary engineering goal: improving safety behavior under a prescribed observation-noise protocol using a plug-in module that leaves the underlying model-free learner unchanged.

A common and practical plug-in mechanism is reward shaping. Among many options, potential-based reward shaping (PBRS) is attractive because it integrates cleanly with existing pipelines and can preserve the set of optimal policies under standard assumptions [14]; recent robotics work also continues to build on potential-based shaping formulations for structured reward design [2]. However, a potential that is reasonable in nominal conditions can become over-optimistic under uncertainty, particularly near obstacles. If the shaping signal systematically overestimates safety, the learned policy may adopt risky maneuvers that are only weakly corrected by sparse failure feedback.

Motivated by these considerations, this paper explores a robust fuzzy-guided shaping strategy for safe UAV obstacle avoidance. Our key observation is that robustness in this setting is not equivalent to taking a pointwise minimum over perturbations. For safety-critical learning, the robustification should be (i) *operationally conservative* in near-obstacle regimes and (ii) *interpretable*, so that one can identify which perturbations dominate the shaping signal. To achieve these properties without explicit min–max optimization, we use a λ-fuzzy measure and ordered Choquet aggregation over a fixed perturbation set. The fuzzy measure provides a compact representation of non-additive interactions among perturbations, which directly aligns with coupled uncertainty: when uncertainty sources interact, a purely additive aggregation can misrepresent their joint impact on safety. The ordered Choquet form yields Choquet weights, providing a direct and auditable account of how individual perturbations influence the robust potential.

A critical nuance is that a larger λ does not necessarily imply a more conservative safety bias. We therefore define conservatism operationally and calibrate it offline: in near-obstacle segments, the robust potential is encouraged to avoid overly high values that would underestimate risk. This design keeps the method practical and avoids claiming monotonicity properties that may not hold for a fixed parameterization.

Our approach is intended as an engineering robustification rather than a worst-case min–max procedure. We do not provide formal guarantees for all admissible uncertainties. Instead, we make the perturbation set explicit and the resulting Choquet weights auditable, which yields a shaping signal that is conservative under the observation-noise protocol used in our experiments.

Accordingly, the resulting potential is designed to be pessimistic in safety-critical regimes rather than universally optimal under an abstract uncertainty set. This emphasis matches pilot-scale robotics learning, where transparency and tunability are often preferred to substantially more complex robust-optimization pipelines.

Contributions. We summarize our contributions as follows:

- **A plug-in shaping module for model-free safe navigation.** We develop a potential-based reward shaping module that can be layered onto standard model-free algorithms without changing the underlying optimization objective. In the present study, all reported experiments use Soft Actor–Critic (SAC) [9] as the base learner, while the formulation itself remains compatible with other standard model-free algorithms such as Proximal Policy Optimization (PPO) [17].
- **Non-additive robustification with interpretable Choquet weights.** We robustify a bounded safety potential by scoring a fixed perturbation set and aggregating the resulting values with an ordered Choquet integral induced by a λ-fuzzy measure, which captures non-additive interactions and exposes an auditable Choquet weight spectrum.
- **Operational conservatism via offline calibration.** We calibrate the fuzzy-measure parameters offline with a near-obstacle objective that penalizes overly optimistic potentials, avoiding monotonic assumptions about λ while keeping training unchanged.

2 Related Work

Fuzzy logic, fuzzy measures, and Choquet aggregation. Fuzzy logic provides a principled language for handling imprecision in decision systems, and it has been widely used as a building block for learning-based control. Beyond classical fuzzy inference, fuzzy measures and the Choquet integral [4] offer a compact way to model non-additive interactions among multiple criteria, which is useful when uncertainty sources interact in a coupled manner [7,8]. FLINS and ISKE papers have studied fuzzy UAV flight control and fuzzy navigation in embodied robotic systems [16,19]. Recent work has started to bring such non-additive aggregation into reinforcement learning. In particular, Fuz-RL uses Choquet-based aggregation for robust value estimation under uncertainty, providing a fuzzy-guided alternative to explicit min–max training [20]. Our work follows this non-additive-robustification line but targets a different locus. Instead of modifying the value operator or the policy-optimization objective, we robustify the *shaping signal* via a plug-in safety potential and keep the underlying model-free learner unchanged, emphasizing modularity and interpretability.

Learning-based UAV navigation and uncertainty handling. Recent UAV-navigation studies have used deep reinforcement learning for autonomous path planning, while agile-flight and multi-drone-racing studies further underscore that perception quality, adaptation, and closed-loop robustness can materially affect flight performance [5,10,12]. Our task is simpler and more local-reactive than those settings, but it addresses the same engineering question: whether an uncertainty-aware add-on can improve robustness without replacing the base model-free learner.

Safe and robust reinforcement learning under uncertainty. Safe reinforcement learning (RL) has been studied from both risk-sensitive criteria and constraint-based perspectives [6], and more recent surveys review safe learning methods in robotics settings with an emphasis on practical deployment [3]. Representative constraint-based methods include Constrained Policy Optimization (CPO) [1]. Classical robust RL formulations build on robust Markov decision process (MDP) models [11,15], while recent fuzzy-guided alternatives such as Fuz-RL revisit uncertainty handling through non-additive aggregation rather than explicit inner maximization [20]. In contrast, our goal is not to replace robust RL objectives, but to provide an engineering-friendly, uncertainty-aware enhancement that can be layered onto standard model-free training procedures.

Reward shaping for accelerating learning. Potential-based reward shaping is a classical technique for accelerating RL while preserving optimal policies under standard conditions [14]. We adopt the potential-based shaping form for ease of integration. Rather than designing an ad-hoc shaping bonus, we use a bounded potential built from aligned safety cues, make it conservative by aggregating finitely many perturbed scores with a λ-fuzzy-measure-induced ordered Choquet integral, and then tune the aggregation parameters offline so that the potential does not become overly optimistic in near-obstacle segments. Compared with fuzzy-guided robust value estimation methods such as Fuz-RL [20], this shaping-based approach emphasizes modularity and interpretability of the resulting Choquet weights, at the expense of not directly optimizing a robust return objective.

3 Preliminaries

MDP and objective. We consider UAV obstacle avoidance as a discounted Markov decision process (MDP) $\langle \mathcal{S}, \mathcal{A}, P, r, \gamma \rangle$, where $\mathcal{S}$ and $\mathcal{A}$ denote the state and action spaces, $P(s_{t+1} \mid s_t, a_t)$ is the transition kernel, $r(s_t, a_t, s_{t+1})$ is the immediate reward, and $\gamma \in (0,1)$ is the discount factor. A policy $\pi(a \mid s)$ specifies a distribution over actions conditioned on state s and induces a trajectory distribution. The discounted return is $J(\pi) = \mathbb{E}_\pi[\sum_{t=0}^{\infty} \gamma^t r_t]$, where r_t denotes the reward collected at step t. Safety is evaluated using collision-related metrics, such as the collision rate, rather than enforced as a hard constraint. This choice supports a plug-in enhancement that remains compatible with standard model-free RL pipelines.

Plug-in shaping setup. We keep the base RL objective $J(\pi)$ unchanged and introduce an additional shaping term computed from a bounded safety potential $\Phi(s) \in [0, 1]$. The shaping is applied in a potential-based form (PBRS) as specified in Sect. 4, so the module can be layered onto standard model-free pipelines.

Fuzzy measure and discrete Choquet aggregation. Let $\Omega = \{1, \dots, K\}$ index K perturbation samples. A (normalized) fuzzy measure is a monotone set function $m : 2^{\Omega} \to [0, 1]$ with $m(\emptyset) = 0$ and $m(\Omega) = 1$. Given K scalar scores $\{\phi_k\}_{k=1}^{K}$ and their sorted order $\phi_{(1)} \leq \cdots \leq \phi_{(K)}$, define nested prefixes $A_i = \{(1), \dots, (i)\}$, with $A_0 = \emptyset$. The discrete ordered Choquet aggregation is

$$\mathcal{C}_m(\phi) = \sum_{i=1}^{K} \phi_{(i)} \big(m(A_i) - m(A_{i-1})\big), \tag{1}$$

which reduces to a weighted sum with data-dependent ordering. Section 4 instantiates $m(\cdot)$ via a λ-fuzzy measure and uses (1) to build a conservative robust potential. *Remark (notation).* We reserve ϕ for the unclipped scores, such as ϕ_{nom} and ϕ_{rob}, and use $\Phi = \text{clip}(\phi, 0, 1)$ when PBRS requires a bounded safety potential.

4 Method

4.1 Overview: Plug-in Potential Shaping and Robust Potential Construction

We use potential-based reward shaping as a drop-in safety bias. It adds a dense, directional signal without changing the base algorithm. Figure 1 provides a visual summary of the module before we detail each component.

$$r'_t = r_t^{\text{base}} + w_{\text{shape}} \big(\gamma_{\text{sh}} \Phi(s_{t+1}) - \Phi(s_t)\big), \tag{2}$$

where r'_t is the shaped reward used by the learner, r_t^{base} is the base environment reward before shaping, $\Phi(\cdot) \in [0, 1]$ is a safety potential (higher is safer), $w_{\text{shape}} > 0$ is a shaping coefficient, and $\gamma_{\text{sh}} \in (0, 1]$ is the shaping discount factor (distinct from the MDP discount γ in Sect. 3). Under standard assumptions, including deterministic state-based potentials, PBRS preserves the set of optimal policies while improving learning signals [14]. In our implementation, Φ is deterministic given the augmented state. The perturbation set is fixed and the EMA variable can be treated as part of the state. We do not claim strict policy-invariance guarantees once clipping and terminal handling are introduced. In our experiments, we fix $w_{\text{shape}} = 1.0$ and set $\gamma_{\text{sh}} = 0.99$ unless otherwise stated. At episode termination (including success), we set $\Phi(s_T) = 0$; consequently, the last-step shaping term equals $-\Phi(s_{T-1})$ and can introduce a negative bias in the shaped return at termination.

The remainder of this section specifies $\Phi(s_t)$: we define aligned features and a nominal score, then robustify it by aggregating finitely many perturbed scores with a λ-fuzzy-measure-induced Choquet integral.

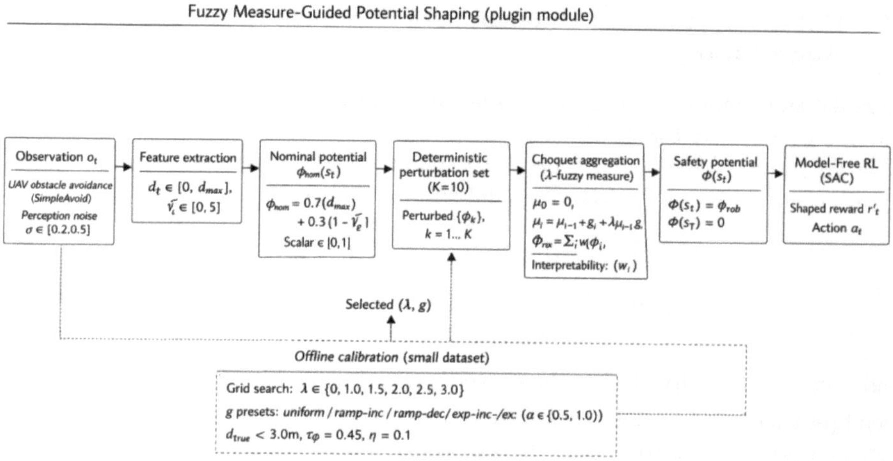

Fig. 1. Framework overview of the fuzzy measure-guided potential shaping module. The robust potential is computed from aligned distance and closing-speed features, evaluated on a fixed perturbation set, and aggregated by a λ-fuzzy-measure-induced Choquet integral. The resulting safety potential is used for potential-based reward shaping.

4.2 Safety Features and Nominal Potential

To keep comparisons fair, all shaping variants use the same two inputs. The shaper uses the perceived minimum distance d_t^{noisy}; d_t^{true} exists only in simulation and is used offline for calibration and evaluation.

$$d_t^{\text{noisy}} \leftarrow \text{clip}(d_t^{\text{noisy}}, 0, d_{\max}), \tag{3}$$

This distance is paired with a smoothed closing speed derived from its change:

$$\begin{aligned} \dot{d}_t &\approx \frac{d_t^{\text{noisy}} - d_{t-1}^{\text{noisy}}}{\Delta t}, \\ v_t &= \max(0, -\dot{d}_t), \\ \bar{v}_t &= \beta \bar{v}_{t-1} + (1-\beta) v_t, \\ \tilde{v}_t &= \text{clip}(\bar{v}_t, 0, v_{\max}). \end{aligned} \tag{4}$$

Unless otherwise stated, $d_{\max} = 15$ and $v_{\max} = 5$. We use EMA coefficient $\beta = 0.3$. In our implementation, Δt defaults to 0.1; for numerical stability, we clamp Δt to a small positive constant when needed.

The nominal potential is a simple "higher-is-safer" score:

$$\phi_{\text{nom}}(s_t) = w_d \frac{d_t^{\text{noisy}}}{d_{\max}} + w_v \left(1 - \frac{\tilde{v}_t}{v_{\max}}\right), \qquad w_d + w_v = 1, \tag{5}$$

with defaults $w_d = 0.7$ and $w_v = 0.3$.

4.3 Robustification via Deterministic Perturbations and Choquet Aggregation

To make the shaper conservative under observation noise, we evaluate the nominal score on a fixed perturbation set of size $K = 10$. The set is sampled once with a fixed seed and then reused, so the shaper stays reproducible. For each $k \in \{1, \ldots, K\}$, we construct the following perturbed features:

$$\begin{aligned}\tilde{d}_t^{(k)} &= \text{clip}\Big(d_t^{\text{noisy}} + \sigma_k z_d^{(k)},\, 0.1,\, d_{\max}\Big),\\ \tilde{v}_t^{(k)} &= \text{clip}\Big(\tilde{v}_t + \rho\,\sigma_k z_v^{(k)},\, 0,\, v_{\max}\Big).\end{aligned} \tag{6}$$

Here $\{\sigma_k\}_{k=1}^{K}$ is a fixed linearly spaced grid on $[0, 0.5]$, and $\rho = 0.5$ scales the speed perturbation relative to distance. The directions $(z_d^{(k)}, z_v^{(k)})$ are drawn once from a clipped standard normal distribution using seed 0 and then fixed, enabling interpretability through stable Choquet weights. The lower bound 0.1 acts as a small numerical floor for the perturbed distance and matches the training-time implementation summarized in Algorithm 1. For the main quantitative noise levels $\sigma \in \{0.2, 0.35, 0.5\}$, the perturbation grid upper bound remains fixed at $\sigma_{\max} = 0.5$. For each sample we compute

$$\phi_k = \phi_{\text{nom}}(\tilde{d}_t^{(k)}, \tilde{v}_t^{(k)}), \tag{7}$$

and sort them in ascending order:

$$\phi_{(1)} \le \phi_{(2)} \le \cdots \le \phi_{(K)}. \tag{8}$$

We then aggregate the sorted scores; this is where we encode non-additive interactions among coupled uncertainties. Let $g = (g_1, \ldots, g_K)$ be a normalized density vector with $g_i > 0$ and $\sum_i g_i = 1$. Given interaction parameter λ and density g, we compute cumulative prefix measures on the sorted index order by a stable recursion:

$$\begin{aligned}\mu_0 &= 0,\\ \mu_i &= \mu_{i-1} + g_i + \lambda\,\mu_{i-1} g_i, \quad i = 1, \ldots, K.\end{aligned} \tag{9}$$

followed by normalization $\mu_i \leftarrow \mu_i / \mu_K$. This recursion induces Choquet weights on the ordered samples:

$$w_i = \mu_i - \mu_{i-1}, \qquad \sum_{i=1}^{K} w_i = 1, \tag{10}$$

and the robust score is

$$\phi_{\text{rob}}(s_t) = \sum_{i=1}^{K} w_i\, \phi_{(i)}. \tag{11}$$

Intuitively, placing more mass on smaller $\phi_{(i)}$ makes the aggregation more conservative under the "higher-is-safer" semantics.

Algorithm 1. Robust potential shaping module (per environment step).

Require: Noisy minimum distance d_t^{noisy}, timestep Δt, optional noise level σ; previous $(d_{t-1}^{\text{noisy}}, \bar{v}_{t-1})$.

Require: Fixed perturbations $\{z_d^{(k)}, z_v^{(k)}\}_{k=1}^{K}$ drawn once from clipped $\mathcal{N}(0,1)$ with seed 0.

Time-step safeguard: clamp Δt to a small positive constant if needed

$v_t \leftarrow \max\left(0, -\frac{d_t^{\text{noisy}} - d_{t-1}^{\text{noisy}}}{\Delta t}\right)$

$\bar{v}_t \leftarrow \beta\,\bar{v}_{t-1} + (1-\beta)\,v_t$

$\tilde{v}_t \leftarrow \text{clip}(\bar{v}_t, 0, v_{\max})$

$\phi_{\text{nom}} \leftarrow w_d(d_t^{\text{noisy}}/d_{\max}) + w_v(1 - \tilde{v}_t/v_{\max})$

Build $\{\sigma_k\}_{k=1}^{K}$ linearly from 0 to $\sigma_{\max} = 0.5$

for $k = 1$ to K **do**

    $d^{(k)} \leftarrow \text{clip}(d_t^{\text{noisy}} + \sigma_k z_d^{(k)}, 0.1, d_{\max})$

    $v^{(k)} \leftarrow \text{clip}(\tilde{v}_t + \rho\,\sigma_k z_v^{(k)}, 0, v_{\max})$

    $\phi_k \leftarrow w_d(d^{(k)}/d_{\max}) + w_v\left(1 - v^{(k)}/v_{\max}\right)$

end for

Sort $\{\phi_k\}$ ascending to obtain $\phi_{(1)} \le \cdots \le \phi_{(K)}$

$\mu_0 \leftarrow 0$

for $i = 1$ to K **do**

    $\mu_i \leftarrow \mu_{i-1} + g_i + \lambda\,\mu_{i-1} g_i$

end for

Normalize $\mu_i \leftarrow \mu_i/\mu_K$; $w_i \leftarrow \mu_i - \mu_{i-1}$

$\phi_{\text{rob}} \leftarrow \sum_{i=1}^{K} w_i\,\phi_{(i)}$; $\Phi(s_t) \leftarrow \text{clip}(\phi_{\text{rob}}, 0, 1)$

return $\Phi(s_t)$ and Choquet weights $\{w_i\}_{i=1}^{K}$

Configuration. The pair (λ, g) is selected via offline calibration (Sect. 5). We use $\lambda = 2.5$ and a decreasing ramp preset for g as defaults. The special case $\lambda = 0$ reduces the recursion to an additive aggregation under the same density preset, which serves as an interpretive reference for the role of non-additivity.

Algorithmic summary. Algorithm 1 provides a step-by-step view of the robust potential computation used at training time, including the deterministic perturbation set and the λ-measure prefix recursion.

5 Offline Calibration

Our robust potential ϕ_{rob} is designed to be conservative in near-obstacle regions, where overly optimistic safety estimates can lead to risk underestimation. Accordingly, the interaction parameter λ and density g are tuned *offline* using a small set of representative trajectories, rather than introducing an additional on-policy loop. This design reflects an engineering trade-off. Calibration is executed once as a lightweight pre-processing step, while the subsequent model-free training procedure and its hyperparameters remain unchanged. Moreover, since the Choquet aggregation yields an explicit Choquet weight spectrum in Sect. 4, calibration provides a practical way to control conservatism under the "higher-is-safer" semantics without hard-coding a worst-case minimum.

Calibration data. The calibration dataset consists of 10–20 successful trajectories with randomized start and goal configurations collected under mild noise, for example $\sigma \in [0.1, 0.2]$ m, to improve coverage.[1] Let $\mathcal{I} = \{t \mid d_t^{\text{true}} < d_{\text{thr}}\}$ denote the near-obstacle index set, where $d_{\text{thr}} = 3.0$ m by default. We focus on $\mathcal{I}$ because it corresponds to safety-critical segments where optimistic potentials are most harmful to learning and evaluation.

Objective: penalize high potential near obstacles. We directly penalize robust potentials that exceed an upper tolerance τ_ϕ inside $\mathcal{I}$, reflecting the requirement that near obstacles the safety potential should not be overly optimistic:

$$\mathcal{L}_{\text{under}} = \frac{1}{|\mathcal{I}|} \sum_{t \in \mathcal{I}} \max\big(0, \phi_{\text{rob}}(s_t) - \tau_\phi\big)^2, \qquad \tau_\phi = 0.45. \tag{12}$$

The hinge-squared form focuses the penalty on *overly high* potentials; it does not enforce a strict monotone mapping between distance and potential, which would be brittle under perception errors. We also consider a mild auxiliary term that encourages ϕ_{rob} to roughly follow the distance trend. This term can stabilize calibration without changing the primary conservatism objective:

$$\mathcal{L}_{\text{mono}} = \text{MSE}\big(\phi_{\text{rob}}(s_t), \text{clip}(d_t^{\text{true}}/d_{\text{max}}, 0, 1)\big)\,. \tag{13}$$

The final calibration loss is

$$\min_{\lambda, g} \ \mathcal{L}_{\text{under}} + \eta\, \mathcal{L}_{\text{mono}}, \qquad \eta = 0.1. \tag{14}$$

This operational definition of conservatism avoids assuming that increasing λ monotonically increases pessimism; in practice, conservatism is determined jointly by (λ, g) through the induced Choquet weight spectrum.

Grid-search procedure. Algorithm 2 summarizes the offline grid search used to select (λ, g) from a small set of successful trajectories.

Search space and defaults. We perform a grid search over $\lambda \in \{0, 1.0, 1.5, 2.0, 2.5, 3.0\}$ (with $\lambda = 0$ giving the additive special case), and several predefined density presets {`uniform`, `ramp-inc`, `ramp-dec`, `exp-inc`, `exp-dec`} with exponential scale $\alpha \in \{0.5, 1.0\}$ for the exp variants. The search space is intentionally small and structured to keep calibration lightweight and reproducible, and to facilitate interpretation of the resulting Choquet weight spectrum. Unless otherwise stated, we use the best-performing calibration setting (default: $\lambda = 2.5$ with a ramp-decreasing density, $\alpha = 1.0$) in all main experiments.

[1] Ground-truth distance d_t^{true} is available in simulation and is used only for offline calibration, not as a training input. In real systems, a proxy signal could be substituted, such as a high-confidence range estimate, a map-based clearance estimate, or a learned collision-probability predictor.

Algorithm 2. Offline grid search for calibrating (λ, g).

Require: Successful trajectories with $\{d_t^{\text{noisy}}, d_t^{\text{true}}\}$; candidate set Λ; density presets $\mathcal{G}$.
Require: $d_{\text{thr}} = 3.0$, $\tau_\phi = 0.45$, $\eta = 0.1$.
for each $\lambda \in \Lambda$ **do**
  for each $g \in \mathcal{G}$ **do**
    Instantiate the shaper with (λ, g); compute $\{\Phi_t\}$ from $\{d_t^{\text{noisy}}\}$ using Algorithm 1
    $\mathcal{I} \leftarrow \{t \mid d_t^{\text{true}} < d_{\text{thr}}\}$
    $L_{\text{under}} \leftarrow \frac{1}{|\mathcal{I}|} \sum_{t \in \mathcal{I}} \max(0, \Phi_t - \tau_\phi)^2$
    $L_{\text{mono}} \leftarrow \text{MSE}(\Phi_t, \text{clip}(d_t^{\text{true}}/d_{\max}, 0, 1))$
    $L \leftarrow L_{\text{under}} + \eta\, L_{\text{mono}}$
  end for
end for
return $(\lambda^*, g^*) \leftarrow \arg\min_{\lambda \in \Lambda,\, g \in \mathcal{G}} L$

6 Experiments and Results

6.1 Task, Environment, and Compared Variants

The experiments use an AirSim-based UAV obstacle-avoidance task, `SimpleAvoid`, instantiated in the AirSim simulator. We cite Shah et al. [18] for AirSim itself and He et al. [10] for a published deep-RL UAV path-planning study that uses the same `SimpleAvoid` task configuration. In this task, the agent must reach a target while avoiding collisions. The comparison includes four shaping variants under an otherwise identical SAC-based model-free training pipeline: **None**, with no shaping; **Nominal**, potential-based reward shaping (PBRS) with ϕ_{nom} [14]; **Robust**, PBRS with ϕ_{rob}; and **Mamdani-FIS**, a Mamdani-style fuzzy-inference baseline using the same two inputs [13]. All variants share the same base reward, algorithm configuration, and noise protocol, and differ only in how $\Phi(\cdot)$ is constructed. Unless otherwise stated, all reported training curves, tables, and trajectory examples in this paper are obtained with SAC as the sole base learner.

6.2 Task Specification and Environment Interface

Arena Layout (Map). Fig. 2 shows the obstacle layout used in our `SimpleAvoid` arena. Since both training and evaluation use the same arena, we include both a top-down view and an oblique view to make the task geometry explicit and visually convey obstacle height and spacing.

Episode Initialization and Goal Sampling. At the beginning of each episode, the UAV is reset to a fixed start position. A goal is sampled at a fixed radial distance (goal radius 50 m) with a uniformly random angular direction in the horizontal plane. An episode is declared *successful* when the UAV enters an acceptance ball of radius $r_{\text{acc}} = 2$ m centered at the goal.

Fig. 2. Multi-view visualization of the AirSim-based `SimpleAvoid` arena used for training and evaluation. Left: top-down obstacle layout. Right: an oblique view highlighting the 3D appearance of cuboid (walls/blocks) and cylindrical obstacles. The UAV starts from a fixed initial pose and must reach a randomly sampled goal while avoiding collisions.

Observation (Depth Image With a Lightweight State Overlay). Our main experiments use a depth-image observation modality with a minimal goal-directed overlay. Let $D_t \in \mathbb{R}^{H\times W}$ denote the depth map in meters (after resizing), with H=60, W=90, and a maximum sensing range $d_{\max}$=15 m. We clip depth values to $[0, d_{\max}]$ and convert them to an 8-bit intensity:

$$I_t = 255 - 255 \cdot \frac{\text{clip}(D_t, 0, d_{\max})}{d_{\max}}, \tag{15}$$

so that nearer obstacles correspond to higher intensity. To provide minimal goal context, the observation includes a second channel that encodes two low-dimensional state features, normalized to $[0, 255]$, in the first row of an otherwise zero matrix: (i) planar distance-to-goal and (ii) relative yaw, which represents the heading error to the goal direction. The final observation is a two-channel image $o_t \in \{0, \dots, 255\}^{H\times W\times 2}$, which motivates a CNN-based policy.

Action Space and Control Rate. We consider horizontal-plane navigation in the main setting (2D navigation). The continuous action is

$$a_t = [v_{xy}, \dot{\psi}], \tag{16}$$

where v_{xy} is the commanded planar speed and $\dot{\psi}$ is the yaw-rate command. In our configuration, $v_{xy} \in [0.5, 5.0]$ m/s and $\dot{\psi} \in [-30°/\text{s}, 30°/\text{s}]$. Actions are applied at a fixed control period $\Delta t = 0.1$ s.

Termination (Safety and Workspace Constraints). An episode terminates when any of the following conditions is met:

- **Goal reached:** $\|p_t - p_{\text{goal}}\|_2 < r_{\text{acc}}$.
- **Collision / unsafe proximity:** AirSim reports a collision event, or the true minimum obstacle distance falls below the crash threshold $d_{\text{crash}} = 2$ m.

- **Out of workspace:** the UAV leaves a pre-defined workspace bounding box.
- **Time limit:** a truncation occurs when the episode length reaches $T_{\max} = 400$ steps.

Safety-related termination checks use the clean, pre-noise geometry for consistency, while the policy is trained on noisy observations.

Base Reward (Task Reward). The base reward encourages goal progress while penalizing unsafe proximity, excessive turning, and yaw misalignment. Terminal rewards are assigned as: goal +10, collision −20, outside-workspace −10, and timeout 0 (default configuration). All shaping variants share the *same* base reward; they only differ in the augmentation term.

Augmentation and Coupled Perceptual Uncertainty. Our shaping module uses *perceived* near-obstacle signals computed from the noisy observation, such as the minimum depth after noise injection, to form $\Phi(\cdot)$. Therefore, uncertainty couples directly into the potential computation, while success/collision outcomes remain grounded in the true environment dynamics and safety events.

6.3 Noise Protocol

For the main quantitative comparison, we consider three non-zero noise levels, $\sigma \in \{0.2, 0.35, 0.5\}$. The clean setting $\sigma = 0.0$ is retained only for the qualitative trajectory panel in Sect. 6.6. When $\sigma = 0.0$, noise injection is disabled; otherwise the observation noise standard deviation is fixed to the selected σ for the entire run. Unless otherwise stated, the noise is applied to the raw depth map before resizing and clipping, and a fresh noise sample is drawn at each environment step. This protocol keeps the underlying task dynamics unchanged and isolates the effect of perceptual corruption on both the policy input and the shaping potential, as described in Sect. 4.

6.4 Training Budget, Seeds, and Metrics

All configurations use the same training budget and evaluation protocol. Each configuration is trained for $T = 100\text{k}$ environment steps, and we report results for a single random seed (seed 0). These results are intended as pilot-scale findings. We report two primary episode-level metrics: **success rate** (fraction of evaluation episodes that reach the goal without collision) and **collision rate** (fraction of evaluation episodes that terminate by collision or unsafe proximity). For each trained policy, the metrics are computed from episode termination types and summarized at the end of training. When multiple seeds are available, we additionally report standard error over seeds. Learning curves are included as diagnostic plots when available (Fig. 3).

Table 1. Final success and collision rates on `SimpleAvoid` after $T = 100\text{k}$ environment steps under the three non-zero noise levels $\sigma \in \{0.2, 0.35, 0.5\}$.

Variant	$\sigma = 0.2$		$\sigma = 0.35$		$\sigma = 0.5$	
	Success ↑	Collision ↓	Success ↑	Collision ↓	Success ↑	Collision ↓
None	0.406	0.251	0.236	0.498	0.033	0.704
Nominal	0.471	0.260	0.294	0.428	0.116	0.595
Robust (ours)	0.616	0.245	0.397	0.394	0.208	0.534
Mamdani FIS	0.571	0.300	0.342	0.432	0.113	0.564

Note: Values are end-of-training evaluation metrics for seed 0 after $T = 100\text{k}$ environment steps; success and collision rates are computed from episode termination types

6.5 Quantitative Robustness Under Non-Zero Noise

This section reports the main quantitative robustness results under the three non-zero noise settings, $\sigma \in \{0.2, 0.35, 0.5\}$. The clean setting ($\sigma = 0.0$) is not used as a main quantitative reference point here; instead, it is reserved for the qualitative trajectory examples in Sect. 6.6. Table 1 reports the exact end-point success and collision rates for these non-zero noise settings, and Fig. 3 provides the corresponding training-time diagnostics.

At $\sigma = 0.2$, all shaping variants outperform **None**, and the proposed **Robust** shaping gives the strongest pilot-run result, with 0.616 success and 0.245 collision. Relative to **Nominal** (0.471/0.260) and **Mamdani FIS** (0.571/0.300), the robust variant improves task completion while keeping collisions slightly lower.

At the intermediate noise level $\sigma = 0.35$, the same ordering is largely preserved. **Robust** reaches 0.397 success with 0.394 collision, compared with 0.342/0.432 for **Mamdani FIS**, 0.294/0.428 for **Nominal**, and 0.236/0.498 for **None**.

At $\sigma = 0.5$, all methods degrade further, but **Robust** again remains the strongest among the compared methods in this pilot run, with 0.208 success and 0.534 collision. Across the three non-zero noise levels, the robust variant is associated with the highest success rate and the lowest collision rate among the compared methods, although the evidence remains preliminary under the current single-seed protocol.

As σ increases from 0.2 to 0.5, all compared variants show lower end-point success and higher collision rates. Under the current pilot-scale, single-seed protocol, we treat this as a descriptive trend rather than as a basis for stronger statistical claims.

6.6 Representative Trajectories Under Clean and Moderate Noise

To complement the non-zero-noise aggregate metrics, Fig. 4 shows representative trajectories for the same start–goal layout under $\sigma = 0.0$ and $\sigma = 0.2$, comparing the unshaped baseline (**None**) with the proposed **Robust** shaping.

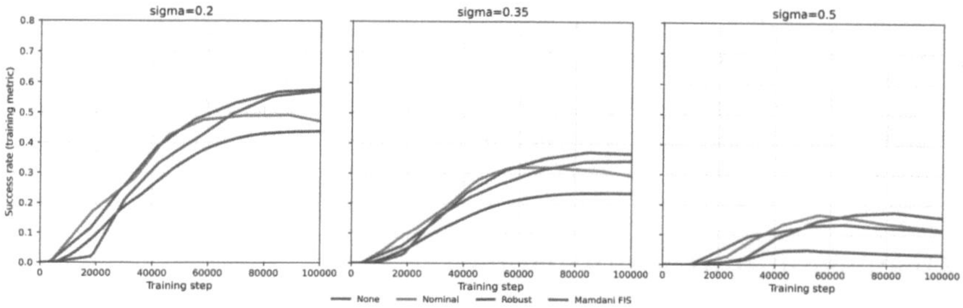

Fig. 3. Training success-rate curves (seed=0) under the three non-zero noise settings $\sigma \in \{0.2, 0.35, 0.5\}$ for the four compared variants.

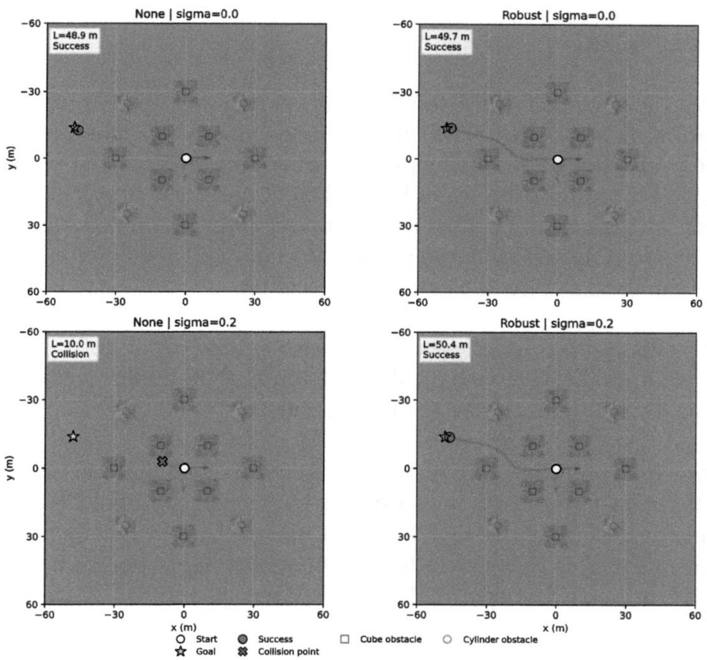

Fig. 4. Representative trajectories under clean and moderate observation noise. Top row: under $\sigma = 0.0$, both **None** and **Robust** reach the goal, while the robust variant follows a slightly more conservative route. Bottom row: under $\sigma = 0.2$, **None** terminates in collision, whereas **Robust** still reaches the goal on the matched start–goal layout.

These examples are intended as qualitative case studies rather than statistical evidence, and the clean-setting row is included only as a visual baseline rather than a main quantitative reference point.

The clean-setting row mainly illustrates route style under accurate observations: both methods reach the goal on this matched layout, while the robust variant follows a slightly more conservative path. The moderate-noise row is more diagnostic for robustness under uncertainty: once perceptual corruption is introduced, the unshaped policy can commit to an unsafe path early, whereas the robust potential still guides the agent through the cluttered region and preserves task completion.

7 Discussion and Limitations

We interpret the empirical findings in light of the design goal, which is a plug-in safety enhancement for model-free training under coupled perceptual uncertainty, and summarize key limitations. Given the pilot-scale, single-seed evaluation reported here, the discussion focuses on qualitative trends rather than statistical significance.

Main Take-Away Under Uncertainty. Under non-zero perceptual noise, the robust shaping achieves the best success rates among the compared methods at all three evaluated noise settings in this pilot run, as reported in Table 1 and supported by the training curves in Fig. 3. Under our noise protocol, this pattern is consistent with improved resilience to coupled observation noise relative to the unshaped baseline, nominal shaping, and Mamdani-FIS shaping. The evidence remains preliminary due to the single-seed setting.

Role of the clean-setting examples. The clean setting ($\sigma = 0.0$) is retained in Fig. 4 only as a qualitative visualization of route style under accurate observations, not as a main quantitative ranking result.

Mechanistic Interpretation. Relative to nominal shaping and Mamdani-FIS shaping, the robust potential is computed by scoring an explicit perturbation set and then aggregating the scores with a non-additive Choquet integral. One plausible mechanism is that the induced Choquet weight spectrum assigns more mass to adverse perturbations, making the potential less prone to overestimate safety in near-obstacle regimes. The offline calibration objective complements this effect by directly penalizing overly high potentials when the ground-truth distance is small, as described in Sect. 5. Finally, since the shaping is added in a potential-based form, as described in Sect. 3, the method remains compatible with standard model-free training pipelines.

Terminal Shaping Bias. We set the terminal potential to $\Phi(s_T) = 0$ for all termination types, including success. As a consequence, the last-step shaping term in (2) becomes $-\Phi(s_{T-1})$ when an episode ends, which introduces a negative bias in the cumulative shaped return for successful episodes. This choice keeps Φ bounded and simplifies implementation, but it may slightly discourage terminating transitions when the agent is already in a safe state. A refined design could condition the terminal potential on termination type or apply an explicit terminal correction; we leave such variants for future work.

Other Limitations and Practical Considerations. This study is pilot-scale: it uses one environment, `SimpleAvoid`, three quantitative non-zero noise settings, and one random seed, seed 0, with a training budget of $T = 100$k steps; the clean setting is used only for the qualitative trajectory panel in Fig. 4. We therefore avoid statistical claims. In addition, the perturbation set is discrete and protocol-specific rather than a continuous uncertainty set, and the offline calibration uses simulator ground truth d_t^{true}, which may not be available in real deployments. Finally, the potential is built from two aligned features (minimum distance and closing speed), so richer scenes may require additional cues. Taken together, these choices make the current results best viewed as a transparent proof of concept for uncertainty-aware shaping rather than a broadly validated robust-control solution.

8 Conclusion

We proposed a plug-in fuzzy measure-guided potential shaping module for safe UAV obstacle avoidance under coupled perceptual uncertainty. The module builds a bounded safety potential from aligned distance and closing-speed cues, evaluates it on a fixed perturbation set, and aggregates the resulting scores via an ordered Choquet integral induced by a λ-fuzzy measure. To obtain an operational safety bias, we calibrate the interaction parameter offline using a near-obstacle objective.

In the main quantitative comparison on `SimpleAvoid` with a limited training budget, the robust shaping was associated with higher task completion across the three non-zero noise settings while keeping collision rates lower than the other compared methods in this pilot run (Table 1).

Several limitations remain, most notably the pilot-scale evaluation and the reliance on simulator ground truth for offline calibration. Future work will expand the empirical study to additional environments, noise sources, and random seeds, and will explore calibration signals that do not require access to ground-truth distances, as well as improved terminal handling.

Disclosure of Interests. The authors have no competing interests to declare that are relevant to the content of this article.

References

1. Achiam, J., Held, D., Tamar, A., Abbeel, P.: Constrained policy optimization. In: Proceedings of the 34th International Conference on Machine Learning. Proceedings of Machine Learning Research, vol. 70, pp. 22–31 (2017)
2. Berducci, L., Aguilar, E.A., Ničković, D., Grosu, R.: HPRS: hierarchical potential-based reward shaping from task specifications. Front. Robot. AI **11**, 1444188 (2025). https://doi.org/10.3389/frobt.2024.1444188
3. Brunke, L., et al.: Safe learning in robotics: from learning-based control to safe reinforcement learning. Annual Rev. Control Robot. Autonom. Syst. **5**, 411–444 (2022). https://doi.org/10.1146/annurev-control-042920-020211

4. Choquet, G.: Theory of capacities. Annales de l'Institut Fourier **5**, 131–295 (1953)
5. Di, J., Li, K., Kang, Y., Dong, Q., Chen, S., Lv, W.: Semi-adaptive spectrally normalized identifier based model uncertainty online compensation for racing drones. SCIENCE CHINA Technol. Sci. **68**(7), 1720404 (2025). https://doi.org/10.1007/s11431-024-2914-2
6. García, J., Fernández, F.: A comprehensive survey on safe reinforcement learning. J. Mach. Learn. Res. **16**(42), 1437–1480 (2015)
7. Grabisch, M.: The application of fuzzy integrals in multicriteria decision making. Eur. J. Oper. Res. **89**(3), 445–456 (1996). https://doi.org/10.1016/0377-2217(95)00175-X
8. Grabisch, M., Marichal, J.L., Mesiar, R., Pap, E.: Aggregation Functions. Cambridge University Press (2009)
9. Haarnoja, T., Zhou, A., Abbeel, P., Levine, S.: Soft actor-critic: off-policy maximum entropy deep reinforcement learning with a stochastic actor. In: ICML, pp. 1861–1870 (2018)
10. He, L., Aouf, N., Song, B.: Explainable deep reinforcement learning for UAV autonomous path planning. Aerosp. Sci. Technol. **118**, 107052 (2021). https://doi.org/10.1016/j.ast.2021.107052
11. Iyengar, G.N.: Robust dynamic programming. Math. Oper. Res. **30**(2), 257–280 (2005). https://doi.org/10.1287/moor.1040.0129
12. Kang, Y., Di, J., Li, M., Zhao, Y., Wang, Y.: Autonomous multi-drone racing method based on deep reinforcement learning. SCIENCE CHINA Inf. Sci. **67**(8), 180203 (2024). https://doi.org/10.1007/s11432-023-4029-9
13. Mamdani, E.H., Assilian, S.H.: An experiment in linguistic synthesis with a fuzzy logic controller. Int. J. Man Mach. Stud. **7**(1), 1–13 (1975). https://doi.org/10.1016/S0020-7373(75)80002-2
14. Ng, A.Y., Harada, D., Russell, S.: Policy invariance under reward transformations: theory and application to reward shaping. In: Proceedings of the Sixteenth International Conference on Machine Learning, pp. 278–287. Morgan Kaufmann (1999)
15. Nilim, A., El Ghaoui, L.: Robust control of Markov decision processes with uncertain transition matrices. Oper. Res. **53**(5), 780–798 (2005). https://doi.org/10.1287/opre.1050.0216
16. Rojas-Ramírez, E., Benítez-Read, J.S.: Multilevel fuzzy navigation control scheme applied to a monitoring mobile robot. In: ISKE, pp. 617–621 (2010). https://doi.org/10.1109/ISKE.2010.5680764
17. Schulman, J., Wolski, F., Dhariwal, P., Radford, A., Klimov, O.: Proximal policy optimization algorithms. arXiv preprint arXiv:1707.06347 (2017)
18. Shah, S., Dey, D., Lovett, C., Kapoor, A.: AirSim: high-fidelity visual and physical simulation for autonomous vehicles. In: Field and Service Robotics, Springer Proceedings in Advanced Robotics, vol. 5, pp. 621–635. Springer (2018). https://doi.org/10.1007/978-3-319-67361-5_40
19. van der Wal, A.J.: Fuzzy logic assisted helicopter UAV flight control. In: FLINS 2012, pp. 1257–1262 (2012). https://doi.org/10.1142/9789814417747_0202
20. Wan, X., Yang, C., Yang, C., Song, J., Sun, M.: FUZ-RL: a fuzzy-guided robust framework for safe reinforcement learning under uncertainty. In: Advances in Neural Information Processing Systems (NeurIPS) (2025)

Wind Power Data Missing Value Completion Method Based on DBO-LRTC

Tiefeng Zhang[1(✉)], Weiye Wang[1], Jian Qian[2,3], Hao Sun[2,3], Botao Sun[1], Xiaozhi Li[1], Wei Jia[2,3], and Shengyu Xu[1]

[1] Department of Electronic and Communication Engineering, North China Electric Power University, Baoding 071003, China
zhangtiefeng@ncepu.edu.cn
[2] Longyuan Damao Wind Power Generation Co., Ltd., Baotou 014500, China
[3] Inner Mongolia Longyuan New Energy Development Co., Ltd., Hohhot 010000, China

Abstract. Wind power data often suffer from missing values due to communication failures and equipment maintenance, which severely impacts the reliability of wind farm operations. This paper proposes a novel multivariate collaborative tensor completion framework for high-accuracy wind power data recovery. To overcome the limitation of traditional low-rank tensor completion (LRTC) where weight parameters are fixed by empirical experience, we propose the DBO-LRTC method. This method integrates the Dung Beetle Optimization (DBO) algorithm to adaptively determine the optimal weight combinations for different tensor modes, enabling the model to capture the intrinsic low-rank features of wind power data more robustly. Furthermore, an Adaptive Alternating Direction Method of Multipliers (Ada-ADMM) is developed to solve the model, which significantly accelerates convergence speed by dynamically adjusting penalty parameters. Extensive experimental results under random and continuous missing scenarios demonstrate that DBO-LRTC consistently outperforms benchmark methods in terms of RMSE, MAPE, and RSE metrics. The proposed approach offers superior accuracy and computational efficiency, providing a robust tool for intelligent wind farm data preprocessing and maintenance.

Keywords: DBO algorithm · data missing · LRTC · Ada-ADMM method

1 Introduction

Under the background of the "dual-carbon" goals and the transformation of the energy structure, wind energy, as a crucial component of clean and renewable energy, has been playing an increasingly significant role in the global energy system [1]. The rapid growth of installed wind power capacity has led to a continuous expansion in the scale of operational data generated by wind farms [2].

J. Lu et al. (Eds.): FLINS-ISKE 2026, LNCS 16752, pp. 195–214, 2027.
https://doi.org/10.1007/978-981-92-2369-5_14

High-quality data support has consequently become a fundamental prerequisite for accurate wind power forecasting, optimal power grid dispatching, and intelligent operation and maintenance of wind farms [3]. Among various types of operational data, wind power generation data serve as core indicators reflecting the operational status of wind farms. Their completeness and accuracy directly affect the reliability of subsequent analysis and decision-making processes. However, in practical wind farm operations, wind power data frequently suffer from random or consecutive missing values due to communication link failures, equipment maintenance, sensor malfunctions, and data acquisition system faults. Data missingness not only degrades overall data quality but also adversely impacts the training performance of power forecasting models, and may even result in biased dispatching decisions. Therefore, the development of high-precision and robust wind power data imputation methods is of significant theoretical importance and practical engineering value.

In the field of missing data recovery, Reference [4] proposed a wind power data imputation method based on similarity measurement. This approach exploits data characteristics and employs similarity metrics such as Euclidean distance, dynamic time warping (DTW), and longest common subsequence (LCSS) to identify the most similar equal-length sequences to complete missing values. Reference [5] introduced an interpolation method that combines horizontal and vertical interpolation based on the K-nearest neighbors (KNN) algorithm, enabling accurate reconstruction of missing entries. Similarly, reference [6] addressed the data missing problem using a hybrid strategy integrating the KNN algorithm with linear interpolation. Reference [7] developed a dual-channel generative adversarial network (GAN) to complete missing load data, while reference [8] proposed an improved GAIN-based data imputation algorithm. By incorporating the Wasserstein discriminator, the authors constructed a GAIN-W model for time-series missing data imputation. Although these methods are computationally efficient and relatively straightforward to implement, they often fail to fully exploit the multi-dimensional correlation structures inherent in wind power data. Consequently, their performance becomes limited under conditions of high missing rates or long consecutive missing intervals. With the rapid development of machine learning and low-rank theory, matrix completion and tensor completion methods have gradually emerged as research hotspots, providing new perspectives and effective tools for the recovery of multi-dimensional wind power data.

In recent years, numerous data imputation methods have been proposed to address the issue of missing data. Matrix Completion (MC) methods based on the low-rank assumption recover missing entries by minimizing the nuclear norm and have achieved remarkable success in image inpainting and recommender systems. Some studies model wind power data in the form of two-dimensional matrices and exploit their low-rank properties for data reconstruction. Reference [9] proposed a phase sequence identification algorithm for low-voltage distribution networks based on quadratic matrix completion, targeting scenarios with insufficient data integrity. Reference [10] introduced a method that integrates matrix completion with channel optimization techniques, employing the alternating direction method of multipliers (ADMM) to address missing data in the original dataset. Reference [11] presented a matrix completion approach combining alternating minimization with moving average to analyze and reconstruct distribution network power flow data under limited measurement conditions. However, wind power data inherently exhibit multi-dimensional structural characteristics, including temporal, spatial, and variable dimensions. Directly reducing such data to a matrix form may destroy the intrinsic multi-dimensional structural information, thereby limiting the performance of data completion methods.

To better explore the multi-dimensional correlations within data, Tensor Completion (TC) methods have gradually gained attention. Tensor models are capable of capturing the intrinsic low-rank characteristics of data while preserving its multi-dimensional structure, using techniques such as CP decomposition, Tucker decomposition, and tensor singular value decomposition (t-SVD) for imputation. In particular, methods based on Low-Rank Tensor Completion (LRTC), which minimize the multiple nuclear norms, have effectively improved imputation accuracy by applying nuclear norm constraints to the unfoldings of different modes of the tensor. Reference [12] proposed a method for completing missing data in low-voltage load profiles based on multi-characteristic sample factors. Reference [13] introduced an effective fourth-order dimension-preserved tensor completion (FDPTC) model with temporal constraints. Reference [14] developed a tensor completion method for measurement data imputation based on CP decomposition and regularized low-rank tensor completion. This algorithm overcomes the limitations of traditional single-dimensional data processing by constructing a three-way observation tensor for Non-Intrusive Load Monitoring (NILM) multi-dimensional measurement data, thereby utilizing the internal temporal dependencies and electrical correlations between parameter dimensions for regularized low-rank tensor completion. Reference [15] proposed a method for recovering missing electrical measurement data based on low-rank tensor completion decomposition (LRTC-TCP). Reference [16] incorporated tensor nuclear norm into an audio signal completion algorithm, utilizing the tensor nuclear norm-based imputation method to restore multi-channel audio signals with missing data. These tensor-based approaches offer significant advantages by leveraging the inherent multi-dimensional structures in the data, leading to enhanced imputation performance, particularly when dealing with complex, high-dimensional datasets such as wind power data.

In recent years, the multi-nuclear norm minimization tensor completion model has become a mainstream approach. This method improves computational efficiency by leveraging the weighted sum of the core nuclear norms of different modes, while preserving the multi-dimensional structure of the data [17]. Although existing mainstream methods improve computational efficiency by weighting the core norms of each mode, the weight parameters are typically set based on empirical experience, lacking theoretical guarantees, which can lead to a decline in model performance. Recently, the Dung Beetle Optimization (DBO) algorithm has shown remarkable performance due to its unique optimization mechanism [18], particularly in balancing the exploration and exploitation of the solution space. It has demonstrated advantages in fields such as image processing, path planning, and microgrid scheduling [19]. Compared to particle swarm optimization and genetic algorithms, DBO can quickly escape local optima, requires fewer parameters to adjust, converges faster, and possesses strong global optimization capabilities, while maintaining stable performance. To address the issues mentioned above, this paper combines the DBO algorithm with the multi-nuclear norm minimization tensor completion model and proposes the DBO-LRTC method. The core theoretical motivation for selecting DBO lies in its unique multi-strategy search mechanism, which enables the model to efficiently escape local optima when navigating the complex, non-convex solution space of weight parameters. Unlike traditional LRTC methods that rely on fixed empirical weights—which often neglect the intrinsic structural disparities across temporal, spatial, and variable modes—our method adaptively determines the optimal weight combination. This'self-evolutionary' weighting process ensures that the model precisely captures the low-rank features most consistent with the underlying data structure. The core advantage and distinction of the DBO-LRTC method lie in its ability to cleverly solve the problem of adaptively determining the optimal combination of nuclear norm weights for different modes in low-rank tensor completion. Existing studies typically set fixed weights using heuristic or empirical methods (such as equal distribution), overlooking the intrinsic differences in the low-rank features of each dimension across different datasets. Additionally, the Alternating Direction Method of Multipliers (ADMM) algorithm is used to solve the model. To overcome the slow convergence of ADMM, an Adaptive Alternating Direction Method of Multipliers (Ada-ADMM) is proposed. Furthermore, to resolve the convergence bottlenecks of the standard Alternating Direction Method of Multipliers (ADMM) in large-scale tensor problems, we introduce an Ada-ADMM solver. By dynamically adjusting the penalty parameter based on primal and dual residuals, the proposed strategy significantly enhances computational efficiency and robustness. Experimental results show that the proposed method effectively completes data under various missing data scenarios and missing rates, and that the Ada-ADMM method significantly improves solving speed, demonstrating the effectiveness of this approach.

In summary, missing wind power data caused by communication failures, equipment maintenance, and sensor abnormalities seriously affects the quality of wind farm operation data and the reliability of subsequent forecasting and dispatching tasks. To address this issue, this paper proposes a hybrid missing-data recovery method that combines Dung Beetle Optimization (DBO), Low-Rank Tensor Completion (LRTC), and an adaptive alternating direction multiplier method (Ada-ADMM). The proposed method effectively improves the accuracy and efficiency of wind power data recovery, and has important practical value for wind farm data preprocessing and intelligent operation. The main contributions of this paper are summarized as follows.

(a) A multi-variable collaborative tensor modeling framework is constructed for wind power missing data recovery. By organizing wind power data into a tensor form and exploiting its multi-dimensional structural characteristics, the proposed method preserves temporal and spatial correlations more effectively than traditional matrix-based or interpolation-based approaches, providing a stronger foundation for missing data completion.

(b) A novel DBO-LRTC framework is proposed. By formulating weight determination as a global optimization problem, DBO is employed to adaptively optimize parameters for different tensor modes. This overcomes the bias of empirical settings and provides a more rigorous methodology for multi-dimensional data recovery.

(c) An Ada-ADMM solution strategy is introduced to improve the scalability and efficiency of the tensor completion process. The dynamic parameter adjustment mechanism accelerates convergence and ensures stable performance under varying missing rates.

The rest of this paper is organized as follows: Sect. 2 introduces the tensor definition and the tensor completion problem. Section 3 presents the proposed multi-kernel norm minimization tensor completion model based on the dung beetle optimization algorithm, including the Ada-ADMM solution method and the overall DBO-LRTC workflow. Section 4 provides case studies under random and continuous missing scenarios, together with comparative experiments and computational performance analysis. Finally, Sect. 5 concludes the paper by summarizing the main findings and contributions of this research.

2 Tensor Definition and Its Completion Problem Description

2.1 Tensor Definition

A tensor is a multi-dimensional array, which is a higher-order generalization of data based on vector representation and matrix representation. The order of tensors is what is commonly known as tensor dimensions and tensor modalities. A schematic diagram of tensors of each order is shown in Fig. 1.

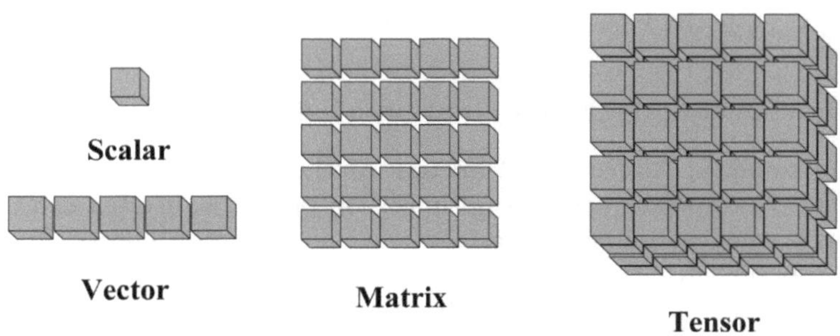

Fig. 1. Schematic diagram of tensors in each order.

2.2 Tensor Completion Problem Description

Tensors are multi-dimensional extensions of matrices. Tensor completion problems can be regarded as higher-order matrix completion problems. Tensor completion problems can be expressed as:

$$\begin{cases} \min y_1 = rank(\mathcal{X}) \\ \text{s.t.} : \mathcal{P}_\Omega(\mathcal{M}) = \mathcal{P}_\Omega(\mathcal{X}) \end{cases} \tag{1}$$

where $\mathcal{X}$ is the completed tensor, $\mathcal{M}$ is the tensor to be completed, y_1 is the rank of the tensor, Ω is the index set of observation data in $\mathcal{M}$, $\mathcal{P}_\Omega$ is the projection operator.

3 Multi-kernel Norm Minimization Tensor Data Completion Method Based on Dung Beetle Optimization Algorithm

3.1 Tensor Completion Model Based on Multi-kernel Norm Minimization

Since the above minimization of low-rank tensor completion is an NP-hard problem and difficult to calculate. In order to deal with this problem, the multi-core norm is added to the model and the rank function is replaced by the multi-core norm, the objective function can be expressed as:

$$\begin{aligned} &\min_x \sum_{n=1}^{N} \lambda_n \parallel \mathbf{X}_{(n)} \|_* \\ &s.t. \mathcal{P}_\Omega(\mathcal{M}) = \mathcal{P}_\Omega(\mathcal{X}) \end{aligned} \tag{2}$$

where $\lambda_n \geq 0,\ n = 1, 2, \cdots, N$ is the weight parameter, $\mathbf{X}_{(n)},\ n = 1, 2, \cdots, N$ is the expansion matrix for each mode of tensor $\mathcal{X}$, and its nuclear norm is defined as:

$$\mathcal{X}_{(n)*} = \sum_{k=1}^{r} \sigma_k(\mathbf{X}_{(n)}) \tag{3}$$

where $\sigma_k(\mathbf{X}_{(n)})$ is the k singular value of the expanded matrix.

3.2 Ada-ADMM Algorithm Solution Model

While the standard Alternating Direction Method of Multipliers (ADMM) is effective for decomposing complex optimization problems, it often exhibits slow convergence when handling high-dimensional wind power tensors. To address this, we propose an Adaptive ADMM (Ada-ADMM) solver. By introducing auxiliary variables $\{\mathcal{T}_n\}$, the augmented Lagrangian function is formulated to decouple the nuclear norm terms. The updates for $\mathcal{X}$ and the Lagrange multipliers $\mathcal{Y}_n$ follow standard iterative procedures. The core refinement of our method lies in the Adaptive Step Size Mechanism for the penalty parameter ρ. Instead of using a fixed value, $\rho^{(k+1)}$ is dynamically adjusted based on the primal residual e_1 and dual residual e_2:

$$\begin{cases} e_1 = \| \mathcal{X}^{(k+1)} - \mathcal{M}^{(k+1)} \|_F \\ e_2 = \rho^{(k)} \| \mathcal{X}^{(k+1)} - \mathcal{X}^{(k)} \|_F \end{cases} \tag{4}$$

where e_1 is the original residual and e_2 is the dual residual. The update formula for obtaining step size ρ can be expressed as:

$$\rho^{(k+1)} = \begin{cases} \tau_{\text{inc}}\rho^{(k)} & \text{if} \quad e_1 > \mu e_2 \\ \tau_{\text{dec}}\rho^{(k)} & \text{if} \quad e_2 > \mu e_1 \\ \rho^{(k)} & \text{otherwise} \end{cases} \tag{5}$$

where τ_{inc} is the multiplication increasing coefficient, τ_{dec} is the multiplication decreasing coefficient, and μ is the balance coefficient of dividing the original residual by the dual residual. This mechanism ensures a robust balance between primal and dual feasibility during the iteration process. By increasing ρ when the primal residual dominates and decreasing it when the dual residual is larger, the Ada-ADMM significantly accelerates convergence and reduces the overall computation time compared to conventional solvers.

3.3 DBO Algorithm

Dung beetle optimizer (DBO) is a swarm intelligence optimization algorithm proposed by Professor Shen Bo's team in 2022. DBO divides the dung beetle population into four populations, each of which plays a different role. The theoretical rationale for employing DBO over traditional meta-heuristics like PSO or GA lies in its structured multi-strategy search mechanism. The optimization of weight parameters λ_n is a high-dimensional, non-convex problem with a simplex constraint. Traditional algorithms often suffer from"premature convergence" due to a lack of population diversity.In contrast, DBO ensures a robust balance between global exploration and local refinement through its four-tier population structure: the rolling ball beetles utilize simulated light intensity to explore the broader search space; the breeding and foraging beetles establish dynamic, shrinking boundaries to concentrate search efforts in high-potential regions; and the stealing beetles introduce stochastic perturbations to effectively escape local optima. This synergistic approach allows for a more rigorous and efficient search for the optimal λ_n, ensuring that the resulting tensor structure accurately reflects the underlying multi-dimensional correlations of wind power data. These four types of roles correspond to different search strategies:

(1) Population 1 (Rolling Ball Dung Beetle)

When there are no obstacles, the rolling forward position update formula of the dung beetle can be expressed as:

$$\begin{gathered} x_i(t+1) = x_i(t) + \alpha \times k \times x_i(t-1) + b \times \Delta x, \\ \Delta x = \mid x_i(t) - X^w \mid \end{gathered} \tag{6}$$

where t is the number of iterations, $x_i(t)$ is the position information of the i dung beetle in the t iteration, $k \in (0, 0.2]$ is a constant, $b \in (0, 1)$ is a fixed value, α is 1 or -1, X^w is the global worst position, and Δx is the change in simulated light intensity.

When there are obstacles, you need to dance to reposition yourself with the goal of obtaining a new route. The position update formula can be expressed as:

$$x_i(t+1) = x_i(t) + \tan(\theta) \mid x_i(t) - x_i(t-1) \mid \tag{7}$$

where $\mid x_i\,(t) - x_i\,(t+1) \mid$ is the difference between the position of the i dung beetle at the t iteration and its position at the $t-1$ iteration, $\theta \in [0, \pi]$ is the deflection angle.

(2) Population 2 (breeding dung beetles)

Breeding dung beetles roll dung balls into suitable areas to lay their eggs. The area boundaries are defined as:

$$\begin{aligned} Lb^* &= \max(X^* \times (t/T_{\max}), Lb) \\ Ub^* &= \min(X^* \times (2 - t/T_{\max}), Ub) \end{aligned} \tag{8}$$

where X^* is the current local optimal position, Lb^* and Ub^* are the lower bounds and upper bounds of the spawning area, $T_{\max}$ the maximum number of iterations, Lb and Ub are the lower bounds and upper bounds of the optimization problem.

The female dung beetle determines the egg laying area and only 1 egg is laid in each iteration. The egg ball position update formula can be expressed as:

$$\begin{aligned} B_i(t+1) &= X^* + b_1 \times (B_i(t) - Lb^*) \\ &+ b_2 \times (B_i(t) - Ub^*) \end{aligned} \tag{9}$$

where $B_i(t)$ is the position information of the i egg ball in the iteration, b_1 and b_2 are two independent random vectors of $1 \times D$, and D is the dimension of the optimization problem.

(3) Population 3 (foraging dung beetle)

To feed dung beetles to find food, it is necessary to establish the optimal dung beetles area, and the boundaries of the area are defined as:

$$\begin{aligned} Lb^b &= \max\left(X^b \times (t/T_{\max}), Lb\right) \\ Ub^b &= \min\left(X^b \times (2 - t/T_{\max}), Ub\right) \end{aligned} \tag{10}$$

where X^b is the global optimal feeding position, Lb^b and Ub^b is the lower and upper bounds of the region. The position update formula can be expressed as:

$$\begin{aligned} x_i(t+1) &= x_i(t) + C_1 \times \left(x_i(t) - Lb^b\right) \\ &+ C_2 \times (x_i(t) - Ub^b) \end{aligned} \tag{11}$$

where $x_i(t)$ is the position information of the i dung beetle in the t iteration, C_1 is a random number obeying Gaussian distribution, $C_2 \in (0, 1)$ is a random vector.

(4) Population four (stealing dung beetles)

Stealing dung beetles will steal food. Assuming X^b is the global best feeding position, the position update formula can be expressed as:

$$x_i(t+1) = X^b + S \times g \times (\mid x_i(t) - X^* \mid + \mid x_i(t) - X^b \mid) \tag{12}$$

where $x_i(t)$ is the position information of the i dung beetle in the t iteration, g is a random vector of dimension $1 \times D$, obeys the Gaussian distribution, S is a fixed value.

3.4 The DBO-LRTC Completion Method

The selection of weight parameters λ_n is critical for the multi-nuclear norm minimization model. Unlike existing studies that typically assign fixed weights based on heuristic rules, our DBO-LRTC method treats weight determination as a constrained global optimization problem. The theoretical motivation for this integration is to eliminate human bias and ensure that the low-rank constraints are dynamically aligned with the intrinsic structural disparities of the specific wind power dataset, the objective function can be expressed as:

$$\begin{aligned} \min_{\lambda} \quad & \mathcal{F}(\lambda) = \tfrac{1}{|\overline{\Omega}|} \parallel \mathcal{P}_{\overline{\Omega}}(\mathcal{X}(\lambda)) - \mathcal{P}_{\overline{\Omega}}(\mathcal{O}) \parallel_F^2 \\ \text{s.t.} \quad & \lambda_n \geq 0, \quad \sum_{n=1}^{N} \lambda_n = 1 \end{aligned} \tag{13}$$

where $\mathcal{X}(\lambda)$ is the tensor after completion under the current weight λ. The DBO algorithm uses minimizing tensor completion error as the objective function to find a set of optimal ones that enables the multi-core norm minimization model to mine the low-rank features that are most consistent with the real underlying structure of the data, thereby obtaining higher completion accuracy.

This paper proposes the DBO-LRTC algorithm to recover missing values, and combines the dung beetle optimization algorithm and the data completion method process of multiple kernel norm minimization tensors (DBO-LRTC) as shown in Fig 2.

The DBO-LRTC method flow is as follows.

(1) Data preprocessing: organize the data into third-order tensors to obtain a complete data set, and randomly missing the data set is processed to obtain the missing data set $\mathcal{M}$, and identify the location of the non-missed data to generate the observation index set Ω.

(2) Parameter initialization: Set the parameters of the low-rank tensor completion algorithm, the initial value of the weight parameter λ, the maximum number of iterations and the convergence condition. The convergence condition is that the original residual and dual residual are less than the preset tolerance ϵ.

(3) Optimizing the weight parameters: According to the initialized parameters, call the LRTC function, use the ADMM algorithm to alternately update $\mathcal{X}$,

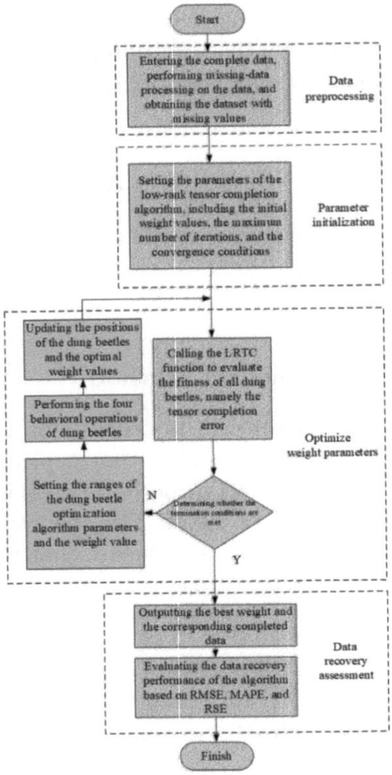

Fig. 2. DBO-LRTC method flow chart.

$\mathcal{T}$ and $\mathcal{Y}_n$, calculate the original residual and dual residual. When the residual meets the convergence conditions, output the optimal weight parameter λ and the completed tensor. When the convergence conditions are not met, set the DBO algorithm parameters and the value range of the weight parameters, call the DBO algorithm, and optimize the weight parameters until the iteration ends or the convergence conditions are met.

(4) Data completion and evaluation: This paper uses RMSE,MAPE and RSE to compare the similarity of data before and after completion and the effectiveness of the completion method. The calculation method can be expressed as:

$$RMSE = \sqrt{\frac{1}{n}\sum_{i=1}^{n}(x_i - \hat{x}_i)^2} \quad (14)$$

$$MAPE = \frac{1}{n}\sum_{i=1}^{n}\left|\frac{x_i - \hat{x}_i}{x_i}\right| \times 100\% \quad (15)$$

$$RSE = \frac{\|\mathcal{P}_\Omega(\mathcal{X} - \mathcal{W})\|_\mathrm{F}}{\|\mathcal{P}_\Omega(\mathcal{W})\|_\mathrm{F}} \times 100\% \tag{16}$$

where x_i is the simulated value, $\hat{x}_i$ is the actual observed value, $\mathcal{W}$ is the original complete tensor, $\mathcal{X}$ is the complete tensor after completion, $\| \cdot \|_F$ is the Frobenius norm. The smaller the RMSE value, the closer it is to the original value. The smaller the values of RMSE and MAPE are, the closer they are to the original values, indicating better performance. The same goes for RSE. The smaller the value, the closer it is to the original one, and it is used to assist in evaluation.

4 Case Analysis

4.1 Source and Preprocessing of Experimental Data

The data studied in this paper is the real-time data of a wind turbine cluster in a certain wind farm in Longyuan, Inner Mongolia for 20 consecutive days, including 9 wind turbines. The data sampling interval is 30 min, with a total of 8640 data points. The data was collected through the SCADA system, and the data source is reliable and rich, which can provide sufficient data support for the experiment.

4.2 Analysis of Results in Random Missing Scenarios

To verify the superiority of the DBO-LRTC algorithm, missing value recovery algorithms such as KNN, TRPCA, LRTC, and GAIN are introduced for comprehensive comparison. In the experiment, a scenario with random missing data was selected as the experimental environment. The wind power data matrix to be completed was constructed with a data missing rate of 5% to 30% by random sampling with a step size of 5%. Considering that the location of random data loss in each simulation is different, 10 repeated completion experiments are conducted for each lost data, and the error of each time is taken as the average of the 10 experiments. Subsequently, the completion effects of the LRTC, TRPCA, KNN, GAIN, and DBO-LRTC methods were analyzed respectively, and their corresponding RMSE, MAPE, and RSE values were calculated. The error values of each method are shown in Table 1,Table 2 and Table 3, and the error line graphs are shown in Fig 3,Fig 4 and Fig 5.

Table 1. The RMSE(MW) values of different completion methods.

Missing rate (%)	5%	10%	15%	20%	25%	30%
KNN	0.0753	0.1347	0.3149	0.6273	0.9473	1.28
TRPCA	0.0627	0.1154	0.2973	0.5864	0.9213	1.13
GAIN	0.0634	0.1124	0.2853	0.5795	0.9210	1.08
LRTC	0.0589	0.1003	0.2634	0.5604	0.8641	0.9874
DBO-LRTC	0.0542	0.0923	0.2244	0.5197	0.8187	0.9453

Table 2. The MAPE(%) values of different completion methods.

Missing rate (%)	5%	10%	15%	20%	25%	30%
KNN	1.34	6.43	10.97	13.27	18.94	23.5
TRPCA	1.25	5.92	9.54	11.85	17.23	21.3
GAIN	1.19	5.89	9.23	11.27	16.89	19.5
LRTC	1.06	5.72	8.85	10.74	15.63	19.1
DBO-LRTC	0.92	5.13	8.25	10.04	14.37	16.3

Table 3. The RSE(%) values of different completion methods.

Missing rate (%)	5%	10%	15%	20%	25%	30%
KNN	5.37	11.32	17.84	25.47	34.59	41.23
TRPCA	4.64	10.27	14.39	20.91	26.45	37.28
GAIN	4.78	10.34	14.54	20.32	25.47	32.97
LRTC	4.24	9.43	12.37	17.29	21.53	26.67
DBO-LRTC	3.72	8.84	11.42	15.77	18.35	21.3

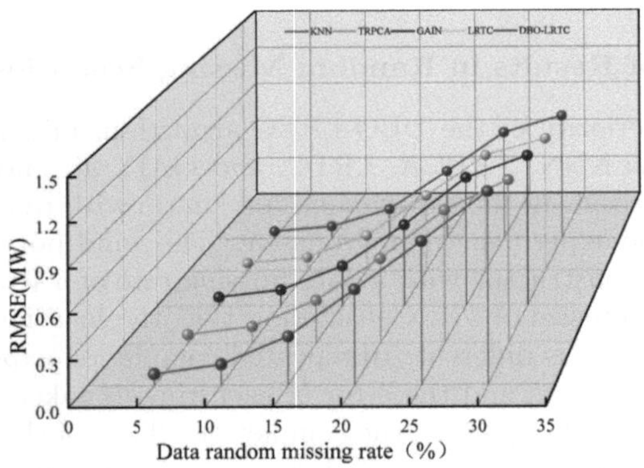

Fig. 3. RMSE(MW) line chart with different completion methods.

The Fig 3,Fig 4 and Fig 5 reveal the performance of several completion algorithms when dealing with different degrees of missing. With the increase of the degree of data loss, the RMSE, MAPE and RSE of several algorithms all show varying degrees of increase, and the effectiveness of the completion algorithm gradually decreases. The reason for this phenomenon lies in the fact that as the amount of missing data increases, the information available to the algorithm decreases, thereby affecting its completion accuracy.

It can be seen from the analysis of the results in the chart that the RMSE of the DBO-LRTC algorithm is approximately 0.0542MW when the missing rate is 5% low. It is approximately 0.9453MW when the high loss rate is 30%.

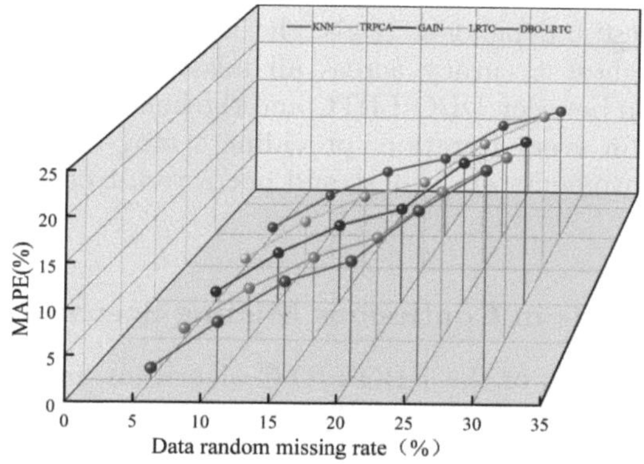

Fig. 4. MAPE(%) line chart with different completion methods.

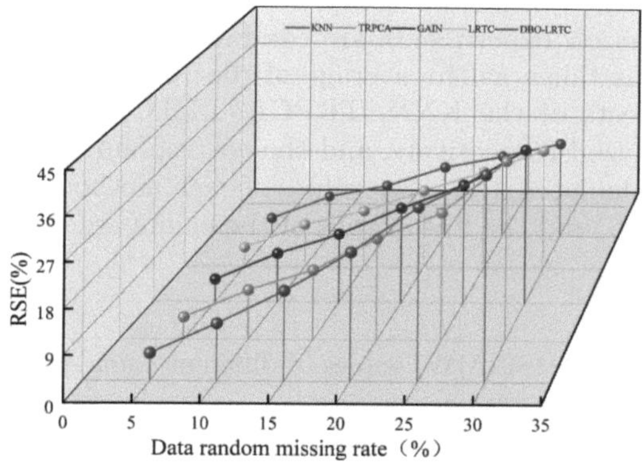

Fig. 5. RSE(%) line chart with different completion methods.

Compared with the four comparison algorithms, DBO-LRTC has the smallest error. The MAPE of the DBO-LRTC algorithm is approximately 0.92% at a low missing rate of 5%, which is 0.42% points lower than KNN, 0.33% points lower than TRPCA, 0.27% points lower than GAIN, and 0.14% points lower than LRTC. At a high deletion rate of 30%, it is approximately 16.3%, which is 7.2% points lower than KNN, 5% points lower than TRPCA, 3.2% points lower than GAIN, and 2.8% points lower than LRTC. The RSE of the DBO-LRTC algorithm is approximately 3.72% when the missing rate is 5%, which is 1.65% points lower than that of KNN, 0.92% points lower than that of TRPCA, 1.06% points lower than that of GAIN, and 0.52% points lower than that of LRTC. At a high deletion rate of 30%, it was approximately 21.3%, which was 19.93% points lower than KNN, 15.98% points lower than TRPCA, 11.67% points lower than GAIN, and 4.97% points lower than LRTC. Therefore, whether in low-data loss scenarios or high-data loss scenarios, the accuracy of random data loss recovery of the DBO-LRTC method is higher than that of other methods.

Throughout these independent trials, the DBO-LRTC method consistently maintained the highest accuracy across all missing scenarios and rates. The performance margin between DBO-LRTC and the benchmark methods remained remarkably stable in every iteration, providing strong empirical evidence that the observed improvements are robust and not a result of stochastic variations in data sampling.

4.3 Result Analysis in Continuous Missing Scenario

To verify the superiority of the DBO-LRTC algorithm, missing value recovery algorithms such as KNN, TRPCA, LRTC, and GAIN are introduced for comprehensive comparison. In the experiment, consecutive missing scenarios were respectively selected as the experimental environment, and the segment missing data matrix was constructed from the data matrix to be completed. Considering that the errors obtained from each completion experiment may have deviations, 10 repeated completion experiments are conducted for each lost data, and the error of each time is taken as the average of the 10 experiments. Subsequently, the completion effects of the KNN, TRPCA, LRTC, GAIN and DBO-LRTC methods were analyzed respectively, and the corresponding RMSE, MAPE and RSE values were calculated.The error values of each method are shown in Table 4 , Table 5 and Table 6, and the line graphs of the errors are shown in Fig 6 , Fig 7 and Fig 8.

Table 4. The RMSE(MW) values of different completion methods.

Missing time (h)	2	4	6	8	10
KNN	0.0347	0.0454	0.0689	0.0941	0.1157
TRPCA	0.0302	0.0431	0.0643	0.0894	0.1121
GAIN	0.0297	0.0421	0.0597	0.0823	0.1101
LRTC	0.0234	0.0325	0.0486	0.0714	0.1043
DBO-LRTC	0.0216	0.0297	0.0434	0.0637	0.0966

Table 5. The MAPE(%) values of different completion methods.

Missing time (h)	2	4	6	8	10
KNN	0.62	0.84	1.25	1.72	2.34
TRPCA	0.57	0.78	1.13	1.45	1.97
GAIN	0.48	0.73	1.05	1.27	1.74
LRTC	0.43	0.64	0.94	1.19	1.53
DBO-LRTC	0.41	0.54	0.79	1.13	1.42

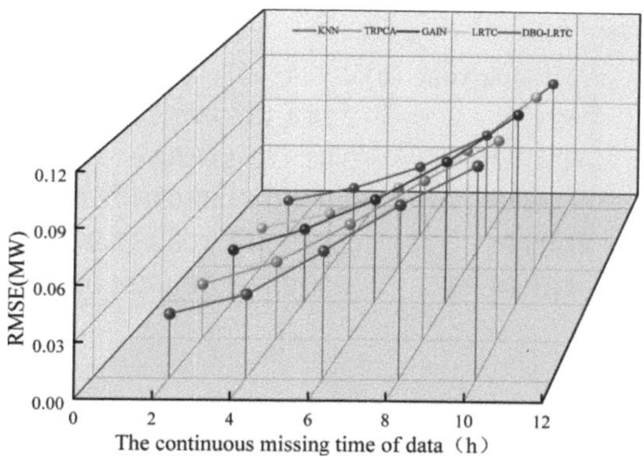

Fig. 6. RMSE(MW) line chart with different completion methods.

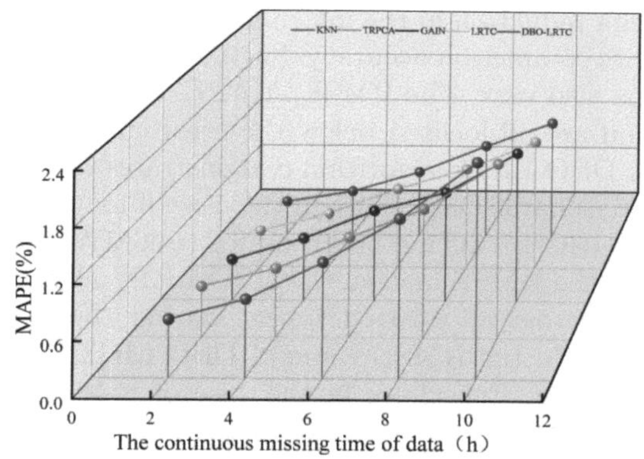

Fig. 7. MAPE(%) line chart with different completion methods.

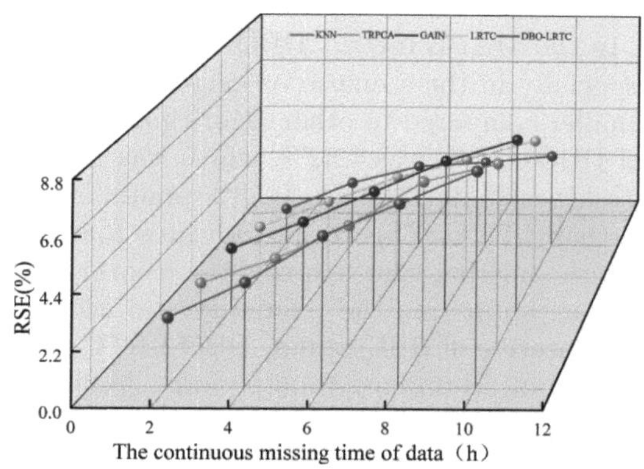

Fig. 8. RSE(%) line chart with different completion methods.

Table 6. The RSE(%) values of different completion methods.

Missing time (h)	2	4	6	8	10
KNN	2.74	4.13	5.97	7.24	8.53
TRPCA	2.62	3.63	4.98	6.78	7.54
GAIN	2.6	3.74	5.02	6.34	7.26
LRTC	2.13	3.24	4.31	5.13	5.97
DBO-LRTC	1.56	2.74	3.51	3.72	4.02

As can be seen from the results of the chart, on the whole, with the increase of the degree of data loss, the RMSE, MAPE and RSE of several methods all show varying degrees of increase, and the effectiveness of the completion algorithm gradually decreases. The reason for this phenomenon lies in that as the number of continuously missing data increases, the temporal correlation of the data is disrupted, resulting in a reduction in the information that the algorithm can utilize and thus affecting its completion accuracy. Furthermore, the data filling errors of different algorithms also vary. The RMSE, MAPE, and RSE error curves of the DBO-LRTC method are all located below the error curves of other algorithms, indicating that the DBO-LRTC algorithm demonstrates relatively excellent performance in all test scenarios, especially in the case of longer missing times. The RMSE value of the DBO-LRTC method is only 0.0966MW. From a detailed perspective, in terms of the RMSE metric, the DBO-LRTC algorithm demonstrates the minimum error across all missing times. During the missing time period ranging from 2 h to 10 h, the RMSE value of DBO-LRTC is always lower than that of other algorithms. Especially in the case of a 10-hour missing period, the RMSE of DBO-LRTC is 0.0966MW. Compared with KNN, it decreased by 0.0191MW; compared with TRPCA, it decreased by 0.0155MW; compared with GAIN, it decreased by 0.0077MW; compared with LRTC, it decreased by 0.0155MW. This result indicates that DBO-LRTC outperforms other methods in the recovery effect of different missing time lengths and can effectively reduce the recovery error. In the MAPE metric, DBO-LRTC also demonstrated a relatively low error, especially in the scenario of long-term absence, where its error was significantly smaller compared to other algorithms. At the 10-hour absence, the MAPE value of DBO-LRTC was 1.42%, which was 1.01% points lower than KNN, 0.55% points lower than TRPCA, 0.32% points lower than GAIN, and 0.11% points lower than LRTC. This result indicates that DBO-LRTC can not only effectively fill in the missing values during the recovery process, but also has a smaller relative error, improving the reliability and accuracy of the recovery results. From the perspective of RSE values, DBO-LRTC maintains a relatively low error value throughout all missing time periods. At the 2-hour absence, the RSE of DBO-LRTC was 1.56%, significantly lower than that of KNN (2.74%), TRPCA (2.62%), GAIN (2.6%), and LRTC (2.13%). With the increase of the deletion time, the advantage of DBO-LRTC remains, especially at the 10-hour deletion, its RSE is 4.02%, which is significantly better than other methods. This

Table 7. Data recovery time when random missing.

Random missing rate (%)	This method takes time (s)	Time-consuming to use ADMM method(s)	LRTC method takes time (s)
5	36	43	89
10	41	62	105
15	56	74	124
20	61	95	146
25	79	103	173
30	95	132	221

Table 8. Data recovery time when continuous missing.

Continuous missing period (h)	This method takes time (s)	Time-consuming to use ADMM method(s)	LRTC method takes time (s)
2	47	76	97
4	55	92	127
6	67	115	164
8	85	134	198
10	94	178	241

result indicates that DBO-LRTC demonstrates a strong control ability in error accumulation and can effectively reduce the error growth during the recovery process in the case of long-term absence.

Observational data from multiple runs confirm that the superiority of DBO-LRTC is highly repeatable. The consistent lead in RMSE, MAPE, and RSE metrics over other algorithms demonstrates a high level of statistical stability, effectively addressing the requirement for reliability in wind farm data recovery.

4.4 Ada-ADMM Method Calculation Performance Verification

In the process of solving the model, this paper uses the alternating direction multiplier method (ADMM) to iteratively solve the model, decomposing the large-scale optimization model into multiple single-variable optimization models for joint optimization, which effectively reduces the difficulty of solving the model. In order to solve the problem of slow convergence of the alternating direction multiplier method, an adaptive step mechanism is added, and the step size is dynamically adjusted with the residual to accelerate convergence. In the experiment, this paper conducted statistics on the data recovery time of LRTC, DBO-LRTC solved using the ADMM method, and data recovery time solved using the Ada-ADMM method. The results are shown in Table 7 and Table 8.

From the analysis of the results in the table, it can be seen that compared with several comparison algorithms, the DBO-LRTC method solved by the Ada-ADMM method has the shortest recovery time. It shows that the adaptive alternating direction multiplier method greatly improves the convergence speed and saves time and cost.

4.5 Parameter Sensitivity Analysis

To address the impact of key hyper-parameters on the imputation performance, we conducted a sensitivity analysis for the DBO population size N_{pop} and the weights λ in the scenario of a 20% random absence.The results are shown in Table 9 and Table 10.

Table 9. Impact of DBO Population Size N_{pop}

Population Size	RMSE(MW)	Iterations	Time(s)
10	0.5642	82	24
20	0.5315	64	48
30	0.5193	52	61
50	0.5144	48	105

Table 10. Performance comparison of different Weight λ strategies.

Weight Strategy	λ_1 (Time)	λ_2 (Space)	λ_3 (Physical quantity)	RMSE(MW)
LRTC	0.333	0.333	0.333	0.5612
DBO-LRTC	0.621	0.234	0.145	0.5193

Table 9 illustrates the impact of population size on convergence and efficiency. A smaller population (N_{pop} = 10) leads to higher errors due to insufficient coverage of the search space. When the population size is increased to 30, the RMSE decreases significantly and the convergence speed accelerates. Although N_{pop} = 50 can slightly further improve the accuracy, the computational time cost per iteration increases substantially; therefore, 30 is selected as the optimal setting to balance both accuracy and efficiency.

Table 10 provides a performance comparison across various weight allocation strategies. Traditional equal weighting ignores the inherent disparities in correlation strength among the different modes of wind power data. Through the global optimization of DBO, the model adaptively assigns a higher weight to the temporal mode ($\lambda_1 = 0.621$), which is consistent with the pronounced temporal periodicity inherent in wind power sequences. Compared with the equal-weight setting, this adaptive weighting mechanism reduces the RMSE by approximately 7.4%, demonstrating the theoretical superiority of DBO-LRTC in mining the intrinsic low-rank structure of the data.

5 Conclusion

Aiming at the problem of missing wind power data, this paper proposes a novel recovery method based on the Dung Beetle Optimization algorithm and multiple nuclear norm minimization tensor completion (DBO-LRTC). The main conclusions are as follows:

(a) The proposed method utilizes the DBO algorithm to adaptively optimize weight parameters, effectively overcoming the theoretical limitations of fixed empirical weights in traditional LRTC. By formulating weight determination as a global optimization problem, the model can precisely capture the intrinsic multi-dimensional structural features and spatial-temporal correlations of wind power data.

(b) To enhance computational efficiency, an Ada-ADMM solver is introduced. By dynamically adjusting the penalty parameter based on primal and dual residuals, the solver achieves faster convergence and superior scalability compared to standard ADMM, significantly reducing the time cost for large-scale data repair.

(c) Experimental results on real-world datasets demonstrate that DBO-LRTC consistently outperforms benchmark methods like KNN, TRPCA, and GAIN. Parameter sensitivity analysis further confirm that our method provides robust and reliable performance under both random and continuous missing scenarios.The high repeatability and consistent superiority observed across multiple independent runs provide sufficient evidence of the method's practical engineering value and statistical stability, even in the absence of traditional variance analysis.

Future research will focus on improving the model's robustness against measurement noise and validating its generalization capabilities across diverse renewable energy domains, such as solar power and regional microgrids.

Acknowledgments. The authors would like to express their sincere gratitude to the technical and operational staff at Longyuan Wind Farm for their assistance in data collection and field support. We also appreciate the constructive feedback from colleagues during the preparation of this manuscript. The authors have thoroughly reviewed, revised, and validated all content and assume full responsibility for the work presented in this publication.

Disclosure of Interests. The authors declare no conflicts of interest. The funders (China Longyuan Power Group) had no role in the design of the study; in the collection, analyzes, or interpretation of data; in the writing of the manuscript; or in the decision to publish the results.

Statement of Funding Sources. This research was funded by China Longyuan Power Group, grant number No. LYX-2024-17 and The APC was funded by China Longyuan Power Group.

References

1. Wu, Y., Wang, Y., Xu, H., et al.: Review of wind power output forecasting technology. Comput. Sci. Explor. **16**(12), 2653–2677 (2022)
2. Li, G., et al.: Research on short-term wind power forecasting based on VMD-IDBO-SVM. Electronics **15**(3), 533 (2026)

3. Jia, H., Marti-Puig, P., Caiafa, C. F., Serra-Serra, M., Sun, Z., Solé-Casals, J.: Exploring tensor completion for missing data estimation in wind farms. IEEE Sensors Lett. **8**(12), 1–4 (2024). Art. no. 7006604
4. Ren,Z., Liu, J.: Wind power data imputation method based on similarity measurement," Electric Eng. Technol. **178**(20), 53–55 (2022)
5. Xie, K., Wang, J., Liu, C., et al.: Study of electricity data processing based on KNN interpolation and wavelet soft-threshold denoising methods. IAENG Int. J. Compu. Sci. **52**(9) (2025)
6. Li, N., Liu, J., Lai, X., et al.: Ultra-short-term multi-step photovoltaic power forecasting based on time-series neural hierarchical interpolation model. Smart Power **52**(4), 69–77 (2024)
7. Liu, Z., Tao, Y., Liu, H., et al.: Urban load missing data completion method based on dual-channel generative adversarial network. Automat. Electric Power Syst. **48**(17), 161–170 (2024)
8. A, R., Liu, L.: Research and implementation of missing value imputation method for time-series data. Comput. Simulat. **41**(9), 520–525 (2024)
9. Hong, W., Li, Q., Zhang, Y., et al.: Phase sequence identification algorithm for low-voltage distribution network based on quadratic matrix completion. Electric Power Automat. Equip. **42**(9), 133–138 (2022)
10. Jin, T., Wang, W., Liu, Y., et al.: Advanced electricity theft detection in smart metering systems via channel-correlation enhanced hierarchical kernel networks and matrix completion. Measurement **253**(117768) (2025)
11. Jiang, R., Yu, X., Jin, X., et al.: Spatiotemporal measurement data completion method for distribution network based on alternating minimization matrix completion and moving average. Electric Power Automat. Equip. **45**(8), 20–27 (2025)
12. Wang, G., Xu, W., Gu, X., et al.: Missing data completion method for low-voltage load based on multi-characteristic sample factors. Energy China **47**(4), 84–92 (2025)
13. Chen, H., Lin, M., Zhao, L., Xu, Z., Luo, X.: Fourth-order dimension preserved tensor completion with temporal constraint for missing traffic data imputation. IEEE Trans. Intell. Transp. Syst. **26**(5), 6734–6748 (2025)
14. Yang, T., Ye, Z., Xu, J., et al.: Missing data recovery method for non-intrusive load monitoring based on low-rank tensor completion. Power Syst. Technol. **48**(1), 394–405 (2024)
15. Yang, T., Yang, Z., Pen, H.: Electricity measurement missing data recovery method based on low-rank tensor completion with triangular canonical polyadic decomposition. IEEE Trans. Instrument. Measur. **74**(9008713), 1–13 (2025)
16. Zhu, L., Yang, L., Guo, Y., et al.: Multi-channel audio completion algorithm based on tensor nuclear norm. Electronics **13**(9) (2024)
17. Miyata, T.: Traffic matrix completion by weighted tensor nuclear norm minimization and time slicing: special section on recent progress in nonlinear theory and its applications. Nonlinear Theory Appl. IEICE **15**(2), 311–323 (2024)
18. Xiao, P., Fang, Z., Jin, K., et al.: Study of the road feel model of a steer-by-wire system based on an improved dung beetle optimization algorithm. Electrical Engineering, 1–12 (2025)
19. Du, W., Wang, Y., Meng, G., Guo, Y.: Privacy-preserving vertical federated KNN feature imputation method. Electronics **13**(381) (2024)

FuzzyRules: A Fuzzy Rule-based System for Artificial Intelligence package for R

Ronei M. Moraes[1], Jodavid A. Ferreira[2], Sergio C. Santos[3], and Liliane S. Machado[3](✉)

[1] Department of Statistics, Federal University of Paraiba, João Pessoa, Brazil
ronei@de.ufpb.br
[2] Department of Statistics, Federal University of Pernambuco, Recife, Brazil
jodavid.ferreira@ufpe.br
[3] Department of Computer Science, Federal University of Paraiba, João Pessoa, Brazil
sergio.caua.dos.santos@academico.ufpb.br, liliane@di.ufpb.br

Abstract. This paper introduces FuzzyRules, an R package for the systematic development, analysis, and evaluation of fuzzy rule-based systems. The framework provides a unified and extensible architecture that supports the full modeling pipeline of interpretable fuzzy inference systems, with emphasis on Artificial Intelligence and classification tasks. The package implements a comprehensive set of tools for the creation and manipulation of fuzzy sets, including set-theoretic operations based on t-norms and dual t-conorms (intersection, union, and complement), as well as structural descriptors such as support, kernel, core, height, degree of membership, overlap functions and indices, grouping functions and indices, and the Type-1 Jaccard fuzzy similarity index. It incorporates a broad collection of membership functions and classical t-norms and t-conorms. For inference, the framework implements Mamdani-type fuzzy systems with multiple defuzzification strategies, including the classical center of gravity, and nine other methods. Methodological foundations and an illustrative example are provided to facilitate reproducible experimentation and practical adoption. A brief comparison with other software is also presented.

Keywords: Fuzzy rule-based systems · R software · 'FuzzyRules'

1 Introduction

Fuzzy Logic (FL), as defined by Zadeh [22], is grounded in formal principles of approximate reasoning. The foundational concept of fuzzy logic control was initially proposed in notes from 1968 [18] and 1972 [19], and was more thoroughly elaborated in 1973 [20]. The first practical implementation was carried out by Mamdani and Assilian in 1974 [6], in the context of controlling a steam engine.

Several implementations of programs for operations over fuzzy sets and fuzzy logic (particularly systems based on fuzzy rules) can be found. Examples are the

J. Lu et al. (Eds.): FLINS-ISKE 2026, LNCS 16752, pp. 215–231, 2027.
https://doi.org/10.1007/978-981-92-2369-5_15

"Fuzzy Logic Toolbox for Matlab" [9] and "Fuzzy Logic Toolkit for Octave" [10]. Regarding programming languages, developments can be found in Java, as the "Juzzy" [16], and in Python, as "fuzzylogic" [11] and "scikit-fuzzy" [12]. In particular, there are implementations for R Software, as the "frbs" [13], "FuzzyR" [2] and "sets" [15] packages. However, "frbs" received its last update in 2019; "FuzzyR" package in 2021 and "sets" in 2023. From a didactic perspective, "frbs" must be programmed using encodings not suitable for novice users. The "FuzzyR" package has a graphical user interface (GUI), and it is easy for novices, but the code generated by its GUI is difficult to understand, and its interfaces require more memory to run. The "sets" package has more intuitive programming and may be suitable for teaching fuzzy rule-based systems (however, it is not easy to understand the rules compilation).

In an attempt to create a more intuitive programming package for students and professionals, a new package for R was developed, called `FuzzyRules` [8]. `FuzzyRules` implements the creation of fuzzy sets and basic operations on them using various methods available in the literature. It also implements Type-1 Fuzzy Logic using the Mamdani type and ten different defuzzification methods. This new implementation aims to be efficient and usable on computers with low processing power and memory, making it accessible specially to students, but also to professionals. `FuzzyRules` is the first package for R to implement some fuzzy set operations, as well as some defuzzification methods described below. Thus, `FuzzyRules` offers significant advantages over its free and paid international counterparts.

2 Fuzzy Sets and Fuzzy Logic

The Theory of Fuzzy Sets was proposed by Zadeh [17]. Formally, a fuzzy subset A of a universe of discourse X is characterized by a membership function $\mu_A : X \rightarrow [0,1]$, where for each element $x \in X$ associates a number $\mu_A(x)$ in the interval $[0,1]$ which assigns the degree of membership of x in A.

Formally, let X and Y be universes of discourse. The intersection of two fuzzy sets A and B, with membership functions $\mu_A : X \rightarrow [0,1]$, and $\mu_B : Y \rightarrow [0,1]$, results in a set C, whose membership function μ_C is given by:

$$\mu_C(x,y) = \wedge\{\mu_A(x), \mu_B(y)\}, \quad x \in X \text{ and } y \in Y, \tag{1}$$

where $\wedge$ is given by a t-norm. Table 1 presents four t-norms defined for $a, b \in [0,1]$.

The union of two fuzzy sets A and B, with membership functions μ_A and μ_B results in a set C, whose membership function μ_C is given by:

$$\mu_C(x,y) = \vee\{\mu_A(x), \mu_B(y)\}, \quad x \in X \text{ and } y \in Y, \tag{2}$$

where $\vee$ is given by a t-conorm (or s-norm). Table 2 presents four t-conorms defined for $a, b \in [0,1]$.

Table 1. Some t-norms used in Fuzzy Sets/Fuzzy Logic.

t-conorm	Equation
Zadeh	$\wedge_{\texttt{Zadeh}}(a,b) = \min(a,b)$
Probabilistic	$\wedge_{\texttt{Prob}}(a,b) = ab$
Lukasiewicz	$\wedge_{\texttt{Luk}}(a,b) = \max(a+b-1,\ 0)$
Weber	$\wedge_{\texttt{Web}}(a,b) = \max\left(0, \frac{a+b-1+\lambda ab}{1+\lambda}\right), \lambda > -1$

Table 2. Some t-conorms used in Fuzzy Sets/Fuzzy Logic.

t-conorm	Equation
Zadeh	$\vee_{\texttt{Zadeh}}(a,b) = \max(a,b)$
Probabilistic	$\vee_{\texttt{Prob}}(a,b) = a+b-ab$
Lukasiewicz	$\vee_{\texttt{Luk}}(a,b) = \min(a+b,\ 1)$
Weber	$\vee_{\texttt{Web}}(a,b) = \min\left(1, \frac{a+b+\lambda ab}{1+\lambda}\right), \lambda > -1$

The complement of a fuzzy set A, denoted by $\bar{A}$, is given by:

$$\mu_{\bar{A}}(x) = 1 - \mu_A(x), \quad x \in X. \tag{3}$$

The support of a fuzzy set A is defined as the set of all elements of the universe of discourse X that have a nonzero degree of membership in A [17]:

$$\text{supp}(A) = \{x \in X \mid \mu_A(x) > 0\}. \tag{4}$$

The kernel of a fuzzy set A is defined as the set of all elements of the universe of discourse X that have a degree of membership equal to 1 in A:

$$\ker(A) = \{x \in X \mid \mu_A(x) = 1\}. \tag{5}$$

The core of a fuzzy set A is defined as the set of all elements of the universe of discourse X that have the largest degree of membership in A:

$$\text{core}(A) = \{x \in X \mid \mu_A(x) = \sup_{x \in X} \mu_A(x)\}. \tag{6}$$

Note that when A is a normal fuzzy set, i.e., $\sup_{x \in X} \mu_A(x) = 1$, the core coincides with the kernel, that is, $\text{core}(A) = \ker(A)$.

The height of a fuzzy set A is defined as the supremum of its membership function over the universe of discourse X [17]:

$$\text{height}(A) = \sup_{x \in X} \mu_A(x). \tag{7}$$

A fuzzy set A is said to be normal if and only if $\text{height}(A) = 1$.

A bi-dimensional function $o : [0,1]^2 \to [0,1]$ is an overlap function if it satisfies the following conditions [1]:

(o1) o is symmetric, i.e. $o(x, y) = o(y, x) \forall x, y \in [0, 1]$;
(o2) $o(x, y) = 0$ if and only if $xy = 0$;
(o3) $o(x, y) = 1$ if and only if $xy = 1$;
(o4) o is non-decreasing;
(o5) o is continuous.

A bi-dimensional function $g : [0, 1]^2 \to [0, 1]$ is said to be a grouping function if it satisfies the following conditions [1]:

(g1) g is symmetric, i.e. $g(x, y) = g(y, x) \forall x, y \in [0, 1]$;
(g2) $g(x, y) = 0$ if and only if $x = y = 0$;
(g3) $g(x, y) = 1$ if and only if $x = 1$ or $y = 1$;
(g4) g is non-decreasing;
(g5) g is continuous.

Some examples of mathematical functions of Overlap and Grouping are presented in the Table 3.

Table 3. Some Overlap and Grouping functions.

Function	Equation
Overlap	$(min(fs1, fs2))/length(U)$
Overlap	$(fs1^p * fs2^p), p > 1$
Overlap	$(min(fs1, fs2))^p, p > 0$
Overlap	$1 - (1 - min(fs1, fs2))^p, p > 0$
Grouping	$(pmax(fs1, fs2))/length(U)$
Grouping	$(1 - (1 - fs1)^p * (1 - fs2)^p), p > 1$
Grouping	$1 - (1 - max(fs1, fs2))^p, p > 0$
Grouping	$(max(fs1, fs2))^p, p > 0$

Let U be the universe of discourse and FS(U) be the space of all fuzzy sets defined over U. An Overlap index is a mapping $O : FS(U) \times FS(U) \to [0, 1]$ such that [3]:

O1) O(A, B) = O(B, A);
O2) $O(A, B) = 0$ if and only if A and B are completely disjoint; i.e., $A(u_i)B(u_i) = 0, \forall u_i \in U$;
O3) $O(A, B) = 1$ if $\exists u_i \in U | A(u_i) = B(u_i) = 1$
O4) If $B \leq C$, then $O(A, B) \leq O(A, C)$.

Let U be the universe of discourse and FS(U) be the space of all fuzzy sets defined over U. A Grouping index G is a mapping $G : FS(U) \times FS(U) \to [0, 1]$ if, for all $A, B \in FS(U)$, it satisfies:

(G1) $G(A, B) = G(B, A)$

(G2) $G(A, B) = 0$ if and only if $A = \emptyset$ and $B = \emptyset$
(G3) $G(A, B) = 1 \iff A \cup B = U$
(G4) If $A \leq A'$ and $B \leq B'$ then $G(A, B) \leq G(A', B')$.

Overlap indices are measures used to quantify the degree of similarity between fuzzy sets. The Grouping index provide complementary information of notion of overlap index, w.r.t. a negation function [14]. Grouping indices measures to quantify how close the union of two fuzzy sets is to the universe set U.

The Type-1 Jaccard index provides a direct and interpretable measure of similarity between two fuzzy sets A and B, defined over a universe of discourse X, and it is given by:

$$J(A, B) = \frac{\sum_{x \in X} \min(\mu_A(x), \mu_B(x))}{\sum_{x \in X} \max(\mu_A(x), \mu_B(x))}. \tag{8}$$

The result $J = 1$ is achieved when the two sets are identical, that is, for all $x \in X$, the membership degree in set A and in set B is the same. On the other hand, when $J = 0$, the index is interpreted as totally disjoint sets. The numerator represents the overlap between the two sets, while the denominator represents the grouping function. As the *min* and *max* operations are applied independently to each element of the universe of discourse, the computational complexity scales linearly with the size of X.

Fuzzy Logic uses degrees of truth in order to evaluate a statement "x is A", as for instance, "Temperature is High". However, it is necessary to have a value for the variable temperature in order to evaluate that statement, i.e., a crisp input (25 °C, for example) in order to determine its degree of truth. The statement "x is A AND y is B" represents the intersection of two fuzzy sets, A and B, where A is a fuzzy set defined on the universe of discourse X and B is defined on the universe of discourse Y. The connective AND evaluates the degree of truth for which a crisp input satisfies both conditions ("x is A" and "y is B") simultaneously.

Mamdani and Assilian [6] proposed in 1974 the first implementation using fuzzy logic. According to Mamdani and Assilian [7], the assessment of rules (or logical expressions) with conjunctions (connective AND) and disjunctions (connective OR) correspond to the operations of t-norm minimum and t-conorm (or s-norm) maximum on fuzzy sets, respectively. The negation (NOT) of a fuzzy set is performed by the complement of that set.

In general, it is possible to use the dual notion of t-norm, which is a t-conorm in order to perform the connectives AND and OR. For instance, the ones proposed by Zadeh, Lukasiewicz, Probabilistic or Weber can be used. Mamdani and Assilian [7] proposed using t-norm min for modeling connective AND and t-conorm max for connective OR.

2.1 Fuzzy Rule-Based Systems

A typical Fuzzy Rule-based System receives as input crisp values. Those values are transformed by a Fuzzifier, which assigns a degree of membership of each input according to each linguistic term. For example, a person aged 17 could belong to the "very young" fuzzy set with degree of membership 0.35 and also could belong to the "young" fuzzy set with degree of membership 0.55.

After Fuzzifier, an inference engine receives the fuzzified inputs and according to the type (in the case of actual version of `FuzzyRules` package: Mamdani type), the rules in Rulebase are processed. A rule in Rulebase for a Mamdani type is:

$$R_k : \textit{IF } N_1 \text{ is } N_{k,1} \textit{ AND } ... \textit{ AND } N_n \text{ is } N_{k,n} \tag{9}$$

$$\textit{THEN } O_k \text{ is } O_{k,n}, \; k = \{1, ..., K\}, \tag{10}$$

where K is the number of rules in the database, $N_{k,n}$ is a linguistic term in the kth rule for the variable X_i, $i = 1, ..., n$. The first part of the rule R_k (Eq. 4) is the antecedent and the second part (Eq. 5) is the consequent. In order to compute the consequent part of a rule, an implication operator must be used. An implication operator on two fuzzy sets A and B, with membership functions μ_A and μ_B results in a fuzzy set C, whose membership function μ_C is given by:

$$\mu_C(x, y) = \nabla\{\mu_A(x), \mu_B(y)\}, \quad x \in X \quad \text{and} \quad y \in Y, \tag{11}$$

where ∇ is an implication operator. For instance, it is possible to use those proposed by Kleene-Diemes, Lukasiewicz, Zadeh-Wilmott, Mamdani, Larsen, among others. When the Mamdani type is used, the implication operator is given by t-norm min (Mamdani's implication).

After all rules are computed, an aggregation operator must be used in order to aggregate all partial conclusions. An aggregation operator on two fuzzy sets A and B, with membership functions μ_A and μ_B results in a fuzzy set C, whose membership function μ_C is given by:

$$\mu_C(x, y) = \Diamond\{\mu_A(x), \mu_B(y)\}, \quad x \in X \quad \text{and} \quad y \in Y, \tag{12}$$

where $\Diamond$ is an aggregation operator.

Some aggregation operators that can be used include the minimum, maximum, average, weighted average, product, among others. When the Mamdani type is used, the aggregation operator is given by t-conorm max.

The fuzzy set resulting from aggregation passes through one of the defuzzification methods to produce a final crisp value. Several methods have been presented in the literature for this purpose [5]. Ten of these methods are reported below:

- **Centroid or Center of Gravity (CoG):** The Centroid method determines the defuzzified value as the X coordinate corresponding to the center of gravity of the fuzzy set A [4]. Formally, it is computed as:

$$\text{CoG}(A) = \frac{\sum_{i=1}^{n} x_i \mu_A(x_i)}{\sum_{i=1}^{n} \mu_A(x_i)}, \tag{13}$$

where x_i are the elements of the universe of discourse X and $\mu_A(x_i)$ is the membership degree of x_i in A.

- **Center of Sums (CoS):** The CoS method computes the defuzzified value as a weighted mean of the geometric centers of the individual output fuzzy sets A_{c_i}, $i = 1, \ldots, n$ [4]:

$$\mathrm{CoS}(A) = \frac{\sum_{i=1}^{n} \bar{x}_i \mathrm{Area}(A_{c_i})}{\sum_{i=1}^{n} \mathrm{Area}(A_{c_i})}, \tag{14}$$

where $\bar{x}_i$ is the midpoint between the extremes of the support of A_{c_i}, and $\mathrm{Area}(A_{c_i})$ is approximated via Monte Carlo integration.
- **Center of Largest Area (CoLA):** The CoLA method determines the crisp output of all rules computing the centroid of the single fuzzy set A_{c_k} that has the maximum area [4]

$$\mathrm{CoLA}(A) = \frac{\sum_{i=1}^{m} x_i \mu_{A_{c_k}}(x_i)}{\sum_{i=1}^{m} \mu_{A_{c_k}}(x_i)}, \tag{15}$$

where A_{c_k} is the output fuzzy set with the largest area, $x_i \in X$ are the discrete elements of the universe of discourse, and $\mu_{A_{c_k}}(x_i)$ is the membership degree of x_i in A_{c_k}.
- **First of Maxima (FoM):** The FoM method computes the defuzzified value as the smallest element of the universe of discourse X at which the membership function of the aggregated fuzzy set A achieves its maximum value [5]:

$$\mathrm{FoM}(A) = \min\{x_i \in X \mid \mu_A(x_i) = \max_{x_j \in X} \mu_A(x_j)\}, \tag{16}$$

where x_i are the elements of the universe of discourse X and $\mu_A(x_i)$ is the membership degree of x_i in A.
- **Last of Maxima (LoM):** The LoM method computes the defuzzified value as the largest x_i in the universe of discourse X that achieves the highest membership degree within the aggregation fuzzy set A [5]:

$$\mathrm{LoM}(A) = \max\{x_i \in X \mid \mu_A(x_i) = \max_{x_j \in X} \mu_A(x_j)\}, \tag{17}$$

where $\mu_A(x_i)$ is the membership degree of x_i in A.
- **Mean of Maxima (MoM):** The MoM method computes the defuzzified value as the arithmetic mean of all x_i in the universe of discourse X that present the maximum membership degree in the fuzzy set A [5]:

$$\mathrm{MoM}(A) = \frac{\sum_{x_i \in \mathcal{M}} x_i}{|\mathcal{M}|}, \tag{18}$$

where $\mathcal{M} = \{x_i \in X \mid \mu_A(x_i) = \max_{x_j \in X} \mu_A(x_j)\}$ is the set of all elements that maximize the membership function, and $|\mathcal{M}|$ denotes the total number of such elements.
- **First of Support (FoS):** The FoS method computes the defuzzified value as the smallest element of the support of the set A that is strictly positive [5]:

$$\mathrm{FoS}(A) = \min\{x_i \in X \mid \mu_A(x_i) > 0\}, \tag{19}$$

where x_i are the elements of the universe of discourse X and $\mu_A(x_i)$ is the membership degree of x_i in A.
- **Last of Support (LoS):** The LoS method computes the defuzzified value as the largest element of the support of the fuzzy set A at which the membership function assumes a strictly positive value [5]:

$$\text{LoS}(A) = \max\{x_i \in X \mid \mu_A(x_i) > 0\}, \tag{20}$$

where x_i are the elements of the universe of discourse X and $\mu_A(x_i)$ is the membership degree of x_i in A.
- **Mean of Support (MoS):** Selecting an extreme point, the MoS method takes a more balanced approach by computing the arithmetic mean of all elements belonging to the support of the aggregated fuzzy set A, that is, the average of all points where the membership function is strictly positive [5]:

$$\text{MoS}(A) = \frac{\sum_{x_i \in \mathcal{S}} x_i}{|\mathcal{S}|}, \tag{21}$$

where $\mathcal{S} = \{x_i \in X \mid \mu_A(x_i) > 0\}$ is the support of A and $|\mathcal{S}|$ denotes its cardinality.
- **Basic Defuzzification Distribution (BADD):** The BADD method computes the defuzzified value as a weighted mean of the elements of the universe of discourse X, where the weights are given by a power v of the membership function of the aggregated fuzzy set A [5]:

$$\text{BADD}(A) = \frac{\sum_{i=1}^{n} x_i [\mu_A(x_i)]^v}{\sum_{i=1}^{n} [\mu_A(x_i)]^v}, \tag{22}$$

where x_i are the elements of the universe of discourse X, $\mu_A(x_i)$ is the membership degree of x_i in A, and $v \geq 0$ is a parameter that controls the strength of the membership degrees on the defuzzified value. Note that the BADD method generalizes several other defuzzification methods as special cases: when $v = 0$, the MoS method is obtained; when $v = 1$, the standard CoG method is obtained; and finally, when $v \to \infty$, it converges to the MoM method.

3 The `FuzzyRules` package

The `FuzzyRules` package provides numerous membership functions, such as Triangular, Trapezoidal, Gaussian, Cauchy, Generalized Bell, Pi, Constant, and Singleton, are available in the `FuzzyRules` package. Each membership function implemented returns a list to the user containing four information (name, parameters, membership function and alphacuts) that can be accessed directly by the user. Plots for membership functions of a Fuzzy Set as well as the alpha-cuts for that fuzzy set were implemented. Additionally, it supports common operations on fuzzy sets, including the fuzzy set's intersection, union, complement, support, kernel, core, and height. Overlap functions and indices, as well as grouping functions and indices for two fuzzy sets are available. It also computes the Jaccard fuzzy index and the degree of membership of an element x (see Sect. 2).

The package includes the Zadeh, Probabilistic, Lukasiewicz, and Weber t-norms and dual t-conorms. The logical operators that are available are aggregation, implication (Mamdani), AND/OR (for multiple fuzzy sets), and NOT. Ten defuzzification methods have been implemented by the `FuzzyRules` package: CoG; CoS; CoLA; FoM; LoM; MoM; FoS; LoS; MoS and BADD. These methods were described in the Sect. 2.1.

Table 4 presents a comparison among `FuzzyRules` package and other R packages, as well as Matlab and Scikit-fuzzy (Python). `FuzzyRules` package allows direct access to functions assign with '*'. Some of these features are not available in other software. Although tools such as Matlab allow users to define these methods when creating or customizing a Fuzzy Inference System, direct access to these functionalities is provided only by `FuzzyRules`.

Table 4. Some differences among software which implement Fuzzy Logic.

Available functions	Fuzzy Rules	FuzzyR	Sets	Matlab	Scikit-fuzzy
Fuzzy Sets					
Union of two Fuzzy Sets	X	X	X	X	X
Intersection of two Fuzzy Sets	X	X	X	X	X
Complement of a Fuzzy Set	X		X		X
Support of a Fuzzy Set	X		X		X
Kernel of a Fuzzy Set	X				
Core of a Fuzzy Set	X		X		
Height of a Fuzzy Set	X				
Degree of membership for x	X	X	X		X
Jaccard fuzzy index of two sets	X	X	X		
Plot a fuzzy set	X	X	X	X	
Plot alpha-cuts of a Fuzzy Set*	X				
Overlap functions*	X				
Overlap indices*	X				
Grouping functions*	X				
Grouping indices*	X				
t-norms and dual t-conorms					
Zadeh	X	X	X	X	
Probabilistic	X	X	X		
Lukasiewicz	X		X		
Weber*	X				

(continued)

Table 4. (*continued*)

Available functions	Fuzzy Rules	FuzzyR	Sets	Matlab	Scikit-fuzzy
Membership functions					
Triangular	X	X	X	X	X
Trapezoidal	X	X	X	X	X
Rectangular*	X				
Constant*	X				
Sigmoidal	X		X	X	X
Singleton	X	X			
two Gaussians			X	X	X
Gaussian	X	X	X	X	X
Cauchy	X				
Generalized Bell	X	X	X	X	X
Pi-shaped	X		X	X	X
S-shaped	X			X	X
Z-shaped	X			X	X
Custom		X	X		
Type-1 Fuzzy Logic (Mamdani)					
AND	X	X	X		X
OR	X	X	X		X
NOT	X				X
Implication	X	X	X		
Aggregation	X	X	X		
Defuzzification					
Centroid (CoG)	X	X	X		X
Center of Sums (CoS)	X	X			
CoLA	X				
FoM/LoM/MoM	X	X	X		X
FoS/LoS/MoS*	X				
Basic Defuzz. Dist. (BADD)*	X				
Bisector		X			X

* Available only in `FuzzyRules` package for R, using a user-directed access function.

Beyond its technical capabilities, the package serves not merely as a computational tool, but as a pedagogical resource. It is useful for both those who already master the process of creating fuzzy rules and those encountering Fuzzy Rule-Based Systems for the first time.

By making the rule creation process clearer, it enables the user to interact at intermediate stages, from the definition of the universe of discourse, through the construction of membership functions, the application of implication and aggregation operators, to the defuzzification into crisp values. Consequently, the package promotes methodological transparency and helps users understand, in a clearer and more structured way, the theoretical principles underlying fuzzy inference.

4 Examples of Application: Tipping

This is an illustrative example of a tipping problem. The universe of discourse for the input variables (service and food quality) and the output variable (tip) is defined over the interval $[0, 25]$. Three fuzzy linguistic variables are considered: Service (Poor, Good, Excellent), Food (Rancid, Delicious), and Tip (Cheap, Average, Generous).

```
> library(FuzzyRules)
> ## Universe X - evaluation by the consumer in [0,25]
> minX = 0
> maxX = 25
> X <- Set.Universe(minX, maxX)
> FRBS = "Mamdani"
```

The linguistic variable *Service* is modeled by three Gaussian fuzzy sets centered at 0, 5, and 10, respectively, each with standard deviation $\sigma = 1.5$. The term *Excellent* is assigned a reduced maximum membership degree of 0.3 to reflect a more restrictive evaluation.

```
> ## Linguistic variable "Service" with three Gaussian-type terms
> service.poor <- GauFS(X, 0.0, 1.5)
> service.good <- GauFS(X, 5.0, 1.5)
> service.excellent <- GauFS(X, 10, 1.5, alphamax = 1.0)
```

The linguistic variable *Food* is modeled by two trapezoidal fuzzy sets. The term *Rancid* is defined over the interval $[0, 4]$, with a flat top between 0 and 2, reflecting full membership for very low quality scores. The term *Delicious* is defined over the interval $[7, 10]$, with a flat top between 9 and 10, capturing the highest quality evaluations.

```
> ## Linguistic variable "Food" with two trapezoidal terms
> food.rancid <- TraFS(X, 0, 0, 2, 4)
> food.delicious <- TraFS(X, 7, 9, 10, 10)
```

The information returned from the membership functions can be accessed either by using its position in the list or by the parameter name (for instance, using the sintax `food.rancid[[1]]` or `food.rancid[['name']]`).

The output linguistic variable *Tip* is modeled by three triangular fuzzy sets. The term *Cheap* is centered at 5, with support over $[0, 10]$. The term *Average* is centered at 12.5, with support over $[7.5, 17.5]$. The term *Generous* is centered at 20, with support over $[15, 25]$. The symmetric and evenly spaced distribution of these terms reflects a uniform partition of the output universe of discourse.

```
> ## Output linguistic variable "Tip" with three triangular
> ## fuzzy terms
> tip.cheap <- TriFS(X, 0, 5, 10)
> tip.average <- TriFS(X, 7.5, 12.5, 17.5)
> tip.generous <- TriFS(X, 15, 20, 25)
```

The union and intersection operations are computed under the default Zadeh method, Probabilistic method, Lukasiewicz method, or Weber method with distinct values of the parameter λ (see Tables 1 and 2). The complement operation can be applied to each term of the linguistic variable *Service*, yielding the fuzzy sets $\overline{Poor}$, $\overline{Good}$, and $\overline{Excellent}$, respectively.

The fuzzy inference process is illustrated by setting the input values to *service* = 3 and *food* = 8, representing a specific consumer evaluation. The first rule, *IF service is Poor OR food is Rancid THEN tip is Cheap*, is evaluated by computing the membership degrees of the input values in the respective antecedent fuzzy sets via the `dmFS` function, combining them through the disjunction operator `OR`, and applying the implication to obtain the output fuzzy set fr_1, which is then stored in the output list.

```
> ## Set the input: Evaluation by the consumer (Facts)
> service = 3
> food = 8
> ## Rules -- Specify rules and add outputs in a list
> ## A list is created in order to store the output of each rule
> ## Out_Rules_Lst <- list()
> ## 1) IF service is poor OR food is rancid THEN tip is cheap
> ## antecedent of the rule
> temp <- OR(FRBS, dmFS(service,X,service.poor),
> dmFS(food,X,food.rancid))
> ## consequent of the rule
> fr1 <- Implication(FRBS, temp, X, tip.cheap)
> ## Create a list with the output
> Out_Rules_Lst <- listfuzzy(fr1)
```

The second rule, *IF service is Good THEN tip is Average*, is evaluated by computing the membership degree of the input value in the antecedent fuzzy set *Good*, and applying the implication to obtain the output fuzzy set fr_2, which is then appended to the output list.

```
> ## 2) IF service is good THEN tip is average
> temp <- dmFS(service,X,service.good)
> fr2 <- Implication(FRBS, temp, X, tip.average)
> Out_Rules_Lst <- listfuzzy(Out_Rules_Lst,fr2)
```

The third rule, *IF service is Excellent OR food is Delicious THEN tip is Generous*, is evaluated by computing the membership degrees of the input values in the antecedent fuzzy sets *Excellent* and *Delicious*.

```
> ## 3) IF service is excellent OR food is delicious THEN
> tip is generous
> temp <- OR(FRBS, dmFS(service,X,service.excellent),
> dmFS(food,X,food.delicious))
> fr3 <- Implication(FRBS, temp, X, tip.generous)
> ## Compute the aggregation of all previous Rules
> Out_Rules_Lst <- listfuzzy(Out_Rules_Lst,fr3)
```

Figure 1 presents a comprehensive visualization of the inference process, arranged in a 4×3 grid. Each row corresponds to one fuzzy rule, displaying the antecedent fuzzy sets with the respective input values marked by vertical lines, and the resulting implication output. The final panel shows the aggregated output fuzzy set OutFS, obtained by combining the outputs of all three rules.

```
> # Plot all the rules and Aggregation
> par(mfrow=c(4,3))
> plotFS(X,service.poor, col = "blue", main = "service.poor")
> abline(v=service)
> plotFS(X,food.rancid, col = "blue", main = "food.rancid")
> abline(v=food)
> plotFS(X,fr1, col = "blue", main = "Implication")
> plotFS(X,service.good, col = "red", main = "service.good")
> abline(v=service)
> plot(x=0,y=0, axes=F, xlab="", ylab="", type="n")
> plotFS(X,fr2, col = "red", main = "Implication")
> plotFS(X,service.excellent, col = "green",
> main = "service.excellent")
> abline(v=service)
> plotFS(X,food.delicious, col = "green",
> main = "food.delicious")
> abline(v=food)
> plotFS(X,fr3, col = "green", main = "Implication")
> plot(x=0,y=0, axes=F, xlab="", ylab="", type="n")
> plot(x=0,y=0, axes=F, xlab="", ylab="", type="n")
> plotFS(X,OutFS, main = "Aggregation", col="black")
```

The aggregated output fuzzy set OutFS is then submitted to the defuzzification step, where the ten methods available in the FuzzyRules package are applied in order to obtain a final crisp output value.

```
> ## Compute Defuzzification in order provide final decision
> FinalDecision1 <- defuzzFS(X,OutFS,Out_Rules_Lst,method="CoG")
> cat("Tip result using CoG is approx. ", FinalDecision1, "")
> FinalDecision2 <- defuzzFS(X,OutFS,Out_Rules_Lst,method="CoS")
> cat("Tip result using CoS is approx. ", FinalDecision2, "")
> FinalDecision3 <- defuzzFS(X,OutFS,Out_Rules_Lst,method="CoLA")
```

```
> cat("Tip result using CoLA is approx. ", FinalDecision3, "")
> FinalDecision4 <- defuzzFS(X,OutFS,Out_Rules_Lst,method="FoM")
> cat("Tip result using FoM is approx. ", FinalDecision4, "")
> FinalDecision5 <- defuzzFS(X,OutFS,Out_Rules_Lst,method="LoM")
> cat("Tip result using LoM is approx. ", FinalDecision5, "")
> FinalDecision6 <- defuzzFS(X,OutFS,Out_Rules_Lst,method="MoM")
```

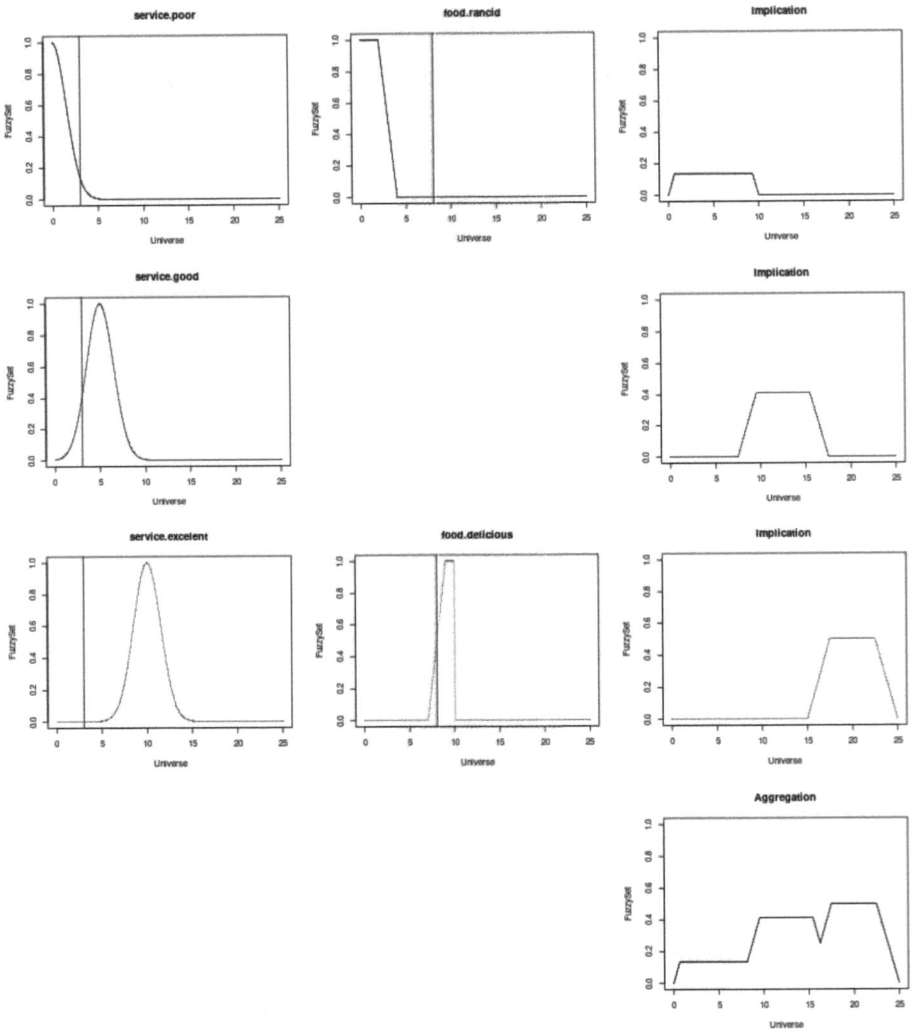

Fig. 1. Visualization of the fuzzy inference process. Each row of the figure corresponds to the fuzzy rules, displaying the fuzzy sets with their respective input values. The aggregated output of the three rules is presented on the last line of figure.

```
> cat("Tip result using MoM is approx. ", FinalDecision6, "")
> FinalDecision7 <- defuzzFS(X,OutFS,Out_Rules_Lst,method="FoS")
> cat("Tip result using FoS is approx. ", FinalDecision7, "\n")
> FinalDecision8 <- defuzzFS(X,OutFS,Out_Rules_Lst,method="LoS")
> cat("Tip result using LoS is approx. ", FinalDecision8, "\n")
> FinalDecision9 <- defuzzFS(X,OutFS,Out_Rules_Lst,method="MoS")
> cat("Tip result using MoS is approx. ", FinalDecision9, "\n")
> FinalDecision10 <- defuzzFS(X,OutFS,Out_Rules_Lst,method="BADD",
> v = 2)
> cat("Tip result using BADD is approx. ", FinalDecision10, "\n")
```

```
Tip result using CoG is approx.   14.8859
Tip result using CoS is approx.  14.75458
Tip result using CoLA is approx.  20
Tip result using FoM is approx.   17.55
Tip result using LoM is approx.   22.45
Tip result using MoM is approx.   20
Tip result using FoS is approx.   0
Tip result using LoS is approx.   25
Tip result using MoS is approx.   12.5
Tip result using BADD is approx.   16.24806
```

5 Comparison with Other Software

Benchmark tests were conducted to evaluate how quickly FuzzyRules is able to run a Fuzzy Inference System, w.r.t. other free software. The "bench" package was used to perform tests on R packages (FuzzyRules, "FuzzyR" and "sets") and the "time" package was used for testing the Python software package "scikit-fuzzy". The computer used had the following configuration: an Intel i5 9400f (9th generation) processor at 2.9GHz, 16GB of RAM, and the Ubuntu Linux 24.04 LTS operating system. Currently, this can be considered a typical system, given that Intel processors are now in their 14th generation.

The Tipping example (presented in the Sect. 4) was the task executed by each package mentioned above. Table 5 presents the execution times related to 10 runnings of this task on the computer cited before. The benchmarks indicate that the FuzzyRules package has a competitive execution time compared to other established solutions.

Table 5. Execution times (s) for the Tipping task using free fuzzy logic packages.

Package	Best performance	Worst performance	Average performance	Standard Diviation
`FuzzyRules`	0.014396	0.017210	0.015326	0.001186
FuzzyR	0.003230	0.003512	0.003300	0.000086
sets	0.021351	0.023899	0.022052	0.000970
Scikit-fuzzy	0.000353	0.000616	000416	0.000076

6 Conclusions and Future Works

This paper presented the free and open source package for R software, named `FuzzyRules`. It implements functions for manipulating fuzzy sets and building fuzzy rule-based systems. Application areas include artificial intelligence, data classification, pattern recognition, and data science.

Although freeware, shareware, and commercial software implementing the methods available in `FuzzyRules` exist, these methods are not typically found together within a single software. A particular feature of the package is the implementation of functions for Overlap and Grouping functions and indices, plot for alpha-cuts and ten different defuzzification methods. Currently, these methods are not jointly available in other software. Furthermore, `FuzzyRules` demonstrates competitive performance even when executed on standard computing systems with limited processing power.

New improvements in progress (to be available shortly) include: 1) a Takagi-Sugeno-Kang Type-1 fuzzy rule-based system; 2) a Mamdani type-2 interval fuzzy sets; and 3) functions and indices for type-2 interval fuzzy sets.

Acknowledgments. This work was supported by the National Council for Scientific and Technological Development (CNPq) grants 301644/2025-0, 315298/2018-9 and 465586/2014-4. The authors would like to thank the reviewers for their valuable comments, which helped improve this paper.

Disclosure of Interests. The authors have no competing interests to declare that are relevant to the content of this article.

References

1. Bedregal, B., Dimuro, G.P., Bustince, H., Barrenechea, E.: New results on overlap and grouping functions. Inf. Sci. **249**, 148–170 (2013)
2. Chen, C., Razak, T.R. and Garibaldi, J.M.: FuzzyR: an extended fuzzy logic toolbox for the r programming language. In: 2020 IEEE International Conference on Fuzzy Systems (FUZZ-IEEE), Glasgow, pp. 1–8 (2020)
3. Garcia-Jimenez, S., Bustince, H., Hüllermeier, E., Mesiar, R., Pal, N.R.: Overlap indices: construction of and application to interpolative fuzzy systems. IEEE Trans. on Fuzzy Systems **234**, 1259–1273 (2015)

4. Jain, D., Sharma, S.K., Dhiman, P.: Comparative analysis of defuzzification techniques for fuzzy output. J. Algebraic Stat. **13**(2), 874–882 (2022)
5. Liu, X.: Parameterized defuzzification with maximum entropy weighting function–Another view of the weighting function expectation method. Math. Comput. Model. **45**(1–2), 177–188 (2007)
6. Mamdani, E.H., Assilian, S.: A case study on the application of fuzzy set theory to automatic control. In: Proceedings of the IFAC Stochastic Control Symposium Budapest (1974)
7. Mamdani, E.H., Assilian, S.: An experiment in linguistic synthesis with a fuzzy logic controller. Int. J. Man Mach. Stud. **7**, 1–13 (1975)
8. FuzzyRules Package. https://github.com/leapigufpb/FuzzyRules. Accessed 26 Apr 2026
9. Fuzzy Logic Toolbox for Matlab. https://www.mathworks.com/products/fuzzy-logic.html. Accessed 27 Feb 2026
10. Fuzzy Logic Toolkit for Octave. https://octave.sourceforge.io/fuzzy-logic-toolkit/index.html. Accessed 27 Feb 2026
11. PyPI.org: fuzzylogic. https://pypi.org/project/fuzzylogic/. Accessed 27 Feb 2026
12. PyPI.org: scikit-fuzzy. https://pypi.org/project/scikit-fuzzy/. Accessed 27 Feb 2026
13. Riza, L.S., Bergmeir, C., Herrera, F., Benítez, J.M.: Fuzzy rule-based systems for classification and regression in R. J. Stat. Softw. **65**(6), 1–30 (2015)
14. Roldán López de Hierro, A.F., et al.: From type-(2,k) grouping indices to type-(2,k) Jaccard indices. Fuzzy Sets and Syst. **501**, 109216 (2025)
15. Package sets. https://cran.r-project.org/web/packages/sets/index.html. Accessed 27 Feb 2026
16. Wagner, C.: Juzzy - a Java based toolkit for type-2 fuzzy logic. In: IEEE Symposium Series on Computational Intelligence, Singapore, pp. 45–52 (2013)
17. Zadeh, L.A.: Fuzzy sets. Inf. Control **8**, 338–353 (1965)
18. Zadeh, L.A.: Fuzzy algorithms. Inf. Control **12**, 94–102 (1968)
19. Zadeh, L.A.: A rationale for fuzzy control. J. Dyn. Syst. Measur. Control **94**(G), 3–4 (1972)
20. Zadeh, L.A.: Outline a New Approach to the Analysis of Complex Systems and Decision Process. IEEE Trans. Systems, Man and Cybernetics, **3**, 28–44 (1973)
21. Zadeh, L.A.: The concept of a linguistic variable and its application to approximate reasoning–I. Inf. Sci. **8**(3), 199–249 (1975)
22. Zadeh, L.A.: Fuzzy logic. IEEE Comput. **21**, 83–93 (1988)

A New Fuzzy Weibull-Based Naive Bayes Network for User Assessment in Virtual Reality Simulations

Ronei M. Moraes[1](✉), Jodavid A. Ferreira[2], Isaac S. L. Araújo[3], and Liliane S. Machado[3]

[1] Department of Statistics, Federal University of Paraíba, João Pessoa, PB, Brazil
ronei@de.ufpb.br

[2] Department of Statistics, Federal University of Pernambuco, Recife, PE, Brazil
jodavid.ferreira@ufpe.br

[3] Department of Informatics, Federal University of Paraíba, João Pessoa, PB, Brazil
isaac.sebastian@academico.ufpb.br, liliane@di.ufpb.br

Abstract. User performance assessment is a fundamental component of virtual reality (VR) training systems, particularly in health-related simulators where objective and real-time skill evaluation is required. Single User Assessment Systems (SUAS) rely on probabilistic inference models to estimate user proficiency from interaction data, with temporal variables playing a central role in the assessment process. Since time-based metrics in VR environments often exhibit asymmetric and heavy-tailed behavior, selecting an appropriate statistical distribution is critical for reliable classification. This paper proposes a Fuzzy Weibull Naive Bayes Network (FWNB) as the core inference model of a SUAS, combining the flexibility of the Weibull distribution for modeling continuous time-to-event variables with fuzzy membership functions to handle uncertainty and variability in performance data. The model was evaluated through one thousand Monte Carlo simulation replicas, considering three performance classes, and compared with traditional Naive Bayes (NB) and Weibull Naive Bayes (WNB) approaches. The proposed SUAS achieved 91.11% accuracy and a Kappa coefficient of 86.19%, outperforming the alternative models and demonstrating greater robustness and discriminative capability. Results indicate that incorporating the Weibull distribution within a fuzzy Bayesian framework improves probabilistic estimation and reduces information loss associated with discretization, for cases similar to this study. Although motivated by SUAS applications in VR simulators, the proposed framework is generalizable to broader classification and machine learning scenarios involving asymmetric continuous and time-to-event data.

Keywords: Weibull distribution · fuzzy Naive Bayes · probabilistic modeling · time-to-event data · classification

J. Lu et al. (Eds.): FLINS-ISKE 2026, LNCS 16752, pp. 232–246, 2027.
https://doi.org/10.1007/978-981-92-2369-5_16

1 Introduction

User performance assessment is a fundamental component of educational and training systems, providing objective indicators of a user's ability to perform specific tasks. Beyond measuring proficiency, performance assessment supports adaptive difficulty adjustment and promotes objective performance awareness [2]. In virtual reality (VR) training systems, online assessment (less than 1 s response) is particularly critical, as it enables rapid and data-driven estimation of user skill based on interaction data generated during simulation. This paradigm has been adopted in VR-based surgical training and objective psychomotor skill assessment [25].

For individual tasks, these assessment systems are referred to as Single User Assessment Systems (SUAS) and must operate efficiently to provide real-time results. The concept of SUAS was formally introduced by Moraes and Machado in early work on intelligent assessment for VR-based training, where the term described real-time automated assessment models embedded within VR simulators [21]. In simplified terms, a SUAS is trained using expert data, embedded into the VR simulator, and uses data from the users̋ simulation to provide a performance outcome (Fig. 1).

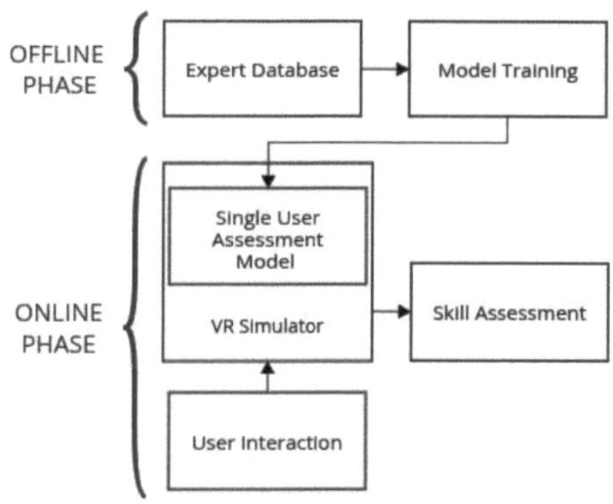

Fig. 1. Overview of a SUAS embedded within a virtual reality simulator.

Several SUAS have already been proposed, indicating their potential and assessment capability across different VR training contexts and different statistical data distributions [6,17,19,31]. The core of a SUAS often relies on a probabilistic inference model, which highlights the importance of selecting a statistical distribution compatible with the underlying performance data. Since probabilistic classifiers estimate likelihoods based on assumed parametric forms, any mismatch between the empirical data distribution and the assumed model may lead to biased probability estimates, reduced discriminative power, and degraded classification accuracy. Consequently, the adequacy of the statistical distribution directly impacts the reliability, robustness, and real-time performance of the SUAS.

Examples of VR simulators that integrate a SUAS include BMHSim and SITEG. BMHSim was proposed to provide a training environment for pediatric bone marrow harvesting [15]. In this procedure, the surgeon must insert a needle into the patient's iliac crest, reach the correct internal region of the bone, and perform bone marrow aspiration. The procedure involves repeated executions of these steps, and its success depends on the perception of tactile properties along the needle trajectory, requiring a high level of manual dexterity. Within a SUAS framework, variables such as trajectory, depth, and force patterns are analyzed together with temporal metrics, including task completion times and time intervals between events, which respectively represent the total time required to complete the procedure and the intervals between relevant task milestones. These variables compose the evidence set used for automatic inference of the user's skill level.

SITEG is another example of VR simulator that integrates a SUAS. Designed for a gynecological examination training system in which the physician must perceive visual and tactile characteristics of the vulvar and intravaginal regions, as well as collect cervical samples using spatulas and swabs [16]. Similarly, the SUAS associated with SITEG employs quantitative metrics related to instrument manipulation, task completion times, and intervals between critical events (e.g., insertion, sampling, and removal), enabling the modeling of not only accuracy but also execution fluency [31]. In both cases, the tasks clearly illustrate the need to complete the procedure within an appropriate time frame, whether due to anesthesia duration or the requirement to ensure patient safety and comfort. Therefore, time is a relevant variable within the SUAS statistical model, directly influencing the probabilistic inference of user's skill level.

Given the central role of temporal variables in SUAS modeling, selecting an appropriate statistical distribution for time-based data becomes a critical design decision. The Weibull distribution represents a versatile two-parameter family widely adopted for modeling lifetime, reliability, and skewed continuous data due to its flexibility in modeling increasing, decreasing, or constant hazard rates, depending on the value of the shape parameter. It is particularly valued for its ability to characterize increasing, constant, or decreasing hazard rates—governed by its shape parameter—making it a cornerstone of survival analysis [10,26]. Beyond its classical foundations, the mathematical properties and numerous generalizations of the Weibull family have been extensively reviewed, highlighting its utility in fitting complex and non-monotonic datasets [1,9,11]. In applied statistics, the Weibull model is employed within accelerated failure-time (AFT) frameworks, offering interpretable regression formulations for various time-to-event outcomes [13]. Recent interdisciplinary research has further demonstrated that Weibull parameters can succinctly capture structural image statistics that correlate with human visual processing, with significant implications for rendering and perceptual modeling in virtual environments [30]. Furthermore, time-based and kinematic data extracted from VR interactions—such as response latencies and locomotion intervals—are highly amenable to Weibull-type modeling. Utilizing either the standard form or extended mixture formulations can significantly improve goodness-of-fit and uncertainty quantification in predictive tasks and classification workflows for VR-based datasets [4,29]. Consequently, the Weibull family serves as a robust analytical framework for modeling the asymmetric and heavy-tailed measurements common in modern VR and human-computer interaction studies.

The present work contributes by proposing a methodology based on a Fuzzy Bayesian Network that leverages the Weibull distribution for data modeling. This approach exploits the inherent flexibility of the Weibull family to characterize continuous performance metrics—such as task completion times, time intervals between two events and reliability—which are crucial for precise skill quantification in virtual environments. The following sections describe the theoretical basis for this new SUAS, the results obtained through its application, and a comparative analysis with alternative SUAS based on different Naive Bayes architectures.

2 Statistical Modeling

Let X be a random variable following a Weibull distribution with shape parameter $\alpha > 0$ and scale parameter $\lambda > 0$, denoted by $X \sim \text{Weibull}(\alpha, \lambda)$. Its probability density function is given by [14]

$$f(x \mid \alpha, \lambda) = \frac{\alpha x^{\alpha-1}}{\lambda^{\alpha}} \exp\left[-\left(\frac{x}{\lambda}\right)^{\alpha}\right], \quad x > 0. \tag{1}$$

The expected value $\mathbb{E}$ and variance Var of X are given by:

$$\mathbb{E}[X] = \lambda \Gamma\left(1 + \frac{1}{\alpha}\right), \tag{2}$$

and

$$\text{Var}(X) = \lambda^2 \Gamma\left(1 + \frac{2}{\alpha}\right) - \left[\lambda \Gamma\left(1 + \frac{1}{\alpha}\right)\right]^2, \tag{3}$$

where $\Gamma(\cdot)$ denotes the Gamma function [8].

Let $X_1, X_2, \ldots, X_n$ be n independent random samples from a Weibull distribution with shape parameter $\alpha > 0$ and scale parameter $\lambda > 0$. In this paper, the parameters are estimated using the method of moments.

Let $\bar{X}$ denote the sample mean and s_m^2 the method of moments estimator of the variance, defined as

$$\bar{X} = \frac{1}{n} \sum_{i=1}^{n} X_i, \tag{4}$$

and

$$S_m^2 = \frac{1}{n} \sum_{i=1}^{n} (X_i - \bar{X})^2. \tag{5}$$

The method of moments estimator of the shape parameter, denoted by $\hat{\alpha}_{\text{mme}}$, is obtained by solving the nonlinear equation

$$\frac{S_m}{\bar{X}} = \left\{ \frac{\Gamma\left(\dfrac{\hat{\alpha}_{\text{mme}}+2}{\hat{\alpha}_{\text{mme}}}\right)}{\left[\Gamma\left(\dfrac{\hat{\alpha}_{\text{mme}}+1}{\hat{\alpha}_{\text{mme}}}\right)\right]^2} - 1 \right\}^{1/2}, \tag{6}$$

where $\Gamma(\cdot)$ denotes the Gamma function.

Once $\hat{\alpha}_{\text{mme}}$ is obtained, the method of moments estimator of the scale parameter is given by

$$\hat{\lambda}_{\text{mme}} = \frac{\bar{X}}{\Gamma\left(\dfrac{\hat{\alpha}_{\text{mme}}+1}{\hat{\alpha}_{\text{mme}}}\right)}. \tag{7}$$

The method of moments estimators based on the unbiased estimator of variance follow the same procedure described above, except that the estimator of the variance in (5) is replaced by the unbiased sample variance, defined as

$$S^2 = \frac{1}{n-1}\sum_{i=1}^{n}(X_i - \bar{X})^2. \tag{8}$$

Accordingly, the estimator $\hat{\alpha}_{\text{mmue}}$ is obtained by replacing s_m with s in Eq. (6), while the estimator of the scale parameter $\hat{\beta}_{\text{mmue}}$ is computed using Eq. (7). In both cases, the resulting equation for the shape parameter does not admit a closed-form solution and must be solved numerically.

3 The Naive Bayes Network

The Naive Bayes network is a probabilistic learning algorithm grounded on Bayes' theorem and widely used in machine learning applications [24]. It is referred to as "naive" because it relies on the assumption that the predictor variables are statistically independent of each other [28].

Consider a dataset composed of n observations and p predictor variables (or attributes), represented by the data matrix $\mathbf{X} = [x_{ij}]_{n\times p}$, where x_{ij} denotes the value of the j-th variable for the i-th observation. For a single observation $\mathbf{x}_i = (x_{i1}, x_{i2}, \ldots, x_{ip})$, the goal is to determine its class membership among a finite set of possible classes $C = \{c_1, c_2, \ldots, c_K\}$.

The objective of the Naive Bayes network is to identify the most probable class c_k associated with the observation $\mathbf{x}_i$, that is,

$$\hat{c} = \arg\max_{c_k \in C} P(c_k \mid \mathbf{x}_i). \tag{9}$$

According to Bayes' theorem, the posterior probability can be expressed as

$$P(c_k \mid \mathbf{x}_i) = \frac{P(\mathbf{x}_i \mid c_k)\, P(c_k)}{P(\mathbf{x}_i)}. \tag{10}$$

Since $P(\mathbf{x}_i)$ is constant across all classes, it can be omitted from the classification rule. Under the conditional independence assumption inherent to the Naive Bayes network, the likelihood function $P(\mathbf{x}_i \mid c_k)$ can be factorized as the product of individual conditional probabilities:

$$P(\mathbf{x}_i \mid c_k) = \prod_{j=1}^{p} P(x_{ij} \mid c_k). \tag{11}$$

Consequently, the final decision rule of the Naive Bayes network classifier is given by

$$\hat{c}_i = \arg\max_{c_k \in C} \left[P(c_k) \prod_{j=1}^{p} P(x_{ij} \mid c_k) \right]. \tag{12}$$

where $P(x_{ij} \mid c_k)$ denotes the conditional probability of the value of the j-th variable for the i-th observation, given that it belongs to class c_k.

The Naive Bayes network assumes multinomial distribution, but it can be used for continuous variables, after discretizing the information. However, this procedure can lead to loss of information [20].

3.1 The Weibull Naive Bayes Network

In this study, we assume that, conditional on the class label c_k, each continuous predictor follows a Weibull distribution, whose probability density function was defined in Eq. (1).

Under this assumption, the conditional probability $P(x_{ij} \mid c_k)$ is replaced by the corresponding Weibull probability density function, parameterized by a shape parameter α_{jc_k} and a scale parameter λ_{jc_k}, both depending on the predictor variable j and the class c_k. Thus, the Eq. (11) can be written as

$$P(\mathbf{x}_i \mid c_k) = \prod_{j=1}^{p} f(x_{ij} \mid \alpha_{jc_k}, \lambda_{jc_k}), \tag{13}$$

where $f(\cdot)$ denotes the Weibull probability density function, as presented in the Eq. (1). After mathematical manipulations, using Eq. (1) on the Eq. (10) and (13) a discriminating function g is given by:

$$\begin{aligned} g(c_k|\mathbf{x}_i) &= \log P(c_k|\mathbf{x}_i) \\ &= \log P(c_k) + \sum_{j=1}^{p} \log \alpha_{jc_k} + \sum_{j=1}^{p} (\alpha_{jc_k} - 1) \log x_{ij} - \\ &- \sum_{j=1}^{p} \alpha_{jc_k} \log \lambda_{jc_k} - \sum_{j=1}^{p} \left(\frac{x_{ij}}{\lambda_{jc_k}} \right)^{\alpha_{jc_k}} \end{aligned} \tag{14}$$

The classification rule of the Naive Bayes classifier reduces to maximizing the product of the class-conditional Weibull densities. Therefore, the decision rule is given by

$$\hat{c}_i = \arg\max_{c_k \in C} \left[g(c_k|\mathbf{x}_i) \right]. \tag{15}$$

The parameters α_{jc_k} and λ_{jc_k} are estimated from the training data separately for each predictor variable and each class c_k, using Eq. (6) and Eq. (7).

3.2 The New Fuzzy Weibull Naive Bayes Network

When the predictor variables are continuous and subject to uncertainty or vagueness, the classical Naive Bayes framework can be extended using fuzzy probability theory [3]. In this study, we adopt Zadeh's definition of the probability of fuzzy events [32] to construct a Fuzzy Naive Bayes classifier with Weibull-distributed predictors.

Let $(\mathbb{R}^p, \mathcal{B}, P)$ be a probability space, where $\mathcal{B}$ denotes the σ-algebra of Borel subsets of $\mathbb{R}^p$ and P is a probability measure. Let F_{jc_k} be a fuzzy event associated with the j-th predictor of the c_k-th class, characterized by a membership function

$$\mu_{jc_k} : \mathbb{R} \to [0, 1]. \tag{16}$$

Consider an observation $\mathbf{x}_i = (x_{i1}, x_{i2}, \ldots, x_{ip})$ and a set of possible classes $C = \{c_1, c_2, \ldots, c_K\}$. Under the Naive Bayes conditional independence assumption, and assuming equal prior probabilities for all classes, the fuzzy posterior probability of class c_k given $\mathbf{x}_i$ can be expressed as

$$\mathcal{P}(c_k \mid \mathbf{x}_i) \propto \prod_{j=1}^{p} \mathcal{P}(\mathbf{x}_{ij} \mid c_k). \tag{17}$$

In this work, we assume that, conditional on class c_k, each continuous predictor follows a Weibull distribution. Consequently, the fuzzy class-conditional probability is defined as

$$\mathcal{P}(x_{ij} \mid c_k) = \mu_{jc_k}(x_{ij})\, f(x_{ij} \mid \alpha_{jc_k}, \beta_{jc_k}) \tag{18}$$

where $f(\cdot \mid \alpha_{jc_k}, \beta_{jc_k})$ denotes the Weibull probability density function with shape parameter α_{jc_k} and scale parameter λ_{jc_k}, both specific to predictor j and class c_k (see Eq. (1)). After mathematical manipulations, using Eq. (18) on the Eqs. (10) and (13), a discriminating function g_f emerges:

$$\begin{aligned} g_f(c_k|\mathbf{x}_i) &= \log P(c_k|\mathbf{x}_i) \\ &= \log P(c_k) + \sum_{j=1}^{p} \mu_{jc_k}(x_{ij}) + \sum_{j=1}^{p} \log \alpha_{jc_k} + \sum_{j=1}^{p} (\alpha_{jc_k} - 1) \log x_{ij} - \\ &\quad - \sum_{j=1}^{p} \alpha_{jc_k} \log \lambda_{jc_k} - \sum_{j=1}^{p} \left(\frac{x_{ij}}{\lambda_{jc_k}} \right)^{\alpha_{jc_k}} \end{aligned} \tag{19}$$

Thus, the decision rule of the Fuzzy Weibull Naive Bayes classifier is given by

$$\hat{c}_i = \arg\max_{c_k \in C} \left[g_f(c_k|\mathbf{x}_i) \right].$$

The Weibull parameters α_{jc_k} and λ_{jc_k} are estimated by Eq. (6) and Eq. (7) too, while the membership functions $\mu_{ij}(\cdot)$ encode the degree of compatibility between observed values and the corresponding class of performance $c_k, k = 1, ..., K$. Those membership functions are estimated from normalized histogram of the data [7,18]. Figure 2 presents an illustrative example of a normalized histogram with is used to estimate a membership function, according to Eq. (16).

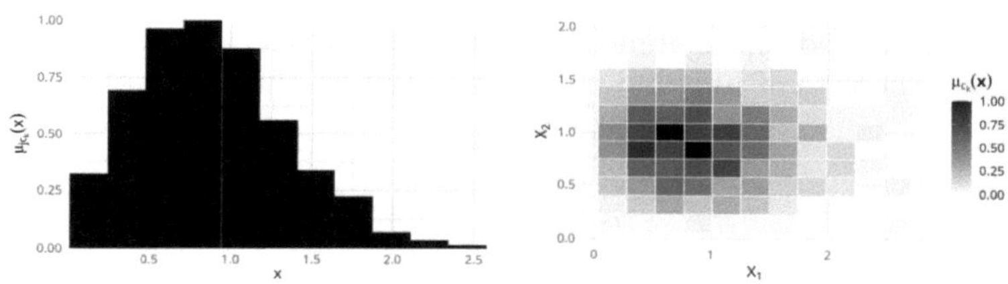

(a) 1-D membership function (b) 2-D membership function

Fig. 2. Example of membership fucntions for 1- and 2-Dimensionals. (a) used $X \sim$ Weibull$(\alpha = 2, \lambda = 1)$ and (b) used $X_1 \sim$ Weibull$(\alpha = 2, \lambda = 1)$ and $X_2 \sim$ Weibull$(\alpha = 3, \lambda = 1)$.

4 Criteria for Comparison

To evaluate and compare the methods that form the foundation of SUAS, several metrics derived from confusion matrices were employed, including the Accuracy index, Kappa Coefficient, Tau Coefficient, and their respective variances.

A confusion matrix of size $M \times M$ illustrates the alignment and discrepancies between two sources of information, such as two human experts or a computational model compared against a human expert. When evaluating model performance, the main diagonal of the matrix (m_{ii}, where $i = 1, ..., M$) represents agreements, where the model's prediction matches the expert's ground truth. Conversely, the off-diagonal elements (m_{ij}, where $i, j = 1, ..., M$ and $i \neq j$) represent disagreements or classification errors.

4.1 Accuracy Index

The Accuracy (A) index [19] is a common measure of agreement. Despite its widespread use, it is often considered overly optimistic by scholars as it does not account for the specific distribution of errors within the assessment. This index is defined as:

$$A = \frac{\sum_{i=1}^{M} m_{ii}}{N}, \quad (20)$$

where m_{ii} denotes the elements along the main diagonal, M is the number of classes, and N represents the total number of decisions. The variance of the Accuracy index is given by:

$$\sigma_A^2 = \frac{A(1-A)}{N}. \quad (21)$$

4.2 Kappa Coefficient

Cohen [5] introduced the Kappa Coefficient (κ) to measure agreement while adjusting for the rate of hits and errors expected by chance. It is calculated as follows:

$$\kappa = \frac{P_0 - P_c}{1 - P_c}, \quad (22)$$

where P_0 and P_c are defined as:

$$P_0 = \frac{\sum_{i=1}^{M} m_{ii}}{N} \quad \text{and} \quad P_c = \frac{\sum_{i=1}^{M} m_{i+} m_{+i}}{N^2}, \quad (23)$$

here, m_{i+} is the sum of row i and m_{+i} is the sum of column i. The variance of the Kappa Coefficient (σ_κ^2) is described by [23] as:

$$\begin{aligned} \sigma_\kappa^2 = & \frac{P_0(1-P_0)}{N(1-P_c)^2} + \frac{2(1-P_0)(2P_0P_c - \theta_1)}{N(1-P_c)^3} \\ & + \frac{(1-P_0)^2(\theta_2 - 4P_c^2)}{N(1-P_c)^4}, \end{aligned} \quad (24)$$

where the intermediate terms θ_1 and θ_2 are:

$$\theta_1 = \frac{\sum_{i=1}^{M} m_{ii}(m_{i+} + m_{+i})}{N^2} \quad \text{and} \quad \theta_2 = \frac{\sum_{i=1}^{M} m_{ii}(m_{i+} + m_{+i})^2}{N^3}. \quad (25)$$

According to the benchmarks by Landis and Koch [12], higher percentage values indicate a stronger degree of agreement, as summarized in Table 1.

4.3 Tau Coefficient

The Tau Coefficient (T) is an alternative metric that, like Kappa, accounts for the error rates inherent in the assessment method. It is defined as [22]:

$$T = \frac{P_0 - 1/M}{1 - 1/M}, \quad (26)$$

Table 1. Interpretation of the Kappa Coefficient agreement levels.

Kappa Coefficient (%)	Degree of Agreement
< 0.00	Poor
0.00–20.00	Slight
21.00–40.00	Fair
41.00–60.00	Moderate
61.00–80.00	Substantial
81.00–100.00	Almost Perfect

where P_0 is the observed agreement. The variance of the Tau Coefficient (σ_T^2) is expressed as:

$$\sigma_T^2 = \frac{P_0(1 - P_0)}{N(1 - 1/M)^2}. \tag{27}$$

5 Results

Monte Carlo simulation is a widely used due to its ability to provide a good understanding of the real world problems. It is used to analyze complex systems, providing different scenarios, including rare but important ones that must be considered in real-world analysis. The method uses statistical distributions in order to estimate the probability of different results [27]. Therefore, the Monte Carlo method was used to evaluate the new Fuzzy Naive Bayes Weibull Network proposed, as the base of a SUAS. It was performed by 1000 replicas in a Monte Carlo simulation in order to obtain the best results. In each replica, two samples were generated with 300 observations each and 100 observations per class. There are 3 Classes of performance, representing three real possible assessments for a procedure: Class 1: "the procedure was performed well", Class 2: "the user needs more training" or Class 3: "the user needs much more training".

The same samples were used to compare the SUAS based on the proposed method with two other SUAS based on Naive Bayes and on Weibull Naive Bayes networks. Figure 3 presents the attribute 2 non-normalizated histograms for all three performance Classes. It is worth noting the large overlap among the Classes. This was done intentionally to simulate the worst-case scenario. The histograms were normalized in the next step.

In the Table 2 are presented the results of assessments. The best result obtained using the SUAS based on Fuzzy Weibull Naive Bayes Network (SUAS-FWNB), according to Kappa Coefficient, with $\kappa = 86.19\%$ with variance 2.2×10^{-3}. Furthermore, the Accuracy resulted in 91.11% with variance 3.1×10^{-3} and that SUAS made mistakes, on average, in 8 cases. The Tau Coefficient for SUAS-FWNB achieved 86.67% with variance 2.0×10^{-2}.

The same procedure was used in order to evaluate the SUAS based on Naive Bayes Network (SUAS-NB) and the other based on Weibull Naive Bayes Network

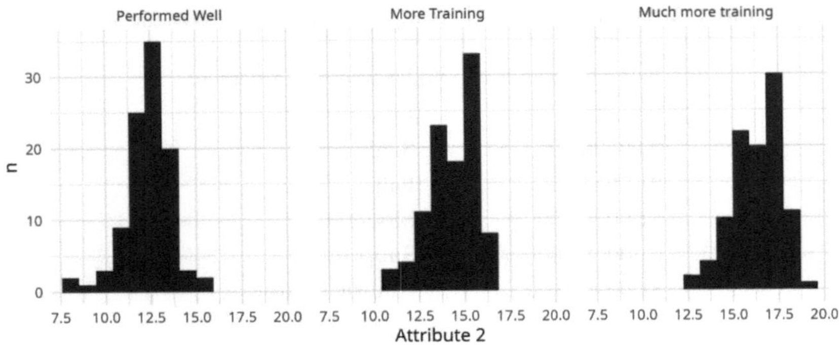

Fig. 3. Histogram for Classes 1 (the procedure was performed well), 2 (the user needs more training) and 3 (the user needs much more training), respectively, according to attribute 2, where n is the frequency.

(SUAS-WNB), i.e., the same samples of training were used to obtain the parameters for all of them, and the same testing samples were used for a controlled and impartial comparison among them.

The SUAS-NB was able to achieve the Accuracy 84.44%, with variance 1.5×10^{-2} and providing, on average, 14 mistakes. Its Kappa Coefficient was 76.32% with variance 3.3×10^{-3} and its Tau Coefficient resulted 76.76% with variance 3.3×10^{-2}. The SUAS-WNB produced Kappa Coefficient $\kappa = 79.59\%$ with variance 3.1×10^{-3}. This SUAS incorrectly assessed the user, on average, 12 times and obtained the Accuracy 86.67% with variance 1.3×10^{-2}. The Tau Coefficient for SUAS-WNB achieved 80.00% with variance 2.9×10^{-2}.

Table 2. Simulation results by Monte Carlo, where A is accuracy, κ is kappa coefficient, σ^2_κ is kappa coefficient variance and Misclassif. refers to errors made by SUAS in user's assessment.

Methods	A (%)	σ^2_A	κ (%)	σ^2_κ	T (%)	σ^2_T
Naive Bayes	84.44	1.5×10^{-2}	76.32	3.3×10^{-3}	76.67	3.3×10^{-2}
Weibull Naive Bayes	86.67	1.3×10^{-2}	79.59	3.1×10^{-3}	80.00	2.9×10^{-2}
Fuzzy Weibull Naive Bayes	91.11	3.1×10^{-3}	86.19	2.2×10^{-3}	86.67	2.0×10^{-2}

It is worth noticing, from the Kappa coefficients and Accuracy, that the performance of the SUAS-FWNB is better than both other SUAS in this comparison. The SUAS-WNB provide the second most accurate results with respect to Accuracy and Kappa coefficient. The SUAS-NB provide the worst results when compared to the previous ones. The reasons are that NB must discretize all variables before computing their probabilities and perform the classification task. As mentioned before, the discretization procedure of variables leads to loss of information. However, the SUAS-NB remains a competitive approach, as its performance was at most 3.33% lower than that achieved by the SUAS-WNB, according to the Tau Coefficient.

A higher accuracy (as well as Kappa and Tau coefficients) is important when considering the assessment of future professionals for health area in training simulators. Due to their generic nature, SUAS can be used for other kinds of professional or trainee assessments, in addition to those ones presented in this paper. Examples include: pilot training in aviation, train drivers, racing drivers, deep-sea divers, among others. The Fuzzy Weibull Naive Bayes Network itself proposed in this paper can also be used in classification tasks (data, signal, image, etc.) and machine learning applications, when data follow a Weibull distribution.

The source code for Fuzzy Weibull Naive Bayes will be available in the Fuzzy-Class package for the R software [7], the current version of which can be accessed at: <https://cran.r-project.org/web/packages/FuzzyClass/index.html>.

6 Conclusions

This paper introduced a Fuzzy Weibull Naive Bayes Network (FWNB) as the inference model for Single User Assessment Systems (SUAS) in virtual reality training environments. The proposed approach combines the probabilistic structure of Naive Bayes, the flexibility of the Weibull distribution for modeling asymmetric continuous variables, and fuzzy membership functions to represent uncertainty in performance measurements.

Monte Carlo simulations demonstrated that the proposed SUAS-FWNB achieved superior performance when compared with classical Naive Bayes and Weibull Naive Bayes models. The method reached an Accuracy of 91.11% and a Kappa coefficient of 86.19%. These results show that incorporating both Weibull-based likelihood modeling and fuzzy compatibility functions improves the discriminative capability of the classifier. The improvement is mainly explained by two factors. First, the Weibull distribution provides a more appropriate model for time-related variables frequently observed in VR simulations, such as task completion times and intervals between events. Second, the fuzzy formulation introduces a smoother representation of uncertainty, allowing the classifier to handle variability in user performance more robustly.

From a practical perspective, the proposed model preserves the computational simplicity of Naive Bayes while allowing continuous Weibull probabilistic modeling without discretization. This characteristic makes the method suitable for real-time assessment systems embedded in VR simulators. Although motivated by medical training simulators such as BMHSim and SITEG, the proposed FWNB network is not restricted to SUAS applications. It can also be applied to broader classification and machine learning problems involving asymmetric continuous variables or time-to-event data.

Future work includes validation using real interaction data from VR simulators, investigation of alternative fuzzy membership structures, and extensions of the model to more complex probabilistic architectures, as multivariate Weibull mixtures.

Acknowledgments. This work was supported by the National Council for Scientific and Technological Development (CNPq) grants 301644/2025-0, 315298/2018-9 and 408940/2024-7. The authors would like to thank the reviewers for their valuable feedback, which helped improve this paper.

Disclosure of Interests. The authors have no competing interests to declare that are relevant to the content of this article.

References

1. Almalki, S.J., Nadarajah, S.: Modifications of the weibull distribution: a review. Reliab. Eng. Syst. Safety **124**, 32–55 (2014)
2. Andrade, J.R.B., et al.: Virtual simulations for health education: how are user skills assessed? Revista Brasileira de Educação Médica **46**(4), e130 (2022). https://doi.org/10.1590/1981-5271v46.4-20210389.ING
3. Araujo, I., Ferreira, J., Machado, L., Moraes, R.: A new fuzzy Hypergeometric Naive Bayes network and its application for user's assessment in virtual reality simulators. In: Kerre, E., Lu, J., Martinez, L., Li, T., Montero, J., Flores-Vidal, P. (eds.) Intelligent Management of Data and Information in Decision Making, vol. 14, pp. 533–540. World Scientific, Singapure (2024). https://doi.org/10.1142/9789811294631_0066
4. Barak Ventura, R., Stewart Hughes, K., Nov, O., Raghavan, P., Ruiz Marín, M., Porfiri, M.: Data-driven classification of human movements in virtual reality-based serious games: preclinical rehabilitation study in citizen science. JMIR Serious Games **10**(1), e27597 (2022)
5. Cohen, J.: A coefficient of agreement for nominal scales. Educ. Psychol. Measur. **20**, 37–46 (1960). https://doi.org/10.1177/001316446002000104
6. Ferreira, J., Machado, L., Moraes, R.: A new double weighted fuzzy hypergeometric naive bayes network and its application for user's assessment in virtual reality simulators. Int. J. Comput. Intell. Syst. **18**, 90 (2025). https://doi.org/10.1007/s44196-025-00816-5
7. Ferreira, J., Moraes, R.: Fuzzyclass: A family of fuzzy and non-fuzzy probabilistic-based classifiers. J. Open Source Softw. **8**, 5613 (2023). https://doi.org/10.21105/joss.05613
8. Forbes, C., Evans, M., Hastings, N., Peacock, B.: Statistical Distributions. John Wiley & Sons, New Jersey (2011)
9. Gómez, Y.M., Gallardo, D.I., Marchant, C., Sánchez, L., Bourguignon, M.: An in-depth review of the weibull model with a focus on various parameterizations. Mathematics **12**(1), 56 (2023)
10. Hallinan, A.J., Jr.: A review of the Weibull distribution. J. Qual. Technol. **25**(2), 85–93 (1993)
11. Lai, C., Murthy, D., Xie, M.: Weibull distributions. Wiley Interdisc. Rev. Comput. Stat. **3**(3), 282–287 (2011)
12. Landis, J., Koch, G.: The measurement of observer agreement for categorical data. Biometrics **33**, 159–174 (1977). https://doi.org/10.2307/2529310
13. Liu, E., Liu, R.Y., Lim, K.: Using the Weibull accelerated failure time regression model to predict time to health events. Appl. Sci. **13**(24), 13041 (2023)
14. Luko, S.N.: A review of the weibull distribution and selected engineering applications. SAE 1984 Trans. **108**(section 2), 398–412 (1999)

15. Machado, L., Mello, A., Odone Fo, V., Zuffo, M.: A virtual reality simulator for bone marrow harvest for pediatric transplant. In: Westwood, J., Hoffman, H., Mogel, G., Stredney, R., Robb, R. (eds.) Medicine Meets Virtual Reality 2001. vol. 81, pp. 293–297. IOS Press, Amsterdam (2001). https://doi.org/10.3233/978-1-60750-925-7-293
16. Machado, L.S., Moraes, R.M.: Vr-based simulation for the learning of gynaecological examination. Lect. Notes Comput. Sci. **4282**, 97–104 (2006)
17. Machado, L.S., Valdek, M.C.O., Moraes, R.M.: On-line assessment system for a training system with multiple stages based on virtual reality. J. Multiple-Valued Logic Soft Comput. **14**(3–5), 511–524 (2008)
18. Mehran Azimbagirad, M., Murta Junior, L.O.: Tsallis generalized entropy for gaussian mixture model parameter estimation on brain segmentation application. Neurosci. Inf. **1**, 100002 (2021). https://doi.org/10.1016/j.neuri.2021.100002
19. Moraes, R.M., Ferreira, J.A., Machado, L.S.: A new bayesian network based on gaussian naive bayes with fuzzy parameters for training assessment in virtual simulators. Int. J. Fuzzy Syst. **23**(3), 849–861 (2020). https://doi.org/10.1007/s40815-020-00936-4
20. Moraes, R., Machado, L.: Gaussian naive bayes for online training assessment in virtual reality-based simulators. Mathware Soft Comput. **16**(2), 123–132 (2009)
21. Moraes, R., Machado, L.: Online assessment in medical simulators based on virtual reality using fuzzy Gaussian Naive Bayes. J. Multiple-valued Logic Soft Comput. **18**, 479–492 (2012)
22. Moraes, R., Soares, E., Machado, L.: A double weighted fuzzy Gamma Naive Bayes classifier. J. Intell. Fuzzy Syst. **38**, 577–588 (2021). https://doi.org/10.3233/JIFS-179431
23. Moraes, R.M., Machado, L.S.: Psychomotor skills assessment in medical training based on virtual reality using a weighted possibilistic approach. Knowl.-Based Syst. **70**, 97–102 (2014)
24. Moraes, R.M.d., Ferreira, R., Machado, L.: A new bayesian network based on gaussian naive bayes with fuzzy parameters for training assessment in virtual simulators. Mathematics **8**(5), 1–12 (2020)
25. Oropesa, I., Sanchez-Gonzalez, P., Lamata, P., Sanchez-Margallo, J., Sanchez-Margallo, F., Gomez, E.: Methods and tools for objective assessment of psychomotor skills in laparoscopic surgery. J. Surgical Reas. **171**(1), E81–E95 (2011)
26. Quinn, J.B., Quinn, G.D.: A practical and systematic review of weibull statistics for reporting strengths of dental materials. Dent. Mater. **26**(2), 135–147 (2010)
27. Robert, C.P., Casella, G.: Monte Carlo Statistical Methods, 2nd edn. Springer, New York (2004)
28. Rodrigues, A.K.G., Batista, T.V.V., Moraes, R.M., Machado, L.S.: A new exponential naive bayes classifier with fuzzy parameters. In: 2016 IEEE International Conference on Fuzzy Systems (FUZZ), pp. 1188–1193. IEEE (2016). proposes ENB-FP classifier using fuzzy exponential distribution
29. RRolff, T., Schmidt, S., Steinicke, F., Frintrop, S.: A deep learning architecture for egocentric time-to-saccade prediction using weibull mixture-models and historic priors. In: Proceedings of the 2023 Symposium on Eye Tracking Research and Applications, pp. 1–8 (2023)
30. Scholte, H.S., Ghebreab, S., Waldorp, L., Smeulders, A., Lamme, V.: Brain responses strongly correlate with weibull image statistics when processing natural images. J. Vision **9**(4), 29.1–15 (2009). https://doi.org/10.1167/9.4.29

31. Soares, E., Moraes, R.: Fusion of online assessment methods for gynecological examination training: A feasibility study. Trends Comput. Appl. Math. **19**(3), 423–436 (2018). https://doi.org/10.5540/tema.2018.019.03.423
32. Zadeh, L.: Probability measure of fuzzy events. J. Math. Anal. Appl. **22**, 421–427 (1968). https://doi.org/10.1016/0022-247X(68)90078-4

Multi-view Masked ECG Representation Learning with Hierarchical Fuzzy Memory

Tianyi Zhou[1], Yuepeng Chen[1], Wei Zhang[1], Fan Fu[1], and Weiping Ding[1,2](✉)

[1] School of Artificial Intelligence and Computer Science, Nantong University, Jiangsu, China

[2] Faculty of Data Science, City University of Macau, Macau, China

dwp9988@163.com

Abstract. Learning robust representations from electrocardiogram (ECG) signals remains challenging due to heterogeneous temporal dynamics and overlapping diagnostic patterns. In this paper, we propose a multi-view ECG representation learning framework that integrates masked autoencoding with a hierarchical fuzzy memory mechanism. The model jointly captures complementary information from time-domain morphology and time–frequency rhythmic structures through two masked representation learning branches. To aggregate heterogeneous latent tokens, a hierarchical fuzzy memory encoder is introduced to compress multi-view representations into a compact set of learnable prototypes using soft membership assignments. This design forms a structured semantic bottleneck that supports both masked signal reconstruction and multi-label ECG diagnosis. Extensive experiments on three public ECG datasets demonstrate that the proposed method achieves strong and consistent performance across multiple evaluation metrics. Additional analyses on reconstruction under severe masking and ablation studies further confirm the effectiveness of multi-view modeling and fuzzy prototype aggregation. These results indicate that the proposed framework provides a compact and robust representation for ECG analysis and offers a promising direction for scalable ECG representation learning.

Keywords: Masked Autoencoder · Fuzzy Memory · ECG Representation Learning

1 Introduction

Electrocardiogram analysis plays a fundamental role in cardiovascular screening and diagnosis due to its noninvasive nature, low cost, and wide availability in clinical practice. Recent advances in deep learning have shown that ECG signals contain substantially richer diagnostic information than traditional rule-based interpretation, enabling automated detection of diverse cardiac abnormalities and high-risk conditions [1]. Large-scale ECG-AI studies further demonstrate

J. Lu et al. (Eds.): FLINS-ISKE 2026, LNCS 16752, pp. 247–262, 2027.
https://doi.org/10.1007/978-981-92-2369-5_17

the value of representation learning for risk stratification and outcome prediction from routine ECG recordings [2]. Despite these advances, learning robust and clinically meaningful representations from multi-lead ECG signals remains challenging due to inter-patient variability, noise contamination, heterogeneous acquisition conditions, and ambiguous diagnostic annotations [3].

A key difficulty arises from the intrinsically heterogeneous temporal characteristics of ECG signals. Clinically relevant information appears at multiple temporal scales. Local waveform morphology, such as the shape and duration of P waves, QRS complexes, and ST segments, provides important cues for conduction abnormalities and structural disorders [4]. In contrast, rhythm-related conditions are often reflected in long-range and quasi-periodic temporal dynamics that only become reliable over extended observation horizons [5]. Capturing both fine-grained morphological patterns and long-range rhythm dynamics within a unified representation therefore remains a central challenge in ECG representation learning [6].

Most existing ECG modeling approaches rely on a single dominant representation. Time-domain models based on convolutional or Transformer architectures mainly focus on local waveform patterns and short-term temporal dependencies [3]. Although effective for many tasks, these approaches may struggle to fully capture global periodic structures associated with rhythm disorders [6]. Alternatively, time–frequency representations explicitly expose periodic components of ECG signals, but traditional transforms are usually fixed and rarely optimized end-to-end within modern deep learning frameworks. Recently, self-supervised learning has emerged as a scalable solution for ECG representation learning by leveraging large amounts of unlabeled data [7]. In particular, masked autoencoders (MAE) have shown promising results by reconstructing masked signal segments and improving label efficiency [8]. However, most existing ECG MAE models operate under a single-view assumption and produce large numbers of latent tokens, which can introduce redundancy and limit the extraction of compact diagnostic representations.

These observations suggest that a single homogeneous latent space may be insufficient for modeling the heterogeneous temporal structure of ECG signals. Multi-view representation learning provides a natural solution by allowing different encoders to capture complementary aspects of the data [9]. However, multi-view encoders often generate large sets of tokens, raising an additional challenge: how to aggregate heterogeneous representations into compact and semantically meaningful abstractions. In clinical practice, ECG interpretation relies on recognizing a small number of recurring diagnostic patterns that often overlap and exhibit fuzzy boundaries rather than mutually exclusive categories [10].

To address these challenges, we propose a unified ECG representation learning framework that combines multi-view masked autoencoding with a hierarchical fuzzy memory mechanism. The proposed framework models ECG signals from both time-domain and time–frequency perspectives, enabling simultaneous learning of local morphology and long-range rhythm patterns. A hierarchical fuzzy memory encoder is introduced to aggregate multi-view latent tokens into a small set of interpretable prototypes through soft membership assignments. This design forms a compact semantic bottleneck that supports both masked signal reconstruction and multi-label ECG diagnosis.

The main contributions of this work are summarized as follows: 1) We propose a multi-view ECG masked autoencoder that jointly models complementary time-domain and time–frequency representations to capture both morphological and rhythmic characteristics of ECG signals. 2) We introduce a hierarchical fuzzy memory encoder that compresses multi-view latent tokens into a compact set of fuzzy prototypes, enabling structured semantic abstraction and efficient representation aggregation. 3) Extensive experiments on multiple ECG datasets demonstrate the effectiveness and robustness of the proposed framework for multi-label ECG classification.

2 Related Work

2.1 Masked Modeling for ECG Representation Learning

Recent ECG modeling research has increasingly shifted from fully supervised learning toward self-supervised pretraining due to limited annotations, label noise, and the need for robust cross-cohort generalization. Among various self-supervised objectives, masked modeling and masked autoencoding have proven particularly effective, as they formulate representation learning as a generative reconstruction task and demonstrate strong scalability in sequential data [8].

In the ECG domain, masked learning strategies have been adapted to account for the multi-lead structure and waveform characteristics of ECG signals. Representative approaches include spatio-temporal masked modeling that captures dependencies across leads and time [11], masking strategies guided by waveform priors [12], and ECG-specific MAE architectures designed for large-scale pretraining and downstream transfer [13]. Recent ECG foundation models further show that large self-supervised Transformers can learn transferable ECG representations across multiple diagnostic tasks [14].

However, most existing masked ECG models operate under a single-view representation, typically based on time-domain patches. These approaches often produce long sequences of latent tokens that must be aggregated through attention or pooling mechanisms. While effective for reconstruction, such token-heavy representations provide limited inductive bias toward compact diagnostic abstractions, particularly for long ECG recordings where relevant information is distributed across multiple temporal scales [14].

2.2 Multi-view and Time–Frequency Representations

Modeling heterogeneous temporal dynamics has motivated the use of multi-view representation learning in time-series analysis. Multi-view approaches introduce complementary inductive biases by modeling different aspects of the data, rather than relying on a single homogeneous latent space [15]. For physiological signals such as ECG, this perspective is particularly relevant because waveform morphology and rhythm structure exhibit distinct statistical properties across temporal scales [10].

Recent work has explored combining multiple signal representations through cross-view reconstruction or alignment objectives. For example, the Cross Reconstruction Transformer enforces consistency between time-domain and frequency-domain representations through cross-domain masking and reconstruction [9]. Related studies also demonstrate that different temporal views capture complementary semantics in long time-series modeling [16]. In addition, long-sequence and multi-resolution Transformer architectures emphasize hierarchical or patch-based representations for capturing long-range dependencies [17].

Nevertheless, most existing multi-view approaches rely on token-level fusion, such as concatenation followed by self-attention. Such mechanisms implicitly assume that heterogeneous views can be reconciled within a shared token space, which may lead to redundant representations and limited semantic structure [18]. This limitation motivates structured aggregation mechanisms capable of distilling heterogeneous multi-view representations into compact and interpretable abstractions.

2.3 Prototype-Based Abstraction and Fuzzy Memory

Beyond multi-view modeling, several studies have explored reducing the computational cost of Transformer architectures through token compression techniques such as token pruning or merging [19]. Although these methods improve efficiency, they primarily target redundancy reduction and typically do not enforce semantic abstraction in the resulting representations [20].

Prototype- and memory-based representation learning provides a more structured alternative by introducing a small set of learnable prototypes or memory slots that summarize variable-length inputs [20]. These mechanisms have been widely studied in clustering-based learning, vector quantization, and memory-augmented neural networks, where they serve as bottlenecks that encourage concept-level representation learning [21].

However, most existing prototype mechanisms rely on competitive or exclusive assignments, typically implemented via nearest-neighbor matching or softmax-based attention. Such winner-take-most behavior is often unsuitable for domains characterized by overlapping or ambiguous concepts. In ECG analysis, diagnostic patterns frequently exhibit partial membership and co-occurrence across multiple conditions [22]. Fuzzy assignment mechanisms address this limitation by allowing elements to belong to multiple prototypes with graded memberships [23]. Despite their theoretical advantages, fuzzy prototype aggregation has rarely been integrated into Transformer-based architectures or multi-view representation learning frameworks. This gap motivates the hierarchical fuzzy memory mechanism proposed in this work.

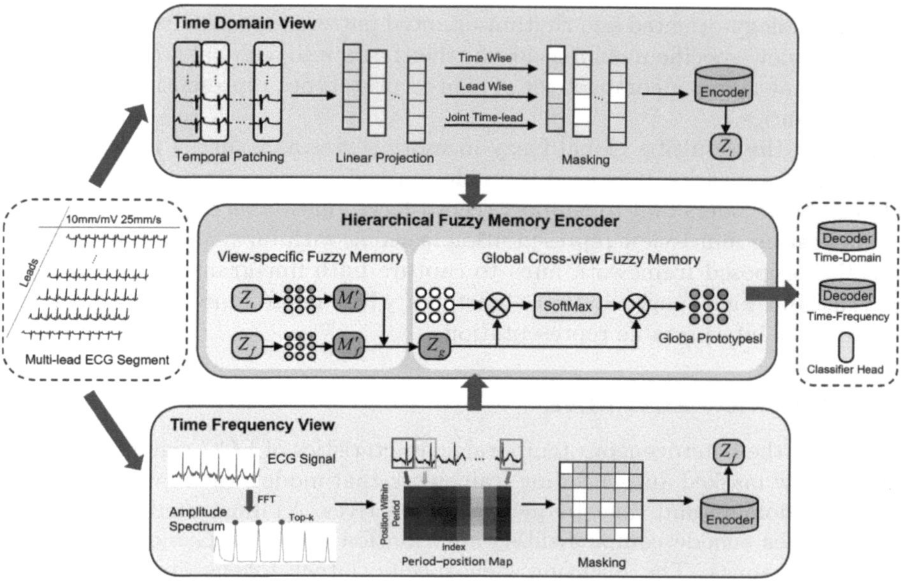

Fig. 1. Architecture of the proposed multi-view ECG masked autoencoder with hierarchical fuzzy memory. Time-domain and time–frequency representations are jointly encoded and compressed into global fuzzy prototypes for reconstruction and diagnosis.

3 Proposed Method

3.1 Framework Overview

The overall architecture of the proposed framework is illustrated in Fig. 1. Given a multi-lead ECG signal, our model learns a compact representation by combining multi-view masked representation learning with hierarchical fuzzy memory aggregation.

First, the input ECG is processed through two complementary views: a *time-domain view* and a *time–frequency view*. The time-domain branch captures local waveform morphology by segmenting the signal into temporal patches, projecting them into embedding tokens, and learning contextual representations through a Transformer encoder under random masking. In parallel, the time–frequency branch extracts periodic structures of ECG signals by identifying dominant frequencies through Fourier analysis and constructing period–position maps, which are subsequently tokenized and encoded to capture rhythm-related dynamics. These two branches produce view-specific latent tokens denoted as Z_t and Z_f, respectively.

Second, the latent tokens from both views are aggregated through a *hierarchical fuzzy memory encoder*. In the first stage, view-specific fuzzy memory modules summarize tokens within each representation view by assigning them to a set of learnable prototypes with soft memberships. These prototypes capture morphology-oriented and rhythm-oriented patterns separately. In the second stage, the view-specific memories are further fused into a global cross-view fuzzy memory that learns a compact set of shared prototypes representing the overall ECG semantics.

Finally, the resulting global fuzzy memory serves as a shared representation for downstream tasks. It is used to guide masked signal reconstruction through lightweight decoders and to support record-level multi-label ECG classification. By integrating multi-view representation learning with fuzzy prototype abstraction, the proposed framework aims to capture both fine-grained morphological patterns and long-range rhythm structures while maintaining a compact and semantically interpretable representation.

3.2 Multi-view ECG MAE

To capture the heterogeneous temporal characteristics of ECG signals, we adopt a multi-view masked autoencoding framework that models the input signal from both time-domain and time–frequency perspectives. As illustrated in Fig. 1, the two branches encode complementary information about ECG morphology and rhythm dynamics. The resulting view-specific latent tokens are subsequently aggregated by the hierarchical fuzzy memory encoder described in Sect. 3.3.

Let an ECG segment be denoted as

$$x = [x_1, \ldots, x_T] \in \mathbb{R}^{T \times C}, \tag{1}$$

where T represents the number of time samples and C is the number of ECG leads. The model learns two complementary representations of x through a time-domain encoder and a time–frequency encoder under masked autoencoding.

Time-Domain View. The time-domain view focuses on local waveform morphology and short-term temporal dependencies. The input ECG signal is first divided into non-overlapping temporal patches, which are linearly projected into embedding tokens and processed by a Transformer encoder. During pretraining, a subset of patches is randomly masked to form a masked autoencoding objective.

Given the visible token sequence H_t^v, the time-domain encoder E_t produces latent representations

$$Z_t = E_t(H_t^v) \in \mathbb{R}^{|\Omega_t| \times d}, \tag{2}$$

where $|\Omega_t|$ denotes the number of visible patches and d is the embedding dimension. These latent tokens mainly capture local ECG morphology, including waveform shapes and short-term temporal context.

Time–Frequency View. While the time-domain view emphasizes local morphology, many ECG abnormalities are reflected in long-range rhythmic patterns. To capture such periodic structures, we construct a complementary time–frequency representation.

Specifically, we apply a Fourier transform along the temporal dimension and compute the amplitude spectrum

$$A(f) = \frac{1}{C}\sum_{c=1}^{C} |FFT(x)(f,c)|. \tag{3}$$

The dominant frequencies are selected to estimate representative periods, which are then used to reshape the ECG signal into period–position maps that explicitly organize repeated cardiac cycles. Similar to the time-domain branch, these maps are partitioned into patches and encoded by a Transformer under random masking.

The time–frequency encoder E_f generates the corresponding latent tokens

$$Z_f = E_f(H_f^v) \in \mathbb{R}^{|\Omega_f| \times d}. \tag{4}$$

Because each token aggregates information across cardiac cycles and within-cycle positions, the resulting representation emphasizes long-range periodic structure and rhythm variability.

The latent tokens from the two views provide complementary information: Z_t captures fine-grained waveform morphology, while Z_f models global rhythmic patterns. These tokens are subsequently aggregated by the hierarchical fuzzy memory encoder to produce a compact representation for reconstruction and downstream ECG diagnosis.

3.3 Hierarchical Fuzzy Memory

The multi-view encoders described in Sect. 3.2 produce two sets of latent tokens, Z_t and Z_f, which capture complementary ECG information: local waveform morphology and long-range rhythm patterns. However, directly processing these tokens using full self-attention can be computationally expensive and may obscure higher-level diagnostic structure.

To address this issue, we introduce a hierarchical fuzzy memory encoder that compresses multi-view tokens into a compact set of learnable prototypes. The proposed module performs two stages of aggregation: (1) view-specific fuzzy memory that summarizes tokens within each representation view, and (2) global cross-view fuzzy memory that integrates information across views to produce a shared representation.

View-Specific Fuzzy Memory. Let $Z_t \in \mathbb{R}^{L_t \times d}$ and $Z_f \in \mathbb{R}^{L_f \times d}$ denote the latent tokens from the time-domain and time–frequency encoders, respectively. For each view $v \in \{t, f\}$, we introduce a set of learnable memory slots M_v that represent view-specific prototypes.

To assign tokens to prototypes, we compute fuzzy membership weights using an attention-based mechanism

$$A_v = \text{softmax}\left(\frac{Z_v M_v^\top}{\sqrt{d}}\right), \tag{5}$$

where each row of A_v represents the soft membership of a token to different prototypes. The view-specific prototypes are then obtained through membership-weighted aggregation

$$M'_v = A_v^\top Z_v. \tag{6}$$

This operation summarizes a large number of tokens into a small set of prototypes that capture dominant morphology patterns in the time domain and rhythm patterns in the time–frequency domain.

Global Cross-View Fuzzy Memory. To integrate information from both views, the view-specific memories are concatenated

$$Z_g = [M'_t; M'_f], \tag{7}$$

and further compressed into a shared global fuzzy memory. Similar to the previous stage, a small set of global memory slots aggregates the stacked prototypes through fuzzy memberships, producing a compact set of global prototypes that jointly represent morphology and rhythm information.

To stabilize training and encourage meaningful prototype usage, we apply a lightweight regularization that promotes balanced utilization and diversity among memory slots.

The resulting global fuzzy memory serves as a shared representation for downstream tasks. For masked autoencoding, the global prototypes guide reconstruction through cross-attention in the decoder. For ECG diagnosis, the pooled global memory representation is used as the input to a classification head for record-level multi-label prediction. In this way, both reconstruction and diagnosis are driven by the same compact prototype-based representation.

3.4 Training Objectives

The proposed framework is trained with a multi-task objective that combines masked signal reconstruction and record-level multi-label classification. All learning objectives operate on the shared fuzzy memory representation introduced in Sect. 3.3.

Masked Reconstruction. For self-supervised pretraining, the model reconstructs masked patches in both the time-domain and time–frequency views. Let $\bar{\Omega}_t$ and $\bar{\Omega}_f$ denote the sets of masked patches in the two views. The reconstruction losses are defined as

$$\mathcal{L}_{rec}^{t} = \frac{1}{|\bar{\Omega}_t|} \sum_{\ell \in \bar{\Omega}_t} \left\| x_t^{(\ell)} - \hat{x}_t^{(\ell)} \right\|_1, \tag{8}$$

$$\mathcal{L}_{rec}^{f} = \frac{1}{|\bar{\Omega}_f|} \sum_{m \in \bar{\Omega}_f} \left\| x_f^{(m)} - \hat{x}_f^{(m)} \right\|_1. \tag{9}$$

To encourage the two views to learn compatible representations, we further introduce a simple cross-view alignment term

$$\mathcal{L}_{align} = 1 - \frac{z_t^\top z_f}{\|z_t\|_2 \|z_f\|_2}, \tag{10}$$

where z_t and z_f denote the pooled representations from the two views.

Multi-label Classification. For ECG diagnosis, the global fuzzy prototypes are pooled to obtain a compact representation g, which is fed into a linear classifier to predict the presence of multiple cardiac conditions. The model is trained using the standard binary cross-entropy loss

$$\mathcal{L}_{cls} = -\frac{1}{K}\sum_{k=1}^{K}[y_k \log \hat{y}_k + (1 - y_k)\log(1 - \hat{y}_k)]. \tag{11}$$

Overall Objective. We adopt a two-stage training strategy. During self-supervised pretraining, the model is optimized using reconstruction and alignment losses together with fuzzy memory regularization:

$$\mathcal{L}_{pre} = \lambda_t \mathcal{L}_{rec}^t + \lambda_f \mathcal{L}_{rec}^f + \lambda_a \mathcal{L}_{align} + \lambda_{fuzzy}\mathcal{L}_{fuzzy}. \tag{12}$$

During supervised fine-tuning, the classification loss becomes the primary objective while the reconstruction and regularization terms are retained as auxiliary constraints

$$\mathcal{L}_{fine} = \mathcal{L}_{cls} + \beta_t \mathcal{L}_{rec}^t + \beta_f \mathcal{L}_{rec}^f + \beta_a \mathcal{L}_{align} + \beta_{fuzzy}\mathcal{L}_{fuzzy}. \tag{13}$$

This multi-task training strategy enables the model to learn general ECG representations from unlabeled data while preserving task-specific discriminative information during supervised learning.

4 Experiments

4.1 Experimental Setup

Datasets. We evaluate the proposed method on three publicly available ECG datasets: PTB-XL [24], CPSC2018 [25], and Chapman [26]. PTB-XL is a large-scale 12-lead clinical ECG dataset containing 21,837 recordings with multi-label diagnostic annotations. CPSC2018 focuses on arrhythmia detection with variable-length ECG recordings. Chapman consists of long-duration ambulatory ECG signals annotated with rhythm-related diagnoses. Together, these datasets cover diverse clinical scenarios and recording conditions, enabling comprehensive evaluation of ECG representation learning.

Evaluation Metrics. We evaluate the model on record-level multi-label ECG classification using three commonly used metrics: macro-AUROC, macro-AUPRC, and macro-F1. These metrics provide complementary evaluation under class imbalance and label co-occurrence. In addition, reconstruction quality during pretraining is measured using mean absolute error.

Baselines. We compare our approach with several representative ECG and time-series models, including self-supervised ECG pretraining methods (MTAE [13], STMEM [11]), Transformer-based time-series models (iTransformer [27], PatchTST [28], Crossformer [29]), and architectures with strong temporal inductive biases (ModernTCN [30], Medformer [31]). We also include a knowledge-guided feature fusion model (KFFusion [32]) as a domain-aware baseline.

Implementation Details. All ECG signals are resampled to 250 Hz and segmented into 10-second windows when necessary. Each segment is normalized using per-record z-score normalization. Models are trained using the Adam optimizer with a learning rate of 1×10^{-4} and batch size 64. Self-supervised pretraining is performed for 100 epochs followed by 50 epochs of supervised fine-tuning. The numbers of view-specific and global fuzzy memory slots are set to 16 and 8, respectively.

Table 1. Main results on multi-label ECG classification across three datasets.

Model	PTB-XL			CPSC2018			Chapman		
	AUROC	AUPRC	F1	AUROC	AUPRC	F1	AUROC	AUPRC	F1
Autoencoding and self-supervised ECG baselines									
MTAE	89.14 ± 0.48	61.63 ± 1.73	56.48 ± 3.57	94.15 ± 0.35	69.43 ± 2.33	65.57 ± 2.84	91.66 ± 0.70	58.74 ± 2.05	**57.19 ± 3.16**
STMEM	87.73 ± 0.64	56.85 ± 2.75	48.26 ± 3.51	93.67 ± 0.46	63.06 ± 1.68	58.33 ± 3.60	89.92 ± 0.41	52.19 ± 2.3	49.48 ± 4.37
Long-sequence and patch-based Transformers									
iTransformer	75.06 ± 0.77	38.65 ± 3.31	35.44 ± 5.1	81.16 ± 0.53	54.75 ± 2.44	48.03 ± 4.4	78.29 ± 0.79	43.7 ± 3.37	36.52 ± 5.57
PatchTST	72.17 ± 0.67	40.26 ± 3.26	35.72 ± 4.92	78.93 ± 0.54	42.56 ± 1.58	30.54 ± 4.85	68.79 ± 0.40	37.58 ± 3.90	35.11 ± 4.68
Crossformer	82.37 ± 0.56	35.14 ± 3.75	31.82 ± 5.61	85.23 ± 0.33	49.78 ± 1.80	42.23 ± 2.15	83.59 ± 0.67	42.91 ± 2.21	37.00 ± 3.27
Models with strong inductive biases for time-series									
ModernTCN	90.81 ± 0.37	45.88 ± 1.42	42.44 ± 2.16	82.41 ± 0.44	42.18 ± 1.52	38.73 ± 3.73	92.38 ± 0.54	59.11 ± 2.26	54.35 ± 2.50
Medformer	82.41 ± 0.44	42.18 ± 1.52	38.73 ± 3.73	87.70 ± 0.29	65.46 ± 1.85	55.81 ± 3.50	88.21 ± 0.62	45.49 ± 3.13	42.76 ± 3.05
Knowledge-guided and feature-fusion baselines									
KFFusion	88.55 ± 0.35	59.29 ± 2.55	52.32 ± 2.41	93.86 ± 0.25	**74.36 ± 1.08**	63.38 ± 2.83	92.45 ± 0.58	61.86 ± 2.19	54.01 ± 3.90
Proposed method									
Ours	**91.94 ± 0.26**	**66.90 ± 1.02**	**59.21 ± 3.39**	**94.36 ± 0.27**	72.66 ± 1.26	**69.12 ± 2.87**	**94.27 ± 0.37**	**62.25 ± 1.92**	56.87 ± 2.21

4.2 Main Results on ECG Classification

Table 1 reports the main multi-label ECG classification results on the PTB-XL, CPSC2018, and Chapman datasets. Overall, the proposed method consistently achieves the best or competitive performance across most evaluation metrics, demonstrating the effectiveness of combining multi-view masked representation learning with hierarchical fuzzy memory aggregation.

On the large-scale PTB-XL dataset, our method achieves the highest scores across all metrics, reaching an AUROC of 91.94, an AUPRC of 66.90, and an F1 score of 59.21. Compared with the strongest baseline MTAE, the proposed approach improves AUPRC by more than 5%, indicating better precision–recall trade-offs under significant label imbalance and co-occurrence.

On the CPSC2018 dataset, our model achieves the best AUROC (94.36) and F1 score (69.12), while maintaining competitive AUPRC performance. The improvement in F1 score suggests that the learned representations provide more balanced classification performance across arrhythmia categories.

On the Chapman dataset, which contains long-duration ambulatory ECG recordings, the proposed model achieves the best AUROC (94.27) and AUPRC (62.25), demonstrating strong robustness for rhythm-oriented diagnosis tasks. These results highlight the advantage of explicitly modeling long-range periodic structures through the time–frequency view.

Across all datasets, models originally designed for generic time-series forecasting perform noticeably worse than ECG-specific methods, suggesting that domain-aware representation learning remains important for clinical ECG analysis. Furthermore, compared with existing self-supervised ECG models such as MTAE and STMEM, the proposed approach consistently achieves stronger performance, indicating that the combination of multi-view modeling and fuzzy prototype aggregation provides a more effective representation of ECG signals.

Overall, these results demonstrate that the proposed framework can learn robust and clinically meaningful representations across diverse ECG datasets and diagnostic scenarios.

Table 2. Reconstruction performance under different masking ratios on PTB-XL. Lower mean absolute error indicates better reconstruction.

Method	50% mask	75% mask	90% mask
Single-view MAE	0.32	0.49	0.72
Ours w/o Fuzzy	0.30	0.45	0.66
Ours	**0.27**	**0.39**	**0.54**

4.3 Reconstruction Under Severe Masking

To further examine the quality of the learned representations, we analyze the reconstruction behavior under severe masking conditions. Masked reconstruction evaluates whether the model captures globally coherent ECG structure rather than relying on local interpolation.

We conduct experiments on PTB-XL with masking ratios of 50%, 75%, and 90%. Higher masking ratios correspond to more challenging settings where only a small portion of the signal is visible. Reconstruction quality is measured using mean absolute error between the reconstructed and original signals.

Table 2 reports the results. Under moderate masking (50%), the differences between methods are relatively small. However, as the masking ratio increases, the advantage of the proposed approach becomes more evident. At 75% masking, our model reduces mean absolute error from 0.49 (single-view MAE) and 0.45 (without fuzzy memory) to 0.39. The gap further enlarges at 90% masking, where

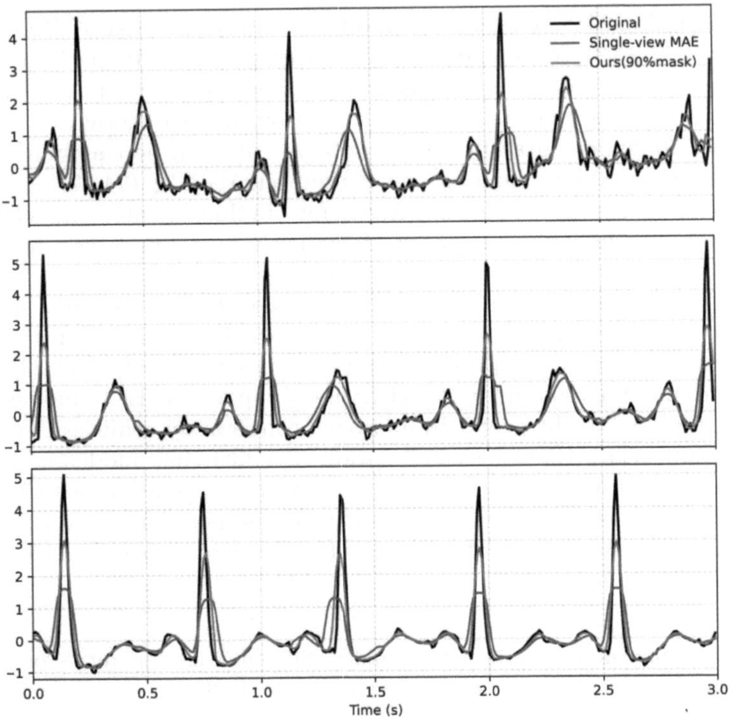

Fig. 2. Qualitative examples of ECG reconstruction under 90% masking on the PTB-XL dataset.

the proposed method achieves a mean absolute error of 0.54 compared with 0.66 and 0.72 for the two baselines.

Figure 2 shows qualitative reconstruction examples under a masking ratio of 90 Compared with the single-view MAE baseline, the proposed model better preserves ECG waveform morphology and rhythm continuity.

These results indicate that the proposed multi-view fuzzy representation better preserves global ECG structure, enabling more reliable reconstruction even under extremely sparse observations.

4.4 Ablation Study

To evaluate the contribution of each component in the proposed framework, we conduct ablation studies by progressively removing key modules from the full model. The results on PTB-XL and Chapman are summarized in Table 3.

Removing the multi-view design leads to the largest performance degradation across both datasets, indicating that modeling complementary time-domain and time–frequency representations provides essential information for ECG diagnosis. Eliminating the fuzzy memory module also causes noticeable performance drops, demonstrating the importance of prototype-based aggregation for capturing overlapping diagnostic patterns.

Replacing the hierarchical memory structure with a single-level design further reduces performance, suggesting that separating view-specific and global prototypes helps improve semantic abstraction. In addition, removing the consistency regularization results in a modest but consistent decrease in performance, indicating its role in encouraging alignment between different representation views.

Overall, the full model consistently achieves the best performance across all metrics. These results confirm that each component contributes to the final performance, and that the improvement arises from the synergistic integration of multi-view representation learning and hierarchical fuzzy memory modeling.

Table 3. Ablation study on core components of the proposed method. Results are reported as mean ± standard deviation over five runs. The best performance in each column is highlighted in bold.

Variant	PTB-XL			Chapman		
	AUROC	AUPRC	F1	AUROC	AUPRC	F1
w/o fuzzy memory	91.03 ± 0.24	64.78 ± 1.07	57.85 ± 3.29	93.57 ± 0.37	60.77 ± 1.80	55.48 ± 2.30
w/o hierarchy (single-level memory)	91.22 ± 0.33	65.41 ± 0.93	58.42 ± 3.02	93.74 ± 0.49	61.32 ± 1.80	55.95 ± 2.27
w/o multi-view (time-domain only)	90.78 ± 0.38	63.82 ± 1.15	56.89 ± 3.17	93.25 ± 0.44	59.84 ± 1.79	54.92 ± 2.35
w/o consistency regularization	91.52 ± 0.26	66.01 ± 1.07	58.63 ± 3.28	94.06 ± 0.32	61.75 ± 1.96	56.25 ± 2.27
Ours (full)	**91.94 ± 0.26**	**66.90 ± 1.02**	**59.21 ± 3.39**	**94.27 ± 0.37**	**62.25 ± 1.92**	**56.87 ± 2.21**

4.5 Representation Analysis

To further analyze the quality of the learned representations, we visualize the record-level ECG embeddings using t-SNE. Figure 3 shows the distribution of samples in the latent space across three datasets.

As shown in the figure, samples associated with similar diagnostic categories tend to form compact local clusters, indicating that the learned representations

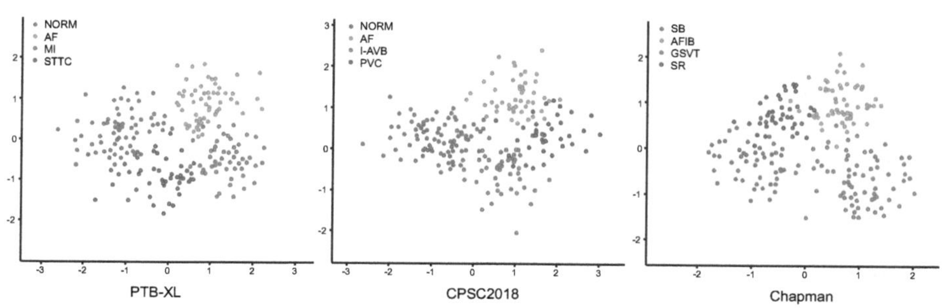

Fig. 3. t-SNE visualization of record-level ECG representations learned by the proposed method on three datasets. Different colors indicate different diagnostic categories.

capture meaningful similarities among ECG signals. At the same time, different categories exhibit distinguishable distributions in the embedding space, suggesting that the model preserves discriminative information for downstream classification tasks.

These observations demonstrate that the proposed framework is able to organize heterogeneous ECG patterns into a structured representation space.

5 Conclusion

This paper presented a multi-view ECG representation learning framework that integrates masked autoencoding with hierarchical fuzzy memory. By jointly modeling time-domain morphology and time–frequency rhythmic structures, the proposed approach enables the learning of complementary ECG representations across multiple temporal scales.

The hierarchical fuzzy memory mechanism further compresses heterogeneous multi-view tokens into a compact set of prototype-based representations. This structured aggregation encourages semantic abstraction while maintaining efficient representation capacity, which proves beneficial for both masked reconstruction and downstream ECG classification tasks.

Extensive experiments on three public ECG datasets demonstrate that the proposed framework consistently achieves strong performance across multiple evaluation metrics. These results suggest that combining multi-view representation learning with fuzzy prototype abstraction provides an effective and scalable solution for ECG representation learning. Future work will explore extending this framework to larger clinical datasets and investigating its potential for broader physiological signal modeling.

Acknowledgments. This work was supported in part by the National Key Research and Development Program of China under Grant 2024YFE0202700, the National Natural Science Foundation of China under Grants U2433216, 62576178, and 62506176, the Natural Science Foundation of Jiangsu Province under Grant BK20231337 and BK20250960, the Modern Agricultural Machinery Equipment and Technology Extension Project of Jiangsu Province under Grant NJ2024-06, the Joint Program on Health Science and Technology Innovation of Hainan Province under Grant WSJK2026ZD285, and the Natural Science Foundation of the Jiangsu Higher Education Institutions of China under Grant 25KJB520041.

Disclosure of Interests. The authors have no competing interests to declare that are relevant to the content of this article.

References

1. Poterucha, T.J., et al.: Detecting structural heart disease from electrocardiograms using ai. Nature **644**(8075), 221–230 (2025)
2. Hughes, J.W., et al.: A deep learning-based electrocardiogram risk score for long term cardiovascular death and disease, NPJ Digit. Med. **6**(1), 169 (2023)

3. Ribeiro, A.H., et al.: Automatic diagnosis of the 12-lead ecg using a deep neural network. Nat. Commun. **11**(1), 1760 (2020)
4. Liu, X., Wang, H., Li, Z., Qin, L.: Deep learning in ECG diagnosis: a review. Knowl.-Based Syst. **227**, 107187 (2021)
5. Tohyama, T., et al.: Deep learning of ecg for the prediction of postoperative atrial fibrillation. Circ. Arrhythmia Electrophysiol. **16**(2), e011579 (2023)
6. Moody, J.B., et al.: A foundation transformer model with self-supervised learning for ECG-based assessment of cardiac and coronary function. NEJM AI **2**(12), AIoa2500164 (2025)
7. Xu, P., et al.: Multi-stage temporal and cross-view contrastive learning for self-supervised multi-lead ECG classification. Biomed. Signal Process. Control **114**, 109317 (2026)
8. He, K., Chen, X., Xie, S., Li, Y., Dollár, P., Girshick, R.: Masked autoencoders are scalable vision learners. In: Proceedings of the IEEE/CVF Conference on Computer Vision and Pattern Recognition, pp. 16000–16009 (2022)
9. Zhang, W., Yang, L., Geng, S., Hong, S.: Self-supervised time series representation learning via cross reconstruction transformer. IEEE Trans. Neural Netw. Learn. Syst. (2023)
10. Fu, F., et al.: ECG adaptive synthesis with multigranular fuzzy easnn: Enhancing wearable cardiac monitoring, IEEE Trans. Fuzzy Syst. (2025)
11. Na, Y., Park, M., Tae, Y., Joo, S.: Guiding masked representation learning to capture spatio-temporal relationship of electrocardiogram. In: The Twelfth International Conference on Learning Representations (2024). https://openreview.net/forum?id=WcOohbsF4H
12. Wang, L., Wang, X., Tao, R.: Wave masked autoencoder: an electrocardiogram signal diagnosis model based on wave making strategy. Inf. Sci. **690**, 121516 (2025)
13. Zhang, H., et al.: Maefe: Masked autoencoders family of electrocardiogram for self-supervised pretraining and transfer learning. IEEE Trans. Instrum. Meas. **72**, 1–15 (2022)
14. Wu, H., Hu, T., Liu, Y., Zhou, H., Wang, J., Long, M.: Timesnet: Temporal 2D-variation modeling for general time series analysis. In: International Conference on Learning Representations (2023)
15. Yue, Z., et al.: Ts2vec: Towards universal representation of time series. In: Proceedings of the AAAI conference on artificial intelligence, Vol. 36, pp. 8980–8987 (2022)
16. Eldele, E., et al.: Time-series representation learning via temporal and contextual contrasting. In: Proceedings of the Thirtieth International Joint Conference on Artificial Intelligence, IJCAI-21, pp. 2352–2359 (2022)
17. Nie, Y., Nguyen, N.H., Sinthong, P., Kalagnanam, J.: A time series is worth 64 words: Long-term forecasting with transformers. In: The Eleventh International Conference on Learning Representations (2023). https://openreview.net/forum?id=Jbdc0vTOcol
18. Zhou, H., et al.: Informer: Beyond efficient transformer for long sequence time-series forecasting. In: Proceedings of the AAAI Conference on Artificial Intelligence, Vol. 35, pp. 11106–11115 (2021)
19. Bolya, D., Fu, C.-Y., Dai, X., Zhang, P., Feichtenhofer, C., Hoffman, J.: Token merging: Your ViT but faster. In: International Conference on Learning Representations (2023)
20. Chen, Z., Shen, Y., Song, Y., Wan, X.: Cross-modal memory networks for radiology report generation. In: Proceedings of the 59th Annual Meeting of the Association

for Computational Linguistics and the 11th International Joint Conference on Natural Language Processing (volume 1: long papers), pp. 5904–5914 (2021)
21. Ghods, A., Cook, D.J.: Pip: Pictorial interpretable prototype learning for time series classification. IEEE Comput. Intell. Mag. **17**(1), 34–45 (2022)
22. Karunaratne, G., et al.: Robust high-dimensional memory-augmented neural networks. Nat. Commun. **12**(1), 2468 (2021)
23. Ding, W., Zhou, T., Huang, J., Jiang, S., Hou, T., Lin, C.-T.: Fmdnn: a fuzzy-guided multigranular deep neural network for histopathological image classification. IEEE Trans. Fuzzy Syst. **32**(8), 4709–4723 (2024)
24. Wagner, P., et al.: Ptb-xl, a large publicly available electrocardiography dataset. Sci. data **7**(1), 1–15 (2020)
25. Liu, F., et al.: An open access database for evaluating the algorithms of electrocardiogram rhythm and morphology abnormality detection. J. Med. Imaging Health Inf. **8**(7), 1368–1373 (2018)
26. Zheng, J., Zhang, J., Danioko, S., Yao, H., Guo, H., Rakovski, C.: A 12-lead electrocardiogram database for arrhythmia research covering more than 10,000 patients. Sci. Data **7**(1), 48 (2020)
27. Liu, Y., et al.: itransformer: Inverted transformers are effective for time series forecasting, arXiv preprint arXiv:2310.06625 (2023)
28. Nie, Y.: A time series is worth 64words: Long-term forecasting with transformers, arXiv preprint arXiv:2211.14730 (2022)
29. Zhang, Y., Yan, J.: Crossformer: Transformer utilizing cross-dimension dependency for multivariate time series forecasting. In: The Eleventh International Conference on Learning Representations (2023)
30. Luo, D., Wang, X.: Moderntcn: A modern pure convolution structure for general time series analysis. In: The Twelfth International Conference on Learning Representations, pp. 1–43 (2024)
31. Wang, Y., Huang, N., Li, T., Yan, Y., Zhang, X.: Medformer: a multi-granularity patching transformer for medical time-series classification. Adv. Neural. Inf. Process. Syst. **37**, 36314–36341 (2024)
32. Chen, Y., Gao, S., Yang, J., Zhou, T., Fu, F., Ding, W.: A knowledge-guided feature fusion and fuzzy decision integration model for ecg-based cardiovascular disease diagnosis. IEEE Trans. Fuzzy Syst. (2026)

Fuzzy Control With Feedforward for the Secondary-Side-Parallel CLLC Resonant Converter

Li Lai, Bo Zhang(✉), and Jingyang Lin

South China University of Technology, Wushan Road 381, Guangzhou, China
epbzhang@scut.edu.cn

Abstract. A secondary parallel CLLC resonant converter is proposed to address the issue of traditional CLLC resonant converters lacking current source output characteristics. The proposed converter has two working modes of constant current and constant voltage, making it highly suitable for applications such as battery charging. In constant current mode, the proposed converter adjusts the output current by changing the phase shift angle of the full bridge inverter; In constant voltage mode, the voltage gain of the resonant network is adjusted by changing the switching frequency of the full bridge inverter. Feedforward fuzzy control is proposed to address the drawbacks of PID control, such as difficulty in parameter tuning without a model, mediocre control effect on strongly nonlinear systems, and sensitivity to noise. The proposed control strategy has the advantages of rapid response, small overshoot, fast convergence, and strong robustness. To verify the feasibility of the proposed converter and control strategy, a simulation model was built using PLECS software for validation. The results showed that the relevant characteristics of the proposed converter and control strategy were in line with theoretical analysis.

Keywords: Fuzzy control · CLLC · resonant converter

1 Introduction

The resonant converter is a power electronic converter which can perform electrical energy conversion processes such as step-up, step-down, and electrical isolation. Compared to other types of power electronic converters, resonant converters offer distinct advantages including high efficiency, low electromagnetic radiation, a wide output power range, and reduced component stress [1,2]. Accordingly, resonant converters are extensively employed in diverse scenarios requiring electrical power conversion. From low-power computer power supplies to medium-power server power supplies, and extending to high-power electric vehicle on-board chargers, all rely on high-performance resonant converters [3,4].

Resonant converters were first proposed in the 1970s [5,6]. Over more than five decades of rapid development, diverse resonant topologies have emerged.

J. Lu et al. (Eds.): FLINS-ISKE 2026, LNCS 16752, pp. 263–276, 2027.
https://doi.org/10.1007/978-981-92-2369-5_18

Based on the characteristics of the resonant network, these can be categorised as series resonant converters, parallel resonant converters, series-parallel resonant converters, LLC resonant converters, and CLLC resonant converters [7].

At present, the control strategies for resonant converters primarily include proportional-integral-derivative (PID) control. PID control stands as one of the most prevalent and widely applied control strategies in power electronic converters, characterised by its straightforward mathematical modelling, robust performance, and minimal computational overhead. However, PID control also exhibits numerous drawbacks, such as difficulty in parameter tuning without a model of the controlled object, suboptimal control performance, generally poor control effectiveness for strongly non-linear systems, and sensitivity to noise in the error signal. Consequently, in applications demanding high dynamic performance characteristics such as response time and convergence speed, the effectiveness of PID control often proves unsatisfactory.

Fuzzy control is a model-free nonlinear control strategy based on fuzzy set theory, fuzzy linguistic variables, and fuzzy reasoning. Common fuzzy control architectures comprise four components: the fuzzy encoder, the knowledge base, the fuzzy reasoner, and the defuzzifier. Therefore, fuzzy control is widely applied in non-linear systems with imprecise models. Following decades of development, the application of fuzzy control have expanded from water temperature control systems to encompass advanced and complex fields, such as domestic appliances [8], the chemical industry [9], biomedicine [10], unmanned aerial vehicles [11], and robotics [12,13]. In power electronic converters, fuzzy control is typically combined with PID control or disturbance observation techniques, primarily employed to achieve objectives such as reducing line-frequency ripple [14], enhancing dynamic performance [15], and improving system robustness [16,17].

As introduced earlier, the existing resonant converter topologies and control strategies still exhibit shortcomings. Therefore, in terms of topology, this paper proposes a secondary-side parallel-connected CLLC resonant converter with current-source characteristics; in terms of control strategy, this paper proposes feedforward fuzzy control. The main contributions of this paper include:

1. A secondary-side parallel CLLC resonant converter is proposed, its operating principle is elucidated, and its characteristics are analyzed in detail.
2. A feedforward fuzzy control strategy is introduced, the controller architecture is described, and a design procedure is provided.
3. The novel topology and control strategy are validated through simulation, with the results subjected to detailed analysis.

2 Secondary-Side-Parallel CLLC Resonant Converter

2.1 Working Principle

As shown in Fig. 1, the secondary-side-parallel CLLC resonant converter comprises a full-bridge inverter (Q_1–Q_4), a resonant network, a full-bridge rectifier (D_1–D_4), and an output filter (L_o, C_o).

The resonant network features an asymmetric structure. The primary-side resonant capacitor C_p is arranged in series, while the secondary-side resonant capacitor C_s is connected in parallel. L_{r1} and L_{r2} constitute series resonant inductors, with L_m and n representing the excitation inductance and turn ratio of transformer T respectively.

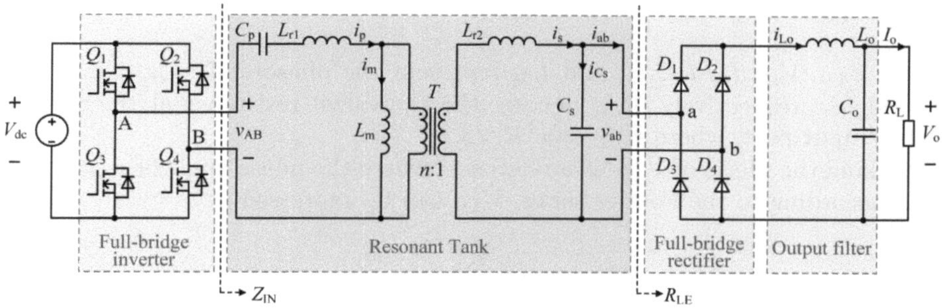

Fig. 1. A Schematic Diagram of a Secondary-Side-Parallel CLLC Resonant Converter.

The resonant tank parameters are designed such that C_p resonates with L_{r1} and C_s resonates with L_{r2}, with both resonant frequencies being equal, as follows:

$$\omega_0 = 2\pi f_0 = \frac{1}{\sqrt{L_{r1}C_p}} = \frac{1}{\sqrt{L_{r2}C_s}} \tag{1}$$

where ω_0 and f_0 denote the resonant angular frequency and resonant frequency respectively.

V_{dc} is the input DC voltage for the converter. The full-bridge inverter employing phase-shift modulation outputs a square-wave voltage v_{AB} with an amplitude of $\pm V_{dc}$, a frequency of f_{SW}, and a duty cycle of α/π. Following frequency selection and gain amplification by the resonant tank, a sinusoidal voltage v_{ab} is output. This voltage is subsequently rectified and filtered through a full-bridge rectifier and output filter, yielding a DC voltage V_o which is applied to the load resistor R_L.

2.2 Characteristic Analysis

The currents i_p, i_s, and i_{Cs} are all approximated as sinusoidal waves. Consequently, fundamental analysis can be employed, which involves neglecting higher harmonics and considering only the fundamental components of each voltage and current. According to Kirchhoff's laws and the phasor method, the voltage and current relationships in each branch of the converter can be expressed as:

$$\begin{cases} \dot{V}_{\mathrm{AB}} = \dot{I}_{\mathrm{p}}\left(j\omega_{\mathrm{s}}L_{\mathrm{r1}} + \dfrac{1}{j\omega_{\mathrm{s}}C_{\mathrm{p}}}\right) + \dot{I}_{\mathrm{m}}j\omega_{\mathrm{s}}L_{\mathrm{m}}, \quad \dot{I}_{\mathrm{p}} = \dot{I}_{\mathrm{m}} + \dfrac{1}{n}\dot{I}_{\mathrm{s}} \\ \dot{I}_{\mathrm{m}}j\omega_{\mathrm{s}}L_{\mathrm{m}} = n\left(\dot{I}_{\mathrm{s}}j\omega_{\mathrm{s}}L_{\mathrm{r2}} + \dot{I}_{\mathrm{Cs}}\dfrac{1}{j\omega_{\mathrm{s}}C_{\mathrm{s}}}\right), \quad \dot{I}_{\mathrm{Cs}}\dfrac{1}{j\omega_{\mathrm{s}}C_{\mathrm{s}}} = \dot{I}_{\mathrm{ab}}R_{\mathrm{LE}} \\ \dot{I}_{\mathrm{s}} = \dot{I}_{\mathrm{Cs}} + \dot{I}_{\mathrm{ab}}, \quad \dot{V}_{\mathrm{ab}} = \dot{I}_{\mathrm{ab}}R_{\mathrm{LE}} \end{cases} \tag{2}$$

where $\dot{V}_{AB}$, $\dot{V}_{ab}$, $\dot{I}_p$, $\dot{I}_m$, $\dot{I}_s$, and $\dot{I}_{Cs}$ represent the phasors of v_{AB}, v_{ab}, i_p, i_m, i_s, and i_{Cs}, respectively. R_{LE} denotes the equivalent resistance at the rectifier bridge input port, where $R_{LE} = \pi^2 R_L/8$.

Taking the phase of $\dot{V}_{AB}$ as a reference, where the phase angle of v_{AB} equals zero, according to the Fourier series, $\dot{V}_{AB}$ can be expressed as:

$$\dot{V}_{\mathrm{AB}} = \frac{2\sqrt{2}}{\pi}V_{\mathrm{dc}}\sin\left(\frac{\alpha}{2}\right) \tag{3}$$

where α is the phase-shift angle of the full-bridge inverter.

The proposed converter features multi-mode operation, operating in both fixed-frequency and variable-frequency modes. The detailed analysis is as follows:

1) Fixed-frequency mode

In this mode, the converter operates at a frequency identical to the resonant frequency f_0. Substituting $\omega_s = 2\pi f_0$ into (1), (2), and (3) yields the root-mean-square values of the node voltages and currents as follows:

$$\begin{cases} I_{\mathrm{p}} = \dfrac{2\sqrt{2}V_{\mathrm{dc}}}{\pi n^2\omega_0 L_{\mathrm{r2}}}\sin\left(\dfrac{\alpha}{2}\right)\sqrt{\dfrac{(n^2L_{\mathrm{r2}}+L_{\mathrm{m}})^2}{L_{\mathrm{m}}^2} + \dfrac{R_{\mathrm{LE}}^2}{\omega_0^2L_{\mathrm{r2}}^2}}, \\ I_{\mathrm{s}} = \dfrac{2\sqrt{2}V_{\mathrm{dc}}}{\pi n\omega_0^2L_{\mathrm{r2}}^2}\sin\left(\dfrac{\alpha}{2}\right)\sqrt{\omega_0^2L_{\mathrm{r2}}^2 + R_{\mathrm{LE}}^2} \\ I_{\mathrm{m}} = \dfrac{2\sqrt{2}V_{\mathrm{dc}}}{\pi\omega_0 L_{\mathrm{m}}}\sin\left(\dfrac{\alpha}{2}\right), \quad I_{\mathrm{Cs}} = \dfrac{2\sqrt{2}V_{\mathrm{dc}}R_{\mathrm{LE}}}{\pi n\omega_0^2L_{\mathrm{r2}}^2}\sin\left(\dfrac{\alpha}{2}\right) \\ I_{\mathrm{ab}} = \dfrac{2\sqrt{2}V_{\mathrm{dc}}}{\pi n\omega_0 L_{\mathrm{r2}}}\sin\left(\dfrac{\alpha}{2}\right), \quad V_{\mathrm{ab}} = \dfrac{2\sqrt{2}V_{\mathrm{dc}}R_{\mathrm{LE}}}{\pi n\omega_0 L_{\mathrm{r2}}}\sin\left(\dfrac{\alpha}{2}\right) \end{cases} \tag{4}$$

The output direct current I_o is the rectified and filtered i_{ab}. The average of the absolute values of i_{ab} can be expressed as:

$$I_{\mathrm{o}} = \frac{1}{T_{\mathrm{s}}}\int_0^{T_{\mathrm{s}}}|i_{\mathrm{ab}}(t)|\,\mathrm{d}t = \frac{8V_{\mathrm{dc}}}{\pi^2 n\omega_0 L_{\mathrm{r2}}}\sin\left(\frac{\alpha}{2}\right) \tag{5}$$

Therefore, in fixed-frequency mode, I_o is independent of the load, indicating the proposed converter exhibits current-source output characteristics. Simultaneously, the magnitude of I_o can be adjusted by modifying α.

2) Variable-frequency mode

The parameters of the components in (2) can be expressed in impedance form, yielding the simplified expression:

$$\begin{cases} \dot{V}_{\mathrm{AB}} = \dot{I}_{\mathrm{p}} Z_1 + \dot{I}_{\mathrm{m}} Z_2, & \dot{I}_{\mathrm{p}} = \dot{I}_{\mathrm{m}} + \dfrac{1}{n} \dot{I}_{\mathrm{s}} \\ \dot{I}_{\mathrm{m}} Z_{\mathrm{m}} = n \dot{I}_{\mathrm{s}} (Z_3 + Z_4), & \dot{V}_{\mathrm{ab}} = \dot{I}_{\mathrm{s}} Z_4 \end{cases} \tag{6}$$

where $Z_1 = Z_{L_{r1}} + Z_{C_p}$, $Z_2 = Z_{\mathrm{m}}$, $Z_3 = Z_{L_{r2}}$, and $Z_4 = \frac{Z_{C_s} R_{\mathrm{LE}}}{Z_{C_s} + R_{\mathrm{LE}}}$.

Solving (6) yields the output voltage V_{ab} of the resonant tank as:

$$V_{\mathrm{ab}} = \frac{2\sqrt{2} n Z_2 Z_4 V_{\mathrm{dc}}}{\pi \left[Z_1 Z_2 + n^2 (Z_1 Z_3 + Z_1 Z_4 + Z_2 Z_3 + Z_2 Z_4)\right]} \tag{7}$$

The DC voltage gain G_{dc} can be expressed as:

$$G_{\mathrm{dc}} = \frac{V_{\mathrm{o}}}{V_{\mathrm{dc}}} = \frac{8 n Z_2 Z_4}{\pi^2 \left[Z_1 Z_2 + n^2 \left(Z_1 Z_3 + Z_1 Z_4 + Z_2 Z_3 + Z_2 Z_4\right)\right]} \tag{8}$$

The characteristic curve of G_{dc} versus normalized f_s when the turn ratio $n = 1$ is shown in Fig. 2. $R_{LE,trans}$ denotes the minimum load resistance in the variable-frequency conversion mode.

Obviously, when the resonant tank is in an over-resonant state, that is, when the switching frequency f_s is greater than the resonant frequency f_0, the voltage gain monotonically decreases with the increase of f_s, which is beneficial for the design of the controller.

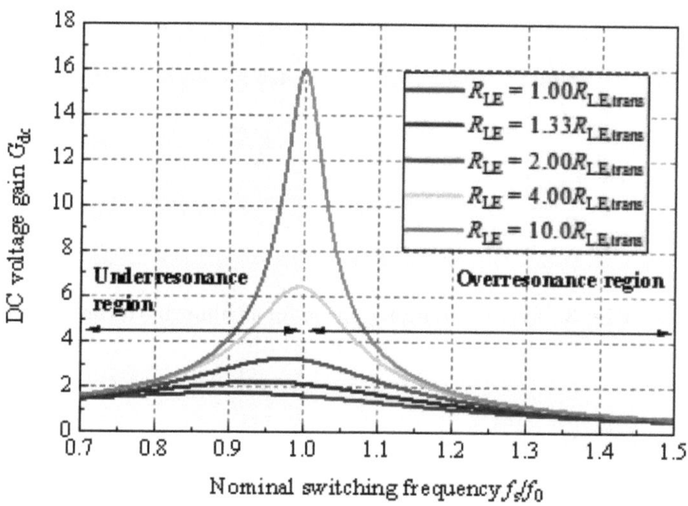

Fig. 2. Relationship curve between voltage gain and switching frequency.

In summary, the proposed converter exhibits current-source characteristics in fixed-frequency mode, where the output current is independent of the load and can be adjusted by altering the phase angle α. In variable-frequency mode, the voltage gain can be modified by changing the switching frequency f_s, thereby regulating the output voltage.

2.3 Battery Charging Application

Figure 3 depicts a typical voltage-current curve for lithium-ion battery charging. During the initial charging phase, the battery undergoes constant-current charging, requiring the converter to deliver a steady direct current output while the voltage progressively increases. During the final stage of charging, the battery enters a constant-voltage charging phase. In this state, the converter must output a steady voltage while the current gradually decreases.

Consequently, the proposed converter is highly suitable for battery charging applications. During the constant-current charging phase, a fixed-frequency operating mode can be employed, with the charging current adjustable via α according to different battery types. In the constant-voltage charging phase, a variable-frequency mode can be adopted, maintaining a constant output voltage by adjusting f_s.

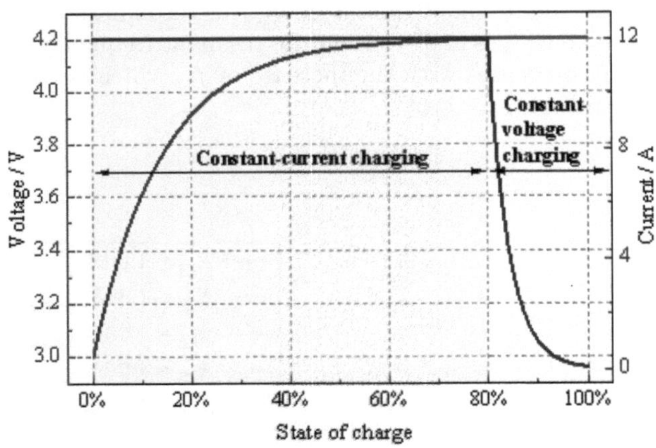

Fig. 3. Typical charging curve for lithium batteries.

2.4 Parameter Design

For the application scenario of electric vehicle charging, with a charging power of 5 kW, the total battery voltage range is 300V 420V, the maximum charging current is 12 A, and the converter input DC voltage range is 380V 420V. The design process for the relevant parameters is outlined below.

1) Operating frequency f_s
Typically, constrained by the losses of switching devices, the switching frequency should be less than 150 kHz. Constrained by the soft-switching range and the monotonicity of gain, the switching frequency should be greater than the resonant frequency. As shown in Fig. 2, a substantial gain variation can be achieved by varying the switching frequency from the resonance point to 1.5 times the resonance frequency. Therefore, a resonance frequency of $f_0 = 100\text{kHz}$ and a maximum switching frequency of $f_{s,max} = 150\text{kHz}$ can be adopted.

2) Excitation Inductance L_m
The design of excitation inductors requires comprehensive consideration of factors such as excitation current, transformer turns, and maximum magnetic flux density of the core. Where turns and maximum flux density allow, the excitation inductor should be maximized to minimize excitation current. Considering the above factors, take the excitation inductance $L_m = 500\mu\text{H}$.

3) Turns ratio n, resonant inductance L_{r1}, resonant inductance L_{r2}
The turns ratio n and resonant inductance L_{r2} determine the maximum output DC current in fixed-frequency operation mode, while the turns ratio n, resonant inductance L_{r1}, and resonant inductance L_{r2} collectively determine the DC voltage gain in variable-frequency mode.

Based on the converter characteristics in fixed-frequency mode and considering a 15% design margin, substituting the minimum input voltage $V_{dc,min}$ and maximum output DC current $I_{o,max}$ into the (5) yields the following constraint:

$$\frac{8V_{dc,min}}{\pi^2\omega_0 n L_{r2}} \geq 1.15 I_{o,max} = 13.8 \tag{9}$$

The operating condition with the lowest input voltage and minimum load resistance should ensure that the maximum DC voltage gain at $f_s = f_0$ is no less than $1.15V_{bat,max}/V_{dc,min}$. Under the operating condition with the lowest input voltage and highest load resistance, the minimum DC voltage gain at $f_s = 1.5f_0$ should be less than or equal to $0.85V_{bat,max}/V_{dc,min}$. Therefore, substituting into (9) yields the following constraints:

$$\begin{cases} G_{dc}(R_{LE,trans}, f_0) \geq 1.15\dfrac{V_{bat,max}}{V_{dc,min}} = 1.271 \\ G_{dc}(10R_{LE,trans}, 1.5f_0) \leq 0.85\dfrac{V_{bat,max}}{V_{dc,max}} = 0.85 \end{cases} \tag{10}$$

Taking $n = 1$, $L_{r1} = 10.0\mu\text{H}$, and $L_{r2} = 35.52\mu\text{H}$, substitution into (9) and (10) verifies that the conditions are satisfied.

4) Resonant capacitor C_p, resonant capacitor C_s
Substituting ω_0, L_{r1}, and L_{r2} into (1) yields $C_p = 253.3\text{nF}$ and $C_s = 71.31\text{nF}$.

The design specifications of the proposed converter are summarized in Table 1.

Table 1. The design specifications of the proposed converter.

Parameters	Value
Input DC voltage range $[V_{\mathrm{dc,min}}, V_{\mathrm{dc,max}}]$	[380 V, 420 V]
Output DC voltage range $[V_{\mathrm{o,min}}, V_{\mathrm{o,max}}]$	[300 V, 420 V]
Maximum output current $I_{\mathrm{o,max}}$	12.0A
Switching point load resistor $R_{\mathrm{L,trans}}$	35.0Ω
Resonant frequency f_0	100kHz
Maximum switching frequency $f_{\mathrm{s,max}}$	150kHz
Excitation inductance L_{m}	500μH
Transformer turns ratio n	1
Resonant inductor L_{r1} / L_{r2}	10.00μH / 35.52μH
Resonant capacitor C_{p} / C_{s}	253.3nF / 71.31nF
Filter inductors and capacitors L_{o} / C_{o}	470μH / 2.0μF

3 Fuzzy Controller Design

3.1 Fuzzy Control Logic

As analysed previously, in the application scenario of lithium-ion battery charging, the control objective during the constant-current charging phase is the output direct current I_{o}, with the control variable being the phase shift angle α of the full-bridge inverter. During the constant-voltage charging phase, the control objective is the output direct voltage V_{o}, with the control variable being the switching frequency f_{s}. The two modes are determined and switched based on the battery voltage.

The logic of the proposed converter fuzzy control is as follows:

1. When V_{o} is less than the maximum charging voltage $V_{\mathrm{bat,max}}$ of the lithium battery, the system enters fixed-frequency operation mode. The switching frequency f_{s} is set to the resonant frequency f_0. First, the error $E_I(t)$ between I_{o} and its reference value I_{o}^*, along with the rate of change $dE_I(t)/dt$, are calculated. These are fuzzyfied to yield the corresponding fuzzy quantities A_I and B_I. The fuzzy inference module calculates the fuzzyfied control quantity C_I, which is defuzzified to obtain the phase shift increment $\Delta\alpha$. α is accumulated over each control cycle, then combined with the phase shift angle feedforward value SD_α to yield the phase shift angle α.
2. When V_{o} is greater than the maximum charging voltage $V_{\mathrm{bat,max}}$ of the lithium battery, the variable frequency operation mode is entered, and the phase shift angle α is set to π. First, determine the error $E_V(t)$ between V_{o} and its reference value V_{o}^*, along with the rate of change of this error $dE_V(t)/dt$. Subsequently, apply fuzzyfying to obtain the corresponding fuzzy quantities A_V and B_V. The fuzzy inference module then calculates the fuzzy control variable C_V, which is defuzzified to yield the switching frequency increment Δf_{s}. The cumulative value SD_{fs} is obtained for each control cycle, which is

then combined with the switch frequency feedforward value f_{FW} to yield the switch frequency f_{s}.

3.2 Defining Fuzzy Concepts

1) Universe

The error ranges E_I and E_V can set to $\pm 10\%$ of the target value, namely [-1.2, 1.2] and [-42, 42]. The error change rates $dE_I(t)/dt$ and $dE_V(t)/dt$, like the output voltage and current, are constrained by the output filter and may be set to $[-6.23\times 10^4, 6.23\times 10^4]$ and $[-2.18\times 10^6, 2.18\times 10^6]$, respectively. The phase shift increment range is $[-\pi/200, \pi/200]$, and the switching frequency increment Δf_{s} ranges from $[-2.5\times 10^4, 2.5\times 10^4]$.

To reduce the complexity of controller design, the fuzzy domains for all measured variables and control variables are uniformly set to $U = \{-4, -3, -2, -1, 0, 1, 2, 3, 4\}$. Consequently, the quantisation factors for each measured variable and control variable are obtained as follows:

$$\begin{cases} k_{\mathrm{EI}} = 1.2/4 = 0.3, & k_{\mathrm{dEI}} = 6.23\times 10^4/4 = 1.5575\times 10^4 \\ k_{\mathrm{EV}} = 42/4 = 10.5, & k_{\mathrm{dEV}} = 2.18\times 10^6/4 = 5.45\times 10^5 \\ k_{\Delta\alpha} = \pi/800, & k_{\Delta f_{\mathrm{s}}} = 2.5\times 10^4/4 = 6.25\times 10^3 \end{cases} \tag{11}$$

2) Fuzzy language

For the fuzzy measurement variables A_I, B_I, A_V, B_V, and the control variables C_I, C_V, the following fuzzy linguistic sets are employed: negative large (NB), negative medium (NM), negative small (NS), zero (Z), positive small (PS), positive medium (PM), and positive large (PB).

3) Membership function

Using Gaussian distribution membership functions that conform to human brain thinking to describe membership degrees, the fuzzy languages can be expressed as:

$$\begin{aligned} & f_{\mathrm{NB}}(u) = \begin{cases} 1, & u < -3 \\ \mathrm{e}^{-\left(\frac{u+3}{0.6}\right)^2}, & u \geq -3 \end{cases}, \quad f_{\mathrm{PB}}(u) = \begin{cases} \mathrm{e}^{-\left(\frac{u-3}{0.6}\right)^2}, & u \leq 3 \\ 1, & u > 3 \end{cases} \\ & f_{\mathrm{NM}}(u) = \mathrm{e}^{-\left(\frac{u+2}{0.9}\right)^2}, \quad f_{\mathrm{NS}}(u) = \mathrm{e}^{-\left(\frac{u+0.5}{0.6}\right)^2}, \quad f_{\mathrm{ZO}}(u) = \mathrm{e}^{-\left(\frac{u}{0.1}\right)^2} \\ & f_{\mathrm{PS}}(u) = \mathrm{e}^{-\left(\frac{u-0.5}{0.6}\right)^2}, \quad f_{\mathrm{PM}}(u) = \mathrm{e}^{-\left(\frac{u-2}{0.9}\right)^2} \end{aligned} \tag{12}$$

The corresponding membership function curve is shown in Fig. 4, and the corresponding discretized membership degree table is shown in Table 2.

3.3 Fuzzy Control Rules

Due to the inconsistent dynamic characteristics of output voltage and output current, corresponding fuzzy control rules should be designed separately.

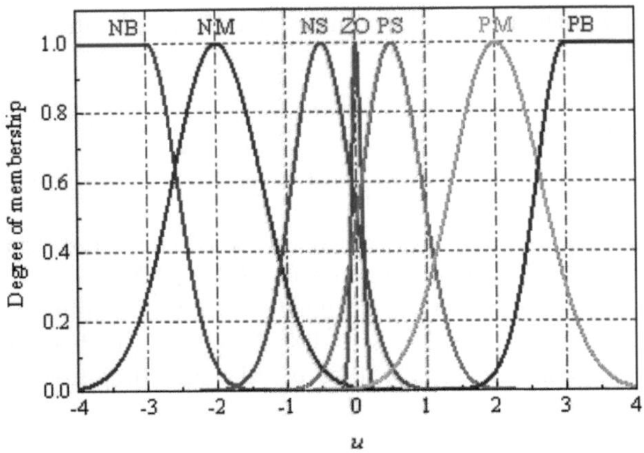

Fig. 4. Membership function curve.

Table 2. Membership degree table.

F \ U	−4	−3	−2	−1	0	1	2	3	4
NB	1	1	0.06	0	0	0	0	0	0
NM	0	0.29	1	0.29	0	0	0	0	0
NS	0	0	0	0.5	0.5	0	0	0	0
ZO	0	0	0	0	1	0	0	0	0
PS	0	0	0	0	0.5	0.5	0	0	0
PM	0	0	0	0	0	0.29	1	0.29	0
PB	0	0	0	0	0	0	0.06	1	1

1) Fuzzy control rules for output current
Based on the design experience of the converter, the fuzzy control rules for the output current can be obtained as shown in Table 3. The total fuzzy relationship can be obtained from the synthesis theorem of fuzzy relationships as:

$$\mathbf{R}_{\mathrm{I}} = \cup_l \mathbf{R}_{\mathrm{I},l} \tag{13}$$

where, $R_{I,l}$ can be obtained by $R_{I,l} = A_{I,i} \times B_{I,j} \times C_{I,k} = D_{I,m}^T \times C_{I,k}$, where D_I^T is the "flattening" of the result of $A_I \times B_I$ and then transposed, so that the two inputs can be equivalent to a single input.

After obtaining the fuzzy relationship matrix R_I, the fuzzy control quantity can be expressed as:

$$C_{\mathrm{I}}' = (A_{\mathrm{I}}' \times B_{\mathrm{I}}') \circ R_{\mathrm{I}} \tag{14}$$

Employing the weighted average method for defuzzification, $\Delta\alpha$ is expressed as:

$$\Delta\alpha = \frac{\sum_{i=1}^{7} U(i)C'_{\mathrm{I}}(i)}{\sum_{i=1}^{7} C'_{\mathrm{I}}(i)} \tag{15}$$

Table 3. Fuzzy Control Rules for Output DC Current I_{o}.

$A_I \setminus B_I$	NB	NM	NS	ZO	PS	PM	PB
NB	NB	NB	NB	NM	NM	NS	ZO
NM	NB	NB	NB	NS	NS	ZO	PS
NS	NB	NB	NM	ZO	ZO	PS	PM
ZO	NB	NM	NS	ZO	PS	PM	PB
PS	NM	NS	ZO	ZO	PM	PB	PB
PM	NS	ZO	PS	PS	PB	PB	PB
PB	ZO	PS	PM	PM	PB	PB	PB

2) Fuzzy control rules for output voltage

Based on the design experience of the converter, the fuzzy control rules for the output voltage can be obtained as shown in Table 4. Similar to the design method of the current controller, the fuzzy output can be expressed as:

$$C'_{\mathrm{V}} = (A'_{\mathrm{V}} \times B'_{\mathrm{V}}) \circ R_{\mathrm{V}} \tag{16}$$

Similarly, employing the weighted average method for defuzzification, the switching frequency increment Δf_{s} can be expressed as:

$$\Delta f_{\mathrm{s}} = \frac{\sum_{i=1}^{7} U(i)C'_{\mathrm{V}}(i)}{\sum_{i=1}^{7} C'_{\mathrm{V}}(i)} \tag{17}$$

Table 4. Fuzzy Control Rules for Output DC Voltage V_{o}.

$A_V \setminus B_V$	NB	NM	NS	ZO	PS	PM	PB
NB	PB	PB	PB	PB	PM	PS	ZO
NM	PB	PB	PM	PM	PS	ZO	NS
NS	PB	PM	PM	PM	NS	NS	NM
ZO	PB	PM	PS	ZO	NS	NM	NB
PS	PM	PS	PS	NS	NM	NB	NB
PM	PM	PS	NS	NS	NB	NB	NB
PB	PS	NS	NM	NM	NB	NB	NB

4 Simulation Verification

To validate the feasibility of the proposed resonant converter topology and control strategy, a simulation model was constructed within the PLECS simulation software.

The charging current for the constant current stage is set at 12A, and the charging voltage for the constant voltage stage is set at 420V. In practical applications, to prevent impact on the switching devices and battery, the charging current is not applied as a step input but rather through a soft-start measure, introduced in a ramped manner. Consequently, the simulation model adopts the same operational approach.

The simulated waveform of the charging process is shown in Fig. 5. At the commencement of the constant-current phase, the output direct current I_o is observed to accurately track the reference ramp signal. Upon reaching steady state, no oscillations occur, and the steady-state error is negligible. When the battery voltage rises to 420V, the constant-voltage charging phase commences. The switching response is rapid with minimal overshoot, and the steady-state error remains below three per thousand, demonstrating excellent control precision.

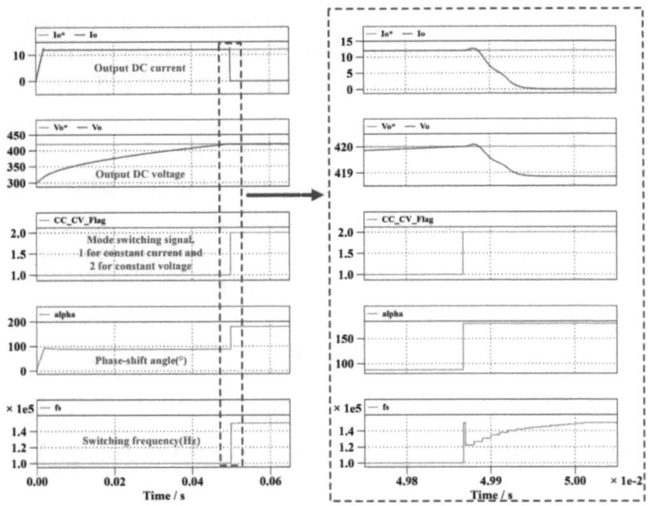

Fig. 5. Charging process.

The steady-state waveforms of the resonant tank input voltage v_1, output voltage v_2, input current i_1, and secondary inductance current i_{r2} are shown in Fig. 6. It can be observed that the current waveform is nearly sinusoidal with low harmonic content, demonstrating the feasibility of the proposed topology and the validity of the analytical approach.

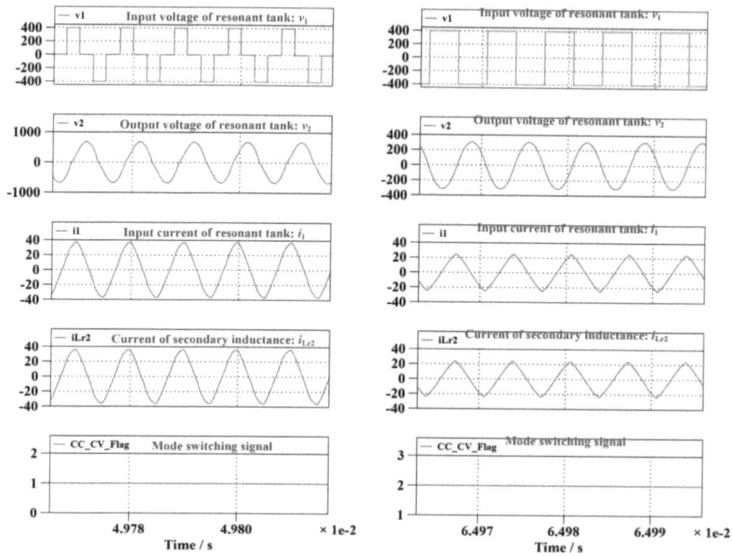

(a) Constant current mode (b) Constant Voltage Mode

Fig. 6. Resonant tank voltage and current waveform: (a) Constant current mode; (b) Constant Voltage Mode.

5 Conclusion

This paper proposes a secondary-side-parallel CLLC resonant converter by replacing the series capacitor in the conventional CLLC resonant network with a parallel capacitor. By switching between different modulation schemes, the proposed converter exhibits both current-source and voltage-source characteristics, making it highly suitable for battery charging applications. Furthermore, a feedforward fuzzy control strategy tailored to the converter characteristics is introduced, offering the advantages of rapid response and minimal overshoot.

Acknowledgments. This study was funded by the Key Program of National Natural Science Foundation of China under Grant 52130705.

References

1. Esteve, V., Jordán, J., Dede, E., et al.: Optimal LLC inverter design with SiC MOSFETs and phase shift control for induction heating applications. IEEE Trans. Industr. Electron. **69**(11), 11100–11111 (2022)
2. Li, X., Ma, H., Ren, S., et al.: A novel LCL resonant converter with inherent CC-CV output for on-board chargers of plug-in electric vehicles. IEEE Trans. Power Electron. **38**(4), 4212–4217 (2023)

3. Mukherjee, S., Ruiz, J., Barbosa, P.: A high power density wide range DC–DC converter for universal electric vehicle charging. IEEE Trans. Power Electron. **38**(2), 1998–2012 (2023)
4. Li, L., Xu, G., Sha, D., et al.: Review of dual-active-bridge converters with topological modifications. IEEE Trans. Power Electron. **38**(7), 9046–9076 (2023)
5. Miller, E.: The Use of Resonant Circuits in Power Conditioning Equipment. In: IEEE Power Electronics Specialists Conference, pp. 94–100. IEEE, Pasadena (1971)
6. Miller, E.: Resonant Switching Power Conversion. In: IEEE Power Electronics Specialists Conference, pp. 206–211. IEEE, Cleveland (1976)
7. Mudiyanselage, G., Keshmiri, N., Emadi, A.: A review of DC-DC resonant converter topologies and control techniques for electric vehicle applications. IEEE Open J. Power Electron. **4**, 945–964 (2023)
8. Gao, F.: A survey on analysis and design of model-based fuzzy control systems. IEEE Trans. Fuzzy Syst. **14**(5), 676–697 (2006)
9. Fery, C., Kuntze, H.: A neuro-fuzzy supervisory control system for industrial batch processes. IEEE Trans. Fuzzy Syst. **9**(4), 570–577 (2001)
10. Seker, H., Odetayo, M., Petrovic, D., et al.: A fuzzy logic based-method for prognostic decision making in breast and prostate cancers. IEEE Trans. Inf Technol. Biomed. **7**(2), 114–122 (2003)
11. Kadmiry, B., Driankov, D.: A fuzzy gain-scheduler for the attitude control of an unmanned helicopter. IEEE Trans. Fuzzy Syst. **12**(4), 502–515 (2004)
12. Bai, A., Zhuang, H., Roth, Z.: Fuzzy logic control to suppress noises and coupling effects in a laser tracking system. IEEE Trans. Control Syst. Technol. **13**(1), 113–121 (2005)
13. Hagras, H.: A hierarchical type-2 fuzzy logic control architecture for autonomous mobile robots. IEEE Trans. Fuzzy Syst. **12**(4), 524–539 (2004)
14. Osorio, R., Alonso, J., Vazquez, N., et al.: Fuzzy logic control with an improved algorithm for integrated LED drivers. IEEE Trans. Industr. Electron. **65**(9), 6994–7003 (2018)
15. Elkhateb, A., Rahim, N., Selvaraj, J., et al.: Fuzzy-logic-controller-based SEPIC converter for maximum power point tracking. IEEE Trans. Ind. Appl. **50**(4), 2349–2358 (2014)
16. Nabulsi, A., Dhaouadi, R.: Efficiency optimization of a DSP-based standalone PV system using fuzzy logic and dual-MPPT control. IEEE Trans. Industr. Inf. **8**(3), 573–584 (2012)
17. Wu, T., Chang, C., Chen, Y.: A fuzzy-logic-controlled single-stage converter for PV-powered lighting system applications. IEEE Trans. Industr. Electron. **47**(2), 287–296 (2000)

A Hybrid Knowledge-Guided Training-Free Decision Framework for Engineering Construction Progress Monitoring

Yumeng Kang[1], Hongbo Zhou[2], Kefeng Huang[3], Zhenyu Liu[4], Baozhu Zhang[4], Zefeng Qian[1], and Chongyang Zhang[1](✉)

[1] Shanghai Jiao Tong University, Shanghai, China
{kangyumeng,Zefeng_Qian,sunny_zhang}@sjtu.edu.cn
[2] Shanghai Research Institute of Building Sciences Group Co., Ltd., Shanghai, China
zhouhongbo@sribs.com
[3] Shanghai Jianke Engineering Consulting Co., Ltd., Shanghai, China
[4] Ningbo HTVision Digital Technology Co., Ltd., Ningbo 315000, China

Abstract. Engineering construction progress monitoring plays a pivotal role in schedule control and resource optimization of engineering projects. Traditional manual inspection methods for tracking engineering construction progress are inefficient and lack traceability. Existing computer vision-based approaches typically rely on large-scale annotated datasets or task-specific model training, which restricts their scalability and practical deployment in complex construction scenarios. To address this issue, we propose a novel training-free decision framework that integrates a general-purpose vision-language model (VLM) with a task-specific visual module and explicit domain knowledge, thereby balancing efficiency and accuracy. Specifically, one existing pre-trained object detector is adopted to extract explicit visual cues (e.g., the presence of workers and spatial locations of construction components), which are then fused with textual prompts as domain knowledge inputs for VLM-based construction stage inference. Furthermore, a dynamic sliding window post-processing strategy is introduced to mitigate ambiguity during stage transitions and reduce frame-level noise. Experimental results verify that the proposed framework can achieve reliable engineering construction progress recognition, providing an efficient and scalable solution for intelligent construction decision support.

Keywords: Construction progress monitoring · Prompt Engineering · Concrete pouring · Vision–Language Model · Intelligent construction decision

1 Introduction

Engineering construction progress monitoring plays a pivotal role in schedule control and resource optimization of engineering projects. For example, concrete

J. Lu et al. (Eds.): FLINS-ISKE 2026, LNCS 16752, pp. 277–308, 2027.
https://doi.org/10.1007/978-981-92-2369-5_19

pouring construction is a fundamental and continuous process in construction engineering, consisting of four stages: formwork installation, rebar binding, concrete pouring, and concrete curing. Accurately identifying the construction phase is crucial for tracking the progress and quality control of engineering projects. However, in practice, the complexity of the construction environment, the flexibility of worker actions, the diversity of workpiece processes in each construction stage, and the gradual transition and coexistence of various actions present significant challenges to automation and reliable stage identification.

To improve engineering construction monitoring efficiency, computer vision techniques have been increasingly applied to automate construction action and progress analysis. Early studies [9] focused on detecting construction equipment and key components to infer construction states and productivity. More recent approaches [35] adopt deep learning–based action recognition and semantic segmentation for finer-grained monitoring. Although these methods achieve certain effectiveness, they all assume relatively stable visual patterns and predefined operation procedures, which do not align with the high dynamics of construction sites in the real world. Furthermore, the general action recognition methods [5, 22] and the visual scene understanding method [39] have achieved remarkable success on their respective benchmarks, but they struggle to be transferred and generalized in variable actual construction scenarios and actions. This is because they usually require large-scale manually annotated datasets and specific task model training. This dependency limits their scalability and adaptability, especially in complex and dynamic construction environments where actions in different construction stages vary significantly in visual appearance, work processes, and on-site layout. This makes it difficult and costly to train a single model that can accurately recognize all stages.

In parallel, vision–language model (VLM) have demonstrated revolutionary visual understanding and language reasoning capabilities through multimodal pretraining on large-scale visual and textual data [23]. By utilizing natural language prompts to achieve flexible scene understanding, VLM provides a promising alternative for untrained engineering stage recognition, especially in situations where labeled data is scarce or acquisition costs are high. Nevertheless, directly applying VLM to construction imagery poses two key challenges. First, VLM may produce hallucinated predictions when there is a lack of sufficient visual constraints or domain-specific semantics are insufficiently grounded, leading to unreliable stage reasoning. Second, utilizing VLM to process the engineering construction video input will result in significant computational resource consumption, while utilizing VLM at the frame level of the video is sensitive to visual noise and transitional stages between stages, leading to frequent and unstable stage switching in continuous video streams. These issues hinder the direct deployment of VLM in actual construction monitoring systems.

An emerging paradigm enhances large language models (LLMs) by integrating task-specific small models (SMs) [36], which provide domain knowledge and factual cues to mitigate hallucinations and domain limitations. Motivated by this paradigm, we propose a hybrid Knowledge-guided training-free decision framework to suppress hallucinations and realize efficient and stable monitoring during the construction phase. The process begins with dynamic frame sampling from the construction site video stream for subsequent frame-level image processing to avoid excessive computational overhead of directly processing the video. We introduce the well trained object detection model YOLO-World [6] to detect workers in the images. The resulting bounding boxes are then overlaid onto the original images to provide explicit visual cues for the VLM, reducing ambiguity in analyzing complex construction scenes. The VLM processes these enhanced images, utilizing a hybrid prompt composed of prior textual knowledge of each construction stage's characteristics and the visual evidence of the highlighted human regions. While inferring the construction phase, VLM also provides corresponding confidence scores and reasons to ensure interpretability. Furthermore, to ensure stable and reliable decision-making over time, we design a dynamic sliding-window post-processing strategy that integrates the VLM's outputs. This strategy adaptively adjusts the window length based on stage dynamics and prediction confidence. In contrast to prior learning-centric approaches, this work explores a training-free decision framework which does not require any form of model fine-tuning for engineering construction progress monitoring task, including full parameter fine-tuning and efficient parameter tuning. In this work, we take the concrete pouring construction monitoring as an example to verify the effectiveness of the proposed framework, which achieves task adaptation only through pre-trained VLM and prompt engineering. Experimental results demonstrate that the proposed framework effectively suppresses frame-level misclassifications among the four standard concrete pouring stages during process monitoring, alleviates ambiguous transitional states, and produces clear and well-defined stage boundaries for calculating the corresponding construction durations.

2 Related Works

2.1 Engineering Construction Stage Recognition

The research on engineering construction monitoring explored automated methods to replace manual inspections. Sensor-based approaches have been widely used to capture operational signals during the pouring process. Moon et al. [19] utilized RFID to track the movement of concrete trucks and achieved transportation-level progress monitoring. Kim et al. [13] combined acoustic sensing with deep learning technology to detect the state of concrete pouring, while Danel et al. [7] utilized embedded sensors on cranes to monitor the pouring operation and analyze delays caused by resources. Although these sensor-driven methods have improved the time resolution at each stage, they typically provide limited scene-level context and have difficulty linking equipment operations with the overall construction progress.

Computer vision-based methods are increasingly being adopted to achieve more granular and scene-aware monitoring. Gong et al. [9] proposed a visual-based framework that identifies the state of concrete pouring by detecting buckets and mapping them to predefined operation areas. Wang et al. [30] further simulated the interaction between concrete buckets and crane cables to accurately identify the pouring cycle in the dam construction scene with high precision. Zhang et al. [38] utilized object detect model to detect the appearance and position changes of construction equipment such as pump pipes and extract key construction events. These methods mainly focus on detecting machines or parts during construction, which have high accuracy but lack representation of the overall construction process.

To better utilize the information in complex construction environments, Pfitzner et al. [20] proposed a concrete pouring progress monitoring method that combines a knowledge graph. By detecting relevant objects such as workers and concrete buckets and associating them with BIM information to construct a spatial-temporal knowledge graph, and then using rules and GNN to infer the pouring state, it is more universal than traditional image recognition. Zhang et al. [37] proposed a pouring progress tracking framework based on semantic segmentation using U-Net [25], combining pixel-level recognition and work status identification algorithms, which can identify production efficiency indicators such as delays and stops. Yang et al. [35] further applied semantic segmentation and multi-camera semantic fusion to achieve pixel-level modeling of the concrete pouring area. However, these methods are still sensitive to extreme occlusion and camera frame loss issues, and rely on a large amount of annotation, project-specific parameters, or fixed assumptions about the layout and continuity of the work process, which limits their reliability and generalization in complex and variable construction environments.

2.2 Collaboration Between Large Models and Small Models in Industrial Monitoring Applications

With the rapid development of LLMs, their potential for application in industrial scenarios is gradually emerging. Under the reasonable design of prompt engineering, LLMs can, with their powerful semantic understanding and reasoning capabilities, achieve high-level comprehension of complex construction environments, abstract engineering semantics, and multi-source information, demonstrating strong generalization performance [3]. The introduction of multimodal large models has incorporated the visual modality, enabling a more comprehensive capture of key information in the construction environment through visual knowledge, providing new ideas for automated engineering supervision [8,21,29].

However, in actual industrial deployment, large models still face issues such as hallucination risks, high inference costs, and insufficient fine-grained perception capabilities [24], making them difficult to independently handle tasks with high requirements for accuracy and real-time performance. On the other hand, specialized small models have accuracy and efficiency, so combining traditional computer vision models or lightweight small models to build a framework where large and small models work collaboratively, leveraging the advantages of both, has gradually become an important paradigm in industrial intelligent systems [17,31].

In the industrial operation and maintenance monitoring scenarios, Yuan et al. [36] proposed a collaborative integration framework that combines the cross-domain knowledge integration capability of large models with the advantages of small models in precise data modeling. In the case of energy-saving in communication machine rooms, the large model, combined with industry knowledge and LSTM models, analyzes energy consumption time series data and formulates dynamic air conditioning adjustment strategies; in the case of operation and maintenance of energy storage stations, the large model queries and calls the small model for fault detection based on the health status, and the two work collaboratively to complete complex operation and maintenance decision-making tasks, demonstrating the feasibility of the collaborative framework in real industrial systems. Wang et al. [32] proposed a dynamic selection strategy framework CoLLM. For complex inputs, it routes to the large model to obtain more reliable results; for low-complexity inputs, the small model is responsible for rapid output, balancing efficiency and accuracy. However, the design of the small model is sensitive and requires domain experts to fine-tune.

In the application of visual small models, Jeoung et al. [12] achieved idle state and operation task recognition by conducting target detection and tracking on the equipment and combining multimodal model reasoning, without the need for additional training data to adapt to building scenarios; however, its adaptability to multi-device collaborative scenarios is limited, and it cannot achieve precise time event recognition. This indicates that there are still deficiencies in addressing construction monitoring in continuous multi-stage complex environments at present. This prompts us to need a more robust and systematic large and small model collaboration paradigm, which explicitly considers temporal coherence and stage-level semantics in construction process monitoring.

3 Method

3.1 Overview

Given continuous video streams with timestamps captured from engineering construction sites, the objective of the task is to automatically identify the construction stages corresponding to each time point, accurately determine stage transitions to derive the duration of each construction stage, and output structured temporal records suitable for construction management.

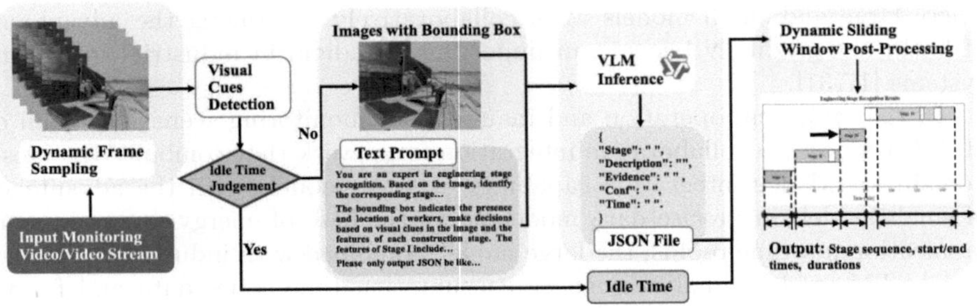

Fig. 1. Overall framework of the proposed method. This framework integrates dynamic frame sampling, visual cues detection via exiting trained object detection, VLM inference with domain knowledge prompts, and dynamic sliding-window post-processing. The idle time judgment mechanism conditionally controls the invocation of VLM. When workers are continuously detected, the VLM performs inference and outputs a structured JSON log (stage, descriptions, etc.) to achieve reliable post-processing and robustly infer the sequence of construction stages and their time boundaries.

The proposed framework is illustrated in Fig. 1. Input on-site monitoring video streams or recorded construction videos, the system performs five consecutive steps:

1. Dynamic frame sampling from the input video, reducing redundant computation while preserving critical temporal information.
2. Visual cues detection via exiting well-trained human detection, utilizing the YOLO-World model [6] to detect workers, providing VLM explicit and task-relevant visual constraints.
3. Idle time judgment mechanism: If there is no detection of workers for a continuous period and the SSIM approximation value of adjacent frames is 1, skip the VLM inference to reduce computational overhead, and mark this time period as "idle".
4. Visual-language model inference, where the VLM infers construction stages based on combined visual cues and text-based knowledge prompts, and records stages, descriptions, evidence, and confidence levels as structured JSON logs.
5. Dynamic sliding window post-processing, which analyzes the generated logs to improve the accuracy of decision-making for stage transition boundary construction.

The framework outputs standardized JSON files and visualizations that present the start time, end time, and duration of each construction stage, achieving reliable and interpretable construction progress monitoring through the cooperation of task-specific small models for visual grounding and VLM for semantic reasoning.

3.2 Dynamic Frame Sampling Strategy

The on-site monitoring video stream is continuously received. Fixed-interval sampling often leads to redundant frames during static phases and may miss critical events during rapidly evolving stages. To address this issue, a content-aware dynamic frame sampling strategy is introduced, with the initial sampling interval set at 2 min and the corresponding frame timestamps recorded. The upper bound of the sampling interval was set at one frame every 10 min, and the lower bound was set at one frame every 30 s. The sampling rate was dynamically adjusted within the upper and lower bounds through adaptive similarity-based interval modulation and stage-based interval modulation to achieve controllable and dynamic regulation. The complete dynamic frame sampling workflow, integrating both similarity-based and stage-based modulation mechanisms, is illustrated in Fig. 2.

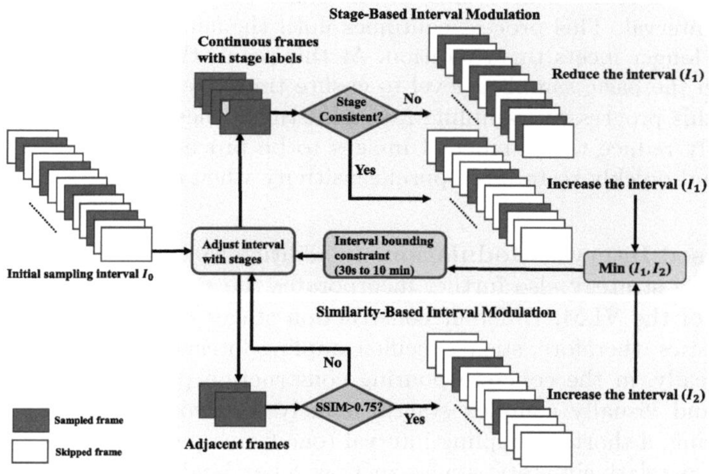

Fig. 2. Dynamic frame sampling strategy. The stage-based modulation adjusts intervals based on predicted stage consistency. The similarity-based modulation detects visual changes. Combine two modules to dynamically adjust the sampling interval.

Similarity-Based Interval Modulation. To ensure that critical temporal changes can be captured promptly while reducing the redundant image processing overhead, an adaptive similarity-based interval modulation mechanism is introduced. The structural similarity index measure (SSIM) [34] is used to measure the degree of scene change between adjacent frames, and its value is highly sensitive to the content changes of the images, but relatively insensitive to image noise and slight blurring, making it more suitable for long-term monitoring scenarios in complex construction environments.

For two adjacent frames I_{t_k} and $I_{t_{k+1}}$, sampled at timestamps t_k and t_{k+1} respectively, the value of SSIM is computed as:

$$\mathrm{SSIM}(I_{t_k}, I_{t_{k+1}}) = \frac{(2\mu_{t_k}\mu_{t_{k+1}} + C_1)(2\sigma_{t_k,t_{k+1}} + C_2)}{(\mu_{t_k}^2 + \mu_{t_{k+1}}^2 + C_1)(\sigma_{t_k}^2 + \sigma_{t_{k+1}}^2 + C_2)} \tag{1}$$

where μ, σ^2, and $\sigma_{t_k,t_{k+1}}$ denote the mean, variance, and covariance of local image patches, respectively, and C_1, C_2 are stability constants. The value of SSIM is between 0 and 1, with closer values to 1 indicating higher similarity between images.

When the similarity between adjacent frames is higher than the preset threshold 0.75, it indicates that the current construction scene is changing relatively slowly in time. At this time, increase the sampling interval to twice the previous value. If the similarity between multiple adjacent frames continues to be higher than this threshold in the subsequent sampling process, the sampling frequency will be further reduced according to the same rule, gradually expanding the sampling interval. This process continues until the similarity between adjacent frames no longer meets this condition. At this point, the sampling frequency is restored to the basic sampling level to ensure timely response to scene changes. Through this progressive sampling rate adjustment mechanism, the system can significantly reduce the number of images to be processed during stable scene periods, and quickly restore temporal sensitivity when changes occur.

Stage-Based Interval Modulation. In addition to low-level visual dynamics, the sampling strategy also further incorporates the stage information from the reasoning of the VLM. Different construction stages exhibit distinct temporal characteristics; therefore, stage-specific sampling intervals are assigned.

Specifically, in the concrete pouring construction process, for the dynamic evolving and visually complex stages, namely formwork installation and concrete pouring, a shorter sampling interval (one frame per minute) is adopted. In contrast, for relatively static stages such as rebar binding and concrete curing, a longer interval (one frame every three minutes) is used.

Furthermore, the consistency in stage prediction during the stages is utilized to improve the sampling process. When VLM generates consistent stage predictions over consecutive frames, the sampling interval gradually increases to reduce redundancy. Conversely, when the stage prediction fluctuates, indicating potential transitions or ambiguous states, the interval is shortened to enhance temporal resolution and capture fine-grained changes.

Compare the current intervals of the two modulation strategies, and select the smaller one to ensure that no critical information is lost. The finally adjusted interval will be fed back to the system to form a closed-loop adaptive sampling. The extracted images are named with timestamps and saved. They are then subjected to unified object detection processing and encoding to meet the input format requirements for subsequent model inference.

3.3 Visual Cues Detection

The presence and spatial configuration of workers plays a crucial role in construction stage recognition, as workers' behaviors and their interactions with construction objects can represent the underlying process state. To explicitly capture these cues, we employ YOLO-World [6], a well trained object detection model that performs open-vocabulary detection without task-specific fine-tuning and demonstrates strong generalization to common category of persons, enabling direct deployment in unseen construction scenarios.

Given an input image, YOLO-World is queried with the category "person" to identify visible workers. Only bounding boxes with confidence scores greater than 0.6 are retained. This threshold mitigates missed detections caused by overly strict filtering while avoiding inaccurate regions introduced by low-confidence predictions. The retained bounding boxes will be superimposed on the original image, effectively guiding the VLM to focus on this area while maintaining the overall image context [26]. Then, the image is encoded in Base64 format and input into the VLM, where the bounding boxes highlight the presence and spatial location of the staff members and serve as useful visual cues for visual-language reasoning.

3.4 Idle Time Judgment

Due to the presence of a significant amount of idle time in actual building monitoring, no construction activities occur during this period. Directly invoking VLM on each sampling frame would waste computing resources and fail to provide useful stage information.

We solve this problem by utilizing a lightweight idle time judgment module, which evaluates two conditions before VLM inference. Firstly, it checks if there are workers present by using the YOLO-World detection results; a situation where there are no workers for more than 10 consecutive frames is assumed to indicate no active construction behavior during the corresponding time period. Secondly, it calculates the value of SSIM between adjacent frames to verify the stability of the scene; if the value is approximately 1, it indicates that the scene is truly stationary rather than experiencing a brief occlusion. Only when both conditions are met will VLM inference be skipped, and the stage label remains unchanged, marking it as idle time. This mechanism enables the use of a lightweight object detection model to filter out meaningless frames, concentrating computing resources on meaningful construction activities rather than redundant idle scenes.

3.5 Visual–Language Model Inference

Engineering construction typically constitutes a long-cycle engineering process, with each stage lasting several days, concrete pouring construction is a typical example; therefore, real-time inference is not a strict requirement. However, complex backgrounds, frequent occlusions, the diversity of visual features within the same stage, and strong visual similarity between adjacent stages make it difficult for a single supervised visual model to generalize across all stages effectively.

Equipped with extensive prior knowledge and strong multimodal reasoning capabilities, VLMs are well suited for such scenarios where high accuracy is required but strict real-time performance is not. In this work, we adopt the Qwen3-VL-plus [1] multimodal model, which provides high-resolution image understanding and robust reasoning under complex industrial conditions, including challenging lighting and occlusions. Preliminary analysis suggests that, for typical concrete construction scenes, the model can accurately identify construction stages utilizing only a small amount of textual prompts, demonstrating strong reasoning capability.

Nevertheless, the hallucination of VLM is prone to inducing misjudgment of construction stages. To improve reliability, we design diversified and fine-grained prompts. Specifically, object detection models provide visual cues regarding the presence and locations of workers, while stage-specific visual characteristics are summarized and incorporated as textual prompts. In addition, the model is explicitly required to distinguish non-construction or non-concrete scenes to enhance robustness against abnormal inputs.

Rather than performing simple classification, the prompt instructs VLM to infer based on visual evidence, including detected worker activities, construction materials, and scene context, and to output structured results in JSON format, consisting of the predicted stage, a brief scene description, supporting visual evidence, confidence score, and timestamp. This reasoning-oriented design improves interpretability, mitigates hallucination, and supports traceable and auditable decision-making.

The carefully designed prompts based on the above requirements are shown in Table 1, which clearly specify visual features related to the task, including whether workers are present, visible construction tools, construction environment, and observable material status. These clear clues are helpful in establishing the reasoning process of the visual language model based on specific visual evidence. Combined with specific domain prior knowledge, they further emphasize the different characteristics of concrete construction materials at different stages (as shown in Table 2), in order to solve the frequent confusion problems that occur between adjacent construction stages.

Furthermore, to enable reliable post-processing and temporal reasoning, the VLM is instructed to generate outputs in a standardized format, thereby preventing uncontrolled variations in natural language responses. In addition, within the dynamic frame sampling strategy, extracted frames are saved using timestamp-based filenames, ensuring a one-to-one correspondence between each frame and its capture time. Accordingly, a dedicated "time" field is incorporated into the JSON representation to explicitly associate the VLM's inference results with the corresponding images. Ultimately, the VLM outputs are stored as structured log files in this unified format, which facilitates dynamic sliding-window post-processing for construction stage duration monitoring while also supporting data traceability.

Table 1. Template for concrete pouring construction stage recognition prompt.

System Prompt

You are an expert in identifying the stages of concrete pouring construction and are skilled at determining the construction stage based on images.

User Prompt

- <Input Image> Analyzes and infers the worker's activities based on the image, and determines the corresponding stage: *formwork installation*, *rebar binding*, *concrete pouring*, *concrete curing*, or *non-concrete pouring scenarios*.

- The red bounding boxs in the image serve as a reminder to indicate the presence and distribution of the workers, but they cannot be the sole basis for determining the stage.

- Attention should also be paid to structural features, material conditions, visible construction traces, and the overall scene background, and a rigorous judgment should be made by combining prior knowledge in areas such as the characteristics of each stage.

Output Format

Please only output JSON be like:{
"Stage": "The stage at which the image was derived from reasoning",
"Description": "Summarize the content of the image",
"Evidence": "Explain the key visible features supporting the stage decision",
"Conf": "A confidence score between 0 and 1. Scores above 0.8 indicate high confidence; scores below 0.8 indicate lower confidence."}

3.6 Dynamic Sliding Window Post-processing

Based on the structured JSON logs generated by VLM, time analysis is conducted to infer the complete construction cycle and the start and end times of each stage. Since frame-level prediction is prone to transient misclassification, ambiguous stage transitions, and environmental noise, direct aggregation usually leads to fragmented stage sequences and inaccurate boundaries. To address these issues and improve the robustness of stage identification and boundary localization of stage transitions, a dynamic sliding window post-processing strategy is proposed. The core principle is to perform stage decisions at the time window level. This design does not preset a fixed optimal window for all scenarios. Instead, it adaptively adjusts the window length by considering the dynamic characteristics of different construction stages and the confidence of VLM reasoning, combined with a fusion strategy, to achieve reliable state discrimination and stable transition detection. The strategy for dynamically adjusting the window length is described as follows.

Firstly, set the basic window length to 10 frames as the benchmark for adaptive adjustment, and in actual operation, dynamically adjust it based on the perception of stage characteristics and confidence level.

Stage Perception. Different construction stages exhibit different time characteristics. In order to better capture the beginning and end of each stage, the window length is set according to the time characteristics of each construction

Table 2. Construction stage characteristics and discriminative visual features.

Stage	Key Visual Features
Formwork Installation	• Visible formwork (wooden/steel plates, support systems) • Template unclosed or being installed • No concrete body visible • No rebar cage formation • No pump pipes
Rebar Binding	• Dense steel bars (horizontal and vertical arrangement) • Stirrups, reinforcement cages visible • Workers may be tying rebars • No formwork closure • No wet/flowing concrete
Concrete Pouring	• Concrete pump and delivery pipes present • Gray, flowing concrete visible • Wet slurry surface texture • Fresh concrete inside formwork
Concrete Curing	• Hardened concrete surface • Maintenance measures (watering, covering, insulation) • No pump pipes • No flowing concrete
Non-Construction	• Not a construction site • Completely unrelated to the four stages above

activity. The rapidly changing stage requires a shorter window to maintain time sensitivity, while the relatively static stage uses a longer window to suppress accidental interference.

For the concrete pouring construction scenario, the configuration of the stage perception window adopted in this paper is summarized in Table 3. The characteristics of formwork installation and rebar binding are moderately dynamic and continuous operation, therefore 8 frames of medium length windows have been allocated. Concrete pouring involves sudden transitions and rapid scene changes, such as pump activation and termination, and is assigned a minimum window length of 5 frames to avoid missing critical boundary events. In contrast, the solidification stage is visually static and more susceptible to sporadic interference, such as temporary worker passages or auxiliary equipment. Therefore, a maximum window of 15 frames is used to ensure stable recognition. For non architectural scenes that do not exhibit meaningful temporal patterns, the basic window length of 10 frames is retained because only rough distinctions are needed.

Table 3. Stage perception sliding window length configuration for different construction scenarios.

Construction Stage	Characteristics	Window Length (Frames)
Formwork installation	Moderate dynamic	8
Rebar binding	Moderate dynamic	5
Concrete pouring	Highly dynamic	5
Concrete curing	Highly static	15
Non-concrete construction	No fixed dynamic patterns	10

Confidence Perception. In addition to stage perception adaptation, the window length is further adjusted by the confidence value output by VLM. For scenes with frequent changes, the model output is easily affected, and the single frame inference confidence of VLM reflects the reliability of the judgment. Therefore, additional confidence parameters are used to control the length of the sliding window. For dynamically changing scenes with low confidence, filter the output results of noisy scenes, and use the continuous high confidence analysis results as the final monitoring results.

The threshold is designed to balance the stability of stage prediction and the requirements for temporal smoothing. In the prompt, we stipulate that values above 0.8 indicate a relatively certain accuracy of the inference results, which is regarded as a medium-high confidence interval. Values below 0.8 represent a low confidence interval. Specifically, if several consecutive frames are within the medium-high confidence interval (conf ≥ 0.8), the model has a clear judgment for the current stage, with low noise interference, usually corresponding to the stable interval within the stage, and can maintain the default window length of the current stage; conversely, if there are frames with a low confidence interval (conf < 0.8), the model may have a blurry judgment due to lighting, occlusion, and scene transitions, usually occurring during stage transitions or when visual information is unclear. They are regarded as noise frames and are filtered out; at the same time, on the basis of the default length of the current stage, moderately expand the window length to integrate more contextual information from the time series samples, in order to reduce the probability of misjudgment and effectively alleviate noise and unstable predictions near the stage boundaries. When several consecutive frames are within the medium-high confidence interval, the default window length is restored.

The process of adjusting the window length according to the confidence level can be formulated as:

$$L = \begin{cases} L_0, & c_t \geq 0.8, \\ (1+\alpha)L_0, & c_t < 0.8. \end{cases} \tag{2}$$

where L is the current window length and L_0 denotes the initialized window length; c_t is the stage confidence predicted by the VLM at frame t. When the confidence is no less than 0.8, the window size is preserved; otherwise, it will expand on the basis of L_0. Specifically, α is dynamically determined through linear interpolation based on the proportion of low-confidence frames within the current window:

$$\alpha = \alpha_0 + \Delta \times \frac{N_{\text{low}}}{N_{\text{total}}} \tag{3}$$

where α_0 denotes the base expansion coefficient, and Δ controls the sensitivity of the window expansion to the noise level. N_{low} denotes the number of frames with confidence below 0.8, and N_{total} represents the total number of frames in the window. This adaptive mechanism will expand the window accordingly based on the observed noise rate. The larger the proportion of low-confidence frames, the greater the extent of window expansion. Specifically, α_0 is set to $1/L_0$ to ensure that the window will at least increase by one frame even under low noise conditions. The upper limit of α is set by requiring that the length of the extended window does not exceed twice the length of the current stage window:

$$L_{\max} = (1 + \alpha_{\max})L_0 \leq 2L_0 \tag{4}$$

Algorithm 1. Dynamic sliding window post-processing algorithm

Input: frame number N, JSON log $\mathcal{L} = \{(t_i, s_i, c_i)\}_{i=1}^{N}$, base window lengths $L_{\text{base}}[\cdot]$
Initialize: load $L_{\text{base}}[\cdot]$ per stage; set $L_0 \leftarrow 10$, $i \leftarrow 1$
for $i = 1$ **to** N **do**
  # stage perception
  if $s_i = s_{i+1} = \cdots = s_{i+4}$ **then**
    $L_0 \leftarrow L_{\text{base}}[s_i]$ # trigger stage-aware adjustment
  end if
  # confidence perception
  $\mathcal{W} \leftarrow \{(s_j, c_j)\}_{j=i}^{i+L_0-1}$
  $N_{\text{low}} \leftarrow |\{j \in \mathcal{W} : c_j < 0.8\}|$, $N_{\text{total}} \leftarrow |\mathcal{W}|$
  if $N_{\text{low}} > 0$ **then**
    $\alpha \leftarrow \alpha_0 + \Delta \times (N_{\text{low}}/N_{\text{total}})$ # Eq. (3)
    $L \leftarrow (1+\alpha) \cdot L_0$
    $\mathcal{W} \leftarrow \{(s_j, c_j)\}_{j=i}^{i+L-1}$ # Eq. (2)
  else
    $L \leftarrow L_0$
  end if
  # sliding window voting
  for each label k in $\mathcal{W}$ **do**
    $\text{vote}(k) \leftarrow \frac{|\{j \in \mathcal{W}: s_j = k, c_j \geq 0.8\}|}{|\mathcal{W}|}$
  end for
  if $\max_k \text{vote}(k) \geq \theta_c$ **then**
    $s_i^* \leftarrow \arg\max_k \text{vote}(k)$
    if $s_i^* \neq s_{i-1}^*$ **and** $\text{vote}(s_i^*) \geq \theta_t$ **then**
      $t_{\text{trans}} \leftarrow \min\{t_j : c_j \geq 0.8,\ s_j = s_i^*,\ j \in \mathcal{W}\}$
      record $(s_{i-1}^* \rightarrow s_i^*,\ t_{\text{trans}})$ into $\mathcal{T}$
    else if $s_i^* \neq s_{i-1}^*$ **and** $\text{vote}(s_i^*) < \theta_t$ **then**
      $s_i^* \leftarrow s_{i-1}^*$ # transition not validated
    end if
  end if
  $i \leftarrow i + 1$ # dynamic update
end for
return smoothed stage sequence $\mathcal{S}^* = \{s_i^*\}_{i=1}^{N}$, transition boundaries $\mathcal{T}$

where $L_{\max}$ denotes the upper bound of the expanded window. This approach limits the maximum time context range introduced by window expansion to an appropriate range corresponding to the basic window, thereby preventing over-smoothing and avoiding blurring of the actual stage transition boundaries. In the scenario of concrete pouring construction, based on the preset basic window values according to the characteristics of each stage, it is calculated that α_0 is 0.2 and Δ is 0.8.

In each window, the stage labels are fused through a sliding window voting mechanism. Specifically, the voting proportion vote(k) of each stage label within the current window is calculated, and two decision thresholds are introduced: the confirmation threshold θ_c and the transition threshold θ_t. The constraint condition $\theta_c > \theta_t$ is set, meaning the voting proportion required for confirming a stable stage should be higher than that for triggering a transition, to prevent interference from inaccurate predictions while maintaining sensitivity to true stage changes. When the voting proportion of a certain stage exceeds θ_c, that stage is confirmed. Only when the voting proportion of the new stage in the window continuously exceeds θ_t will the stage transition be verified. The transition point is the first occurrence time of consecutive high-confidence predictions, thereby achieving smooth and stable positioning of the stage boundaries. This dual-threshold design is similar to the lag threshold in signal processing. The high threshold θ_c is used to confirm stable states, while the lower threshold θ_t is used to trigger state transitions. This lag structure can effectively suppress false triggers in noisy environments.

By integrating stage perception and confidence perception through window adaptation, the proposed post-processing strategy can reduce frame level noise and transient misclassification while maintaining sensitivity to real construction stage transitions.

4 Experiments

4.1 Dataset

Fig. 3. Representative samples from four concrete construction stages in the dataset, covering a variety of lighting conditions, viewpoints, and occlusions.

To evaluate the effectiveness of the proposed method, experiments were conducted utilizing real-world construction videos and images collected from public online sources. The constructed dataset was designed to cover various practical variations commonly found in real concrete construction environments. We collected a total of 60 videos related to concrete construction, sampling frames uniformly at intervals of 4 s, and collected some images of concrete construction as supplements. After screening, we finally obtained 1675 image frames. The collected samples cover different lighting conditions, including daytime, nighttime, and cloudy scenes, different degrees of occlusion, including no occlusion and partial occlusion, and different camera perspectives, including remote and close-up perspectives. This reflects the high visual variability and complexity of real-world construction scenes. In this dataset, in addition to including four categories of concrete construction stages such as formwork installation, rebar binding, concrete pouring, and concrete curing, there are also 350 images of non-concrete construction scenarios, including open spaces, office areas, living areas, and road construction scenes, for evaluating the robustness of the proposed method in distinguishing target construction stages from visually similar but unrelated scenes. Representative samples from these four construction stages are shown in Fig. 3.

Moreover, to ensure the reliability of the ground truth annotations, each image was independently annotated by two annotators with knowledge in the field of construction engineering. In case of disagreement, a third expert would act as an arbitrator to resolve the inconsistency and ultimately determine the labels. This effectively reduces subjective bias and improves the consistency and accuracy of the annotations. Our dataset can be obtained from https://github.com/aslearning/Concrete-Construction-Recognition-Dataset.

All experiments utilizing our method on this dataset were conducted in an inference-only manner. No task-specific fine-tuning or additional training was performed for either the object detector or the visual–language model.

4.2 Performance Evaluation of Hybrid Knowledge-Based Inference

To evaluate the reasoning performance of the VLM that integrates visual enhancement and fine-grained text prompts, standard classification metrics were adopted, including the overall classification accuracy, Precision (P), Recall (R), Average Precision (AP) for all categories at each stage and mean Average Precision (mAP). Utilizing the Qwen3-VL-plus model [1], the system achieved an overall accuracy of 88.3% and an mAP of 80.7%, with an average inference time of 5.4 s per frame. Table 4 shows that the proposed framework has performed excellently in all stages of construction.

We further applied vision–language models (VLMs) with different model scales to the construction stage recognition task to validate the robustness of the proposed method as well as its deployment potential on edge devices. For this purpose, three representative VLMs were selected, namely gpt-4o-mini as a lightweight model, Qwen3-VL-plus as a medium-scale model, and Gemini-3-pro-preview as a heavyweight model, thereby covering a broad spectrum of

Table 4. Performance evaluation of construction stage recognition.

Stage	Precision (%)	Recall (%)	AP (%)
Formwork installation	95.0	78.0	72.7
Rebar binding	77.4	93.3	78.7
Concrete pouring	94.7	93.2	89.1
Concrete curing	90.9	88.2	81.8
Non-concrete construction	94.2	89.1	81.2

lightweight, medium-scale, and large-scale VLMs. Table 5 presents the performance evaluation results, systematically reporting key metrics including accuracy, mean average precision (mAP), and inference efficiency (inference time per frame). These results provide practical guidance for selecting appropriate VLMs in construction stage recognition tasks. Overall, all evaluated VLMs achieved performance levels close to or exceeding 80%, indicating that the proposed visual cues and refined textual prompts can function reliably across different VLM architectures and capacity levels, demonstrating strong model independence and robustness. Specifically, Gemini-3-pro-preview outperformed the other two models in terms of accuracy and mAP, reflecting its stronger ability to capture fine-grained semantic and visual features of building scenes. However, its average inference time is considerably longer, reaching 23.1 s per frame, which limits its applicability in real-time or resource-constrained scenarios. In contrast, Qwen3-VL-plus achieved the fastest inference speed while maintaining competitive recognition performance, representing a favorable balance between accuracy and efficiency and making it more suitable for practical deployment. Meanwhile, gpt-4.1-mini, as a lightweight VLM, still maintained competitive accuracy and mAP with moderate inference efficiency, highlighting the potential of the proposed framework for deployment on edge devices or low-resource platforms.

Table 5. Performance comparison of different VLMs.

VLM	Accuracy (%)	mAP (%)	Inference time (s/frame)
GPT-4.1-mini	84.9	77.4	9.1
Qwen3-VL-plus	88.3	80.7	**5.4**
Gemini-3-pro-preview	**89.2**	**85.5**	23.1

The above experiments have demonstrated that under the guidance of the combined knowledge of visual enhancement and fine-grained text prompts, VLM exhibits significantly enhanced reasoning capabilities. This further validates the roles of each module in improving the recognition performance during the construction stage. Table 6 conducts a systematic comparative analysis of the combined effects of different prompt strategies and the visual enhancement module.

Among them, the basic prompt only provides the stage classification instructions to the model, without any specific construction stage semantic descriptions or structured prior information. From the results, it can be seen that when using only the basic prompt, the model achieves an accuracy rate of 79.4% and mAP of 72.8%, indicating that VLM has a certain stage understanding ability, but its reasoning still mainly relies on general visual semantics, and is difficult to fully distinguish visually similar or transitional stages. After introducing the fine-grained prompt, the accuracy rate increases to 85.6%, and mAP increases to 76.4%, indicating that by clearly injecting stage definitions, key actions, and context constraints, the model can effectively focus on more discriminative semantic cues, thereby significantly improving the stage recognition performance. After introducing the visual enhancement module under the condition of the basic prompt, the model performance also improves, with an accuracy rate and mAP of 83.9% and 74.0% respectively. This shows that explicit visual cues such as the presence and location of workers can compensate for the visual perception deficiencies of pure text prompts, enhancing the model's understanding of key entities and spatial relationships in construction scenarios. Furthermore, when the fine-grained prompt, visual cues detection and idle time judgment module are utilized simultaneously, the performance is optimal, with the minimum number of VLM calls and the lowest cost. This result indicates that the structured semantic constraints at the text level and the explicit information at the visual level have a significant collaborative effect. They can jointly enhance the reasoning stability and discrimination ability of VLM in complex construction scenarios, and idle time judgment module can significantly reduce the overall resource consumption without affecting the accuracy of stage monitoring.

In terms of inference efficiency, the visual enhancement module does not significantly increase the inference time, while introducing the fine-grained prompt inevitably increases the inference time per frame. However, for the long-term construction process, the overall remains within an acceptable range, indicating that the proposed mixed knowledge-guided framework achieves a reasonable balance between performance improvement and computational overhead. The ablation experiment results (conducted utilizing the Qwen3-VL-plus model) further validate the effectiveness and practicality of the proposed hybrid knowledge-guided framework in the construction stage recognition task.

To comprehensively evaluate the effectiveness of the proposed hybrid knowledge-guided unsupervised training framework, we designed a comparative experiment on the same dataset and compared it with representative traditional methods and zero-shot methods. The benchmark methods include: the framework based on semantic segmentation [35], the knowledge graph-enhanced computer vision method [20] the classic behavior recognition method TSN [33], TSM [16], these four training-required methods, as well as the zero-shot CLIP [23] and methods based on VLM inference that does not require extra training. The core ideas of these methods were applied to the tasks in this paper.

Specifically, the dataset was divided into training and test sets in a 7:3 ratio. For the semantic segmentation method proposed by Yang et al. [35], a

Table 6. Ablation study on prompt design, visual cues detection and idle time judgement for construction stage recognition utilizing Qwen3-VL-plus. Accuracy and mAP are reported in %, inference time is measured in seconds per frame, and VLM Calls denotes the number of model invocations per hour.

Module	Accuracy	mAP	Inference Time	VLM Calls
Basic prompt	79.4	72.8	**2.8**	95
Fine-grained prompt	85.2	76.4	5.3	95
Basic prompt + Visual cues	83.4	74.5	2.9	95
Fine-grained prompt + Visual cues	**88.3**	**80.7**	5.4	95
Fine-grained prompt + Visual cues + Idle time judgment	88.0	80.6	5.4	**72**

DeepLabv3+ model with a MobileNetV2 architecture was used, and it was trained for 100 epochs on our dataset. This framework performs pixel-level semantic segmentation, classifying each pixel into different categories: concrete, rebar, formwork, background, etc. Stage recognition is achieved through the analysis of semantic composition based on area proportion: when the proportion of the template area is greater than 60% and there is no concrete, it recognizes the installation of the formwork; when the proportion of steel bars is greater than 50%, it recognizes the binding of steel bars; when the proportion of concrete pixels is between 20% and 80%, it detects the pouring of concrete; when the area of concrete is stable and there is a film or water column, it recognizes the curing of concrete. For the method proposed by Pfitzner et al. [20], we used YOLOv5m to detect objects in the construction scene, including workers, concrete buckets, concrete mixers, etc., and combined spatio-temporal rules to construct a knowledge graph. Through graph neural network inference on the knowledge graph structure, the concrete construction stage is inferred.

For the video action recognition methods, we evaluated architectures based on CNN and Transformer. TSN [33] and TSM [16] utilizd pre-trained ResNet50 on ImageNet as the backbone network and were fine-tuned on the construction dataset. Long videos were cropped into multiple segments, and the network was trained on the training set through sparse time sampling. For transformer based methods, Video Swin Transformer [18] adopts a hierarchical architecture with shifted window attention to capture local spatio-temporal patterns, while TimeSformer [2] relies on global space-time self-attention for feature modeling. Both adopt the pre-trained weights of Kinetics-400. Input clips were uniformly sampled from videos, and standard data augmentation strategies were applied during training.

For the training-free method, the zero-shot classification ability of CLIP [23] (ViT-B/32) was utilized, and text prompts were designed for each concrete construction stage. Stage prediction is obtained by calculating the cosine similarity between the image and text embeddings. Furthermore, we also explored a zero-shot method based on Video-LLaVA [15], where the sampled frames are used as input, and natural language prompts are employed to process the sampled frames to generate stage predictions, without the need for training on our dataset. For the methods based on VLM reasoning, we utilized the GPT-4.1-mini and Gemini-3-pro-preview models and conducted the reasoning solely through the basic prompt. Since concrete construction is a long-term process, when using the above methods, frames were processed at 30-second intervals to balance computational efficiency, and then the stages were classified at the frame level. Finally, after basic time smoothing, a majority voting method with a 10-frame window was used to determine the stage transition points to adapt to the tasks of this paper.

Table 7. Comprehensive performance comparison with baseline methods on concrete construction stage recognition. "*" denotes that the corresponding VLM performs inference utilizing only a basic prompt. The best results are highlighted in bold and the second-best results are underlined.

Method	Training Required	Accuracy (%)	mAP (%)	FTR (%)
Yang et al. [35]	Yes	67.4	61.2	14.5
Pfitzner et al. [20]	Yes	44.6	41.0	44.8
TSN [33]	Yes	46.8	41.8	43.9
TSM [16]	Yes	52.6	45.3	42.6
TimeSformer [2]	Yes	38.2	34.4	51.9
Video Swin Transformer [18]	Yes	45.3	40.7	41.2
Zero-shot CLIP [23]	No	48.6	44.7	43.4
Video-LLaVA [15]	No	47.1	36.7	48.1
GPT-4.1-mini*	No	77.5	68.6	11.3
Gemini-3-pro-preview*	No	<u>84.9</u>	<u>79.2</u>	<u>6.9</u>
Ours	No	**88.3**	**80.7**	**3.4**

The comprehensive comparison results of the above methods are shown in Table 7. The evaluation indicators include whether training is required, accuracy, mean average precision (mAP), and False Transition Rate (FTR), which is the ratio of the number of false transitions to the total number of predicted transitions, used to measure the stability of the method and the number of false alarms generated. The experimental results show that the proposed method without training achieved the highest accuracy and mAP of 88.3% and 80.7%, respectively, and the lowest FTR, significantly outperforming all benchmark methods.

The reason for the poor performance of the training-based methods may be that traditional supervised learning heavily relies on large-scale labeled data, and construction scenarios are diverse and complex. Under the condition of limited training data, the model has difficulty learning the subtle visual differences between different construction stages. In contrast, the zero-shot CLIP method, although without training, pays more attention to global feature matching and lacks understanding of local details, ignoring the local discriminative cues in the construction scenarios, resulting in poor performance. Furthermore, although the method of VLM for reasoning based on the basic prompt has a relatively high accuracy per frame, it only uses simple post-processing rules and has limited effect in suppressing incorrect conversions. Additionally, although the reasoning time of these methods (excluding VLM inference) is at the millisecond level, lower than the proposed method in this paper, due to the shortest sampling interval of 30 s in the concrete construction stage recognition task, the inference speed of all methods is much faster than the sampling frequency, and the practical significance of the speed advantage is limited. However, our method can achieve superior practical performance without any training. By combining the clear visual cues in object detection with specific domain knowledge prompts, VLM performs more reliably in stage distinction. The mixed knowledge input effectively builds the reasoning of VLM on observable visual evidence, reduces hallucination phenomena, and improves the credibility of predictions. The dynamic sliding window post-processing strategy can adaptively adjust the window length according to stage features and prediction confidence, thereby significantly reducing frame-level noise and false transitions, achieving the best performance.

4.3 Hyperparameter Setting and Sensitivity Analysis

After performing VLM inference, clear stage transition boundaries and the start and end times as well as durations of each stage are obtained through dynamic sliding post-processing. To determine the hyperparameters in the dynamic sliding window post-processing strategy, namely the confirmation threshold θ_c and the transition threshold θ_t, and to quantify their effects on system performance, a small-scale validation set is constructed based on the dataset presented in Sect. 4.1 for sensitivity analysis. The validation subset is established from 12 consecutive construction monitoring video clips, covering multiple groups of four sequential concrete construction stages with complete transition processes for each stage. This subset contains 273 annotated frames. First, frame-level VLM inference is performed on these frames. Subsequently, the sliding window post-processing operation is conducted with different threshold configurations. After stage confirmation, inconsistent misclassified frames within the sliding window are rectified. Finally, calculate and analyze the corrected frame-level accuracy and False Transition Rate (FTR).

The relationship between accuracy and θ_c, and that between FTR and θ_t, are shown in Fig. 4. As shown in Fig. 4(a), accuracy increases as θ_c rises from 0.60 to 0.80, since a higher confirmation threshold suppresses noisy or transient misclassifications. When θ_c exceeds 0.85, accuracy declines because overly strict criteria delay stage confirmation and induce false transitions. Therefore, 0.80 is selected as the optimal value.

As shown in Fig. 4(b), a small θ_t leads to frequent false transitions caused by short-term voting fluctuations. As θ_t increases to 0.70, FTR decreases to 3.1%, which indicates that moderately strict conversion criteria can effectively distinguish genuine stage changes from noise. However, when θ_t further approaches

θ_c, the reduced hysteresis margin weakens state stability and causes oscillatory switching near true boundaries. Therefore, 0.70 is chosen as the optimal transition threshold.

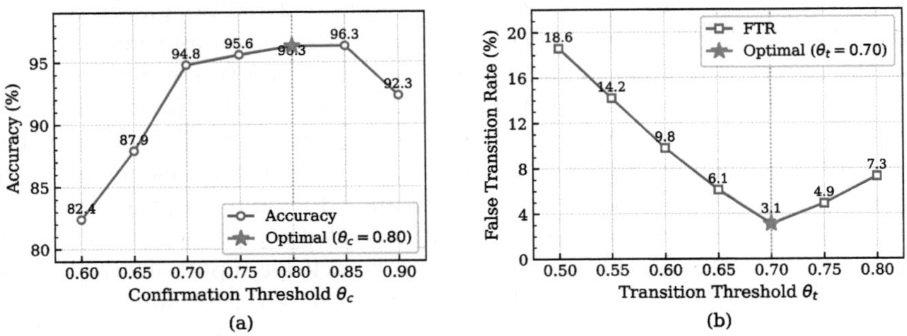

Fig. 4. Sensitivity analysis of the confirmation threshold θ_c on Accuracy (a) and the transition threshold θ_t on False Transition Rate (b).

4.4 Visualization of Post-processing Results

To further demonstrate the practical effectiveness of the selected hyperparameter configuration, visualization experiments were conducted on representative construction monitoring videos. By comparing the raw VLM predictions with the outputs after dynamic sliding window post-processing, the temporal smoothing capability and transition correction performance of the proposed method can be more intuitively illustrated. Figure 5 shows the impact of the proposed dynamic sliding window post-processing module on the recognition of construction stages over time. The top plot presents the original frame-level prediction results generated by VLM, where frequent short-term fluctuations and sudden stage transitions can be observed, especially during stage transition periods and long-duration stages. These brief misclassifications are mainly caused by visual blurring, intermittent occlusion, and short-term scene changes, resulting in noisy and temporally inconsistent prediction results.

In contrast, the bottom plot shows the processing results after applying the dynamic sliding window strategy. By dynamically adjusting the window length based on stage features and confidence distribution, the proposed method can effectively integrate temporal context and suppress isolated erroneous predictions. Therefore, false stage transitions are largely eliminated, and the predicted stages exhibit significantly improved temporal continuity and stability. Moreover, the post-processing module can generate clearer and more consistent stage boundaries, which are more in line with the actual construction process. This indicates that the proposed dynamic sliding window not only reduces frame-level noise but also improves the reliability of stage duration estimation, which is crucial for subsequent progress analysis and decision support in actual construction monitoring scenarios.

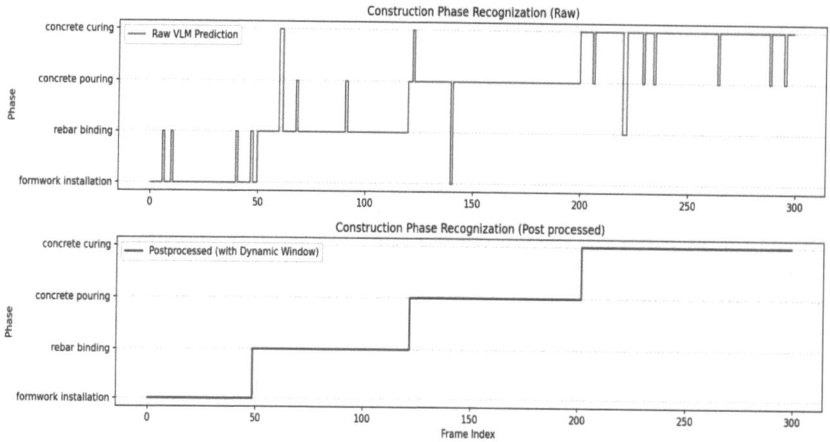

Fig. 5. Comparison between construction phase recognition over time before and after post-processing. The top plot shows raw VLM predictions with frequent, noisy phase switches. The bottom plot shows results after applying a dynamic sliding window, producing smoother and more stable transitions between the four stages

To simulate real-site monitoring videos, the sampled frames are combined and certain idle state interference frames are inserted to verify whether the proposed framework can distinguish the construction stage while identifying the idle stage. Figure 6 shows a representative example. The horizontal axis represents time, and the vertical axis corresponds to the four standard concrete construction stages. The colored bars represent the time intervals of each stage obtained after processing through the dynamic sliding window analysis, while the white segments embedded within the colored bars represent the idle time periods where no construction activities were observed. This framework can accurately identify the sequence of processes such as formwork installation, rebar binding, concrete pouring, and concrete curing. More importantly, the short idle periods that occur within certain stages are clearly retained rather than being wrongly classified as stage transitions, indicating that this framework can distinguish temporary inactive states from true stage changes.

By excluding these idle periods when estimating the effective working time, the proposed method can provide a more realistic and management-oriented representation of the construction schedule. Moreover, the structured JSON output, including visual evidence and confidence ratings, also enhances interpretability and promotes traceable decision-making in actual construction monitoring scenarios.

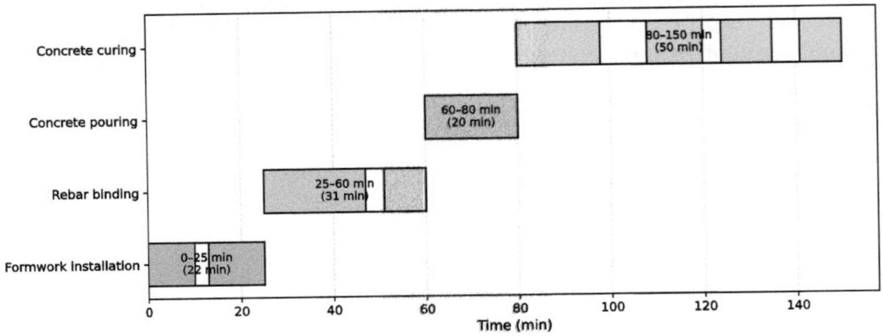

Fig. 6. Temporal construction stage recognition and duration estimation results. Each colored bar represents a recognized construction stage along the timeline. The white segments embedded within the colored bars correspond to idle periods, during which no active construction operations are detected. The annotated intervals indicate the estimated start time, end time, and effective duration of each stage after excluding idle periods.

4.5 Generalizability Evaluation Across Multiple Application Domains

To assess the generalizability of the proposed hybrid knowledge-guided training-free framework beyond concrete construction stage recognition, we extend its application to three representative domains: construction safety inspection, general action recognition, and temporal action localization. For each scenario, the core framework components remain unchanged. All utilize YOLO-World for object detection and then undergo inference utilizing the mixed knowledge-guided VLM. Task adaptation is achieved solely through prompt engineering and post-processing modes of the output, eliminating the need for any task-specific retraining. These application scenarios collectively demonstrate the generality and scalability of the proposed paradigm, and are applicable to a variety of visual reasoning tasks.

Construction Safety Inspection. The construction safety inspection was conducted on the ConstructionSite10k dataset [4], which contains 10,013 images captured from different construction sites. From these images, 1,500 were selected as a subset, which included images that had safety violation annotations and those that did not violate safety rules, covering the four types of safety rule violations defined in the dataset as well as the images that did not violate safety rules. Specifically, the following four safety rules were stipulated: (1) Use of basic personal protective equipment when walking at construction sites; (2) Use of safety harness when working from a height of three meters and the edges are without any edge protection; (3) Adoption of edge protection or edge warning including guardrails, fences, for underground projects three meters in depth with steep retaining walls and for humans to stand; (4) Appearance of workers in the blind spots of the operator and within the operation radius of operating excavators, or excavators with operators inside. In Table 8, Rule ID 0 represents no safety violation, while Rule IDs 1 to 4 correspond to the four safety rules listed above, respectively. Unlike the stage recognition task, construction safety detection does not involve time sequence and is a multi-label reasoning problem, requiring independent evaluation of each image. That is, in one image, one or more types of safety rule violations may occur simultaneously, and the identification of these violations requires detailed reasoning about the state of each worker and the spatial relationship between the worker and the surrounding dangerous environment.

Consistent with the process described in Sect. 3.3, YOLO-World detected workers with confidence over 0.6, and bounding boxes were overlaid to guide VLM attention to high-risk areas. These boxes provide spatial anchors and semantic context about worker-hazard relationships, reducing VLM hallucinations.In Addition, a structured prompt instructed the VLM to utilize bounding boxes as positioning cues rather than behavior evidence, enabling detailed safety compliance checks. VLM outputs structured JSON containing violation categories, evidence descriptions, and confidence scores, with no fine-tuning required.

Performance was evaluated using multi-label precision and recall metrics, and the VLM employs Qwen3-VL-plus. The results are shown in Table 8, where the 'Human upper bound' reference is adopted directly from the Construction Site 10k baseline [4]. This human performance ceiling was derived by having civil engineering graduate students evaluate the exact same image-prompt combinations as those processed by the models, thereby offering a reliable human-level standard for precision and recall in safety violation detection. And VLM with object detection (OD+VLM) achieves high accuracy across all violation categories, approaching or exceeding human-level performance. Critically, object detection reduces VLM inference time from 18.9 s to 11.2 s per frame (40.7% reduction) by pre-localizing workers, allowing the VLM encoder to focus on relevant regions and avoid exhaustive scene exploration.

Table 8. The precision (P) and recall (R) (in %) of the safety rule violation detection on the ConstructionSite10k subset.

Rule ID	0		1		2		3		4	
	P	R	P	R	P	R	P	R	P	R
VLM-only	82.3	78.2	60.9	60.0	58.9	84.3	35.7	74.3	48.2	71.4
OD+VLM	92.8	79.6	83.4	69.7	77.4	84.6	41.2	81.6	66.1	79.4
Human upper bound	95.3	98.3	95.6	66.6	92.0	92.0	59.4	60.3	81.0	65.4

General Action Recognition. To evaluate the applicability of this framework in general human action recognition, we selected representative subsets from three public action recognition benchmark datasets for experiments. The HMDB51 dataset [14] contains 51 action categories collected from various video sources, while UCF101 [28] covers 101 action categories covering various real activities. SSv2 [10] is a time-sensitive dataset containing 174 fine-grained human-object interaction actions that are difficult to reliably infer from static appearances. For each benchmark dataset, 20 videos from each action category were randomly selected to construct subsets. Action recognition is defined as a single-label classification task at the frame level: given one or more frames of images extracted from an action video segment, the model needs to predict the most likely action category from the corresponding predefined label set.

For action recognition, the spatial position and posture attributes of the human subject are the main distinguishing features. YOLO-World will be applied to each extracted frame to detect all visible humans, and the generated bounding boxes will be superimposed on the original image to provide clear human-centered visual cues before VLM inference. In addition, a specific structured text prompt is designed for each dataset. This prompt provides a complete list of candidate action categories for the target subset, guiding the model to reason about observable human postures, limb movements, object interactions, and scene backgrounds, and requiring the output of a single predicted action label along with a brief evidence description. To leverage the powerful prior knowledge of VLM, the category names will be directly provided without adding additional visual examples, thus achieving training-free inference. For the SSv2 dataset, its action semantics are mainly defined by object-oriented action basic elements rather than scene backgrounds, so this prompt also emphasizes the reasoning of the relationship between the visible human subjects and the target objects in the scene.

Utilizing Qwen3-VL-plus model, the recognition accuracy of the proposed framework on the three benchmark subsets is shown in Table 9. Overall, the hybrid knowledge-guided VLM inference method achieves competitive recognition performance on all three datasets. For the two action category benchmarks with low temporal dependence, such as HMDB51 and UCF101, the accuracy of VLM under single-frame inference is 56.8% and 86.4% respectively. When using two equally spaced sampled frames as input, the accuracy increases to 92.2% and 94.1% respectively, indicating that even in sparse multi-frame input, when the temporal context has information value, it can significantly improve the recognition effect. For SSv2, it requires fine temporal reasoning of object interactions, and the categories in the dataset are difficult to infer from a single-frame appearance, so two or three equally spaced frames are used as the input to VLM. The performance under all configurations is significantly lower, which is consistent with the inherent higher temporal dependence of the action categories in SSv2. The slight improvement from 2 frames to 3 frames indicates that increasing the frame number provides certain benefits, but capturing fine motion dynamics from sparse sampling still poses challenges. Moreover, after adding the visual enhancement module, all baselines with only text prompts have achieved consistent accuracy improvements, verifying that clear human positioning cues enable the model to focus on relevant human regions, reduce the influence of clutter or irrelevant background on the model, and help with more accurate posture and behavior reasoning.

Table 9. Action recognition accuracy (%) on subsets of HMDB51, UCF101, and SSv2-Small under different frame sampling configurations, and the method utilizing only VLM is compared with the object detection enhanced VLM (OD+VLM).

Method	HMDB51		UCF101		SSv2-Small	
	1-frame	2-frame	1-frame	2-frame	2-frame	3-frame
VLM-only	56.8	92.2	86.4	94.1	40.3	56.7
OD+VLM	58.2	93.6	88.5	94.7	41.2	56.9

Temporal Action Localization. Temporal Action Localization (TAL) not only requires classifying the action categories in the video sequence, but also precisely determining the time boundaries of each action instance, namely the start and end timestamps. In this problem, the continuous video stream must be segmented into time-coherent stage intervals, which is similar to the stage recognition task. Inspired by this structural correspondence, we apply the proposed framework to the TAL task to further verify its generalization in temporal reasoning outside the construction domain. Given a video sequence containing one or more action instances, after dynamic frame sampling, VLM inference is performed, and then a post-processing strategy is adopted to generate action category predictions as well as the associated time boundaries and valid duration.

The experiments are evaluated on subsets of two widely used temporal action localization benchmark datasets. Charades [27] contains 9,848 daily indoor activity videos with an average duration of 30 s, covering 157 action categories, and videos usually contain multiple concurrent or consecutive action instances. THUMOS'15 [11] covers videos of 20 sports action categories with an average duration of 2-3 min, and the video duration and action density vary greatly. Both datasets provide time segments annotations for evaluating the localization accuracy as a benchmark. For each dataset, 10 videos from each action category video are randomly selected to construct a subset.

First, dynamic frame sampling is performed on the video, and the initial sampling rate is 1 frame per second. When a potential action start is detected, the sampling rate is dynamically adjusted to 0.5 frames per second to capture the fine-grained temporal changes during the action execution until the action ends. This adaptive sampling mechanism significantly reduces the processing overhead of redundant frames while ensuring the coverage of critical moments. For each frame, YOLO-World detects persons and overlays bounding boxes as visual cues. The input for each VLM inference is two temporally adjacent frames, using the temporal continuity between frames to provide context information for the action evolution. VLM is instructed to output structured JSON format, including action category, action stage labels, where 0 indicates the start of the action, 1 indicates the action is in progress, 2 indicates the end of the action, and brief visual evidence. By explicitly outputting the temporal evolution stage of the action, it provides fine-grained semantic cues for subsequent temporal localization.

To aggregate the frame-level predictions of VLM into accurate and coherent action instances and infer their time boundaries, a set of rule-based post-processing strategies is designed. This strategy is based on the temporal consistency assumption: the true action instance should exhibit stable category continuity and clear stage transition patterns over time. The specific rules are as follows:

(1) When one of the following conditions is met, it is considered that the action has started: three consecutive outputs of the same action category, and the sequence begins with a stage label of 0; or two consecutive frames output the same action category and both are stage labels of 1. The start time of the action is defined as the moment when the first stage label 0 appears; if there is no label 0 in the sequence, the start time is defined as the moment when the first label 1 appears.
(2) The trigger condition for the end of the action is that the output action category changes and a stage label 2 is detected, explicitly indicating the completion of the action. The end time of the action is defined as the moment when the last output of the action category occurs and the stage label is 2.
(3) Only when the action instance contains a clear end signal, that is, the appearance of stage label 2, is the instance considered an effective action. Incomplete action fragments, such as only having the start but no end, are filtered out.

Utilizing standard temporal action localization evaluation metrics, including the mean average precision (mAP) under the time intersection-over-union (tIoU) thresholds of 0.5 and 0.7, these metrics comprehensively assess the performance of both action recognition and temporal localization. All experiments were not fine-tuned for specific tasks. Utilizing the Qwen3-VL-plus model, the experimental results are shown in Table 10. The proposed framework achieved mAP@tIoU=0.5 of 60.5% and 78.2% on the Charades and THUMOS'15 datasets, although the performance declined under the stricter 0.7 IoU threshold. Furthermore, there is a significant performance gap between THUMOS'15. This difference can be attributed to the inherent characteristics of the datasets and the core assumptions of our proposed framework. Our post-processing strategy largely relies on the sequential stage transformation assumption. However, THUMOS'15 mainly contains sequential motion actions, which are highly consistent with the continuous stage-based transition assumption of the framework. In contrast, Charades includes complex, multi-label indoor activities, which often occur simultaneously or overlap, exhibit higher temporal fragmentation, shorter action instances, and frequent context switching. These concurrent activities can cause certain disruption to the strict stage transformation constraints of the framework, resulting in a decrease in prediction performance. However, the overall results indicate that through reasonable stage modeling and temporal reasoning post-processing strategies, the framework originally designed for construction phase monitoring can effectively be applied to general time action localization, further verifying the scalability and cross-domain applicability of the proposed framework.

Table 10. Temporal action localization performance on Charades and THUMOS'15 datasets under different IoU thresholds.

Dataset	mAP@IoU=0.5 (%)	mAP@IoU=0.7 (%)
Charades	60.5	48.7
THUMOS'15	78.2	63.2

Among the above three application scenarios, the proposed framework based on hybrid knowledge guidance and without the need for training consistently demonstrates robust performance. Through the visual cues detection module, the structured hybrid prompt design, and the post-processing rules tailored for specific application scenarios, the scalability of the proposed paradigm has been verified.

5 Discussion

In the proposed framework, small visual models and VLMs are flexibly called and combined. VLM performs visual-constrained reasoning based on the image visual cues extracted by the small visual model, thereby improving the accuracy and interpretability of the construction stage discrimination, and achieving robust performance through its complementary error correction mechanism in three stages. When object detection fails to extract visual cues due to occlusion, VLM compensates through overall scene understanding; when frame-level errors occur in VLM, dynamic post-processing will enforce temporal consistency. This design builds a fault-tolerant system, ensuring reliable stage recognition under various conditions. Furthermore, by jointly leveraging detection results and idle time judgment, the framework adaptively reduces VLM invocation frequency during inactive periods, significantly lowering computational costs without sacrificing monitoring accuracy.

Moreover, the proposed training-free decision framework is inherently generalizable to other long-cycle construction tasks, such as road paving and structural assembly. For different application scenarios, task-specific lightweight visual models (e.g., detectors focusing on specific tools, machinery, or materials) can be flexibly selected and integrated with a general-purpose VLM to form an effective perception–reasoning pipeline. Since the framework does not rely on training, adapting it to new construction processes primarily involves redefining stage descriptions and visual cues within the prompt, rather than collecting additional annotated datasets. This modular and knowledge-guided design highlights the potential of our decision frameworks as scalable and practical solutions for real-world construction monitoring and intelligent project management.

6 Conclusion

This paper proposes a hybrid knowledge-guided training-free decision framework for automatically monitoring engineering construction progress and key temporal points. The framework constructs hybrid knowledge guided by visual cues and domain-specific industrial knowledge, integrating existing pre-trained human detection model with large visual–language model, thereby improving both the accuracy and interpretability of VLM inference. In addition, the dynamic frame sampling strategy and dynamic sliding-window post-processing module are introduced to further enhance computational efficiency and robustness. Experimental results on real-world construction videos demonstrate that the proposed framework achieves accurate and stable stage recognition across diverse scenarios while effectively suppressing noise and spurious transitions, which provides a methodological reference for monitoring engineering projects.

Acknowledgments. This work was supported in part by the Shanghai Jiao Tong University AI for Engineering Initiative (No. WH410263001/001), and the Science and Technology Commission of Shanghai Municipality (No. 22DZ2229005).

References

1. Bai, S., et al.: Qwen3-vl technical report. arXiv preprint arXiv:2511.21631 (2025)
2. Bertasius, G., Wang, H., Torresani, L.: Is space-time attention all you need for video understanding? In: Proceedings of the International Conference on Machine Learning (ICML), pp. 813–824 (2021)
3. Chaudhary, N., Uddin, S., Chandra, S.S., Ovid, A., Albert, A.: Prompt to protection: A comparative study of multimodal LLMs in construction hazard recognition. arXiv preprint arXiv:2506.07436 (2025)
4. Chen, X., Zou, Z.: Are large pre-trained vision language models effective construction safety inspectors? arXiv preprint arXiv:2508.11011 (2025)
5. Chen, Y., Li, Y., Zhang, C., Zhou, H., Luo, Y., Hu, C.: Informed patch enhanced hypergcn for skeleton-based action recognition. Inf. Process. Manag. 102950 (2022)
6. Cheng, T., Song, L., Ge, Y., Liu, W., Wang, X., Shan, Y.: Yolo-world: Real-time open-vocabulary object detection. In: Proceedings of the IEEE/CVF Conference on Computer Vision and Pattern Recognition, pp. 16901–16911 (2024)
7. Danel, T., Lafhaj, Z., Puppala, A., BuHamdan, S., Lienard, S., Richard, P.: Identifying tower crane activities with data: the case of the concrete pouring. Eng. Constr. Archit. Manag. **31**(2), 939–956 (2024)

8. Ersoz, A.B.: Demystifying the potential of chatgpt-4 vision for construction progress monitoring. arXiv preprint arXiv:2412.16108 (2024)
9. Gong, J., Caldas, C.H.: Computer vision-based video interpretation model for automated productivity analysis of construction operations. J. Comput. Civ. Eng. **24**(3), 252–263 (2010)
10. Goyal, R., et al.: The "something something" video database for learning and evaluating visual common sense. In: Proceedings of the IEEE International Conference on Computer Vision, pp. 5842–5850 (2017)
11. Idrees, H., et al.: The thumos challenge on action recognition for videos "in the wild". Comput. Vision Image Understanding **155**, 1–23 (2017)
12. Jeoung, J., Jung, S., Hong, T.: Zero-shot framework for construction equipment task monitoring. Comput.-Aided Civil Infrastruct. Eng. (2025)
13. Kim, I., Kim, Y., Chin, S.: Deep-learning-based sound classification model for concrete pouring work monitoring at a construction site. Appl. Sci. **13**(8), 4789 (2023)
14. Kuehne, H., Jhuang, H., Garrote, E., Poggio, T., Serre, T.: Hmdb: a large video database for human motion recognition. In: 2011 International Conference on Computer Vision, pp. 2556–2563. IEEE (2011)
15. Lin, B., et al.: Video-llava: Learning united visual representation by alignment before projection. In: Proceedings of the 2024 Conference on Empirical Methods in Natural Language Processing, pp. 5971–5984 (2024)
16. Lin, J., Gan, C., Han, S.: Tsm: Temporal shift module for efficient video understanding. In: Proceedings of the IEEE International Conference on Computer Vision (2019)
17. Liu, Y., et al.: Towards harnessing the collaborative power of large and small models for domain tasks. arXiv preprint arXiv:2504.17421 (2025)
18. Liu, Z., et al.: Video swin transformer. In: Proceedings of the IEEE/CVF Conference on Computer Vision and Pattern Recognition, pp. 3202–3211 (2022)
19. Moon, S., Yang, B.: Effective monitoring of the concrete pouring operation in an rfid-based environment. J. Comput. Civ. Eng. **24**(1), 108–116 (2010)
20. Pfitzner, F., Hu, S., Braun, A., Borrmann, A., Fang, Y.: Monitoring concrete pouring with knowledge graph-enhanced computer vision: A case study from Munich, Germany. In: EC3 Conference 2024, vol. 5, pp. 0–0. European Council on Computing in Construction (2024)
21. Pu, H., Yang, X., Shi, Z., Jin, N.: Automated inspection report generation using multimodal large language models and set-of-mark prompting. In: ISARC. Proceedings of the International Symposium on Automation and Robotics in Construction, vol. 41, pp. 1003–1009. IAARC Publications (2024)
22. Qian, Z., Yao, X., Huang, Y., Zhang, C., Ying, J., Sun, H.: Beyond label semantics: Language-guided action anatomy for few-shot action recognition. In: Proceedings of the IEEE/CVF International Conference on Computer Vision, pp. 10421–10431 (2025)
23. Radford, A., et al.: Learning transferable visual models from natural language supervision. In: International Conference on Machine Learning, pp. 8748–8763. PmLR (2021)
24. Raza, M., Jahangir, Z., Riaz, M.B., Saeed, M.J., Sattar, M.A.: Industrial applications of large language models. Sci. Rep. **15**(1), 13755 (2025)
25. Ronneberger, O., Fischer, P., Brox, T.: U-Net: Convolutional Networks for Biomedical Image Segmentation. In: Navab, N., Hornegger, J., Wells, W.M., Frangi, A.F. (eds.) MICCAI 2015. LNCS, vol. 9351, pp. 234–241. Springer, Cham (2015). https://doi.org/10.1007/978-3-319-24574-4_28

26. Shtedritski, A., Rupprecht, C., Vedaldi, A.: What does clip know about a red circle? visual prompt engineering for vlms. In: Proceedings of the IEEE/CVF International Conference on Computer Vision, pp. 11987–11997 (2023)
27. Sigurdsson, G.A., Varol, G., Wang, X., Farhadi, A., Laptev, I., Gupta, A.: Hollywood in Homes: Crowdsourcing Data Collection for Activity Understanding. In: Leibe, B., Matas, J., Sebe, N., Welling, M. (eds.) ECCV 2016. LNCS, vol. 9905, pp. 510–526. Springer, Cham (2016). https://doi.org/10.1007/978-3-319-46448-0_31
28. Soomro, K., Zamir, A.R., Shah, M.: Ucf101: A dataset of 101 human actions classes from videos in the wild. arXiv preprint arXiv:1212.0402 (2012)
29. Wang, C., Shamsabadi, E.A., Chen, Z., Shen, L., Fini, A.A.F., Dias-da Costa, D.: Automating construction safety inspections using a multi-modal vision-language rag framework. arXiv preprint arXiv:2510.04145 (2025)
30. Wang, D., et al.: Vision-based productivity analysis of cable crane transportation using augmented reality-based synthetic image. J. Comput. Civ. Eng. **36**(1), 04021030 (2022)
31. Wang, F., et al.: A survey on collaborating small and large language models for performance, cost-effectiveness, cloud-edge privacy, and trustworthiness. arXiv preprint arXiv:2510.13890 (2025)
32. Wang, H., Ren, L., Zhao, T., Jiao, L.: Collm: Industrial large-small model collaboration with fuzzy decision-making agent and self-reflection. IEEE Trans. Fuzzy Syst. (2025)
33. Wang, L., Xiong, Y., Wang, Z., Qiao, Yu., Lin, D., Tang, X., Van Gool, L.: Temporal Segment Networks: Towards Good Practices for Deep Action Recognition. In: Leibe, B., Matas, J., Sebe, N., Welling, M. (eds.) ECCV 2016. LNCS, vol. 9912, pp. 20–36. Springer, Cham (2016). https://doi.org/10.1007/978-3-319-46484-8_2
34. Wang, Z., Bovik, A.C., Sheikh, H.R., Simoncelli, E.P.: Image quality assessment: from error visibility to structural similarity. IEEE Trans. Image Process. **13**(4), 600–612 (2004)
35. Yang, B., Gao, B., Han, Y., Zhang, B.: Semantic segmentation-based framework for concrete pouring progress monitoring by using multiple surveillance cameras. Dev. Built Environ. **16**, 100283 (2023)
36. Yuan, X., et al.: Towards next-generation intelligent maintenance: Collaborative fusion of large and small models. arXiv preprint arXiv:2506.05854 (2025)
37. Zhang, R., Deng, R., Zhang, Z., Mao, Y.: Vision-based real-time progress tracking and productivity analysis of the concrete pouring process. Dev. Built Environ. **21**, 100609 (2025)
38. Zhang, Y., Liu, J., Zhao, Y., Xu, J., Wu, W.: Crack detection on concrete composite slab using you only look once version 8 based lightweight system. J. Build. Eng. 112974 (2025)
39. Zhou, H., Zhang, C., Hu, C.: Visual relationship detection with relative location mining. In: Proceedings of the 27th ACM International Conference on Multimedia, pp. 30–38 (2019)

Fuzzy-Guided Agentic Memory for LLM-Based Recommendation

Di Zhang(✉), Jie Lu, and Guangquan Zhang

University of Technology Sydney, Sydney, Australia
di.zhang-7@student.uts.edu.au, {jie.lu,guangquan.zhang}@uts.edu.au

Abstract. Language models (LLMs) are promising for recommender systems due to their parametric knowledge and natural language reasoning over user preferences. However, direct LLM prompting incurs high inference costs and context collapse, where long user histories inflate prompts—driving latency and hallucinations—while yielding shallow, pattern-matching recommendations. We introduce a multi-agent memory framework that distills long and short-term user interactions plus item audience signals into dual-format profiles: structured features paired with natural-language summaries. A Mamdani fuzzy inference engine aggregates signals across memory dimensions to derive compatibility scores, constraining LLM context and offloading scoring. This design further enables incremental real-time updates to user/item representations, obviating full-dataset retraining. Crucially, how fuzzy scores are presented to the LLM matters as much as the scores themselves: a fuzzy-first prompt strategy—where the LLM defaults to the top-ranked candidate and overrides only with concrete evidence—outperforms both standalone fuzzy scoring and naive score injection. On MovieLens 1M, the combined system attains 60.5% Hit@1 (random negatives) and 48.5% (hard negatives)—up to 3× over direct prompting—while reducing token consumption ∼80% and inference time ∼3.6× across LLM backbones.

Keywords: Recommender Systems · Large Language Models · Fuzzy Logic · Multi-Agent Systems · Agentic Memory

1 Introduction

Recommender systems are a canonical instance of data-driven decision-making, where large volumes of interaction data are distilled into personalised choices under uncertainty [1]. Large language models have emerged as a compelling paradigm for this setting. As LLMs are pre-trained on internet-scale datasets encompassing reviews, synopses, forum discussions, and metadata, they internalize rich item knowledge as a byproduct of language modeling. As a result, the need for tedious manual feature engineering that traditional content-based systems require is significantly eased. Natural language reasoning and real-world knowledge give LLM-based recommender systems the ability to handle cold-start scenarios where collaborative filtering data is sparse [6]. Unlike traditional systems relying on opaque learned embeddings, LLMs can explicitly reason about thematic resonance, directorial style, cultural context, and narrative structure.

J. Lu et al. (Eds.): FLINS-ISKE 2026, LNCS 16752, pp. 309–325, 2027.
https://doi.org/10.1007/978-981-92-2369-5_20

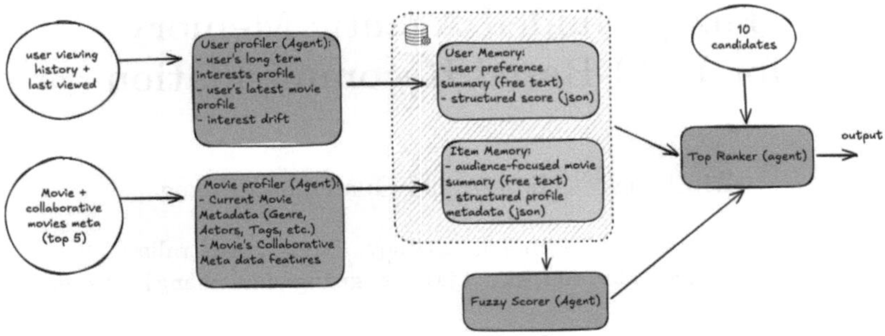

Fig. 1. System architecture. Four independent agents communicate through a shared database memory store. The User Profiler and Movie Profiler generate dual-format memory profiles that are consumed by both the Recommender Agent (LLM-only) and the Fuzzy Recommender Agent (fuzzy pre-ranking + LLM).

However, LLM-based recommenders face challenges that their original LLM training objectives were not designed to handle. First, even moderately active users' viewing histories span hundreds of items, exceeding practical context windows and triggering the "lost-in-the-middle" phenomenon [7], where mid-sequence signals are ignored due to positional attention degradation. Second, LLMs exhibit systematic biases in candidate selection: they favour surface-level genre similarity, show popularity and prestige biases toward high-profile films, and struggle with non-obvious audience fits where preference profiles diverge from genre norms—yielding shallow pattern-matching over nuanced compatibility. Third, unconstrained LLM generation frequently produces hallucinated or out-of-domain item features or recommended output, as the model conflates world knowledge with the bounded item space of a recommendation corpus. Fourth, auto-regressive scoring of even a modest candidate pool demands repeated full-sequence generation, making real-time serving economically and latency-wise impractical compared to traditional scoring pipelines. The above challenges make the LLM recommender system's production use rather limited.

To address these challenges, we propose a framework with two key innovations (Fig. 1):

(1) Agentic Memory. A multi-agent pipeline compresses user viewing histories and movie metadata into compact, dual-format memory profiles. Each user profile sequentially captures long-term patterns, short-term deviations, then a recommendation strategy into free-text (for LLM reasoning) and structured JSON (as extracted features for quantitative scoring). Movie profiles similarly encode audience-fit signals, collaborative neighborhood data, and content characteristics. This compression reduces context requirements from hundreds of raw ratings to a concise profile, while preserving the signals most relevant for recommendation. Because memory generation is a one-time cost amortized across all subsequent recommendations, this architecture also yields significant inference speed and token savings at recommendation time. As model updates and new interactions occur, only the affected user or movie profiles need regeneration, enabling real-time adaptability without retraining or fine-tuning the entire model.

(2) Fuzzy Logic as LLM Guardrail. A Mamdani fuzzy inference system computes multi-signal compatibility scores across user/item feature dimensions, using membership rules empirically calibrated from training data analysis. The fuzzy scorer is designed to *saturate* genre influence (preventing the genre trap), *boost* underweighted signals like appeal breadth fit, and *penalize* content conflicts the LLM misses. Further, we find that the integration strategy matters as much as the scorer itself: a **fuzzy-first prompt design**—where the LLM defaults to the fuzzy top-ranked candidate and only overrides with specific, concrete evidence—outperforms both naive fuzzy injection and standalone fuzzy scoring.

We evaluate our approach on the MovieLens 1M dataset with 200/200 users split for training and test, using Hit@1 as the primary metric across two difficulty levels of test sets (random and hard negatives) and three LLMs (GPT-5.2, DeepSeek-V3.2, Llama-4-Scout). Our research questions are:

- **RQ1**: Does agentic memory improve recommendation accuracy and inference efficiency over direct LLM prompting?
- **RQ2**: Can fuzzy logic correct systematic LLM biases, particularly the genre trap?
- **RQ3**: How should fuzzy scoring and LLM reasoning be combined for optimal results?
- **RQ4**: How does the choice of LLM foundation model affect memory quality and downstream recommendation?

The remainder of this paper is organized as follows. Section 2 reviews related work. Section 3 details our methodology. Section 4 presents the experimental evaluation. Section 5 concludes with a discussion of limitations and future directions.

2 Related Work

2.1 Foundations of Recommender Systems

Collaborative filtering and content-based recommender systems (RS) are the foundations of modern RS research. Matrix factorisation methods [2] decompose user-item interaction matrices into latent factor representations, enabling scalable prediction of missing ratings. Bayesian Personalised Ranking (BPR) [4] use implicit-feedback as pairwise learning. Neural collaborative filtering [5] replaced the fixed inner product of matrix factorisation with learned nonlinear interaction functions, achieving stronger expressiveness on benchmark comparison. Hybrid approaches extend these ideas by unifying multiple information sources; for example, Shambour and Lu [3] combine user and item trust information for B2B recommendation, mitigating sparsity and cold-start by propagating trust through the interaction graph.

These methods operate on user-item interactions and encode preferences as embeddings. While the approach is simple and effective, only requires user-item pairs. They struggle with cold-start scenarios and sparse data, and they are unable to effectively incorporate the rich, unstructured information (reviews, synopses, cultural context, directorial style) that often drives human preference. Large language models address this gap: pre-trained on internet-scale text, they internalise broad item knowledge and can perform natural language reasoning over qualitative preference signals that embedding-based methods cannot represent. This motivates the growing body of work applying LLMs to recommendation.

2.2 LLM-Based Recommendation

The application of LLMs to recommendation has progressed rapidly, as surveyed by Wu et al. [6]. Early approaches, such as P5 [9] treated recommendation as text-to-text generation, fine-tuning language models on formatted recommendation data. Chat-REC [10] leveraged chat-tuned models to elicit preferences through dialogue. More recently, zero-shot approaches have shown that LLMs can serve as rankers without task-specific training: Hou et al. [8] demonstrated competitive zero-shot ranking by presenting candidate items as text, though they also revealed systematic position and popularity biases and difficulty with long sequential histories.

Several works address integrating collaborative signals into language understanding. CLLM4Rec [11] internalises collaborative knowledge by embedding user/item ID tokens directly into the LLM, using a prediction head to avoid hallucination. RLMRec [12] uses LLMs to build rich textual user/item profiles and aligns them with collaborative embeddings—an approach close in spirit to our dual-format profiles, though scoring remains within neural encoders. PURE [13] is arguably the closest existing work to our memory design: it builds evolving user profiles by extracting preferences from reviews, continuously compressing them to stay within token limits. However, PURE still relies on the LLM for final ranking, whereas we externalise scoring to a fuzzy engine.

Our work differs from these approaches in three respects: we decouple memory generation from recommendations for independent optimisation, use a *single-call* structured output design that guarantees bounded latency, and offload compatibility scoring to a non-parametric Mamdani engine.

2.3 LLM Agentic Architecture and Memory Management

Agent-based frameworks decompose recommendations into specialised modules with explicit memory and action capabilities. Agent4Rec [14] uses LLM-powered generative agents with profile, memory, and action modules to simulate users for evaluation. RecAgent [15] models users as persistent LLM agents with memory for simulation studies, while InteRecAgent [16] integrates traditional recommendation tools within an LLM agent loop. MASRank [17] proposes specialised LLM agents with sequence understanding, content analysis, and bias detection that collaborate to mitigate ranking biases. demonstrating that multi-agent decomposition can improve ranking quality and robustness.

On the memory side, MemoryBank [18] introduces an external memory system with a forgetting mechanism inspired by the Ebbinghaus curve, enabling long-term user interaction with LLMs—a conceptual precedent for our long/short-term memory split. Recursive summarisation approaches [19] maintain long-term dialogue coherence by progressively compressing the interaction history, which supports our use of natural-language summaries for preference distillation.

Our framework moves from simulation and conversational companions to be more focused on inference optimisation. The dual memory format not only effectively compacts a long context window, but it also extracts signals that power the fuzzy scorer, with the LLM constrained to a single reasoning call over compact profiles.

2.4 Fuzzy Logic and Neuro-Symbolic Recommendation

Fuzzy logic has a substantial history in recommender systems and decision-making. Yera and Martínez [20] survey fuzzy tools across RS domains, covering fuzzy similarity measures, fuzzy clustering, and fuzzy rule-based preference modelling. The original Mamdani fuzzy controller [21]—which aggregates multiple input dimensions through linguistic rules and defuzzification—provides the methodological foundation for our compatibility scoring. In the movie domain specifically, Calderón-Vilca et al. [22] used fuzzy logic over audience scores, critic scores, and metadata attributes for Netflix movie recommendation.

More recently, differentiable fuzzy neural networks [23] learn fuzzy logic rules over human-readable atoms to provide transparent recommendations, evaluated on MovieLens 1M. This is methodologically closest to our fuzzy component; the key difference is that they integrate fuzzy rules inside a differentiable model for end-to-end training, whereas we use an explicit Mamdani engine with rules *empirically calibrated from LLM failure analysis*—each rule targeting a specific, observed bias in LLM recommendation behaviour rather than being learned from user-item interactions.

2.5 Efficiency and LLM Hallucinations

Deploying LLMs for recommendation raises practical concerns around inference cost, context budget, and output quality. Systematic biases compound these issues: LLMs exhibit bias favouring early candidates and popularity bias toward better-known items [8].

Setwise prompting approaches [24] provide quantitative evidence of the trade-off between ranking quality and token consumption, motivating architectures that minimise LLM involvement in tight scoring loops.

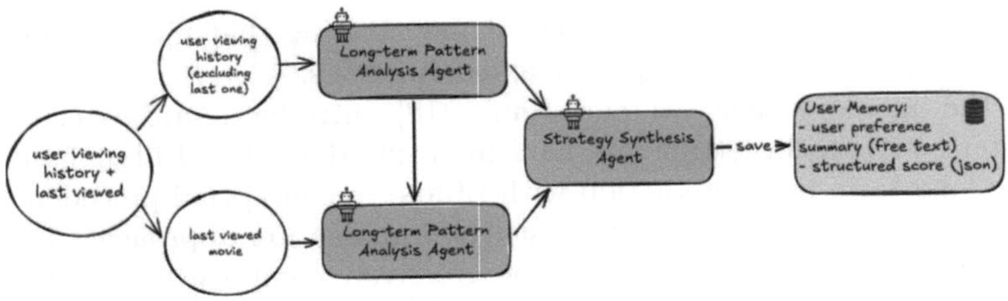

Fig. 2. User Profiling Agent workflow. Three sequential sub-agents progressively refine user preferences: long-term pattern analysis identifies stable viewing habits, short-term deviation analysis classifies recent behavior, and strategy synthesis produces the final dual-format memory profile (free text + structured JSON).

3 Methodology

3.1 System Overview

Our Framework consists of four main independent agents connected through a shared memory store (Fig. 1):

1. **User Profiler Agent**: Generates user preference profiles from viewing history via three sequential sub-agents calls:
   - **Long-term Pattern Analysis Agent**: Analyzes long-term viewing patterns to identify stable preferences and rating behaviors.
   - **Short-term Deviation Analysis Agent**: Evaluates recent viewing behavior for mood/context preference consistency check.
   - **Strategy Synthesis Agent**: Combines long-term and short-term analyses into a final dual-format memory profile.
2. **Movie Profiler Agent**: Generates audience-fit profiles for movies, cleanly extracting free-form metadata values into a well-formatted structured output.
3. **Fuzzy Scorer Agent**: Read pre-process user and item (movie) memories, apply membership scoring against each candidates.
4. **Ranker Agent**: Making final decision based on the Fuzzy score quality and user/item memory context.

All agents produce **dual-format memory**: a free-text narrative for LLM reasoning and a structured JSON profile for quantitative scoring. This design allows the same memory to serve both the LLM's natural language understanding and the fuzzy scorer's numerical computations.

3.2 User Profiling Agent

The User Profiling Agent, acts more like a User Profiling Team. (Fig. 2) It generates a comprehensive preference profile from a user's viewing history through three sub-agents, each building on the output of the previous one:

Long-term Pattern Analysis Agent. The agent receives the user's enriched viewing history (movies with genres, actors, directors, tags, and ratings) and identifies consistent patterns across genre preferences, rating behavior, thematic preferences, era preferences, tone/vibe preferences, and pacing preferences.

Short-term Deviation Analysis Agent. Using the long-term analysis and detailed metadata for the most recently watched movie, the agent classifies the recent viewing as CONSISTENT, EXPLORATION, or MOOD-DRIVEN, and determines how it should influence recommendations.

Strategy Synthesis Agent. The agent combines both analyses into the final dual-format memory:

- **Free text**: A 3–5 paragraph natural language strategy describing core preferences, rating patterns, current mood, and recommendation guidance.
- **Structured profile output** : A json object with fields for:
  - `genre_preferences`: 18 canonical MovieLens genres, scored 0–1
  - `era_preference`: classic/modern_classic/contemporary
  - `vibe_tags`: from a canonical list of 26 descriptors
  - `pacing_preference`: slow/medium/fast
  - `avoid`: genres, subgenres, tones, and content types to avoid mood/rating metadata

This three-step design decomposes the complex profiling task into manageable subtasks while keeping context focused. The sequential architecture ensures that the strategy synthesis has access to both the broad patterns and the recent-viewing context.

3.3 Movie Profiling Agent

The Movie Profiler Agent generates audience-fit profiles describing *who* tends to enjoy a movie and *why*, rather than summarizing the plot. It uses a single LLM call with structured output via Pydantic schema enforcement.

The agent receives comprehensive movie data including metadata (genres, cast, directors, year), tag-genome relevance scores, collaborative neighbors (similar movies based on rating patterns), and aggregate rating statistics. From these signals, it produces:

- **Free text**: A 2–4 sentence audience-focused description of the ideal viewer.
- **Structured profile output**: Structured fields including:
  - `genre_profile`, `vibe_tags`, `pacing`,
  - `era` (5-bucket: classic through contemporary),
  - `audience_fit` (target description and appeal breadth from 0 for niche to 1 for mainstream),
  - `collaborative_signals` (average rating, rating count, genre concentration, neighbor genre distribution), and
  - `avoid_for` (audience types poorly served by this movie).

The data set we used comes from different sources, besides the original MovieLens dataset, we also use IMDb and Tag Genome project to enrich the movie metadata, Multiple data sources means inconsistent formatting and noise are common. The LLM's role is to synthesize this heterogeneous data into a coherent profile features. Besides, effectively controlling effective context window for downstream recommender agents, these 2 profilers also serve as a critical *signal distillation* step, extracting/cleaning the most relevant features and filtering out noise. Deterministic structure output against a well-defined schema ensures that the resulting profiles are consistent and machine-readable for downstream fuzzy scoring, while the free-text description provides rich context for the LLM's qualitative reasoning during recommendation.

3.4 Fuzzy Compatibility Scorer

The fuzzy compatibility scorer computes a compatibility score for each (user, movie) pair using membership functions and Mamdani-style rules empirically calibrated from training-set failure analysis.

Input Dimensions and Fuzzification. From the dual-format memory profiles, we extract eight compatibility dimensions for each candidate: genre similarity, appeal breadth fit, era match, collaborative overlap, tag-genome similarity, cast/director affinity, avoid violation (binary), and pacing distance. These dimensions bridge the rich natural-language profiles and interpretable numerical scoring.

Each continuous dimension is fuzzified into linguistic terms (low/medium/high) via trapezoidal membership functions whose shapes are calibrated to *reshape* dimensional influence and counter known LLM biases. The most critical design choice is **genre saturation**: even a strong genre match is capped at a moderate consequent, preventing surface-level genre similarity from dominating the score—the "genre trap" responsible for 20.2% of observed failures (Table 2). Conversely, appeal breadth fit is deliberately **boosted** to reflect its empirical importance as a robust audience-alignment signal.

Cross-Dimensional Rules. The fuzzified terms are evaluated against a Mamdani rule base whose strength lies in cross-dimensional interactions: *genre trap detection* penalizes high genre similarity lacking collaborative or audience support; *hidden gem rescue* provides alternate scoring pathways for low-genre-similarity candidates with exceptional audience and era fit; and *fine-grained tiebreaking* uses cast affinity or tag-genome matches to resolve close cases.

Defuzzification and Scoring. To obtain a single scalar score suitable for ranking, we aggregate the activated rules using standard Mamdani defuzzification as a weighted average of rule consequents:

$$s_{\text{base}} = \frac{\sum_{i=1}^{N} w_i \cdot c_i}{\sum_{i=1}^{N} w_i} \tag{1}$$

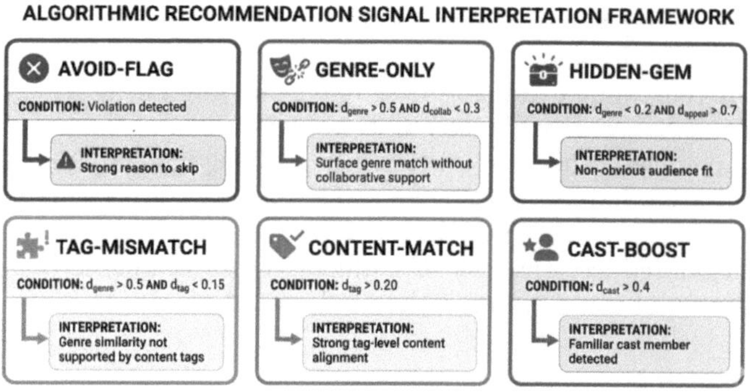

Fig. 3. Diagnostic flags generated by the fuzzy scorer for each candidate. Flags surface rule activations and scoring rationale to the LLM, enabling it to respect the fuzzy ranking while applying qualitative judgment for edge cases.

where $x_1, \ldots, x_k$ are the fuzzified input values for the k compatibility dimensions introduced in Sect. 3.4 (genre similarity, appeal breadth fit, era match, collaborative overlap, tag-genome similarity, cast/director affinity, avoid violation, pacing distance), $A_1, \ldots, A_k$ are the linguistic fuzzy sets (low/medium/high) defined by the trapezoidal membership functions μ_{A_j}, $w_i = \min(\mu_{A_1}(x_1), \ldots, \mu_{A_k}(x_k))$ is the firing strength of rule i under the minimum T-norm for conjunctive rules, and $c_i \in [0, 1]$ is its consequent score. This step compresses the full fuzzy inference process into a single compatibility score $s_{\text{base}} \in [0, 1]$ for each (user, movie) pair.

Hard constraints are applied as deterministic post-hoc penalties: avoid-list violations incur a ~45% reduction, and severe pacing mismatches a ~10% reduction, ensuring non-negotiable conflicts are not averaged away by otherwise high compatibility.

Beyond producing a numerical score, the scorer exposes human-readable **diagnostic flags** (e.g., *genre-only*, *hidden-gem*, *avoid-flag*) that are injected into the LLM prompt (Fig. 3), surfacing *why* a candidate scored well or poorly and enabling the LLM to apply qualitative judgment where the fuzzy rules are insufficient.

3.5 Recommendation Agent

The Recommender Agent selects the best movie for a user from a candidate set. The agent receives:

1. The user's memory profile (free text + JSON) from the User Profiler.
2. Candidate movie cards, each combining factual metadata (genres, cast, directors, tags) with the audience-fit profile from the Movie Profiler.
3. Fuzzy compatibility scores and diagnostic flags for each candidate from the Fuzzy Scorer.

The system prompt instructs the LLM to follow an *elimination-then-selection* strategy: first scan all candidates and rule out obvious mismatches, then select the single best remaining candidate based on genre alignment, vibe/tone match, pacing/era fit, collaborative signals, and avoid-for warnings.

This design anchors the LLM on the fuzzy ranking while preserving its ability to apply qualitative judgment for edge cases that fuzzy rules cannot capture.

The LLM returns a structured output with the selected movie ID and reasoning. Such output reduces the risk of hallucinated recommendations and forces llm to validate its choice as part of the final output.

4 Experiment

4.1 Setup

We evaluate on the **MovieLens 1M** dataset [25], which contains approximately 1 million ratings from 6,040 users on 3,952 movies. The movie metadata is enriched with actor, director, and tag-genome data from IMDb and the Tag Genome project.

We then randomly select 400 users with at least 10 historical ratings for profiling and evaluation. For each user, we construct 10-candidate test sets with one ground-truth positive (a movie the user rated ≥ 4) and nine negatives. Finally, we split the 400 users into 200 for training (used for memory generation and fuzzy rule calibration) and 200 for testing (used for final evaluation of recommendation performance).

4.2 Test Sets

Two test set difficulty levels evaluate different aspects of recommendation quality:

Random Negatives. Nine movies sampled uniformly at random from the user's non-interacted items. This tests basic preference matching—can the system distinguish a liked movie from unrelated alternatives?

Hard Negatives. Nine movies selected to share surface characteristics with the positive: three same-genre movies, two same-era movies, two movies with high tag overlap, and two popular movies from the dataset. This tests the system's ability to make fine-grained distinctions when candidates are superficially similar to the ground truth.

4.3 LLM Foundation Models

We evaluate three LLMs as foundation models accessed via API:

- **GPT-5.2** (OpenAI): Proprietary model.
- **DeepSeek-V3.2** (DeepSeek): Open-weight model (37B).
- **Llama-4-Scout** (Meta): Open-weight model (17B).

All agents share the same prompts with default parameters.

4.4 Configurations

We compare four recommendation configurations:

1. **Direct LLM**: (baseline): The LLM receives the user's raw viewing history and candidate metadata directly, with no memory compression or fuzzy scoring.
2. **Profiled Memory**: The LLM receives the user's agentic memory profile (from the User Profiler) and movie audience profiles (from the Movie Profiler), but no fuzzy scores.
3. **Fuzzy on Memory**: No LLM in the recommendation loop. The fuzzy scorer ranks candidates using pre-generated memory profiles, and the top-scored candidate is selected deterministically.
4. **Ranker on FM (Fuzzified Memory)**: The fuzzy scorer pre-ranks candidates, and the Ranker Agent makes the final selection using the fuzzy-first prompt strategy described in Sect. 4.6.

We also conducted experiments variations:

1. **Integration prompt comparison**: as mentioned in Sect. 4.6 to validate the importance of integration strategy.
2. **Mixed-memory ablation**: using 1 llm model for memories, and with another LLM for ranking, to test memory generalizability and the impact of memory quality on fuzzy scoring and LLM reasoning.

4.5 Results

Table 1. Hit@1 (%) across four configurations, three LLMs, and two test set difficulties. R = random negatives, H = hard negatives. Best result per column in **bold**

Configuration	GPT-5.2		DeepSeek-V3.2		Llama-4-Scout	
	R	H	R	H	R	H
Direct LLM	38.0	18.5	41.5	18.0	22.5	13.5
Profiled Memory	39.5	23.0	35.5	21.0	30.5	17.0
Fuzzy on Memory	58.0	43.0	53.0	36.0	52.0	34.0
Ranker on FM	**60.5**	**48.5**	**57.0**	**46.0**	**58.0**	**42.0**

Table 1 presents the Hit@1 results across all configurations, models, and test sets.

4.6 Analysis

Memory Provides Gains on Accuracy, and a Significant Boost on Inference Efficiency (RQ1). User/Item Profiled Memory is the fundamental building block of our approach. When it directly used without fuzzy scoring, it yields

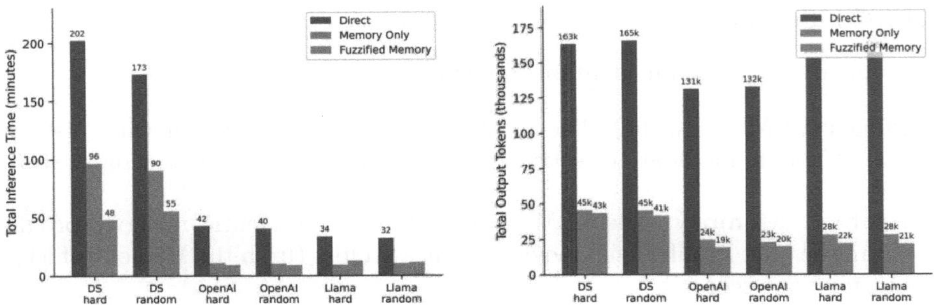

Fig. 4. Inference time (left) and output token counts (right) for the recommendation phase across configurations.

modest accuracy improvements over Direct LLM prompting, particularly on hard negatives where it provides a 4.5–5.5pp lift across models. This suggests that the distilled memory profiles help the LLM make better use of the user's viewing history and movie metadata, especially when fine-grained distinctions are required. The context compression from hundreds of raw ratings to a concise profile providing more help for harder tasks than easy ones. The structured format of the memory profiles also enables the fuzzy scorer to operate effectively, which is critical for the larger gains seen in the Fuzzy on Memory and Ranker on FM configurations.

Beyond accuracy, agentic memory yields substantial efficiency gains at recommendation time (Fig. 4). Because user and movie profiles are pre-computed and stored in the database, the recommender prompt contains only the compact memory rather than the full viewing history, dramatically reducing per-request token consumption and latency. As shown in Fig. 4, memory-based approaches reduce token consumption by approximately 80% and inference time by approximately 3.6× compared to Direct LLM, demonstrating the compound efficiency benefits of memory compression and fuzzy pre-ranking.

Fuzzy Logic Corrects Systematic Biases (RQ2). The Fuzzy on Memory configuration demonstrates that fuzzy scoring alone can substantially improve recommendation on random test set, and modest gains on the hard test set. Across all models, Fuzzy on Memory achieves 58.0–52.0% Hit@1 on random negatives and 43.0–34.0% on hard negatives, representing a 14.5–20.0pp lift over Direct LLM on random negatives and a 11.0–16.0pp lift on hard negatives.

This validates our hypothesis that the genre trap is a dominant failure mode during the Profiled Memory LLM inference: by saturating genre influence and boosting appeal breadth fit, the fuzzy scorer avoids the systematic bias that undermines LLM-only approaches. The failure mode distribution from our analysis of randomly selected false negatives (Table 2) also confirms the genre trap as the primary issue.

Table 2. Failure mode distribution from false-negative LLM recommendations. Categories are not mutually exclusive.

Failure Mode	Percentage
Marginal difference	35.6%
Appeal breadth mismatch	20.2%
Genre mismatch (genre trap)	20.2%
Pacing mismatch	16.8%
Collaborative signal ignored	15.8%
Avoid violation on selected	15.8%
Era mismatch	5.5%
Vibe mismatch	3.4%

Integration Strategy Is Critical (RQ3). A key finding from the interaction between fuzzy scoring and LLM reasoning. Three observations emerge:

First, Ranker on FM achieves the best results on random negatives across all models (57–60.5%), substantially outperforming both Fuzzy on Memory (43–53%) and Profiled Memory (30.5–39.5%). This confirms that combining fuzzy pre-ranking with LLM reasoning produces synergistic gains.

Second, *how* fuzzy scores are presented to the LLM matters as much as the scores themselves. The fuzzy-first prompt design is essential.

Our earlier experiments with naive fuzzy-score injection—simply appending the numeric score to each candidate—found that the LLM frequently *overrode* the fuzzy ranking in the wrong direction (47 score-guided damages vs. 16 rescues on the training set), making the combination *worse* than fuzzy scoring alone. The fuzzy-first prompt redesign reverses this rescue-to-damage ratio. Candidates are presented in score-descending order with explicit rank labels (#1, #2, etc.), and the system prompt instructs the LLM to:

1. start from the top-ranked candidate as the default recommendation;
2. override only if a specific, concrete conflict is identified; and
3. justify any deviation by naming both the mismatch that disqualifies the fuzzy top pick and the specific reason the alternative is better.

With this framing, the LLM rescues more cases than it damages, making the combination beneficial overall.

Third, we observe an additional bias not fully captured by our fuzzy rules: *prestige bias*. When fuzzy scores are close, LLMs consistently default to the more critically acclaimed or widely known film. For example, *Bio-Dome* (1996) loses to *Almost Famous* (2000), and *Night of the Creeps* (1986) loses to *L.A. Confidential* (1997), even when the user's profile suggests a preference for the less prestigious alternative. This bias is difficult to address through prompting alone and represents an opportunity for future work.

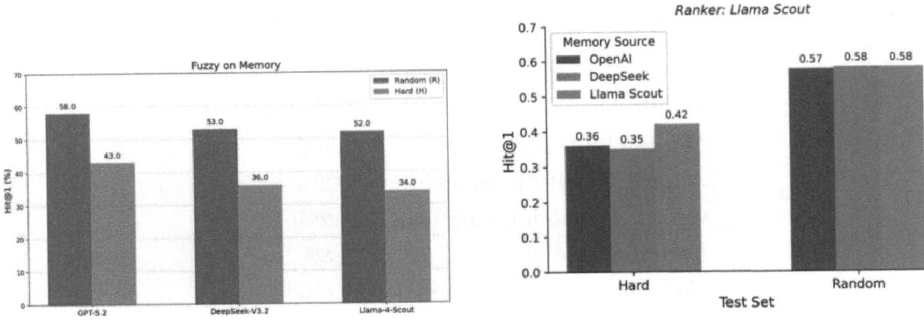

Fig. 5. (Left) Hit@1 performance across LLM models for Fuzzy on Memory, showing consistent improvement with model capacity on both random (R) and hard (H) negatives—suggesting stronger LLMs produce higher-quality memory profiles that feed the fuzzy scorer. (Right) Mixed-memory ablation: Llama Scout memory achieves the highest Hit@1 using its own memory profile.

Impact of LLM Foundation Model Is Complicated (RQ4). An interesting observation from our limited model comparison is shown in Fig. 5: Fuzzy Scoring Hit@1 has a direct correlation with LLM parameter count—the more powerful the LLM, the better the fuzzy scoring performs, even when no LLM is involved at recommendation time. A possible explanation is that more powerful LLMs generate higher-quality memory profiles, which in turn provide more accurate and informative inputs for the fuzzy scorer.

However, does "better" memory quality mean we can use a less powerful LLM as the ranker? To test this, we conducted a mixed-memory ablation where one LLM is used for memory generation and another LLM is used for ranking, to decouple the impact of memory quality from the ranking model's reasoning ability. The right panel of Fig. 5 shows the results of this ablation.

To our surprise, the best performance is achieved by Llama-4-Scout when using its own memory for ranking. It is almost as if the best person at understanding someone's notes is the one who wrote them: even readers with stronger comprehension may not interpret the notes as faithfully as the author does.

4.7 Limitations and Generalization

Fuzzy Rule Generalization. Our rules were calibrated from a human-in-the-loop analyst reviewing MovieLens failures. The protocol is domain-agnostic, but the analyst and several dimensions (era, pacing, cast affinity) are movie-native. Scaling to open domains will require replacing the analyst with an autonomous hypothesize-test-refine loop that proposes new dimensions, not only re-weights existing ones.

Cold-Start. Cold-start is handled asymmetrically. Cold *items* are supported: the Movie Profiler uses metadata, tag-genome, and collaborative neighbors, none of which require user interactions with the target item. Cold *users* are not:

long-term pattern analysis presupposes rating history. Plausible bootstraps—onboarding ratings or demographic priors routed to cluster-level memory—are left to future work.

Domain-Specific Preferences. The biases our rules correct (genre trap, appeal-breadth underweighting, prestige bias) and several dimensions (pacing, era, cast) are movie-specific. Porting the framework preserves the Mamdani skeleton (saturation, boosting, hard penalties) and the analyst-in-the-loop protocol, but requires re-deriving dimensions from target-domain failure analysis—e.g., in books, pacing becomes narrative density and cast affinity becomes author loyalty. Whether rescue rates transfer is an open empirical question.

5 Conclusion

We have presented a fuzzy-guided agentic memory framework for LLM-based movie recommendation that addresses the genre trap—a systematic bias where LLMs over-index on surface-level genre similarity, identified as a leading failure mode in our analysis (20.2% of false negatives, Table 2). Our framework combines multi-agent memory compression (reducing raw viewing histories to dual-format profiles) with a Mamdani fuzzy inference system that operates as a pre-ranking guardrail across eight compatibility dimensions.

Four key findings emerge from our evaluation. First, agentic memory provides modest but consistent improvements on hard negatives (4.5–5.5pp lift), confirming that context compression helps with nuanced discrimination. Second, fuzzy logic is highly effective at correcting systematic LLM biases: the Fuzzy on Memory baseline yields 14.5–20.0pp gains on random negatives and up to 3$\times$ improvement on hard negatives over direct LLM prompting, by saturating genre influence and boosting underweighted signals. Third, the *integration strategy* between fuzzy scoring and LLM reasoning is critical: naive score injection causes net regression, while our fuzzy-first prompt strategy—where the LLM defaults to the fuzzy top pick and overrides only with concrete evidence—achieves the best overall results (60.5% Hit@1 on random, 48.5% on hard negatives). Fourth, the memory architecture delivers a 3.6$\times$ inference speedup and $\sim$80% token reduction at recommendation time, making the system practical for production deployment.

Future Work. First, ablation studies across pipeline stages—profiling, memory compression, and recommendation—to clarify the specific role the LLM serves at each step. Explore whether smaller, distilled, or domain-adapted models can replace large general-purpose LLMs, reducing inference cost, latency, and hardware requirements without sacrificing recommendation quality. Second, extension to cross-domain recommendation settings, testing how effective this framework can help with cold start problem when user coming from a different domains. Third, closing the loop on fuzzy rule calibration by building an agentic pipeline in which agents analyse recommendation failures, propose rule adjustments, and refine membership functions and rule weights automatically—with human-in-the-loop oversight to validate changes—replacing the current manual calibration process with a semi-automated, iterative workflow.

Acknowledgments. This work is supported by the Australian Research Council under grant DP220102635.

References

1. Lu, J., Yan, Z., Han, J., Zhang, G.: Data-Driven Decision-Making (D3M): framework, methodology, and directions. IEEE Trans. Emerging Top. Comput. Intell. **3**(4), 286–296 (2019)
2. Koren, Y., Bell, R., Volinsky, C.: Matrix factorization techniques for recommender systems. Computer **42**(8), 30–37 (2009)
3. Shambour, Q., Lu, J.: An effective recommender system by unifying user and item trust information for B2B applications. J. Comput. Syst. Sci. **81**(7), 1110–1126 (2015)
4. Rendle, S., Freudenthaler, C., Gantner, Z., Schmidt-Thieme, L.: BPR: Bayesian Personalized Ranking from Implicit Feedback. In: UAI 2009, pp. 452–461. AUAI Press (2009)
5. He, X., Liao, L., Zhang, H., Nie, L., Hu, X., Chua, T.S.: Neural collaborative filtering. In: WWW 2017, pp. 173–182. ACM (2017)
6. Wu, L., et al.: A survey on large language models for recommendation. World Wide Web **27**(5), 60 (2024)
7. Liu, N.F., et al.: Lost in the Middle: how language models use long contexts. Trans. ACL **12**, 157–173 (2024)
8. Hou, Y., Zhang, J., Lin, Z., Lu, H., Xie, R., McAuley, J., Zhao, W.X.: Large language models are zero-shot rankers for recommender systems. In: ECIR 2024, LNCS, vol. 14609, pp. 364–381. Springer, Cham (2024). https://doi.org/10.1007/978-3-031-56060-6_24
9. Geng, S., Liu, S., Fu, Z., Ge, Y., Zhang, Y.: Recommendation as Language Processing (RLP): a unified pretrain, personalized prompt & predict paradigm (P5). In: RecSys 2022, pp. 299–315. ACM (2022)
10. Gao, Y., Sheng, T., Xiang, Y., Xiong, Y., Wang, H., Zhang, J.: Chat-REC: towards interactive and explainable LLMs-augmented recommender system. arXiv:2303.14524 (2023)
11. Zhu, Y., Wu, L., Guo, Q., Hong, L., Li, J.: Collaborative large language model for recommender systems. In: WWW 2024, pp. 3162–3172. ACM (2024). https://doi.org/10.1145/3589334.3645347
12. Ren, X., et al.: Representation learning with large language models for recommendation. In: TheWebConf 2024, pp. 3464–3475. ACM (2024). https://doi.org/10.1145/3589334.3645458
13. Bang, S., Song, H.: LLM-based user profile management for recommender system. In: GenNext Workshop at SIGIR 2025 (2025). arXiv:2502.14541
14. Zhang, A., Chen, Y., Sheng, L., Wang, X., Chua, T.S.: On generative agents in recommendation. In: SIGIR 2024, pp. 1807–1817. ACM (2024). https://doi.org/10.1145/3626772.3657844
15. Wang, L., et al.: User behavior simulation with large language model based agents. ACM Trans. Inf. Syst. (2025). https://doi.org/10.1145/3708985

16. Huang, Y., et al.: InteRecAgent: recommendation as language agent. arXiv:2308.16505 (2023)
17. Ma, B., Li, H., Hu, Z., Liu, Y., Liu, L., Gui, X., Lau, S.: MASRank: multi-agent collaborative large language models for zero-shot ranking in recommender systems. In: 2025 6th International Conference on Intelligent Computing and Human-Computer Interaction (ICICHCI). IEEE (2025). https://doi.org/10.1109/ICICHCI63033.2025.11291733
18. Zhong, W., Guo, L., Gao, Q., Ye, H., Wang, Y.: MemoryBank: enhancing large language models with long-term memory. In: AAAI 2024, vol. 38(17), pp. 19724–19731 (2024). https://doi.org/10.1609/aaai.v38i17.29946
19. Wang, Q., Fu, Y., Cao, Y., Wang, S., Tian, Z., Ding, L.: Recursively summarizing enables long-term dialogue memory in large language models. Neurocomputing (2025). arXiv:2308.15022
20. Yera, R., Martínez, L.: Fuzzy tools in recommender systems: a survey. Int. J. Comput. Intell. Syst. **10**(1), 776–803 (2017)
21. Mamdani, E.H., Assilian, S.: An experiment in linguistic synthesis with a fuzzy logic controller. Int. J. Man Mach. Stud. **7**(1), 1–13 (1975)
22. Calderón-Vilca, H.D., Aragón-Tirado, J.W., Canahuire-Cabello, S.R., Ticona-Herrera, R.: Netflix movie recommendation using fuzzy logic. In: FRUCT 2021, vol. 30, pp. 17–25 (2021)
23. Bartl, S., Innerebner, K., Lex, E.: Differentiable fuzzy neural networks for recommender systems. In: HyPer Workshop at ACM UMAP 2025. arXiv:2505.06000 (2025)
24. Zhuang, S., Zhuang, H., Koopman, B., Zuccon, G.: A setwise approach for effective and highly efficient zero-shot ranking with large language models. In: SIGIR 2024. ACM (2024). https://doi.org/10.1145/3626772.3657813
25. Harper, F.M., Konstan, J.A.: The MovieLens Datasets: history and context. ACM Trans. Interact. Intell. Syst. **5**(4), 19:1–19:19 (2016). https://doi.org/10.1145/2827872

Adaptive Hybrid PSO with Q-Learning-Based Algorithm Selection

Mohamed El Khoukhi and Li Zhang(✉)

Department of Computer Science, Royal Holloway, University of London, Surrey TW20 0EX, UK

Mohamed.ElkhoukhiJaadari.2023@live.rhul.ac.uk, li.zhang@rhul.ac.uk

Abstract. Swarm-based algorithms are powerful in solving diverse optimisation problems. However it is still a challenging task for any single metaheuristic optimisation algorithm to perform optimally across all benchmark landscapes. As an example, while Particle Swarm Optimisation (PSO) often converges quickly on smooth unimodal functions, it may suffer from premature convergence on multimodal landscapes with many deceptive local optima. Therefore, this research aims to propose an adaptive hybrid optimisation algorithm in which a Q-Learning method selects complementary optimisation algorithms. Specifically, the proposed model uses a PSO warm-start phase followed by a tabular Q-Learning controller that dynamically selects between Firefly (FA) and Genetic Algorithm (GA). Algorithm selection is determined by population-level characteristics such as normalised diversity and normalised fitness spread. This RL-based search action controller is trained offline to learn an interpretable algorithm-selection policy and is applied with a dynamic Q-table during evaluation. Evaluated using various continuous and complex benchmark functions over multiple independent runs, the proposed algorithm outperforms standalone PSO, GA, FA and Simulated Annealing across diverse landscapes. Statistical validation using the Wilcoxon rank-sum test confirms significant improvements in most comparisons.

Keywords: Hybrid Metaheuristics · Particle Swarm Optimisation · Q-Learning · Adaptive Optimisation · Reinforcement Learning

1 Introduction

Metaheuristic optimisation algorithms are widely used for solving complex optimisation problems where exact solutions are computationally infeasible. Such problems arise in engineering design, scheduling, logistics and other areas where objective functions are often non-linear, non-convex, and expensive to evaluate.

Particle Swarm Optimisation (PSO) [1] is a popular population-based method due to its simplicity, low parameterisation and fast convergence. However PSO can converge prematurely when swarm diversity collapses too early

J. Lu et al. (Eds.): FLINS-ISKE 2026, LNCS 16752, pp. 326–341, 2027.
https://doi.org/10.1007/978-981-92-2369-5_21

on multimodal functions containing many local minima. This is a common challenge in continuous optimisation, where search dynamics change over time, i.e. strong exploration is emphasized in early search stages, while local exploitation becomes more important at a later search process.

The No-Free-Lunch theorem motivates hybrid approaches that combine complementary behaviours, owing to the fact that no algorithm is best for all problem landscapes. Existing hybrid metaheuristics often rely on fixed schedules or hand-designed switching rules. While these approaches can work well, they are not adaptive to the observed search states. In practice, the same fixed search operation switching rule may be beneficial on one function and harmful on another.

This research therefore proposes a hybrid algorithm where reinforcement learning (RL) is used at a meta-level stage to select suitable optimisation operators to apply, based on population-level properties computed during run time. Instead of tuning PSO parameters, the RL-based controller chooses between distinct search operators, allowing the algorithm to react to changes in diversity and fitness distribution.

The main contributions are summarised as follows.

- A hybrid PSO with a Q-Learning (QL)-based algorithm selection is proposed to adapt to exploration of benchmark functions with diverse complex landscapes.
- A modular implementation integrating PSO warm-start with Firefly Algorithm (FA) and Genetic Algorithm (GA) operators is adopted to increase search diversity. The Q-Learning method is used to select either FA or GA based on swarm distribution and fitness improvement status over iterations.
- Empirical evaluation on a number of benchmark functions with nonparametric statistical testing is performed to indicate effectiveness of the proposed strategies.

2 Related Work

Hybrid metaheuristics combining PSO with complementary optimisation strategies have been widely investigated in the literature. A common approach is deterministic hybridisation, where PSO is combined with another algorithm such as FA, Differential Evolution (DE), or GA using predefined switching schedules or stagnation-triggered rules [12]. For example, hybrid PSOFirefly variants integrate attraction-based local exploitation with swarm-based exploration, often activating the secondary operator when convergence slows or diversity drops below a threshold [3]. While effective in specific settings, these switching conditions are manually designed and problem-dependent.

Another line of research focuses on adaptive parameter control within PSO. Reinforcement learning and fuzzy control techniques have been used to dynamically adjust inertia weight, acceleration coefficients, mutation strength, or exemplar learning strategies [4,5]. In these approaches, adaptation occurs at the parameter level, while the underlying optimisation process remains PSO-based.

Such methods improve flexibility but do not fundamentally change the search operator being applied. In another word, such methods rely on purely the original PSO operation without any other search methods adopted.

The proposed work in this research differs from prior research in two key respects. First, reinforcement learning is applied at a meta-level above PSO, enabling the controller to select between distinct optimisation algorithms rather than adjusting PSO parameters. Second, the state representation in RL is based explicitly on interpretable population-level signals (diversity and fitness spread), allowing the learned policy to remain compact and transparent.

This separation between base optimiser and meta-controller yields a modular framework in which optimisation operators can be replaced or extended without modifying the reinforcement learning structure. The resulting hybrid system therefore combines interpretability, adaptability, and extensibility, distinguishing it from both deterministic hybrid approaches and parameter-level adaptation methods.

3 Key Algorithms Used in the Proposed Model

We first introduce all key algorithms used in our proposed hybrid PSO model, including the original PSO, FA, GA and the RL Q-Learning method.

3.1 PSO

PSO was introduced by Kennedy and Ebehart [1] and is inspired by social behaviour such as flocking and schooling. PSO maintains a swarm of N particles in a d-dimensional search space. Each particle i has a position $x_i^t \in \mathbb{R}^d$, velocity $v_i^t \in \mathbb{R}^d$, and a personal best position p_i^t, while the swarm maintains a global best g_i^t.

The velocity update is commonly defined as:

$$v_i^{t+1} = wv_i^t + c_1 r_1 (p_i^t - x_i^t) + c_2 r_2 (g^t - x_i^t), \tag{1}$$

where w is the inertia weight, c_1 and c_2 are acceleration coefficients, and $r_1, r_2 \sim U(0,1)$ are independent random variables. Particle positions are then updated by:

$$x_i^{t+1} = x_i^t + v_i^{t+1}. \tag{2}$$

The inertia component encourages exploration by preserving momentum, while the cognitive and social terms move particles toward personal and global attractors, increasing exploration and exploitation, respectively. In many practical cases, PSO converges quickly at the start of the search process, but can stall if the swarm becomes too concentrated around a sub-optimal solution [13,16–18,26,27].

In this research, PSO is used as a warm-start operator to rapidly locate promising regions before the RL controller begins selecting complementary operators, i.e. FA or GA, to diversify the population.

3.2 FA

FA [8,9,38] is a population-based metaheuristic method, inspired by the flashing behaviour of fireflies, where individuals are attracted to others based on perceived brightness. In the context of optimisation, brightness is associated with the quality of a solution as measured by the objective function, and attraction decreases with distances. This mechanism allows the algorithm to naturally balance between local exploitation and global exploration, particularly in multimodal search spaces.

Moreover, each firefly represents a candidate solution in the search space. The attractiveness between two fireflies is proportional to their relative brightness and inversely related to the distance between them. Each firefly is guided by all neighbouring brighter solutions.

Also, as distance increases, the attractiveness term diminishes, causing fireflies to behave more independently and reducing the risk of premature convergence. As a result, FA is particularly effective at exploring highly multimodal landscapes with many local optima [34].

3.3 GA

GA [10,11,21,33] is another population-based optimisation method inspired by the principles of natural selection and evolutionary biology. Rather than explicitly modelling particle trajectories or attraction dynamics, GA evolves a population of candidate solutions through iterative application of selection and variation operators, encouraging the survival and propagation of higher quality solutions over successive generations.

In the GA, each candidate solution is referred to as a chromosome. At each generation, chromosomes are evaluated using a fitness function, and a subset of population is selected for reproduction based on their relative fitness. Selection mechanisms such as tournament selection (the one used in this research) or fitness proportionate selection increase the likelihood that high-quality solutions contribute genetic material to the next generation, while still allowing weaker chromosomes to be selected occasionally in order to preserve diversity.

New chromosomes are generated through crossover and mutation operators. The crossover operation combines information from two parent chromosomes to produce offspring solutions, typically by exchanging segments of their representations. Mutation introduces random perturbations to solutions, ensuring continued exploration of the search space and reducing the risk of premature convergence.

Unlike PSO, which relies on collective social learning, GA explores the search space through recombination of partial solutions. This makes them particularly effective at discovering diverse solution structures and escaping local optima. However, GA often requires careful tuning and may converge more slowly than swarm-based methods like PSO and FA [14,39].

3.4 Reinforcement Learning and Tabular Q-Learning

Reinforcement learning (RL) provides a principled framework for sequential decision-making under uncertainty, in which an agent learns a policy through interaction with an environment. At each discrete time step, the agent observes a state, selects an action, and receives a scalar reward signal that reflects the quality of the action taken. The objective is to learn a policy that maximises the expected cumulative discounted reward over time [2,15,17,19,20].

In the context of the proposed hybrid optimisation framework, the reinforcement learning agent operates at a meta-level. The environment corresponds to the optimisation process itself, while the state summarises current population behaviour. The action corresponds to selecting an optimisation operator, and the reward reflects relative improvement in the objective function value.

Tabular Q-Learning, introduced by Watkins [2], is an off-policy temporal-difference learning algorithm that estimates the optimal action-value function $Q(s, a)$. For discrete state and action spaces, the update rule, also known as Bellman Equation, is given by:

$$Q(s_t, a_t) \leftarrow Q(s_t, a_t) + \alpha\Big[r_t + \gamma \max_{a'} Q(s_{t+1}, a') - Q(s_t, a_t)\Big], \tag{3}$$

where $\alpha \in (0, 1]$ is the learning rate and $\gamma \in [0, 1)$ is the discount factor. The term $\max_{a'} Q(s_{t+1}, a')$ represents the estimated optimal future value, making Q-Learning an off-policy method.

Action selection follows an ϵ-greedy strategy:

$$a_t = \begin{cases} \text{random action}, & \text{with probability } \epsilon, \\ \arg\max_a Q(s_t, a), & \text{otherwise.} \end{cases} \tag{4}$$

During training, ϵ is gradually decayed to balance exploration and exploitation, allowing sufficient sampling of the stateaction space early in training while increasingly exploiting learned preferences as learning progresses [22].

A practical advantage of tabular Q-Learning in this work is interpretability. Because the state and action spaces are small and explicitly defined, the learned Q-table directly reveals operator preferences under specific population conditions.

4 Proposed Hybrid PSO-RL Algorithm

We introduce the proposed hybrid PSO model in detail below. It includes PSO warm-start, RL-based search action selection, and FA and GA-based swarm enhancement.

4.1 Overall Architecture

The pseudo-code of the proposed model is shown in Fig. 1.

Each key process is explained below.

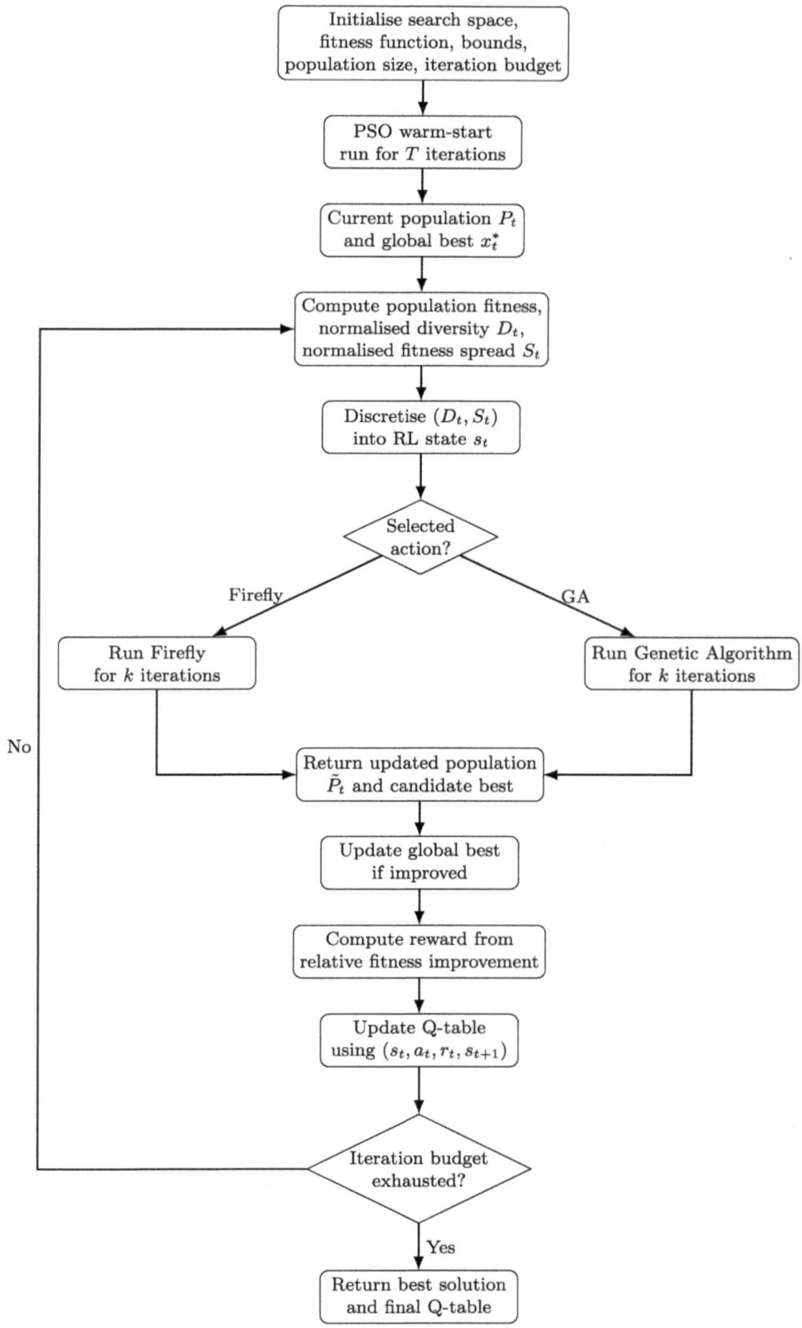

Fig. 1. Dataflow of the proposed hybrid optimisation framework

1. **PSO warm-start:** PSO runs first to produce a non-random, reasonably good population [35,36].
2. **Meta-level RL-based controller:** a tabular Q-Learning agent observes population signals such as population diversity and spread, and selects an operator (FA or GA).
3. **Operators:** The selected search operation, i.e. FA (exploitation) or GA (diversification), is executed for a fixed budget to increase search diversity.

At each decision step, the Q-Learning controller computes the population state, chooses an action (FA or GA), runs the selected operator for a bounded number of iterations, and updates the Q-table using the observed reward.

4.2 Population Signals and State Discretisation

Two population-level signals are computed from the current population P_t:

- **Normalised diversity** (mean distance from centroid, scaled by search range):

$$D_t = \frac{\frac{1}{N}\sum_{i=1}^{N} \|x_i^t - \bar{x}^t\|_2}{(x_{\max} - x_{\min}) + \epsilon}, \tag{5}$$

  where $\bar{x}^t$ is the population centroid. $x_{\max}$ and $x_{\min}$ indicate the upper and lower position boundaries, respectively.
- **Normalised fitness spread** (fitness standard deviation scaled by best fitness magnitude):

$$S_t = \frac{\sigma\big(f(P_t)\big)}{|f_{\text{best}}^t| + \epsilon}. \tag{6}$$

where $\sigma\big(f(P_t)\big)$ represents the standard deviation of the fitness scores of the overall swarm and f_{best} denotes the fitness score of the swarm leader.

These are discretised into 8 states. First, a binary indicator I_t identifies a "multimodal-like" regime:

$$I_t = \begin{cases} 1, & S_t > \tau_S, \\ 0, & \text{otherwise}, \end{cases} \tag{7}$$

Then a base state is determined by diversity and spread thresholds (high/moderate/low diversity and stagnation). The final state index is:

$$s_t = \text{base}(D_t, S_t) + 4I_t. \tag{8}$$

This design allows the same diversity categories to be represented separately for unimodal and multimodal-like regimes.

4.3 Search Actions

The Q-Learning method is used to select either FA or SA based on the swarm diversity and fitness spread signals defined above. Specifically, Q-Learning maintains a Q-table throughout the search process for optimal search action recommendation. The Q-table employs the action space as follows.

$$\mathcal{A} = \{\text{Firefly Algorithm}, \text{Genetic Algorithm}\}.$$

Firefly is treated as the exploitation operator, while GA provides diversification through crossover and mutation. The Q-Learning agent is initialised with a randomly selected search action, with optimal search actions recommended by the Q-table which maximises the overall long-term reward. Precisely, after performing a selected search action, the Bellman Equation shown in Eq. 3 is used to update the Q-Learning table for the corresponding state-action pair. The immediate reward generation is provided below.

4.4 Reward Generation in Q-Learning

The immediate reward in the Q-table update is based on relative improvement in best fitness between consecutive decision steps:

$$\Delta_t = \frac{f_{\text{best}}^t - f_{\text{best}}^{t+1}}{|f_{\text{best}}^t| + \epsilon}. \tag{9}$$

A piecewise immediate reward is applied to stabilise learning across functions based on the following strategies. Such an immediate reward along with the future reward is used to update Q-table as indicated in Eq. 3.

- if $\Delta_t > 10^{-2}$: $r_t = 1.0$
- else if $\Delta_t > 10^{-4}$: $r_t = 0.3$
- else if $\Delta_t \geq 0$: $r_t = -0.2$
- else: $r_t = -0.5$

The above RL Q-Learning based search action selection is integrated with the original PSO method for optimal search strategy selection. E.g. it adopts either FA or GA based on swarm diversity and fitness improvement status throughout the search process to tackle stagnation and increase diversification. The empirical studies indicate its effectiveness in balancing between local and global search operations.

5 Experimental Setup

The proposed hybrid PSO and baseline search methods are evaluated on six continuous benchmark functions [23–25,28–30].

- **Michalewicz:** highly multimodal with steep valleys and inter-variable dependencies.
- **Rastrigin:** regularly structured multimodal function with many local minima.
- **Rosenbrock:** narrow curved valley, testing stable convergence and local search.
- **Levy:** flat regions with sharp drops near the optimum.
- **Powell:** function with coupled terms and sensitivity to search stability.
- **Schwefel:** deceptive landscape with many local optima far from the global optimum.

Example benchmark functions are also visualised in Figs. 2, 3 and 4 [6,7]. To account for the stochastic nature of all considered optimisation algorithms, each method was evaluated over 20 independent runs on each benchmark function [31–33,37]. The search algorithms compared include, PSO, FA, GA, Simulated Annealing (SA) and the proposed hybrid PSO model integrated with RL-based action selection.

For each algorithm and benchmark, the mean, standard deviation, best, and worst final fitness values were recorded. This multi-run evaluation provides a more reliable comparison than single-run experiments and allows assessment of both optimisation performance and robustness. The reinforcement learning component of the proposed hybrid PSO algorithm was trained offline on a subset of benchmark functions (Michalewicz, Rastrigin and Rosenbrock). The learned Q-table was then reused unchanged during evaluation, allowing the generalisation capability of the learned algorithm selection policy to be assessed.

6 Results and Discussions

We present the detailed evaluation results for all selected benchmark functions in this section. Tables 1, 2, 3, 4, 5 and 6 illustrate the outcomes of the numerical functions for all search methods. Overall, the proposed hybrid algorithm outperforms FA and SA, and shows competitive robustness against PSO and GA for most landscapes. On multimodal landscapes such as Michalewicz and Rastrigin, the proposed hybrid method is able to mitigate premature convergence and improve search diversity in comparsion with several baselines. On functions where PSO is already extremely strong (e.g. Levy), the proposed hybrid model shows even better optimal solutions with competitive performance.

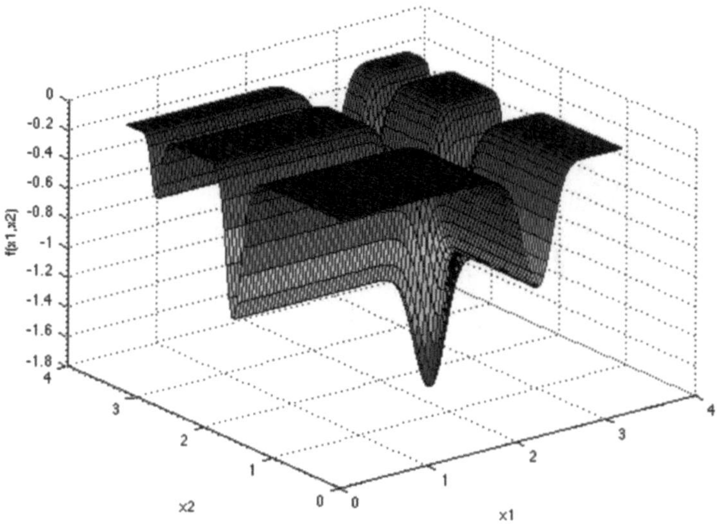

Fig. 2. Visualization of the Michalewicz benchmark function [6]

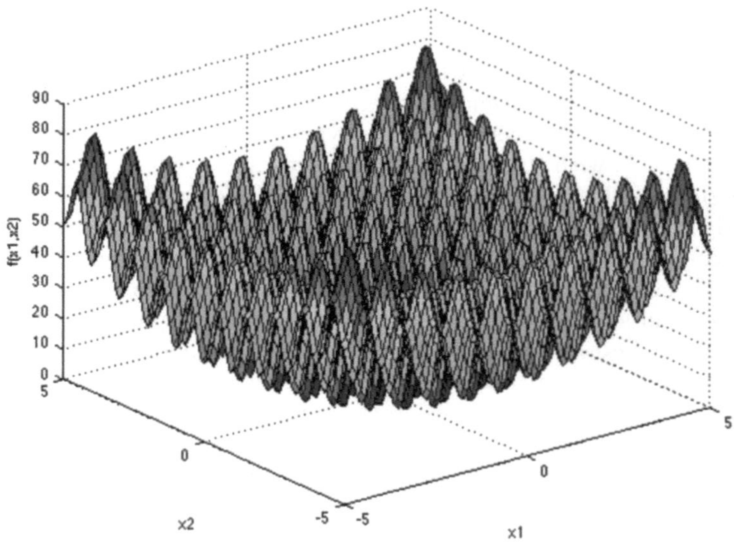

Fig. 3. Visualization of the Rastrigin benchmark function [6]

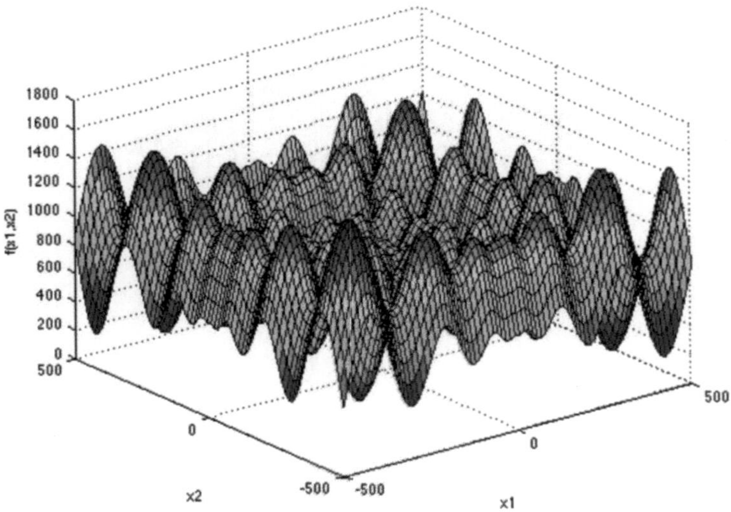

Fig. 4. Visualization of the Schwefel benchmark function [6]

Across all benchmark functions considered, i.e. Schwefel, Rastrigin, Rosenbrock, Powell, Michalewicz, and Levy, the proposed hybrid optimisation algorithm demonstrated robust and competitive performance when compared to standalone metaheuristic methods. The hybrid approach consistently outperformed SA and FA, with statistically significant improvements observed in most cases. These algorithms showed high variance and poor worst-case performance, particularly on complex multimodal functions.

When compared to the original PSO, the proposed hybrid algorithm achieved better performance on complex and challenging multimodal problems along with comparable results on simple unimodal landscapes (e.g. Sphere and Levy). The GA achieved the best mean performance on some multimodal functions, whereas our proposed hybrid algorithm remained competitive and exhibited greater stability, avoiding the extreme failures observed in some GA runs.

Overall, these results support our central hypothesis, i.e. the RL-based algorithm selection improves robustness and adaptability in metaheuristic optimisation, particularly when the characteristics of the optimisation landscapes are unknown with complex narrow curved valleys and peaks with inter-variable dependencies.

The significance of the proposed algorithm is also further ascertained by statistical test functions, i.e. the Wilcoxon rank-sum test. As indicated in Table 7, our proposed model is statistically better than all the baseline methods in most test cases.

Table 1. Results on the Michalewicz function (20 runs).

Algorithm	Mean	Std	Best	Worst
Prop. hybrid	-4.775957	0.160375	-4.943974	-4.205392
PSO	-4.408701	0.429657	-4.747226	-2.897554
Firefly	-2.454612	0.650078	-3.657034	-1.549219
GA	-4.759008	0.073868	-4.907420	-4.663982
SA	-2.717618	1.011036	-4.648076	-0.648979

Table 2. Results on the Rastrigin function (20 runs).

Algorithm	Mean	Std	Best	Worst
Prop. hybrid	0.052574	0.216414	7.78e-07	0.995295
PSO	2.636640	2.208601	0.000000	10.944519
Firefly	10.745535	5.798124	0.994959	20.894080
GA	0.216328	0.394998	6.95e-11	0.996830
SA	34.574649	14.052684	5.969779	62.682006

Table 3. Results on the Rosenbrock function (20 runs).

Algorithm	Mean	Std	Best	Worst
Prop. hybrid	0.420596	0.814891	0.002590	3.950010
PSO	1.031647	1.302133	0.090606	4.305046
Firefly	19.548390	48.614100	0.290844	177.616388
GA	13.255975	33.443308	0.026077	124.390645
SA	1.979715	1.642299	0.136421	6.226919

Table 4. Results on the Levy function (20 runs).

Algorithm	Mean	Std	Best	Worst
Prop. hybrid	4.94e-20	7.38e-20	3.51e-23	3.29e-19
PSO	6.49e-11	1.58e-10	9.61e-14	6.45e-10
Firefly	8.32e-1	1.19e-0	1.34e-15	4.56e-0
GA	1.49e-04	3.33e-04	8.92e-08	1.424073e-03
SA	4.13e+01	1.91e+01	1.07e+01	7.98e+01

Table 5. Results on the Powell function (20 runs).

Algorithm	Mean	Std	Best	Worst
Prop. hybrid	1.00e-06	1.00e-06	6.50e-10	4.00e-06
PSO	5.00e-06	7.00e-06	2.74e-07	3.00e-05
Firefly	8.94e-01	2.52e+00	1.76e-06	1.12e+01
GA	2.02e-02	1.47e-02	6.02e-04	6.13e-02
SA	5.08e-04	2.84e-04	3.49e-05	1.23e-03

Table 6. Results on the Schwefel function (20 runs).

Algorithm	Mean	Std	Best	Worst
Prop. hybrid	1660.180411	1.170692e-02	1660.156268	1660.196566
PSO	1660.196566	2.273737e-13	1660.196566	1660.196566
Firefly	1661.822373	1.391874e+00	1660.196566	1665.347167
GA	1659.150983	4.381068e+00	1640.054376	1660.181598
SA	1660.196566	2.273737e-13	1660.196566	1660.196566

Table 7. Wilcoxon Rank-Sum Test Results (Hybrid vs Baselines, $\alpha = 0.05$)

Test Function	Comparison	p-value	Significant
Michalewicz	Hybrid vs PSO	5.52e-06	Yes
	Hybrid vs FA	3.40e-08	Yes
	Hybrid vs GA	8.69e-01	No
	Hybrid vs SA	3.40e-08	Yes
Rastrigin	Hybrid vs PSO	1.74e-06	Yes
	Hybrid vs FA	6.16e-08	Yes
	Hybrid vs GA	4.81e-02	Yes
	Hybrid vs SA	3.40e-08	Yes
Rosenbrock	Hybrid vs PSO	2.99e-03	Yes
	Hybrid vs FA	1.60e-05	Yes
	Hybrid vs GA	2.83e-02	Yes
	Hybrid vs SA	1.67e-03	Yes
Levy	Hybrid vs PSO	1.35e-04	Yes
	Hybrid vs FA	6.00e-07	Yes
	Hybrid vs GA	3.40e-08	Yes
	Hybrid vs SA	3.40e-08	Yes
Powell	Hybrid vs PSO	9.57e-04	Yes
	Hybrid vs FA	1.03e-06	Yes
	Hybrid vs GA	1.97e-07	Yes
	Hybrid vs SA	1.49e-07	Yes
Schwefel	Hybrid vs PSO	2.27e-03	Yes
	Hybrid vs FA	2.00e-08	Yes
	Hybrid vs GA	1.00e+00	No
	Hybrid vs SA	2.27e-03	Yes

7 Policy Analysis

The learned RL Q-table shows structured operator preferences across the discretised population states. In practice, FA tends to be preferred in states associated with moderate diversity where exploitation can still refine promising regions, while GA is favoured when diversity drops and diversification is required to escape stagnation. Because the state space is small and explicitly defined using population metrics, the resulting policy remains interpretable and easy to inspect.

8 Conclusion

This research presents an adaptive hybrid PSO algorithm integrating with reinforcement learning-based search operator selection. The approach combines a PSO warm-start phase with a tabular Q-Learning controller that dynamically selects between FA and GA based on population diversity and fitness spread signals. Experimental evaluation across six benchmark functions demonstrates that the proposed algorithm improves robustness across diverse optimisation landscapes and frequently outperforms standalone metaheuristics, particularly in multimodal settings where premature convergence along with local optima traps is common.

Future work will investigate extending the action space to include additional population-based optimisation algorithms, enabling richer meta-level operator selection. Furthermore, alternative state representations based on different diversity measures, convergence indicators, or landscape-aware metrics will be explored to enhance the adaptability and generalisation capability of the learned policy.

References

1. Kennedy, J., Eberhart, R.: Particle swarm optimization. In: Proceedings of IEEE International Conference on Neural Networks, pp. 1942–1948 (1995)
2. Watkins, C.J.C.H.: Learning from Delayed Rewards. PhD Thesis, King's College, Cambridge (1989)
3. Karaçizmeli, İ.H., Aydilek, İ.B., Gümüşçü, A., Tenekeci, M.E., Kaya, S.: Evaluating solution performance of hybrid firefly and particle swarm optimization algorithm in flow shop scheduling problems. In: Proceedings Book, p. 126 (2019)
4. Yin, S., et al.: Reinforcement-learning-based parameter adaptation method for particle swarm optimization. Complex Intell. Syst. **9**(5), 5585–5609 (2023)
5. Qiu, H., Xue, B., Xiao, Q. and Niu, B.: Q-learning based particle swarm optimization with multi-exemplar and elite learning. In: International Conference on Intelligent Computing, pp. 310–321. Springer Nature Singapore, Singapore (2023)
6. Andrei, N.: An unconstrained optimization test functions collection. Adv. Model. Optim **10**(1), 147–161 (2008)
7. Sharma, P., Raju, S.: Metaheuristic optimization algorithms: a comprehensive overview and classification of benchmark test functions: P. Sharma, S. Raju. Soft Comput. **28**(4), pp. 3123–3186 (2024)
8. Yang, X.S., Slowik, A.: Firefly algorithm. In: Swarm Intelligence Algorithms, pp. 163–174. CRC Press (2020)
9. Zhang, L., et al.: Semantic segmentation using firefly algorithm-based evolving ensemble deep neural networks. Knowl.-Based Syst. **277**, 110828 (2023)

10. Forrest, S.: Genetic algorithms. ACM Comput. Surv. (CSUR) **28**(1), 77–80 (1996)
11. Lambora, A., Gupta, K., Chopra, K.: Genetic algorithm-a literature review. In: 2019 International Conference on Machine Learning, Big Data, Cloud and Parallel Computing (COMITCon), pp. 380–384. IEEE (2019)
12. Zhang, L., Lim, C.P., Yu, Y., Jiang, M.: Sound classification using evolving ensemble models and particle swarm optimization. Appl. Soft Comput. **116**, 108322 (2022)
13. Slade, S., Zhang, L., Yu, Y., Lim, C.P.: An evolving ensemble model of multi-stream convolutional neural networks for human action recognition in still images. Neural Comput. Appl. **34**(11), 9205–9231 (2022)
14. Lim, T.Y., Tan, C.J., Kerk, Y.W., Zhang, L., Lim, C.P.: Beta distribution-based monogamous pairs genetic algorithm for knowledge transfer in many-task optimization. Knowl.-Based Syst. **316**, 113361 (2025)
15. Sedov, V., Zhang, L.: Auxiliary generative adversarial networks with iliustration2vec and q-learning based hyperparameter optimisation for anime image synthesis. In: 2024 IEEE International Conference on Systems, Man, and Cybernetics (SMC), pp. 4396–4403. IEEE (2024)
16. Cunha, L., Zhang, L., Sowan, B., Lim, C.P., Kong, Y.: Video deepfake detection using particle swarm optimization improved deep neural networks. Neural Comput. Appl. **36**(15), 8417–8453 (2024)
17. Zhang, L., Lim, C.P., Liu, C.: Enhanced bare-bones particle swarm optimization based evolving deep neural networks. Expert Syst. Appl. **230**, 120642 (2023)
18. Slade, S., et al.: Neural inference search for multiloss segmentation models. IEEE Trans. Neural Netw. Learn. Syst. **35**(11), 15113–15127 (2023)
19. Kondrotas, K., Zhang, L., Lim, C.P., Asadi, H., Yu, Y.: Audio-visual emotion classification using reinforcement learning-enhanced particle swarm optimisation. IEEE Trans. Affect. Comput. (2025)
20. Slade, S., et al.: Cluster search optimisation of deep neural networks for audio emotion classification. Knowl.-Based Syst. **314**, 113223 (2025)
21. Woodcock, A., Zhang, L.: Genetic algorithm with reinforcement learning based parameter optimisation. In: 2024 IEEE International Conference on Systems, Man, and Cybernetics (SMC), pp. 4833–4839. IEEE (2024)
22. Zhang, L., et al.: Video deepfake classification using particle swarm optimization-based evolving ensemble models. Knowl.-Based Syst. **289**, 111461 (2024)
23. Zhang, L., Srisukkham, W., Neoh, S.C., Lim, C.P., Pandit, D.: Classifier ensemble reduction using a modified firefly algorithm: An empirical evaluation. Expert Syst. Appl. **93**, 395–422 (2018)
24. Zhang, Y., Zhang, L., Hossain, M.A.: Adaptive 3D facial action intensity estimation and emotion recognition. Expert Syst. Appl. **42**(3), 1446–1464 (2015)
25. Pandit, D., Zhang, L., Chattopadhyay, S., Lim, C.P., Liu, C.: A scattering and repulsive swarm intelligence algorithm for solving global optimization problems. Knowl.-Based Syst. **156**, 12–42 (2018)
26. Zhang, L., Lim, C.P., Yu, Y.: Intelligent human action recognition using an ensemble model of evolving deep networks with swarm-based optimization. Knowl.-Based Syst. **220**, 106918 (2021)
27. Lawrence, T., Zhang, L., Lim, C.P., Phillips, E.J.: Particle swarm optimization for automatically evolving convolutional neural networks for image classification. IEEE Access **9**, 14369–14386 (2021)
28. Xie, H., Zhang, L., Lim, C.P., Yu, Y., Liu, H.: Feature selection using enhanced particle swarm optimisation for classification models. Sensors **21**(5), 1816 (2021)

29. Tan, C.J., et al.: Application of an evolutionary algorithm-based ensemble model to job-shop scheduling. J. Intell. Manuf. **30**(2), 879–890 (2019)
30. Zhang, L., Lim, C.P.: Intelligent optic disc segmentation using improved particle swarm optimization and evolving ensemble models. Appl. Soft Comput. **92**, 106328 (2020)
31. Fielding, B., Zhang, L.: Evolving image classification architectures with enhanced particle swarm optimisation. IEEE Access **6**, 68560–68575 (2018)
32. Zhang, Y., Zhang, L., Neoh, S.C., Mistry, K., Hossain, M.A.: Intelligent affect regression for bodily expressions using hybrid particle swarm optimization and adaptive ensembles. Expert Syst. Appl. **42**(22), 8678–8697 (2015)
33. Chin Neoh, S., et al.: An intelligent decision support system for Leukaemia diagnosis using microscopic blood images. Sci. Rep. **5**(1), 14938 (2015)
34. Xie, H., et al.: Improving K-means clustering with enhanced Firefly Algorithms. Appl. Soft Comput. **84**, 105763 (2019)
35. Srisukkham, W., Zhang, L., Neoh, S.C., Todryk, S., Lim, C.P.: Intelligent Leukaemia diagnosis with bare-bones PSO based feature optimization. Appl. Soft Comput. **56**, 405–419 (2017)
36. Tan, T.Y., Zhang, L., Neoh, S.C., Lim, C.P.: Intelligent skin cancer detection using enhanced particle swarm optimization. Knowl.-Based Syst. **158**, 118–135 (2018)
37. Xie, H., Zhang, L., Lim, C.P.: Evolving CNN-LSTM models for time series prediction using enhanced grey wolf optimizer. IEEE Access **8**, 161519–161541 (2020)
38. Mistry, K., Zhang, L., Neoh, S.C., Lim, C.P., Fielding, B.: A micro-GA embedded PSO feature selection approach to intelligent facial emotion recognition. IEEE Trans. Cybern. **47**(6), 1496–1509 (2016)
39. Neoh, S.C., et al.: Intelligent facial emotion recognition using a layered encoding cascade optimization model. Appl. Soft Comput. **34**, 72–93 (2015)

A Multi-attribute Group Decision-Making Approach Based on Linguistic Truth-Valued Lattice Implication Algebra and Prospect Theory

Tianyu Pang[1], Kuo Pang[1(✉)], Pengbo Liu[1], Tao Zhang[1], and Li Zou[2]

[1] School of Information Science and Engineering, Yanshan University, Qinhuangdao 066004, China
kuopang@ysu.edu.cn

[2] School of Computer Science and Technology, Shandong Jianzhu University, Jinan 250102, China

Abstract. In uncertain linguistic environments, multi-attribute group decision-making (MAGDM) problems face the fuzziness of linguistic evaluations and the dynamic nature of decision-makers' psychological behaviors. This paper proposes an improved VIKOR MAGDM approach that integrates linguistic truth-valued lattice implication algebra (LTV-LIA) with prospect theory. LTV-LIA establishes operational rules for linguistic terms with strict logical semantics to address evaluation fuzziness. Prospect theory constructs a prospect value matrix that captures decision-makers' psychological behaviors. A Home purchasing case study and comparative analysis with VIKOR, TOPSIS and PROMETHEE validate the feasibility and effectiveness of the proposed approach.

Keywords: Multi-attribute group decision-making · Linguistic truth-valued lattice implication algebra · Prospect theory · VIKOR

1 Introduction

Multi-attribute group decision-making (MAGDM) has been extensively applied in economic management, engineering technology, and social sciences. Expert evaluations typically exhibit fuzziness, uncertainty, and discrete linguistic expressions. Traditional numerical approaches cause information loss and fail to accurately reflect experts' true intentions. Three key challenges motivate this work.

Challenge 1: Processing Comparable and Incomparable Linguistic Information. Existing linguistic representation models [1–3] handle comparable assessments effectively, yet experts often express inherently incomparable judgments such as "very good" or "somewhat poor". The LTV-LIA is proposed by Xu et al. [4,5], characterizes the internal logical structure of linguistic information and supports rigorous mathematical operations, providing a robust foundation for MAGDM [6,7].

J. Lu et al. (Eds.): FLINS-ISKE 2026, LNCS 16752, pp. 342–354, 2027.
https://doi.org/10.1007/978-981-92-2369-5_22

Challenge 2: Incorporating Decision-Makers' Psychological Behavior. Classical decision models assume complete rationality, yet behavioral decision theory shows that risk preferences and reference-point dependence substantially affect choices. Kahneman and Tversky's prospect theory [8] captures these behaviors: decision-makers are risk-averse over gains and risk-seeking over losses. Integrating prospect theory into MAGDM yields more realistic outcomes [9–11].

Challenge 3: Selecting an Efficient Compromise Ranking Approach. The VIKOR approach [12] ranks alternatives by computing compromise solutions and has been widely applied [13,14]. Its simultaneous consideration of group utility and individual regret makes it particularly suitable here.

This paper combines LTV-LIA, prospect theory, and VIKOR into an integrated framework whose core contributions are: (i) using LTV-LIA to process both comparable and incomparable linguistic evaluations without information loss; (ii) integrating prospect theory to model decision-makers' risk preferences; (iii) applying an improved VIKOR procedure on the resulting prospect value matrix; and (iv) validating the approach on a Home purchasing case syudy with comparative analysis against TOPSIS and PROMETHEE.

2 Preliminaries

2.1 LTV-LIA

Definition 1 ([15]). *Let $AD_n = \{h_1, h_2, \ldots, h_n\}$ denote a set of n linguistic hedges satisfying $h_1 < h_2 < \cdots < h_n$, and $MT = \{t, f\}$ denote meta-linguistic truth values, such that $f < t$. Let $L_{V(n\times2)} = AD_n \times MT$. Define the mapping $g\colon L_{V(n\times2)} \to L_n \times L_2$ and induce operations $\vee, \wedge, ', \to$ on $L_{V(n\times2)}$ via g. The resulting structure $(L_{V(n\times2)}, \vee, \wedge, ', \to)$ is called the* LTV-LIA .

The four operations in $L_{V(n\times2)}$ are given explicitly as follows.

$$\begin{aligned}\vee:\quad &(h_i,t)\vee(h_j,f) = (h_{\max\{i,\,n-j+1\}},t),\\ &(h_i,t)\vee(h_j,t) = (h_{\max\{i,j\}},t),\quad (h_i,f)\vee(h_j,f) = (h_{\min\{i,j\}},f).\end{aligned} \tag{2-1}$$

$$\begin{aligned}\wedge:\quad &(h_i,t)\wedge(h_j,f) = (h_{\max\{j,\,n-i+1\}},f),\\ &(h_i,t)\wedge(h_j,t) = (h_{\min\{i,j\}},t),\quad (h_i,f)\wedge(h_j,f) = (h_{\max\{i,j\}},f).\end{aligned} \tag{2-2}$$

$$':\quad (h_i,t)' = (h_i,f),\qquad (h_i,f)' = (h_i,t). \tag{2-3}$$

$$\begin{aligned}\to:\quad &(h_i,t)\to(h_j,t) = (h_{\min\{n-i+j,\,n\}},t),\\ &(h_i,f)\to(h_j,f) = (h_{\min\{n-j+i,\,n\}},t),\\ &(h_i,t)\to(h_j,f) = (h_{\max\{0,\,i+j-n\}},f),\\ &(h_i,f)\to(h_j,t) = (h_{\min\{n,\,i+j\}},t).\end{aligned} \tag{2-4}$$

Definition 2 ([16]). *For* $A = (h_i, c_{j_1})$, $B = (h_j, c_{j_2}) \in L_{V(n\times 2)}$, *the* shortest distance $d_{LI}(A, B)$ *is defined by*

$$d_{LI}(A,B) = \begin{cases} |i-j|, & j_1 = j_2, \\ i+j-n, & j_1 \neq j_2 \text{ and } i+j \geq n+1, \\ n+2-(i+j), & j_1 \neq j_2 \text{ and } i+j < n+1. \end{cases} \tag{2-5}$$

Definition 3. *Let* $\Im$ *be an evaluation set which is a LTV-LIA. A mapping* $LVWA : \Im^m \to \Im$ *is called a linguistic-valued weighted aggregation (LVWA) operator, if*

$$LVWA_W(\Im_1, \Im_2, \ldots, \Im_m) = \bigwedge_{i=1}^{m} (h_i \to \Im_i), \tag{2-6}$$

where $H = (h_1, h_2, \cdots, h_m)^{\mathrm{T}}$ *is the linguistic weight vector with* $h_i \in H$, $\Im_i \in \Im$ $(i = 1, 2, \ldots, m)$ *are the linguistic-valued arguments to be aggregated,* $\to$ *is the implication operation in* $\Im$, *and* $\bigwedge$ *is the meet operation in* $\Im$.

2.2 VIKOR Approach

VIKOR simultaneously maximizes group utility and minimizes individual regret. For alternative x_i, define

$$S_i = \sum_{j=1}^{s} w_j \frac{v_j^+ - v_{ij}}{v_j^+ - v_j^-}, \qquad R_i = \max_{1\leq j\leq s} w_j \frac{v_j^+ - v_{ij}}{v_j^+ - v_j^-}, \tag{2-7}$$

where v_j^+ and v_j^- are the positive and negative ideal values of attribute j. The compromise ranking index is

$$Q_i = k\,\frac{S_i - S^*}{S^- - S^*} + (1-k)\,\frac{R_i - R^*}{R^- - R^*}, \tag{2-8}$$

where $S^* = \min_i S_i$, $S^- = \max_i S_i$, $R^* = \min_i R_i$, $R^- = \max_i R_i$, and $k \in [0, 1]$ is the compromise weight.

3 Methodology

We employ a ten-element LTV-LIA $L_{V(5\times 2)}$ with hedge set $H = \{h_1 = \text{slightly}, h_2 = \text{somewhat}, h_3 = \text{moderately}, h_4 = \text{very}, h_5 = \text{extremely}\}$ and truth set $C = \{c_1 = \text{poor}, c_2 = \text{good}\}$.

3.1 Linguistic Truth-Valued Entropy Weight Approach

In linguistic MAGDM problems, accurately determining the entropy weight coefficients of each indicator is crucial. While numerous approaches exist for determining entropy weight coefficients, this paper proposes the linguistic truth-valued entropy weight evaluation approach, considering the characteristic that the indicator data contains linguistic information. The fundamental principle of this approach is to calculate the entropy weight of each indicator by measuring the amount of information contained within the indicator data of the evaluation system. The specific steps are as follows:

Step 1 – Normalize the Aggregated Linguistic Evaluation Matrix. For a benefit attribute:

$$x_{ij} = d_{LI}(f_{ij}, \min_i f_{ij})/d_{LI}(\max_i f_{ij}, \min_i f_{ij}). \tag{3-1}$$

For a cost attribute:

$$x_{ij} = d_{LI}(\max_i f_{ij}, f_{ij})/d_{LI}(\max_i f_{ij}, \min_i f_{ij}). \tag{3-2}$$

where x_{ij} denotes the normalized value of alternative i on attribute j, d_{LI} is the shortest distance function based on LTV-LIA (definition 2), f_{ij} is the original linguistic evaluation of alternative i on attribute j, $\min_i f_{ij}$ and $\max_i f_{ij}$ represent the minimum and maximum linguistic evaluations of attribute j across all alternatives, respectively.

Step 2 – Evaluation Proportion
The resulting evaluation proportion matrix is $Y_{r\times s} = [y_{ij}]_{r\times s}$. To prevent the expression $\ln y_{ij}$ from being undefined, the calculation formula for the evaluation proportion y_{ij} is defined as follows:

$$y_{ij} = (x_{ij}+1)\Big/\sum_{i=1}^{r}(x_{ij}+1) \tag{3-3}$$

Step 3 – Criterion Entropy

$$m_j = (\ln r)^{-1}\sum_{i=1}^{r}(-y_{ij}\ln y_{ij}). \tag{3-4}$$

where m_j is the entropy value of attribute j (ranging from 0 to 1), reflecting the dispersion degree of evaluations on attribute j, and $\ln r$ is the normalization coefficient to ensure the maximum entropy is 1.

Step 4 – Entropy Weights

$$w_j = \frac{1-m_j}{\sum_{j=1}^{s}(1-m_j)}. \tag{3-5}$$

where w_j is the entropy weight of attribute j (satisfying $\sum_{j=1}^{s} w_j = 1$), $1 - m_j$ represents the discrimination degree of attribute j, and s is the total number of attributes.

Remark 1. The normalization Equations ((3-1)–(3-2)) are not merely numerical scaling, but are established on the LTV-LIA defined in Sect. 2 Specifically, the distance function $d_{L_I}(\cdot,\cdot)$ constitutes a lattice-valued metric that preserves the logical implication structure and partial order relations of linguistic truth values. The minimum and maximum operations are determined based on the Hasse diagram of the linguistic truth-valued set $L_{V(5\times 2)}$. Consequently, the normalized $x_{ij} \in [0,1]$ is not an arbitrary numerical substitution, but a lattice embedding of linguistic evaluations that retains the logical relationships between linguistic assessments. In this regard, the entropy weights computed in subsequent steps reflect the intrinsic information content of linguistic evaluations themselves, rather than merely that of their numerical encoding.

3.2 A LT-PT-VIKOR Approach for MAGDM

Step 1. Aggregate the k decision-makers' linguistic matrices into a unified matrix $F = (f_{ij})_{r\times s}$ using the LVWA operator (definition 3), where k is the number of decision-makers, $F = (f_{ij})_{r\times s}$ is the aggregated linguistic evaluation matrix.

Step 2. Apply the entropy weight approach (Sect. 3.1) to obtain $W = (w_1,\ldots,w_s)$. Then compute the secondary normalized matrix $B = (b_{ij})_{r\times s}$:

$$b_{ij} = \begin{cases} x_{ij}/\max_i x_{ij}, & \text{benefit attribute,} \\ 1 - x_{ij}/\max_i x_{ij}, & \text{cost attribute.} \end{cases} \tag{3-6}$$

Select the column-median of B as the prospect theory reference point b_0, where $B = (b_{ij})_{r\times s}$ is the secondary normalized matrix, b_{ij} is the secondary normalized value of alternative i on attribute j, and b_0 is the reference point of prospect theory reflecting the psychological benchmark of decision-makers.

Step 3. Construct the prospect value matrix $V = (V_{ij})_{r\times s}$. Let $\Delta b_{ij} = b_{ij} - b_0$; with risk-attitude coefficients $\alpha, \beta \in (0,1]$ and loss-aversion coefficient $\lambda > 1$:

$$V_{ij} = \begin{cases} (\Delta b_{ij})^{\alpha}, & \Delta b_{ij} \geq 0, \\ -\lambda(-\Delta b_{ij})^{\beta}, & \Delta b_{ij} < 0. \end{cases} \tag{3-7}$$

Step 4. Determine the prospect positive ideal solution (PIS) and negative ideal solution (NIS):

$$v_j^+ = \max_{1\leq i\leq r} V_{ij}, \qquad v_j^- = \min_{1\leq i\leq r} V_{ij}. \tag{3-8}$$

where v_j^+ and v_j^- are the positive and negative ideal solutions of attribute j, representing the optimal and worst prospect values of attribute j, respectively.

Step 5. Compute the prospect group utility S_i and individual regret R_i via Eq. (2-7), replacing v_{ij} with V_{ij}, where S_i is the group utility value of alternative

i (ranging from 0 to 1), reflecting the overall performance of the alternative, and R_i is the individual regret value of alternative i (ranging from 0 to 1), reflecting the worst attribute performance of the alternative.

Step 6. Compute the compromise index Q_i via Eq. (2-8), where Q_i is the compromise ranking index of alternative i , $S^* = \min_i S_i$, $S^- = \max_i S_i$, $R^* = \min_i R_i$, $R^- = \max_i R_i$, and $k \in [0, 1]$ is the compromise weight .

Step 7. Rank all alternatives in ascending order of Q, S, and R, where a smaller value of Q_i, S_i, or R_i indicates a better alternative.

Step 8. Determine the compromise solution using two acceptance conditions:

(C1) *Acceptable advantage*: $Q(A^{(2)}) - Q(A^{(1)}) \geq \frac{1}{r-1}$. (C2) *Acceptable stability*: $A^{(1)}$ also ranks first in S or R. If both (C1) and (C2) hold, $A^{(1)}$ is the unique compromise solution. If (C1) fails, the set $\{A^{(1)}, \ldots, A^{(m)}\}$ with $Q(A^{(m)}) - Q(A^{(1)}) < 1/(r-1)$ constitutes the compromise set. If only (C2) fails, both $A^{(1)}$ and $A^{(2)}$ are proposed as solutions, where $A^{(1)}$ and $A^{(2)}$ are the first and second ranked alternatives by Q, $A^{(m)}$ is the last alternative in the compromise set, and $1/(r-1)$ is the threshold of acceptable advantage to ensure a significant difference between the optimal and suboptimal alternatives.

3.3 Algorithm

Algorithm 1. LT-PT-VIKOR Multi-Attribute Group Decision-Making

Require: Linguistic matrix $F[r][s]$; weight vector W; $k = 0.5$; α, β, λ; reference point b_0

Ensure: Ranked alternatives and compromise solution BEST

for each $f_{ij} \in F$ **do**
  Compute x_{ij} by linguistic entropy normalisation (Sec. 3.1)
end for
for each x_{ij} **do**
  Compute b_{ij} (Eq. 3-6); $\Delta b_{ij} \leftarrow b_{ij} - b_0$; V_{ij} (Eq. 3-7)
end for
Compute v^+, v^- (Eq. 3-8); S, R (Eq. 2-7); Q (Eq. 2-8)
Sort alternatives in ascending order of Q; sort S and R likewise
if $Q(A^{(2)}) - Q(A^{(1)}) \geq 1/(r-1)$ **and** $A^{(1)}$ ranks first in S or R **then**
  BEST $\leftarrow A^{(1)}$
else if $Q(A^{(2)}) - Q(A^{(1)}) \geq 1/(r-1)$ **and** $A^{(1)}$ does *not* rank first **then**
  BEST $\leftarrow \{A^{(1)}, A^{(2)}\}$
else
  BEST $\leftarrow \{A^{(1)}, \ldots, A^{(m)}\}$ where $Q(A^{(m)}) - Q(A^{(1)}) < 1/(r-1)$
end if
return BEST

The dominant cost is sorting r alternatives, giving overall time complexity $O(r * s + r \log r)$, where $O(r * s)$ is the time complexity of basic operations such as normalization and prospect value calculation, and $O(r \log r)$ is the time complexity of sorting r alternatives, resulting in an overall linear logarithmic complexity suitable for large-scale decision problems.

4 Home Purchase Decision-Making

4.1 Background

Home purchasing is among the most critical economic decisions for households, shaped by the interplay of multiple variables including property value, living experience, and housing price [17]. The three primary decision factors are:

(1) Property Value Appreciation Potential. Transportation networks, commercial facilities, educational and medical resources all impact long-term asset value and regional attractiveness.

(2) Living Space Comfort. Sunlight exposure, ventilation, orientation, building spacing, and layout design collectively determine quality of life.

(3) Housing Price. As the core cost factor, price exerts a significant negative influence on the purchasing decision, requiring consumers to trade off cost against positive-valued attributes.

4.2 Case Study

This case study investigates a typical household home-purchasing decision under a multi-attribute and multi-decision-maker environment. The selected alternatives represent four candidate residential properties, denoted as $X = \{x_1, x_2, x_3, x_4\}$, which may correspond to four real estate projects or four housing units with different locations and quality profiles. The three evaluation attributes are defined as y_1: Housing Price (cost-type), y_2: Living Space Comfort (benefit-type), and y_3: Location Value Appreciation Potential (benefit-type).

The decision-making information was collected from three consumers $Z = \{z_1, z_2, z_3\}$ who jointly participated in the house selection process. Their linguistic assessments reflect each decision-maker's subjective preferences and judgment on the housing alternatives with respect to the three attributes y_1, y_2, and y_3.

The linguistic evaluation data are presented in Tables 1, 2 and 3. Each table records the preference of one consumer over the four alternatives under the three attributes y_1, y_2, and y_3.

Step 1 – Aggregation. Applying LVWA with equal decision-maker weights yields the aggregated matrix:

$$F = \begin{pmatrix} (h_3, c_2) & (h_3, c_2) & (h_4, c_2) \\ (h_4, c_2) & (h_3, c_1) & (h_3, c_2) \\ (h_4, c_1) & (h_4, c_2) & (h_3, c_1) \\ (h_3, c_2) & (h_2, c_1) & (h_3, c_2) \end{pmatrix}.$$

Table 1. Evaluation matrix provided by consumer z_1

	x_1	x_2	x_3	x_4
y_1	(very, good)	(mod., poor)	(som., good)	(sli., good)
y_2	(mod., poor)	(sli., good)	(ext., poor)	(som., good)
y_3	(ext., good)	(very, poor)	(mod., poor)	(sli., good)

Table 2. Evaluation matrix provided by consumer z_2

	x_1	x_2	x_3	x_4
y_1	(mod., poor)	(very, good)	(ext., poor)	(mod., good)
y_2	(som., good)	(mod., poor)	(ext., good)	(sli., poor)
y_3	(very, good)	(mod., poor)	(ext., poor)	(sli., good)

Step 2 – Normalization and entropy weights. ; The standardized matrix X and normalized matrix B are:

$$X_{4\times3} = \begin{pmatrix} \frac{1}{3} & \frac{1}{2} & 1 \\ 0 & 0 & \frac{1}{2} \\ 1 & 1 & 0 \\ \frac{1}{3} & \frac{1}{2} & \frac{1}{2} \end{pmatrix}, \qquad B_{4\times3} = \begin{pmatrix} \frac{2}{3} & \frac{1}{2} & 1 \\ 1 & 0 & \frac{1}{2} \\ 0 & 1 & 0 \\ \frac{2}{3} & \frac{1}{2} & \frac{1}{2} \end{pmatrix}.$$

Using column medians as the reference point gives $b_0 = \left(\frac{2}{3}, \frac{1}{2}, \frac{1}{2}\right)$, and the entropy weights are $W = (0.3596,\ 0.3202,\ 0.3202)$.

Step 3 – Prospect value matrix. With $\alpha = 0.88$, $\beta = 0.9$, $\lambda = 2.25$:

$$V = \begin{pmatrix} 0 & 0 & 0.5434 \\ 0.3803 & -1.2057 & 0 \\ -1.5621 & 0.5434 & -1.2057 \\ 0 & 0 & 0 \end{pmatrix}.$$

Table 3. Evaluation matrix provided by consumer z_3

	x_1	x_2	x_3	x_4
y_1	(very, good)	(mod., poor)	(ext., poor)	(mod., good)
y_2	(som., good)	(sli., poor)	(mod., good)	(mod., poor)
y_3	(ext., poor)	(very, good)	(mod., poor)	(sli., good)

Steps 4–6 – Ideal solutions, utility, regret, and compromise index.

$$v^+ = (0.3803,\ 0.5434,\ 0.5434),\quad v^- = (-1.5621,\ -1.2057,\ -1.2057).$$
$$S = (0.1699,\ 0.4197,\ 0.6798,\ 0.2694),\quad R = (0.0995,\ 0.3202,\ 0.3596,\ 0.0995).$$
$$Q = (0,\ 0.6695,\ 1,\ 0.0975)\quad (k = 0.5).$$

Steps 7–8 – Ranking and compromise solution.

$$S: x_1 \prec x_4 \prec x_2 \prec x_3,\quad R:\ x_1 = x_4 \prec x_2 \prec x_3,\quad Q:\ x_1 \prec x_4 \prec x_2 \prec x_3.$$

Since $Q(x_4) - Q(x_1) = 0.0975 < 1/(4-1) = 0.333$, condition1 is not satisfied; both x_1 and x_4 form the compromise set. Because S_1 and R_1 are both minimum, $\boldsymbol{x_1}$ is identified as the *optimal solution.*

4.3 Parameter Sensitivity Analysis

Tables 4, 5, 6 and 7 report the effect of varying each prospect-theory parameter while keeping the others at their baseline values ($\alpha = 0.88$, $\beta = 0.9$, $\lambda = 2.25$, $k = 0.5$).

Table 4. Effect of risk preference coefficient α on Q and ranking

α	Q_1	$Q_2/Q_3/Q_4$	Ranking
0.2	0	0.697 / 0.399 / 1	$x_1 \prec x_4 \prec x_2 \prec x_3$
0.4	0	0.717 / 0.394 / 1	$x_1 \prec x_4 \prec x_2 \prec x_3$
0.6	0	0.731 / 0.389 / 1	$x_1 \prec x_4 \prec x_2 \prec x_3$
0.8	0	0.740 / 0.384 / 1	$x_1 \prec x_4 \prec x_2 \prec x_3$

Table 5. Effect of risk aversion coefficient β on Q and ranking

β	Q_1	$Q_2/Q_3/Q_4$	Ranking
0.2	0	0.713 / 0.631 / 1	$x_1 \prec x_4 \prec x_2 \prec x_3$
0.4	0	0.721 / 0.533 / 1	$x_1 \prec x_4 \prec x_2 \prec x_3$
0.6	0	0.729 / 0.459 / 1	$x_1 \prec x_4 \prec x_2 \prec x_3$
0.8	0	0.738 / 0.404 / 1	$x_1 \prec x_4 \prec x_2 \prec x_3$

Across all four parameters and every tested value, the ranking $x_1 \prec x_4 \prec x_2 \prec x_3$ remains unchanged and x_1 is consistently the optimal alternative, confirming the robustness of the proposed approach.

Table 6. Effect of loss aversion coefficient λ on Q and ranking

λ	Q_1	$Q_2/Q_3/Q_4$	Ranking
1.2	0	0.755 / 0.421 / 1	$x_1 \prec x_4 \prec x_2 \prec x_3$
1.6	0	0.751 / 0.403 / 1	$x_1 \prec x_4 \prec x_2 \prec x_3$
2.0	0	0.746 / 0.390 / 1	$x_1 \prec x_4 \prec x_2 \prec x_3$
2.4	0	0.741 / 0.379 / 1	$x_1 \prec x_4 \prec x_2 \prec x_3$

Table 7. Effect of compromise coefficient k on Q and ranking

k	Q_1	$Q_2/Q_3/Q_4$	Ranking
0.0	0	0.852 / 0.393 / 1	$x_1 \prec x_4 \prec x_2 \prec x_3$
0.2	0	0.808 / 0.389 / 1	$x_1 \prec x_4 \prec x_2 \prec x_3$
0.6	0	0.721 / 0.380 / 1	$x_1 \prec x_4 \prec x_2 \prec x_3$
0.8	0	0.677 / 0.376 / 1	$x_1 \prec x_4 \prec x_2 \prec x_3$
1.0	0	0.634 / 0.372 / 1	$x_1 \prec x_4 \prec x_2 \prec x_3$

4.4 Comparative Analysis

Table 8 compares the rankings of the proposed approach with classical VIKOR, TOPSIS [18], and PROMETHEE [19] on the same house-purchasing data.

Table 8. Ranking comparison across four methods

Method	Rank 1	Rank 2	Rank 3	Rank 4
Proposed (LT-PT-VIKOR)	x_1	x_4	x_2	x_3
VIKOR	x_1	x_4	x_2	x_3
TOPSIS	x_1	x_4	x_3	x_2
PROMETHEE	x_1	x_4	x_2	x_3

All four methods unanimously identify x_1 as optimal and x_4 as second-best. The minor difference in lower ranks (TOPSIS swaps x_2 and x_3) reflects the distinct evaluation logic of that method. The proposed approach is the only one that simultaneously handles incomparable linguistic terms via LTV-LIA and explicitly models risk attitudes via prospect theory, while yielding consistent top-level results.

Beyond ranking consistency, the four methods differ substantially in their underlying assumptions and capabilities. Table 9 provides a structured comparison across six methodological dimensions.

Table 9. Methodological feature comparison of the four approaches

Feature	Proposed	VIKOR	TOPSIS	PROMETHEE
Linguistic model	LTV-LIA	Crisp/fuzzy	Crisp/fuzzy	Crisp/fuzzy
Incomparable terms	✓	×	×	×
Behavioral theory	Prospect Theory	None	None	None
Rationality assumption	Bounded	Complete	Complete	Complete
Reference point	✓	×	×	×

As shown in Table 9, the proposed approach is the only method that simultaneously satisfies all six dimensions. Classical VIKOR, TOPSIS, and PROMETHEE all rely on crisp or standard fuzzy numbers and cannot natively process incomparable linguistic expressions; any such evaluation must first be forcibly converted into a scalar, inevitably causing semantic loss. By contrast, LTV-LIA preserves the internal logical structure of linguistic terms through its lattice operations, so both comparable and incomparable evaluations enter the model on equal footing. Furthermore, while the three classical methods implicitly assume that decision-makers are fully rational and indifferent to the framing of outcomes, the prospect-theory component of the proposed approach captures asymmetric sensitivity to gains and losses via the value function (Eq. (3-7)) and explicitly models loss aversion through the coefficient λ. These behavioral features are particularly important in high-stakes consumer decisions such as home purchasing, where psychological factors demonstrably influence final choices.

5 Conclusion

Validated through a Home purchasing case study and comparison with existing approaches, this approach yields consistent optimal solutions withthe existing decision-making approaches, which verifies its feasibility, effectiveness and robustness. Furthermore, it is adaptable to diverse consumer psychological characteristics, and can be applied to complex practical decision-making scenarios that balance objective data and subjective preferences, thereby providing more scientific and effective decision-making solutions for such scenarios

Limitations: First, prospect theory parameters remain challenging to determine; current parameter values lack systematic calibration, affecting modeling accuracy. Second, computational complexity increases substantially with more alternatives and criteria, limiting real-time or large-scale applications. Third, the approach assumes consistent linguistic assessments across criteria, which often fails in practice; furthermore, while capturing loss aversion and probability distortion, it overlooks other behavioral phenomena such as regret theory, emotions, and dynamic preference changes.

Future work will focus on: (1) machine-learning-based calibration of prospect-theory parameters; (2) extension to cumulative prospect theory and regret theory; (3) handling dynamic decision environments; and (4) integration with natural language processing for intelligent decision support.

Acknowledgment. This work was supported in part by the Natural Science Foundation of Hebei Province under Grant F2025203025, and in part by the Basic Innovative Research Cultivation Program of Yanshan University under Grant 2024LGQN009

Disclosure of Interests. The authors have no competing interests to declare that are relevant to the content of this article.

References

1. Yi, Z. H.: Decision-making based on probabilistic linguistic term sets without loss of information. Complex Intell. Syst. **8**(3), 1–15 (2022)
2. Chen, B., Cai, Q.,Wei, G. W., Mo, Z. W.: Novel Aczel-Alsina operations-based linguistic Z-number aggregation operators and their applications in multi-attribute group decision-making process. Eng. Appl. Artif. Intell. **124** (2023)
3. Wang, F., Pan, Y., Xu, G.L., Wei, Q.: Probabilistic linguistic multi-attribute group decision making method considering the important degrees of experts and attributes. Expert Syst. Appl. **265**, 125899 (2025)
4. Xu, Y., Chen, S. W., Ma, J.: Linguistic truth-valued lattice implication algebra and its properties. In: Proceedings of the Multiconference on Computational Engineering in Systems Applications, IEEE, pp. 1413–1418 (2006)
5. Xu, Y., Chen, S., Liu, J., Ruan, D.: Weak completeness of resolution in a linguistic truth-valued propositional logic. In: Theoretical Advances and Applications of Fuzzy Logic and Soft Computing, Springer, pp. 358–366 (2007)
6. Pang, K., Lu, Y.F., Martínez, L., Pedrycz, W., Zou, L., Lu, M.Y.: A conceptual clustering method for large-scale group decision-making with linguistic truth-valued lattice implication algebra. Appl. Soft Comput. **155**, 111418 (2024)
7. Yang, L., Wang, Y.H., Li, H.X.: Research on the disease intelligent diagnosis model based on linguistic truth-valued concept lattice. Complexity **2021**, 6630077 (2021)
8. Kahneman, D., Tversky, A.: Prospect theory: an analysis of decision under risk. Econometrica **47**(2), 363–391 (1979)
9. Gao, J. et al.: Multi-attribute decision-making for PPP projects. J. Manag. (2020)
10. Chai, N. J., Zhou, W. L., Jiang, Z. G.: Sustainable supplier selection using an intuitionistic and interval-valued fuzzy MCDM approach based on cumulative prospect theory. Info. Sci. **626**, 710–737 (2023)
11. Zhang, H. Y., Wei, G. W ., Chen, X. D.: SF-GRA method based on cumulative prospect theory for multiple attribute group decision making and its application to emergency supplies supplier selection. Eng. Appl. Artif. Intell. **110**, 104679 (2022)
12. Opricovic, S.: Multicriteria optimization of civil engineering systems. Faculty Civil Eng. Belgrade **2**(1), 5–21 (1998)
13. Zhao, C., Liu, R., Xu, X., He, J.: Fuzzy heterogeneous multi-attribute group decision with dual interaction of attributes and alternatives. Expert Syst. Appl. 125532 (2025)
14. Ruan, C. Y., Yan, L., Zeng, S. Z.: Multi-attribute decision-making with extended VlseKriterijuska Optimizacija I Komoromisno Resenje method based on generalized Fermatean probabilistic hesitant fuzzy distance measure and cross-entropy. Eng. Appl. Artif. Intell.**159**(PC), 111684–111684 (2025)

15. Zou, L., Ruan, D., Pei, Z., Xu, Y.: A linguistic-valued lattice implication algebra approach for risk analysis. J. Multiple-Valued Logic Soft Computi. **17**(4), 293–303 (2011)
16. Li, X., Ruan, D., Liu, J., Xu, Y.: A linguistic-valued weighted aggregation operator to multiple attribute group decision making with quantitative and qualitative information. Int. J. Comput. Intell. Syst. **1**(3), 274–284 (2008)
17. Hassan, M.M., Ahmad, N., Hashim, A.H.: The conceptual framework of housing purchase decision-making process. Int. J. Academ. Res. Bus. Soc. Sci **11**(11), 1673–1690 (2021)
18. Han, F., Alkhawaji, N.R., Shafieezadeh, M.M.: Evaluating sustainable water management strategies using TOPSIS and fuzzy TOPSIS methods. Appl. Water Sci. **15**(1), 4 (2024)
19. Deng, J., Zhang, J. M., Wu, W. Z.: A ranking method with a preference relation based on the PROMETHEE method in incomplete multi-scale information systems. Info. Sci. **608**, 1261–1282 (2022)

From Assistants to Decision-Makers: Identifying and Tracing the Evolution of Knowledge-Related Roles Among Outstanding Female Scientists (16th-21st Century)

Yanmin Du[1], Qiuxia Li[1], Tanyao Zhang[1], Huizhen Fu[2], and Zhixiang Wu[1](✉)

[1] Nanjing Tech University, Nanjing 210000, JS, China
cnwzx2012@njtech.edu.cn
[2] Zhejiang University, Hangzhou 310000, ZJ, China

Abstract. Although the participation and activity levels of female scientists in academia have increased markedly, existing studies have largely attributed this change to external institutional and cultural drivers, portraying women as passive beneficiaries while overlooking their agentic roles and substantive contributions in knowledge-creation practices. To address this gap, this study introduces the concept of **Knowledge-Related Roles (KRRs)** and establishes a role classification framework comprising three major categories and nine subcategories, aiming to examine the evolution of female scientists' roles and the underlying nature of their rising status from the perspective of knowledge production. The study develops a role identification system based on large language model (LLM) driven text mining to systematically extract and quantify the specific roles performed by 1,190 outstanding female scientists, along with their corresponding textual and lexical evidence. By tracing the dynamic changes of these roles across historical periods, the study reveals that the core driver of female scientists' status advancement lies in the gradual transformation of their knowledge-related roles from "knowledge assistants" to "knowledge leaders". This finding demonstrates both consistency and meaningful heterogeneity across different disciplines and regional-cultural contexts. Theoretically, this research offers a knowledge-centered perspective for understanding changes in female's status in science, and it provides profound and insightful implications for the cultivation and development of female talent.

Keywords: Female scientists · Knowledge-related roles · Role evolution · Large language model

1 Introduction

In recent years, as the scientific contributions of female scientists have become increasingly visible, related research has gradually emerged as an important topic

J. Lu et al. (Eds.): FLINS-ISKE 2026, LNCS 16752, pp. 355–377, 2027.
https://doi.org/10.1007/978-981-92-2369-5_23

of scholarly attention. Interdisciplinary evidence indicates that over the past few decades, the proportion of active female scientists has risen from approximately 12% to more than 35% [1]. This growth has occurred not only in fields that have traditionally exhibited higher levels of female participation [2], but also in disciplines long dominated by men and characterized by pronounced gender stereotypes [3,4]. From the perspectives of stability, volatility, and persistence in knowledge production, performance differences between female and male researchers have continued to narrow over time [5]. Moreover, no significant difference has been observed in average citation counts per paper [6]. On the contrary, increased female participation often enhances the innovative capacity of research teams, and teams with more gender-balanced structures tend to exhibit higher citation impact [7,8].

However, alongside this positive trend, the knowledge contributions of female scientists have long been underestimated or rendered invisible, with their roles frequently described as those of "invisible technicians" [9]. The resulting Matilda Effect highlights the recurrent neglect or misattribution of women's contributions to scientific knowledge creation [10]. Such narratives not only weaken recognition of the actual roles played by female scientists in research practice, but also obscure the concrete pathways through which they advance scientific progress.

From a historical perspective, since the late nineteenth and early twentieth centuries, when women gradually entered universities and received scientific education, female researchers have consistently been present in laboratories. Although their numbers were limited, they were no longer regarded as mere "exceptions" [11]. Second-wave feminism posed systematic challenges to the sociocultural and institutional structures that legitimized gender inequality, enabling the contributions of female scientists to gain broader social recognition and pushing related research from the margins of academia toward the center [12]. Against this backdrop, female scientists as a collective gradually became a significant object of study. Over recent decades, fields such as gender studies, the history of science, and science of science have continuously expanded their analytical perspectives, focusing primarily on the gender biases faced by female scientists in research activities and the mechanisms underlying changes in their status.

Existing studies generally argue that gender bias encountered by female scientists in the research process manifests mainly at two levels: scientific cultural narratives and research practices. In cultural narratives, the image of the scientist has long been associated with masculine traits, while female scientists are constructed as "anomalies" rather than as normative members of the scientific community [13,14]. Such narratives provide a cultural foundation for institutional arrangements and subtly reinforce the social perception that "women are not suited for high-intensity scientific research" [15]. In research practice, these cultural biases are further translated into measurable inequalities, including gendered divisions of labor, disparities in resource allocation, and disputes over authorship credit [16,17]. Empirical studies show that female scientists continue to face systematic disadvantages in areas such as invitations to academic presentations, access to

STEM (Science, Technology, Engineering and Mathematics) fields, and the sustainability of academic careers [18, 19]

Nevertheless, in recent years the overall academic status of female scientists has shown a clear upward trend. This change is commonly attributed to the combined effects of external factors such as expanded educational access, institutional reforms, and improvements in the research ecosystem, including the continued rise in female participation in higher education [20], the implementation of gender equality policies [21], and the deepening recognition within the scientific community of the relationship between team diversity and innovation [22]. At the same time, the growing presence of female scientific role models and the diffusion of gender equality norms have, to some extent, improved the social perception environment for female researchers [23].

However, existing studies still tend to interpret the improved status of female scientists primarily as a passive outcome of institutional and cultural change, overlooking the agency, creativity, and substantive impact of female scientists as knowledge producers within research practice. To address this limitation, this paper adopts an internal perspective on knowledge production and introduces the concept of Knowledge-Related Roles (KRRs). Using large language model–driven (LLM-driven) text mining methods, it systematically analyzes the role evolution of female scientists across different historical periods, disciplines, and regional cultural contexts, and explores the internal mechanisms underlying their status advancement.

Specifically, this study seeks to answer the following three research questions:

RQ1: What key KRRs do female scientists assume in research activities, and what are their core characteristics?

RQ2: How have these roles evolved dynamically over historical time?

RQ3: What forms of heterogeneity characterize the role evolution of female scientists across different disciplinary and regional cultural contexts?

By addressing these questions, the main contributions of this paper are threefold. First, at the theoretical level, it proposes the concept of knowledge-related roles, providing a new analytical framework centered on knowledge creation for understanding changes in the status of female scientists. Second, at the methodological level, it constructs a role identification system based on large language model–driven text mining, enabling the systematic extraction and quantitative analysis of the knowledge roles of female scientists across disciplines and historical periods. Third, at the empirical level, it reveals a transformation in the roles of female scientists from "knowledge assistants" to "knowledge leaders", demonstrating that their status advancement fundamentally stems from their continuously expanding knowledge contributions, thereby offering new theoretical insights for future research.

The rest of this paper is organized as follows: Sect. 2 presents the data and methodology. Section 3 presents the experimental results. Section 4 discusses the implications and limitations of this work.

2 Data and Methodology

The overall research framework of this study is illustrated in Fig. 1 and consists of the following main components. First, at the data level, Wikipedia serves as the primary data source. Four core lists of female scientists spanning the sixteenth to the twenty-first centuries are integrated, and after rigorous data cleaning, a final dataset of 1,190 valid records is constructed. Second, in terms of research design, drawing on the concepts of the division of scientific labor and social role theory, this study proposes the concept of KRRs and predefines a role system comprising three major categories and nine subcategories. ChatGPT-5.1 is employed as the core extraction model, and through systematic prompt design, roles and their corresponding textual evidence are accurately identified, with their validity further assessed through small-scale manual verification. Third, in the result-processing stage, Python is used to conduct text preprocessing, including tokenization and lemmatization, followed by word frequency analysis and the construction of co-occurrence matrices from both temporal and overall perspectives. Subsequently, a set of multidimensional key indicators covering roles, time, regions, and disciplines is designed to establish a standardized data analysis framework, thereby providing methodological support for subsequent empirical analyses.

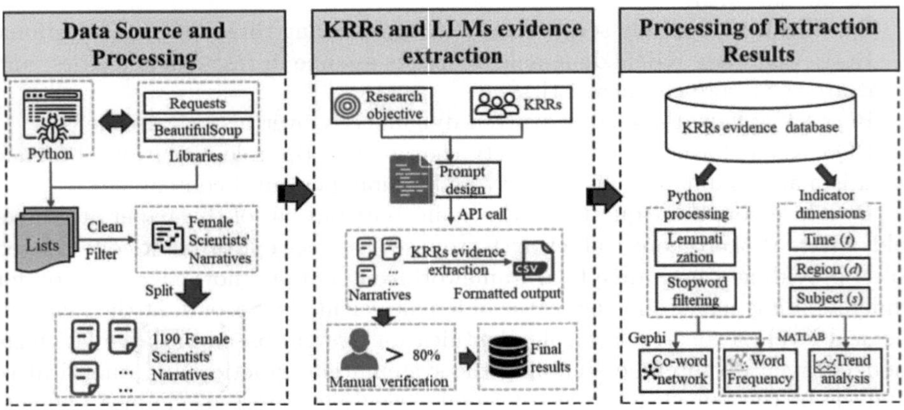

Fig. 1. Overall research framework.

2.1 Data Source and Processing

The study uses Wikipedia as its primary data source and employs Python programs for data collection. To construct a foundational dataset with both historical depth and broad representativeness, four core lists are utilized: Timeline of women in science, List of female scientists before the 20th century, List of female scientists in the 20th century, and List of female scientists in the 21st century. The reasons for this selection are as follows.

First, the complementarity of multiple lists ensures completeness along the temporal dimension. Together, these lists cover approximately five centuries of scientific development from the sixteenth to the twenty-first century, ensuring that female scientists from different historical periods are included. Second, validation against key figures confirms the authority of the dataset. By cross-checking with rosters of prestigious awards, the dataset includes female scholars who have received top honors such as the Nobel Prize, the Turing Award, and the Fields Medal, demonstrating its ability to accurately represent highly influential members of the scientific community. Third, Wikipedia's textual data consist of short biographies of female scientists that document key milestones and significant events in their life and academic careers, thereby providing sufficient information for the effective identification of KRRs.

After obtaining the raw data, rigorous data cleaning was conducted through manual inspection. Records with insufficient biographical information or duplicate entries referring to the same individual were removed. The final dataset comprises 1,190 valid biographical entries of female scientists. The distributions across time periods, disciplines, and regions are presented in Fig. 2.

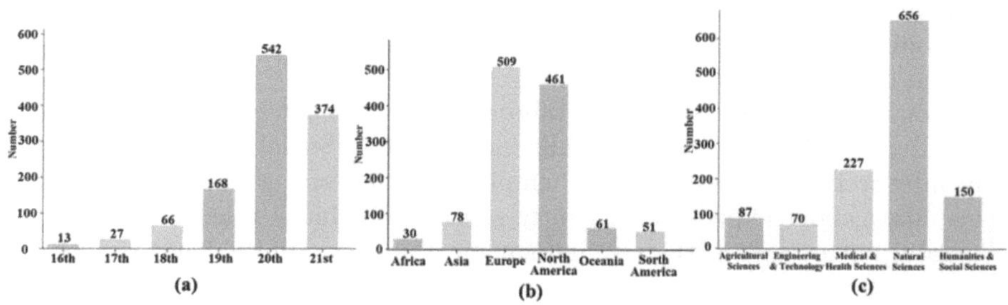

Fig. 2. Distribution of female scientists from the 16th to the 21st century.

Temporally, more than 77% of the individuals are concentrated in the twentieth and twenty-first centuries, reflecting the rapid growth of female scientists in the context of the modern scientific revolution and the expansion of higher education. In terms of geographic distribution, approximately 82% of the scientists come from Europe and North America, highlighting the regional concentration of global science and scholarship. With regard to disciplinary distribution, the dataset is dominated by the natural sciences (656 individuals) and the medical and health sciences (227 individuals), which is consistent with the tendency for female scientists to be more prevalent in traditionally advantaged fields. Overall, the distributional characteristics of this dataset align with existing literature on historical, regional, and disciplinary imbalances among female scientists, thereby providing a solid foundation for subsequent analyses.

2.2 Definition and Identification of KRRs

KRRs Framework Design. The Division of Labor (DOL) in science refers not only to the allocation and execution of research tasks among team members [24], but also constitutes a theoretical category within science and technology studies (STS) for describing knowledge production, skill differentiation, and power configurations. Larivière et al. argue that author contribution statements, by decomposing the research production process into clearly identifiable task units, reveal the types of labor that are crucial to scientific outputs [16]. Accordingly, existing empirical studies on scientists' roles have largely taken the division of scientific labor as their point of departure, examining role allocation and collaboration patterns from the perspective of research practice or workflow [25]. A typical example is the use of the CRediT (Contributor Roles Taxonomy) standard to quantify the division of labor in team-based research, breaking down research activities into multiple contribution roles such as conceptualization, data collection, and methodological design [16]. Although this classification approach refines the operational-level division of labor, its perspective remains relatively limited and is insufficient to fully reveal the positioning and interconnections of scientists as agents of knowledge production within the broader construction and circulation of knowledge systems. To address this limitation, this paper introduces the concept of Knowledge-Related Roles (KRRs), defined as the specific functional roles that individuals assume at different stages of scientific knowledge production and dissemination, reflecting their positions within knowledge flows and collaborative processes.

To construct a KRRs framework tailored to female scientists, this study draws on the division of scientific labor while incorporating Social Role Theory (SRT) as an interpretive perspective. SRT emphasizes that social structures, cultural norms, and gender expectations jointly shape individuals' role positioning and behavioral patterns [26], thereby influencing the opportunities and constraints faced by different groups within the division of labor and career trajectories. Historically, women were expected to assume the roles of mothers and caregivers; accordingly, within the division of scientific labor, male household heads acted as the public representatives of scientific careers, while female relatives served as assistants [27]. For example, Rosalind Franklin conducted the critical recording and analysis of X-ray crystallography data in the discovery of the DNA double helix, effectively undertaking core technical support work; however, constrained by prevailing gender stereotypes and institutional biases, she did not receive academic recognition equivalent to that of her male counterparts [28].

In contemporary society, with the awakening of gender equality consciousness, social expectations of women's roles have undergone a fundamental transformation. Women are no longer regarded as dependent subordinate groups, but rather as independent knowledge producers and decision-makers who participate deeply in the development of national scientific endeavors [29]. This shift has enabled female scientists to move beyond a singular auxiliary positioning in the division of research labor and to extend into more diversified roles. At the same time, studies have shown that knowledge brokers play a critical role in environmental science and policy translation and are key practitioners within the research process [30]. Moreover, compared with men, science communication is considered "a field where women feel 'at home' " [31], making women more likely to assume communicator roles.

On the basis of these studies, the present research classifies the functional roles of female scientists in the research process from a knowledge-related role perspective into three major categories: knowledge assistants (with an emphasis on technical and data support), knowledge leaders (with an emphasis on innovation and decision-making), and knowledge brokers (with an emphasis on communication and connection). Together, these categories constitute a complete ecosystem of research activities and encompass nine specific sub-roles. The detailed classification is presented in Table 1.

LLM-Driven Identification of KRRs. This study leverages the technical strengths of ChatGPT-5.1 in deep English semantic parsing, contextual inference, and complex information filtering, employing it as the core processing model. With the primary objective of guiding the model to accurately recognize role-specific semantics, prompts are systematically designed to enable the extraction of KRRs and their corresponding textual evidence based on these prompts, as illustrated in Fig. 3. In total, 10,888 evidence instances are extracted.

After the extraction process, a small-scale manual verification is conducted to assess the overall reliability of the system. The verification results show that the model achieves a composite accuracy of 81.87% in role classification and evidence matching. This outcome confirms the effectiveness of the prompt design strategy in combination with the model's capabilities, and further demonstrates that this approach exhibits strong applicability and scalability for KRRs identification tasks in textual corpora.

2.3 Methods for Interpreting and Tracing the Evolution of KRRs

KRRs Meanings Based on Co-Word Network Construction. For the role evidence extracted by the LLM, this study first employs Python to perform text segmentation and preprocessing, including lemmatization, stop-word removal, and the cleaning of invalid symbols, in order to ensure semantic consistency of terms and computational accuracy in statistical analyses. Subsequently, the frequencies of role-related terms are calculated and co-occurrence matrices are constructed from both a time-segmented perspective and an overall perspective. During the co-occurrence analysis, low-weight edges that appear only once are removed to enhance the interpretability of the network structure (see Table 2).

For example, the sentence "Become assistant to her husband" is identified as evidence for the role of Knowledge Facilitator. After the above processing, the representative word pair with high co-occurrence strength is "Assistant – Husband," where the hyphen "–" indicates that the two terms co-occur within the same sentence. By constructing co-word networks, the meanings of different roles can be further delineated, along with the representative vocabularies that capture the core connotations of each role.

Table 1. KRRs framework and definitions of female scientists.

Primary Classification	Secondary Classification	Definition
Knowledge Assistant	Knowledge Facilitator [27]	Provide essential technical and operational support facilitating knowledge production.
	Knowledge Inheritor [32]	Acquire knowledge and skills from predecessors through mechanisms such as apprenticeship, familial transmission, or community sharing, and ensure their preservation and continuation through teaching or mentorship.
	Knowledge Recorder [27]	Document empirical information, such as natural phenomena or species data, etc.
Knowledge Leader	Knowledge Creator [33]	Generate knowledge through original discoveries, innovative methodologies, theoretical advancements, or technological inventions; their contributions typically include authoring original research articles, monographs, or scholarly books.
	Knowledge Reconstructor	Pioneers of emerging research fields who establish foundational methodologies, guide scientific inquiry directions, and fundamentally reshape disciplines, thereby driving transformative progress.
	Knowledge Decision-Maker [34]	Scientists occupying leadership or managerial roles within research institutions or academic organizations, exercising authority over research agendas, resource distribution, policy development, and project oversight.
Knowledge Broker	Knowledge Applier [35]	Implement existing knowledge, theories, or technologies in practical, applied settings, translating scientific insights into real-world applications.
	Knowledge Communicator [31]	Disseminating scientific knowledge to diverse audiences, including the general public, students, policymakers, and interdisciplinary communities.
	Knowledge Symbolizer [36]	Scientists whose scientific achievements or public prominence elevate them to symbolic status, embodying broader cultural, ideological, or societal values.

Table 2. Representative terms of KRRs and their evidence (partial).

KRRs	Representative co- words (From evidence)	Evidence (Obtained from the original text)	Scientists' names	Century
Knowledge Facilitator	Assistant - Husband	Become assistant to her husband	Maria Kirch	18th
Knowledge Inheritor	Father - Taught	Her father taught her the basics of science in his laboratory	Astrid Cleve	20th
Knowledge Recorder	Record - Observation	She recorded in a weather observation diary	Margaretha Kirch	18th
Knowledge Creator	Publish - Discover	Published her discovery of white dwarf stars	Williamina Fleming	19th
Knowledge Reconstructor	Pioneer - Field	Pioneer in the field of garden archaeology	Wilhelmina Feemster Jashemski	20th
Knowledge Decision-Maker	Director – Study	She is director of clinical hepatology on a study for trials	Anna Suk-Fong Lok	20th
Knowledge Applier	Application – Study	Application of her expertise to the monitoring of surface ozone as part of air pollution studies	Edith Farkas	20th
Knowledge Communicator	Deliver - Lecture	Planned to deliver lectures to the public	Margaret Bryan	18th
Knowledge Symbolizer	Award - Women	An award for young women scientists would be named the Anneke Levelt-Senger Prize	Anneke Levelt Sengers	20th

Indicator Design for Tracing the Evolution of KRRs. To systematically characterize the distributional patterns of different KRRs across historical periods, regions, and disciplines, this study designs a set of key indicators. First, KRRs are classified into nine specific roles (r) and analyzed along the dimensions of time period (t), region (d), and discipline (s). For each dimension, the number of evidence instances associated with each role within a given time segment is counted, and the relative proportion of each role within the total body

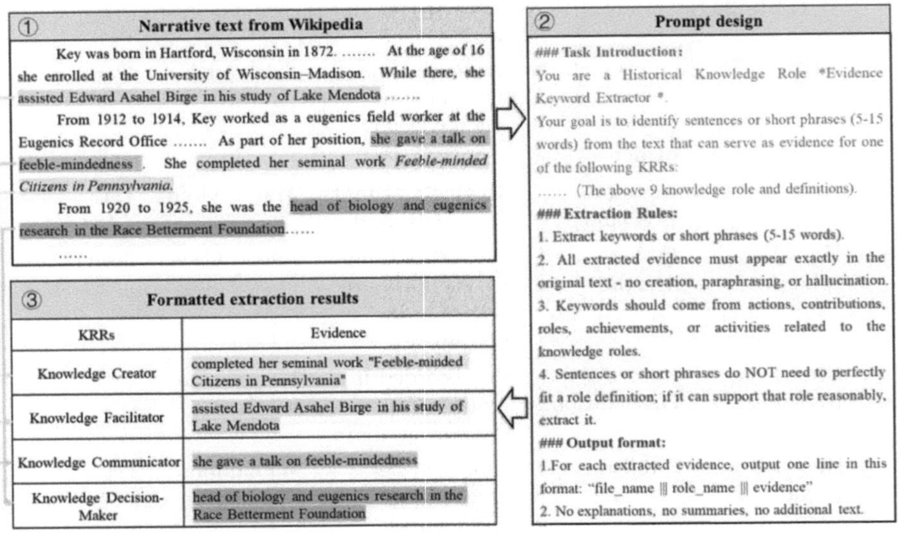

Fig. 3. Prompt design and examples of partial extraction results (Wilhelmine Key as an example).

of evidence is then calculated. Specifically, Table 3 provides a systematic definition of all variables and indicators, including their mathematical expressions and corresponding interpretations, thereby laying a solid foundation for subsequent data analysis and result interpretation.

Table 3. Key variables and indicators with explanations.

Variable	Mathematical Definition	Interpretation
r	$R = \{r_1, r_2 \ldots, r_9\}$	Set of nine knowledge roles
t	$t = \{t_1, t_2 \ldots, t_5\}$	Historical period used in analysis
$E_{t,r}$	-	Evidence count for role r in time t
E_t	$E_t = \sum_{r \in R} E_{t,r}$	Total evidence in time t
$p_{t,r}$	$p_{t,r} = \frac{E_{t,r}}{E_t}$	Relative prominence of role r in period t
d	$d = \{d_1, d_2, \ldots, d_6\}$	Region set
$E_{t,d,r}$	-	Evidence count of role r in region d during time t
$E_{t,d}$	$E_{t,d} = \sum_{r \in R} E_{t,d,r}$	Total evidence for all roles in region d during time t
$p_{t,d,r}$	$p_{t,d,r} = \frac{E_{t,d,r}}{E_{t,d}}$	Proportion of role r evidence in region d at time t
s	$s = \{s_1, s_2, \ldots, s_5\}$	Set of scientific disciplines
$E_{t,s,r}$	-	Count of role r occurrences in discipline s at time t
$E_{t,s}$	$E_{t,s} = \sum_{r \in R} E_{t,s,r}$	Total evidence for all roles in subject s at time t
$p_{t,s,r}$	$p_{t,s,r} = \frac{E_{t,s,r}}{E_{t,s}}$	Proportion of role r evidence in subject s at time t

3 Results and Analysis

3.1 Temporal Evolution Trends of KRRs

The study employs the indicator $p_(t, r)$ to measure the proportion of specific KRRs across different historical periods. Based on the calculated values of this indicator, trend charts illustrating the temporal evolution of female scientists' KRRs, as well as separate subplots for Knowledge Assistant, Knowledge Leader, and Knowledge Broker, are constructed, as shown in Fig. 4. Overall, the KRRs of female scientists exhibit a structural evolutionary trend, shifting from early roles centered on recording, assistance, and transmission toward core roles involving knowledge creation, reconstruction, and decision-making.

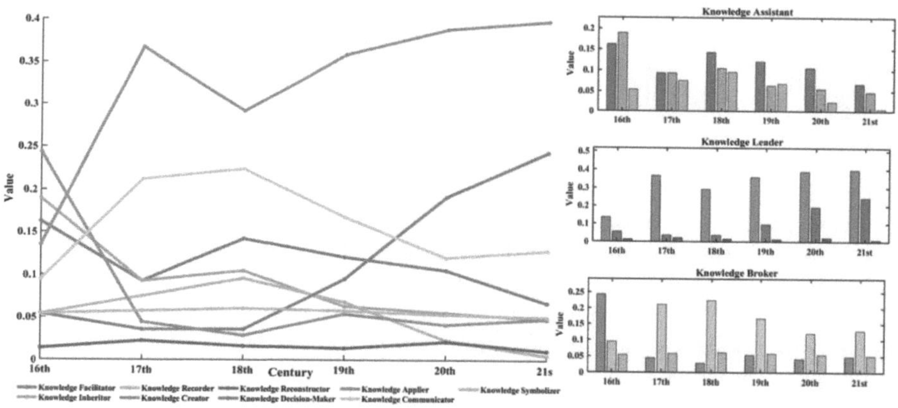

Fig. 4. Temporal evolution of female scientists' KRRs.

First, the long-term decline in the Knowledge Assistant indicator indicates that women initially entered the scientific system in roles focused on recording, inheritance, and assistance. Quantitative analysis shows that this group underwent a prolonged and systematic contraction, with its overall proportion decreasing significantly from 40.5% in the sixteenth century to 12.1% in the twenty-first century. This contraction is particularly evident in the near disappearance of the Knowledge Recorder role. However, this decline does not imply that women exited the scientific domain as assistants; rather, it reflects a systematic shift in their role structure. For example, in the sixteenth century, Sophia Brahe, although restricted to participating in astronomical observations as an assistant, performed work that provided indispensable data for Kepler's laws, fully demonstrating her expertise in core scientific labor. Thus, the decline of this role group serves as quantitative evidence of women successfully transitioning from peripheral knowledge activities to more professional and independent core knowledge activities through accumulated experience and skill development.

Second, the continuous rise of Knowledge Leader serves as a core indicator of the elevation of women's scientific status. The data show that this group's proportion increased from 20.3% in the sixteenth century to 65.3% in the twenty-first century, solidifying women's position in modern science. First, Knowledge Creator represents the cornerstone for women entering the core layer. As early as the seventeenth century, this role accounted for 36.7% of female scientists and remained stable at a high level of 35%-40% over the following centuries (39.8% in the twenty-first century), indicating that women had already established their identity as Knowledge Creators through knowledge production. Second, the delayed breakthrough of Knowledge Decision-Maker is noteworthy. This role remained stagnant at around 3.5% until the eighteenth century, but surged to 19.2% in the twentieth century and reached 24.4% in the twenty-first century. This rapid growth essentially reflects that female scientists, after accumulating sufficient innovative achievements, successfully transitioned from Knowledge Creators to Knowledge Decision-Makers, marking the true leap in women's scientific status.

Third, women's roles as Knowledge Brokers exhibit dynamic changes across different historical periods. Data show that Knowledge Applicators accounted for a relatively high proportion in the sixteenth century (approximately 24.3%) but subsequently declined and remained at a low level, reflecting women's early involvement in the practical application of knowledge. Knowledge Communicators peaked during the seventeenth to eighteenth centuries (around 21%-22%), and although their proportion declined after the twentieth century, it remained relatively stable, highlighting women's significant role in science communication and education. Although the proportion of Knowledge Symbolizers is small, its presence indicates that women began to influence public perception as symbolic representatives of scientific achievements. Overall, women in brokerage roles acted as a bridge connecting scientific production and societal application, serving as a crucial link between the scientific community and the public.

3.2 Analysis of KRRs' Meanings and Evolution

Using Gephi, co-word networks for each KRRs were constructed to visually present the semantic association structures and knowledge distribution characteristics of different roles, as shown in Fig. 5. By analyzing the core vocabulary of these knowledge graphs, it was found that the central terms corresponded well with the nine predefined KRRs, indirectly validating the accuracy of the LLM-based extraction. For example, in the co-word network for Knowledge Facilitator, the term work serves as the central node. Differences in node size and edge density highlight its foundational position within Knowledge Facilitator practice scenarios. Around this core, secondary nodes such as assistant, research, and help form associated clusters, anchoring the action attributes of knowledge facilitators, participating assistively in knowledge production and practice processes, while also clarifying their functional orientation: facilitating the flow and implementation of knowledge through supportive activities.

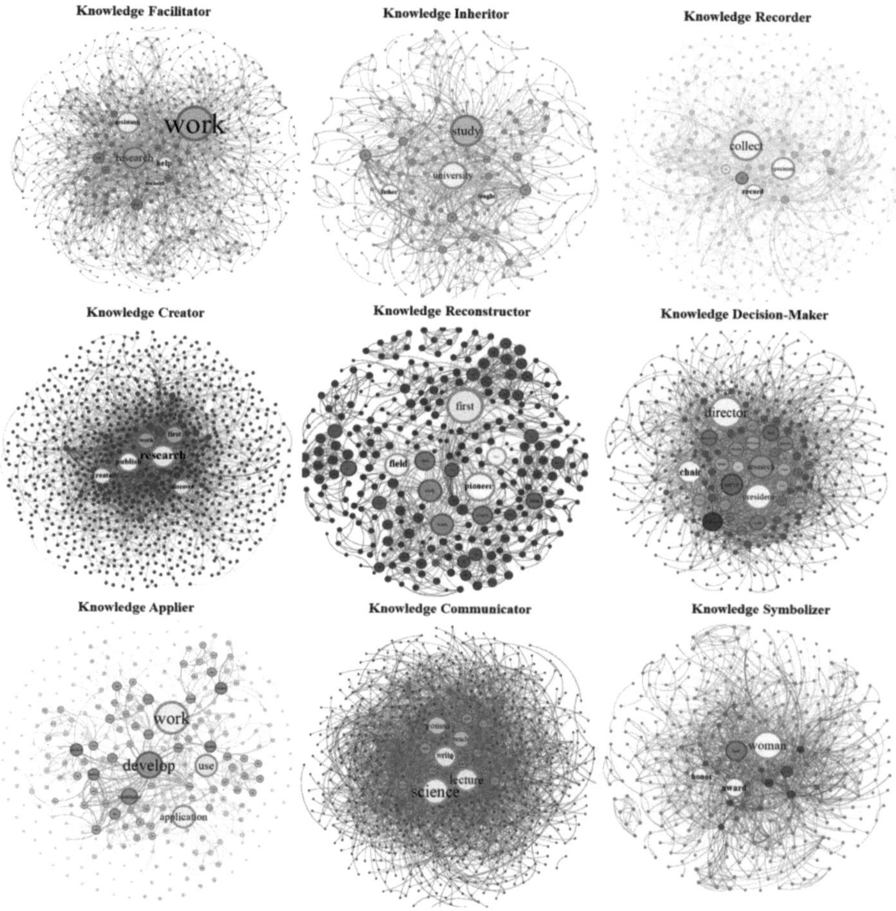

Fig. 5. Co-word networks of core vocabulary for each KRRs.

Based on the co-word networks of KRRs, high-weight feature terms representing the core meanings of each role were further selected. These feature terms were then used to construct word frequency sequences across temporal windows, presenting the dynamic semantic evolution of roles over historical periods. Ultimately, evolution charts of feature term frequencies over time were created (Fig. 6), revealing the long-term semantic trajectories of different KRRs.

Observations from the co-word networks and the temporal evolution of core vocabulary indicate the following:

(1) Knowledge Assistants: Assistive terms show a significant decline, such as "help", "record", and "assistant". Two interesting patterns emerge: within the Knowledge Facilitator role, the early "assistant" function of female scientists was deeply tied to their husbands, relying on intimate relationships for sup-

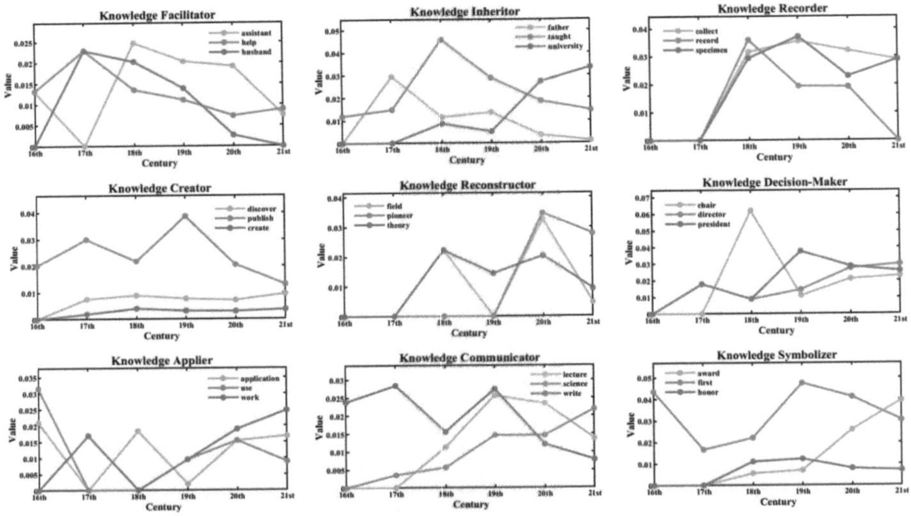

Fig. 6. Cross-century trends of core vocabulary for each KRRs.

port, whereas later it shifted toward an independent knowledge assistance positioning. In the Knowledge Inheritor role, knowledge acquisition pathways evolved from the early familial model of father-to-child transmission to institutionalized channels centered on universities, marking a transition from private to public education systems for knowledge inheritance.

(2) Knowledge Leaders: Leadership-related vocabulary gradually increases, such as "discover", "create", and "director". In the Knowledge Creator role, "publish" consistently shows a high proportion, indicating that the core output of this role is primarily reflected through publication achievements. In the Knowledge Decision-Maker role, terms like "director", "chair", and "head" become prominent. Over time, their frequencies steadily increase, collectively forming the core leadership identity of women in knowledge decision-making and reflecting the growing diversity and balanced representation of women in leadership roles within this domain.

(3) Knowledge Brokers: In the Knowledge Communicator role, "lecture", "science", and "write" are core dimensions. After the eighteenth century, the value of "lecture" and "science" rises significantly, while "write" initially increases and then declines, indicating that knowledge dissemination gradually expanded from traditional written outputs to more public-oriented paths, such as lectures and science popularization. In the Knowledge Symbolizer role, "award", "first", and "honor" are central dimensions. After the nineteenth century, the value of "first" experiences explosive growth, and "award" also rises noticeably in later periods. This shift directly reflects that female scientists' academic achievements increasingly overcame traditional

recognition limits, gaining broad societal acknowledgment and institutional validation.

3.3 Evolution of KRRs Across Different Disciplines

The study uses the indicator $p_(t, s, r)$ to measure, within a specific time period t and discipline s, the proportion of evidence for a particular role r relative to the total evidence of all knowledge roles in that discipline. Based on the calculated values of this indicator, trend charts depicting the temporal proportion changes of each KRRs across different disciplines were plotted, as shown in Fig. 7.

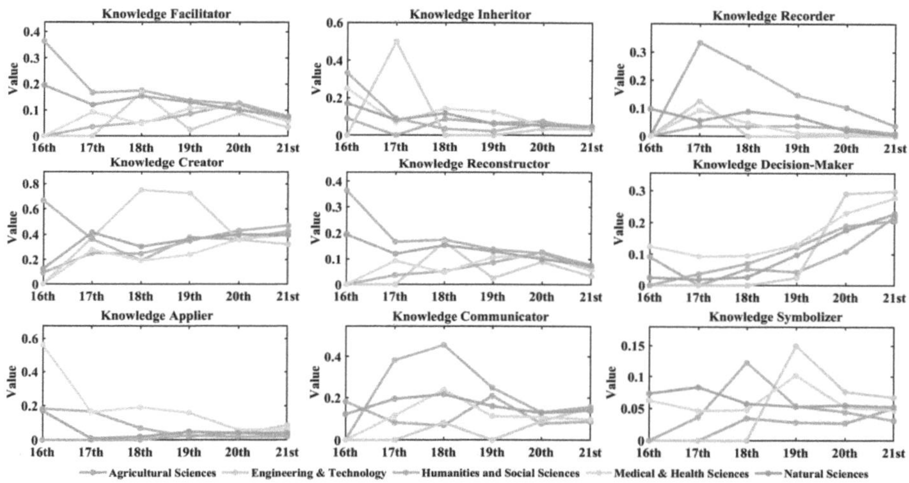

Fig. 7. Temporal trends of different KRRs across disciplines.

Analysis of the results reveals the following insights:

(1) Universality of knowledge role evolution: Although different disciplines vary in pace and magnitude, they generally follow a consistent evolutionary direction: transitioning from low-threshold roles focused on recording, inheritance, and assistance toward core roles characterized by creation, reconstruction, and decision-making. This consistency indicates that the historical rise of women's scientific status is not an isolated event within certain regions or disciplines, but rather a general pattern driven by internal dynamics of the knowledge system. As knowledge accumulation, educational opportunities, and research experience increase, women progressively move from peripheral participation to core contributions.

(2) Medicine and health sciences as a typical pathway for women entering knowledge leadership: Medicine and health sciences have long been a key domain through which women assume leadership roles in knowledge production,

reflecting the strong inclusivity of this discipline. The relatively high proportion of female Nobel laureates in biomedical fields further supports this trend. From the perspective of social expectation theory, traditional gender divisions associate rationality and authority with "masculine" traits and sensibility and care with "feminine" traits. Medicine aligns closely with feminine traits, making it a discipline more accessible and socially recognized for women [37]. This cognitive alignment creates path dependency and a "Matthew effect," whereby the increasing number of women further reinforces gender stereotyping in medicine and promotes their expansion into leadership knowledge roles. Specifically, women's core leadership roles in medicine show significant growth: Knowledge Creator steadily rises from the sixteenth century to nearly 40%, and Knowledge Decision-Maker increases from 12.5% to 27.7%. Meanwhile, auxiliary roles such as Knowledge Inheritor and Knowledge Facilitator decline markedly, reflecting women's gradual transition from supportive roles to innovation and decision-making. Overall, the sustained growth of women in medicine, in interaction with social recognition, has jointly facilitated their elevation in leadership knowledge roles.

(3) Disciplinary differences in women's knowledge roles: In the humanities, women account for a high proportion among Knowledge Creators, but a low proportion among Knowledge Leaders. In the natural sciences, women have a significantly higher proportion among Knowledge Reconstructors than in the humanities. This pattern results from the interplay of three mechanisms: First, the "lightweight" and inclusive nature of knowledge production in the humanities allows women to fully realize their academic potential in low-gender-barrier environments, becoming important knowledge creators. Second, knowledge leadership relies on resource allocation and discourse power, but the humanities' relative resource disadvantage and implicit gender biases in research constrain women's advancement into leadership roles. Third, in the natural sciences, the objective evaluation system centered on empirical data reduces the influence of gender bias on academic recognition, allowing women's innovative contributions to be more easily acknowledged through quantifiable metrics, thereby enhancing their knowledge reconstruction capability.

These results not only confirm the stage-wise progress of gender equality in academia but also reveal the dual influence of disciplinary attributes and social mechanisms on the academic development of women.

3.4 Evolution of KRRs Across Different Regions

The study uses the indicator $p_(t, d, r)$ to measure, within a specific time period t and region d, the proportion of evidence for a particular knowledge role r relative to the total evidence of all roles in that region. Based on the calculated values of this indicator, trend charts depicting the temporal proportion changes of each knowledge role across different regions were plotted, as shown in Fig. 8.

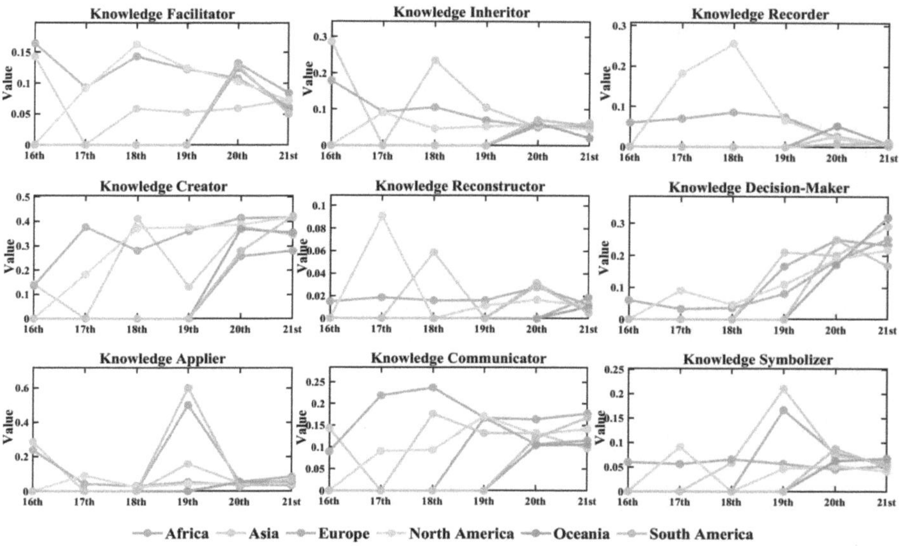

Fig. 8. Temporal trends of different KRRs across regions.

Analysis reveals the following insights:

(1) Structural consistency in regional evolution: The historical changes of female scientists across the nine KRRs display an overall pattern of "structural consistency." All regions generally experienced a migration from knowledge-assistive roles toward core knowledge leadership roles. Auxiliary roles such as Recorder and Facilitator gradually declined over time, while core roles such as Creator and Decision-Maker steadily increased globally, reflecting the historical elevation of women's overall scientific status.

(2) Regional differences in the pace of role upgrading: Economically and scientifically advanced regions tend to exhibit richer role structures, with roles emerging earlier and growing more steadily. In culturally and socially advanced Europe, all nine KRRs appeared relatively early. Over time, North America's scientific capacity strengthened, and the full set of nine KRRs emerged. Developed regions also achieved higher proportions of Creator and Decision-Maker roles earlier. In contrast, regions such as Africa and South America experienced significant delays. This indicates that the advancement of female scientists' roles is strongly dependent on regional economic and social development.

(3) The uniqueness of Oceania: Entering the twenty-first century, Oceania, represented by Australia, showed a notable rise in Knowledge Decision-Maker roles. This distinct pattern stems from dual structural and institutional adjustments in the regional research system. On one hand, Australia's national research assessment mechanisms, exemplified by the Excellence in Research for Australia (ERA) initiative, concentrated research

resources into high-impact and strategic fields, institutionally strengthening the demand for decision-making and management-oriented roles. On the other hand, gender equity policies in Oceania, such as the Science in Australia Gender Equity (SAGE) program, increased female researchers' participation in research governance structures, facilitating their entry into core positions with organizational authority, resource allocation, and strategic decision-making functions. Moreover, Australia's highly internationalized research system, characterized by large-scale collaborations, cross-institutional alliances, and expanding national strategic research projects, has significantly increased the organization of knowledge production, further boosting the demand for knowledge decision-makers capable of coordination, integration, and strategic judgment. These combined institutional, structural, and policy factors have driven Oceania's growth in knowledge decision-maker roles faster than in other regions, creating its unique trajectory in the global evolution of female scientists' roles.

Overall, the elevation of female scientists' status is both a global long-term trend and heavily influenced by regional economic conditions and the maturity of research systems. Their role transitions exhibit path-dependent characteristics, progressing gradually from lower-tier to mid-tier to upper-tier roles. These features collectively provide key evidence of the historical evolution of women's scientific status.

4 Discussion and Conclusion

4.1 Progress of Female Scientists as Knowledge Creators and Leaders

Based on the KRRs framework, this study systematically traces the role evolution of female scientists from the sixteenth to the twenty-first century, revealing a structural shift from early knowledge-assisting roles toward core roles as knowledge creators and decision-makers.

First, grounded in Division of Labor (DOL) and Social Role Theory (SRT), and drawing on existing literature in the history of science and gender studies, this study constructs a KRRs system consisting of three major categories and nine subcategories. The KRRs framework classifies the roles of female scientists into knowledge assistants, knowledge leaders, and knowledge brokers. Using LLM-based text mining techniques, these roles, as well as their supporting evidence texts and keywords, are identified from short biographical entries of eminent female scientists. By analyzing the extracted evidence and constructing semantic networks of keywords, the substantive meanings of each role are made explicit. This addresses RQ1.

Second, the study develops quantitative indicators to trace the dynamics of KRRs over time. The results show that female scientists' KRRs follow a "bottom-up and inside-out" evolutionary pathway: roles centered on assistance and inheritance gradually contract, while core roles related to knowledge creation and decision-making steadily expand, forming a long-term and stable developmental trend. Through the production of breakthrough knowledge, female scientists progressively influence academic norms and social perceptions, thereby gaining higher authority within the scientific community. Marie Curie provides a representative example: early in her career, when she and her husband were invited to give lectures, only Pierre Curie was allowed to speak; as her scientific achievements gained global recognition, she was not only permitted to lecture publicly but was also regarded as a knowledge leader who "defined an international standard for radioactive emissions" [38]. This addresses RQ2.

Third, the study finds that the evolution of women scientists' KRRs exhibits both consistency and heterogeneity. At the disciplinary level, women in medicine and health sciences account for a relatively high proportion of knowledge leadership roles; in the natural sciences, women are disproportionately represented among knowledge reconstructors; and in the humanities, women are more prominent as knowledge creators but underrepresented as leaders. At the regional level, developed regions such as Europe and North America display a richer repertoire of roles and more stable upgrading trajectories, whereas less developed regions lag behind. These patterns indicate that economic development, the maturity of research systems, and policy support significantly facilitate the advancement of female scientists' knowledge roles. This addresses RQ3.

Overall, the progress of female scientists as knowledge creators and leaders results from the combined effects of individual agency and external institutional and sociocultural conditions. Through the sustained accumulation of original contributions, female scientists have not only advanced their own scientific standing but have also driven structural transformations in scientific norms, social perceptions, and academic authority within the scientific community.

4.2 Theoretical and Practical Implications

The core theoretical contribution of this study lies in providing an analytical pathway grounded in the internal structure of knowledge production to understand the academic status and role transformations of female scientists. The KRRs framework no longer takes positions, awards, or institutional identities as its primary indicators; instead, it focuses on the concrete functions that women actually perform in the process of knowledge production, such as recording, assisting, inheriting, creating, reconstructing, deciding, communicating, symbolizing, and brokering. In doing so, it captures their real roles and sources of authority within the knowledge chain.

This shift in perspective carries three main theoretical implications. First, it bridges the explanatory gap between macro-level institutional studies and individual biographical narratives, enabling the construction of a clear analytical chain linking "concrete research practices–functional divisions of knowledge labor–social perceptions of roles". Second, KRRs locates gender inequality within the internal processes of knowledge production, rather than treating it solely as the outcome of external discrimination or institutional exclusion, thereby revealing the endogenous mechanisms through which the academic status of female scientists improves. Third, the framework demonstrates strong transferability: it is applicable not only to the study of female scientists but also serves as a general analytical tool for understanding the role positioning and evolution of other marginalized groups within knowledge systems.

From a theoretical standpoint, this study suggests that the advancement of women's status in science is not merely a matter of "identity acquisition" or "institutional compensation", but a structural transformation realized through the reconfiguration of knowledge roles. This finding pushes gender and science studies beyond the question of "who is recognized" toward deeper inquiries into "who creates knowledge, how knowledge is created, and how knowledge authority is established."

The study also offers the following practical implications:

First, in the cultivation of female scholars, openness and equality in opportunities and task allocation should be maintained so that they can naturally participate in, and fully realize, their agency and creative potential in knowledge creation. The stages of graduate training, postdoctoral development, and early-career scholarship are critical periods during which research interests, capabilities, and academic styles take shape. During these stages, supervisors, research groups, and teams can adopt more open approaches to task allocation and collaboration, allowing women scholars, without being constrained by predefined role expectations, to engage in core activities such as formulating research questions, designing research pathways, and leading manuscript writing. In the daily operation of laboratories and research groups, practices such as rotating the chairing of group meetings, jointly leading subprojects, and supporting appropriate authorship arrangements can ensure that women and men have equal access to participation along the knowledge production chain. A training environment oriented toward freedom of choice and capability development helps women, according to their own interests and strengths, to form stable and diverse pathways of knowledge creation through research practice.

Second, in the design of research organizations and institutional arrangements, more inclusive and flexible mechanisms can be adopted to provide feasible space for women's participation in knowledge leadership and research governance. Women's engagement in knowledge decision-making and organizational roles often builds on their long-term accumulation of research experience and professional judgment. Accordingly, in the operation of research projects and academic organizations, more open selection mechanisms for leadership positions and team structures can enable women to assume roles such as project leaders, topic coordinators, or organizers of collaborative networks when conditions are appropriate. By supporting opportunities for female scholars to chair academic conferences, lead academic organizational affairs, or participate in research advisory activities, they can engage more fully in the practice of knowledge governance and academic decision-making, without being compelled toward predefined leadership trajectories, thus contributing to a more diverse, open, and sustainable structure of scientific leadership.

4.3 Limitations

This study has several limitations. First, the data used in this research are primarily drawn from Wikipedia biographical texts, which remain uneven in terms of regional distribution and disciplinary coverage. This imbalance may affect the overall depiction of the role distribution of female scientists. Second, the study mainly interprets the historical evolution of female scientists' KRRs at the level of textual semantics and does not yet fully incorporate structural factors such as institutional environments and the allocation of research resources for more comprehensive cross-validation. Future research may expand data sources by incorporating multilingual and multi-regional historical materials, introduce heterogeneous data from multiple sources, such as research institutions, scientific collaboration networks, and funding systems, and further apply and test the KRRs framework in interdisciplinary contexts, thereby providing more comprehensive theoretical explanations and stronger empirical support.

Author contributions. Yanmin Du: Software, Investigation, Data curation, Experiment, Result analysis, Writing-original draft. **Qiuxia Li**: Investigation, Data curation, Result analysis. **Tanyao Zhang**: Software, Investigation, Data curation, Experiment. **Huizhen Fu**: Conceptualization, Writing-review & editing, offering timely and critical suggestions to improve the quality of the paper. **Zhixiang Wu**: Conceptualization, Methodology, Formal analysis, Writing-original draft, Funding acquisition, Supervision.

Funding Information. This study was supported by Social Science Foundation Youth Program of China (No. 23CTQ027).

Declarations. The authors declare no competing interests.

References

1. Huang, J., Gates, A.J., Sinatra, R., Barabási, A.-L.: Historical comparison of gender inequality in scientific careers across countries and disciplines. Proc. Natl. Acad. Sci. **117**(9), 4609–4616 (2020)
2. Ceci, S.J., Ginther, D.K., Kahn, S., Williams, W.M.: Women in academic science: a changing landscape. Psychol. Sci. Public Int. **15**(3), 75–141 (2014)
3. Teich, E. G., et al.: Citation inequity and gendered citation practices in contemporary physics. Nat. Phys. **18**(10), 1161–1170 (2022)
4. Jaramillo, A. M., Macedo, M., Oliveira, M., Karimi, F., Menezes, R.: Systematic comparison of gender inequality in scientific rankings across disciplines (2025). arXiv: 2501.13061
5. Zheng, B., Yang, C., Hou, J.: Beyond productivity gaps: temporal patterns of gender differences in scientific knowledge creation. Data Sci. Info. **5**(2), 100084 (2025)
6. Hildrun, K., Alexander, P., Johannes, S.: Research evaluation. Part II: gender effects of evaluation: are men more productive and more cited than women? Scientometrics **93**(1), 17–30 (2012)
7. Shi, W., Wu, R.: Women's strength in science: exploring the influence of female participation on research impact and innovation. Scientometrics **129**(7), 4529–4551 (2024)

8. Sofi-Mahmudi, A., et al.: Cochrane reviews' authorship has become more gender-diverse but remains geographically concentrated: a meta-research study. J. Clinic. Epidemiol. **181**, 111719 (2025)
9. Jalloh, M. B., Vernooij, E., Street, A.: Invisible and undervalued: a qualitative study of laboratory workers' experiences and perceptions of laboratory strengthening in Sierra Leone. African J. Laborat. Med. **13**(1), 11 (2024)
10. Goyanes, M., Demeter, M., Simeunović Bajić, N., Gil de Zúñiga, H.: Gender disparities in first authorship: examining the matilda effect across communication, political science, and sociology. Scientometrics **130**(5), 2947–2961 (2025)
11. Jones, C.G.: 'all your dreadful scientific things': women, science and education in the years around 1900. Hist. Educ. **46**(2), 162–175 (2017)
12. Foster, J.E.: In keeping with family tradition: American second-wave feminists and the social construction of political legacies. Qual. Sociol. Rev. **14**(1), 6–28 (2018)
13. Blagdanic, S., Kadijevic, G.M., Kovacevic, Z.: Gender stereotypes in preschoolers' image of scientists. Eur. Early Child. Educ. Res. J. **27**(2), 272–284 (2019)
14. Owens, T.: Madame curie above the fold: divergent perspectives on curie's visit to the United States in the American press. Sci. Commun. **33**(1), 98–119 (2011)
15. Dasgupta, N., Stout, J.G.: Girls and women in science, technology, engineering, and mathematics: STEMing the tide and broadening participation in STEM careers. Policy Insights Behav. Brain Sci. **1**(1), 21–29 (2014)
16. Larivière, V., Pontille, D., Sugimoto, C. R.: Investigating the division of scientific labor using the contributor roles taxonomy (CRediT). Quantit. Sci. Stud. **2**(1), 111–128 (2021)
17. Lawson, M.A., Martin, A.E., Huda, I., Matz, S.C.: Hiring women into senior leadership positions is associated with a reduction in gender stereotypes in organizational language. Proc. Natl. Acad. Sci. U.S.A. **119**(9), e2026443119 (2022)
18. Koletsi, D., Papadopoulou, A.K.: Gender disparity in speakers of the European orthodontic society conferences held between 2015 and 2020: a cross-sectional report. Eur. J. Orthod. **44**(6), 614–621 (2022)
19. Spoon, K., et al.: Gender and retention patterns among U.S. faculty. Sci. Adv. **10**(14), eadi2205 (2024)
20. Correa, A., Glas, M.G., Opara, J.: Females in higher education and leadership: insights from a multi-method approach. Front. Educat, **9**, 1464416 (2025)
21. Jörg M., Rachel P.: Pathways to gender equality in academia. National factors shaping women's early career progression across Europe, Science and Public Policy scaf035 (2025)
22. Yang, Y., Tian, T.Y., Woodruff, T.K., Jones, B.F., Uzzi, B.: Gender-diverse teams produce more novel and higher-impact scientific ideas. Proc. Natl. Acad. Sci. U.S.A. **119**(36), e2200841119 (2022)
23. Buckley, C., Farrell, L., Tyndall, I.: Brief stories of successful female role models in science help counter gender stereotypes regarding intellectual ability among young girls: a pilot study. Early Educ. Dev. **33**(7), 1196–1211 (2021)
24. Lu, C., Zhang, Y., Ahn, Y.-Y., Ding, Y., Zhang, C., Ma, D.: Co-contributorship network and division of labor in individual scientific collaborations. J. Am. Soc. Inf. Sci. **71**(10), 1162–1178 (2020)
25. Hofmann, B., Salomon, H., Hoffmann, S.: Roles of researchers in inter-and trans-disciplinary sustainability research: a reflection tool. Sustain. Sci. **20**(3), 777–792 (2025)
26. Lesch, L., Scharfenkamp, K., Wicker, P.: The perceived role fit of women and men academics: evidence from the social sports sciences. Front. Psychol. **14**, 239944 (2023)

27. Yale, E.: Teaching women's work and thought in undergraduate history of science courses. History Compass **21**(9), e12780 (2023)
28. Cobb, M., Comfort, N.: What Rosalind Franklin truly contributed to the discovery of DNA's structure. Nature **616**(7958), 657–660 (2023)
29. Khairullah, K.: Changes in women's social roles and functions in gender dichotomy resistance. PERSPEKTIF **11**(2), 990–996 (2022)
30. Cvitanovic, C., Karcher, D., Breen, J., Badullovich, N., Cairney, P., Dalla Pozza, R., Posner, S.: Knowledge brokers at the interface of environmental science and policy: a review of knowledge and research needs. Environm. Sci. Policy **163**, 103973 (2025)
31. Wilkinson, C., Milani, E., Ridgway, A., Weitkamp, E.: Roles, incentives, training and audiences for science communication: perspectives from female science communicators. J. Sci. Commun. **21**(4), A04 (2022)
32. Kammermeier, M., Muckenthaler, M., Weiß, S., Kiel, E.: Feminization of teaching: gender and motivational factors of choosing teaching as a career. Front. Educat. **10**, 1472945 (2025)
33. Schuhmacher, A., Germann, P.-G., Trill, H., Gassmann, O.: Models for open innovation in the pharmaceutical industry. Drug Discovery Today **18**(23–24), 1133–1137 (2013)
34. Buss, M., Andler, S., Tiberius, V.: Female leadership: an integrative review and research framework. Leadersh. Q. **36**(3), 101858 (2025)
35. Wood, C.V.: Knowledge practices, institutional strategies, and external influences in the making of an interdisciplinary field: insights from the case of women"s and gender studies. Am. Behav. Sci. **56**(10), 1301–1325 (2012)
36. Alaimo, L., Chambers, L., Puddephatt, A.: Images of Marie Curie: how reputational entrepreneurs shape iconic identities. Engag. Sci. Technol. Soc. **4**, 8–21 (2018)
37. Steiner-Hofbauer, V., Katz, H.W., Grundnig, J.S., Holzinger, A.: Female participation or "feminization" of medicine. Wien. Med. Wochenschr. **173**(5–6), 125–130 (2023)
38. Leone, M., Robotti, N.: Constructing a Nobel Prize: the case of madame curie. Phys. Perspect. **26**(3–4), 132–179 (2024)

Personalized Services Through Lifecycle Tracking: A Probabilistic Graphical Approach to Modeling Consumer Psychological Dynamics

Yao Mu(✉) and Baojun Ma

Key Laboratory of Brain-Machine Intelligence for Information Behavior (Ministry of Education and Shanghai), School of Business and Management, Shanghai International Studies University, Shanghai 201620, China
muy@shisu.edu.cn

Abstract. As e-commerce consumer acquisition costs soar, maximizing long-term consumer value through lifecycle management has become a strategic priority. Consumers typically shift between distinct psychological periods (e.g., introduction, growth, maturation, dormancy, and churn), making the precise tracking of these dynamics essential for personalized services. However, existing methods encounter significant challenges in decoding continuous psychological evolution, integrating long- and short-term behavioral features, and accounting for submarket competition. This study proposes a lifecycle probabilistic graphical model (LC-PGM) to address these gaps. By employing sequential learning, the LC-PGM approach uncovers psychological shifting patterns across multiple shopping journeys and explicitly decodes specific lifecycle periods. Furthermore, a shopping journey-oriented separation strategy is introduced to organically fuse long- and short-term behavioral features. To enhance predictive accuracy, the model incorporates multi-merchant interaction data within submarkets to synthetically capture consumer psychology in the competitive environment. Real-world data experiments and multi-view post-hoc analysis manifest the superior performance of the proposed LC-PGM approach. This research provides merchants with interpretable intelligent decision support, facilitating precision tracking and refined consumer relationship management.

Keywords: Consumer lifecycle management · Probabilistic graphical models · Psychological dynamics · Personalized tracking services · Submarket competition

1 Introduction

In today's digital business ecosystem, e-commerce merchants face the dual pressures of escalating consumer acquisition costs and the waning of "traffic dividends." Consequently, maintaining existing consumers and maximizing their long-term value has become critical for corporate survival and growth. Research

J. Lu et al. (Eds.): FLINS-ISKE 2026, LNCS 16752, pp. 378–396, 2027.
https://doi.org/10.1007/978-981-92-2369-5_24

indicates that the cost of acquiring a new consumer is five to six times higher than retaining an existing one [1]. Moreover, loyal consumers not only contribute stable revenue but also generate derivative value through mechanisms such as word-of-mouth and brand interaction [2]. Therefore, there is an urgent need for merchants to provide personalized tracking services and achieve high-value retention through precise intervention.

However, the high degree of information interconnectivity has significantly lowered consumers' switching costs between merchants, making the maintenance of long-term relationships increasingly challenging. Behavioral trajectories accumulated on platforms can reflect consumers' psychological dynamics that shift in response to personal circumstances and market trends [3]. Timely detection of such psychological evolution based on observed behavioral records is an essential prerequisite for merchants to formulate precise tracking strategies and secure a sustained competitive advantage.

From a lifecycle perspective, interactions between consumers and merchants can be categorized into sequential psychological periods, including introduction, growth, maturation, dormancy, and churn [4,5]. Each period corresponds to distinct demand patterns and value contributions. While merchants strive to keep consumers in the high-value maturation period, psychological dynamics are often non-linear, influenced by various internal and external factors [6]. Thus, it is imperative to accurately locate a consumer's current period based on historical behavior to implement appropriate strategies, either driving them toward maturation or initiating active intervention before churn occurs [7].

Regarding merchant service strategies, existing intelligent decision-making methods focus primarily on short-term preferences, often overlooking long-term psychological patterns [8]. Although recent approaches have attempted to analyze long-term behavior, several challenges remain in continuously tracking psychological dynamics, integrating long- and short-term modeling, and incorporating submarket competition. First, capturing the continuity and shifts of long-term psychological periods is difficult. Current methods largely focus on static detection of key periods [9], making it hard to decode the underlying psychological evolution patterns. Second, an organic fusion of long- and short-term psychological dynamics remains elusive. While long-term psychology is essentially the accumulation of short-term experiences, existing models often treat the two in isolation [10]. Third, competitive factors within submarkets have not been fully integrated into the analysis. Consumer psychological periods are significantly influenced by interactions with competing merchants [11], yet few studies have performed collaborative modeling in this regard.

To address these challenges, this study proposes a novel approach named lifecycle probabilistic graphical model (LC-PGM) to support merchant tracking services. This approach innovatively models long-term psychological dynamics alongside observable behavioral features, uncovers interpretable period shifting patterns, and enhances prediction through short-term features and submarket competition. The primary contributions are threefold.

- We construct a probabilistic graphical model for the sequential learning of consumer lifecycle behavior. Through a constrained sampling strategy, we

explicitly decode psychological periods, providing interpretable support for merchants' intelligent decision-making.
- We introduce a shopping journey-oriented sequence separation technique to organically fuse long- and short-term features, thereby uncovering psychological evolution patterns across multiple shopping journeys.
- From an industry competition perspective, we incorporate consumer interactions with multiple merchants within a submarket. By collaboratively learning psychological period shifts across different merchants, we account for competitive dynamics to improve the performance of tracking services.

The remaining sections of this paper are organized as follows. Section 2 summarizes the literature on intelligent methods for tracking services and the consumer lifecycle. Section 3 provides details of our methodological design. In Sect. 4, extensive experimental results on real-world data are analyzed to illustrate the validity of the proposed approach. Section 5 finally concludes this work and discusses its theoretical and practical implications.

2 Related Work

2.1 Intelligent Methods for Tracking Services

To provide effective lifecycle tracking services, intelligent methods must accurately characterize the long-term psychological dynamics and behavioral patterns of consumers. However, existing studies predominantly focus on short-term behaviors, often overlooking the dominant role of long-term demands in consumers' decision-making, which leads to biased service strategies [8]. Research has found that short-term consumer purchasing behavior is frequently influenced by trending events and represents a hybrid of inherent needs and external market trends [12]. Consequently, relying solely on short-term interests provides insufficient support for effective long-term services.

In addressing long-term dynamics, one branch of research employs reinforcement learning to adjust decisions based on continuous consumer feedback, with typical applications in interactive recommender systems [13]. Policy-based methods learn optimal actions for given states, while value-based approaches evaluate action quality through value functions, with some hybrid models integrating both perspectives [14]. Another research stream models long-term and short-term psychology separately, for instance, by combining matrix factorization with recurrent neural networks (RNNs) [15] or designing variations of long short-term memory (LSTM) with asymmetric singular value decomposition [10]. Nevertheless, these methods often lack supervision during the interest representation phase, leading to inconsistent performance and limited interpretability.

From a consumer lifecycle management perspective, existing studies utilize classification frameworks to predict critical junctures such as conversion and churn. These models incorporate features such as demographics and recency, frequency, and monetary metrics [16], employing algorithms ranging from logistic regression and decision trees to neural networks [9].

Despite these advancements, three primary limitations persist. First, consumer psychological evolution is a continuous process. Thus, static or cumulative features fail to capture granular dynamic shifts, necessitating sequence learning methods suitable for low-interaction scenarios. Second, tracking services must encompass the entire lifecycle rather than isolated junctures. Interpretable long-term detection of the shifting periods would offer significantly higher value for differentiated retention strategies. Third, market competition has not been adequately integrated. Consumer psychological dynamics is significantly influenced by their interactions with competing merchants, and ignoring this factor limits the outcome of tracking services.

2.2 The Consumer Lifecycle

From a long-term perspective, the repeated interactions between consumers and merchants reflect a complete progression from initial contact and peak activity to decline and eventual churn, defined as the consumer lifecycle. Grounded in consumer behavior theory, this process is categorized into sequential psychological periods, including introduction, growth, maturation, dormancy, and churn [4,5]. In the introduction period, consumers have not yet established a deep connection, showing only preliminary intent. The growth period is marked by initial purchases and increasing interaction, fostering early-stage loyalty. Consumers in the maturation period exhibit high purchase frequency and may even become brand advocates. Conversely, the dormancy period sees a decline in activity, while the churn period is characterized by a prolonged lack of interaction, making reactivation difficult.

Prior studies have found that consumer value typically follows an inverted U-shaped trend across the lifecycle, peaking during the maturation period [17]. Merchants need to detect a consumer's current period and evolving demands based on behavioral data to implement differentiated strategies. For consumers in the introduction and growth periods, merchants should facilitate conversion through brand promotion, content marketing, and discount incentives. For those in the dormancy period, active retention efforts, such as exclusive benefits, new product trials, and loyalty programs, are required to underscore brand indispensability before churn occurs [7].

From a lifecycle management standpoint, the necessity for targeted services across different periods suggests that intelligent methods should change from event prediction to sequential mining. By learning generalizable patterns of psychological period shifts, models can form a comprehensive understanding of consumers' psychological dynamics in varying shopping journeys, thereby providing richer practical insights for consumer management.

3 The LC-PGM Approach

Inspired by consumer behavior theory, this study proposes a novel approach named lifecycle probabilistic graphical model (LC-PGM) to mine and decode the psychological dynamic patterns of consumers throughout their entire lifecycle. The overall framework is illustrated in Fig. 1. The approach begins by preprocessing long-term consumer behavioral records from the e-commerce platform and separating them into individual shopping journeys. Then, a dynamic Bayesian network (DBN) [18] is trained to characterize short-term psychological and behavioral dynamics for each journey. Psychological features are extracted from the stage transition paths identified via Viterbi-like inference, which, alongside behavioral sequence features, serve as inputs for the PGM. To learn period shifting patterns from integrated long-term sequences, the PGM is trained using a constrained sampling strategy and the expectation-maximization (EM) algorithm. Finally, the decoded psychological periods are used to predict consumer behavior and assist merchants in personalized tracking services.

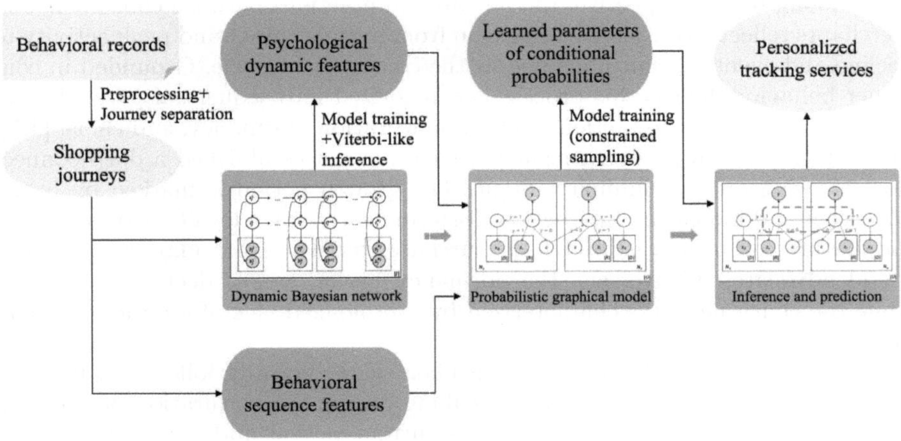

Fig. 1. The LC-PGM approach framework.

3.1 Short-Term Feature Extraction

Since long-term psychological dynamics result from the accumulation of short-term experiences, LC-PGM starts by analyzing individual shopping journeys to detect lifecycle periods. Unlike existing methods that use fixed-time statistics and assume psychological stability within periods [16], LC-PGM treats shopping journey as the fundamental unit. This allows for flexibility in time duration and behavior volume, better capturing the inherent heterogeneity of psychological dynamics. To simplify the model, we control for product category, assuming a consumer's value to a target merchant is primarily influenced by competitors within the same submarket rather than across disparate categories.

Given the highly variable length of shopping journeys, LC-PGM avoids modeling raw sequences directly. To maintain computational efficiency, it trains a well-developed DBN model [18] via the EM algorithm to identify psychological stage transitions. The model features a multistage DBN framework to extract generalizable regularity of consumer psychological dynamics during shopping journeys, two latent layers to depict variability in consumers' interest shifts across multiple stages, and the identification strategies that dynamically detect the invisible stages and interests. Utilizing the identified psychological dynamics, LC-PGM extracts nine key features for each journey, including the proportions of stage advancement, recession, and maintenance, the average transition span, and the probability distribution of stays at each stage. Here, advancement is defined as moving forward along the "awareness–consideration–preference–action" trajectory of psychological stages. Additionally, five behavioral sequence features are extracted to supplement the psychological data, that is, time duration, number of behaviors, purchase occurrence, repeat purchase, and interaction ratio with the target merchant. Thus, consumer shopping journey can be jointly characterized by a series of short-term psychological dynamic and behavioral sequence features, serving as observable input information for mining long-term period shifting patterns.

3.2 Lifecycle Probabilistic Graphical Model

In an increasingly competitive e-commerce landscape, LC-PGM considers a consumer's psychological period regarding both the target merchant and competitors through collaborative learning. We define an observable binary variable y to represent merchant choice. $y = 1$ if the consumer primarily interacts with the target merchant, and $y = 0$ otherwise (determined by purchase outcome or interaction volume). By focusing only on the primary merchant for each journey, the model shares information across competitors via collaborative learning, efficiently capturing generalizable patterns of psychological period shifts.

The LC-PGM approach integrates three graphical modeling paradigms to represent the consumer lifecycle. That is, latent Dirichlet allocation (LDA) to generate discrete data (e.g., merchant choice, purchase occurrence) and detect latent psychological periods, Gaussian mixture model (GMM) to handle continuous data (e.g., stage transition proportions) and identify short-term psychological dynamic types, and hidden Markov model (HMM) to capture temporal dependencies and shifting patterns between lifecycle periods.

The model structure is illustrated in Fig. 2. We assume a consumer set U in the preprocessed dataset, containing a total of $|U|$ consumers. For a specific consumer $u \in U$, the number of shopping journeys within the target merchant's product category over a given period is N_u. Each shopping journey j of consumer u is defined by a variable $y_{u,j}$ (merchant choice) and described by $|D|$ psychological dynamic features $x_{d,u,j}$ and $|R|$ behavioral sequence features $z_{r,u,j}$. Beyond these observable variables, the consumer's shopping journey involves three latent attributes, including the psychological period corresponding to the target merchant $t_{u,j}$, the period corresponding to competing merchants $c_{u,j}$, and the short-term psychological dynamic type $e_{u,j}$.

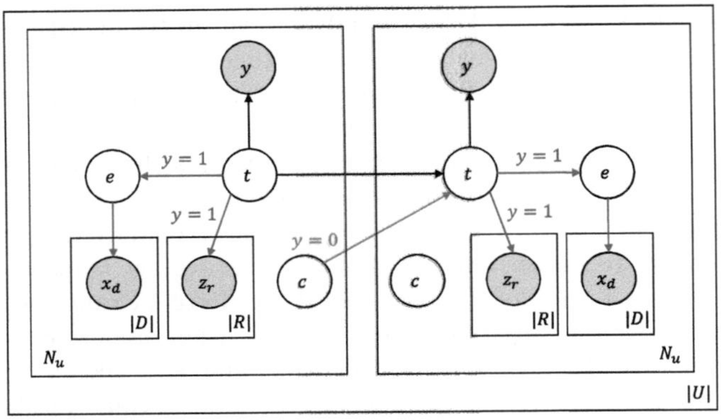

Fig. 2. Probabilistic graphical model depicting consumer lifecycle.

As shown in the figure, the model describes a consumer's psychological period shifts throughout lifecycle from the perspective of the target merchant. The latent psychological period of the next journey depends on the current period, denoted as $P(t_{u,j+1}|t_{u,j})$. Furthermore, considering that a consumer's psychological dynamics are easily influenced by recent interactions with competitors, the model posits that if a consumer chooses a competitor in the current journey ($y_{u,j} = 0$), the target merchant's psychological period for the next journey will be jointly determined by $t_{u,j}$ and $c_{u,j}$, expressed as $P(t_{u,j+1}|t_{u,j}, c_{u,j})$. Specifically, since multiple competitors naturally exist within a submarket, the model selects the psychological period with the highest consumer value among all competitors to serve as $c_{u,j}$. This strategy effectively captures submarket competition while reducing model complexity, based on the premise that a consumer's value to the target merchant is mainly influenced by the strongest competitor. The simplification aligns with classical economic theory, which suggests that in the face of a limited consumer group, businesses primarily compete for resources with the strongest competitors [19,20].

The model also characterizes the influence of unobservable psychological dynamics on observable variables. On the one hand, consumers make different merchant choices based on their current psychological period. Thus, the model captures the conditional dependency of merchant choice on psychological period corresponding to the target merchant, that is, $P(y_{u,j}|t_{u,j})$. Generally, consumers in the maturation period are more likely to choose the target merchant, while those in the introduction or dormancy periods are more likely to interact with competitors. On the other hand, a consumer's psychological period affects their shopping behavior and mental state, manifesting as a series of observable psychological and behavioral features. The model utilizes a latent variable e representing short-term psychological dynamic types to summarize the former. Specifically, the short-term psychological type and behavioral sequence features are conditionally dependent on the latent lifecycle period, that is, $P(e_{u,j}|t_{u,j})$

and $P(z_{r,u,j}|t_{u,j})$. The psychological dynamic features are, in turn, conditionally dependent on the latent type, denoted as $P(x_{d,u,j}|t_{u,j})$.

Based on these variable relationships, the generation process of consumer lifecycle derived from the PGM is shown in the following Algorithm 1.

3.3 Parameter Estimation and Decoding

The LC-PGM approach estimates model parameters, including θ, λ, δ, and η, primarily through Gibbs sampling [21]. This process iteratively samples latent variables (i.e., the psychological period $t_{u,j}$ and short-term dynamic type $e_{u,j}$ for each shopping journey j of consumer u) based on conditional probabilities until the updated parameters converge. Under the collaborative learning framework, latent variables $c_{u,j}$ representing competitor-related psychological periods are derived directly from the corresponding merchant's $t_{u,j}$, enabling information sharing without redundant sampling. Furthermore, continuous variables in the psychological dynamic and behavioral sequence features are assumed to follow Gaussian distributions, and their mean value μ and standard deviation σ are estimated using the EM algorithm.

The complexity analysis shows that the primary computational burden of LC-PGM's parameter estimation lies in the random sampling of latent variables. For $|U|$ consumers with an average of $\overline{N_u}$ shopping journeys, the complexity per iteration is approximately $O(\overline{N_u}|U|(|T| + |E|))$, where $|T|$ and $|E|$ are the total number of psychological periods and short-term psychological dynamic types, respectively. Since $|T|$ and $|E|$ are predefined hyperparameters, the training time scales linearly with the number of consumers and shopping journeys per consumer, ensuring high scalability for large-scale e-commerce data.

To ensure the learned patterns are interpretable for targeted marketing decisions, we incorporate two constrained sampling strategies in the general Gibbs algorithm to explicitly decode consumers' psychological periods. First, considering consumer interactions with the target merchant, psychological period sampling is constrained by (repeat) purchase probabilities. Second, considering mixed interactions within a competitive submarket, the sampling is further constrained by merchant choice inclination.

Specifically, the proposed strategies constrain the learning and optimization of conditional probabilities based on established theoretical findings, that is, higher-value consumers are more likely to purchase and prefer the target merchant [11,17]. In light of the findings, we may classify consumers' lifecycle period shifts into "uphill" (moving toward maturity) and "downhill" (moving toward dormancy or churn). LC-PGM initializes conditional probabilities to reflect the purchase or merchant choice trend and apply constraints during Gibbs sampling. For example, if a consumer's value increases (e.g., manifested in increasing (repeat) purchase probability or choice inclination to the target merchant), the probability of a "downhill" shift is constrained to zero. After each iteration, period indices are reordered to maintain a clear correspondence with the lifecycle progression, balancing predictive accuracy with model interpretability.

Algorithm 1: Consumer lifecycle generation by LC-PGM

input : probabilistic priors α, γ, π, and τ
output: consumer lifecycle $t_{u,j}$, $y_{u,j}$, $e_{u,j}$, $x_{d,u,j}$, and $z_{r,u,j}$

for *psychological period* $t \leftarrow 1$ **to** $|T|$ **do**
  draw distribution of merchant choices $\theta_t \sim Beta(\alpha)$;
  draw distribution of short-term psychological dynamic types $\lambda_t \sim Dirichlet(\gamma)$;
  for *competitor period* $c \leftarrow 1$ **to** $|T|$ **do**
    draw distribution of period shifts $\delta_{c,t} \sim Dirichlet(\pi_c)$;
  end
  for *behavioral sequence feature* $r \leftarrow 1$ **to** $|R|$ **do**
    draw distribution of values $\eta_{r,t} \sim Beta(\tau_r)$ or $\mu_{r,t}, \sigma_{r,t}$;
  end
end
for *psychological dynamic type* $e \leftarrow 1$ **to** $|E|$ **do**
  for *psychological dynamic feature* $d \leftarrow 1$ **to** $|D|$ **do**
    draw distribution of values $\mu_{d,e}, \sigma_{d,e}$;
  end
end
for *consumer* $u \leftarrow 1$ **to** $|U|$ **do**
  draw initial psychological period $t_{u,1} \sim Mult(\delta_{random})$;
  draw initial merchant choice $y_{u,1} \sim Bernoulli(\theta_{t_{u,1}})$;
  if $y_{u,1} = 1$ **then**
    draw initial short-term psychological dynamic type $e_{u,1} \sim Mult(\lambda_{t_{u,1}})$;
    for *psychological dynamic feature* $d \leftarrow 1$ **to** $|D|$ **do**
      draw value $x_{d,u,1} \sim N(\mu_{d,e_{u,1}}, \sigma^2_{d,e_{u,1}})$;
    end
    for *behavioral sequence feature* $r \leftarrow 1$ **to** $|R|$ **do**
      draw value $z_{r,u,1} \sim Bernoulli(\eta_{r,t_{u,1}})$ or $z_{r,u,1} \sim N(\mu_{r,t_{u,1}}, \sigma^2_{r,t_{u,1}})$;
    end
  end
  for *shopping journey* $j \leftarrow 2$ **to** N_u **do**
    if $y_{u,j-1} = 1$ **then** draw psychological period $t_{u,j} \sim Mult(\delta_{t_{u,j-1}})$;
    else draw psychological period $t_{u,j} \sim Mult(\delta_{c_{u,j-1},t_{u,j-1}})$;
    draw merchant choice $y_{u,j} \sim Bernoulli(\theta_{t_{u,j}})$;
    if $y_{u,j} = 1$ **then**
      draw short-term psychological dynamic type $e_{u,j} \sim Mult(\lambda_{t_{u,j}})$;
      for *psychological dynamic feature* $d \leftarrow 1$ **to** $|D|$ **do**
        draw value $x_{d,u,j} \sim N(\mu_{d,e_{u,j}}, \sigma^2_{d,e_{u,j}})$;
      end
      for *behavioral sequence feature* $r \leftarrow 1$ **to** $|R|$ **do**
        draw value $z_{r,u,j} \sim Bernoulli(\eta_{r,t_{u,j}})$ or $z_{r,u,j} \sim N(\mu_{r,t_{u,j}}, \sigma^2_{r,t_{u,j}})$;
      end
    end
  end
end

3.4 Prediction of Consumer Behavior and Merchant Choice

Once the model parameters are learned, LC-PGM performs inference on test consumers to identify their current psychological period and predict future actions. To this end, LC-PGM first employs a Viterbi-like inference strategy. The strategy detects the most probable latent psychological periods of consumer u for the 1st through $(N_u - 1)$-th shopping journeys based on observable variables, including merchant choice, short-term psychological dynamic features, and behavioral sequence features. Let $\phi^j_{u,t}$ denote the maximum probability that consumer u is in period t during journey j while exhibiting observations $y_{u,j}$, $x_{d,u,j}$, and $z_{r,u,j}$. Based on the conditional dependencies defined by the PGM, a recursive formula from $\phi^1_{u,t}$ to $\phi^{N_u-1}_{u,t}$ can be derived via probabilistic calculation. Through recursive backtracking as follows, the model reconstructs the inferred sequence of all psychological periods $t_{u,j} (j = 1, \cdots, N_u - 2)$ experienced by the consumer across multiple journeys. That is,

$$\hat{t_{u,N_u-1}} = \arg\max_t \phi^{N_u-1}_{u,t}, \tag{1}$$

$$\hat{t_{u,j}} = \arg\max_t \phi^j_{u,t} P(\hat{t_{u,j+1}}|t)^{y_{u,j+1}} P(\hat{t_{u,j+1}}|c_{u,j}, t)^{1-y_{u,j+1}}. \tag{2}$$

Subsequently, LC-PGM employs a probabilistic prediction strategy, which predicts the probability of (repeat) purchase behavior in consumer u's future N_u-th shopping journey, as well as the choice of the target merchant and the corresponding interaction proportion. Specifically, for the N_u-th shopping journey, although Viterbi-like inference can decode the most likely psychological period, there remains a possibility that the consumer experiences other periods. To provide accurate and robust behavioral prediction and competitive analysis, LC-PGM considers all possible values of t_{u,N_u-1}. Simultaneously, it prunes paths where the consumer reached t_{u,N_u-1} with a low probability. By calculating the maximum probability of the consumer being in each psychological period during the $(N_u - 1)$-th journey, the model effectively accounts for uncertainty and noise in the data. Finally, based on the integrated probabilities across all t_{u,N_u-1}, the model outputs probabilistic predictions regarding consumer u's (repeat) purchase behavior $P(z_{r,u,N_u})$ and merchant choice inclination $P(y_{u,N_u})$ for the N_u-th shopping journey, as follows.

$$P(z_{r,u,N_u}) = \sum_{t_{u,N_u-1}} \sum_{t_{u,N_u}} \frac{\phi^{N_u-1}_{u,t}}{\sum_t \phi^{N_u-1}_{u,t}} P(t_{u,N_u}|t_{u,N_u-1}) \times P(y_{u,N_u} = 1|t_{u,N_u}) P(z_{r,u,N_u}|t_{u,N_u}), \tag{3}$$

$$P(y_{u,N_u}) = \sum_{t_{u,N_u-1}} \sum_{t_{u,N_u}} \frac{\phi^{N_u-1}_{u,t}}{\sum_t \phi^{N_u-1}_{u,t}} P(t_{u,N_u}|t_{u,N_u-1}) P(y_{u,N_u}|t_{u,N_u}). \tag{4}$$

In the inference and prediction phases, the computational cost of LC-PGM primarily stems from the calculation of $\phi^j_{u,t}$. Given a predefined total of $|T|$ psychological periods, the complexity of calculating $\phi^j_{u,t}$ for consumer u with N_u

shopping journeys is $O(N_u|T|)$. By controlling the maximum value of N_u, the time required to infer and predict the sequence of psychological periods for each consumer remains at a constant scale. Consequently, once the PGM parameters are learned from the training data, LC-PGM can achieve high-efficiency behavioral prediction and competitive analysis for each newly arriving consumer in real-world tracking service applications.

4 Real-World Data Experiments

4.1 Dataset and Settings

The experiments utilize a large-scale real-world dataset from an online cosmetics store on the REES46 platform[1], spanning from October 2019 to February 2020. Each record contains information such as consumer ID, product ID, brand, behavior type, and timestamp. The dataset focuses on similar product categories and a limited number of brands, which can effectively capture consumer lifecycle and submarket competition, making it highly suitable for our research context.

During preprocessing, the data was first partitioned by consumer ID. Records were then segmented into multiple shopping journey sequences, using either a purchase event or a time interval exceeding one week as the boundary. Following common practices in the literature [22], we retained only products with at least 10 interactions and shopping journeys containing at least 10 behavioral records. Furthermore, only consumers who generated at least five shopping journeys during the collection period were included. The final processed dataset comprises 1,037,029 behavioral records, covering 5,922 consumers and 22,424 products. While purchase records are sparse, consumers interacted with a limited number of brands. The average interaction proportion with the target merchant within a single shopping journey reached 0.70. This indicates stable competitive relationships within the submarket, providing a feasible basis for the collaborative learning framework of LC-PGM.

The experiment first trains the multi-stage dynamic Bayesian network (MS-DBN) model [18] to identify the psychological stage transition paths during consumer shopping journeys, and then extract short-term psychological dynamic features as inputs for LC-PGM. Subsequently, we conduct 10-fold cross-validation. For the training set, all historical behaviors are utilized. In the testing phase, LC-PGM predicts the (repeat) purchase probability and merchant choice inclination for a consumer's final shopping journey. The method performance is evaluated based on the average metrics across the ten folds.

The hyperparameters for LC-PGM include the number of psychological periods $|T|$, the number of short-term psychological dynamic types $|E|$, and the priors $\alpha, \gamma, \pi, \tau$. Since consumers in the churn period rarely interact with the target merchant, we focus only on periods prior to churn. According to the literature, $|T|$ is set to 4 (introduction, growth, maturation, and dormancy) [5].

[1] https://www.kaggle.com/datasets/mkechinov/ecommerce-events-history-in-cosmetics-shop.

Moreover, $|E|$ is tuned within the interval $[3, 8]$, and priors are selected from $\{0.01, 0.05, 0.1, 0.2\}$ based on predictive performance. To manage model complexity and mitigate the impact of sparse data, we retain the top 10 brands with the highest interaction proportions for each consumer, merging all remaining brands into an "Others" one.

4.2 Prediction of Consumer Behavior

To evaluate the consumer behavior prediction performance of LC-PGM, we compare it against seven baseline methods. The baselines are categorized into three groups. (1) Standard classification methods, including random forest [23], XGBoost [24], and multi-layer perceptron (MLP) [25]; (2) Dynamic modeling methods, including deep interest network (DIN) [26] and recurrent neural network (RNN) [27]; (3) Ablated models, including no psychology (excluding short-term psychological dynamic features) and no competition (excluding submarket collaborative learning). To ensure a fair comparison, all methods are provided with the same feature set based on the unified shopping journey separation, and hyperparameters for each are tuned to achieve optimal predictive performance.

Using the dataset described in Sect. 4.1, we predict the purchase and repurchase probabilities for each consumer's final shopping journey. Given the observed purchase and repurchase behavior as the ground truth, the experiment utilizes AUC (area under the curve) as the primary evaluation metric. As shown in Fig. 3, following 10-fold cross-validation, LC-PGM achieves an AUC of 0.8531 for purchase prediction and 0.7852 for repurchase prediction, significantly outperforming all baseline models.

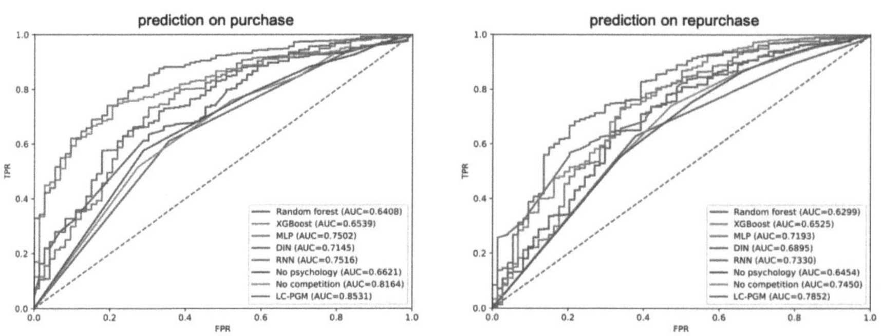

Fig. 3. Method performance on consumer behavior prediction.

The results yield the following insights. First, all methods perform better at predicting purchases than repurchases. The latter presents a greater learning challenge due to the inherent sparsity of repeat transaction data. Second, deep learning-based methods generally outperform shallow models, demonstrating the superior capability of neural networks in capturing complex non-linear relationships. Third, LC-PGM surpasses deep learning baselines such as MLP, DIN, and RNN. This is attributed to its ability to extract long-term psychological patterns across multiple shopping journeys by integrating short-term features to characterize distinctive lifecycle periods. This approach improves prediction accuracy while maintaining a relatively concise model architecture. Fourth, LC-PGM outperform both ablated models, confirming that both short-term psychological dynamics and submarket competitive factors are critical modeling components. Ignoring either element prevents a comprehensive capture of consumer psychological evolution and demand shifts, thereby compromising the accuracy of conversion behavior predictions.

4.3 Prediction of Merchant Choice

To further evaluate the performance of each method in submarket competitive analysis, the experiment conducts two merchant choice prediction tasks using the test set. The first task involves predicting consumers' merchant choices in their final shopping journey to generate a list of the top-K most likely brands, and HR@K (hit ratio) and NDCG@K (normalized discounted cumulative gain) are used to evaluate accuracy and ranking quality. The second task is predicting the consumer's interaction proportion with the target merchant during the final journey, using MSE (mean squared error) and RMSE (root mean squared error) to measure the prediction error.

Setting $K = 5$, the cross-validated performance of each method across these metrics is summarized in Table 1. We conduct paired t-tests between LC-PGM and the baseline method with the best average performance for each metric across the ten folds. In the table, *, **, and *** denote statistical significance levels of 0.1, 0.05, and 0.01, respectively.

Table 1. Method performance on merchant choice prediction.

Method	Brand choice		Interaction proportion	
	HR@5	NDCG@5	MSE	RMSE
Random forest	0.4668	0.3187	0.0789	0.2809
XGBoost	0.5559	0.3854	0.0774	0.2783
MLP	0.6324	0.5354	0.0529	0.2300
DIN	0.6224	0.4823	0.0630	0.2511
RNN	0.6756	0.5046	0.0510	0.2259
No psychology	0.6085	0.4706	0.0683	0.2613
No competition	0.7231	0.5689	0.0478	0.2187
LC-PGM	0.7677***	0.5806***	0.0429**	0.2070***

The results indicate that LC-PGM significantly outperforms all baseline methods across all four metrics for both tasks, demonstrating its superiority in submarket competitive analysis. On the one hand, LC-PGM accurately predicts a consumer's merchant choice in a shopping journey, effectively assisting merchants in identifying their primary competitors. On the other hand, LC-PGM predicts the interaction proportion with the target merchant (or brand) with minimal error. This provides valuable insights for granular analysis of competitive landscapes and market share within specific product categories. These findings validate that the proposed LC-PGM approach effectively models the submarket competitive factors that influence psychological dynamics throughout the consumer lifecycle.

4.4 Visualization of Psychological Period Decoding

LC-PGM decodes the psychological periods of consumer shopping journeys using a constrained sampling strategy. To reveal the managerial insights regarding long-term consumer psychological dynamics, we visualize the key learned parameters. The experiment defines four latent psychological periods (i.e., introduction, growth, maturation, and dormancy), whose sequence is jointly determined by (repeat) purchase probabilities and merchant choice inclinations, consistent with their theoretical definitions.

The heatmap in Fig. 4 illustrates the shifting probabilities of a consumer's psychological period corresponding to the target merchant, conditioned on the current period relative to competitors, that is, $P(t_{u,j+1}|t_{u,j}, c_{u,j+1})$. The results indicate that consumers in the later phases of the lifecycle are more inclined to remain in their current period, with minimal shifts observed out of the dormancy period. Consumers' psychological shifts primarily occur between adjacent periods, particularly for growth and maturation. Furthermore, the shifts are flexible and not strictly linear, reflecting the inherent complexity of consumers' psychological dynamics in e-commerce contexts. The impact of submarket competition is also significant. When consumers enter the growth or maturation period of a competitor, the probability of them entering or remaining in the target merchant's maturity period decreases, while the probability of shifting to dormancy increases. Conversely, if the competitor period is dormancy, consumers are more likely to shift from introduction or growth to the maturity period of the target merchant, with an increased probability of staying there. These insights are consistent with the theoretical findings and thus validate the effectiveness of our submarket competition modeling.

The stacked bar chart in the left subfigure of Fig. 5 presents the probabilistic distribution of psychological stage advancement, maintenance, and recession within a single shopping journey across the awareness, consideration, preference, and action stages. That is, $P(x_{d,u,j}|t_{u,j}) = \sum_{e_{u,j}} P(x_{d,u,j}|e_{u,j})P(e_{u,j}|t_{u,j})$. Generally, the probability of staying in the current stage is the highest, followed by advancement, with recession being the lowest. In the introduction period, consumers show the highest overall shifting potential, reflecting varying interests. The growth and maturation periods are characterized by a decrease in psychological stage recession probability, and the maturation period exhibits the highest probability of stage advancement, indicating rapid decision-making.

Moreover, the dormancy period features the lowest advancement and highest recession probabilities, corresponding to a marked decline in conversions. The significant differences in psychological dynamic features across the four lifecycle periods align with our modeling hypotheses and demonstrate the discriminative power of LC-PGM's decoding strategy.

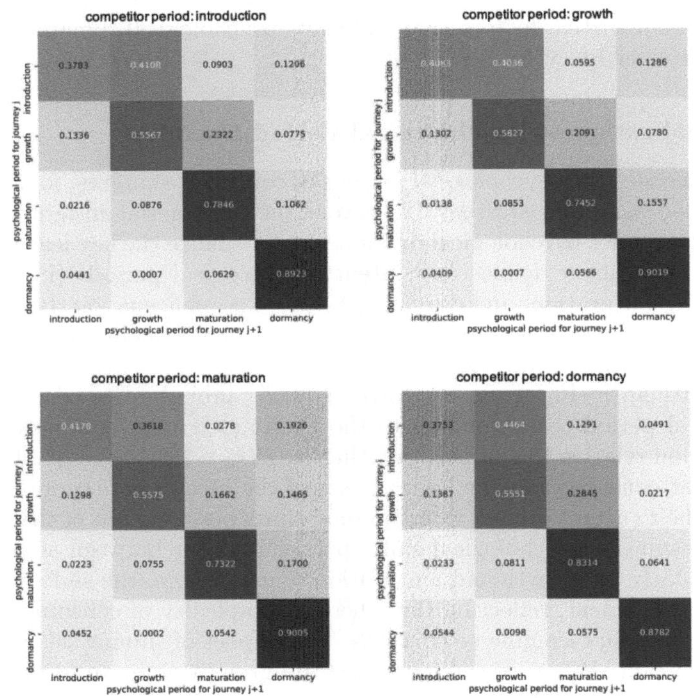

Fig. 4. Learned probability of psychological period shifts.

The line graph in the right subfigure of Fig. 5 displays the purchase and repurchase probabilities, alongside the average interaction proportion with the target merchant for each period, that is, $P(z_{r,u,j}|t_{u,j})$. Specifically, in the introduction period, consumers' purchase probability and interaction proportions remain relatively low. When stepping into growth or maturation, both metrics rise significantly, with interaction proportions exceeding 70%. In the dormancy period, both indicators decline sharply to levels below the introduction period, approaching a state of churn. This trend is highly consistent with the standard lifecycle value curve, further verifying the validity of our decoding approach.

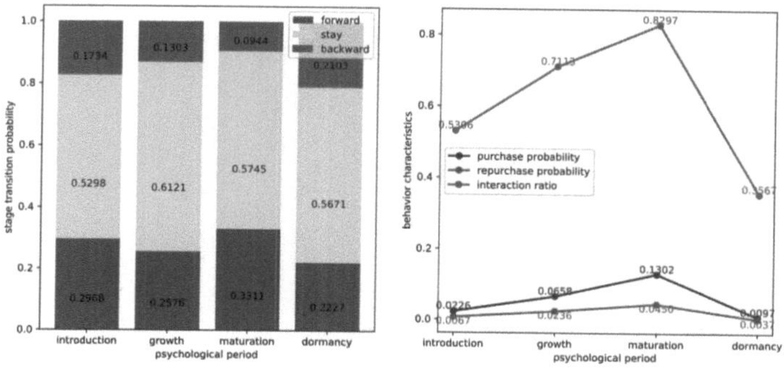

Fig. 5. Learned short-term features of each psychological period.

In summary, LC-PGM effectively learns long-term psychological dynamics and short-term features. The decoding strategy accurately distinguishes latent psychological periods, providing interpretable managerial insights for consumer behavior prediction and competitive analysis.

4.5 Heterogeneity Analysis

To evaluate the performance of LC-PGM across different record lengths, we partition the 5,922 consumers in the test set into three groups of approximately equal frequency based on their number of shopping journeys. That is, short-record group (5 journeys, $n = 1,911$), medium-record group (6–7 journeys, $n = 2,193$), and long-record group (≥ 8 journeys, up to a maximum of 72, $n = 1,818$). Figure 6 summarizes the comparative results between LC-PGM and five baseline methods for behavior prediction and competitive analysis across these groups.

The results indicate that LC-PGM significantly outperforms the baseline methods for all groups, further validating its effectiveness and robustness. Overall, all methods demonstrate better performance in the long-record group, suggesting that a higher number of shopping journeys provides substantial information gain. Notably, the advantage of LC-PGM is most pronounced in the long-record group. This is because the approach can fully leverage accumulated shopping journeys to collaboratively learn long- and short-term psychological dynamics and submarket competitive factors. In contrast, the baselines show a limited capacity to comprehensively utilize long-sequence information, with their performance gains diminishing as the record length increases.

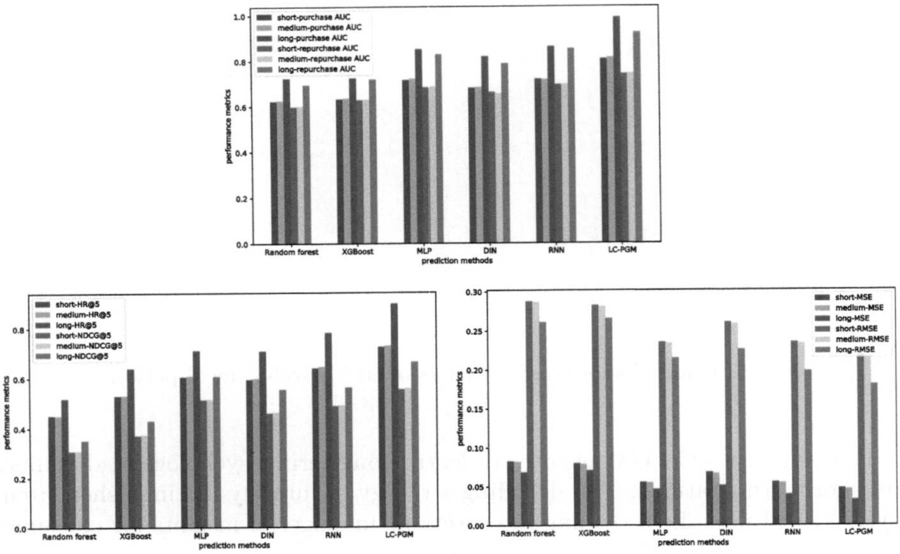

Fig. 6. Method performance across different record lengths.

5 Conclusion

To support e-commerce merchants in providing long-term tracking services, this study proposes the LC-PGM approach. By modeling shifts of psychological periods across multiple consumer shopping journeys, the approach characterizes the underlying mechanisms by which psychological dynamics drive observable behaviors. LC-PGM integrates short-term behavioral features with submarket competitive factors while employing constrained sampling, Viterbi-like inference, and probabilistic prediction strategies. This ensures its high predictive accuracy while maintaining intrinsic interpretability. Experiments on real-world datasets demonstrate that LC-PGM outperforms baseline methods in large-scale lifecycle tracking applications, particularly in long-record scenarios, thereby effectively supporting merchants' service decision-making.

The contributions of this study are reflected in three primary aspects. First, LC-PGM tracks and decodes long-term consumer psychological dynamics. By extracting generalizable shifting patterns, the approach achieves precise predictions even in sparse interaction scenarios, empowering managerial decisions in an interpretable manner. Second, LC-PGM organically fuses long- and short-term shopping dynamics. It leverages short-term features from individual journeys to assist in mining long-term patterns, thereby flexibly capturing psychological periods under varying behavioral rhythms and enhancing overall predictive performance. Third, from the perspective of competitive intelligence, LC-PGM incorporates submarket competition factors. By employing a collaborative learning framework to integrate multi-merchant interaction data, the approach comprehensively detects consumers' psychological shifts, improving learning outcomes while providing practical insights for platform-level competitive analysis.

Future research may explore in the following directions. First, integrate multimodal information, such as review text and images, to more accurately analyze psychological shifts and market competition. Second, expand the temporal span of data to fully encompass the entire consumer lifecycle, thereby further strengthening the model's learning efficacy. Third, explore adaptive refinements of the methodology to suit a broader range of diverse tracking service tasks.

Acknowledgments. This work was supported by the National Natural Science Foundation of China [Grants 72302153 and 72172092], Shanghai Key Laboratory of Brain-Machine Intelligence for Information Behavior [Grant 22dz2261100], and Shanghai Chenguang Program [Grant 24cga34].

Disclosure of Interests. The authors have no competing interests to declare that are relevant to the content of this article.

References

1. Son, Y., Oh, W.: Digitalization of loyalty: impacts of mobile technology on reward redemption and engagement level. J. Manag. Inf. Syst. **40**(4), 1301–1327 (2023)
2. Lee, J., Lee, J., Feick, L.: Incorporating word-of-mouth effects in estimating customer lifetime value. J. Database Market. Custom. Strat. Manage. **14**(1), 29–39 (2006)
3. Ali, N., Shabn, O.S.: Customer lifetime value (CLV) insights for strategic marketing success and its impact on organizational financial performance. Cogent Bus. Manage. **11**(1), 2361321 (2024)
4. Yu, X., Guo, S., Guo, J., Huang, X.: An extended support vector machine forecasting framework for customer churn in e-commerce. Expert Syst. Appl. **38**, 1425–1430 (2011)
5. Li, Q.: Design of customer churn early warning system based on mobile communication technology based on data mining. J. Electr. Comput. Eng. (1), 9701349 (2022)
6. Fader, P., Hardie, B.: How to project customer retention. Harv. Bus. Rev. **99**(1), 130–136 (2021)
7. Garcia, D.L., Nebot, A., Vellido, A.: Intelligent data analysis approaches to churn as a business problem: a survey. Knowl. Inf. Syst. **51**, 719–774 (2017)
8. Duan, S., Ouyang, M., Wang, R., Li, Q., Xiao, Y.: Let long-term interests talk: a disentangled learning model for recommendation based on short-term interests generation. Inform. Process. Manage. **62**(2), 103997 (2025)
9. Verbeke, W., Martens, D., Mues, C., Baesens, B.: Building comprehensible customer churn prediction models with advanced rule induction techniques. Expert Syst. Appl. **38**, 2354–2364 (2011)
10. Yu, Z., Lian, J., Mahmoody, A., Liu, G., Xie, X.: Adaptive user modeling with long and short-term preferences for personalized recommendation. In: Proceedings of the 28th International Joint Conference on Artificial Intelligence, pp. 4213–4219. AAAI Press, Macao, China (2019)

11. Bayer, J., Servan-Schreiber, E.: Gaining competitive advantage through the analysis of customers' social networks. J. Direct Data Digit. Mark. Pract. **13**(2), 106–118 (2011)
12. Taylor, O., Rossi, M., Lee, J., Hernandez, M.: Behavioral economics in consumer decision-making: analyzing the impact of cognitive biases. Int. J. Manage., Bus., Econom. **1**(1), 15–17 (2024)
13. Lee, H., Hwang, D., Min, K., Choo, J.: Towards validating long-term user feedbacks in interactive recommendation systems. In: Proceedings of the 45th International ACM SIGIR Conference on Research and Development in Information Retrieval, pp. 2607–2611. ACM, Madrid, Spain (2022)
14. Xin, X., Karatzoglou, A., Arapakis, I., Jose, J. M.: Self-supervised reinforcement learning for recommender systems. In: Proceedings of the 43rd International ACM SIGIR Conference on Research and Development in Information Retrieval, pp. 931–940. ACM, Virtual Event, China (2020)
15. Zhao, W., Wang, B., Ye, J., Gao, Y., Yang, M., Chen, X.: PLASTIC: prioritize long and short-term information in top-n recommendation using adversarial training. In: Proceedings of the 27th International Joint Conference on Artificial Intelligence and the 23rd European Conference on Artificial Intelligence, pp. 3676–3682. AAAI Press, Stockholm, Sweden (2018)
16. Johny, C.P., Mathai, P.P.: Customer churn prediction: a survey. Int. J. Adv. Res. Comput. Sci. **8**(5), 2178–2181 (2017)
17. Mosaddegh, A., Albadvi, A., Sepehri, M.M., Teimourpour, B.: Dynamics of customer segments: a predictor of customer lifetime value. Expert Syst. Appl. **172**, 114606 (2021)
18. Wei, Q., Mu, Y., Guo, X., Jiang, W., Chen, G.: Dynamic Bayesian network-based product recommendation considering consumers' multistage shopping journeys: a marketing funnel perspective. Inf. Syst. Res. **35**(3), 1382–1402 (2024)
19. Clark, J.B.: The limits of competition. Political Sci. Quart. **2**(1), 45–61 (1887)
20. Vance, R.R.: Interference competition and the coexistence of two competitors on a single limiting resource. Ecology **65**(5), 1349–1357 (1984)
21. Heinrich, G.: Technical Report: Parameter Estimation for Text Analysis. University of Leipzig, Germany (2008)
22. Sahoo, N., Singh, P.V., Mukhopadhyay, T.: A hidden Markov model for collaborative filtering. MIS Q. **36**(4), 1329–1356 (2012)
23. Breiman, L.: Random forests. Mach. Learn. **45**, 5–32 (2001)
24. Chen, T., Guestrin, C.: Xgboost: a scalable tree boosting system. In: Proceedings of the 22nd ACM SIGKDD International Conference on Knowledge Discovery and Data Mining, pp. 785–794. ACM, San Francisco, CA, USA (2016)
25. Popescu, M.C., Balas, V.E., Perescu-Popescu, L., Mastorakis, N.: Multilayer perceptron and neural networks. WSEAS Trans. Circ. Syst. **8**(7), 579–588 (2009)
26. Zhou, G., et al.: Deep interest evolution network for click-through rate prediction. In: Proceedings of the 33rd AAAI Conference on Artificial Intelligence, pp. 5941–5948. AAAI Press, Honolulu, Hawaii, USA (2019)
27. Hidasi, B., Karatzoglou, A., Baltrunas, L., Tikk, D.: Session-based recommendations with recurrent neural networks. arXiv preprint arXiv:1511.06939 (2015)

A Modular and Reproducible Z-Number Fuzzy AHP–TOPSIS Framework in R

Gizem Intepe[1(✉)], Irem Ucal Sari[2], and Chng Wei Lau[1]

[1] Western Sydney University, Rydalmere, NSW 2116, Australia
{g.intepe,c.lau}@westernsydney.edu.au
[2] Baskent University, 06790 Ankara, Turkey
iremucals@baskent.edu.tr

Abstract. Fuzzy multi-criteria decision-making (MCDM) methods such as AHP and TOPSIS are widely used to evaluate complex alternatives under uncertainty. The introduction of Z-numbers, incorporating both fuzzy restriction and reliability information, has further extended the expressive capability of these models. However, despite their growing application, computational realisations of Z-number fuzzy AHP–TOPSIS are often presented as procedural workflows in which modelling assumptions remain implicit and difficult to isolate.

This paper presents an R-based computational implementation and framework prototype that formalises established triangular Z-number AHP–TOPSIS procedures within an explicit, script-controlled architecture. The framework preserves the conventional mathematical structure while exposing key modelling components - linguistic scale specification, reduction of Z-numbers, weight defuzzification, and ranking configuration - as selectable parameters.

The approach is demonstrated through the computational replication of a published Z-number fuzzy AHP–TOPSIS study, with all data and code provided in an online repository to enable independent execution of the complete decision pipeline. By separating representation, transformation, aggregation, and ranking stages, the framework improves transparency and analytical traceability without altering the theoretical foundations of Z-number fuzzy decision-making.

Keywords: Z-numbers · fuzzy AHP · fuzzy TOPSIS · reproducible research · computational framework · R

1 Introduction

Multi-criteria decision-making (MCDM) methods provide systematic tools for evaluating alternatives in the presence of competing criteria and uncertainty. Classical techniques such as the Analytic Hierarchy Process (AHP) and Technique for Order Preference by Similarity to Ideal Solution (TOPSIS) have become ubiquitous in engineering, management, and logistics applications due to their intuitive structure and ability to combine subjective judgements with

J. Lu et al. (Eds.): FLINS-ISKE 2026, LNCS 16752, pp. 397–415, 2027.
https://doi.org/10.1007/978-981-92-2369-5_25

quantitative analysis. To capture linguistic assessments and imperfect information, fuzzy set extensions of these methods - particularly fuzzy AHP and fuzzy TOPSIS - have been widely adopted, allowing decision makers to express preferences using fuzzy numbers rather than single crisp values [18,28].

The concept of Z-numbers, first introduced by Zadeh [38], extends fuzzy numbers by pairing a fuzzy restriction with a fuzzy measure of reliability, thereby enabling a richer representation of both uncertainty and confidence in expert judgements. Z-number based extensions have been integrated into numerous MCDM methods - including AHP, TOPSIS, VIKOR, and hybrid approaches - demonstrating enhanced descriptive power in decision contexts where reliability matters [6,31,33,35]. Despite the growing methodological interest, the translation of Z-number based decision models into reusable computational tools remains fragmentary.

In parallel, the computational support for fuzzy MCDM within the R ecosystem displays an uneven landscape. While crisp MCDM techniques and foundational fuzzy arithmetic utilities are available through packages such as `RMCDA`, `topsis`, `FuzzyAHP`, and `FuzzyNumbers` [5,13,24,36], these tools tend to be isolated in scope or provide only partial workflow integration. Moreover, several earlier packages offering multi-method fuzzy functionality have been archived, highlighting challenges in long-term maintenance and reproducibility.

Despite the expressive capability of advanced fuzzy set paradigms such as intuitionistic, Pythagorean, spherical, and Z-number representations, the R ecosystem lacks a modular and reproducible framework supporting an end-to-end fuzzy MCDM pipeline that integrates weighting and ranking within a unified computational architecture. In applied research, such analyses are often implemented in spreadsheets [7,10,12,25,32], which are convenient for matrix construction but poorly suited to script-based reproducibility, automated validation, and scalable experimentation. Establishing an R-based framework therefore strengthens transparency, traceability, and extensibility in fuzzy decision-making research.

To address these gaps, this paper presents a modular R implementation that operationalises established triangular Z-number AHP–TOPSIS procedures within a reproducible, script-controlled framework. The current prototype supports triangular Z-numbers and separates scale definition, Z-reduction, fuzzy aggregation, defuzzification, and ranking into composable modules. Users can declare custom linguistic and reliability scales and execute the full workflow under explicit parameter settings. The implementation is available as a working research prototype in R, together with replication data and source code.

While the present prototype focuses on triangular fuzzy Z-numbers within an AHP–TOPSIS structure, the overall architecture is designed to support future extension to additional fuzzy paradigms. The modular separation between representation, transformation, and aggregation layers is designed to facilitate future integration of alternative fuzzy extensions, including intuitionistic, spherical, and proportional fuzzy sets. In this way, the framework provides both a concrete implementation for triangular Z-number MCDM and a scalable foundation for broader fuzzy decision models within R.

The main contributions of this paper can be summarised as follows:

- A modular, layered computational architecture for Z-number fuzzy AHP–TOPSIS in R, separating representation, transformation, aggregation, and ranking stages.
- Explicit parameterisation of linguistic scales, reliability modelling, and Z-reduction procedures within a reproducible script-based environment.
- Integration of group aggregation and ranking into a unified, end-to-end decision pipeline.
- Demonstration of reproducibility through replication of a published Z-AHP–TOPSIS study within a fully executable workflow.

By formalising the computational realisation of Z-number fuzzy decision models within a transparent and parameter-driven architecture, the proposed framework contributes to methodological standardisation in computational intelligence and reliability-aware decision making.

2 Related Work

R provides a diverse ecosystem of packages supporting MCDM, ranging from general-purpose toolkits to method-specific implementations and foundational libraries for fuzzy representations. However, functionality is distributed across largely disjoint packages, with limited support for integrated fuzzy pipelines and no widely adopted implementation of Z-number decision models. Table 1 summarises representative packages in the current R ecosystem.

A substantial portion of the R MCDM ecosystem focuses on crisp data. For example `RMCDA` offers a broad collection of decision-analysis methods and is representative of general-purpose MCDM libraries in R, providing implementations of widely used techniques such as AHP, TOPSIS, WASPAS, PROMETHEE and MULTIMOORA within a unified interface [24]. Complementing such general toolkits, several packages implement specific crisp methods, including TOPSIS-focused libraries such as `TopSisWM` [21] and the earlier `topsis` package [36], and a dedicated implementation of WASPAS in `waspasR` [1]. Hybrid workflows are also present: `ahptopsis2n` provides a crisp AHP–TOPSIS procedure, illustrating common practice in which AHP-derived criteria weights are combined with TOPSIS ranking [8,9]. For AHP modelling specifically, `AHPtools` provides utilities for structuring and analysing AHP hierarchies [2]. Collectively, these packages demonstrate mature support for crisp MCDM in R, but they generally treat each judgement or performance score as crisp numbers and therefore do not natively represent layered uncertainty or confidence information.

Fuzzy extensions exist, but are typically method-specific and less integrated at the workflow level. The `FuzzyAHP` package implements triangular fuzzy AHP computation and supports fuzzy pairwise comparisons based on standard triangular fuzzy scales [5]. It provides a practical implementation of classical FAHP

Table 1. Selected R packages related to fuzzy MCDM

Package	Source	Year	Author(s)	Fuzzy Types	MCDM Methods
FuzzyAHP	CRAN	2019	Caha	Triangular	Fuzzy AHP
RMCDA	CRAN	2025	Najafi	Triangular; (workflow integration limited)	AHP, TOPSIS, WASPAS (crisp)
IFMCDM	CRAN	2023	Dudek et al.	Intuitionistic	IF-TOPSIS, IF synthetic measures
ipsfs	CRAN	2022	Panda & Nagwani	Intuitionistic, Pythagorean, Spherical	Similarity measures (not full MCDM workflow)
ahptopsis2n	CRAN	2021	Coutinho & Santos	Crisp	Hybrid AHP-TOPSIS
TopSisWM	CRAN	2025	Santos	Crisp	TOPSIS
topsis	CRAN	2013	Yazdi	Crisp	TOPSIS
FuzzyNumbers	CRAN	2021	Gagolewski	Triangular, Trapezoidal	None (fuzzy arithmetic only)
sets	CRAN	2023	Meyer et al.	General fuzzy sets	None (set operations only)

procedures [17]. However, the scale structure and core computational choices (e.g., aggregation and defuzzification) are embedded within the package design rather than exposed as fully modular, user-defined components. As a result, adapting the implementation to alternative linguistic scales or methodological variations may require modification of internal functions rather than simple parameter selection. At a lower level, `FuzzyNumbers` provides tools for defining and manipulating fuzzy numbers in R, enabling custom implementations of fuzzy arithmetic, ranking and defuzzification that are frequently required when building fuzzy MCDM pipelines [13]. In addition, `sets` package offers general data structures and operations for fuzzy sets and generalised set formalisms [23]; its associated methodological paper highlights flexible representations within R that can serve as building blocks for fuzzy modelling [22]. These foundational packages are valuable for algorithm development, but they do not by themselves provide end-to-end fuzzy AHP–TOPSIS workflows.

Earlier multi-method fuzzy MCDM functionality was provided by the `FuzzyMCDM` package, which implemented several triangular fuzzy ranking procedures, including Fuzzy TOPSIS, Fuzzy VIKOR, Fuzzy Multi-MOORA, and Fuzzy WASPAS, together with a MetaRanking function for aggregating results across methods [19]. While this package represented one of the more comprehensive fuzzy MCDM toolkits available in R at the time of its release, its focus was primarily on alternative-ranking procedures and it did not provide fuzzy AHP-based weight derivation or a modular architecture integrating scale definition, transformation, aggregation, and ranking stages. In addition, both `FuzzyMCDM` and the related crisp `MCDM` package have since been archived from CRAN [19,20], highlighting the broader challenge of maintaining reusable research software for fuzzy decision analysis.

R also contains specialised packages targeting alternative fuzzy paradigms. The `IFMCDM` package implements multi-criteria decision-making methods under the intuitionistic fuzzy framework, where evaluations are represented by membership and non-membership degrees. Specifically, it provides the Intuitionistic Fuzzy Synthetic Measure (IFSM) and Intuitionistic Fuzzy TOPSIS (IFTOPSIS) methods for ranking alternatives [11,14,29]. While the package offers dedicated support for intuitionistic fuzzy decision models, it is focused on this specific fuzzy extension and does not provide a unified architecture adaptable across multiple fuzzy paradigms or hybrid AHP–TOPSIS workflows. In parallel, `ipsfs` provides computational support for advanced fuzzy paradigms, including intuitionistic, Pythagorean, and spherical fuzzy sets [26,27]. The package focuses primarily on the computation of similarity and distance measures under these representations, offering tools for handling multi-dimensional fuzzy evaluations. While `ipsfs` expands the representational capabilities of fuzzy modelling in R beyond triangular forms, it does not implement complete MCDM workflows such as AHP or TOPSIS, nor does it provide an integrated decision pipeline. These packages expand the representational capacity beyond standard triangular fuzziness, but they are typically oriented towards particular fuzzy extensions (e.g., intuitionistic or spherical) rather than offering an integrated MCDM workflow across fuzzy types. Furthermore, the `lfl` package contributes processable linguistic fuzzy logic constructs in R, supporting linguistic modelling and operations that are relevant when decision judgements are expressed in verbal terms rather than numeric scales [4].

Overall, the reviewed packages demonstrate that R provides strong support for (i) crisp AHP/TOPSIS/WASPAS methods and hybrids [1,2,8,21,24,36], (ii) triangular fuzzy AHP weighting [5], and (iii) advanced fuzzy representations and similarity measures [4,11,26,27]. However, the ecosystem lacks a standardised and actively maintained computational framework that integrates layered uncertainty modelling with modular, parameter-driven decision pipelines. In particular, decision models requiring two-component judgement structures, such as Z-numbers, are not supported within a unified, extensible architecture. These observations motivate the development of a reproducible and modular R framework for Z-number fuzzy AHP–TOPSIS, designed to support user-defined linguistic and reliability scales while remaining extensible to emerging fuzzy set variants.

While similar implementations may exist in other programming environments such as Python, the present work focuses specifically on the R ecosystem, where integrated support for Z-number fuzzy MCDM workflows remains limited.

3 A Modular Computational Framework for Z-Number Fuzzy AHP–TOPSIS

This work develops a modular R-based computational framework for conducting Z-number fuzzy AHP–TOPSIS analyses in a structured and reproducible manner. The current prototype supports Z-numbers constructed on triangular fuzzy numbers, consistent with the dominant representation adopted in applied fuzzy MCDM studies.

Importantly, the framework formalises the computational realisation of established Z-AHP and Z-TOPSIS procedures within a transparent, parameter-driven environment. The objective is to preserve the familiar algorithmic structure widely used in the literature while replacing manual spreadsheet-based workflows with script-based, auditable, and reproducible implementations.

3.1 Algorithmic Structure

The computational pipeline mirrors the conventional stages of Z-number fuzzy AHP–TOPSIS:

1. Definition of linguistic restriction and reliability scales;
2. Construction of Z-number pairwise comparison matrices;
3. Reduction of Z-numbers to operational triangular fuzzy numbers;
4. Fuzzy AHP weight derivation via geometric mean aggregation;
5. Defuzzification and normalisation of weights;
6. Construction of the weighted decision matrix;
7. Application of fuzzy TOPSIS to obtain final rankings.

Each stage corresponds directly to formulations commonly reported in the Z-number fuzzy MCDM literature. The framework simply encapsulates these steps into clearly defined computational layers.

3.2 Flexible Linguistic and Reliability Modelling

Following Zadeh [38], a Z-number is defined as an ordered pair of fuzzy numbers

$$Z_{ij} = (\tilde{A}_{ij}, \tilde{R}_{ij}),$$

where $\tilde{A}_{ij}$ represents a fuzzy restriction on the value of a variable, and $\tilde{R}_{ij}$ expresses a fuzzy assessment of the reliability associated with that restriction. In Z-number based MCDM applications, $\tilde{A}_{ij}$ typically encodes a pairwise comparison intensity, while $\tilde{R}_{ij}$ captures the decision maker's confidence in that judgement. Both restriction and reliability components are represented in this framework by triangular fuzzy numbers (TFNs).

The triangular components of Z_{ij} are specified as

$$\tilde{A}_{ij} = (l_{ij}, m_{ij}, u_{ij}), \quad \tilde{R}_{ij} = (l'_{ij}, m'_{ij}, u'_{ij}).$$

where each component is represented using a triangular membership function [37].

For a generic triangular fuzzy number $\tilde{a} = (l, m, u)$, the membership function is defined as:

$$\mu_{\tilde{a}}(x) = \begin{cases} \dfrac{x-l}{m-l}, & l \leq x \leq m, \\ \dfrac{u-x}{u-m}, & m \leq x \leq u, \\ 0, & \text{otherwise.} \end{cases}$$

Accordingly, each Z-number used in pairwise comparisons is expressed as

$$Z_{ij} = \left((l_{ij}, m_{ij}, u_{ij}), (l'_{ij}, m'_{ij}, u'_{ij})\right),$$

where (l_{ij}, m_{ij}, u_{ij}) denotes the restriction TFN and $(l'_{ij}, m'_{ij}, u'_{ij})$ denotes the reliability TFN. In typical Z-AHP formulations, reliability terms are defined over the unit interval $[0, 1]$, reflecting degrees of confidence.

Unlike many implementations that embed fixed linguistic scales, the proposed framework allows decision makers to declare their own restriction and reliability scales explicitly as sets of triangular fuzzy numbers;

$$\tilde{A}^{(k)} = (l_k, m_k, u_k), \qquad \tilde{R}^{(s)} = (l^r_s, m^r_s, u^r_s),$$

thereby reproducing exactly the scales adopted in a target study or supporting domain-specific calibration. Judgements are entered using paired linguistic tokens of the form `(restriction term, reliability term)`, for example `(Moderate, Strong)`. Each token pair is mapped programmatically to its corresponding triangular restrictionreliability representation $\left(\tilde{A}^{(k)}, \tilde{R}^{(s)}\right)$ based on the user-declared scales, ensuring that all numerical representations remain explicit and reproducible.

Although the modelling layer is fully parameter-driven, the algebraic structure of fuzzy AHP imposes basic coherence conditions. Each triangular fuzzy number must satisfy admissibility ($l \leq m \leq u$), and pairwise comparison matrices are assumed to form fuzzy positive reciprocal matrices following Buckley [3], that is,

$$\tilde{a}_{ij}^{-1} = \tilde{a}_{ji}.$$

In Z-number formulations, reliability terms act as confidence modifiers in the subsequent reduction stage. While no additional algebraic constraints are imposed on reliability beyond admissibility and boundedness, extreme or inconsistent reliability specifications may distort the effective weighting of restriction terms after reduction.

Accordingly, the framework performs structural validation prior to computation. Triangular admissibility is verified for all declared scale elements, and reciprocal consistency of the fuzzy pairwise comparison matrix is verified before fuzzy aggregation. These safeguards help verify that user-defined linguistic scales conform to the structural assumptions required by fuzzy AHP and Z-number reduction procedures.

3.3 Configurable Z-Reduction and Defuzzification

As introduced above, a Z-number consists of a restriction component $\tilde{A}_{ij}$ and a reliability component $\tilde{R}_{ij}$. Since classical fuzzy AHP–TOPSIS operate on a single-component fuzzy number, a reduction step is required to transform each Z_{ij} into an operational triangular fuzzy number prior to aggregation.

Reliability Defuzzification. In triangular Z-fuzzy AHP implementations, the reliability component $\tilde{R}_{ij}$ is converted into a scalar coefficient α_{ij} prior to restriction scaling. For triangular fuzzy numbers, a commonly used formulation is the arithmetic centroid (centre-of-area) defuzzification [15,16]

$$\alpha_{ij} = \frac{l'_{ij} + m'_{ij} + u'_{ij}}{3}.$$

Modal-weighted formulations are also adopted in the literature. For example, Çekmiş et al. [6], Tüysüz and Şimşek [34], and Sergi et al. [33] apply the weighted triangular mean

$$\alpha_{ij} = \frac{l'_{ij} + 2\,m'_{ij} + u'_{ij}}{4},$$

which assigns greater influence to the modal value.

The proposed framework therefore allows selection of the defuzzification rule used to compute α_{ij}. The resulting coefficient is subsequently used in the Z-number reduction step described below.

Restriction Scaling. After obtaining the scalar reliability coefficient α_{ij}, the restriction component $\tilde{A}_{ij}$ is transformed into a triangular fuzzy number $\tilde{A}'_{ij}$ prior to aggregation.

Following the Z-number conversion approach proposed by Otay and Kahraman [15], the restriction component is adjusted through square-root reliability scaling:

$$\tilde{A}'_{ij} = \sqrt{\alpha_{ij}}\,\tilde{A}_{ij}.$$

This transformation preserves the triangular structure of $\tilde{A}_{ij}$ while allowing the reliability coefficient to attenuate the restriction in a moderated manner. Compared to direct proportional scaling, the square-root formulation reduces the magnitude of the restriction without excessively collapsing its support, thereby maintaining interpretability of the fuzzy comparison.

Defuzzification of Weights. After fuzzy aggregation, triangular fuzzy weights are converted to crisp values for ranking purposes. For triangular fuzzy numbers, this is commonly performed using the arithmetic centroid operator (e.g., [3,16]), although alternative defuzzification and ranking operators exist in the fuzzy decision-making literature. The framework supports selectable weight defuzzification operators; the replication in Sect. 4 uses the weighted triangular mean to match [6].

TOPSIS Ranking. The derived criteria weights are subsequently used within the TOPSIS ranking stage. In the present replication experiment, crisp performance values are used, consistent with the data reported in the original study. In the current implementation, alternative performance values may be provided as crisp or triangular fuzzy numbers. When fuzzy performance values are supplied, vertex-based distance metrics are applied to compute separations from the fuzzy ideal and anti-ideal solutions. Benefit and cost criteria are handled through standard normalisation rules prior to distance computation.

By parameterising the Z-reduction, weight defuzzification, and TOPSIS configuration stages, the framework enables structured sensitivity analysis with respect to modelling assumptions that may influence weight magnitudes and final ranking outcomes in fuzzy MCDM models.

3.4 Layered Computational Architecture

Z-number fuzzy AHP–TOPSIS consists of mathematically distinct stages: (i) linguistic representation, (ii) restrictionreliability integration, (iii) fuzzy aggregation and normalisation, and (iv) alternative ranking. In most existing implementations, these stages are embedded within a single procedural workflow, making modelling assumptions implicit and difficult to isolate.

The proposed framework formalises this methodological decomposition into computationally independent layers. Each layer corresponds to a well-defined mathematical operator in the Z-AHP–TOPSIS pipeline and exposes its assumptions as explicit parameters.

The architecture consists of:

- **Representation layer:** Definition and validation of restriction and reliability scales as triangular fuzzy numbers, ensuring admissibility and reciprocity.
- **Transformation layer:** Integration of Z-number components through a selectable reduction operator.
- **Aggregation layer:** Fuzzy AHP weight derivation via geometric mean computation and normalisation.
- **Ranking layer:** Fuzzy TOPSIS computation of ideal and anti-ideal distances and closeness coefficients.
- **Execution layer:** Parameter-driven orchestration of all modelling choices in a reproducible, script-controlled environment.

This separation localises modelling assumptions to their corresponding mathematical operators. Reliability defuzzification, fuzzy aggregation, and ranking procedures are implemented as independent layers, ensuring that adjustments in one stage do not alter the formal structure of the others. As a result, the computational pipeline mirrors the theoretical decomposition of Z-AHP–TOPSIS and makes each modelling decision explicit and auditable.

Figure 1 illustrates the interaction between these layers and highlights the explicit declaration of modelling choices at each stage of the pipeline.

3.5 Execution Model and Structural Validation

The framework adopts a declarative execution model in which modelling assumptions are specified explicitly prior to computation. Restriction and reliability scales, pairwise comparison matrices, defuzzification operators, normalisation strategies, and distance metrics are declared as workflow parameters rather than embedded within implicit procedural routines.

This design makes the analysis deterministic under identical inputs and parameter settings, enabling version-controlled storage, transparent auditing of modelling assumptions, and replication of the complete decision pipeline.

Prior to numerical computation, the framework performs structural validation of all declared fuzzy elements and comparison matrices. These checks verify that structural assumptions required by the Z-number fuzzy AHP–TOPSIS computational pipeline are satisfied before weight derivation and TOPSIS-based ranking are executed.

Triangular Admissibility. Each triangular fuzzy number must satisfy $l \leq m \leq u$; violations trigger explicit execution errors. Pairwise comparison matrices are also checked for fuzzy reciprocity, and consistency diagnostics are computed using the classical Saaty [30] consistency ratio on the defuzzified operational comparison matrix.

Reciprocal Consistency. Pairwise comparison matrices are verified to satisfy fuzzy reciprocity

$$\tilde{a}_{ji} = \tilde{a}_{ij}^{-1}$$

If reciprocal symmetry is violated, computation is stopped in strict mode and a diagnostic message identifies the inconsistent entries.

Reciprocity is imposed on the restriction component of the pairwise comparison matrix. Reliability terms are interpreted as confidence modifiers attached to the underlying judgement. Consistent with the Z-number conversion approach of Otay and Kahraman [15], the same reliability specification is applied to reciprocal entries. This ensures that after Z-reduction,

$$\tilde{A}'_{ji} = (\tilde{A}'_{ij})^{-1}$$

preserving the structural requirements of fuzzy AHP.

In addition, consistency diagnostics are computed by evaluating the classical Saaty consistency ratio (CR) [30] on the defuzzified operational comparison matrix. This provides a standard coherence check consistent with common practice in fuzzy AHP applications.

4 Reproducibility Demonstration

This section illustrates the application of the proposed R framework to reproduce the Z-number fuzzy AHP–TOPSIS analysis reported by Çekmiş et al. [6]. The

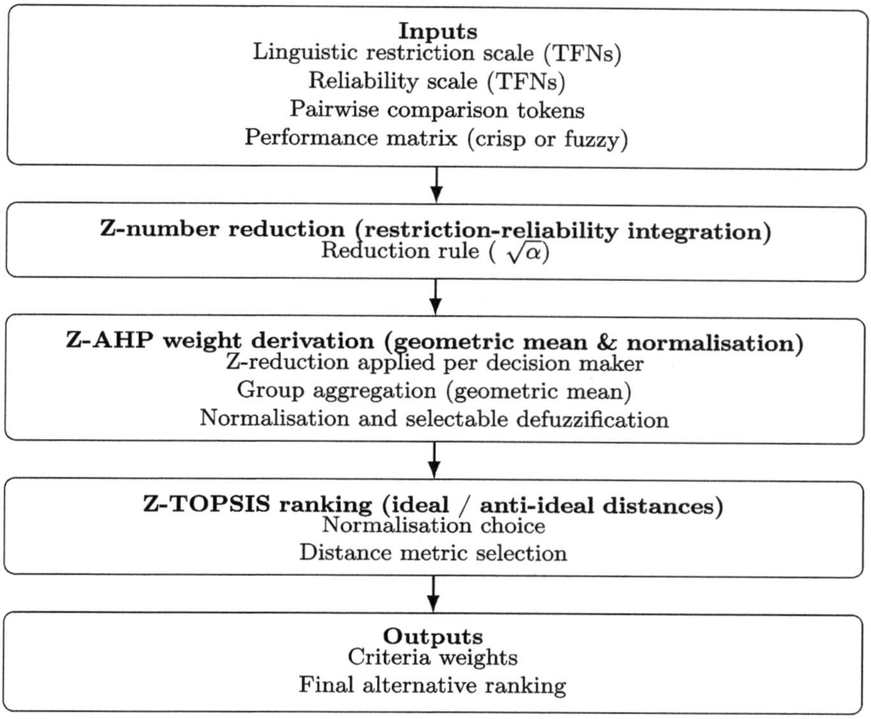

Fig. 1. Workflow of the proposed Z-number fuzzy AHP–TOPSIS framework. Key modelling choices (e.g., reduction and defuzzification strategies) are declared explicitly as parameters.

objective is to demonstrate that their published decision model can be reconstructed within a fully script-driven environment, where linguistic scales, modelling assumptions, and computational operators are declared explicitly rather than embedded in spreadsheet formulas.

4.1 Replication Configuration

To reproduce the results reported in [6], the computational workflow was executed using the same modelling assumptions described in the original study. In particular,

- Reliability coefficients were obtained using the weighted triangular mean $\alpha = (l' + 2m' + u')/4$;
- Z-number restriction components were reduced using the square-root reliability scaling rule;
- Criteria weights were defuzzified using the weighted triangular mean $(l+2m+u)/4$;

– TOPSIS ranking used ratio normalisation and Euclidean distance.

All linguistic scales, decision-maker pairwise comparison matrices, and performance criteria values are taken directly from the published study. Using these inputs, the Z-number fuzzy AHP–TOPSIS computational procedure is reproduced within the proposed R framework.

4.2 End-to-End Workflow Implementation

The reproduction follows a sequence of computational stages reflecting the mathematical components introduced in Sect. 3 and employed in Çekmiş et al. [6]. Each stage is implemented through declared inputs and operator selections, allowing the complete decision pipeline to be executed deterministically under identical configurations.

Prepare Input Files (Decision Makers and Alternatives). Decision makers provide pairwise comparisons using paired linguistic tokens consisting of a restriction term and a reliability term, e.g., `"MB, HR"` or `"1/MB, HR"`. Each decision maker contributes a $k \times k$ matrix stored as a CSV file. Alternative evaluations for TOPSIS are provided in a performance table with one row per alternative and one column per criterion. The current prototype supports either crisp or triangular fuzzy performance values.

Define Restriction and Reliability Scales. The first reproducibility requirement is making the linguistic scales explicit. Instead of relying on an implicit Excel sheet, the restriction and reliability vocabularies are explicitly declared as structured input tables within the workflow, either directly in code or through accompanying input files.

```
# Restriction (linguistic) scale: user-defined terms -> TFNs
ling_df <- data.frame(
  term = c("AM","GM","MM","WM","E","WL","ML","GL","AL"),
  tri  = c("(7,9,9)",
           "(5,7,9)",
           "(3,5,7)",
           "(1,3,5)",
           "(1,1,1)",
           "(1/5,1/3,1)",
           "(1/7,1/5,1/3)",
           "(1/9,1/7,1/5)",
           "(1/9,1/9,1/7)")
)

# Reliability scale: user-defined terms -> TFNs
rel_df <- data.frame(
```

```
  term = c("AR","SR","VHR","HR","FR","WR","VWR","SU","AU"),
  tri  = c("(1,1,1)",
           "(0.7,0.8,0.9)",
           "(0.6,0.7,0.8)",
           "(0.5,0.6,0.7)",
           "(0.4,0.5,0.6)",
           "(0.3,0.4,0.5)",
           "(0.2,0.3,0.4)",
           "(0.1,0.2,0.3)",
           "(0,0.1,0.2)")
)
```

The linguistic vocabularies correspond exactly to the restriction and reliability scales reported in Table II of [6].

Read DM Pairwise Matrices and Aggregate Judgements. Consistent with the group decision-making procedure adopted in [6], the framework reads each decision maker's pairwise comparison matrix and constructs a collective matrix through fuzzy aggregation.

Token entries are first mapped to their corresponding triangular fuzzy numbers and reduced to operational restriction TFNs $\tilde{A}'_{ij}$. Reliability defuzzification follows the weighted triangular mean formulation $\alpha_{ij} = (l'_{ij} + 2m'_{ij} + u'_{ij})/4$ used in [6]. The resulting operational fuzzy numbers are then aggregated across decision makers using the fuzzy geometric mean proposed by Buckley [3].

```
library(readr)

dm_files <- c("DM1_pairwise.csv","DM2_pairwise.csv","DM3_pairwise.csv")

dm_tokens <- lapply(dm_files, function(f) {
  as.matrix(read_csv(f, col_names = FALSE))
})

# Combine DMs:
# - map tokens to TFNs
# - apply Z-reduction (reliability defuzzification selected below)
# - aggregate corresponding cells across DMs (geometric mean)
tokens_group <- aggregate_dm_tokens(
  dm_tokens,
  ling_df = ling_df,
  rel_df  = rel_df,
  reliability_defuzz = "weighted_mean"
)
```

The resulting object `tokens_group` represents the aggregated group comparison matrix expressed in operational triangular fuzzy numbers, which serves as the input for the fuzzy AHP weight derivation stage.

Compute Fuzzy AHP Weights (Scripted and Reproducible). Given the aggregated operational triangular fuzzy comparison matrix, criterion weights are computed using fuzzy geometric mean synthesis, normalisation, and defuzzification. For consistency with [6], weight defuzzification is set to the weighted triangular mean $(l + 2m + u)/4$.

```
w <- compute_fuzzy_weights_from_inputs(
  tokens_mat      = tokens_group,
  row_labels      = c("C1","C2","C3","C4","C5"),
  col_labels      = c("C1","C2","C3","C4","C5"),
  weight_defuzz   = "weighted_mean"
)

# w is a table with fuzzy weights (l,m,u) + crisp weights + ranks
print(w)
```

The consistency ratio (CR), computed on the defuzzified operational comparison matrix, satisfied the acceptable threshold (CR < 0.10), indicating coherent pairwise judgements.

Run TOPSIS Using the Derived Weights. The AHP-derived weights are then passed directly to the TOPSIS procedure. This avoids manual transfer of intermediate results and permits deterministic reproduction of the ranking under identical inputs and parameter settings.

```
perf <- read_csv("Performance_criteria.csv")

topsis_res <- ztopsis_rank(
  performance = perf,              # alternatives x criteria
  weights     = w$weight,          # AHP-derived crisp weights
  criteria    = c("benefit","benefit","cost","benefit","cost"),
  mode        = "crisp",           # or "fuzzy" if TFN performance is provided
  normalise   = "ratio",           # configurable TOPSIS option
  distance    = "euclidean"        # configurable distance option
)

print(topsis_res)
```

The object `topsis_res` reports (i) closeness coefficients and (ii) the final ranking of alternatives. Because the workflow is executed through explicitly declared inputs and multiple operator selections, identical configurations produce identical ranking outcomes.

4.3 Replication Results

The criteria weights and TOPSIS rankings obtained from the R implementation were compared with those reported in [6]. The results show close agreement with the published study, with only minor numerical differences in intermediate values

Table 2. Comparison of TOPSIS closeness coefficients and rankings

Alternative	CC (Published)	CC (R)	Rank (Published)	Rank (R)
A1	0.73	0.74	1	1
A2	0.68	0.69	2	2
A3	0.45	0.45	6	6
A4	0.36	0.35	7	7
A5	0.49	0.52	5	5
A6	0.61	0.59	3	3
A7	0.36	0.37	8	8
A8	0.26	0.28	11	10
A9	0.28	0.27	10	11
A10	0.59	0.58	4	4
A11	0.28	0.30	9	9

and closeness coefficients, attributable to rounding conventions in spreadsheet-based calculations and floating-point precision in R.

Table 2 compares the published and reproduced closeness coefficients and final rankings. The highest-ranked alternatives are reproduced exactly, with A1, A2, and A6 remaining the top three options in both implementations. A minor discrepancy appears only among the lowest-ranked alternatives, where A8 and A9 exchange positions due to very small differences in closeness coefficients.

Overall, the findings indicate that the proposed R implementation successfully reproduces the published ranking structure within a fully script-based and reproducible computational environment.

Reproducible Research Artefacts. All data files (`DM1_pairwise.csv`, `DM2_pairwise.csv`, `DM3_pairwise.csv`, `Performance_criteria.csv`), the linguistic scale definitions, and the analysis script are stored together in a version-controlled repository. This enables independent reproduction of the complete computational pipeline under identical input data and parameter configurations. Sensitivity to modelling assumptions can be examined by modifying parameter selections within the same executable workflow.

The implementation and replication materials are publicly available in the project repository: https://github.com/FuzzyDecisionLab/Z-AHP-TOPSIS. Executing the provided scripts with the accompanying data files under the specified modelling configuration reproduces the alternative ranking reported in [6]. Small numerical differences in intermediate criteria weights may occur due to floating-point precision and rounding differences between spreadsheet-based calculations and R; however, the overall ranking structure is preserved, and the highest-ranked alternatives remain identical.

5 Discussion

The growing body of research on advanced fuzzy MCDM models - including Z-number, intuitionistic, and spherical extensions - has introduced increasingly expressive representations of uncertainty and reliability. However, computational standardisation has evolved more slowly. In practice, Z-number reduction strategies, defuzzification operators, and aggregation procedures are frequently implemented in spreadsheets or isolated scripts, with limited transparency regarding modelling assumptions. This introduces a methodological gap between theoretical development and reproducible computational practice.

The framework presented in this paper addresses this gap by formalising the computational structure of Z-number fuzzy AHP–TOPSIS within a parameter-driven and layered environment. Modelling choices are declared explicitly as configuration parameters rather than embedded implicitly within procedural code or spreadsheets. This is methodologically important, as modelling choices such as reliability defuzzification and crisp weight extraction can influence derived weights and ranking stability. By making these assumptions explicit within the computational pipeline, the framework improves traceability of results and preserves computational coherence.

This is particularly valuable in fuzzy decision-making studies where modelling choices such as scale definitions, reduction operators, or defuzzification rules are not always reported explicitly in published applications. Parameterised execution also supports rapid recalculation under alternative assumptions, facilitating sensitivity analysis and comparative experimentation across Z-number formulations.

Relative to existing R packages, the main contribution is the formalisation of an end-to-end, parameterised and reproducible computational pipeline for Z-AHP–TOPSIS in R, enabling transparent sensitivity analysis and reuse. While current tools provide support for crisp MCDM or isolated fuzzy procedures, they rarely combine Z-number transformation, group aggregation, and ranking in a single extensible workflow. The layered structure adopted here separates scale definition, Z-reduction, fuzzy aggregation, and ranking into interoperable components, enabling future integration of alternative fuzzy paradigms without altering the overall decision architecture.

Several limitations should be acknowledged. The current prototype focuses on triangular Z-number fuzzy AHP with an integrated TOPSIS layer, and does not yet support alternative fuzzy representations such as intuitionistic or spherical extensions. In addition, although multiple reduction and defuzzification options are implemented, systematic benchmarking across alternative modelling configurations has not been performed. Validation is currently limited to the replication of a single published case study, although the framework is designed to support multiple datasets and alternative decision contexts. Future work will include broader empirical evaluation across diverse datasets, comparative analysis of modelling choices, and extension of the framework to additional fuzzy paradigms.

Overall, the results indicate that reliability-aware fuzzy AHP–TOPSIS models can be implemented within a reproducible and extensible R framework without altering their underlying methodological structure.

6 Conclusion

Advanced fuzzy MCDM models, particularly those based on Z-numbers, offer richer representations of uncertainty by modelling both restriction and reliability. Despite their theoretical development, their computational application has remained fragmented, frequently relying on ad hoc scripts or spreadsheet implementations that may reduce reproducibility and obscure explicit documentation of modelling assumptions.

This paper presented a modular R-based framework for Z-number fuzzy AHP–TOPSIS analysis that formalises the associated computational pipeline. By separating scale definition, Z-reduction, aggregation, and ranking into explicitly structured components, the framework makes modelling assumptions visible within an executable workflow. The reproducibility demonstration showed that published Z-AHP–TOPSIS applications can be replicated within this structured environment without altering their underlying methodological formulations.

The current prototype focuses on triangular Z-number fuzzy AHP integrated with TOPSIS ranking. Future work includes broader benchmarking, expanded paradigm support, and full CRAN-compliant release of the package.

Overall, the framework establishes a reproducible computational foundation for reliability-aware fuzzy MCDM models and contributes toward improving methodological transparency and comparability in advanced decision analysis research.

Use of Generative AI. A large language model was used to assist with language polishing. The authors are solely responsible for the scientific content and results.

Disclosure of Interests. The authors have no competing interests to declare that are relevant to the content of this article.

References

1. Barbara, F.: waspasR: Tool Kit to Implement a W.A.S.P.A.S. Based Multi-Criteria Decision Analysis Solution (2023). https://CRAN.R-project.org/package=waspasR. r package version 0.1.5
2. Bose, A.: AHPtools: Tools for Analytic Hierarchy Process (AHP) Modeling in R (2025). https://doi.org/10.32614/CRAN.package.AHPtools. https://CRAN.R-project.org/package=AHPtools. r package version 1.0.1
3. Buckley, J.J.: Fuzzy hierarchical analysis. Fuzzy Sets Syst. **17**(3), 233–247 (1985)
4. Burda, M., Štěpnička, M.: LFL: an r package for linguistic fuzzy logic. Fuzzy Sets Syst. **431**, 1–38 (2022)
5. Caha, J., Drážná, A.: Information about FuzzyAHP package for R (ver.0.9.5) (2019). http://github.com/JanCaha/FuzzyAHP/. r package version 0.9.5

6. Cekmis, A., Intepe, G., Ucal Sari, I., Ozer, S.: A fuzzy z-ahp z-topsis model to rank architectural adaptation of bank branches to new technologies and changing customer profiles. In: Engineering, Construction and Architectural Management, pp. 1–34 (2025)
7. Choudhary, D., Shankar, R.: An steep-fuzzy ahp-topsis framework for evaluation and selection of thermal power plant location: a case study from india. Energy **42**(1), 510–521 (2012)
8. Coutinho, R., dos Santos, M.: ahptopsis2n: Hybrid Method for Multiple Criteria Decision-Making (MCDM) (2021). https://CRAN.R-project.org/package=ahptopsis2n. r package version 0.2.0
9. De Souza, L.P., Gomes, C.F.S., De Barros, A.P.: Implementation of new hybrid ahp-topsis-2n method in sorting and prioritizing of an it capex project portfolio. Int. J. Inf. Technol. Decis. Mak. **17**(04), 977–1005 (2018)
10. Demircan, B.G., Yetilmezsoy, K.: A hybrid fuzzy ahp-topsis approach for implementation of smart sustainable waste management strategies. Sustainability **15**(8), 6526 (2023)
11. Dudek, A., Jefmański, B., Kusterka-Jefmańska, M., Roszkowska, E.: IFMCDM: Intuitionistic Fuzzy Multi-Criteria Decision Making Methods (2023). https://CRAN.R-project.org/package=IFMCDM. r package version 0.1.17
12. Ekmekcioğlu, Ö., Koc, K., Özger, M.: Stakeholder perceptions in flood risk assessment: a hybrid fuzzy ahp-topsis approach for Istanbul, Turkey. Int. J. Disast. Risk Reduct. **60**, 102327 (2021)
13. Gagolewski, M., Caha, J.: FuzzyNumbers Package: Tools to Deal with Fuzzy Numbers in R (2021). https://github.com/gagolews/FuzzyNumbers/
14. Jefmański, B., Roszkowska, E., Kusterka-Jefmańska, M.: Intuitionistic fuzzy synthetic measure on the basis of survey responses and aggregated ordinal data. Entropy **23**(12), 1636 (2021)
15. Kahraman, C., Otay, I.: Solar pv power plant location selection using a z-fuzzy number based ahp. Int. J. Anal. Hierarchy Process **10**(3) (2018)
16. Klir, G., Yuan, B.: Fuzzy Sets and Fuzzy Logic, vol. 4. Prentice hall, New Jersey (1995)
17. Krejčí, J., Pavlačka, O., Talašová, J.: A fuzzy extension of analytic hierarchy process based on the constrained fuzzy arithmetic. Fuzzy Optim. Decis. Making **16**(1), 89–110 (2017)
18. Liu, Y., Eckert, C.M., Earl, C.: A review of fuzzy ahp methods for decision-making with subjective judgements. Expert Syst. Appl. **161**, 113738 (2020)
19. Martin, B.A.C.: FuzzyMCDM: Multi-Criteria Decision Making Methods for Fuzzy Data (2016), https://cran.r-project.org/src/contrib/Archive/FuzzyMCDM/. r package version 1.1, archived from CRAN on 2023-08-18
20. Martin, B.A.C.: MCDM: Multi-Criteria Decision Making Methods for Crisp Data (2016). https://cran.r-project.org/src/contrib/Archive/MCDM/. r package version 1.2, archived from CRAN on 2022-04-05
21. Martins dos Santos, W.: TopSisWM: Multi-Criteria Method for Decision (TOPSIS) (2025). https://CRAN.R-project.org/package=TopSisWM. r package version 1.0.3
22. Meyer, D., Hornik, K.: Generalized and customizable sets in R. J. Stat. Softw. **31**(2), 1–27 (2009). https://doi.org/10.18637/jss.v031.i02
23. Meyer, D., Hornik, K., Buchta, C.: Sets: Sets, Generalized Sets, Customizable Sets and Intervals (2023), https://CRAN.R-project.org/package=sets. r package version 1.0-25
24. Najafi, A.: RMCDA: Multi-Criteria Decision Analysis (2025). https://CRAN.R-project.org/package=RMCDA. r package version 0.3

25. Önüt, S., Kara, S.S., Işik, E.: Long term supplier selection using a combined fuzzy MCDM approach: a case study for a telecommunication company. Expert Syst. Appl. **36**(2), 3887–3895 (2009)
26. Panda, R.R., Nagwani, N.K.: IPSFS: intuitionistic, pythagorean, and spherical fuzzy similarity computation package in r. Softw. Impacts **14**, 100424 (2022)
27. Panda, R.R., Nagwani, N.K.: ipsfs: Intuitionistic, Pythagorean, and Spherical Fuzzy Similarity Measure (2022), https://CRAN.R-project.org/package=ipsfs, r package version 1.0.0
28. Pandey, V., Komal, D.H.: A review on topsis method and its extensions for different applications with recent development. Soft. Comput. **27**(23), 18011–18039 (2023)
29. Roszkowska, E., Kusterka-Jefmańska, M., Jefmański, B.: Intuitionistic fuzzy topsis as a method for assessing socioeconomic phenomena on the basis of survey data. Entropy **23**(5), 563 (2021)
30. Saaty, T.L.: The analytic hierarchy process (1980)
31. Sari, I.U., Tüysüz, N.: Covid-19 risk assessment of occupations using interval type 2 fuzzy z-ahp & topsis methodology. J. Multiple-Valued Logic Soft Comput. **38** (2022)
32. Sequeira, M., Adlemo, A., Hilletofth, P.: A hybrid fuzzy-ahp-topsis model for evaluation of manufacturing relocation decisions. Oper. Manag. Res. **16**(1), 164–191 (2023)
33. Sergi, D., Ucal Sari, I.: Prioritization of public services for digitalization using fuzzy z-ahp and fuzzy z-waspas. Complex Intell. Syst. **7**(2), 841–856 (2021)
34. Tüysüz, N., Kahraman, C.: Codas method using z-fuzzy numbers. J. Intell. Fuzzy Syst. **38**(2), 1649–1662 (2020)
35. Tüysüz, N., Kahraman, C.: An integrated picture fuzzy z-ahp & topsis methodology: application to solar panel selection. Appl. Soft Comput. **149**, 110951 (2023)
36. Yazdi, M.M.: topsis: TOPSIS method for multiple-criteria decision making (MCDM) (2013). https://CRAN.R-project.org/package=topsis. r package version 1.0
37. Zadeh, L.A.: Generalized theory of uncertainty (gtu)–principal concepts and ideas. Comput. Stat. Data Anal. **51**(1), 15–46 (2006)
38. Zadeh, L.A.: A note on z-numbers. Inf. Sci. **181**(14), 2923–2932 (2011)

Dual-Scale Adaptive Fuzzy Rough Sets for Feature Selection

Shimin Yang, Jin Qian(✉), Shaowei Yan, and Dongsheng Zhi

East China Jiaotong University, Nanchang, China
{2024218085405026,qianjin,2023218083500021,2024218085405001}@ecjtu.edu.cn

Abstract. Traditional fuzzy rough sets usually employ distance metrics for feature selection and classification tasks. However, this approach ignores the relationship between samples and decision classes, especially in classification boundary regions. Consequently, traditional fuzzy approximations may not accurately capture the actual classification status thereby affecting feature selection performance. To address the issue, this paper proposes a feature selection method based on dual-scale fuzzy rough sets. First, an overlap degree function is used to effectively filter less important features. Next, a dual-scale fuzzy rough set model is constructed, and sample membership degrees are dynamically adjusted via a fuzzy adaptive neighborhood radius to accommodate different data distributions. The model further redefines the fuzzy approximation operator by considering the degree of separation between a sample and both the target and non-target classes, thus more accurately reflecting the sample's classification status. Finally, a heuristic backward redundant-feature filtering algorithm is designed to handle both low-dimensional and high-dimensional data. Comparative experiments on real-world datasets show that the proposed method effectively removes redundant features and improves the performance of downstream classification tasks.

Keywords: Fuzzy rough set · Overlap degree · Dual-scale fuzzy rough set model · Feature selection

1 Introduction

In today's era of big data, large-scale high-dimensional data are ubiquitous in fields such as text mining, image recognition, and financial analysis. Such data not only increase computational costs but may also exacerbate the curse of dimensionality. In this context, feature selection [1] has become an essential data preprocessing technique. It improves the efficiency and predictive accuracy of learning algorithms, enhances model interpretability, and reduces the risk of overfitting.

Rough set theory [2] relies on lower and upper approximations to handle uncertain and imprecise data without prior information [3–5]. However, it is sensitive to noise and missing data, and continuous attributes typically require

J. Lu et al. (Eds.): FLINS-ISKE 2026, LNCS 16752, pp. 416–432, 2027.
https://doi.org/10.1007/978-981-92-2369-5_26

discretization, which may cause information loss. To address these limitations, various extensions have been proposed, including neighborhood rough sets [6–9], ordinal rough sets [10–13], fuzzy neighborhood rough sets [14,15], and relational information entropy models [16,17]. These models extend rough set theory to better accommodate complex and fuzzy data.

Nowadays, fuzzy rough sets are used to handle uncertain and fuzzy data by combining rough set theory and fuzzy set theory [18]. In recent years, researchers have proposed many extended models based on fuzzy rough sets [19–22]. For instance, Jensen and Shen extended the dependency functions in the traditional rough set model to the fuzzy environment and proposed for the first time a feature reduction method based on fuzzy rough sets [23]. Dai et al. [24] suggested applying fuzzy implicit granularity information to fuzzy rough sets. Wang et al. [25] redefined the fuzzy binary relation and designed a directed fuzzy rough set model. Sang et al. [26] proposed a robust fuzzy dominant rough set model, which combined category separability and attribute correlation metrics to verify its robustness and classification performance in noisy environments. Zhao et al. [27] constructed a variable-precision fuzzy rough set model for handling interference noise. Wang et al. [28] extended rough entropy to the theory of fuzzy rough sets. Yuan et al. [29] proposed a generalized fuzzy rough set without sample labels for unsupervised attribute reduction.

Traditional fuzzy rough set methods mainly measure sample similarity via distance, which may overlook the intrinsic relationship between samples and decision classes. When class boundaries are vague or classes overlap heavily, this can lead to discrimination errors and higher misclassification risk. To address this, we propose dual-scale fuzzy approximation operators that jointly consider a sample's relevance to the target class and its distance to non-target classes. Building on this, we develop a feature selection algorithm that integrates overlap degree and dual-scale fuzzy rough sets. The overlap degree evaluates the importance of individual features and removes redundant ones, while the dual-scale fuzzy rough set model captures inter-feature relationships to select a key subset. Moreover, a fuzzy adaptive neighborhood radius is introduced to dynamically adjust the similarity range, further improving selection performance and flexibility across different data distributions.

The remainder of this paper is organized as follows. Section 2 reviews the basic concepts of fuzzy rough sets. Section 3 presents the overlap degree and the proposed dual-scale adaptive fuzzy approximation operators, and develops a feature selection algorithm based on dual-scale fuzzy rough sets. Section 4 reports the experimental results and analysis. Finally, Sect. 5 summarizes the content of this article.

2 Preliminaries

This section briefly reviews the relevant concepts of the classical fuzzy rough set theory.

Definition 1. *Given a non-empty domain U and the set of conditional attributes C, let $B \subseteq C$. A fuzzy similarity relation induced by B, denoted by R_B, satisfies: (1) reflexivity: $R_B(x,x) = 1$ for all $x \in U$, (2) symmetry: $R_B(x,y) = R_B(y,x)$ for all $x, y \in U$.*

A fuzzy similarity relation is used to characterize the degree of similarity between elements in fuzzy set theory. It allows the similarity value to range from 0 to 1, reflecting the fuzziness and uncertainty between elements.

Definition 2. *Given a sample set $U = \{x_1, x_2, \ldots, x_n\}$, an attribute subset B, and the fuzzy similarity relation R_B induced by B, the fuzzy similarity matrix is defined as*

$$FSM_{R_B} = \begin{pmatrix} r_{B_{11}} & r_{B_{12}} & \cdots & r_{B_{1n}} \\ r_{B_{21}} & r_{B_{22}} & \cdots & r_{B_{2n}} \\ \vdots & \vdots & \ddots & \vdots \\ r_{B_{n1}} & r_{B_{n2}} & \cdots & r_{B_{nn}} \end{pmatrix} \tag{1}$$

where $r_{B_{ij}} = R_B(x_i, x_j)$.

Definition 3. *Let $T = \langle U, C, D \rangle$ be a decision table, where $U = \{x_1, x_2, \ldots, x_n\}$ is a finite nonempty universe and $D = \{d\}$ is the decision attribute. Assume that d takes r distinct values. For each $t \in \{1, 2, \ldots, r\}$, the crisp decision class induced by d is defined as*

$$D_t = \{x_i \in U \mid d(x_i) = t\} \tag{2}$$

Then $U/D = \{D_1, D_2, \ldots, D_r\}$ forms a crisp partition of U.

Definition 4. *Let U be a finite universe. A fuzzy set $\tilde{M}$ on U is a mapping $\tilde{M} : U \to [0, 1]$, where $\tilde{M}(x)$ denotes the membership degree of x in $\tilde{M}$. All fuzzy sets defined on U constitute the set $F(U)$. A collection of fuzzy sets $\{\tilde{D}_1, \tilde{D}_2, \ldots, \tilde{D}_r\}$ is called a fuzzy partition of U if, for every $x \in U$,*

$$\sum_{j=1}^{r} \tilde{D}_j(x) = 1 \tag{3}$$

Given a crisp decision partition $U/D = \{D_1, \ldots, D_r\}$ and a feature subset $B \subseteq C$, the fuzzy decision of a sample $x \in U$ with respect to class D_j is defined by

$$\tilde{D}_j(x) = \frac{|[x]_B \cap D_j|}{|[x]_B|}, \qquad j = 1, 2, \ldots, r \tag{4}$$

where $[x]_B$ denotes the neighborhood of x induced by B. Under this definition, $\{\tilde{D}_1, \ldots, \tilde{D}_r\}$ constitutes a fuzzy partition of U.

Definition 5. *Let R_B be a fuzzy similarity relation defined on U and let $H \in F(U)$ be a fuzzy set. The lower and upper approximations of H under the relation R_B are defined as*

$$\underline{R}_B H(x) = \inf_{y \in U} \max\{1 - R_B(x, y),\ H(y)\} \tag{5}$$

$$\overline{R}_B H(x) = \sup_{y \in U} \min\{R_B(x, y),\ H(y)\} \tag{6}$$

where $\underline{R}_B H(x)$ represents the certainty that element x belongs to H, and $\overline{R}_B H(x)$ represents the possibility that element x belongs to H.

The fuzzy positive region of the fuzzy decision variable D with respect to the feature subset B and the corresponding fuzzy dependency degree function are defined as follows

$$POS_B(D)(x) = \sum_{D_j \in U/D} \underline{R}_B H(x) \tag{7}$$

$$\gamma_B(D) = \frac{1}{|U|} \sum_{x \in U} POS_B(D)(x) \tag{8}$$

The dependency degree function $\gamma_B(D)$ indicates the importance of the feature subset B.

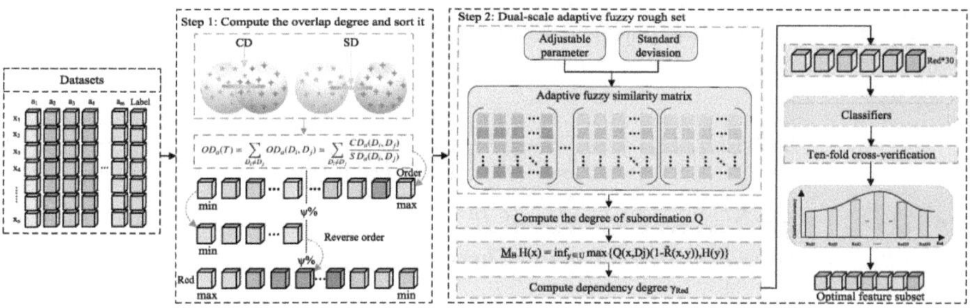

Fig. 1. The framework of the proposed feature selection algorithm.

3 Methodology

In this section, we propose a feature selection algorithm whose overall framework consists of two main stages, as shown in Fig. 1. First, features are ranked by the overlap degree and preliminarily screened. Then, on the resulting candidate subset, the proposed dual-scale adaptive fuzzy approximation operators are introduced and combined with a backward redundancy elimination strategy to further remove redundant features, yielding a better feature subset.

3.1 Feature Ranking Based on Overlap Degree

A ranking and filtering scheme is adopted to select informative features. Features are scored and ranked, and the top ψ features are retained as a candidate subset with low computational cost. Feature importance is measured by the overlap degree, which integrates the coincidence degree and the separation degree. A smaller coincidence degree and a larger separation degree indicate greater importance. Prior to the calculation, each continuous attribute is normalized using min–max scaling to ensure comparability and eliminate scale effects.

Definition 6. *Let $T = \langle U, C, D\rangle$ be a decision table, $a \in C$, and $U/D = \{D_1, \ldots, D_r\}$. For any two distinct decision classes D_i and D_j, the coincidence degree of attribute a is defined as*

$$CD_a(D_i, D_j) = \frac{\left|[m^a_{D_i}, M^a_{D_i}] \cap [m^a_{D_j}, M^a_{D_j}]\right|}{\left|[m^a_{D_i}, M^a_{D_i}] \cup [m^a_{D_j}, M^a_{D_j}]\right| + \xi} \tag{9}$$

Here $m^a_{D_i} = \min_{x\in D_i} f(x,a)$ and $M^a_{D_i} = \max_{x\in D_i} f(x,a)$, and $\xi > 0$ avoids division by zero. The global coincidence degree is

$$CD_a(T) = \sum_{D_i \neq D_j} CD_a(D_i, D_j) \tag{10}$$

Definition 7. *Under the same decision table T, the separation degree of attribute a between two distinct classes D_i and D_j is defined as*

$$SD_a(D_i, D_j) = \frac{|\bar{f}_a(D_i) - \bar{f}_a(D_j)|}{f^a_{\max}(D_i, D_j) - f^a_{\min}(D_i, D_j) + \xi} \tag{11}$$

Here $\bar{f}_a(D_i) = \frac{1}{|D_i|}\sum_{x\in D_i} f(x,a)$, and $f^a_{\max}$ and $f^a_{\min}$ denote the maximum and minimum over $D_i \cup D_j$, respectively. The global separation degree is

$$SD_a(T) = \sum_{D_i \neq D_j} SD_a(D_i, D_j) \tag{12}$$

Definition 8. *Based on the coincidence and separation degrees, the overlap degree of attribute a between two distinct classes D_i and D_j is defined as*

$$OD_a(D_i, D_j) = \frac{CD_a(D_i, D_j)}{SD_a(D_i, D_j) + \xi} \tag{13}$$

The global overlap degree of attribute a on T is defined as

$$OD_a(T) = \sum_{D_i \neq D_j} OD_a(D_i, D_j) \tag{14}$$

A smaller $OD_a(T)$ indicates stronger discriminative capability of attribute a.

3.2 Dual-Scale Adaptive Fuzzy Approximation Operators

Ranking and filtering methods are computationally efficient, but they assess features independently and neglect inter-feature relationships, often yielding redundant subsets. Fuzzy approximation operators derived from fuzzy similarity relations can model the dependency between a feature subset and the decision attribute, and the dependency degree reflects the subsetN's classification capability. However, conventional operators typically emphasize the relationship between a sample and non-target classes while overlooking its association with the target class. In addition, most studies adopt a fixed neighborhood radius, which cannot adequately adapt to varying data distributions across feature granularities. To address these issues, this paper proposes a pair of dual-scale adaptive fuzzy approximation operators. Furthermore, a feature selection algorithm is developed, as presented in Algorithm 1.

Definition 9. *Given a decision table* $T = \langle U, C, D\rangle$ *and* $U/D = \{D_1, D_2, \ldots, D_r\}$*, for any two samples* $x_i, x_j \in U$*, the standardized Euclidean distance is defined as*

$$\widehat{\text{dist}}(x_i, x_j) = \frac{\text{dist}(x_i, x_j)}{\sqrt{m}} = \frac{\sqrt{\sum_{k=1}^{m}(x_{ik} - x_{jk})^2}}{\sqrt{m}} \tag{15}$$

where m *is the number of numerical features, and* x_{ik} *and* x_{jk} *are the values of samples* x_i *and* x_j *on the* k*-th feature, respectively.*

Definition 10. *For each* $x_i \in U$*, the shortest distance from* x_i *to class* $D_j \in U/D$ *is defined as*

$$P(x_i, D_j) = \min\left\{\widehat{\text{dist}}(x_i, x) \;\middle|\; x \in D_j,\ x \neq x_i\right\} \tag{16}$$

Based on $P(x_i, D_j)$*, we define*

$$Q(x_i, D_j) = 1 - P(x_i, D_j) \tag{17}$$

where $Q(x_i, D_j)$ *represents the degree of subordination between sample* x_i *and class* D_j*.*

Remark. *The mapping in (17) transforms* $P(x_i, D_j)$ *to* $(0, 1]$ *under the assumption that* $P(x_i, D_j) \in [0, 1]$*. A smaller distance yields a larger* $Q(x_i, D_j)$*. In particular,* $P(x_i, D_j) = 0$ *implies* $Q(x_i, D_j) = 1$*.*

Definition 11. *Let* $x_i, x_j \in U$ *and let* $f(x_i, a)$ *denote the value of sample* x_i *on attribute* a*. The adaptive fuzzy similarity induced by attribute* a *is defined as*

$$\tilde{R}_a(x_i, x_j) = \begin{cases} 1 - |f(x_i, a) - f(x_j, a)|, & \textit{if } |f(x_i, a) - f(x_j, a)| \le Adp_a \\ 0, & \textit{otherwise} \end{cases} \tag{18}$$

where $Adp_a = \dfrac{\text{std}(a)}{\varepsilon}$ *is the adaptive fuzzy neighborhood radius of attribute* a*,* $\text{std}(a)$ *is the standard deviation of attribute* a *over* U*, and* $\varepsilon > 0$ *is an adjustable parameter controlling the tolerance range.*

After computing $\tilde{R}_a(x_i, x_j)$ for each attribute $a \in B$, the induced fuzzy relation matrix $M_{\tilde{R}_a} \in [0,1]^{n\times n}$ is defined as $M_{\tilde{R}_a} = \left(\tilde{R}_a(x_i, x_j)\right)_{n\times n}$. For an attribute subset $B \subseteq C$, the induced fuzzy relation is obtained by intersection: $\tilde{R}_B(x_i, x_j) = \bigwedge_{a\in B} \tilde{R}_a(x_i, x_j) = \min_{a\in B} \tilde{R}_a(x_i, x_j)$, and the corresponding matrix is $M_{\tilde{R}_B} = \left(\tilde{R}_B(x_i, x_j)\right)_{n\times n}$.

Definition 12. *Consider a decision table $\langle U, C, D\rangle$, an attribute subset $B \subseteq C$, and $U/D = \{D_1, D_2, \ldots, D_r\}$. Let $D_j \in U/D$ be a decision class and let $H \in F(U)$ be a fuzzy set, i.e., $H : U \to [0,1]$. For any $x \in U$, the dual-scale adaptive fuzzy lower and upper approximations of H are defined as*

$$\underline{M}_B H(x) = \inf_{y\in U} \max\left\{Q(x, D_j)\left(1 - \tilde{R}_B(x, y)\right),\ H(y)\right\} \tag{19}$$

$$\overline{M}_B H(x) = \sup_{y\in U} \min\left\{1 - Q(x, D_j)\left(1 - \tilde{R}_B(x, y)\right),\ H(y)\right\} \tag{20}$$

When H is a crisp subset of U, the above operators can be simplified to

$$\underline{M}_B H(x) = \min_{y\notin H}\left\{Q(x, D_j)\left(1 - \tilde{R}_B(x, y)\right)\right\} \tag{21}$$

$$\overline{M}_B H(x) = \max_{y\in H}\left\{1 - Q(x, D_j)\left(1 - \tilde{R}_B(x, y)\right)\right\} \tag{22}$$

Equations (19)–(22) define the dual-scale adaptive fuzzy approximation operators (DSAFAO), which jointly incorporate the inter-sample relation $1 - \tilde{R}_B(x, y)$ and the class association $Q(x, D_j)$. When $Q(x, D_j) = 1$, DSAFAO reduces to the traditional fuzzy approximation operator.

Definition 13. *Let $T = \langle U, C, D\rangle$ be a decision table and let $B \subseteq C$ be a feature subset. For each $x \in U$, the positive degree of regionality of D with respect to B is defined as*

$$POS_B(D)(x) = \sum_{H\in U/D} \underline{M}_B H(x) \tag{23}$$

The degree of dependence of D on B is defined as

$$\gamma_B(D) = \frac{1}{|U|} \sum_{x\in U} POS_B(D)(x) \tag{24}$$

Definition 14. *Let the decision partition be $U/D = \{D_1, D_2, \ldots, D_r\}$. For a given decision equivalence class $D_i \in D$, the discriminant of a sample x in D_i, the overall discriminant of the decision class D_i, and the global discriminant of the decision class set D are respectively defined as*

$$DD(D_i)(x) = \frac{\underline{M}_B(D_i)(x)}{\sum_{D_j\in U/D} \underline{M}_B(D_j)(x)}, \quad \forall x \in D_i \tag{25}$$

$$DD(D_i) = \sum_{x\in D_i} DD(D_i)(x) \tag{26}$$

$$DD(D) = \frac{1}{n} \sum_{D_i\in U/D} DD(D_i) \tag{27}$$

Table 1. Decision table T.

U	a_1	a_2	D
x_1	0.42	0.72	1
x_2	0.25	0.70	1
x_3	0.36	0.80	1
x_4	0.37	0.56	1
x_5	0.45	0.30	2
x_6	0.61	0.54	2
x_7	0.48	0.40	2
x_8	0.60	0.29	2
x_9	0.85	0.60	3
x_{10}	0.90	0.62	3
x_{11}	0.81	0.40	3
x_{12}	0.85	0.66	3

Table 2. Classical fuzzy lower approximation of the decision.

U	$\underline{R}(D_1)$	$\underline{R}(D_2)$	$\underline{R}(D_3)$	D
x_1	0.26	0.10	0.10	1
x_2	0.38	0.15	0.15	1
x_3	0.36	0.10	0.10	1
x_4	0.19	0.17	0.17	1
x_5	0.10	0.27	0.10	2
x_6	0.19	0.24	0.19	2
x_7	0.10	0.19	0.10	2
x_8	0.15	0.24	0.15	2
x_9	0.05	0.05	0.25	3
x_{10}	0.05	0.05	0.30	3
x_{11}	0.20	0.20	0.24	3
x_{12}	0.06	0.06	0.27	3

where $n = |U|$ *is the sample size, and* $\underline{M}_B(D_i)(x)$ *denotes the degree to which sample* x *belongs to the fuzzy lower approximation of decision class* D_i *under* B*. By replacing* $\underline{M}_B(D_i)(x)$ *with a classical fuzzy approximation operator, the corresponding discrimination degree can be obtained.*

Example 1: To demonstrate the advantages of DSAFAO, the following will compare the proposed operator with the classical fuzzy approximation operator through Example 1. Table 1 shows a simple decision table. Table 2 lists the lower approximations calculated using the traditional fuzzy lower approximation operator. As can be seen from Table 1, samples x_4 and x_6 belong

Table 3. Double-scale fuzzy lower approximation of the decision.

U	$\underline{M}(D_1)$	$\underline{M}(D_2)$	$\underline{M}(D_3)$	D
x_1	**0.24**	0.08	0.07	1
x_2	**0.34**	0.11	0.09	1
x_3	**0.33**	0.07	0.06	1
x_4	**0.17**	0.15	0.11	1
x_5	0.08	**0.25**	0.07	2
x_6	0.16	**0.21**	0.16	2
x_7	0.09	**0.18**	0.08	2
x_8	0.11	**0.21**	0.12	2
x_9	0.03	0.04	**0.24**	3
x_{10}	0.03	0.04	**0.29**	3
x_{11}	0.13	0.17	**0.21**	3
x_{12}	0.04	0.05	**0.26**	3

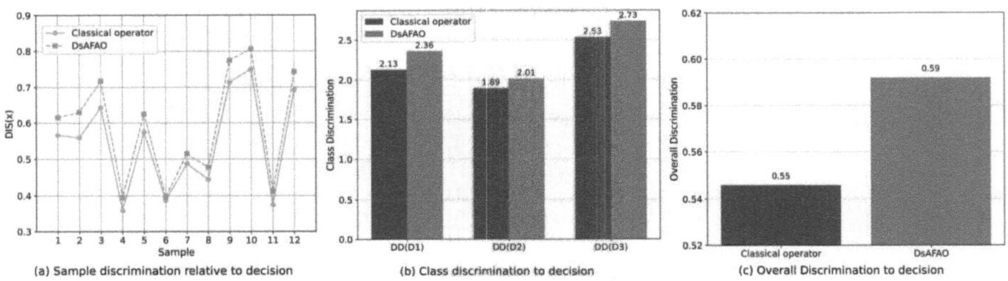

Fig. 2. Degree of discrimination.

to D_1 and D_2 respectively, and sample x_{11} belongs to D_3. But in Table 2, $\underline{R}(D_1)(x_4) = \underline{R}(D_1)(x_6) < \underline{R}(D_1)(x_{11})$. It has been indicated that the membership degree of samples x_4 and x_6 to D_1 is the same, and the membership degree of x_{11} to D_1 is higher than that of x_4. This is obviously unreasonable. $\underline{M}(D_1)$, $\underline{M}(D_2)$, and $\underline{M}(D_3)$ are lower approximations calculated using DSAFAO. Obviously, based on the operators proposed in this paper, $\underline{M}(D_1)(x_4) > \underline{M}(D_1)(x_6)$ and $\underline{M}(D_1)(x_4) > \underline{M}(D_1)(x_{11})$, as presented in Table 3.

To further illustrate the advantages of the dual-scale fuzzy approximation operator over the classical fuzzy approximation operator in terms of sample distinguishability, further experiments will be conducted based on Definition 14 next. Figure 2(a)–(c) compared the discrimination capabilities of the classical fuzzy approximation operators and the dual-scale fuzzy approximation operators from the perspectives of samples, categories, and the whole. The results showed that the dual-scale operator outperformed the classical operator in both the single-sample discrimination curve and the identification ability of each category. This indicates that its discrimination effect is more reasonable and its classification of sample decision categories is more effective. Therefore, it can provide a more reliable basis for feature selection and help obtain a better feature subset.

Algorithm 1. Dual-scale adaptive fuzzy rough sets for feature selection (DsAFRS)

Require: Decision information system $T = \langle U, C \cup D \rangle$, a condition feature $a \in C$, feature retention rate $\psi\%$.
Ensure: Feature subset *Red*.

$Red \leftarrow \varnothing$
Normalize $\langle U, C \cup D \rangle$ by min–max method
for all $(D_i, D_j) \in (U/D) \times (U/D)$ with $D_i \neq D_j$ **do**
    Compute $CD_a(D_i, D_j)$, $SD_a(D_i, D_j)$ according to Eq. (9) and Eq. (11)
    Compute $OD_a(T)$ with respect to Eq. (14)
end for
Sort $OD_a(T)$ in ascending order, select the top $\psi\%$ values, then reverse the order before storing the result into $feature\_rank(K)$
$Red \leftarrow feature\_rank(K)$
Transform crisp labels into fuzzy labels $\tilde{D}_j(x_i)$, obtaining $U/\tilde{D} = \{\tilde{D}_1, \tilde{D}_2, \ldots, \tilde{D}_r\}$
for all $b \in Red$ **do**
    for all $D_j \in U/D$ **do**
        for all $x_i \in U$ **do**
            Compute $P(x_i, D_j)$ and $Q(x_i, D_j)$ according to Eq. (16) and Eq. (17)
            Calculate adaptive fuzzy similarity matrix $M_{\tilde{R}_a}$
            Compute $\underline{M_B}H(x_i)$ according to Eq. (19)
        end for
    end for
end for
Compute γ_{Red} by Eq. (24)
for all $k \in feature\_rank(K)$ **do**
    $SIG \leftarrow \gamma_{Red} - \gamma_{Red-\{k\}}$
    if $|SIG| \leq 10^{-6}$ **then**
        $Red \leftarrow Red \setminus \{k\}$
        $\gamma_{Red} \leftarrow \gamma_{Red-\{k\}}$
    end if
end for
return *Red*

---

4 Experiments

In this section, we validate the effectiveness of the proposed method from several aspects: (1)Whether DsAFRS can improve the classification accuracy of data; (2) Whether DsAFRS is superior to the existing methods; (3) The influence of parameter ε and ψ on the result of feature selection.

Table 4. Basic information of the datasets.

No.	Dataset	Abbr.	$\|U\|$	$\|C\|$	Classes
1	Cervical Cancer Behavior Risk	Cervical	72	19	2
2	Climate Model Simulation Crashes	Climate	540	20	2
3	Connectionist Bench	Connectionist	208	60	2
4	Divorce Predictors data set	Divorce	170	54	2
5	Forest	Forest	325	27	4
6	Image Segmentation	Image	210	19	7
7	Musk (Version 1)	Musk	476	166	2
8	Wisconsin Diagnostic Breast Cancer	WDBC	569	30	2
9	Acute Lymphoblastic Leukemia – Microarray	ALLAML	72	7129	2
10	Prostate	Prostate	102	5966	2
11	WarpPIE10P	warpPIE10P	210	2420	10
12	Yale	Yale	165	1024	15

Table 5. Classification accuracy rates of different algorithms under KNN classifier (%).

Dataset	Baseline	IFSNR	FScNCE	MRNG	WAFS	DsAFRS
Cervical	90.14 ± 0.81	86.28 ± 0.26	90.66 ± 0.88	89.80 ± 1.51	89.26 ± 0.46	**93.85 ± 1.21**
Climate	90.59 ± 0.27	90.42 ± 0.45	91.87 ± 0.27	94.24 ± 0.33	89.92 ± 0.36	**94.27 ± 0.36**
Connectionist	84.62 ± 0.68	69.65 ± 1.66	85.57 ± 0.56	86.26 ± 0.70	85.78 ± 1.25	**86.27 ± 1.04**
Divorce	97.64 ± 0.00	98.22 ± 0.54	98.11 ± 0.25	97.82 ± 0.52	98.17 ± 0.18	**98.23 ± 0.05**
Forest	79.40 ± 0.42	84.99 ± 1.10	80.60 ± 0.77	79.60 ± 0.84	85.30 ± 0.89	**85.53 ± 0.65**
Image	85.00 ± 0.71	89.46 ± 0.88	84.00 ± 0.83	84.76 ± 0.72	89.00 ± 0.85	**89.90 ± 0.72**
Musk	83.59 ± 1.06	71.71 ± 1.18	83.11 ± 0.68	81.20 ± 0.86	83.48 ± 0.62	**84.30 ± 0.74**
WDBC	97.01 ± 0.30	95.80 ± 0.42	96.74 ± 0.29	97.27 ± 0.27	96.99 ± 0.52	**97.30 ± 0.28**
ALLAML	85.17 ± 1.99	98.62 ± 0.22	99.57 ± 0.47	95.37 ± 1.42	96.33 ± 0.54	**100.00 ± 0.00**
Prostate	82.39 ± 1.37	94.31 ± 1.11	92.53 ± 0.58	91.30 ± 1.20	94.99 ± 0.49	**95.22 ± 0.56**
warpPIE10P	95.57 ± 0.30	96.76 ± 0.51	98.19 ± 0.46	97.00 ± 0.64	98.47 ± 0.50	**98.97 ± 0.57**
Yale	61.33 ± 1.66	62.40 ± 1.22	64.92 ± 1.51	42.70 ± 1.94	**65.87 ± 1.26**	64.95 ± 1.42
Average	86.04 ± 0.80	86.55 ± 0.80	88.82 ± 0.63	86.44 ± 0.91	89.46 ± 0.66	**90.73 ± 0.63**

4.1 Dateset

The experiment evaluated 12 datasets from the UCI and ASU databases, all of which have a numerical feature format. The information of these datasets is listed in Table 4.

4.2 Experimental Comparison of Methods and Parameter Settings

To verify the effectiveness of the proposed algorithm, we compared it with a baseline method and also compared it with four representative supervised feature selection methods: IFSNR [30], FScNCE [31], MRNG [32], WAFS [33]. In performance evaluation, we assessed the selected feature subsets using KNN (K=3), Classification Tree (CT), and Gaussian Naive Bayes (GNB), implemented with Scikit-Learn. We adopted 10-fold cross-validation with randomly shuffled data and no fixed seed and repeated each experiment 10 times, reporting the mean and standard deviation of accuracy. For competing methods, the best accuracy over all tuned parameter settings was used. For our method, $\varepsilon \in [0.1, 3]$ with a step size of 0.1 and $\psi \in [0.5, 1]$ with a step size of 0.05. All algorithms were implemented in Python 3.9.

Table 6. Classification accuracy rates of different algorithms under CT classifier (%).

Dataset	Baseline	IFSNR	FScNCE	MRNG	WAFS	DsAFRS
Cervical	83.87 ± 3.11	79.48 ± 0.61	83.25 ± 2.90	83.25 ± 2.23	83.55 ± 0.77	**87.66 ± 2.76**
Climate	91.25 ± 0.81	89.87 ± 0.51	91.31 ± 1.03	92.05 ± 0.84	85.31 ± 0.57	**92.68 ± 0.75**
Connectionist	71.50 ± 2.56	69.54 ± 1.25	76.66 ± 2.24	74.18 ± 2.94	76.71 ± 2.33	**77.39 ± 2.39**
Divorce	97.58 ± 0.71	98.23 ± 0.00	95.82 ± 0.67	96.52 ± 0.71	97.41 ± 0.56	**98.82 ± 0.61**
Forest	75.96 ± 1.13	76.72 ± 1.22	77.78 ± 1.43	78.45 ± 1.30	78.16 ± 1.39	**79.66 ± 1.44**
Image	87.86 ± 1.90	88.57 ± 1.04	88.29 ± 1.35	86.81 ± 1.07	89.68 ± 1.31	**89.69 ± 1.33**
Musk	77.24 ± 1.99	71.97 ± 0.97	80.31 ± 0.83	78.70 ± 0.93	79.58 ± 1.30	**80.37 ± 1.39**
WDBC	92.26 ± 0.73	92.02 ± 0.54	93.05 ± 0.76	92.77 ± 0.69	93.10 ± 0.98	**93.88 ± 0.77**
ALLAML	80.82 ± 2.18	90.92 ± 1.90	91.51 ± 2.34	92.78 ± 2.13	89.91 ± 2.44	**95.12 ± 1.98**
Prostate	81.55 ± 1.40	83.78 ± 1.25	88.87 ± 2.05	86.87 ± 2.16	84.62 ± 2.03	**89.43 ± 1.92**
warpPIE10P	79.47 ± 2.24	80.95 ± 1.83	80.62 ± 1.89	80.52 ± 2.31	81.28 ± 2.00	**81.42 ± 2.19**
Yale	47.40 ± 2.58	54.51 ± 2.61	54.82 ± 2.20	37.54 ± 2.34	**56.33 ± 2.60**	53.14 ± 2.53
Average	80.56 ± 1.78	81.38 ± 1.14	83.52 ± 1.64	81.70 ± 1.64	82.97 ± 1.52	**84.94 ± 1.67**

4.3 Experimental Results

Tables 5 to 7 respectively list the classification accuracy results of the KNN, CT and GNB classifiers, presented in the form of "mean ± standard deviation". The highest accuracy rate is marked in bold. The last row in the table represents the average classification accuracy of these algorithms on the 12 selected datasets. In addition, Table 8 shows the optimal feature subset sizes of different feature selection algorithms under different classifiers.

From Table 5, under the KNN classifier, DsAFRS improves classification accuracy on all datasets and achieves the highest accuracy on 11 datasets. In particular, DsAFRS increases the accuracy on the ALLAML dataset by 17.41%. Moreover, Table 8 shows that although the number of selected features on ALLAML is reduced by 99.80%, the accuracy still reaches 100%. Table 6 indicates that, under the CT classifier, DsAFRS attains the best accuracy on 11 datasets. The ALLAML dataset again exhibits the largest improvement,

Table 7. Classification accuracy rates of different algorithms under GNB classifier (%).

Dataset	Baseline	IFSNR	FScNCE	MRNG	WAFS	DsAFRS
Cervical	90.07 ± 1.73	81.21 ± 0.21	91.20 ± 0.65	89.79 ± 0.61	90.30 ± 0.47	$\mathbf{92.91 \pm 1.03}$
Climate	94.77 ± 0.21	92.51 ± 0.03	94.87 ± 0.26	95.18 ± 0.20	91.48 ± 0.00	$\mathbf{95.35 \pm 0.22}$
Connectionist	67.81 ± 1.35	70.65 ± 0.78	$\mathbf{72.06 \pm 0.77}$	71.26 ± 1.52	71.55 ± 1.39	68.96 ± 0.70
Divorce	97.17 ± 0.23	94.64 ± 0.28	97.11 ± 0.05	98.35 ± 0.30	98.70 ± 0.41	$\mathbf{99.23 \pm 0.38}$
Forest	80.90 ± 0.55	79.83 ± 0.54	81.70 ± 0.63	81.70 ± 0.59	81.14 ± 0.52	$\mathbf{81.74 \pm 0.48}$
Image	69.00 ± 1.16	83.86 ± 0.62	73.24 ± 0.97	85.14 ± 1.00	80.47 ± 1.04	$\mathbf{85.42 \pm 1.02}$
Musk	73.29 ± 0.62	67.89 ± 0.34	75.43 ± 0.42	74.39 ± 0.42	74.91 ± 1.32	$\mathbf{79.22 \pm 0.49}$
WDBC	93.30 ± 0.20	93.79 ± 0.13	93.33 ± 0.22	94.90 ± 0.19	93.60 ± 0.23	$\mathbf{95.30 \pm 0.17}$
ALLAML	98.62 ± 0.08	98.03 ± 0.55	98.89 ± 0.49	92.94 ± 0.96	95.89 ± 0.22	$\mathbf{100.00 \pm 0.00}$
Prostate	59.95 ± 1.86	94.09 ± 0.66	91.34 ± 0.49	89.88 ± 0.75	93.50 ± 0.55	$\mathbf{94.30 \pm 0.94}$
warpPIE10P	90.90 ± 1.24	91.52 ± 1.04	94.76 ± 0.73	87.33 ± 1.05	95.00 ± 0.69	$\mathbf{95.71 \pm 0.84}$
Yale	61.79 ± 1.09	62.11 ± 1.59	61.50 ± 1.63	47.08 ± 1.74	62.77 ± 1.82	$\mathbf{63.73 \pm 1.72}$
Average	81.46 ± 0.86	84.18 ± 0.56	85.45 ± 0.61	84.00 ± 0.78	85.78 ± 0.72	$\mathbf{87.66 \pm 0.67}$

Table 8. Comparison of different models on various datasets.

Dataset	Baseline	IFSNR			FScNCE			MRNG			WAFS			DsAFRS		
		KNN	CT	GNB	KNN	CT	GNB	KNN	CT	GNB	KNN	CT	GNB	KNN	CT	GNB
Cervical	19	3	6	3	18	18	18	7	6	7	17	17	17	6	6	6
Climate	20	2	2	2	19	19	20	11	6	7	1	1	1	6	8	7
Connectionist	60	1	1	1	45	16	45	38	28	28	51	31	50	45	10	7
Divorce	54	1	1	1	36	36	36	8	8	8	5	4	7	4	4	4
Forest	27	18	18	18	24	27	24	26	8	24	18	20	26	10	12	11
Image	19	8	3	12	15	13	8	13	13	6	9	10	9	8	12	7
Musk	166	3	3	3	108	89	108	15	20	15	145	144	128	42	34	17
WDBC	30	11	11	11	25	25	28	23	28	5	29	28	29	21	7	9
ALLAML	7129	232	1	232	32	25	27	6	3	6	18	18	18	7	4	5
Prostate	5966	33	1	33	69	28	34	5	5	5	10	10	10	22	8	10
warpPIE10P	2420	274	617	617	56	80	75	15	15	15	48	64	56	33	87	57
Yale	1024	641	16	641	85	26	85	9	6	8	20	20	96	41	41	41
Average	1411	102	57	131	44	34	42	15	12	11	31	31	37	20	19	15

reaching 17.69%. Table 7 reports the results under the GNB classifier, where DsAFRS ranks first on 11 datasets. Notably, on the prostate dataset, the accuracy increases by 57.30% while the number of selected features decreases by 99.83%. Among the competing methods, FScNCE and WAFS achieve the highest accuracy on 1 and 2 datasets, respectively, under the three classifiers, but their accuracies are close to those of DsAFRS. In addition, as shown in Table 8, our method ranks second in removing redundant attributes, only behind MRNG. However, across all three classifiers, our method consistently achieves higher

classification accuracy than MRNG, further demonstrating its superior overall performance.

4.4 Parameter Sensitivity Analysis

To evaluate robustness under different parameter settings, we conducted sensitivity experiments on the feature retention rate ψ and the control parameter ε. As shown in Fig. 3, retaining a proper proportion of features can improve classification accuracy. Moreover, Fig. 4 shows that as ε increases, the number

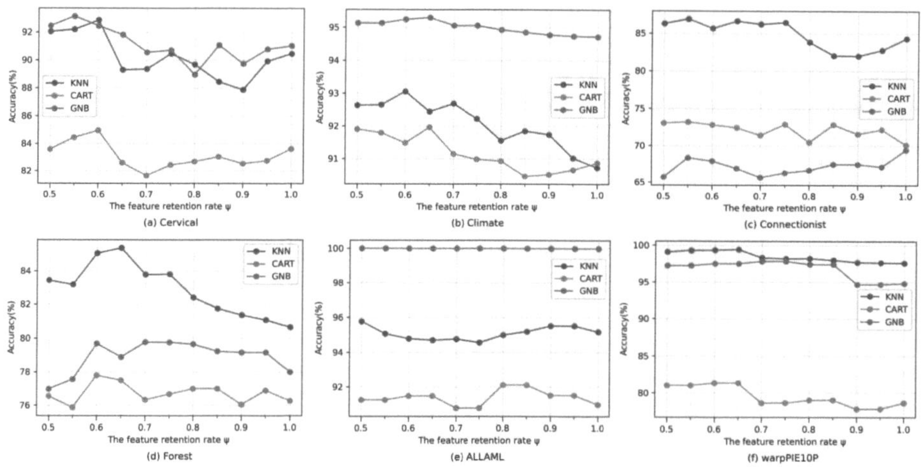

Fig. 3. Classification accuracy of different feature retention rates ψ under KNN, CT, and GNB classifiers.

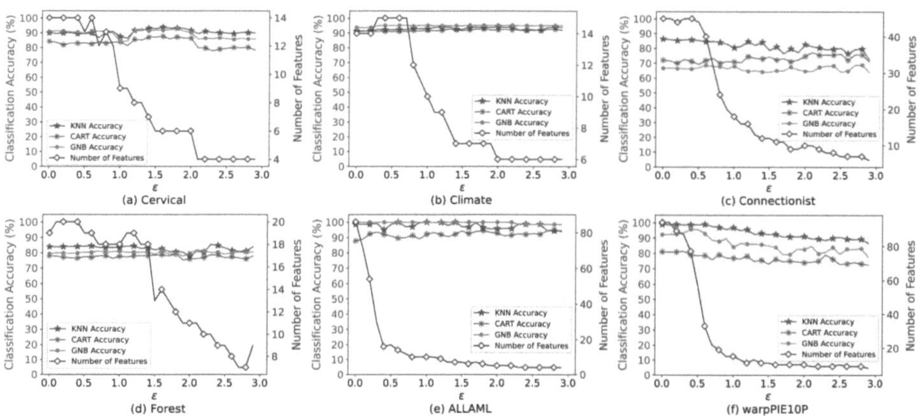

Fig. 4. The variation trend of classification accuracy with the number of features and ε under different classifiers.

of selected features generally decreases and then stabilizes, while the accuracies of KNN, CART, and GNB vary only slightly. These results indicate that the proposed method is insensitive to ε and maintains stable performance with a sparse feature subset.

5 Conclusion

In this paper, we propose a feature selection algorithm based on dual-scale fuzzy rough sets. The method aims to address the limitations of traditional fuzzy rough sets in dealing with fuzziness and uncertainty. First, an evaluation mechanism based on the overlap function is introduced to more accurately measure the discriminative ability of individual features, thereby laying the foundation for subsequent model construction. Then, a dual-scale adaptive fuzzy approximation operator is proposed. Finally, a heuristic backward redundant feature filtering algorithm is designed. It demonstrates strong competitiveness in improving classification accuracy and eliminating redundant features. In conclusion, the method proposed in this study provides a new theoretical framework for the application of fuzzy rough sets. Future research can explore the integration of this method with other fuzzy theories and deep learning techniques. This can further enhance the generalization ability and practicality of the method.

Acknowledgments. The research is supported by the National Natural Science Foundation of China under Grant Nos. 62466017 and Nos. 62066014, Jiangxi Province Natural Science Foundation of China under Grant Nos. 20232ACB202013.

Disclosure of Interests. The authors declare that they have no known competing financial interests or personal relationships that could have appeared to influence the work reported in this paper.

References

1. Das, A.K., Sengupta, S., Bhattacharyya, S.: A group incremental feature selection for classification using rough set theory based genetic algorithm. Appl. Soft Comput. **65**, 400–411 (2018)
2. Pawlak, Z.: Rough sets. Int. J. Comput. Inf. Sci. **11**(5), 341–356 (1982)
3. Pawlak, Z.: Rough set theory and its applications to data analysis. Cybern. Syst. **29**(7), 661–688 (1998)
4. Prasad, M., Tripathi, S., Dahal, K.: An efficient feature selection based Bayesian and rough set approach for intrusion detection. Appl. Soft Comput. **87**, 105980 (2020)
5. Zhang, X., Yao, Y.: Tri-level attribute reduction in rough set theory. Expert Syst. Appl. **190**, 116187 (2022)
6. Chen, H., Li, T., Cai, Y.: Parallel attribute reduction in dominance-based neighborhood rough set. Inf. Sci. **373**, 351–368 (2016)
7. Hu, Q., Yu, D., Liu, J., Wu, C.: Neighborhood rough set based heterogeneous feature subset selection. Inf. Sci. **178**(18), 3577–3594 (2008)

8. Xu, F., Cai, M., Li, Q., Wang, H., Fujita, H.: Shared neighbors rough set model and neighborhood classifiers. Expert Syst. Appl. **244**, 122965 (2024)
9. Zhan, J., Cai, M.: A cost-minimized two-stage three-way dynamic consensus mechanism for social network-large scale group decision-making: Utilizing [formula omitted]-nearest neighbors for incomplete fuzzy preference relations. Expert Syst. Appl. **263**, 125705 (2025)
10. Greco, S., Matarazzo, B., Slowinski, R.: Rough approximation by dominance relations. Int. J. Intell. Syst. **17**(2), 153–171 (2002)
11. Zhang, X., Chen, D., Tsang, E.C.C.: Generalized dominance rough set models for the dominance intuitionistic fuzzy information systems. Inf. Sci. **378**, 1–25 (2017)
12. Kusunoki, Y., Blaszczynski, J., Inuiguchi, M., Slowinski, R.: Empirical risk minimization for dominance-based rough set approaches. Inf. Sci. **567**, 395–417 (2021)
13. Sun, B., Ma, W., Gong, Z.: Dominance-based rough set theory over interval-valued information systems. Expert. Syst. **31**(2), 185–197 (2014)
14. Jiang, J., Zhang, X., Yuan, Z.: Multi-label feature selection using self-information in divergence-based fuzzy neighborhood rough sets. Pattern Recogn. **155**, 110684 (2024)
15. Sun, L., Wang, L., Ding, W., Qian, Y., Xu, J.: Feature selection using fuzzy neighborhood entropy-based uncertainty measures for fuzzy neighborhood multigranulation rough sets. IEEE Trans. Fuzzy Syst. **29**(1), 19–33 (2021)
16. Yan, S., Qian, J., Yu, Y., Ni, Y.: A feature selection method driven by fuzzy implication granularity. Eng. Appl. Artif. Intell. **158**, 111298 (2025)
17. Zhang, X., Fan, Y., Yang, J.: Feature selection based on fuzzy-neighborhood relative decision entropy. Pattern Recogn. Lett. **146**, 100–107 (2021)
18. Dubois, D., Prade, H.: Rough fuzzy sets and fuzzy rough sets. Int. J. Gen. Syst. **17**(2–3), 191–209 (1990)
19. Morsi, N.N., Yakout, M.M.: Axiomatics for fuzzy rough sets. Fuzzy Sets Syst. **100**(1–3), 327–342 (1998)
20. Yue, X., Zhou, J., Yao, Y., Miao, D.: Shadowed neighborhoods based on fuzzy rough transformation for three-way classification. IEEE Trans. Fuzzy Syst. **28**(5), 978–991 (2020)
21. Zhang, X., Mei, C., Li, J., Yang, Y., Qian, T.: Instance and feature selection using fuzzy rough sets: a bi-selection approach for data reduction. IEEE Trans. Fuzzy Syst. **31**(6), 1981–1994 (2022)
22. An, S., Zhao, E., Wang, C., Guo, G., Zhao, S., Li, P.: Relative fuzzy rough approximations for feature selection and classification. IEEE Trans. Cybern. **53**(4), 2200–2210 (2021)
23. Jensen, R., Shen, Q.: Fuzzy–rough attribute reduction with application to Web categorization. Fuzzy Sets Syst. **141**(3), 469–485 (2004)
24. Dai, J., Zhu, Z., Zou, X.: Fuzzy rough attribute reduction based on fuzzy implication granularity information. IEEE Trans. Fuzzy Syst. **32**(6), 3741–3752 (2024)
25. Wang, C., Wang, C., An, S., Ding, W., Qian, Y.: Feature selection and classification based on directed fuzzy rough sets. IEEE Trans. Syst. Man Cybern. Syst. **55**(1), 699–711 (2024)
26. Sang, B., Yang, L., Chen, H., Li, T., Xu, W.: Robust attribute reduction exploring class-separability and attribute-correlation for ordered decision systems. IEEE Trans. Syst. Man Cybern. Syst. **55**(6), 2168–2216 (2025)
27. Zhao, H., Wang, P., Hu, Q., Zhu, P.: Fuzzy rough set based feature selection for large-scale hierarchical classification. IEEE Trans. Fuzzy Syst. **27**(10), 1891–1903 (2019)

28. Wang, Z., Chen, H., Yuan, Z., Yang, X., Zhang, P., Li, T.: Exploiting fuzzy rough mutual information for feature selection. Appl. Soft Comput. **131**, 109769 (2022)
29. Yuan, Z., Chen, H., Li, T., Yu, Z., Sang, B., Luo, C.: Unsupervised attribute reduction for mixed data based on fuzzy rough sets. Inf. Sci. **572**, 67–87 (2021)
30. Zhang, G., Hu, J., Yang, J., Zhang, P.: Interactive streaming feature selection based on neighborhood rough sets. Eng. Appl. Artif. Intell. **139**, 109479 (2025)
31. Zhang, P., Li, T., Yuan, Z., Luo, C., Liu, K., Yang, X.: Heterogeneous feature selection based on neighborhood combination entropy. IEEE Trans. Neural Netw. Learn. Syst. **35**(3), 3514–3527 (2022)
32. Dai, J., Zhu, Z., Li, M., Zou, X., Zhang, C.: Attribute reduction for heterogeneous data based on monotonic relative neighborhood granularity. Int. J. Approx. Reason. **170**, 109210 (2024)
33. Wang, C., Wang, C., Qian, Y., Leng, Q.: Feature selection based on weighted fuzzy rough sets. IEEE Trans. Fuzzy Syst. **32**(7), 4027–4037 (2024)

Time Series Classification via Interval Gray Transformation Frame and Its Application

Liyuan Jiang[1(✉)] and Zhifu Tao[2]

[1] School of Big Data and Statistics, Anhui University, Hefei 230000, Anhui, China
we23101001@stu.ahu.edu.cn

[2] Research Centre of Applied Mathematics, Anhui University, Hefei 230601, Anhui, China
jeff.tao@ahu.edu.cn

Abstract. Uncertainty and observation errors are commonly involved in real-valued time series, which may weaken the performance of conventional time series classification methods based only on point-valued observations. To address this issue, this paper proposes an interval gray transformation framework for time series classification. In the proposed framework, each original point-valued time series is first transformed into a stationary series, and the residual information obtained from an ARIMA model is then used to construct interval gray numbers with known bounds. In this way, the original time series is represented not only by its central values but also by its uncertainty range. Based on this representation, the classical DTW-1NN classifier is extended to the interval gray environment, leading to the proposed IGN-DTW-1NN method. Experiments on several datasets from the UCR time series classification archive demonstrate that the proposed method can effectively reduce classification error rates compared with the conventional DTW-1NN method and several representative classification approaches. The results indicate that incorporating residual-based interval gray information provides a useful and interpretable way to improve point-valued time series classification.

Keywords: Time series classification · interval gray transformation · ARIMA · DTW-1NN · UCR time series classification archive

1 Introduction

Time series data are daily generated in a wide range of fields including finance [35], engineering [9], bioinformatics [25], healthcare, meteorology, and smart manufacturing [11]. Over the years, with the increasing availability of temporal data [36], theoretical and practical research on time series data has gradually become a hot topic, including classification [18,34], forecasting [27] and outlier detection [24].

J. Lu et al. (Eds.): FLINS-ISKE 2026, LNCS 16752, pp. 433–454, 2027.
https://doi.org/10.1007/978-981-92-2369-5_27

Time series classification has been widely applied to many domains including medical diagnosis [2], creation of 2D bifurcation diagrams [17], knowledge extraction [3], and smart manufacturing systems [11]. Furthermore, the diversity of dataset types in the UCR/UEA archives shows different applications of the TSC problem.

In practical time series classification, what is classified is often not merely a sequence of point values, but the dynamic phenomenon reflected by the sequence. Time series generated from the same system often show various similarities, such as similar overall trends or autocorrelation structures, and such similarities may even exist between different categories. Once this phenomenon occurs, it may interfere with the classification process. Moreover, direct observations of time series are often imprecise [47]; measurement errors, environmental disturbances, sensor noise, and model residuals may be involved in the observed data. Recent studies on uncertainty estimation for time series classification also indicate that uncertainty information is important for improving the reliability of classification models [4,31].

The above problems indicate that the information expressed by the original point-valued time series data is uncertain, which provides the possibility of improvement for classification work. Therefore, time series classification is studied from the perspective of interval gray time series via an interval gray transformation frame in this paper, in which classical point-valued time series is analyzed by transforming it into interval gray time series. Specifically, the classical point-valued time series is first transformed into a stationary time series by differencing, and then the residual information obtained from the ARIMA model is used to construct interval gray numbers. The interval constructed in this way contains not only the central information of the observed variable but also the uncertainty range related to residual fluctuations, which makes up for the defect of real-valued data with observation error.

Thus, the DTW-1NN algorithm is used to realize the developed time series classification method via an interval gray transformation frame, namely IGN-DTW-1NN. The main contributions of this paper are summarized as follows:

- An interval gray transformation frame is proposed to represent point-valued time series with uncertainty information.
- Three residual-based interval construction modes are designed by using the residual terms obtained from the ARIMA model.
- The classical DTW-1NN classifier is extended to the interval gray number environment by introducing a distance measure between interval gray numbers.
- Experiments on several datasets from the UCR time series classification archive are conducted to verify the effectiveness of the proposed IGN-DTW-1NN method, and comparisons with representative classification methods are provided.

The remainder of this paper is organized as follows. Section 2 reviews related work on time series classification, interval gray representation, and uncertainty-aware time series analysis. Section 3 introduces the necessary preliminaries, including interval gray numbers, dynamic time warping, and the ARIMA model. Section 4 presents the proposed time series classification method based on the interval gray transformation framework. Section 5 reports the experimental results and comparative analysis. Finally, Sect. 6 concludes this paper and discusses possible future work.

2 Related Work

2.1 Time Series Classification Methods

Up until now, many time series classification methods have been designed and proposed [15,34], which can be mainly partitioned into distance-based methods [1], feature-based methods [41], dictionary-based methods [39], shapelet-based methods [20], ensemble methods [37], and deep learning-based methods [49,51]. Recent survey work has also shown that deep learning has become an important direction for time series classification and extrinsic regression, especially when features are difficult to define manually [34].

For distance-based methods, by introducing distance functions of time series, the similarities among time series can be calculated and further used to derive the classification according to the idea of clustering [14]. One of the most traditional and commonly used methods is to use a nearest neighbor classifier and a distance function [28]. Currently, distance functions including Euclidean distance [5], edit distance [33], longest common subsequence [13], constrained elastic distance [50] and dynamic time warping [9,21,23] have been proposed and used for time series classification. Among various types of distance measures, DTW has shown powerful capability in time series classification [22,38]. Besides, 1-NN has also been used in this field due to its simplicity and ease of implementation [16,40].

On the other hand, when using feature-based methods, statistical or symbolic features [7,8,19] are commonly predefined and then used to classify time series by employing machine learning approaches. Shapelet-based methods have attracted attention because they can provide discriminative subsequences with relatively good interpretability; recent studies further combine shapelet features with neural networks to improve classification performance [20]. Dictionary-based and symbolic methods transform time series into symbolic or word-like representations, and recent segmented word embedding-based methods have also been developed for efficient time series classification [39]. In addition, attention-based and graph-based deep learning methods have been proposed to capture temporal dependencies and multivariate relationships in time series classification [49,51]. Ensemble approaches also remain effective by combining multiple representations or classifiers [37].

Model interpretability is not only important but also necessary [29]. For example, in finance and healthcare, for better monitoring and decision-making, an interpretable model is needed to improve decision support. Due to the interpretability issues of some machine learning approaches, distance-based methods for time series classification have also attracted increasing interest in both industry and academia. Compared with complex deep learning models, DTW-1NN has a simple structure and good interpretability, but it usually works directly on point-valued observations and does not explicitly consider uncertainty information.

2.2 Interval Gray Representation and Uncertainty-Aware Time Series Analysis

In time series research, interval data can express uncertain information more effectively than point-valued data. For this reason, some scholars construct interval data based on point-valued observations. Froelich et al. [12] used an interval representation of the data, where each observation is represented as an interval in which the actual value falls, enabling more efficient forecasting of time series data. Zhou et al. [52] performed data transformation based on point-valued sequences, using high-order difference and interval representation to obtain more detailed information about temporal shape and morphological features for anomaly detection.

Interval gray numbers are a special type of interval data used to describe uncertainties whose bounds are known but whose distribution is not clear. Liu et al. [32] gave the definition of classical gray numbers, developed interval gray numbers on this basis, and defined the kernel, gray level and algorithm of interval gray numbers. In recent years, interval gray number sequences have also been used in research. Ye et al. [48] proposed a prediction model for interval gray number sequences based on the generalized non-decreasing gray axiom. Multivariate interval gray number sequences have also been studied [44,45]. In addition, recent studies on information granules and fuzzy granular modeling have shown that granulated or interval-like representations can be useful for time series analysis, classification, and fault detection tasks [30,53].

All classification problems rely on similarity measures between data. For time series, there are three types of similarity between similar time series: similarity in time domain, similarity in shape, and similarity in change. Differences between time-domain similarity sequences may be caused by noise and phase drift. 1-NN classifiers are well suited to handle such problems, and DTW metrics can mitigate the effects of noise and temporal misalignment. Shape similarity sequences are distinguished by some identical subsequences or shapes. These subsequences or shapes may appear anywhere in the time series, and such cases can be distinguished using methods based on time series features. Change similarity is the least observable similarity. Such similarity appears in sequences with strong autocorrelation and can be introduced into models, such as the HMM model [26] and autoregressive moving average models.

Although the above studies provide important foundations for time series classification and interval-valued representation, most existing distance-based classification methods are still constructed directly on point-valued time series. The residual-based uncertainty information contained in observed time series has not been sufficiently incorporated into the classical DTW-1NN classification framework. This motivates the interval gray transformation frame proposed in this paper, which transforms point-valued time series into interval gray time series and then performs classification under the interval gray DTW-1NN framework.

3 Preliminaries

This section introduces the basic concepts and tools used in the proposed method, including interval gray numbers, dynamic time warping (DTW), and the ARIMA model. These preliminaries are necessary for constructing the interval gray representation of point-valued time series and for measuring the similarity between interval gray time series.

Point-valued time series may contain observation errors, noise, and residual fluctuations. Therefore, a single observed value may not fully describe the uncertainty of the underlying system. To characterize such uncertainty, this paper reconstructs the original point-valued sequence into an interval gray number sequence. Different from a conventional interval sequence, where the values in an interval are often implicitly treated with a certain distribution, an interval gray number only assumes that the possible bounds are known, while the probability distribution inside the interval is unknown. This property makes interval gray numbers suitable for representing the uncertainty contained in observed time series.

Moreover, time series generated from the same system usually have autocorrelation structures, and sequences from different classes may also share similar trends or correlation patterns. The ARIMA model is used to fit each individual sequence and extract the residual term, which is further employed to construct interval gray numbers. DTW is then used to measure the similarity between the transformed interval gray time series. The main notations used in this paper are summarized in Table 1.

3.1 Interval Gray Number

Definition 3.1 (gray number [42]). A gray number, denoted as '$\otimes$', is defined as a number whose certain value is unknown but possible bounds of the true value can be obtained. Let $\mathbf{D}$ and d^* be the information background and true value of given gray number $\otimes$, i.e., $\forall \otimes \Rightarrow d^* \in D = [a, b]$, then

- $\otimes$ is a gray number under information background $\mathbf{D}$.
- $\mathbf{D}$ is a potential value-covered set of $\otimes$.
- d^* is the true value of $\otimes$.

Definition 3.2 (Interval gray number [42]). Given that the lower and upper bounds can be obtained in a continuous value-covered set $\mathbf{D}$, i.e., $\mathbf{D}$ is an interval, then $\tilde{\otimes}$ is called an interval gray number, denoted as $\forall \tilde{\otimes} \Rightarrow d^* \in \mathbf{D}, \mathbf{D} = [\underline{\mathbf{a}}, \bar{\mathbf{a}}]$

Table 1. Main notations used in this paper

Symbol	Meaning
$\{y_t\}$	Original point-valued time series
$\nabla^d y_t$	d-order differenced time series
Y_t	Stationary series after differencing
μ_t	Residual term obtained from the ARIMA model
$\otimes$	Gray number
$\tilde{\otimes}$	Interval gray number
$\tilde{\otimes}G(t)$	Interval gray number at time t
$\underline{a}$, $\bar{a}$	Lower and upper bounds of an interval gray number
$l(\tilde{\otimes})$	Length of an interval gray number
$g(\tilde{\otimes})$	Kernel of an interval gray number
$r(\tilde{\otimes})$	Gray radius of an interval gray number
$d(\tilde{\otimes}_1, \tilde{\otimes}_2)$	Distance between two interval gray numbers
$[X]$, $[Y]$	Two time series used in DTW
W	Warping path in DTW
$\gamma(i, j)$	Cumulative DTW distance at position (i, j)
$\Gamma(i, j)$	Cumulative DTW distance between interval gray time series
$ARIMA(p, d, q)$	ARIMA model with orders p, d, and q
TSS	Time series set

($\tilde{\otimes} = [\underline{a}, \bar{a}]$ for short). Herein, both upper bound $\bar{a}$ and lower bound $\underline{a}$ are crisp numbers.

For an interval gray number $\tilde{\otimes} = [\underline{a}, \bar{a}]$, $l(\tilde{\otimes}) = \bar{a} - \underline{a}$ is named as the length of $\tilde{\otimes}$, $g(\tilde{\otimes}) = (\underline{a} + \bar{a})/2$ and $r(\tilde{\otimes}) = (\bar{a} - \underline{a})/2$ are said to be the kernel and gray radius of $\tilde{\otimes}$. Then, an interval gray number $\tilde{\otimes}$ would degenerate to be a conventional gray number if $l(\tilde{\otimes}) = 0$.

Different from the concept of interval number in fuzzy set theory (Qi et al., 2021), the probability of each point in an interval gray number is unknown but not a uniform distribution.

To handle interval gray numbers, the following definitions are valid:

Definition 3.3 (Operations on interval gray numbers [32]). The algebra operations between any two interval gray numbers $\tilde{\otimes}_1 = [\underline{a}_1, \bar{a}_1]$ and $\tilde{\otimes}_2 = [\underline{a}_2, \bar{a}_2]$ can be defined according to:

$$\tilde{\otimes}_1 \oplus \tilde{\otimes}_2 = [\underline{a}_1 + \underline{a}_2, \bar{a}_1 + \bar{a}_2] \tag{1}$$

$$-\tilde{\otimes}_1 = [-\bar{a}_1, -\underline{a}_1] \tag{2}$$

$$\tilde{\otimes}_1 - \tilde{\otimes}_2 = [\underline{a}_1 - \bar{a}_2, \bar{a}_1 - \underline{a}_2] \tag{3}$$

The set operations on any two interval gray numbers $\tilde{\otimes}_1 = [\underline{a}_1, \bar{a}_1]$ and $\tilde{\otimes}_2 = [\underline{a}_2, \bar{a}_2]$ are expressed as:

$$\tilde{\otimes}_1 \cup \tilde{\otimes}_2 = \{\varsigma | \varsigma \in [a_1, b_1]\, or \varsigma \in [a_2, b_2]\} \tag{4}$$

$$\tilde{\otimes}_1 \cap \tilde{\otimes}_2 = \{\varsigma | \varsigma \in [a_1, b_1]\, and \varsigma \in [a_2, b_2]\} \tag{5}$$

Similar to the distance measure of interval numbers listed in [6], the following distance measure corresponding to interval gray number is valid:

Definition 3.4 [43]. Let $\tilde{\otimes}_1$ and $\tilde{\otimes}_2$ be any two interval gray numbers,

$$\begin{aligned} d(\tilde{\otimes}_1, \tilde{\otimes}_2) &= \textstyle\int_{-0.5}^{0.5} \left\{ \left[g(\tilde{\otimes}_1) + \alpha \left(l \left(\tilde{\otimes}_1 \right) \right) \right] - \left[g(\tilde{\otimes}_2) + \beta \left(l \left(\tilde{\otimes}_2 \right) \right) \right] \right\}^2 d\alpha d\beta \\ &- \textstyle\int_{-0.5}^{0.5} \int_{-0.5}^{0.5} \left\{ \left[\hat{\zeta}_1 + \alpha l(\hat{\zeta}_1) \right] - \left[\hat{\zeta}_1 + \beta l(\hat{\zeta}_1) \right] \right\}^2 d\alpha d\beta \\ &= [g(\tilde{\otimes}_1) - g(\tilde{\otimes}_2)]^2 + \tfrac{1}{12}\{[l(\tilde{\otimes}_1)]^2 + [l(\tilde{\otimes}_2)]^2\} - \tfrac{1}{6}[l(\zeta)]^2 \end{aligned} \tag{6}$$

Then $d(\tilde{\otimes}_1, \tilde{\otimes}_2)$ is defined as the distance measure between $\tilde{\otimes}_1$ and $\tilde{\otimes}_2$, where $\zeta = \tilde{\otimes}_1 \cap \tilde{\otimes}_2$.

Compared with the traditional distance measure that only considers the upper and lower bounds of the interval gray number [46], this measure more fully complements the inherent characteristics of the interval gray number, considers the deviation of every two corresponding points, and eliminates any correspondence in the overlapping area of the two interval gray numbers.

3.2 Dynamic Time Warping

The notion of dynamic time warping [10] was firstly introduced in the real-valued data environment. Suppose that there are two time series, i.e., the time series $[X]$ with lengths n and $[Y]$ with lengths m. DTW aligns the two series by constructing a warping path W, which is a $n \times m$ matrix. The warping path, denoted as $W = \omega_1, \omega_2, \cdots, \omega_K$, where $\omega_k = (i, j), k = 1, 2, \cdots, K$ is the distance between i in $[X]$ and j in $[Y]$ and $\max\{|X|, |Y|\} \leq K \leq |X| + |Y|$.

There are usually several restrictions:

Boundary Conditions: $\omega_1 = (1, 1)$ and $\omega_K = (|X|, |Y|)$, which ensures all positions in $|X|$ and $|Y|$ appear in W.

Monotonic: Let $\omega_k = (i, j)$ and $\omega_{k+1} = (i', j')$, then $i' \geq i$, $j' \geq j$.

Continuity: Let $\omega_k = (i, j)$ and $\omega_{k+1} = (i', j')$, then $i' \leq i + 1$, $j' \leq j + 1$.

The warping path that meets these criteria is growing exponentially, but we are only interested in paths with the lowest warping costs, i.e.,

$$DTW([X], [Y]) = \min \left\{ \sqrt{\sum\nolimits_{k=1}^{K} \omega_k / K} \right\}. \tag{7}$$

Solution of the optimization model can be realized by using the following recursion, which defines cumulative distance d as the minimum value of the distance found in the current cell and the cumulative distance of adjacent elements:

$$\gamma(i, j) = d(q_i, c_j) + \min\{\gamma(i-1, j-1), \gamma(i-1, j), \gamma(i, j-1)\}. \tag{8}$$

The dynamic time warp distance between $[X]$ and $[Y]$ is determined as: $\gamma([X],[Y])$.

3.3 ARIMA Model

Let $\{y_t\}$ be an observed time series, which can be transformed into a stationary time series by d-order difference, then the $ARIMA(p,d,q)$ modeled for $\{y_t\}$ can be listed in the following:

$$\begin{aligned}\nabla^d y_t &= \varphi_1 \nabla^d y_{t-1} + \varphi_2 \nabla^d y_{t-2} + \cdots + \varphi_p \nabla^d y_{t-p} \\ &\quad - \mu_t - \theta_1 \mu_{t-1} - \theta_2 \mu_{t-2} - \cdots - \theta_q \mu_{t-q}.\end{aligned} \tag{9}$$

where p is the number of autoregressive terms and q is the number of moving average terms.

Especially, if $p = 0$, then $ARIMA(p,d,q)$ model degenerates into IMA model. While if $q = 0$, then $ARIMA(p,d,q)$ model will degenerate into ARI model.

4 Time Series Classification via Interval Gray Transformation Frame

4.1 Interval Gray Transformation Frame of Real-Valued Time Series

Point-valued time series may contain uncertainty caused by observation errors, noise, and unexplained residual fluctuations. If only the original point values are used for classification, such uncertainty information is ignored. To address this issue, this subsection introduces an interval gray transformation frame, in which a real-valued time series is transformed into an interval gray time series by using the residual information extracted from an ARIMA model.

Let $\{y_t\}$ be a given real-valued time series, and denote its d-order differenced series $\nabla^d y_t$ as Y_t. By using the ARIMA model introduced in Subsect. 3.3, the residual term μ_t can be obtained. The residual term reflects the part of the sequence that cannot be explained by the fitted autocorrelation structure, and is therefore used to construct the uncertainty range around Y_t.

According to the residual sequence $\{\mu_t\}$, the constructed interval gray time series is denoted as

$$\{\tilde{\otimes} G(t)\} = \{\tilde{\otimes} G(1), \tilde{\otimes} G(2), \cdots, \tilde{\otimes} G(T)\}.$$

In this paper, three interval construction modes are considered as follows.

Mode 1. The upper bound of the interval gray number at time t is defined as Y_t plus the square of μ_t, and the lower bound is defined as Y_t minus the square of μ_t. Thus, the interval gray number at time t is expressed as

$$\tilde{\otimes} G_1(t) = [Y_t - \mu_t^2, Y_t + \mu_t^2], \quad t = 1, 2, \cdots, T. \tag{10}$$

Mode 2. The upper bound of the interval gray number at time t is defined as Y_t plus the absolute value of μ_t, and the lower bound is defined as Y_t minus the absolute value of μ_t. Thus, the interval gray number at time t is expressed as

$$\tilde{\otimes}G_2(t) = [Y_t - |\mu_t|, Y_t + |\mu_t|], \quad t = 1, 2, \cdots, T. \tag{11}$$

Mode 3. The interval is constructed according to the sign of μ_t. When $\mu_t > 0$, the interval is $[Y_t, Y_t + \mu_t]$; when $\mu_t < 0$, the interval is $[Y_t + \mu_t, Y_t]$. Therefore, the interval gray number at time t is expressed as

$$\tilde{\otimes}G_3(t) = \begin{cases} [Y_t, Y_t + \mu_t], \mu_t > 0, \\ [Y_t + \mu_t, Y_t], \mu_t < 0, \\ [Y_t, Y_t], \qquad \mu_t = 0, \end{cases} \quad t = 1, 2, \cdots, T. \tag{12}$$

Herein, Y_t is the stationary series obtained after differencing the original data. In Mode 1 and Mode 2, Y_t is the kernel of the corresponding interval gray number. These three modes transform the point-valued sequence into an interval gray sequence from different perspectives, so that the residual-based uncertainty can be incorporated into the subsequent classification process.

4.2 Distances Between Interval Gray Number Time Series

According to the distance measure between two interval gray numbers defined in Definition 3.4, the DTW algorithm can be extended to measure the distance between two interval gray number time series. Let

$$\{\tilde{\otimes}g_1(t)\} = \{\tilde{\otimes}g_1(1), \tilde{\otimes}g_1(2), \cdots, \tilde{\otimes}g_1(n)\}$$

and

$$\{\tilde{\otimes}g_2(t)\} = \{\tilde{\otimes}g_2(1), \tilde{\otimes}g_2(2), \cdots, \tilde{\otimes}g_2(m)\}$$

be two interval gray number time series. The DTW distance matrix D of the two series is expressed as

$$D = \begin{bmatrix} d(\tilde{\otimes}g_1(1), \tilde{\otimes}g_2(1)) & d(\tilde{\otimes}g_1(2), \tilde{\otimes}g_2(1)) & \cdots & d(\tilde{\otimes}g_1(n), \tilde{\otimes}g_2(1)) \\ d(\tilde{\otimes}g_1(1), \tilde{\otimes}g_2(2)) & d(\tilde{\otimes}g_1(2), \tilde{\otimes}g_2(2)) & \cdots & d(\tilde{\otimes}g_1(n), \tilde{\otimes}g_2(2)) \\ \vdots & \vdots & \ddots & \vdots \\ d(\tilde{\otimes}g_1(1), \tilde{\otimes}g_2(m)) & d(\tilde{\otimes}g_1(2), \tilde{\otimes}g_2(m)) & \cdots & d(\tilde{\otimes}g_1(n), \tilde{\otimes}g_2(m)) \end{bmatrix}.$$

The cumulative distance is calculated according to

$$\Gamma(i,j) = d(\tilde{\otimes}g_1(i), \tilde{\otimes}g_2(j)) + \min\{\Gamma(i-1,j-1), \Gamma(i-1,j), \Gamma(i,j-1)\}. \tag{13}$$

Thus, the DTW distance between the two interval gray time series $\tilde{\otimes}g_1(t)$ and $\tilde{\otimes}g_2(t)$ is determined by $\Gamma(m,n)$.

4.3 Time Series Classification Based on Interval Gray Transformation Frame

By using the interval gray transformation frame, the uncertainty information of observed time series can be represented by interval gray numbers with known bounds. In this subsection, DTW-1NN is generalized to the interval gray number environment. Therefore, classical point-valued time series can be classified from the perspective of interval gray representation. After the interval gray sequence is constructed, the computational complexity of IGN-DTW-1NN remains the same order as that of the classical DTW-1NN framework.

Given an observed time series set TSS, the proposed classification method can be summarized as follows:

Table 2. Algorithm of time series classification based on the interval gray transformation frame

Input: Observed time series set $TSS = \{y_{it} \mid i = 1, 2, \cdots, n\}$
Output: Time series classification results and error indicators

Step 1: Interval gray implicitization module
for each time series y_{it} in TSS **do**
Fit y_{it} by the ARIMA model introduced in Subsection 3.3 and obtain μ_{it}
Construct the interval gray number time series by Eqs. (10)–(12): $\tilde{\otimes} G_{it}$
end for

Step 2: DTW distance calculation among interval gray time series
for any two interval gray time series $\tilde{\otimes} G_l$ and $\tilde{\otimes} G_k$ in TSS **do**
Construct the distance matrix by calculating $d(\tilde{\otimes} G_l(i), \tilde{\otimes} G_k(j))$
Calculate the DTW distance $d(\tilde{\otimes} G_l, \tilde{\otimes} G_k)$ by Eq. (13)
end for

Step 3: DTW-1NN classification of interval gray time series
Partition TSS into the training set and the test set
for each test sample $\tilde{\otimes} G_{\text{test}}$ **do**
Find its nearest neighbor in the training set according to the interval gray DTW distance
Assign the class label of the nearest neighbor to $\tilde{\otimes} G_{\text{test}}$
end for
Calculate the classification error rate

As shown in Table 2, the proposed method consists of three main stages: interval gray implicitization, interval gray DTW distance calculation, and DTW-1NN classification. The three interval construction modes are respectively applied and compared in the experimental section.

5 Experimental Studies

5.1 Data Description

The experimental datasets are selected from the UCR time series classification archive, which contains time series data from various fields, such as finance, biology, medicine, and meteorology. Each time series in these datasets has been assigned a class label. The UCR archive provides predefined training and testing sets, which are directly used in this study to ensure comparability with existing methods. As shown in Table 3, the selected datasets cover different numbers of classes, training samples, testing samples, and sequence lengths. All methods are evaluated on the predefined training/test splits of the UCR archive, and the error rate is used as the evaluation metric.[1]

Table 3. Data description

Name	$Num_{Classes}$	$Size_{Tran.S}$	$Size_{Test.S}$	Length
Beef	5	30	30	470
Car	4	60	60	577
ECG200	2	100	100	96
Fish	7	175	175	463
Lightning-7	7	70	73	319
ShapeletSim	2	20	180	500
Swedish Leaf	15	500	625	128
ProximalPhalanxOutlineAgeGroup	3	400	205	80
ProximalPhalanxTW	6	205	400	80
DistalPhalanxTW	6	139	400	80
DistalPhalanxOutlineAgeGroup	3	139	400	80
MiddlePhalanxOutlineAgeGroup	3	154	400	80
MiddlePhalanxTW	6	154	399	80
WormsTwoClass	2	77	181	900

The classification performance is evaluated by the error rate. For a dataset, if the number of testing samples is k and the number of correctly classified samples is c, the error rate is defined as

$$error\_rate = \frac{k-c}{k}. \tag{14}$$

[1] $Num_{Classes}$, $Size_{Tran.S}$ and $Size_{Test.S}$ represent the number of classes, size of training set and size of testing set, respectively.

5.2 Data Preprocessing

The original point-valued time series are first transformed into interval gray time series according to the proposed interval gray transformation frame.

Taking the Car dataset as an example, after differencing and stationarity testing, two sequences from each of the four classes are randomly selected. It is observed that these eight sequences become stationary after the same order of differencing, i.e., the third-order differencing. The autocorrelation function and partial autocorrelation function values of these eight third-order differenced sequences are shown in Figs. 1 and 2, respectively, where (a) and (b) are from class 1, (c) and (d) are from class 2, (e) and (f) are from class 3, and (g) and (h) are from class 4.

Table 4. Classification error rates under different interval construction methods

Name	DTW-1NN	IGN I-DTW-1NN	IGN II-DTW-1NN	IGN III-DTW-1NN
Beef	0.36667	0.26667(−0.1)	0.33333(−0.03334)	0.33333(−0.03334)
Car	0.26667	0.15(−0.11667)	0.2(−0.06667)	0.15(−0.11667)
ECG200	0.23	0.17(−0.06)	0.17(−0.06)	0.15(−0.08)
Fish	0.17714	0.091429(−0.085711)	0.068571(−0.108569)	0.08(−0.09714)
Lightning-7	0.274	0.21918(−0.05482)	0.23288(0.04112)	0.24658(−0.02742)
ShapeletSim	0.35	0.07222(−0.27778)	0.08333(−0.26667)	0.24444(−0.10556)
Swedish Leaf	0.208	0.1152(−0.0928)	0.1504(−0.0576)	0.1328(−0.0752)
WormsTwoClass	0.337	0.28729(−0.04971)	0.30387(−0.03313)	0.28729(−0.04971)

From Figs. 1 and 2, it can be observed that the autocorrelation patterns of sequences within the same class are similar to those of sequences from different

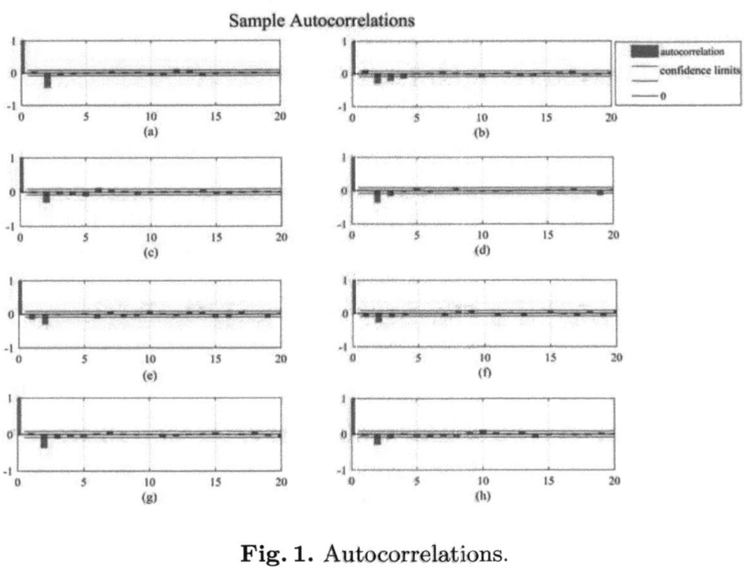

Fig. 1. Autocorrelations.

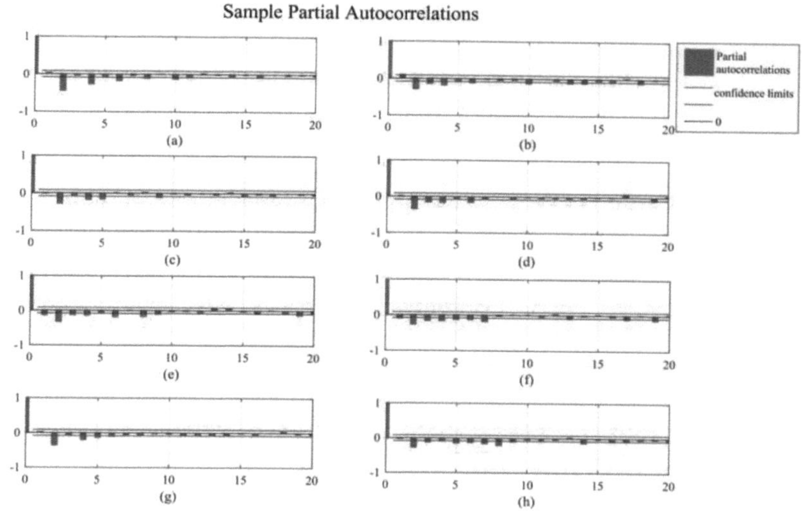

Fig. 2. Partial autocorrelations.

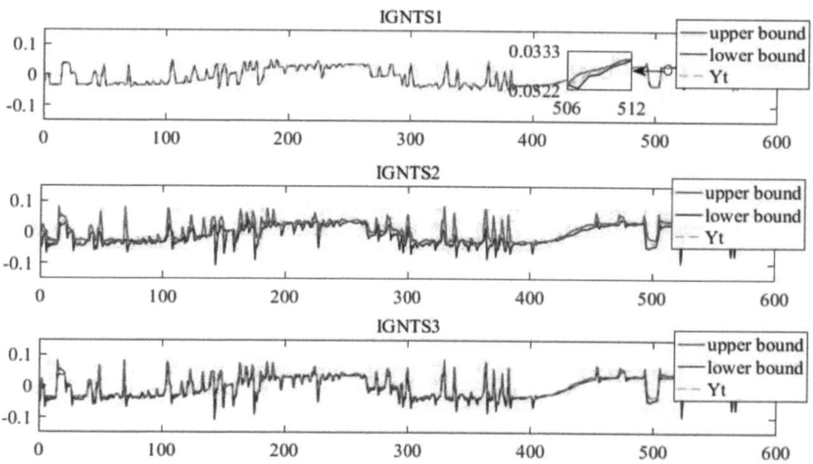

Fig. 3. Lower and upper bounds of transformed interval gray time series.

classes. This phenomenon indicates that similar autocorrelation structures may interfere with direct classification based on point-valued observations. Therefore, each sequence is fitted by an ARIMA model, and the residual fluctuations are extracted to construct interval gray representations.

The maximum likelihood method is used to estimate the parameters of the ARIMA model, and the Akaike information criterion (AIC) is used to select the best model.

After the ARIMA model is established for each sequence, the residual sequence is obtained, and the corresponding interval gray forms can be constructed under the three interval construction modes. The interval gray representations of the selected sequences from the Car dataset are shown in Fig. 3.

In this study, the classification error rates under the three interval construction modes are compared. Table 4 reports the error rates of DTW-1NN and the proposed interval gray DTW-1NN methods. The values in parentheses indicate the changes in error rate compared with the classical DTW-1NN method.

As shown in Table 4, the interval gray transformation frame improves the classification performance on most datasets compared with the classical DTW-1NN method. Among the three construction modes, Mode 1 achieves more stable improvements on the selected datasets. Therefore, Mode 1 is adopted in the following comparative experiments.

5.3 Effectiveness Comparison

Car Dataset. The Car dataset is selected to illustrate the effectiveness of the proposed method for sequences with similar trends and autocorrelation structures.

For the 60 testing sequences, the error rate decreases from 0.26667 under DTW-1NN to 0.15 under the proposed interval gray transformation frame. This means that seven more samples are correctly classified compared with the point-valued DTW-1NN method.

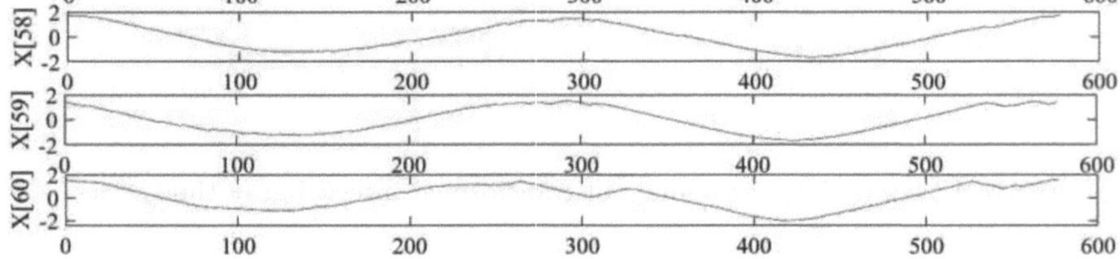

Fig. 4. Point-valued time series.

For the point-valued representation shown in Fig. 4, the classification results of the selected sequences are as follows: Category 1: $x[56]$; Category 2: $x[55]$, $x[58]$; Category 3: $x[59]$; Category 4: $x[57]$, $x[60]$.

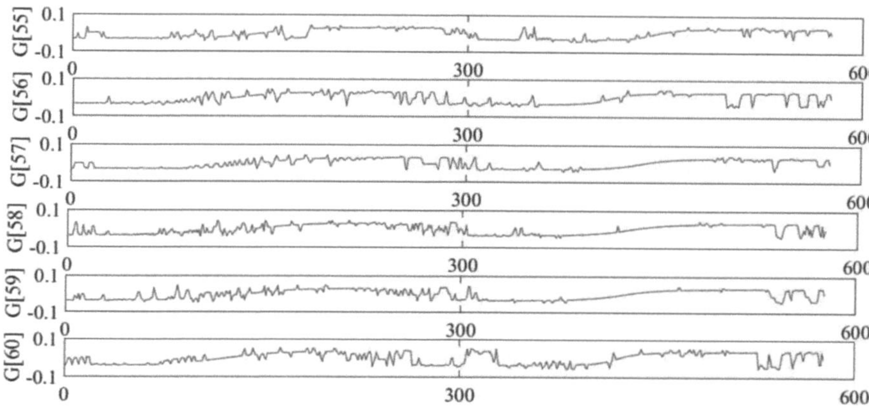

Fig. 5. Interval gray number time series.

For the interval gray representation shown in Fig. 5, the classification results are as follows: Category 1: $G[56]$, $G[57]$; Category 2: $G[55]$; Category 3: $G[58]$, $G[59]$; Category 4: $G[60]$.

Figures 4 and 5 show the classification results under the point-valued representation and the interval gray representation, respectively. The 55th to 60th training sequences are selected for illustration. From direct observation, the overall trends of the six point-valued sequences are very similar, which makes direct classification difficult. By contrast, after the sequences are transformed into interval gray forms, the residual-based uncertainty information enlarges the differences among the sequences. As shown in Fig. 5, the selected sequences can be correctly classified by using the proposed interval gray transformation frame.

ECG200 Dataset. To further examine the effectiveness of the proposed method, the ECG200 dataset is used as another example.

Most sequences in the ECG200 dataset become stationary after the same order of differencing, indicating that their overall trends are similar. However, the autocorrelation patterns are not completely identical across different sequences. This suggests that residual fluctuations may provide useful information for distinguishing sequences with similar global trends.

For the 100 testing sequences, the error rate decreases from 0.23 under DTW-1NN to 0.17 under the interval gray transformation frame. Thus, six more samples are correctly classified compared with the point-valued DTW-1NN method.

Here, three representative sequences are selected. The point-valued classification results are shown in Fig. 7: Category 1: $X[11]$; Category 2: $X[12]$, $X[13]$. After removing the common trend, the stationary sequences and their autocorrelation and partial autocorrelation coefficients are shown in Fig. 6. It can be observed that the residual fluctuations and autocorrelation structures of these sequences are different.

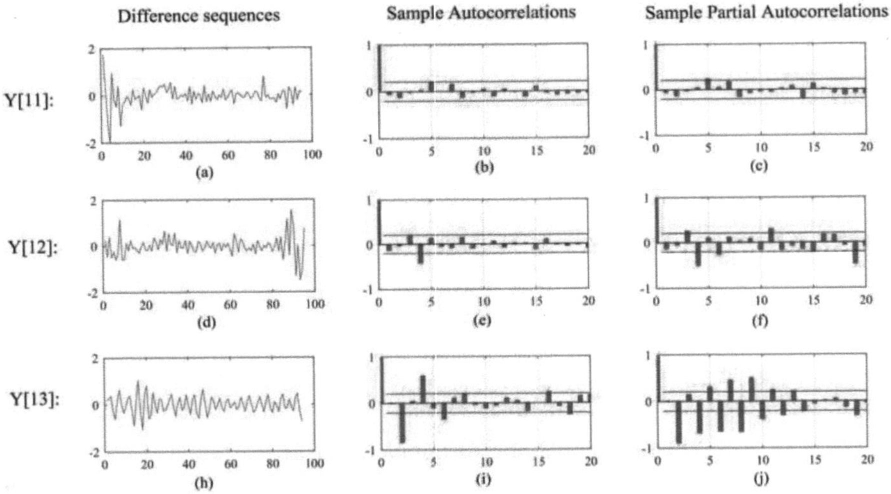

Fig. 6. Autocorrelations.

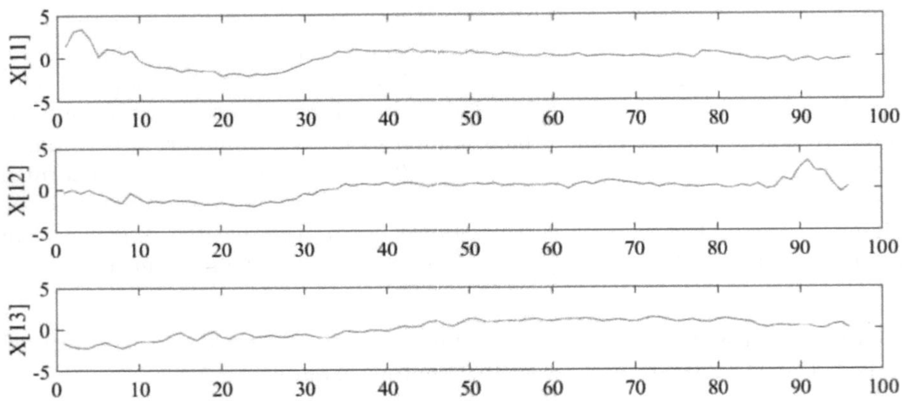

Fig. 7. Point-valued time series.

After interval gray transformation, the classification results are shown in Fig. 8: Category 1: $G[12]$; Category 2: $G[11]$, $G[13]$. Compared with the point-valued representation, the interval gray representation incorporates residual-based uncertainty information. This information helps distinguish sequences that are difficult to separate by using only point-valued observations.

Comparison with Traditional Methods. The above experiments show that IGN-DTW-1NN can effectively improve classification performance for time series with similar trends or autocorrelation structures. In addition, for sequences with

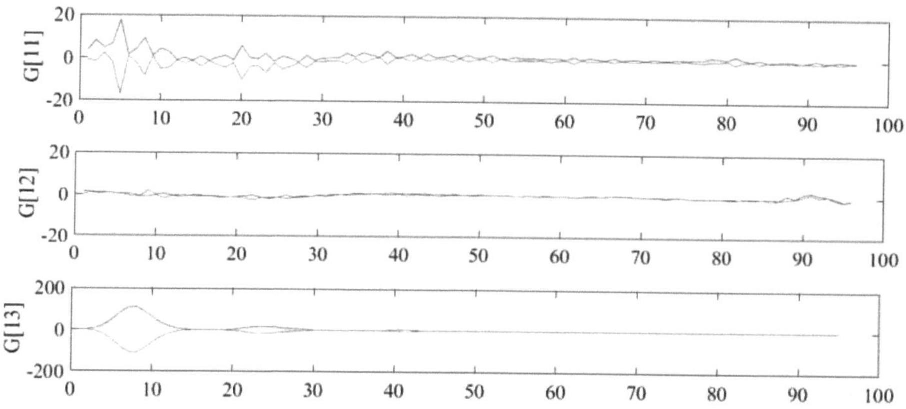

Fig. 8. Interval gray number time series.

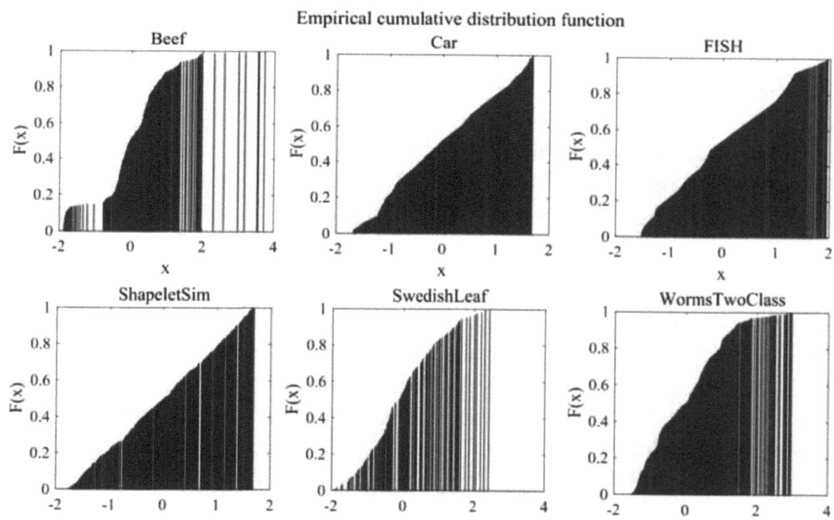

Fig. 9. Empirical cumulative distribution.

relatively uniform value density distributions, as illustrated in Fig. 9, the proposed method also shows obvious improvement over classical DTW-1NN.

To further evaluate the effectiveness of the proposed method, several representative methods are selected for comparison, including DTW-1NN, DDTW-1NN, EE, PF, and BOSS. The comparison results are shown in Table 5.

Table 5. Classification error rates compared with traditional methods

Name	DTW-1NN	IGN-DTW-1NN	DDTW-1NN	EE	PF	BOSS
Beef	0.36667	0.26667	0.4433	0.532	0.28	0.2
Car	0.26667	0.15	0.2685	0.2005	0.1533	0.1667
DisPhOAG	0.208	0.205	0.2667	0.232	0.2691	0.2518
DisPhTW	0.29	0.2875	0.3955	0.3459	0.3403	0.3237
ECG	0.23	0.17	0.1832	0.119	0.091	0.13
Fish	0.17714	0.091429	0.0789	0.0896	0.0651	0.0114
Lightning-7	0.274	0.26027	0.3063	0.2366	0.1781	0.3151
MidPhOAG	0.25	0.285	0.4235	0.3912	0.4377	0.4545
MidPhTW	0.416	0.40602	0.4894	0.4749	0.4708	0.4545
ProPhOAG	0.195	0.2	0.2287	0.1947	0.1537	0.1659
ProPhTW	0.263	0.2425	0.2633	0.2409	0.221	0.2
ShapeletSim	0.350	0.072222	0.3674	0.1727	0.2239	0
Swedish Leaf	0.208	0.1152	0.1039	0.0841	0.0534	0.0784
Worms2	0.337	0.28729	0.2908	0.2826	0.2156	0.1688

Table 5 reports the classification error rates on all selected datasets. The second and third columns compare the classical point-valued DTW-1NN method with the proposed IGN-DTW-1NN method. It can be observed that IGN-DTW-1NN reduces the error rate on most datasets. In particular, the improvement is obvious on datasets such as Beef, Car, ECG200, Fish, ShapeletSim, Swedish Leaf, and WormsTwoClass.

Compared with DDTW-1NN and EE, the proposed method achieves competitive or better results on several datasets while maintaining the simple structure of DTW-1NN. Compared with PF and BOSS, IGN-DTW-1NN does not always obtain the lowest error rate, but it provides a different advantage: it improves the classical DTW-1NN framework by incorporating residual-based uncertainty information while preserving interpretability and ease of implementation.

Therefore, the experimental results demonstrate that the proposed interval gray transformation frame is effective for improving point-valued time series classification, especially when residual fluctuations and uncertainty information are useful for distinguishing sequences with similar overall trends. The comparative results of the existing methods are available at http://timeseriesclassification.com.

6 Conclusion

This paper proposes an interval gray transformation frame for point-valued time series classification, namely IGN-DTW-1NN. In the proposed framework, the original point-valued time series is first transformed into a stationary sequence by differencing. Then, an ARIMA model is used to extract residual information, which is further employed to construct interval gray numbers. Based on this interval gray representation, the classical DTW-1NN classifier is extended to the interval gray number environment.

The main advantage of the proposed method is that it provides a simple and interpretable way to incorporate uncertainty information into point-valued time series classification. Different from conventional DTW-1NN, which only measures the similarity between point-valued observations, IGN-DTW-1NN considers both the central value and the residual-based uncertainty range of each time point. Therefore, it can better distinguish time series with similar overall trends or autocorrelation structures. Moreover, the proposed framework preserves the simplicity and interpretability of DTW-1NN, and does not require complex feature engineering or large-scale model training.

Experimental results on several UCR time series classification datasets demonstrate the effectiveness of the proposed method. Compared with the classical DTW-1NN method, IGN-DTW-1NN reduces the classification error rate on most selected datasets. Comparisons with several representative methods further show that the proposed interval gray transformation frame is feasible and effective for improving point-valued time series classification.

Nevertheless, several aspects deserve further investigation. For example, the quality of interval construction depends on the residuals obtained from the ARIMA model, and the current experiments mainly focus on univariate time series. In future work, more flexible residual extraction models will be considered, and the proposed framework can be further extended to multivariate time series, directly observed interval-valued time series, and other time series analysis tasks, such as anomaly detection and clustering.

Acknowledgments. This work was supported in part by the Humanities and Social Sciences Research Youth Project of the Ministry of Education of China (No. 21YJCZH148), the Natural Science Foundation of Anhui Province (Nos. 2108085MG239, 2108085QG290, 2008085QG334, and 2008085MG226), the National Natural Science Foundation of China (Nos. 72171002, 72001001, 71901001, and 72071001), and the Teacher Project of Anhui Ecology and Economic Development Research Center in 2021 (No. AHST2021002).

Disclosure of Interests. The authors have no competing interests to declare that are relevant to the content of this article.

References

1. Abanda, A., Mori, U., Lozano, J.A.: A review on distance based time series classification. Data Min. Knowl. Disc. **33**(2), 378–412 (2019)

2. Bhaskar, N., Philip, N.Y., Manikandan, S.: Time series classification based correlational neural network with bidirectional LSTM for automated detection of kidney disease. IEEE Sens. J. **21**(4), 4811–4818 (2021)
3. Cabrera, D., Sancho, F., Cerrada, M., et al.: Knowledge extraction from deep convolutional neural networks applied to cyclo-stationary time-series classification. Inf. Sci. **524**, 1–14 (2020)
4. Cádiz-Leyton, M., Cabrera-Vives, G., Protopapas, P., Moreno-Cartagena, D., Donoso-Oliva, C., Becker, I.: Uncertainty estimation for time series classification: exploring predictive uncertainty in transformer-based models for variable stars. Astron. Astrophys. **699**, A168 (2025). https://doi.org/10.1051/0004-6361/202453388
5. Caiado, J., Crato, N., Peña, D.: A periodogram-based metric for time series classification. Comput. Stat. Data Anal. **50**(10), 2668–2684 (2006)
6. Chang, W.J., Zhang, Q., Fu, C., Liu, W.Y., Lu, J.: A cross-domain recommender system through information transfer for medical diagnosis. Decis. Support Syst. **143**, 113489 (2021)
7. Chen, L., Chen, D., Yang, F., et al.: A deep multi-task representation learning method for time series classification and retrieval. Inf. Sci. **555**, 17–32 (2020)
8. Chen, W., Shi, K.: Multi-scale attention convolutional neural network for time series classification. Neural Netw. **136**, 126–140 (2021)
9. Do, D.T., Lee, J., Nguyen-Xuan, H.: Fast evaluation of crack growth path using time series forecasting. Eng. Fract. Mech. **218**, 106567 (2019)
10. D'Urso, P., Giovanni, L., Massari, R.: Trimmed fuzzy clustering of financial time series based on dynamic time warping. Ann. Oper. Res. **299**(1), 1379–1395 (2021)
11. Farahani, M.A., et al.: Time-series pattern recognition in smart manufacturing systems: a literature review and ontology. J. Manuf. Syst. **69**, 208–229 (2023)
12. Fröelich, W., Salmeron, J.L.: Evolutionary learning of fuzzy grey cognitive maps for the forecasting of multivariate, interval-valued time series. Int. J. Approximate Reasoning **55**(6), 1319–1335 (2014)
13. Górecki, T.: Using derivatives in a longest common subsequence dissimilarity measure for time series classification. Pattern Recogn. Lett. **45**, 99–105 (2014)
14. Górecki, T., Piasecki, P.: A comprehensive comparison of distance measures for time series classification. In: Steland, A., Rafajłowicz, E., Okhrin, O. (eds.) SMSA 2019. SPMS, vol. 294, pp. 409–428. Springer, Cham (2019). https://doi.org/10.1007/978-3-030-28665-1_31
15. Gupta, A., Gupta, H.P., Biswas, B., Dutta, T.: Approaches and applications of early classification of time series: a review. IEEE Trans. Artif. Intell. **1**(1), 47–61 (2020)
16. Han, T., Peng, Q.K., Zhu, Z.B., Shen, Y.Q., Abid, N.N.: A pattern representation of stock time series based on DTW. XXPhys. A **550**, 124161 (2020)
17. Hassona, S., Marszalek, W., Sadecki, J.: Time series classification and creation of 2D bifurcation diagrams in nonlinear dynamical systems using supervised machine learning methods. Appl. Soft Comput. **113**, 107874 (2021)
18. Hu, Y., Zhan, P., Xu, Y., Zhao, J., Li, Y., Li, X.: Temporal representation learning for time series classification. Neural Comput. Appl. **33**(8), 3169–3182 (2021)
19. Iwana, B.K., Uchida, S.: Time series classification using local distance-based features in multi-modal fusion networks. Pattern Recogn. **97**, 107024 (2019)
20. Ji, C., Hu, Y., Liu, S., Pan, L., Li, B., Zheng, X.: Fully convolutional networks with shapelet features for time series classification. Inf. Sci. **612**, 835–847 (2022)
21. Kate, R.J.: Using dynamic time warping distances as features for improved time series classification. Data Min. Knowl. Disc. **30**(2), 283–312 (2016)

22. Kowdiki, M., Khaparde, A.: Automatic hand gesture recognition using hybrid meta-heuristic-based feature selection and classification with dynamic time warping. Comput. Sci. Rev. **39**, 100320 (2021)
23. Lahreche, A., Boucheham, B.: A fast and accurate similarity measure for long time series classification based on local extrema and dynamic time warping. Expert Syst. Appl. **168**, 114374 (2020)
24. Li, J., Izakian, H., Pedrycz, W., Jamal, I.: Clustering-based anomaly detection in multivariate time series data. Appl. Soft Comput. **100**, 106919 (2021)
25. Li, Y., Jann, T., Vera-Licona, P.: Benchmarking time-series data discretization on inference methods. Bioinformatics **35**(17), 3102–3109 (2019)
26. Lin, T., Kaminski, N., Bar-Joseph, Z.: Alignment and classification of time series gene expression in clinical studies. Bioinformatics **24**(13), i147–i155 (2008)
27. Lindemann, B., Müller, T., Vietz, H., Jazdi, N., Weyrich, M.: A survey on long short-term memory networks for time series prediction. Procedia CIRP **99**, 650–655 (2021)
28. Lines, J., Bagnall, A.: Time series classification with ensembles of elastic distance measures. Data Min. Knowl. Disc. **29**(3), 565–592 (2015)
29. Lipton, Z.C.: The mythos of model interpretability: in machine learning, the concept of interpretability is both important and slippery. Queue **16**(3), 31–57 (2018)
30. Liu, C., Yuan, Z., Chen, B., Chen, H., Peng, D.: Fuzzy granular anomaly detection using Markov random walk. Inf. Sci. **646**, 119400 (2023). https://doi.org/10.1016/j.ins.2023.119400
31. Liu, M., Yao, P., Cheng, X., Chen, S.: An end-to-end model for time series classification in the presence of missing values. Expert Syst. Appl. **284**, 127954 (2025)
32. Liu, S.F., Fang, Z.G., Xie, N.M.: Algorithm rules of interval gray numbers based on the "kernel" and the degree of grayness of gray numbers. In: 2010 IEEE International Conference on Systems, Man and Cybernetics, pp. 756–760. IEEE (2010)
33. Marteau, P.F., Gibet, S.: On recursive edit distance kernels with application to time series classification. IEEE Trans. Neural Netw. Learn. Syst. **26**(6), 1121–1133 (2017)
34. Mohammadi Foumani, N., Miller, L., Tan, C.W., Webb, G.I., Forestier, G., Salehi, M.: Deep learning for time series classification and extrinsic regression: a current survey. ACM Comput. Surv. **56**(9), 1–45 (2024)
35. Sezer, O.B., Gudelek, M.U., Ozbayoglu, A.M.: Financial time series forecasting with deep learning: a systematic literature review: 2005–2019. Appl. Soft Comput. **90**, 106181 (2020)
36. Silva, D.F., Giusti, R., Keogh, E., Batista, G.E.A.P.A.: Speeding up similarity search under dynamic time warping by pruning unpromising alignments. Data Min. Knowl. Disc. **32**(4), 988–1016 (2018). https://doi.org/10.1007/s10618-018-0557-y
37. Souza, V.M.A., Veiga, P.S., Ribeiro, A.G.: Visemble: a fast ensemble approach for time series classification with multiple visual representations. Knowl.-Based Syst. **309**, 112864 (2025). https://doi.org/10.1016/j.knosys.2024.112864
38. Switonski, A., Josinski, H., Wojciechowski, K.: Dynamic time warping in classification and selection of motion capture data. Multidimension. Syst. Signal Process. **30**(3), 1437–1468 (2019)
39. Tabassum, N., Menon, S., Jastrzebska, A.: Time-series classification with SAFE: simple and fast segmented word embedding-based neural time series classifier. Inf. Process. Manag. **59**(5), 103044 (2022). https://doi.org/10.1016/j.ipm.2022.103044

40. Vinh, V.T., Anh, D.T.: A novel clustering-based 1-nn classification of time series based on MDL principle. In: Król, D., Madeyski, L., Nguyen, N. (eds.) Recent Developments in Intelligent Information and Database Systems. Studies in Computational Intelligence, vol. 642, pp. 29–40. Springer, Cham (2016)
41. Wu, J., Yao, L., Liu, B.: An overview on feature-based classification algorithms for multivariate time series. In: 2018 IEEE 3rd International Conference on Cloud Computing and Big Data Analysis (ICCCBDA), pp. 32–38. IEEE (2018)
42. Xie, N., Liu, S.: Interval grey number sequence prediction by using non-homogenous exponential discrete grey forecasting model. J. Syst. Eng. Electron. **26**(1), 96–102 (2015)
43. Xiong, N.X., Wang, Y.M.: Multi-attribute decision-making method of interval grey number based on prospect theory and evidential reasoning. Comput. Syst. Appl. **28**(9), 33–40 (2019)
44. Xiong, P., He, Z., Chen, S., et al.: A novel GM(1, n) model based on interval gray number and its application to research on smog pollution. Kybernetes **49**(3), 753–778 (2019)
45. Xiong, P.P., Yin, Y., Shi, J., et al.: Nonlinear multivariable GM(1, n) model based on interval gray number sequence. J. Grey Syst. **30**(3), 33–47 (2018)
46. Yan, S.L., Liu, S.F., Fang, Z.G., Zhu, J.J., Wu, L.F.: Dynamic risk grey target decision making method based on cumulative prospect theory. Control Decis. **28**(11), 1655–1666 (2013)
47. Yang, X., Liu, B.: Uncertain time series analysis with imprecise observations. Fuzzy Optim. Decis. Making **18**(3), 263–278 (2019)
48. Ye, J., Dang, Y.G., Ding, S.: Grey prediction model of interval grey numbers based on axion of generalized non-decrease grey degree. Control Decis. **31**(10), 1831–1836 (2016)
49. Younis, R., Hakmeh, A., Ahmadi, Z.: Mts2graph: interpretable multivariate time series classification with temporal evolving graphs. Pattern Recogn. **152**, 110486 (2024)
50. Zg, A., Vk, B., Mi, B., et al.: Weighted kNN and constrained elastic distances for time-series classification. Expert Syst. Appl. **162**, 113829 (2020)
51. Zhao, B., Xing, H., Wang, X., Song, F., Xiao, Z.: Rethinking attention mechanism in time series classification. Inf. Sci. **627**, 97–114 (2023). https://doi.org/10.1016/j.ins.2023.01.093
52. Zhou, Y., Ren, H., Li, Z., Pedrycz, W.: An anomaly detection framework for time series data: an interval-based approach. Knowl.-Based Syst. **228**, 107153 (2021)
53. Zhu, C., Zhang, P., Li, P., Chen, Y., Zheng, X.: Information granules-based long-term forecasting of time series via BPNN under three-way decision framework. Inf. Sci. **634**, 351–372 (2023). https://doi.org/10.1016/j.ins.2023.03.133

Rule-Anchored Semantic Alignment: A Residual TSK Network for Unsupervised Domain Adaptation

Xiu Chen[1], Shengyu Liu[1], Can Ti[1], Yuanpeng Zhang[1(✉)], Saikit Lam[2,3(✉)], and Shitong Wang[4(✉)]

[1] Department of Medical Informatics, Nantong University, Nantong, Jiangsu, China
y.p.zhang@ieee.org

[2] Department of Biomedical Engineering, The Hong Kong Polytechnic University, Hong Kong SAR, China
saikit.lam@polyu.edu.hk

[3] Research Institute for Smart Ageing, The Hong Kong Polytechnic University, Hong Kong SAR, China

[4] School of Computer Science and Artificial Intelligence, Jiangnan University, Wuxi, Jiangsu, China
wxwangst@aliyun.com

Abstract. Existing Deep Unsupervised Domain Adaptation (UDA) methods primarily rely on aligning black-box feature distributions, which often leads to semantic drift and lacks transparency regarding where the alignment occurs. To address these limitations, we propose an Interpretable Residual TSK UDA Network. First, we introduce a Residual TSK Fuzzy Layer that serves as a structural foundation for alignment. We treat the learnable fuzzy rules as semantic anchors that partition the high-dimensional feature space into interpretable fuzzy sub-regions while preserving feature discriminability via a residual connection. Second, leveraging this structured representation, we propose Rule-Class Conditional Alignment. Unlike global alignment strategies, our Rule-Anchored Semantic Alignment treats fuzzy rules as shared coordinate bases to enforce fine-grained consistency between source and target domains, effectively mitigating negative transfer caused by noisy pseudo-labels. We integrate these components into a dual-classifier mutual learning framework to refine decision boundaries. Extensive experiments on VisDA-2017, Office-Home, and ImageCLEF-DA benchmarks demonstrate that our method achieves state-of-the-art performance and offers intuitive visualizations of the adaptation process through rule activation dynamics.

Keywords: Unsupervised Domain Adaptation · TSK Fuzzy Systems · Residual Learning · Deep Neural Networks · Interpretability

J. Lu et al. (Eds.): FLINS-ISKE 2026, LNCS 16752, pp. 455–472, 2027.
https://doi.org/10.1007/978-981-92-2369-5_28

1 Introduction

Deep learning has achieved strong performance in computer vision, yet its reliance on large-scale annotations limits deployment in open-world scenarios. In practice, changes in imaging conditions, sensors, and backgrounds often induce distribution shifts, which may cause significant performance drops on the target domain. Unsupervised domain adaptation (UDA), under the setting where the source domain is labeled while the target domain is unlabeled, transfers knowledge from the source to improve recognition on the target and has been widely used to mitigate the annotation burden and domain shift [24]. Existing UDA methods mainly reduce domain discrepancy, originally focusing on distribution matching based on discrepancy measures and adversarial learning for domain-invariant representations. Recently, the paradigm has shifted towards Transformer-based architectures [12] and masked image consistency learning [10], which achieve remarkable performance benchmarks. However, these state-of-the-art methods typically rely on computationally intensive self-attention mechanisms and massive pre-training, often exacerbating the "black-box" nature of the model. Consequently, adaptation is typically conducted in latent deep feature spaces, making it difficult to observe and explain where and how alignment takes place [3].

Structurally, early UDA approaches often impose alignment in a global feature space, encouraging the overall source and target distributions to be close, while the local structure inside the feature space is rarely modeled explicitly [8,28]. However, domain shift usually comes with changes in class-wise multi-modal structures and reorganization of inter-class geometry. Pure global alignment may reduce the overall discrepancy by compressing or moving clusters across different categories, which can lead to class mismatch and semantic drift [5,21]. To alleviate class mismatch, class-conditional alignment with pseudo labels has been explored, but pseudo-label noise can make the alignment unstable: incorrect pseudo labels may pull samples toward wrong class regions and get further reinforced during training, resulting in negative transfer. In other words, UDA requires fine-grained semantic alignment to avoid class entanglement, while also needing a mechanism to suppress erroneous alignment driven by noisy pseudo supervision. With black-box deep features alone, it is often hard to localize the issue to specific classes or local regions. Without explicit structural anchors, adaptation in the latent space is akin to navigating without landmarks, making the model prone to semantic drift.

Fuzzy logic systems provide a tool for structured representations and interpretable alignment. In particular, the Takagi–Sugeno–Kang (TSK) fuzzy system partitions the space into fuzzy sub-regions via antecedent rules and learns local mappings within these regions. Rule activation strengths indicate the degree of membership to different sub-regions, naturally encoding uncertainty and offering an interpretable handle to track adaptation at the rule level. We argue that TSK fuzzy rules can serve as interpretable anchors. By aligning features relative to these stable rule prototypes rather than unstable global distributions, we can achieve Rule-Anchored Semantic Alignment [18,29].

Motivated by this, we propose a Residual TSK UDA Network. We insert a Residual TSK Fuzzy Layer between the feature extractor and the classifier. The residual pathway preserves discriminative deep features, while the TSK pathway softly partitions the feature space with fuzzy rules, producing interpretable rule activations and fuzzy sub-region representations. Building on this structure, we introduce Rule–Class Conditional Alignment by constructing a rule–class association matrix and enforcing consistency of class distributions conditioned on rule activations across domains, pushing alignment from global statistics to fine-grained semantics. Finally, we integrate this alignment into a dual-classifier mutual learning framework: prediction discrepancy highlights hard samples near the decision boundary and, in turn, encourages the TSK layer to refine rule structures and coverage for better target adaptation, while keeping an interpretable entry for analyzing the adaptation process.

The main contributions are summarized as follows:

- We introduce a residual TSK fuzzy layer that preserves discriminability while providing rule-space representations and interpretable activation signals.
- We propose a rule–class conditional alignment loss that uses rule activations as semantic anchors to align class distributions under rule conditions.
- We combine rule-space alignment with dual-classifier discrepancy learning, using boundary hard-sample signals to facilitate adaptive rule refinement, improving both performance and interpretability.

2 Related Work

2.1 Deep Unsupervised Domain Adaptation

The goal of UDA is to learn representations that remain effective on an unlabeled target domain. Existing methods can be broadly grouped into discrepancy-based distribution matching and adversarial domain-invariant learning. Discrepancy-based approaches align domains by minimizing distribution distances, including kernel mean matching, correlation alignment, and optimal-transport-inspired objectives. They are often stable to optimize, yet marginal alignment may cause class mismatch when data exhibit pronounced multi-modality or when inter-class geometry is reorganized: the overall distributions become closer at the cost of inter-class separability [31]. Adversarial methods employ a minimax game between a feature extractor and a domain discriminator to obtain domain-invariant features. While effective in an end-to-end manner, they may prioritize domain confusion and induce semantic drift or boundary shift when class-conditional discrepancies are large.

To reduce class mismatch, recent studies push alignment from marginal distributions to class-conditional consistency, typically relying on pseudo labels or clustering, including conditional adversarial alignment, joint distribution matching, and prototype/centroid alignment [20]. Such strategies better preserve semantic structure but are sensitive to pseudo-label noise: errors, especially at early stages or near ambiguous regions, may pull samples toward wrong classes and get reinforced across iterations, leading to negative transfer [19]. In practice, UDA often faces a tension: global alignment risks class entanglement, whereas class-conditional alignment can be biased by noisy pseudo supervision.

Boundary- or discrepancy-driven adaptation is closely related to our work. These methods commonly adopt two classifiers and exploit prediction discrepancy to identify target samples near the decision boundary, then optimize the representation to move them toward reliable class regions [6]. To stabilize training, consistency regularization and teacher–student self-training are also widely used to suppress oscillations across iterations [2]. Despite strong empirical performance, most approaches still perform alignment in black-box deep feature spaces, making it hard to localize where negative transfer or class mismatch originates.

Recently, the rapid development of Vision Transformers (ViT) has significantly pushed the boundaries of UDA. Methods such as TVT [27] and CDTrans [26] leverage the self-attention mechanism to capture long-range dependencies, achieving superior transfer performance compared to CNN-based baselines. Furthermore, self-supervised pre-training paradigms, exemplified by MIC [10], improve adaptation by learning context-aware features through masked image consistency. Despite their state-of-the-art accuracy, these approaches often require substantial computational resources and function as complex "black boxes". In contrast, our work focuses on constructing a transparent and interpretable adaptation framework using fuzzy rules, which allows for explicit tracking of semantic drift and rule-level decision dynamics, offering a complementary perspective to these deep black-box models.

2.2 Deep Fuzzy Systems

Deep fuzzy systems provide a path to introduce structured modeling and interpretability into deep networks. Neuro-fuzzy models such as ANFIS parameterize fuzzy inference, making rules learnable while retaining rule-level interpretability. Recently, fuzzy/rule layers have been embedded into deep networks to model uncertainty and local structure through rule activations. In particular, the TSK system defines fuzzy sub-regions via antecedent rules and learns local mappings via consequents, where rule centers and activation strengths serve as observable structured units for analysis [32]. Most existing works focus on supervised learning or robustness; for UDA, it remains less explored how to update rule structures stably without target labels and how to formulate alignment objectives as interpretable rule-level constraints rather than post-hoc visualizations.

In interpretable transfer learning, one line relies on attention or saliency-based explanations, which are intuitive but often weakly coupled with the training objective [1]. Another line adopts structured representations such as prototypes, concepts, or subspaces so that transfer is conducted over more interpretable units. However, in UDA, such units must simultaneously cover target-domain diversity, resist noisy pseudo supervision, and avoid degrading the discriminability of deep features.

Our work uses TSK rules as structured units for adaptation. A residual design preserves discriminative deep features, and alignment is performed in the rule space by enforcing consistency of class distributions conditioned on rule activations, enabling direct inspection and localization of class mismatch and semantic drift via rule activations and rule–class association matrices.

3 Residual TSK UDA Network

3.1 Definition

Given a labeled source-domain dataset $\mathcal{D}_s = \{(\mathbf{x}_i^s, y_i^s)\}_{i=1}^{n_s}$ and an unlabeled target-domain dataset $\mathcal{D}_t = \{\mathbf{x}_j^t\}_{j=1}^{n_t}$. Here, $\mathbf{x}_i^s$ and $\mathbf{x}_j^t$ denote input samples from the source and target domains, respectively, and $y_i^s \in \{1, \dots, C\}$ is the ground-truth class label of a source sample. We consider the closed-set UDA scenario where the two domains share the same C categories, while the input marginal distributions differ due to variations in imaging conditions, acquisition devices, or background styles, i.e., $p_s(\mathbf{x}) \neq p_t(\mathbf{x})$.

During training, we jointly leverage the source supervision $(\mathbf{x}_i^s, y_i^s)$ and the unlabeled target samples $\mathbf{x}_j^t$; the true target labels are unavailable for both training and model selection. During testing, the model is evaluated only on the target domain. Our goal is to achieve strong discriminative performance on the target domain without using any target annotations, while maintaining interpretability so that one can trace which structured units are associated with each class and whether such associations are preserved across domains.

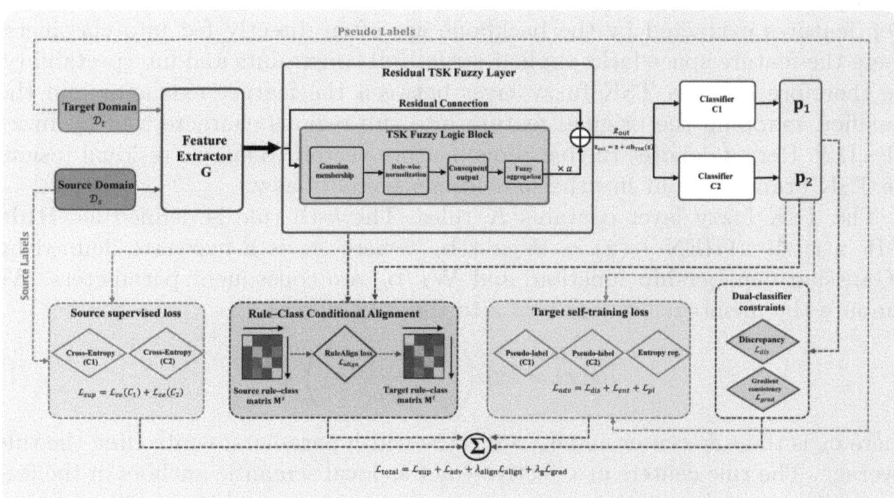

Fig. 1. Overall architecture of the proposed Residual TSK UDA Network.

3.2 Framework and Optimization

The overall architecture of the proposed Residual TSK UDA Network is illustrated in Fig. 1. The framework consists of three main components: a deep feature extractor G, a Residual TSK Fuzzy Layer, and two independent classifiers C_1 and C_2. Input images from both the source domain $\mathcal{D}_s$ and the target domain $\mathcal{D}_t$ are projected into a latent feature space by G. These features are processed by the Residual TSK Fuzzy Layer, which models the local semantic structure using K fuzzy rules and refines the representation via a residual connection $\mathbf{z}_{out} = \mathbf{z} + \alpha h_{\mathrm{TSK}}(\mathbf{z})$. The refined features are subsequently fed into the dual classifiers to obtain the final probability predictions $\mathbf{p}_1$ and $\mathbf{p}_2$.

To achieve robust adaptation, the network is optimized using a composite objective function $\mathcal{L}_{\mathrm{total}}$. As illustrated in the training pipeline, this objective comprises four parts: (1) the source supervised loss $\mathcal{L}_{\mathrm{sup}}$ for discriminative learning on labeled data; (2) the target-side adversarial/self-training objective $\mathcal{L}_{\mathrm{adv}}$, which is instantiated as $\mathcal{L}_{\mathrm{adv}} = \mathcal{L}_{\mathrm{dis}} + \mathcal{L}_{\mathrm{ent}} + \mathcal{L}_{\mathrm{pl}}$, where the dual-classifier discrepancy term $\mathcal{L}_{\mathrm{dis}}$ highlights target samples near decision boundaries, and the entropy regularization $\mathcal{L}_{\mathrm{ent}}$ together with the pseudo-labeling loss $\mathcal{L}_{\mathrm{pl}}$ stabilizes target optimization; (3) the gradient consistency constraint $\mathcal{L}_{\mathrm{grad}}$, which couples the boundary-induced updates with consistent feature/rule adjustment; and (4) the proposed Rule-Anchored Semantic Alignment loss $\mathcal{L}_{\mathrm{align}}$. The alignment module constructs Rule-Anchored Semantic Matrices ($\mathbf{M}^s$ and $\mathbf{M}^t$) to enforce topological consistency of rule-class associations across domains, thereby mitigating semantic drift.

Residual TSK Fuzzy Layer. In conventional dual-classifier UDA frameworks, deep features extracted by the backbone are often directly fed into classifiers, where the feature space lacks explicit structural constraints and interpretability. We therefore insert a TSK fuzzy layer between the feature extractor and the classifier, mapping the original feature into sub-regions characterized by fuzzy rules [22]. Here $\mathbf{z}$ denotes the backbone feature vector; to enable residual fusion, the TSK branch output has the same dimensionality as $\mathbf{z}$.

The TSK fuzzy layer contains K rules. The k-th rule is defined as: **Rule** k: IF $\mathbf{z}$ is A_k, THEN $f_k(\mathbf{z}) = \mathbf{W}_k\mathbf{z} + \mathbf{b}_k$, where A_k is a fuzzy set defined by a Gaussian membership function, and $\mathbf{W}_k, \mathbf{b}_k$ are consequent parameters. We compute the membership degree of $\mathbf{z}$ to the k-th rule as

$$\mu_k(\mathbf{z}) = \exp\left(-\frac{\|\mathbf{z} - \mathbf{c}_k\|^2}{2\delta_k^2}\right) \tag{1}$$

where $\mathbf{c}_k$ is the rule center and $\delta_k > 0$ is the width parameter controlling the rule coverage. The rule centers $\mathbf{c}_k$ effectively act as local semantic anchors in the feature space, capturing representative patterns shared across domains. To compare the relative importance of different rules, we normalize the firing strengths:

$$\tilde{\mu}_k(\mathbf{z}) = \frac{\mu_k(\mathbf{z})}{\sum_{j=1}^{K} \mu_j(\mathbf{z})} \tag{2}$$

The TSK output is the weighted sum of all consequents:

$$\mathbf{h}_{\mathrm{TSK}}(\mathbf{z}) = \sum_{k=1}^{K} \tilde{\mu}_k(\mathbf{z})\,(\mathbf{W}_k\mathbf{z} + \mathbf{b}_k) \tag{3}$$

The rule parameters $\{\mathbf{c}_k, \delta_k, \mathbf{W}_k, \mathbf{b}_k\}$ are learned end-to-end. To stabilize early training, we adopt a short warm-up by freezing the updates of rule centers and widths.

Since fuzzy rules may be unstable at the beginning of training, directly using $\mathbf{h}_{\mathrm{TSK}}(\mathbf{z})$ may disturb the original feature distribution. We adopt a residual design and treat the TSK output as a soft correction [9]:

$$\mathbf{z}_{\mathrm{out}} = \mathbf{z} + \alpha\,\mathbf{h}_{\mathrm{TSK}}(\mathbf{z}) \tag{4}$$

where α balances the correction magnitude, ensuring the model behaves like a standard deep network in early training and gradually incorporates rule-structured information.

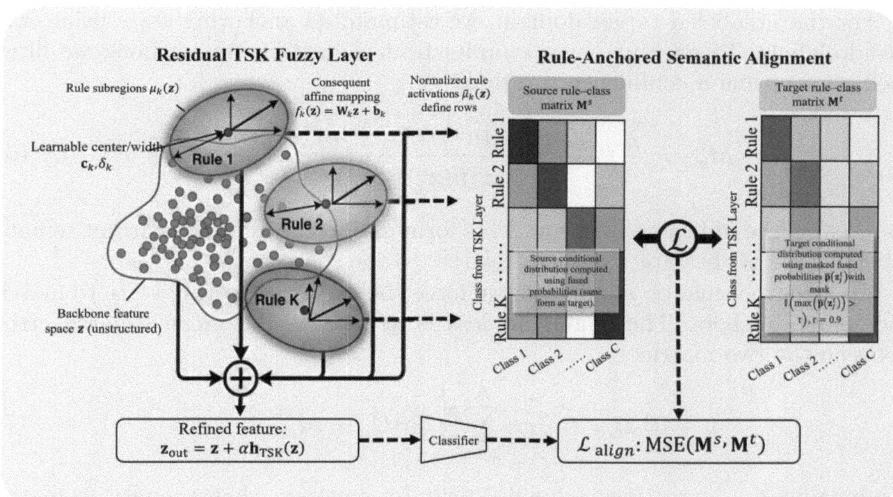

Fig. 2. Principle of the TSK rule space and RuleAlign via rule–class conditional alignment.

Rule-Anchored Semantic Alignment. We propose Rule-Anchored Semantic Alignment to mitigate semantic drift by leveraging TSK rules as stable semantic anchors. Instead of merely aligning global feature distributions, we enforce consistency in the local relationships between fuzzy rules and class semantics.

To quantify this relationship, we construct a Rule-Anchored Semantic Matrix $\mathbf{M} \in \mathbb{R}^{K \times C}$ for each domain, as illustrated in Fig. 2. Unlike a simple response matrix, $\mathbf{M}$ explicitly models the anchoring state of the semantic space: each entry $M_{k,c}$ represents the anchoring strength of class c relative to the k-th fuzzy rule. This provides a structural coordinate system where semantic alignment can be measured precisely.

Let $\mathbf{p}_1(\mathbf{z}), \mathbf{p}_2(\mathbf{z}) \in \mathbb{R}^C$ be the softmax probability vectors from classifiers C_1 and C_2. We define the averaged predictive distribution as $\bar{\mathbf{p}}(\mathbf{z}) = \frac{1}{2}(\mathbf{p}_1(\mathbf{z}) + \mathbf{p}_2(\mathbf{z}))$, where the c-th element is denoted by $\bar{p}_c(\mathbf{z})$.

For the labeled source domain, we compute the anchoring matrix $\mathbf{M}^s$ using the model's own predictive distribution. This captures the intrinsic topology of how the source data is clustered around the fuzzy rules. Using the normalized firing strengths $\tilde{\mu}(\mathbf{z})$, we calculate:

$$M_{k,c}^s = \frac{\sum_{i=1}^{N_b} \tilde{\mu}_k(\mathbf{z}_i^s) \cdot \bar{p}_c(\mathbf{z}_i^s)}{\sum_{i=1}^{N_b} \tilde{\mu}_k(\mathbf{z}_i^s) + \epsilon} \tag{5}$$

where N_b is the batch size and ϵ is a stabilizing constant. Here, $\bar{p}_c(\mathbf{z}_i^s)$ represents the predicted probability of class c for the source sample $\mathbf{z}_i^s$.

For the unlabeled target domain, we estimate its anchoring state using soft pseudo-labels. To prevent noisy samples from distorting the anchors, we filter predictions using a confidence threshold τ:

$$M_{k,c}^t = \frac{\sum_{j=1}^{N_b} \mathbb{I}(\max(\bar{\mathbf{p}}(\mathbf{z}_j^t)) > \tau) \cdot \tilde{\mu}_k(\mathbf{z}_j^t) \cdot \bar{p}_c(\mathbf{z}_j^t)}{\sum_{j=1}^{N_b} \mathbb{I}(\max(\bar{\mathbf{p}}(\mathbf{z}_j^t)) > \tau) \cdot \tilde{\mu}_k(\mathbf{z}_j^t) + \epsilon} \tag{6}$$

Here $\mathbb{I}(\cdot)$ is the indicator function. This formulation captures the current semantic structure of the target domain relative to the shared rules.

To achieve semantic alignment, we force the target anchoring state to match the source anchors. The alignment loss is defined as the mean squared error between the two matrices:

$$\mathcal{L}_{align} = \frac{1}{KC} \sum_{k=1}^{K} \sum_{c=1}^{C} (M_{k,c}^s - M_{k,c}^t)^2 \tag{7}$$

Minimizing $\mathcal{L}_{align}$ rectifies semantic drift by ensuring that samples from the same class in the target domain activate the same set of rule anchors as they do in the source domain.

Dual-Classifier Optimization with Gradient Consistency. To improve robustness near the target decision boundary, we integrate the above TSK fuzzy framework into dual-classifier adversarial learning with two classifiers C_1 and C_2. We use prediction discrepancy to identify target samples away from the source support. The discrepancy loss is defined as the L_1 distance between the two output distributions:

$$\mathcal{L}_{\text{dis}} = \mathbb{E}_{\mathbf{x}^t \sim \mathcal{D}_t} \left[\|\mathbf{p}_1(\mathbf{z}) - \mathbf{p}_2(\mathbf{z})\|_1 \right], \quad \mathbf{z} = G(\mathbf{x}^t) \tag{8}$$

where G denotes the feature extractor (backbone with the residual TSK layer when enabled). With G fixed, we maximize $\mathcal{L}_{\text{dis}}$ to expose boundary samples; with C_1, C_2 fixed, we minimize $\mathcal{L}_{\text{dis}}$ to move features toward regions where the two classifiers agree.

We adopt an alternating optimization strategy with three steps in each iteration: (1) jointly update G and (C_1, C_2) by minimizing the source supervised loss $\mathcal{L}_{\text{sup}}$ together with target-side entropy regularization and pseudo-label supervision; (2) fix G and update (C_1, C_2) to maximize $\mathcal{L}_{\text{dis}}$ under source supervision; (3) fix (C_1, C_2) and update G for multiple inner steps to minimize $\mathcal{L}_{\text{dis}}$ while jointly minimizing $\mathcal{L}_{\text{align}}$ and the gradient constraint $\mathcal{L}_{\text{grad}}$, where $\mathcal{L}_{\text{align}}$ is applied only to high-confidence target samples with $\max(\bar{\mathbf{p}}(\mathbf{z}_j^t)) > \tau$.

For gradient stabilization, instead of explicitly measuring gradient angles, we implement $\mathcal{L}_{\text{grad}}$ using a GMN-style margin-based gradient discrepancy, which suppresses unstable gradient divergence during adaptation and is optimized together with $\mathcal{L}_{\text{dis}}$ when updating G.

Overall, the objective can be summarized as

$$\mathcal{L}_{\text{total}} = \mathcal{L}_{\text{sup}} + \mathcal{L}_{\text{adv}} + \lambda_{\text{align}}\mathcal{L}_{\text{align}} + \lambda_{\text{g}}\mathcal{L}_{\text{grad}} \tag{9}$$

where $\mathcal{L}_{\text{sup}} = \mathcal{L}_{\text{ce}}(C_1, y^s) + \mathcal{L}_{\text{ce}}(C_2, y^s)$, and $\mathcal{L}_{\text{adv}}$ includes $\mathcal{L}_{\text{dis}}$ together with its accompanying entropy regularization and pseudo-label supervision. The alternating scheme enables boundary-sample exploration, feature alignment, and rule-space semantic alignment to be optimized in a unified training pipeline.

4 Experimental Studies

4.1 Experimental Setup

Datasets. We evaluate our method on three standard UDA benchmarks: VisDA-2017 [17], Office-Home [23], and ImageCLEF-DA [4]. VisDA-2017 focuses on the challenging synthetic-to-real shift with 12 categories and large-scale data, and is widely used to test the robustness of UDA methods. Office-Home contains four domains (Artistic, Clipart, Product, Real-World) with 65 categories and substantial cross-domain appearance variations. ImageCLEF-DA includes three domains (C/I/P) with a small yet balanced set of shared categories, which is suitable for assessing robustness under limited data.

Compared Methods. We compare against representative UDA approaches including adversarial alignment methods (e.g., DANN [7], CDAN [14]) and decision-boundary based methods (e.g., MCD [21]). Notably, MCD alternates between maximizing classifier discrepancy to expose boundary target samples and minimizing it by updating the feature generator, which is closely related to our dual-classifier optimization paradigm. To ensure a fair comparison, we strictly evaluate our method against representative ResNet-based approaches. While recent Transformer-based models (e.g., TVT [27]) achieve higher raw accuracy, our focus is on demonstrating the effectiveness and interpretability of the proposed residual TSK layer under the same backbone architecture.

Algorithm 1: Optimization Procedure of Residual TSK UDA Network

Input: Source data $\mathcal{D}_s$, Target data $\mathcal{D}_t$;
Hyper-parameters: Rules K, Residual α, Threshold τ, Weights $\lambda_{align}, \lambda_g$.
Output: Optimized parameters $\theta_G, \theta_{C_1}, \theta_{C_2}, \theta_{TSK}$.
Initialize network parameters and TSK centers c_k, widths δ_k
for *epoch* $= 1$ **to** *MaxEpochs* **do**
  if *epoch* $<$ *WarmUp* **then**
    Freeze updates for TSK centers c_k and widths δ_k
  end
  Generate pseudo-labels $\hat{y}^t$ for $\mathcal{D}_t$ using ensemble predictions
  for *batch* $= 1$ **to** N_{iter} **do**
    Sample $x^s, y^s \sim \mathcal{D}_s$ and $x^t \sim \mathcal{D}_t$
    Compute residual features: $z_{out} = G(x) + \alpha h_{TSK}(G(x))$ (Eq. 4)
    `// Step A: Source Supervision & Self-Training (min-only terms)`
    Update $\theta_G, \theta_{C_1}, \theta_{C_2}, \theta_{TSK}$ to **minimize** $\mathcal{L}_{sup} + \mathcal{L}_{ent} + \mathcal{L}_{pl}$
    `// Step B: Maximize Discrepancy (w.r.t. classifiers)` Fix θ_G, θ_{TSK}. Update $\theta_{C_1}, \theta_{C_2}$ to **maximize** $\mathcal{L}_{dis}$ (Eq. 8)
    `// Step C: Alignment & Generator Update (w.r.t.` G`)` Fix $\theta_{C_1}, \theta_{C_2}$. Initialize the objective for updating (θ_G, θ_{TSK})
      $\mathcal{L}_C = \mathcal{L}_{dis} + \lambda_g \mathcal{L}_{grad} + \mathcal{L}_{ent}$
    if *epoch* $>$ *StartRule* **then**
      `// Rule-Anchored Semantic Alignment` Compute normalized rule activations $\tilde{\mu}_k(z)$ (Eq. 2)
      Construct **Source Anchoring Matrix** M^s (Eq. 5)
      Construct **Target Anchoring Matrix** M^t using confident samples $\mathbb{I}(\max \bar{p} > \tau)$ (Eq. 6)
      Add alignment loss: $\mathcal{L}_C \leftarrow \mathcal{L}_C + \lambda_{align} \mathcal{L}_{align}$ (Eq. 7)
    end
    Update θ_G, θ_{TSK} to **minimize** $\mathcal{L}_C$
  end
end
return Optimized Network

Parameter Settings. Our method is implemented in PyTorch and trained on a single NVIDIA GPU. We adopt different backbones for different benchmarks: ResNet-101 for VisDA-2017, and ResNet-50 for both ImageCLEF-DA and Office-Home. The batch size is fixed to 32 for all experiments. We use mini-batch stochastic gradient descent (SGD) with momentum 0.9 and weight decay 5×10^{-4}. Following the standard UDA practice, we employ a layer-wise learning-rate scheme to balance stability and adaptation: the pre-trained backbone is updated with $0.1 \times$ base_lr, while the TSK fuzzy layer, the bottleneck, and the classifier are updated with $1.0 \times$ base_lr. The base learning rate is set to 3×10^{-4} on VisDA-2017 and 10^{-3} on the other datasets. We train for 10 epochs on VisDA-2017 and 100 epochs on the other two benchmarks.

For hyper-parameters, we set the number of fuzzy rules to $K = 8$ for all datasets, and fix the weight of the rule–class conditional alignment loss to 0.005

throughout the experiments. The residual coefficient α is adjusted according to the data scale and domain gap: we use $\alpha = 2.0$ for the large-scale VisDA-2017 benchmark and $\alpha = 0.25$ for the other tasks. Unless otherwise specified, we fix the confidence threshold τ in Eq. (6) to 0.8 throughout the main experiments. This threshold is used as a common default hyper-parameter across tasks, rather than being selected individually for each target test task according to target-domain test accuracy. A dedicated sensitivity analysis of τ is provided in Sect. 4.4.

Evaluation Metrics. We report Top-1 accuracy on the target domain for all benchmarks. For VisDA-2017, we additionally report per-class accuracy and the mean over 12 classes. For ImageCLEF-DA and Office-Home, we report accuracy for each transfer task and the averaged accuracy across tasks. For ImageCLEF-DA, we repeat experiments with three random seeds and report mean $\pm$ std.

4.2 Performance Evaluation

(1) On VisDA-2017. Table 1 reports the adaptation performance on the Synthesis-to-Real task. Our method achieves a new state-of-the-art mean accuracy of 82.8%, outperforming the strongest competitor STAR [16] while maintaining a significant lead over the ResNet-101 baseline. Detailed inspection reveals that our method dominates in 7 out of 12 categories. A key highlight is the 13.6% improvement in the "truck" class (55.8% vs. 42.2% for STAR). Since trucks and cars share similar visual traits and are frequent sources of confusion in VisDA, this boost suggests that the Residual TSK Layer effectively refines decision boundaries for hard classes. Although this comes with a slight trade-off in the "car" category, the overall stability demonstrates the efficacy of rule-class conditional alignment in large-scale domain shifts.

Table 1. Classification accuracy (%) of different methods on the VisDA-2017 validation set.

Method	plane	bcycl	bus	car	horse	knife	mcycl	person	plant	sktbrd	train	truck	mean
ResNet-101 [9]	55.1	53.3	61.9	59.1	80.6	17.9	79.7	31.2	81.0	26.5	73.5	8.5	52.4
DANN [7]	81.9	77.7	82.8	44.3	81.2	29.5	65.1	28.6	51.9	54.6	82.8	7.8	57.4
DAN [13]	87.1	63.0	76.5	42.0	90.3	42.9	85.9	53.1	49.7	36.3	85.8	20.7	61.1
MCD [21]	87.0	60.9	83.7	64.0	88.9	79.6	84.7	76.9	88.6	40.3	83.0	25.8	71.9
MCC [11]	88.1	80.3	80.5	71.5	90.1	93.2	85.0	71.6	89.4	73.8	85.0	36.9	78.8
STAR [16]	95.0	**84.0**	**84.6**	**73.0**	91.6	91.8	85.9	78.4	**94.4**	84.7	**87.0**	42.2	82.7
Ours	**96.4**	81.5	78.7	56.3	**94.3**	**95.5**	**89.0**	**79.8**	90.4	**90.6**	84.9	**55.8**	**82.8**

(2) On ImageCLEF-DA. To verify robustness on smaller datasets, we conduct experiments on ImageCLEF-DA as shown in Table 2. Our method achieves the highest mean accuracy of 89.6%, surpassing CDAN+E [14] by 1.9%. Notably, we obtain the best performance on 5 out of 6 transfer tasks, demonstrating that the

Table 2. Classification accuracy (%) on ImageCLEF-DA under all transfer tasks.

Method	I→P	P→I	I→C	C→I	C→P	P→C	mean
ResNet-50 [9]	74.8 ± 0.3	83.9 ± 0.1	91.5 ± 0.3	78.0 ± 0.2	65.5 ± 0.3	91.2 ± 0.3	80.7
DANN [7]	75.0 ± 0.6	86.0 ± 0.3	96.2 ± 0.4	87.0 ± 0.5	74.3 ± 0.5	91.5 ± 0.6	85.0
DAN [13]	74.5 ± 0.4	82.2 ± 0.2	92.8 ± 0.2	86.3 ± 0.4	69.2 ± 0.4	89.8 ± 0.4	82.5
JAN [15]	76.8 ± 0.4	88.0 ± 0.2	94.7 ± 0.2	89.5 ± 0.3	74.2 ± 0.3	91.7 ± 0.3	85.8
DMRL [25]	77.3 ± 0.4	90.7 ± 0.3	97.4 ± 0.3	91.8 ± 0.3	76.0 ± 0.5	94.8 ± 0.3	88.0
CDAN+E [14]	77.7 ± 0.3	90.7 ± 0.2	**97.7 ± 0.3**	91.3 ± 0.3	74.2 ± 0.2	94.3 ± 0.3	87.7
Ours	**79.1 ± 0.2**	**92.9 ± 0.5**	96.9 ± 0.5	**93.5 ± 0.3**	**79.4 ± 0.2**	**95.6 ± 0.2**	**89.6**

Table 3. Classification accuracy (%) on Office-Home under all 12 transfer tasks.

Method	Ar→Cl	Ar→Pr	Ar→Rw	Cl→Ar	Cl→Pr	Cl→Rw	Pr→Ar	Pr→Cl	Pr→Rw	Rw→Ar	Rw→Cl	Rw→Pr	mean
ResNet-50 [9]	34.9	50.0	58.0	37.4	41.9	46.2	38.5	31.2	60.4	53.9	41.2	59.9	46.1
DANN [7]	45.6	59.3	70.1	47.0	58.5	60.9	46.1	43.7	68.5	63.2	51.8	76.8	57.6
DAN [13]	43.6	57.0	67.9	45.8	56.5	60.4	44.0	43.6	67.7	63.1	51.5	74.3	56.3
CDAN+E [14]	50.7	70.6	76.0	57.6	70.0	70.0	57.4	50.9	77.3	70.9	56.7	81.6	65.8
MDD [30]	54.9	73.7	**77.8**	**60.0**	71.4	**71.8**	**61.2**	53.6	78.1	**72.5**	**60.2**	82.3	68.1
Ours	**57.6**	**74.1**	76.5	59.8	**72.7**	71.6	61.1	**53.8**	**79.0**	69.2	59.8	**82.9**	**68.2**

proposed fuzzy alignment mechanism generalizes well across different domains and is not limited to specific data scales.

(3) On Office-Home. Finally, we evaluate our method on the Office-Home dataset, which presents a more complex adaptation scenario with 65 classes. As detailed in Table 3, our model achieves a mean accuracy of 68.2%. It not only significantly improves upon the ResNet-50 baseline (46.1%) but also slightly surpasses the competitive MDD method (68.1%). It is worth noting that our method yields performance gains in hard transfer tasks such as Ar→Cl (57.6% vs. 54.9% for MDD) and Pr→Rw (79.0% vs. 78.1% for MDD). While the margin on this dataset is narrower compared to VisDA-2017, the results confirm that the TSK fuzzy layer maintains discriminative stability even when the category space is large and the domain discrepancies are subtle.

Across three diverse benchmarks ranging from large-scale synthetic data (VisDA) to multi-category object recognition (Office-Home), our proposed Residual TSK framework consistently achieves state-of-the-art or comparable performance. The empirical evidence suggests that incorporating interpretable fuzzy rules does not compromise representation learning; rather, it enhances robustness, particularly in scenarios involving ambiguous decision boundaries (e.g., the "truck" class in VisDA and "Art-to-Clipart" in Office-Home).

4.3 Ablation Study

Table 4 reports the ablation results on VisDA-2017. Adding the residual TSK module (+**T**) brings the largest gain, improving the mean accuracy from 79.7 to 84.1 (+4.4), indicating that the structured rule-based mapping is the primary contributor. When further enabling the rule–class conditional alignment

Table 4. Ablation results on VisDA-2017. **+T** denotes adding the residual TSK module, and **+T+A** further includes the rule–class conditional alignment.

Method	plane	bcycl	bus	car	horse	knife	mcycl	person	plant	sktbrd	train	truck	Avg.
Base	94.9	86.5	**81.4**	64.3	93.9	94.9	87.1	43.8	91.3	83.7	81.6	53.2	79.7
+T	95.7	**86.7**	78.8	**66.6**	**94.9**	94.8	87.8	**80.8**	93.0	**91.3**	83.4	**55.8**	**84.1**
+T+A	**96.4**	81.5	78.7	56.3	94.3	**95.5**	**89.0**	79.8	90.4	90.6	**84.9**	**55.8**	82.8

(**+T+A**), the mean accuracy becomes 82.8, which is slightly lower than **+T** but still clearly above the baseline (+3.1). The non-monotonic behavior mainly stems from a few classes with pronounced drops. In particular, *car* decreases from 66.6 to 56.3 (−10.3) and *bicycle* decreases from 86.7 to 81.5 (−5.2), while several other categories continue to improve (e.g., *plane* 95.7→96.4, *knife* 94.8→95.5, *mcycl* 87.8→89.0, and *train* 83.4→84.9). This is consistent with the fact that the conditional alignment term is a stronger structural regularizer: it aligns class distributions conditioned on rule activations and therefore depends on target-domain prediction statistics. For context-sensitive categories with larger appearance gaps, imperfect early predictions may bias these statistics and introduce mild negative transfer under a fixed hyper-parameter setting. Importantly, the full model remains competitive overall and provides a more explicit semantic correspondence at the rule level, which is the intended role of the alignment component.

4.4 Sensitivity Analysis

To evaluate the robustness of the proposed framework, we investigate the sensitivity of four key hyper-parameters on the $C \to I$ task of the ImageCLEF-DA dataset: the number of fuzzy rules K, the residual scale coefficient α, the weight of the rule-class conditional alignment loss λ_{align}, and the confidence threshold τ. The results are illustrated in Fig. 3.

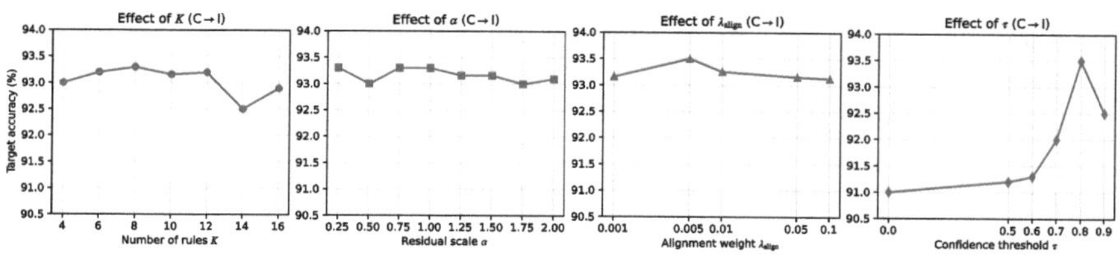

Fig. 3. Parameter sensitivity analysis on the ImageCLEF-DA ($C \to I$) task.

Figure 3(a) reports the classification accuracy with respect to the number of rules K. The performance peaks at $K = 8$. A smaller K limits the capacity to model complex multi-modal distributions, while a larger K degrades performance, likely due to the over-fragmentation of the feature space which hinders the learning of generalized rule-based representations.

The impact of the residual scale α is analyzed in Fig. 3(b). The model maintains high stability within the range $\alpha \in [0.25, 1.0]$, validating the effectiveness of the residual fusion strategy. A slight performance drop is observed at $\alpha = 2.0$, suggesting that an excessive correction signal might interfere with the discriminative features preserved by the backbone network.

Figure 3(c) demonstrates the effect of the alignment weight λ_{align}. The accuracy improves as λ_{align} increases, reaching an optimal value at $\lambda_{align} = 0.005$. Further increasing the weight leads to a decline in performance. This implies that while the rule-class alignment provides necessary semantic regularization, an overly strong constraint may enforce alignment on ambiguous target samples, potentially inducing negative transfer.

Finally, Fig. 3(d) illustrates the effect of the confidence threshold τ, which is utilized to filter noisy pseudo-labels during the construction of the target anchoring matrix. When $\tau = 0.0$ (i.e., no filtering), the accuracy drops to its lowest point, indicating that noisy target predictions disrupt the semantic alignment and cause negative transfer. As τ increases, the performance steadily improves and peaks at $\tau = 0.8$, demonstrating that filtering out uncertain predictions effectively protects the rule anchors from semantic drift. However, a further increase to $\tau = 0.9$ results in a slight performance degradation, likely because an overly strict threshold discards too many correct samples, leaving insufficient target data to estimate robust rule-class statistics.

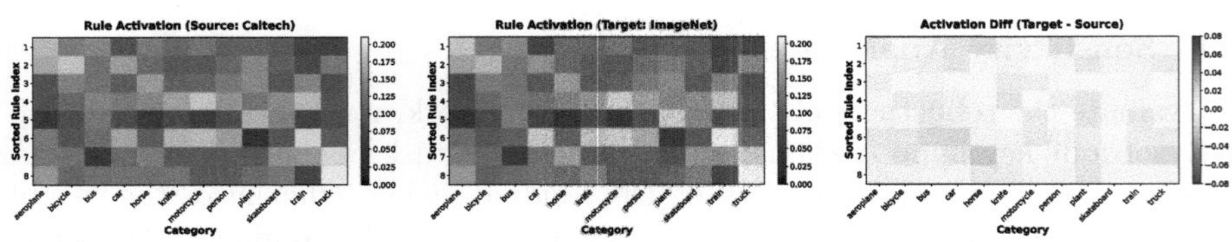

(a) Source Rule Activation (b) Target Rule Activation (c) Activation Difference

Fig. 4. Visualization of TSK rule activations. (a) and (b) show the average rule memberships on source and target domains respectively, while (c) highlights the activation difference.

4.5 Visualization

To intuitively understand how the proposed framework mitigates semantic drift, we visualize the learned rule structures in Fig. 4 and Fig. 5.

Figure 4 displays the class-wise average rule memberships. For both source and target domains, each class consistently activates a specific subset of rules. This stability indicates that the TSK rules effectively function as semantic anchors in the high-dimensional feature space, providing a shared coordinate system that is invariant to domain shifts. The difference map (Fig. 4(c)) remains

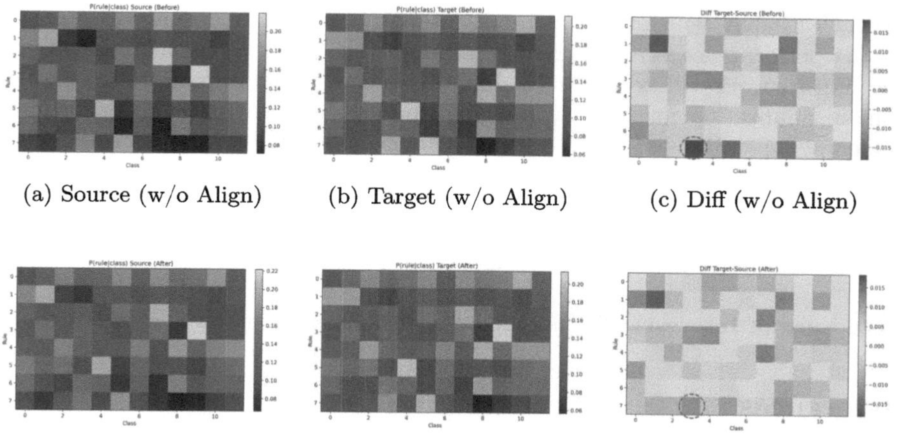

(a) Source (w/o Align) (b) Target (w/o Align) (c) Diff (w/o Align)

(d) Source (w/ Align) (e) Target (w/ Align) (f) Diff (w/ Align)

Fig. 5. Rule-class conditional matrices before (top row) and after (bottom row) alignment. The difference maps in (c) and (f) show that alignment significantly reduces the mismatch between source and target patterns.

sparse, confirming that the anchor functions maintain their selectivity across domains.

Figure 5 further examines the Rule-Anchored Semantic Matrices ($\mathbf{M}^s$ and $\mathbf{M}^t$), which represent the fine-grained logical topology between rules and classes. Without alignment (top row), the target domain's anchoring pattern deviates from the source, leading to prominent "hotspots" in the difference map that signify semantic drift. By enabling our Rule-Anchored Semantic Alignment (bottom row), the target samples are successfully pulled back to their corresponding rule anchors. The resultant difference map becomes remarkably uniform, demonstrating that the proposed method effectively rectifies semantic misalignment by enforcing structural consistency at the rule level.

5 Conclusion

In this paper, we presented a Residual TSK UDA Network to bridge the gap between high-performance adaptation and model interpretability. Unlike conventional black-box methods, our approach explicitly establishes TSK fuzzy rules as learnable semantic anchors. This design transforms the unstructured feature space into a rule-organized topology without compromising discriminative power. Building on this structure, the proposed Rule-Anchored Semantic Alignment mechanism exploits the logical correspondence between rule activations and class semantics. By treating rules as stable reference points, it effectively constrains the target distribution and mitigates semantic drift, even under noisy pseudo-supervision. Extensive experiments on VisDA-2017, Office-Home, and

ImageCLEF-DA benchmarks demonstrate that our method achieves state-of-the-art performance while offering transparent insights into the adaptation process. Future work will explore extending this anchor-based framework to open-set domain adaptation for medical imaging applications, utilizing the uncertainty modeling capabilities of fuzzy rules to enhance reliability when processing data from complex or unseen distributions.

Acknowledgments. This work was supported by the College Students' Innovation and Entrepreneurship Training Program (Grant No. S202510304166).

Disclosure of Interests. The authors have no competing interests to declare.

References

1. Abusitta, A., Li, M.Q., Fung, B.C.: Survey on explainable AI: techniques, challenges and open issues. Expert Syst. Appl. **255**, 124710 (2024)
2. Berthelot, D., Roelofs, R., Sohn, K., Carlini, N., Kurakin, A.: Adamatch: A unified approach to semi-supervised learning and domain adaptation. In: International Conference on Learning Representations (ICLR) (2022). https://openreview.net/forum?id=ccx2PtNpMP
3. Bobek, S., Nowaczyk, S., Pashami, S., Taghiyarrenani, Z., Nalepa, G.J.: Towards explainable deep domain adaptation. In: Proc. European Conference on Artificial Intelligence (ECAI), pp. 101–113. Springer (2023)
4. Caputo, B., et al.: ImageCLEF 2014: Overview and analysis of the results. In: International Conference of the Cross-language Evaluation Forum for European Languages, pp. 192–211. Springer (2014)
5. Deng, Z., Luo, Y., Zhu, J.: Cluster alignment with a teacher for unsupervised domain adaptation. In: Proceedings of the IEEE/CVF International Conference on Computer Vision, pp. 9944–9953 (2019)
6. Du, Z., Li, J., Su, H., Zhu, L., Lu, K.: Cross-domain gradient discrepancy minimization for unsupervised domain adaptation. In: Proceedings of the IEEE/CVF Conference on Computer Vision and Pattern Recognition, pp. 3937–3946 (2021)
7. Ganin, Y., Lempitsky, V.: Unsupervised domain adaptation by backpropagation. In: Proceedings of the 32nd International Conference on Machine Learning (ICML), pp. 1180–1189. PMLR, Lille, France (2015)
8. Gao, J., Ma, X., Xu, C.: Learning transferable conceptual prototypes for interpretable unsupervised domain adaptation. IEEE Trans. Image Process. **33**, 5284–5297 (2024). https://doi.org/10.1109/TIP.2024.3459626
9. He, K., Zhang, X., Ren, S., Sun, J.: Deep residual learning for image recognition. In: Proceedings of the IEEE Conference on Computer Vision and Pattern Recognition, pp. 770–778 (2016)
10. Hoyer, L., Dai, D., Wang, H., Van Gool, L.: MIC: Masked image consistency for context-enhanced domain adaptation. In: Proceedings of the IEEE/CVF Conference on Computer Vision and Pattern Recognition, pp. 11721–11732 (2023)
11. Jin, Y., Wang, X., Long, M., Wang, J.: Minimum class confusion for versatile domain adaptation. In: European Conference on Computer Vision, pp. 464–480. Springer (2020)

12. Li, Y., Zhou, Y., Zhao, J., Du, W.l., Yao, R., Liu, B.: Counterfactual knowledge maintenance for unsupervised domain adaptation. In: Proceedings of the Thirty-Fourth International Joint Conference on Artificial Intelligence, pp. 1476–1484 (2025)
13. Long, M., Cao, Y., Wang, J., Jordan, M.: Learning transferable features with deep adaptation networks. In: International Conference On Machine Learning, pp. 97–105. PMLR (2015)
14. Long, M., Cao, Z., Wang, J., Jordan, M.I.: Conditional adversarial domain adaptation. In: Advances in Neural Information Processing Systems. vol. 31, pp. 1647–1657. Curran Associates, Inc. (2018). https://proceedings.neurips.cc/paper/2018/hash/ab88b15733f543179858600245108dd8-Abstract.html
15. Long, M., Zhu, H., Wang, J., Jordan, M.I.: Deep transfer learning with joint adaptation networks. In: International Conference on Machine Learning, pp. 2208–2217. PMLR (2017)
16. Lu, Z., Yang, Y., Zhu, X., Liu, C., Song, Y.Z., Xiang, T.: Stochastic classifiers for unsupervised domain adaptation. In: Proceedings of the IEEE/CVF Conference on Computer Vision and Pattern Recognition, pp. 9111–9120 (2020)
17. Peng, X., Usman, B., Kaushik, N., Wang, D., Hoffman, J., Saenko, K.: VisDA: a synthetic-to-real benchmark for visual domain adaptation. In: Proceedings of the IEEE Conference on Computer Vision and Pattern Recognition Workshops, pp. 2021–2026 (2018)
18. Pham, D.H., Vu, M.T.: Takagi-Sugeno-Kang fuzzy neural network for nonlinear chaotic systems and its utilization in secure medical image encryption. Mathematics **13**(6), 923 (2025)
19. Prabhu, V., Khare, S., Kartik, D., Hoffman, J.: Sentry: Selective entropy optimization via committee consistency for unsupervised domain adaptation. In: Proceedings of the IEEE/CVF International Conference on Computer Vision, pp. 8558–8567 (2021)
20. Saito, K., Kim, D., Sclaroff, S., Darrell, T., Saenko, K.: Semi-supervised domain adaptation via minimax entropy. In: Proceedings of the IEEE/CVF International Conference on Computer Vision. pp. 8050–8058 (2019)
21. Saito, K., Watanabe, K., Ushiku, Y., Harada, T.: Maximum classifier discrepancy for unsupervised domain adaptation. In: Proceedings of the IEEE Conference on Computer Vision and Pattern Recognition, pp. 3723–3732 (2018)
22. Takagi, T., Sugeno, M.: Fuzzy identification of systems and its applications to modeling and control. IEEE Trans. Syst. Man Cybern. **15**(1), 116–132 (1985). https://doi.org/10.1109/TSMC.1985.6313399
23. Venkateswara, H., Eusebio, J., Chakraborty, S., Panchanathan, S.: Deep hashing network for unsupervised domain adaptation. In: Proceedings of the IEEE Conference on Computer Vision and Pattern Recognition, pp. 5018–5027 (2017)
24. Wang, M., Deng, W.: Deep visual domain adaptation: a survey. Neurocomputing **312**, 135–153 (2018)
25. Wu, Y., Inkpen, D., El-Roby, A.: Dual mixup regularized learning for adversarial domain adaptation. In: European Conference on Computer Vision, pp. 540–555. Springer (2020)
26. Xu, T., Chen, W., Wang, P., Wang, F., Li, H., Jin, R.: CDTrans: Cross-domain transformer for unsupervised domain adaptation. In: International Conference on Learning Representations (ICLR) (2022). https://openreview.net/forum?id=XGzk5OKWFFc

27. Yang, J., Liu, J., Xu, N., Huang, J.: TVT: Transferable vision transformer for unsupervised domain adaptation. In: Proceedings of the IEEE/CVF Winter Conference on Applications Of Computer Vision, pp. 520–530 (2023)
28. Zhang, Y.: A survey of unsupervised domain adaptation for visual recognition. arXiv preprint arXiv:2112.06745 (2021)
29. Zhang, Y., et al.: Takagi-Sugeno-Kang fuzzy system fusion: a survey at hierarchical, wide and stacked levels. Inform. Fusion **101**, 101977 (2024). https://doi.org/10.1016/j.inffus.2023.101977
30. Zhang, Y., Liu, T., Long, M., Jordan, M.: Bridging theory and algorithm for domain adaptation. In: International Conference on Machine Learning, pp. 7404–7413. PMLR (2019)
31. Zhao, H., Des Combes, R.T., Zhang, K., Gordon, G.: On learning invariant representations for domain adaptation. In: International Conference on Machine Learning, pp. 7523–7532. PMLR (2019)
32. Zhou, T., Wang, G., Choi, K.S., Wang, S.: A two-view deep interpretable TSK fuzzy classifier under mutually teachable classification criterion. Inf. Sci. **665**, 120388 (2024)

Receding-Horizon Trajectory Planning for Unmanned Surface Vessel Based on Reachable Set

Yu Luan[1], Hao Sun[1], Qinglin Sun[1(✉)], and Panlong Tan[2]

[1] College of Artificial Intelligence, Nankai University, Tianjin 300350, China
sunql@nankai.edu.cn

[2] Haihe Laboratory of Information Technology Application Innovation, Tianjin 300450, China

Abstract. This paper proposes a reachable-set-based receding-horizon trajectory planning method for unmanned surface vessels (USVs). A dynamic model is established to capture the coupled translational and rotational motions of the USV. Based on this model, forward reachable sets over a finite prediction horizon are constructed to characterize all dynamically feasible future states. A rolling planning framework is then implemented, wherein an optimal state is selected at each planning step by solving a quadratic programming (QP) problem. The designed cost function jointly accounts for position convergence toward the target, heading alignment, and angular velocity smoothness, thereby ensuring both navigation accuracy and maneuvering stability. To further enhance robustness near the target region, a backtracking strategy is incorporated to prevent oscillations and performance degradation during the final approach. Extensive Simulation results under different initial conditions and environmental disturbances demonstrate that the proposed method generates smooth, dynamically feasible trajectories and exhibits strong adaptability and robustness in complex scenarios.

Keywords: Unmanned surface vessel · Trajectory planning · Reachable set · Quadratic programming

1 Introduction

With the continuous development of marine resource exploitation and the increasing complexity of maritime missions, unmanned surface vessels (USVs) have gradually become essential autonomous platforms for applications such as marine environmental monitoring, offshore inspection, search and rescue operations, and maritime defense [1]. Due to their advantages of low operational cost, high safety, and long endurance, USVs are capable of performing tasks in complex, hazardous, and even human-inaccessible marine environments. Consequently, their autonomous navigation and decision-making capabilities have garnered significant attention. Among the key enabling technologies, trajectory

J. Lu et al. (Eds.): FLINS-ISKE 2026, LNCS 16752, pp. 473–489, 2027.
https://doi.org/10.1007/978-981-92-2369-5_29

planning plays a central role in achieving autonomous navigation for USVs, directly impacting navigation safety, feasibility, and operational efficiency [2].

The objective of USV trajectory planning is to guide the vessel smoothly from an initial state to a target position while satisfying system dynamic constraints and control input limitations. Compared to ground or aerial platforms, the motion of USVs is subject to more pronounced environmental disturbances, such as wind, waves, and ocean currents. Furthermore, the dynamic model of a USV typically exhibits strong nonlinearity and coupling characteristics [3]. Consequently, trajectory planning must consider not only geometric feasibility but also strictly ensure dynamic feasibility. If a planned trajectory is dynamically infeasible, it cannot be safely executed on a real system, even if it is geometrically reachable.

Existing USV trajectory planning methods can generally be classified into three categories: sampling-based methods, optimization-based methods, and learning-based methods. Sampling-based motion planning methods generate feasible paths by performing random or heuristic sampling in the state space and then construct smooth trajectories that satisfy constraints through optimization and connection strategies. Han et al. [4] proposed a USV trajectory planning scheme that integrates the multi-scale A* algorithm, ReedsâĂŞShepp curves, and online double-S trajectory planning. Mao et al. [5] proposed spRRT and enhanced spRRT-Informed for USV motion planning. However, sampling-based methods typically enforce kinematic and dynamic constraints in a post-processing manner rather than during the planning process, which makes it difficult to guarantee dynamic feasibility throughout the entire trajectory.

Optimization-based planning methods formulate the trajectory planning problem as a constrained optimization problem and directly compute a performance-optimal trajectory that satisfies kinematic, dynamic, and obstacle-avoidance constraints through numerical solving. Gao et al. [6] proposed a hybrid map that combines topological units and grid maps, along with a feedback mechanism-based genetic algorithm (FGA) for USV path planning. Ntakolia and Lyridis [7] conducted a comparative study on three ACO-based approaches for USV multi-objective path planning ACO-Mamdani ACO-TSK and ACO-RMSE. However, these methods are often computationally intensive and sensitive to modeling accuracy and environmental disturbances, which may affect their robustness in real-time USV applications.

Learning-based planning methods employ data-driven learning models to derive principles of trajectory generation and optimization from historical experience or simulation data, thereby enabling autonomous and efficient trajectory decision-making under complex constraints. Zhai et al. [8] established a horizontal 3-DOF motion model of USV and applied the DQN algorithm to its path planning. Wei et al. [9] proposed an ATCM-SAC deep reinforcement learning framework for USV trajectory planning which constructs a time-varying marine environment model using actual data from the Baltic Sea and integrates BFS-based heuristic guidance with an attention mechanism-based adaptive trajectory control module for multi-source information fusion. However, most learning-based methods lack explicit feasibility guarantees and rely heavily on extensive training data, which limits their applicability in safety-critical marine environments.

In recent years, reachable set theory has provided a novel perspective for feasibility analysis and trajectory planning of complex dynamical systems [10]. A reachable set describes all states that a system can attain within a finite time horizon, given initial conditions and bounded control inputs. Essentially, the reachable set characterizes the dynamically feasible state space of the system and establishes a rigorous feasibility boundary for trajectory planning. However, for nonlinear systems, the exact computation of reachable sets is generally intractable. Consequently, conservative approximations, such as zonotope-based over-approximations, are commonly employed in practical applications [11].

Against this background, combining reachable sets with receding-horizon planning has emerged as an effective strategy that balances feasibility guarantees with real-time performance. By predicting the reachable set over a finite horizon at each planning step and restricting state optimization within this reachable domain, dynamic infeasibility can be fundamentally avoided. Meanwhile, the rolling planning mechanism enables continuous trajectory updates based on the current system state, thereby enhancing robustness against environmental disturbances and modeling errors.

Motivated by the considerations above, this paper proposes a reachable-set-based receding-horizon trajectory planning method for USVs. First, a planar dynamic model of the USV is established, and the forward reachable set over a finite prediction horizon is constructed under bounded control inputs. Then, at each rolling planning step, an optimal state is selected exclusively from the terminal reachable set by solving a quadratic programming problem. The designed cost function jointly considers position error, heading angle alignment, and angular velocity smoothness, guiding the USV toward the target while ensuring smooth heading variation and continuous motion. Additionally, to prevent performance degradation or oscillations near the target region, a backtracking strategy is introduced to enhance the stability and robustness of the planning process.

The remainder of this paper is organized as follows. Section 2 presents the dynamic modeling of the unmanned surface vessel and a qualitative analysis of key parameters. Section 3 details the reachable-set-based receding-horizon trajectory planning algorithm and its optimization formulation. Section 4 provides simulation results and performance analysis under different scenarios. Finally, Sect. 5 concludes the paper and outlines directions for future work.

2 Model Building and Qualitative Analysis

2.1 Model of Unmanned Surface Vessel

As illustrated in Fig. 1, two coordinate frames are adopted to describe the motion of the USV. The ground coordinate frame $\{O_R - X_R Y_R\}$ is defined on the horizontal plane, and the body coordinate frame $\{O_B - X_B Y_B\}$ is attached to the USV.

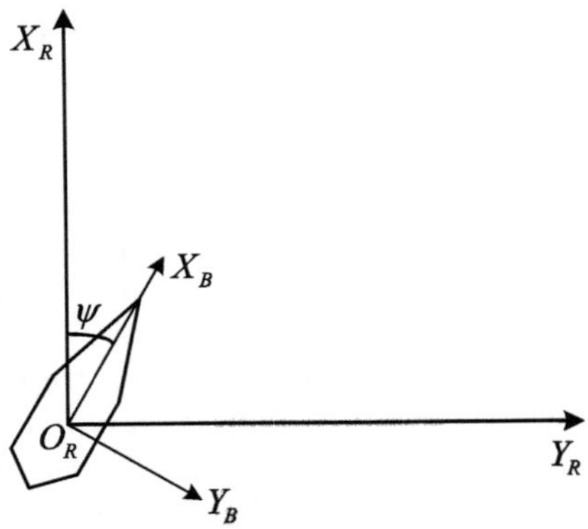

Fig. 1. Schematic diagram of the parafoil system.

The translational and rotational movements of the USV are expressed as follows:

$$\begin{aligned} \dot{\boldsymbol{r}}_R &= \begin{bmatrix} \cos\psi & -\sin\psi \\ \sin\psi & \cos\psi \end{bmatrix} \boldsymbol{v}_B \\ \dot{\psi} &= \omega_B \end{aligned} \tag{1}$$

where $\boldsymbol{r}_R$ refers to the position vector in the ground coordinate frame, ψ denotes the yaw angle, $\boldsymbol{v}_B$ and ω_B correspond to the velocity and angular velocity in the body coordinate frame.

$$\begin{aligned} m\dot{\boldsymbol{v}}_B &= \boldsymbol{F}_D + \boldsymbol{F}_T \\ J\dot{\omega}_B &= M_D + M_T \end{aligned} \tag{2}$$

where m and J denote the mass and rotational inertia of the USV, respectively; $\boldsymbol{F}_D$ and $\boldsymbol{F}_T$ denote the drag force and the thrust, respectively; M_D and M_T denote the damping term and the thrust torque, respectively. The forces and moments of the USV are expressed as follows:

$$\boldsymbol{F}_D = -\frac{1}{2}\rho S_{ref} C_D \left\| \boldsymbol{v}_a \right\| \boldsymbol{v}_a \tag{3}$$

$$\boldsymbol{F}_T = \begin{bmatrix} T\cos\delta_T \\ -T\sin\delta_T \end{bmatrix} \tag{4}$$

$$M_D = -\frac{1}{2}\rho S_{ref} C_n \left| \psi \right| \psi \tag{5}$$

$$M_T = r_T T \sin\delta_T \tag{6}$$

where T and δ_T are the thrust magnitude and the thrust direction angle; r_T denotes the distance from the mass-center to the position of the engine; the relative speed vector $\boldsymbol{v}_a$ is the difference between the USV velocity and the sea wave velocity. The USV parameters are summarized in Table 1.

Table 1. Parameters of the unmanned surface vessel.

Parameter	Value	Parameter	Value
m_s	100000 kg	$T_{\max}$	8000 N
S^s_{ref}	50 m^2	$\delta_T^{\max}$	30°
r_T	15 m	$v^s_{\max}$	16.5 m/s
C^s_D	1.0	C^s_n	0.4
ρ_s	1.025×10^3 kg/m^3	J_s	6.4×10^6 kg · m^2

2.2 Analysis of Key Parameter Effects

The trajectory characteristics of USV are influenced by the coupling of multiple physical parameters. This section employs numerical simulations to qualitatively evaluate the distinct effects of heading angle and angular velocity on the subsequent state of the USV, thereby laying a foundation for the design of objective functions in subsequent trajectory planning.

Effect of Heading Angle on Trajectory. The heading angle is a key attitude parameter in the motion equations of USV. It characterizes the relative orientation between the ground coordinate frame and the body coordinate frame, thereby determining the projection direction of the USV's velocity in the horizontal plane and governing the fundamental evolution direction of its horizontal trajectory. To qualitatively investigate the influence of the heading angle on horizontal motion behavior, the initial position of the USV is set to $(0, 0)$, with initial velocities $v_x = v_y = 8\,\text{m/s}$ and an initial angular velocity of $w = 0\,\text{rad/s}$. Four discrete initial heading angles are considered, namely $\psi \in \{0°, 90°, 180°, 270°\}$. For the control inputs, the thrust magnitude is fixed at 8000 N, while the thrust direction angle is selected from $\delta_T \in \{-30°, 0°, 30°\}$, so as to construct the set of reachable states at the next time step under different control actions.

As illustrated in Fig. 2, the heading angle significantly influences the shape and distribution of the reachable set of horizontal positions at the next time step. Different initial heading angles substantially affect the orientation and spatial coverage of the reachable set in the horizontal plane, thereby constraining the dominant motion directions of the USV under all admissible control inputs. In contrast, the heading angle itself does not directly cause variations in angular velocity; instead, the evolution of angular velocity is primarily governed by the applied control inputs.

Effect of Angular Velocity on Trajectory. Angular velocity is a key state variable that characterizes the turning behavior of an USV. According to the motion equations, angular velocity directly determines the rate of change of the heading angle, namely the USV's turning rate. To qualitatively analyze the influence of angular velocity on the state evolution and horizontal trajectory reachability of the USV, five discrete initial angular velocity values are considered,

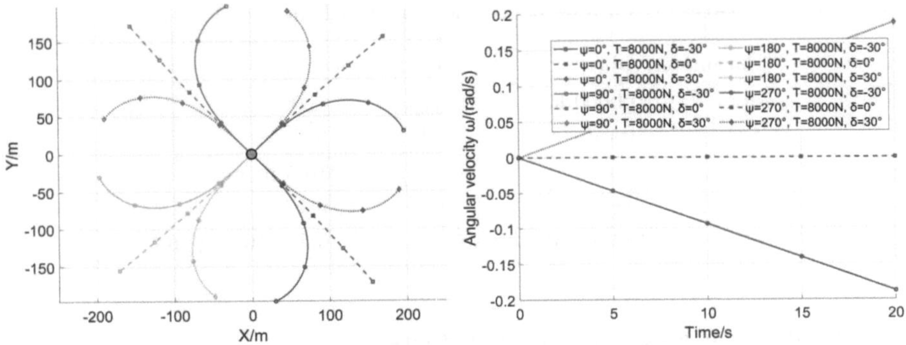

(a) Horizontal trajectory (b) Angular velocity

Fig. 2. Horizontal trajectory and angular velocity with different ψ and δ_T.

namely $w \in \{-0.02, -0.01, 0, 0.01, 0.02\}$ rad/s. The initial state of the USV and the control input settings are kept identical to those in the previous subsection.

As shown in Fig. 3, different initial angular velocities result in different subsequent heading angles. This change in heading directly affects the transformation between the body and ground coordinate frames, which consequently determines the shape and distribution of the reachable set of horizontal positions.

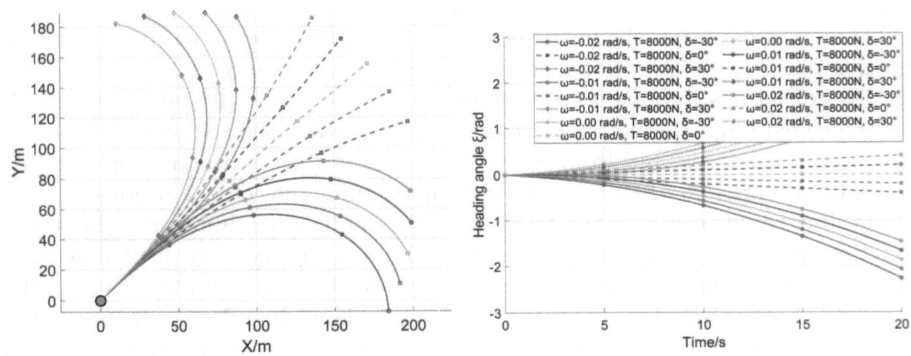

(a) Horizontal trajectory (b) Heading angle

Fig. 3. Horizontal trajectory and heading angle with different ω and δ_T.

3 Design of Rolling Planning Algorithm Based on Reachable Set

3.1 Basic Concept

Reachable Set. This section briefly reviews the concept of reachable sets for general nonlinear systems. Consider a nonlinear dynamical system described by [12,13]

$$\dot{x}(t) = f(x(t), u(t)) \tag{7}$$

where the initial state satisfies $x(0) \in X_0 \subset \mathbb{R}^n$ and the control input is constrained by $u(t) \in U \subset \mathbb{R}^m$. Given the admissible control set U and the initial state set X_0 at time $t = 0$, the exact forward reachable set at time $t = r$ is defined as follows:

$$R^e(r) = \{x(r) \mid x(0) \in X_0,\ u(t) \in U\} \tag{8}$$

which contains all system states that can be reached at time r under allowable control inputs.

In practice, the exact computation of $R^e(r)$ is generally intractable for nonlinear systems. Therefore, reachable sets are usually approximated over finite time horizons, leading to conservative outer approximations. Denote by $R(r)$ an over-approximation of the exact reachable set such that $R(r) \supseteq R^e(r)$. The reachable set over a time interval $[0, r]$ is then defined as the union $R([0, r]) = \bigcup_{t\in[0,r]} R(t)$.

Quadratic Programming Principles. The zonotopic reachable set defines the *feasible state space* for trajectory planning, within which an optimal trajectory point must be selected to minimize deviations from the desired target. To achieve this, the problem is formulated as a QP problem, thereby bridging the reachable-set framework with convex optimization. Its standard form is expressed as follows [14,15]:

$$\begin{aligned} \min_{x} \quad & \frac{1}{2}x^{\mathrm{T}}Hx + f^{\mathrm{T}}x \\ \text{s.t.} \quad & Ax \le b, \\ & A_{\mathrm{eq}}x = b_{\mathrm{eq}}, \\ & x_{\mathrm{lb}} \le x \le x_{\mathrm{ub}} \end{aligned} \tag{9}$$

where x denotes the optimization variable. The matrix H is symmetric and positive semi-definite, ensuring convexity of the objective function. Vectors f, b, and b_{eq} define the linear cost and constraint terms, while matrices A and A_{eq} specify inequality and equality constraints, respectively. The vectors x_{lb} and x_{ub} impose bound constraints on the decision variables.

3.2 Zonotope-Based Reachable Set Computation

To construct the reachable set for the nonlinear USV dynamic model described in Eqs. (1, 2, 3, 4, 5 and 6), a linearization approach is employed over the prediction horizon. Let the USV state vector be explicitly defined as $\mathbf{x} = [x, y, \psi, v_x, v_y, \omega]^\top \in \mathbb{R}^6$ and the control input as $\boldsymbol{u}_c = [T, \delta_T]^\top \in \mathbb{R}^2$. At each sampling interval Δt, the nonlinear system $\dot{\mathbf{x}} = f(\mathbf{x}, \boldsymbol{u}_c)$ is locally linearized using Taylor series expansion around the current operating point $(\mathbf{x}_k, \boldsymbol{u}_{c,k})$, yielding a discrete-time linear time-varying (LTV) system:

$$\mathbf{x}_{k+1} = A_k \mathbf{x}_k + B_k \boldsymbol{u}_c + \boldsymbol{d}_k \tag{10}$$

where $A_k = I + \Delta t \cdot \frac{\partial f}{\partial \mathbf{x}}\Big|_{\mathbf{x}_k, \boldsymbol{u}_{c,k}}$, $B_k = \Delta t \cdot \frac{\partial f}{\partial \boldsymbol{u}_c}\Big|_{\mathbf{x}_k, \boldsymbol{u}_{c,k}}$, and $\boldsymbol{d}_k$ accounts for higher-order truncation errors and environmental disturbances.

Based on this LTV model, the forward reachable set is approximated using zonotopes. A zonotope $Z = \langle c, G \rangle$ is defined by a center vector $c \in \mathbb{R}^n$ and a generator matrix $G \in \mathbb{R}^{n \times m}$. Given the initial state modeled as a zonotope $Z_0 = \langle c_0, G_0 \rangle$ and bounded control inputs modeled as $U = \langle c_u, G_u \rangle$, the reachable set at step $k+1$ is propagated as:

$$Z_{k+1} = A_k Z_k \oplus B_k U \oplus D \tag{11}$$

where $\oplus$ denotes the Minkowski sum and D bounds the linearization errors. After N steps ($h = N\Delta t$), the terminal reachable set is obtained as $Z_N = \{c + G\beta \mid \beta \in [-1, 1]^m\}$, where $c \in \mathbb{R}^6$ is the center and $G \in \mathbb{R}^{6 \times m}$ is the generator matrix. The time step $\Delta t = 0.1\,\mathrm{s}$ is deliberately chosen to balance discrete-time integration accuracy and real-time computational load. Although the zonotope over-approximation introduces a degree of conservativeness, it practically serves as a safety margin for the USV to prevent constraint violations.

3.3 Cost Function Design

To select the optimal state from the predicted reachable set at each rolling planning step, a composite cost function is designed that jointly considers position, heading angle, and angular velocity. This cost function aims to guide the USV toward the target position while ensuring proper orientation alignment and smooth turning behavior.

Position Cost. The position cost penalizes the horizontal distance between a candidate state and the target position. Let $p = [x, y]^\top$ denote the horizontal position of a reachable-set candidate and p_{goal} the target position. To improve numerical robustness and maintain scale consistency during the rolling process, the squared Euclidean distance is normalized by the current maximum distance to the goal, defined as

$$J_{\text{pos}} = w_{\text{pos}} \cdot \frac{\|p - p_{\text{goal}}\|^2}{d_{\max}^2} \tag{12}$$

where w_{pos} is the weight associated with the position cost and $d_{\max} = \max(\|p_{\text{cur}} - p_{\text{goal}}\|, 1)$ is dynamically updated at each iteration.

Heading Angle Cost. The heading angle plays a dominant role in determining the short-term evolution of the USV trajectory, as discussed in Sect. 2.2. A candidate state whose heading angle is aligned with the target direction is more likely to yield favorable future motion under reachable-set constraints. Therefore, a heading angle cost is introduced to penalize the deviation between the candidate heading angle ψ and the desired heading angle ψ_{goal}, defined as follows:

$$\psi_{\text{goal}} = \text{atan2}(y_{\text{goal}} - y,\ x_{\text{goal}} - x) \tag{13}$$

To handle angular periodicity, the heading angle deviation is computed as follows:

$$\Delta\psi = \text{mod}(\psi - \psi_{\text{goal}} + \pi,\ 2\pi) - \pi \tag{14}$$

and the corresponding cost is defined as follows:

$$J_{\psi} = w_{\psi} \cdot \frac{\|\Delta\psi\|^2}{(2\pi)^2} \tag{15}$$

where w_{ψ} denotes the heading angle weight.

Angular Velocity Cost. Angular velocity directly determines the USV's turning rate and thus has a significant impact on maneuvering smoothness and stability. Instead of driving the angular velocity directly to zero, a target angular velocity ω_{goal} is constructed based on the heading angle deviation as follows:

$$\omega_{\text{goal}} = k_{\psi \to \omega} \cdot \Delta\psi \tag{16}$$

where $k_{\psi \to \omega}$ is a proportional coefficient. To reflect physical limitations, ω_{goal} is saturated within the interval $[-\omega_{\max}, \omega_{\max}]$. The angular velocity cost is then defined as follows:

$$J_{\omega} = w_{\omega} \cdot \frac{\|\omega - \omega_{\text{goal}}\|^2}{\omega_{\max}^2} \tag{17}$$

where w_{ω} is the corresponding weight and $\omega_{\max}$ is the normalization factor. The overall cost function is obtained by combining the above components:

$$J = J_{\text{pos}} + J_{\psi} + J_{\omega} \tag{18}$$

3.4 QP-Based Optimization Formulation

At each iteration of the rolling planning process, the reachable set at the prediction horizon is approximated by a zonotopic set, which can be expressed as follows:

$$Z = \{c + G\beta \mid \beta \in [-1, 1]^m\} \tag{19}$$

where $c \in \mathbb{R}^6$ is the zonotope center, $G \in \mathbb{R}^{6 \times m}$ is the generator matrix, and $\beta \in \mathbb{R}^m$ is the decision variable.

By substituting the zonotope representation into the cost function, the optimization problem becomes a quadratic function of β. Based on our explicitly defined state vector $\mathbf{x} = [x, y, \psi, v_x, v_y, \omega]^\top$, the position (x, y) corresponds to indices $1 : 2$, the heading angle ψ to index 3, and the angular velocity ω to index 6. To explicitly link the heading deviation $\Delta\psi$ to the decision variable β, we approximate the terminal heading as $\psi \approx c_3 + G_3\beta$. Additionally, the target angular velocity ω_{goal} is calculated using the heading deviation at the *start* of the current planning step to maintain the convexity of the QP problem.

The resulting objective can be rigorously expressed as follows:

$$\min_{\beta} \left\{ w_{\text{pos}} \cdot \frac{\|c_{1:2} + G_{1:2}\beta - p_{\text{goal}}\|^2}{d_{\max}^2} + w_\psi \cdot \frac{\|\Delta\psi\|^2}{(2\pi)^2} + w_\omega \cdot \frac{\|c_6 + G_6\beta - \omega_{\text{goal}}\|^2}{\omega_{\max}^2} \right\} \tag{20}$$

This optimization problem can be rewritten in the standard QP form:

$$\min_{\beta} \left(\frac{1}{2}\beta^\top H\beta + f^\top\beta \right), \quad \text{s.t.} \quad \beta \in [-1, 1]^m \tag{21}$$

where the Hessian matrix H and gradient vector f are derived as follows:

$$H = 2\left(\frac{w_{\text{pos}}}{d_{\max}^2} G_{1:2}^\top G_{1:2} + \frac{w_\psi}{(2\pi)^2} G_3^\top G_3 + \frac{w_\omega}{\omega_{\max}^2} G_6^\top G_6 \right) \tag{22}$$

$$f = 2\Big(\frac{w_{\text{pos}}}{d_{\max}^2} (c_{1:2} - p_{\text{goal}})^\top G_{1:2} + \frac{w_\psi}{(2\pi)^2} \Delta\psi\, G_3 + \frac{w_\omega}{\omega_{\max}^2} (c_6 - \omega_{\text{goal}})\, G_6 \Big)^\top \tag{23}$$

To ensure the Hessian matrix H is positive semidefinite and numerically stable, a regularization term is added:

$$H \leftarrow H + \lambda I, \quad \lambda = 10^{-8} \tag{24}$$

This regularization effectively prevents algorithmic singularities without noticeably altering the optimization landscape.

The resulting QP problem is efficiently solved using MATLAB's `quadprog` solver. The optimal coefficient vector β^* yields the optimal terminal state within the predicted reachable set, which is then used as the next rolling planning state.

3.5 Iterative Trajectory Generation

This paper proposes a reachable-set-based rolling iterative trajectory generation framework for USV operating under dynamic constraints and in complex environments. Unlike conventional optimization- or sampling-based planners that may generate dynamically infeasible trajectories, the proposed framework explicitly restricts trajectory evolution to the system's reachable set at each iteration, thereby ensuring dynamic feasibility throughout the planning process.

At each iteration, a fixed reachable time horizon h is employed. Starting from the current state, the reachable set over the horizon is computed using the method described in Sect. 3.1, with a fixed discretization step $\Delta t = 0.1$. This yields a sequence of reachable sets $\{R(t_i)\}_{i=1}^{N}$, where $t_i = i\,\Delta t$ and $N = h/\Delta t$. Each reachable set characterizes all admissible states that the USV can attain at the corresponding future time instant under bounded control inputs.

Based on the predicted reachable set sequence, a rolling terminal-state optimization strategy is employed. Specifically, only the reachable set at the terminal time t_N is considered for optimization. Using the QP formulation introduced in Sect. 3.4, an optimal terminal state is selected by minimizing a composite cost function that accounts for position error, heading angle deviation, and angular velocity consistency. This terminal-state selection strategy eliminates the need for frequent global replanning and enables efficient guidance toward the target under dynamic constraints.

After obtaining the optimal terminal state, the system state is updated accordingly, and this optimized terminal state is used as the new initial condition for the next iteration. The reachable set computation and terminal-state QP optimization are then repeated in a rolling-horizon manner. This iterative process continues until the USV reaches a predefined neighborhood around the target position.

To further enhance robustness near the target, a backtracking mechanism is incorporated. When the USV enters a predefined activation radius around the goal, the algorithm monitors the distance-to-goal evolution. If a deterioration trend is detected, the algorithm searches for a closer state within the reachable set sequence of the current horizon. If no improvement is found, the trajectory is rolled back to the historically best state achieved so far. This strategy effectively mitigates oscillations or local degradation near the target while preserving reachability guarantees.

The primary advantage of the proposed rolling iterative framework is its strict adherence to reachable-set constraints at every planning step. By ensuring that each selected state lies within the predicted reachable domain, the method maintains dynamic feasibility throughout the trajectory and exhibits strong robustness against model uncertainties and environmental disturbances. Compared to traditional planning approaches, the proposed method avoids infeasible updates and improves convergence reliability in complex navigation scenarios.

4 Simulation Experiments and Results

4.1 Simulation Setup and Real-Time Performance

The proposed algorithm is implemented in MATLAB. The prediction horizon is set to $h = 1.0\,\mathrm{s}$, with a discretization step of $\Delta t = 0.1\,\mathrm{s}$, resulting in $N = 10$ prediction steps per rolling iteration. The weights in the composite cost function are empirically tuned to balance convergence and smoothness, set as $w_{\mathrm{pos}} = 1.0$, $w_{\psi} = 0.4$, and $w_{\omega} = 0.6$.

As an online receding-horizon planning algorithm, computational efficiency is critical. Tested on a standard PC equipped with an Intel Core i7 processor and 16 GB of RAM, the average execution time-encompassing the zonotopic reachable set propagation and the solving of the QP problem-is approximately 8.5 ms per step. This execution time is well below the control period ($\Delta t = 0.1$ s), thereby directly demonstrating the algorithm's real-time operational capability for deployment on actual USV hardware.

4.2 Simulation with Different Initial Positions

To investigate the influence of different spatial initial conditions on the trajectory planning performance, simulations are carried out with varying initial positions while keeping other state variables identical. Specifically, the initial heading angle, velocity, and target position are fixed, and several representative starting points are selected to cover different spatial distributions in the horizontal plane. The quantitative terminal states and positioning errors for these different initial positions are summarized in Table 2.

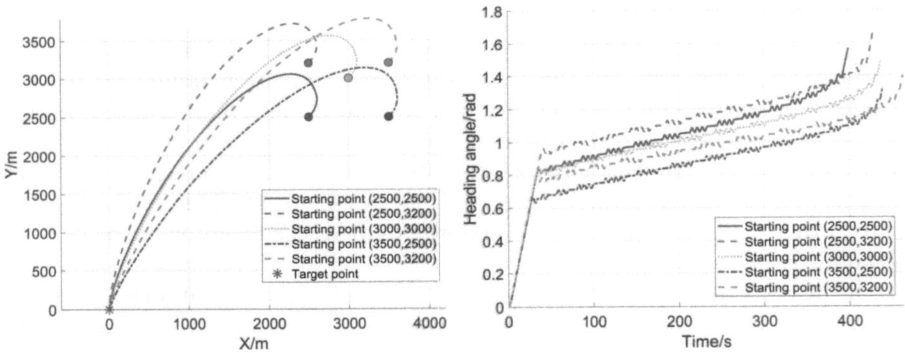

(a) Horizontal trajectory (b) Heading angle

Fig. 4. Horizontal trajectory and heading angle with initial positions.

Figure 4(a) illustrates the horizontal trajectories generated from different initial positions. It can be observed that all trajectories successfully converge to the target point without violating motion constraints, demonstrating the feasibility of the proposed method under diverse initial spatial conditions. Despite significant differences in the initial positions, the planned trajectories exhibit smooth and continuous shapes, indicating that the method can effectively adapt to varying starting locations while maintaining stable convergence behavior.

Figure 4(b) shows the corresponding heading angle profiles over time. Despite differences in the initial positions, the heading angles evolve smoothly without abrupt changes or oscillations. This indicates that the proposed method ensures

Table 2. Terminal states for different initial positions.

Initial State (x, y, ψ)	Terminal Position (m, m)	Terminal Error (m)
$(2500, 2500, 0°)$	$(-4.12, 1.43)$	4.36
$(2500, 3200, 0°)$	$(-0.51, -4.30)$	4.33
$(3000, 3000, 0°)$	$(-2.39, -6.47)$	6.90
$(3500, 2500, 0°)$	$(-1.60, -0.13)$	1.61
$(3500, 3200, 0°)$	$(-1.93, -5.60)$	5.92

gradual heading adjustments and avoids aggressive maneuvers, which is especially important for practical motion execution.

Overall, the simulation results demonstrate that the proposed trajectory planning method exhibits strong adaptability to different initial positions, achieving smooth trajectory generation while accurately guiding the USV toward the target.

4.3 Simulation with Different Initial Headings

To further examine the influence of different initial heading angles on the trajectory planning performance, simulations are conducted with varying initial headings while keeping the initial position, velocity, and target point constant. The initial heading angles are selected as $\psi_0 = 0°, 90°, 180°$, and $270°$, representing a wide range of initial orientation conditions. The specific terminal states and corresponding errors for these heading variations are recorded in Table 3.

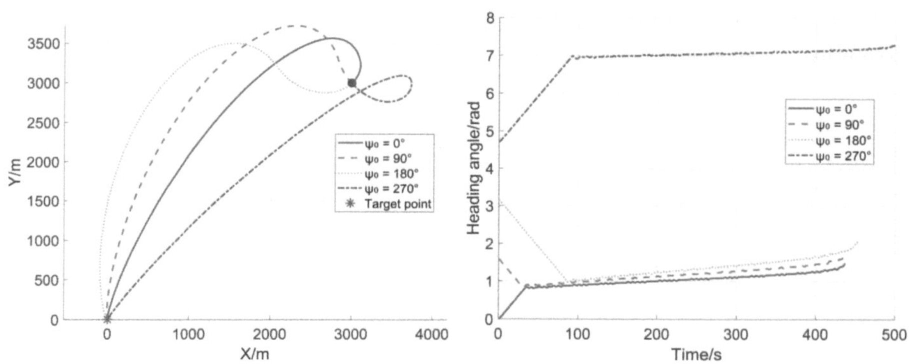

(a) Horizontal trajectory (b) Heading angle

Fig. 5. Horizontal trajectory and heading angle with initial headings.

As shown in Fig. 5(a), the proposed method generates feasible trajectories for all considered initial headings, successfully guiding the USV toward the target

Table 3. Terminal states for different initial headings.

Initial State (x, y, ψ)	Terminal Position (m, m)	Terminal Error (m)
$(3000, 3000, 0°)$	$(-2.39, -6.47)$	6.90
$(3000, 3000, 90°)$	$(-1.85, -0.18)$	1.86
$(3000, 3000, 180°)$	$(-1.32, -1.50)$	2.00
$(3000, 3000, 270°)$	$(-0.53, -0.10)$	0.54

point. Despite significant differences in initial orientations, the planned trajectories remain smooth and continuous, without sharp turns or irregular behavior.

Figure 5(b) shows the evolution of the heading angle over time. It can be observed that the heading angles gradually converge toward a consistent trend, regardless of their initial values. No abrupt changes or excessive oscillations occur during the maneuvering process, even when the initial heading is opposite to the desired direction. This demonstrates that the proposed trajectory planning strategy ensures smooth heading transitions and stable orientation adjustments.

Overall, these results confirm that the proposed method exhibits strong robustness to variations in initial heading angles, enabling reliable and smooth trajectory planning under diverse orientation conditions.

4.4 Simulation with Wind and Wave Disturbances

To evaluate the robustness of the proposed method in realistic marine environments, wind and wave disturbances are incorporated into the system dynamics. Different disturbance intensities are considered to assess the influence of environmental uncertainties on the planned trajectories, while the initial state and target position remain unchanged. The quantitative terminal results under these external perturbations are summarized in Table 4.

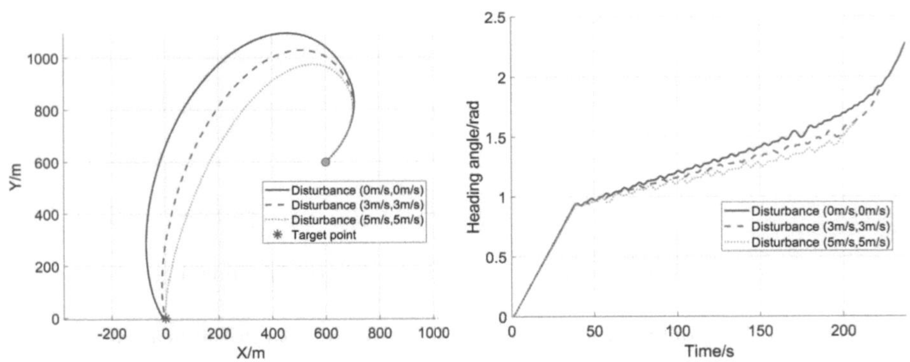

(a) Horizontal trajectory (b) Heading angle

Fig. 6. Horizontal trajectory and heading angle with disturbances.

Table 4. Terminal states under wind and wave disturbances.

Initial State $(x, y), (v_{wx}, v_{wy})$	Terminal Position (m, m)	Terminal Error (m)
$(600, 600), (0, 0)$	$(-3.66, -5.05)$	6.23
$(600, 600), (3, 3)$	$(0.10, -3.34)$	3.34
$(600, 600), (5, 5)$	$(-0.51, -4.87)$	4.90

The horizontal trajectories under different disturbance conditions are shown in Fig. 6(a). It can be observed that, although wind and wave disturbances cause slight deviations from the planned paths, all trajectories still successfully converge to the target point. Compared to the disturbance-free case, trajectories under stronger disturbances exhibit moderate adjustments, indicating that the proposed method effectively compensates for external perturbations while maintaining stable convergence behavior.

Figure 6(b) shows the corresponding heading angle responses under different disturbance intensities. Despite the influence of wind and wave disturbances, the heading angle profiles remain smooth and continuous, without noticeable oscillations or abrupt changes. This indicates that the proposed approach effectively mitigates the impact of environmental disturbances and ensures stable heading control during the trajectory planning process.

In summary, the simulation results demonstrate that the proposed method exhibits strong robustness against wind and wave disturbances, making it suitable for practical applications in uncertain and dynamic marine environments.

5 Conclusions

This paper presents a trajectory planning method for USV, aiming to achieve smooth and reliable motion under various initial conditions and environmental uncertainties. The proposed method integrates system dynamics and motion constraints into the planning process, enabling the generation of feasible trajectories while ensuring stable heading evolution.

To validate the effectiveness of the proposed method, extensive simulations are conducted under different scenarios. The results demonstrate that the method successfully generates smooth trajectories for a wide range of initial positions and heading angles while maintaining consistent convergence behavior. Furthermore, the robustness of the proposed method is verified through simulations incorporating wind and wave disturbances, where stable trajectory planning and smooth heading adjustments are achieved despite external perturbations.

Overall, the simulation results confirm that the proposed trajectory planning strategy exhibits strong adaptability and robustness, making it suitable for practical applications in realistic marine environments. Future work will focus on extending this method to more complex scenarios, such as dynamic obstacles and time-varying environmental conditions, as well as experimental validation on real-world autonomous marine platforms.

Acknowledgments. This work was supported by the National Natural Science Foundation of China under Grant No.62473210 and the Natural Science Foundation of Tianjin under Grant No. 24JCYBJC00240, 2025JCQNJC01350.

Disclosure of Interests. The authors have no competing interests to declare that are relevant to the content of this article.

References

1. Huang, T., Xue, Z., Chen, Z., et al.: Efficient trajectory planning and control for USV with vessel dynamics and differential flatness. In: 2023 IEEE/ASME International Conference on Advanced Intelligent Mechatronics (AIM), pp. 1273–1280. IEEE (2023)
2. Feng, Z., Pan, Z., Chen, W., et al.: USV application scenario expansion based on motion control, path following and velocity planning. Machines **10**(5), 310 (2022)
3. Hao, B., Du, H., Yan, Z.: A path planning approach for unmanned surface vehicles based on dynamic and fast Q-learning. Ocean Eng. **270**, 113632 (2023)
4. Han, X., Zhang, X., Zhang, H.: Trajectory planning of USV: On-Line computation of the double S trajectory based on multi-scale A* algorithm with reeds–shepp curves. J. Marine Sci. Eng. **11**(1), 153 (2023)
5. Mao, S., Yang, P., Gao, D., et al.: A motion planning method for unmanned surface vehicle based on improved RRT algorithm. J. Marine Sci. Eng. **11**(4), 687 (2023)
6. Gao, H., Zhang, T., Zuo, Z., et al.: USV path planning in a hybrid map using a genetic algorithm with a feedback mechanism. J. Marine Sci. Eng. **12**(6), 939 (2024)
7. Ntakolia, C., Lyridis, D.V.: A comparative study on Ant Colony Optimization algorithm approaches for solving multi-objective path planning problems in case of unmanned surface vehicles. Ocean Eng. **255**, 111418 (2022)
8. Zhai, H., Wang, W., Zhang, W., et al.: Path planning algorithms for USVs via deep reinforcement learning. In: 2021 China Automation Congress (CAC), pp. 4281–4286. IEEE (2021)
9. Wei, Y., Liu, D., Ou, J., et al.: Adaptive fusion of multi-source information for USV trajectory planning via deep reinforcement learning. Ocean Eng. **348**, 124076 (2026)
10. Agrawal, D.R., Chen, R., Panagou, D.: gatekeeper: Online safety verification and control for nonlinear systems in dynamic environments. IEEE Trans. Robot. (2024)
11. Huimin, W., Zhen, H., Yuxiao, P.: Iterative learning trajectory control method based on reachable sets for perching maneuvers. Acta Aeronautica et Astronautica Sinica **46** (2025)
12. Althoff, M., Grebenyuk, D., Kochdumper, N.: Implementation of Taylor models in CORA 2018. In: Proceedings of the 5th International Workshop on Applied Verification for Continuous and Hybrid Systems (2018)
13. Feng, B., Zhang, X., Feng, Z.: Reachable set synthesis of delay Markovian jump systems with hidden Markov model and general transition probabilities. Nonlinear Dyn. **111**(3), 2607–2620 (2023)

14. Koung, D., Kermorgant, O., Fantoni, I., et al.: Cooperative multi-robot object transportation system based on hierarchical quadratic programming. IEEE Robot. Autom. Lett. **6**(4), 6466–6472 (2021)
15. Park, S., Park, L., Lee, S., et al.: Whole-body control of redundant hybrid cable-driven robot with manipulator: hierarchical quadratic programming approach. Nonlinear Dyn. **112**(22), 20047–20062 (2024)

Optimize Belief Rule Base Based on BackPropagation

Xinlei Cao[1,2], Jun Liu[1], Bilal Ahmed Lodhi[1], Li Zou[2], and Xin Liu[3(✉)]

[1] Ulster University, Belfast BT15 1AP, Northern Ireland, UK
{cao-x5,j.liu,B.Lodhi}@ulster.ac.uk
[2] Shandong Jianzhu University, Jinan 250101, Shandong, China
[3] Liaoning Normal University, Dalian 116000, Liaoning, China
xinl@lnnu.edu.cn

Abstract. In complex system modeling, the Belief Rule Base (BRB) integrates expert knowledge with data-driven approaches, offering strong interpretability. However, conventional BRB models face challenges such as rule explosion, parameter optimization difficulties, and the trade-off between interpretability and accuracy. Introducing the concept of backpropagation into BRB is of great significance, as combining BRB with gradient descent enables rapid convergence and efficient parameter optimization. This study proposes a backpropagation BRB(BP-BRB) method with parameterized rule center construction. Experiments on 23 public datasets demonstrate that BP-BRB improves training efficiency, enhances generalization ability, and achieves superior performance compared to standard BRB models and existing rule-based reasoning methods, thereby validating the novelty and effectiveness of the proposed approach.

Keywords: Belief Rule Base · backpropagation · gradient descent · rule center · interpretability

1 Introduction

In recent years, Explainable Artificial Intelligence (XAI) has emerged as a prominent research focus, whose principles draw from many advanced AI models, such as deep learning algorithms. Although these models deliver outstanding performance, their internal mechanisms and decision-making processes are often opaque, making it difficult for users to understand and trust their outputs [1]. The rapid emergence of large-scale AI systems, such as ChatGPT [2], has further heightened concerns about the trustworthiness of AI-generated results.

Among various interpretable modeling approaches, the Belief Rule Base (BRB) has attracted significant attention for its strengths in uncertainty modeling and transparent reasoning. Originating from the traditional IF-THEN rule based system [3], BRB not only integrates qualitative knowledge and quantitative information, but also explicitly preserves uncertainty during the reasoning

J. Lu et al. (Eds.): FLINS-ISKE 2026, LNCS 16752, pp. 490–508, 2027.
https://doi.org/10.1007/978-981-92-2369-5_30

process. Its core inference mechanism-the Rule Inference Method using Evidential Reasoning (RIMER) [4]-is built upon Dempster-Shafer (D-S) evidence theory [5,6], decision theory [7], fuzzy theory [8], and conventional IF-THEN rule bases [9]. This approach exhibits strong modeling capabilities for handling data characterized by uncertainty, incompleteness, and nonlinearity. BRB has been successfully applied in diverse domains, including safety assessment [9], graphite composition detection [10], pipeline leak detection [11], medical diagnosis [12], fault detection [13], and product lifetime prediction [14].

However, parameter optimization in traditional BRB models typically relies on heuristic search methods such as genetic algorithms [15] and particle swarm optimization [16]. While these methods offer some global search capability, they suffer from low optimization efficiency and limited parameter update precision in large-scale data scenarios. The reason is that the BRB inference process is non-differentiable, preventing direct end-to-end training using the Backpropagation (BP) algorithm. Originally proposed by Werbos [17], BP was successfully applied by LeCun [18] to handwritten digit recognition, demonstrating its efficiency in parameter optimization and pattern recognition. BP computes the gradient of the loss function with respect to model parameters via the chain rule and iteratively updates them using optimization methods such as gradient descent. It has since become the standard training method for neural networks and has been progressively integrated into partially differentiable symbolic reasoning frameworks.

Although BRB offers significant advantages in interpretability and transparent reasoning, its learning capability is limited, making it difficult to model high-dimensional data in complex systems. In practice, the fixed rule structure is difficult to optimize automatically; moreover, prior rules defined by experts may deviate from reference data, further constraining the model's generalization performance. Therefore, to introduce an efficient parameter learning function that enables a trainable white-box reasoning model, while preserving transparency and interpretability. To address this issue, we propose a Belief Rule Base with Backpropagation (BP-BRB) model. This is the first approach to transform the BRB inference process into a differentiable form, thereby allowing direct application of the BP algorithm for gradient-driven, end-to-end training. Specifically, a Gaussian kernel function is incorporated into the rule matching computation to avoid the non-differentiability of Euclidean distance at zero. Furthermore, the rule centers, rule weights, attribute weights, membership bandwidths, and belief degree distributions are all formulated as trainable parameters. Within the Evidential Reasoning (ER) framework, the proposed method retains the interpretability and transparency of the reasoning process. This design preserves the symbolic reasoning advantages of BRB while equipping it with efficient data-driven learning capability. Compared with existing BRB variants and other white-box models, BP-BRB demonstrates significant improvements in parameter optimization, generalization ability, and computational efficiency.

The main contributions of this study are summarized as follows:

1. Differentiable BRB inference structure: BRB combine integrating backpropagation (BP) and gradient descent optimization to enable end-to-end training. While preserving the original BRB structure and interpretability, the proposed method endows the model with neural network-like learning capability, allowing automatic optimization of rule centers, rule weights, attribute weights, membership bandwidths, and belief degree distributions without relying on manually crafted rules.
2. Label-guided clustering for rule center initialization: We adopt a label-guided clustering strategy to effectively allocate rule center positions, which are then treated as trainable parameters after initialization. This design not only ensures a reasonable rule distribution but also significantly reduces the number of rules, mitigating the combinatorial explosion problem and improving training efficiency.
3. High interpretability with competitive performance: Compared to neural networks with tens of thousands of parameters, BP-BRB requires only a limited number of trainable parameters. The rules retain clear logical expressions, and the inference process remains fully transparent and interpretable.

The remainder of this paper is organized as follows: Sect. 2 reviews the background of BRB and typical rule-based systems. Section 3 describes the construction of the proposed BP-BRB. Section 4 presents comparative experiments and result analysis, while Sect. 5 concludes the paper and outlines future research directions.

2 Related Work

This section first introduces the BRB model and its relevant studies, followed by a discussion of representative rule-based systems for comparison, thereby situating the present work within the broader research landscape.

2.1 BRB Systems

The BRB system extends the conventional IF-THEN rule framework by incorporating a belief distribution into the consequent part of each rule, while also introducing attribute weights and rule weights to quantify the importance of individual attributes and rules [4]. A belief rule in the BRB system can be expressed as follows:

$$\begin{aligned} &R_k : \text{IF } A_1^k \wedge A_2^k \wedge \cdots \wedge A_T^k \text{ THEN } (D_1, \beta_1^k), \ldots, (D_N, \beta_N^k),\ k \in [1, K], \\ &\qquad \text{With rule weight } \theta_k \text{ and attribute weight } \delta_1, \ldots, \delta_T,\ \text{s.t.} \sum_{n=1}^{N} \beta_n^k \leq 1 \end{aligned} \tag{1}$$

where, A_T^k denotes the Tth antecedent attribute of the kth rule, and K is the total number of rules. θ_k represents the weight of the kth rule, while δ_T denotes the weight of the Tth attribute. β_N^k indicates the belief degree associated with the evaluation grade D_N. A larger β_N^k value corresponds to higher decision reliability. If $\sum_{n=1}^{N} \beta_N^k = 1$, the rule is considered complete; otherwise, it is classified as incomplete.

Challenge 1: The weight parameters, including the rule weight θ_k and attribute weight δ_i, are typically assigned by domain experts. This manual specification can introduce bias into the data. Moreover, the BRB framework requires covering all possible combinations of antecedent attributes and their evaluation grades during construction. Consequently, the total number of rules can reach $\prod_{i=1}^{T} J_i$, where T denotes the number of attributes and J_i represents the number of evaluation levels. This combinatorial growth results in a large and potentially unwieldy rule base, forming the first challenge addressed in this study.

Challenge 2: The standard BRB inference process supports only forward propagation, making backpropagation infeasible. This limitation prevents end-to-end learning and forces BRB parameter optimization to rely on heuristic algorithms, which are often inefficient and yield non-differentiable models. Designing a trainable BP-BRB model is the second major challenge of this work.

2.2 Related Work on Addressing BRBS Challenges

To tackle Challenge 1 identified in Sect. 2.1, researchers have made considerable efforts in recent years. These efforts can be broadly categorized into the following two directions.

(1) Yang [4] proposed a belief rule-based (BRB) system derived from Dempster-Shafer (D-S) theory. In BRB systems, combinatorial explosion arises from the large number of antecedent attributes and their evaluation grades. To address this, Hu [21] developed a novel hierarchical BRB structure (HFS-BRB) based on an extreme gradient boosting (XGBoost) feature selection method. By transforming a multi-class problem into multiple binary classification tasks, HFS-BRB can handle an arbitrary number of categories and mitigate the combinatorial explosion issue. Chang [22] introduced the concept of structural learning to tackle this problem, proposing four methods-Grey Target (GT), Multidimensional Scaling (MDS), Isomap, and Principal Component Analysis (PCA) for BRB structure optimization. You [23] proposed an ensemble BRB model within a bagging framework to reduce the size of the rule base. Yang [24] introduced a multi-layer tree structure (MTS) to define the hierarchical MTS-BRB model, effectively reducing rule base size to overcome the combinatorial explosion problem. Zhang [25] constructed a vector based BRB by ranking original attributes according to their contribution rates and transforming them into attribute vectors, enabling the expert system to handle excessive attribute dimensionality.

(2) Although numerous BRB variants [26–28] have been developed, the combinatorial explosion problem remains unresolved. To this end, Liu [19] proposed

the Extended Belief Rule Base (EBRB) system, which uses a belief degree framework for antecedent attributes and generates the system in a data-driven manner. This approach effectively alleviates combinatorial explosion while enhancing the representation of uncertain information. However, EBRB is not without limitations-it suffers from rule boundlessness and rule inconsistency. In early work addressing these issues, Calzada [29] proposed a Dynamic Rule Activation (DRA) method for EBRB to handle classification problems related to incomplete or inconsistent data. Yang [30] incorporated Data Envelopment Analysis (DEA) and the Wang-Mendel rule synthesis mechanism to eliminate invalid or redundant rules in EBRB. Fu [31] applied the Hierarchical Navigable Small World (HNSW) algorithm to build a K-nearest neighbor graph for EBRB, reducing the number of activated rules. Zhang [32] proposed a density-based clustering method for EBRB rule reduction, validated on both regression and classification datasets. Ren [33] developed a dynamic activation factor adjustment method to reduce redundant rules during inference. Gao [34] first constructed a candidate rule set based on rule similarity and then applied a greedy strategy to select optimal rules. Similarly, Bi [35] employed a greedy iterative search to remove similar rules from EBRB. Yang [36] introduced the cumulative belief rule-based (CBRB) system, which uses a nearest-neighbor information fusion strategy to activate consistent cumulative rules, achieving a better balance among interpretability, computational efficiency, and accuracy.

(3) In recent years, researchers have increasingly explored integrating Belief Rule Base (BRB) models with neural networks to combine the data-driven learning capabilities of neural architectures with the symbolic reasoning and interpretability of BRB systems. Neural networks can be employed to predict or adjust BRB parameters-such as attribute weights, rule weights, and belief degree distributions, thereby reducing dependence on expert-defined rules and improving model performance. For example, some studies leverage deep features to model BRB inputs and then optimize BRB parameters through heuristic or semi-automatic approaches, achieving performance gains but still requiring stage-wise training [37]. Another research line uses neural networks purely as feature extractors, with their outputs serving as input conditions for BRB, which then performs transparent, rule-based reasoning. This strategy has shown promising results in tasks such as visual sentiment analysis [38] and air pollution forecasting [39]. However, because the BRB component remains non-differentiable, these methods typically require separate training of the neural network and BRB components, lacking the capability for end-to-end joint optimization. Furthermore, when the number of rules grows large, the combinatorial explosion problem persists [40].

Existing studies indicate that although BRB and its extended variant EBRB have achieved notable progress in balancing performance and interpretability, they are still constrained by non-differentiability, decoupled training, and rule combination explosion. A deeper analysis reveals that these issues originate from the modeling of antecedent attributes in the rule base: conventional approaches rely on discretized evaluation grades, which can cause combinatorial explosion and rule inconsistency, as well as bias due to imprecise grade definitions. To address this limitation, this paper discards the evaluation-grade transformation and instead constructs the rule base directly from raw input data using clustering methods. The resulting rule centers, rule weights, and related parameters are treated as trainable variables, enabling end-to-end optimization via backpropagation. This approach eliminates reliance on expert priors, enhances trainability, and improves generalization capability, thereby providing strong motivation for the differentiable BP-BRB model proposed in this study.

3 Building Backpropagation-BRB

This section introduces BP-BRB optimization and reasoning framework centered on rule prototypes. First, the rule centers are constructed using either the K-Means [41] or fuzzy c-means (FCM) [42] clustering strategy. Second, the evidential reasoning (ER) approach is employed for inference. Finally, a comparative study is conducted to demonstrate the general applicability of the BP-BRB inference procedure.

By generating rules through clustering and integrating gradient descent-based backpropagation with the interpretability of the BRB model, the proposed method addresses the challenge of training white-box models. The detailed formulation of the proposed framework is presented in the following subsections, and its overall architecture is illustrated in Fig. 1.

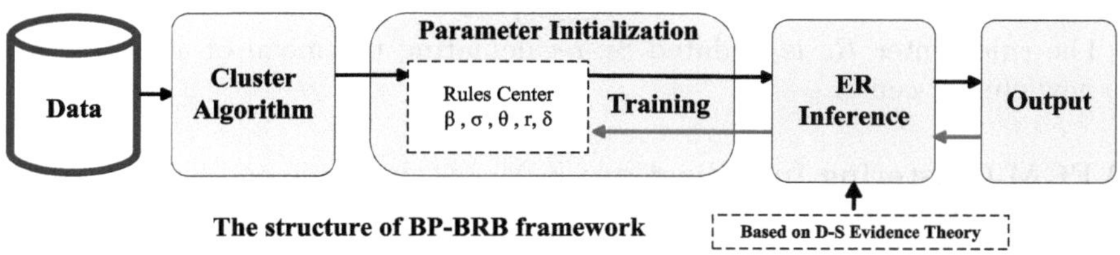

Fig. 1. BP-BRB framework: combining the backpropagation process with the DS evidence reasoning module optimization.

3.1 Construction of Rule Centers

The construction of the rule base is one of the key processes in rule-based systems. In this study, we employed two approaches for initialization: (i) clustering based on class labels using the K-Means algorithm, and (ii) clustering using the Fuzzy C-Means (FCM) algorithm. These methods enable a more reasonable initial distribution of the rule base, facilitating parameterization and accelerating convergence during subsequent training.

(1) K-Means Clustering Initialization: K-Means clustering is a classical hard-clustering algorithm whose core principle is to partition data into several mutually exclusive clusters, where each sample belongs to exactly one cluster center. The method is computationally efficient, converges rapidly, and is well-suited for large-scale datasets. Its simplicity makes it effective for clustering tasks with well-defined structures and clear boundaries.

However, the hard assignment nature of K-Means limits its ability to model complex scenarios, as it cannot represent fuzzy membership or uncertainty in sample assignments. Moreover, it is sensitive to the selection of initial centers, making it prone to local optima, and it performs poorly when dealing with non-spherical or overlapping clusters.

In our implementation, we applied K-Means to cluster all training data into K categories, with each cluster center serving as a rule center (clustering performed according to class labels).

$$J = \sum_{c=0}^{C} \sum_{k=1}^{K_c} \sum_{i=1}^{N} r_{ki}^{c} \left\| x_i - R_k^c \right\|^2 \tag{2}$$

where, J denotes the cost function of K-Means; C is the number of classes; N is the total number of samples; K_c represents the number of rule centers per class (determined using the elbow method); x_i is the ith input sample; R_k denotes the kth rule center; and r_{ik} is an indicator function that equals 1 if x_i belongs to cluster k, and 0 otherwise.

$$R_k = \frac{\sum_{i=1}^{N} r_{ki} x_i}{\sum_{i=1}^{N} r_{ki}} \tag{3}$$

The rule center R_k is updated by recalculating the mean of all samples as the new cluster center.

(2) FCM Clustering Initialization: K-Means clustering and Fuzzy C-Means (FCM) each have distinct advantages. FCM is a soft-clustering algorithm that generalizes K-Means by introducing a fuzzy factor F, which controls the degree of cluster fuzziness. Unlike K-Means, FCM allows each sample to belong to multiple clusters with varying degrees of membership, making it well-suited for modeling real-world problems with ambiguous boundaries or multi-class tendencies.

FCM offers stronger representational capability, particularly in scenarios involving uncertainty modeling or fuzzy rule-based systems. It characterizes the fuzzy relationship between samples and clusters through membership functions, aligning more closely with the requirements of fuzzy classification in practice. However, FCM has higher computational complexity, slower convergence, and greater sensitivity to noise and outliers. Furthermore, its performance is strongly influenced by parameter settings-such as the fuzzy factor-which must be tuned carefully.

$$J = \sum_{c=0}^{C} \sum_{k=1}^{K_c} \sum_{i=1}^{N} u_{ki}^{F} \left\| x_i - R_k^c \right\|^2, \sum_{k=1}^{K_c} u_{ki} = 1, u_{ki} \in [0,1] \tag{4}$$

where, J denotes the cost function of FCM; K_c is the number of cluster centers; N is the total number of samples; u_{ki} represents the membership degree of sample $\mathbf{x}_i$ to cluster center R_k; F is the fuzziness coefficient (typically set to 2) controlling the degree of fuzziness; and C is the number of classes. The updates are:

$$R_k = \frac{\sum_{i=1}^{N} u_{ki} x_i}{\sum_{i=1}^{N} u_{ki}^F}, u_{ki} = \left[\sum_{j=1}^{K_c} \left(\frac{\|x_i - R_k\|}{\|x_i - R_j\|} \right)^{\frac{2}{F-1}} \right]^{-1} \tag{5}$$

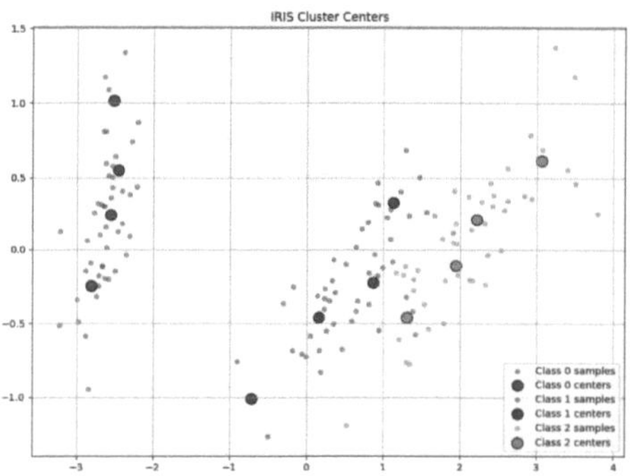

Fig. 2. IRIS cluster center.

Figure 2 presents the visualization results of clustering the IRIS dataset using the Fuzzy C-Means (FCM) method. For ease of visualization, the original four-dimensional feature space was projected onto a two-dimensional plane using a dimensionality reduction technique. In the figure, different colors represent different classes, small circles denote individual samples, and large circles indicate the corresponding cluster centers.

A total of 150 samples produced 12 cluster centers, which are well distributed within the primary regions of each class. This suggests that FCM effectively captures the intra-class distribution characteristics. In particular, the cluster centers for Class 0 and Class 2 exhibit strong spatial aggregation with their corresponding samples, indicating relatively compact internal structures and clear class boundaries. In contrast, the sample distribution of Class 1 appears more dispersed, with its cluster centers positioned closer to the class boundary-likely influenced by neighboring classes-implying some degree of class overlap.

Overall, the clustering results confirm the rationality of the label-guided fuzzy clustering strategy in generating rule centers, thereby supporting the subsequent BP-BRB modeling process.

(3) Label-Guided Clustering Construction: In a parameterized BRB, a rule center serves as the "representative point" of the premise attributes, representing a typical feature combination for a specific class of input samples.

Formally, a rule-based model is represented as:

$$\boldsymbol{R}_K^C : \textbf{IF} \{(\boldsymbol{r}_1^c, \ldots, \boldsymbol{r}_T^c),\ c \in C\} \textbf{ THEN } (\boldsymbol{\beta}_1^c, \ldots, \boldsymbol{\beta}_N^c) \tag{6}$$

with rule weight $\boldsymbol{\theta}_k$ and attribute weight $\{\boldsymbol{\delta}_1, \ldots, \boldsymbol{\delta}_T\}$, s.t. $\sum_{n=1}^{N} \boldsymbol{\beta}_n^c \leq 1$. Where, R_k denotes the kth rule center, and C the total number of classes. The term r_T^c refers to the Tth attribute of the kth rule center for class c. The vector β_N^c represents the belief degree distribution of the kth rule center in class c; higher values indicate greater reliability of the corresponding conclusion or decision. A rule is considered complete if $\sum_{n=1}^{N} \beta_N^p = 1$; otherwise, it is incomplete. The scalar θ_k denotes the rule weight of the kth rule center, while δ_T represents the weight of the Tth attribute. Both are trainable parameters.

In the parameterized BRB framework, these elements are represented as vectors in continuous space. The rule centers are initialized using the results of either K-Means or Fuzzy C-Means (FCM) clustering: $[r_1^c, \ldots, r_T^c, \beta_1^c, \ldots, \beta_N^c] \in \mathbb{R}^K,\ \beta_i \in [0,1]$

Rule weight initialization: $\theta_k \in [0,1],\ k \in K$

Gaussian Kernel Width: $\sigma_j = 1,\ \ j \in K$

Attribute weight initialization: $\delta_i = 1,\ i \in T$

As an example, for the IRIS dataset illustrated in Fig. 2, 12 cluster centers were generated. The corresponding trainable parameters are summarized in Table 1.

Table 1. Number of parameters

Parameter	Rule_Centers	Rule_weight	sigma	Attribute_weight	Belief_degrees	total
Shape	[12,4]	[12]	[12]	[4]	[12,3]	
Num	48	12	12	4	36	112

3.2 Parameterized Rule Base Forward Inference

(1) Similarity Calculation: In traditional BRB or EBRB systems, the degree of match between an input and a rule center is typically computed using the Euclidean distance formula. However, Euclidean distance has a critical limitation-it is non-differentiable at the origin. This poses a severe problem when applying backpropagation to update parameters, as it prevents the computation of gradients at that point.

To address this issue, we replace the Euclidean distance with a Gaussian function to measure the similarity between the input data and the rule centers.

$$S_k(x) = \exp\left(-\frac{\sum_{i=1}^{K}\left(x_i^{\delta_i} - R_{k,i}^{\delta_i}\right)^2}{2\sigma_k^2}\right), R_k \in \mathbb{R}_K \tag{7}$$

where, R_k denotes the kth rule center, σ_k is the bandwidth parameter of the membership, δ_i represents the attribute weight. Both σ_k and δ_i are trainable parameters.

(2) Activation Weight:

$$w_k = \frac{\theta_k S_k}{\sum_{j=1}^{L^*}\theta_j S_j} \tag{8}$$

where, L^* denotes the number of clustering rule center; θ represents the rule weight, S was given by (7).

(3) Rule Inference Using the ER Method: After obtaining the degrees of match and activation weights, the Evidential Reasoning (ER) method is employed to combine the rules and generate the inference output.

$$\beta_s(x) = \frac{\mu\left[\prod_{k=1}^{L^*}\left(w_k(x)\beta_{sk} + 1 - w_k(x)\sum_{s=1}^{N}\beta_{sk}\right) - \prod_{k=1}^{L^*}\left(1 - w_k(x)\sum_{s=1}^{N}\beta_{sk}\right)\right]}{1 - \mu\prod_{k=1}^{L^*}\left(1 - w_k(x)\sum_{s=1}^{N}\beta_{sk}\right)}$$

$$\mu = \left[\sum_{s=1}^{N}\prod_{k=1}^{L^*}\left(w_k(x)\beta_{sk} + 1 - w_k(x)\sum_{s=1}^{N}\beta_{sk}\right) - (N-1)\prod_{k=1}^{L^*}\left(1 - w_k(x)\sum_{s=1}^{N}\beta_{sk}\right)\right]^{-1} \tag{9}$$

where, μ denotes the normalization factor, w_k represents the activation weight of the kth rule, was given by (8).

After obtaining the inference output, the interpretation differs depending on the task type. For regression problems, let $u(D_n)$ denote the confidence of result D_n in attribute D, satisfying $u(D_1) \leq u(D_2) \leq ... \leq u(D_n)$. The inference output is described as:

$$f(x) = \sum_{n=1}^{N} u(D_n)\beta_n \tag{10}$$

For classification problems, D_n represents the nth class, and the inference output is described as (11):

$$f(x) = D_n, \quad n = \arg\left\{\max_{i=1,\ldots,N}[\beta_i]\right\} \tag{11}$$

4 Experiments

4.1 Datasets Description

We employed 23 well-known benchmark classification datasets from the UCI Machine Learning Repository, covering diverse domains such as medical diagnosis (e.g., Diabetes) and image recognition (e.g., Penbased). The classification performance of the proposed BP-BRB model was evaluated through four distinct comparative experiments. The statistical characteristics of these datasets are summarized in Table 2, where Samples denotes the number of instances, Attributes indicates the number of attributes, and Categories refers to the number of classes.

Table 2. Basic information of classification datasets.

Dataset	Samples	Features	Categories	Dataset	Samples	Features	Categories
Tae	151	5	3	Balance	625	4	3
Iris	150	4	3	Twonorm	7400	20	2
Bupa	345	6	2	Phoneme	5404	5	2
Heart	270	13	2	Cleveland	297	13	5
Wdbc	569	30	2	Satimage	6435	36	6
Wine	178	13	3	Penbased	10992	16	10
Glass	214	9	6	Wisconsin	683	9	2
Vowel	990	13	11	Transfusion	748	4	2
Magic	19020	10	2	Page-Blocks	5472	10	5
Pima	768	8	2	Contraceptive	1473	9	3
Yeast	1484	8	10	Mammographic	830	5	2
Diabetes	393	8	2				

4.2 Results and Analysis

For each classification dataset, a BP-BRB model was constructed and evaluated using 10-fold cross-validation (10-CV). In this setting, each dataset was randomly partitioned into ten subsets, with 90% of the data used for training and the remaining 10% for testing. Classification accuracy, one of the most widely used metrics for evaluating classifier performance, was adopted to assess the effectiveness of the BP-BRB model.

Three comparative experiments were conducted, and the experimental reference results are all from: [36,43]. In the first comparison, the proposed BP-BRB model was evaluated against nine improved BRB variants previously reported in the literature. Table 3 presents the basic characteristics of these nine BRB variants, along with BP-BRB, as well as their classification accuracies on selected datasets.

Table 3. Comparison of BP-BRBS with existing BRBSs.

Indicator Dataset	Improved BRBS EBRB	MaSF	CABRA	DBSCAN	SO	BT	DRA	Micro	CBRBS	BP-BRB This Study
Year	2013 [19]	2016 [44]	2018 [45]	2020 [46]	2020 [47]	2020 [48]	2021 [49]	2021 [50]	2022 [36]	-
Challenge 1	Yes	Yes	Yes	Yes	Yes	Yes	Yes	Yes	Yes	Yes
Challenge 2	No	No	No	No	No	No	No	No	No	Yes
Wdbc	94.59(8)	96.30(3)	97.01(2)	95.47(5)	95.22(6)	96.00(4)	94.61(7)	96.49(2)	96.49(2)	97.19(1)
Diabetes	73.39(7)	75.04(5)	76.34(3)	75.98(4)	75.04(5)	-	71.44(8)	74.91(6)	76.84(1)	76.43(2)
Mammographic	77.64(10)	80.61(4)	79.52(8)	79.57(7)	83.95(1)	79.80(6)	78.39(9)	79.88(5)	81.08(3)	81.69(2)
Iris	95.20(6)	95.20(6)	96.00(2)	95.73(4)	97.67(1)	95.99(3)	95.50(5)	-	96.00(2)	96.00(2)
Wine	96.32(7)	96.52(4)	96.63(3)	97.87(1)	96.49(5)	94.76(9)	96.46(6)	95.84(8)	96.63(3)	97.75(2)
Transfusion	76.14(6)	78.65(3)	72.07(7)	57.33(8)	-	-	76.57(4)	76.52(5)	79.14(2)	79.41(1)
Glass	67.85(7)	70.19(4)	72.90(2)	57.07(9)	-	70.01(5)	69.65(6)	63.32(8)	71.03(3)	83.59(1)
Average rank	7.3	4.1	3.9	5.4	3.6	5.4	6.4	5.7	2.3	1.6

From Table 3, it can be observed that, in terms of classification accuracy, BP-BRB achieved the highest performance on 3 datasets and ranked second on 4 datasets, yielding an average rank of 1.6-substantially outperforming all other methods. For example, on the Glass, Transfusion, and Wdbc datasets, BP-BRB attained the highest accuracy, significantly surpassing both the traditional BRB and its variants. Moreover, BP-BRB demonstrated stable and superior performance across diverse data types, including small-sample datasets (e.g., Glass), imbalanced datasets (e.g., Mammographic), and medium-dimensional datasets (e.g., Wdbc). These results further highlight the proposed method's advantages in model expressiveness, generalization capability, and robustness.

To further verify the generality and performance advantages of BP-BRB, a second comparative experiment was conducted on 10 publicly available datasets, benchmarking against eight representative recent BRB variants: EBRB, DEA-EBRB, Micro-EBRB, DBSCAN-EBRB, CBRB, AFDRA-EBRB, RCPL-EBRB, and Optimization-CBRB. These methods collectively represent mainstream advancements in BRB research over recent years, encompassing structural improvements, enhanced inference mechanisms, and optimization algorithm innovations.

From Table 4, it is evident that BP-BRB achieved the highest accuracy on 9 datasets, ranking first on average and exhibiting a significant overall performance advantage. For instance, on structurally complex datasets such as Yeast, Contraceptive, and Page-Blocks, BP-BRB outperformed all other methods with accuracies of 59.02%, 54.04%, and 96.47%, respectively. Notably, on small-sample and hard-to-classify datasets such as Bupa and Tae, BP-BRB also demonstrated strong robustness, with accuracy improvements exceeding 7% points over competing approaches.

Furthermore, considering Tables 3 and 4 together, it can be observed that most existing studies focus solely on addressing Challenge 1, whereas the present work is the first to simultaneously tackle both challenges through the proposed BP-BRB. By mitigating the combinatorial explosion problem and introducing a trainable structure, BP-BRB enables end-to-end optimization, thereby enhancing the model's generalization capability across diverse datasets. In summary, BP-BRB substantially improves modeling capacity and adaptability while pre-

Table 4. Comparison of BP-BRBS with existing BRBSs.

Indicator Dataset	Improved BRBS								BP-BRB
	EBRB	DEA-EBRB	Micro-EBRB	DBSCAN	CBRB	AFDRA	RCPL	Opt-CBRB	This Study
Year	2013 [19]	2017 [30]	2018 [50]	2020 [46]	2022 [36]	2024 [31]	2024 [51]	2025 [43]	-
Challenge 1	Yes	Yes	Yes	Yes	Yes	Yes	Yes	Yes	Yes
Challenge 2	No	No	No	No	No	No	No	No	Yes
Tae	49.01(6)	46.36(8)	44.37(9)	46.39(7)	50.99(2)	49.67(5)	50.97(3)	50.33(4)	56.92(1)
Wisconsin	95.9(5)	95.46(7)	95.46(7)	55.26(8)	95.61(6)	96.19(3)	95.91(4)	96.34(2)	96.48(1)
Bupa	57.97(4)	57.97(4)	57.97(4)	57.97(4)	57.97(4)	57.97(4)	58.26(3)	60.29(2)	67.76(1)
Wine	96.32(4)	80.9(7)	95.84(5)	97.87(1)	96.63(3)	96.63(3)	74.73(8)	93.82(6)	97.75(2)
Page-Blocks	89.88(5)	89.84(6)	89.84(6)	89.88(5)	89.89(4)	89.91(3)	80.6(7)	90.61(2)	96.47(1)
Balance	89.44(3)	84.16(7)	80.32(8)	89.44(3)	84.8(5)	90.56(2)	84.64(6)	85.6(4)	90.57(1)
Contraceptive	47.73(3)	46.23(6)	45.49(7)	43.86(8)	46.37(4)	47.73(3)	46.3(5)	48.34(2)	54.04(1)
Yeast	33.89(6)	38.81(3)	31.33(9)	33.02(7)	32.35(8)	34.43(5)	38.41(4)	42.05(2)	59.02(1)
Mammographic	77.64(9)	80.84(3)	79.88(6)	79.57(8)	79.76(7)	80.48(4)	80.34(5)	81.08(2)	81.69(1)
Phoneme	70.65(4)	70.65(4)	70.65(4)	70.65(4)	82.42(2)	70.65(4)	70.65(4)	75.19(3)	84.51(1)
Average rank	4.9	5.5	6.5	5.5	4.5	3.6	4.9	2.9	1.1

serving interpretability, making it more practical and extensible in real-world applications.

In the third comparative experiment, Table 5 presents the classification accuracy of our two proposed variants-BP-BRB with K-Means initialization (K-BP-BRB) and BP-BRB with FCM initialization (F-BP-BRB)-against other existing approaches on 12 widely used datasets. In this experiment, six classical FRBS-related classifiers were compared with BP-BRB: the Structure Learning Algorithm in a Vague Environment (SLAVE) [52], Fuzzy Hybrid Genetic-Based Machine Learning (FH-GBML) [53], Steady-State Genetic Algorithm for Extracting Fuzzy Classification Rules from Data (SGRED) [54], High-Dimensional Classification based on Fuzzy Association Rules (FARC-HD) [49], Classification based on Fuzzy Association Rules (CFAR) [55], and the Cumulative Belief Rule-Based System (CBRBS) [36]. Among these, SGRED is an improved WM-FRBS designed for classification tasks. Table 5 reports the classification accuracies of these six FRBS-related classifiers across the 12 datasets, where the numbers in parentheses indicate the ranking of each method on the corresponding dataset (with smaller numbers denoting better performance).

From Table 5, it can be observed that F-BP-BRB achieved the best overall performance, ranking first on 7 out of the 12 datasets, with an average rank of 1.5-significantly outperforming other methods. In comparison, K-BP-BRB, while achieving first place on 2 datasets, exhibited slightly lower stability and accuracy overall. This indicates that employing Fuzzy C-Means (FCM) to construct rule centers, and initializing rule weights using the soft membership degrees of samples, effectively enhances the model's representational capacity and classification performance.

Moreover, F-BP-BRB delivered leading results on several high-dimensional or large-scale datasets (e.g., Magic, Satimage, Penbased), demonstrating superior generalization ability. In contrast, traditional methods such as SLAVE, FH-

Table 5. Comparison of CBRBS with classical FRBS-related classifiers.

Dataset	SLAVE	FH-GBML	SGRED	FARC-HD	CFAR	CBRBS	K-BP-BRB	F-BP-BRB
Heart	71.36(8)	75.93(6)	73.21(7)	84.44(1)	82.22(4)	80.37(5)	83.33(3)	83.70(2)
Wdbc	92.33(5)	92.26(6)	90.68(7)	95.25(4)	–	96.49(2)	96.30(3)	97.19(1)
Pima	73.71(6)	75.26(3)	73.37(7)	75.66(2)	65.11(8)	74.09(5)	74.22(4)	77.21(1)
Phoneme	76.41(6)	79.66(5)	75.55(7)	82.14(4)	70.65(8)	82.42(3)	89.01(1)	84.51(2)
Twonorm	86.99(6)	85.97(7)	73.98(8)	95.28(4)	91.66(5)	96.77(3)	97.58(2)	97.78(1)
Magic	73.96(6)	81.30(4)	72.06(7)	84.51(3)	64.84(8)	80.61(5)	86.29(2)	86.32(1)
Iris	94.44(4)	94.00(5)	94.89(2)	96.00(1)	90.67(6)	96.00(1)	94.67(3)	96.00(1)
Wine	89.47(8)	92.61(6)	91.88(7)	94.35(4)	93.24(5)	96.63(3)	97.19(2)	97.75(1)
Cleveland	48.82(8)	53.51(6)	51.59(7)	55.24(4)	53.88(5)	55.89(3)	57.30(1)	56.62(2)
Satimage	81.69(5)	74.72(7)	77.10(6)	87.32(4)	–	90.46(3)	90.92(2)	91.14(1)
Penbased	81.16(5)	50.45(7)	67.93(6)	96.04(4)	36.43(8)	99.14(1)	98.03(3)	99.04(2)
Vowel	71.11(5)	67.07(6)	65.83(7)	71.82(4)	–	98.38(1)	96.97(2)	93.13(3)
Num No.1	0	0	0	2	0	3	2	7
Average rank	6	5.7	6.5	3.25	6.3	2.9	2.3	1.5

GBML, and CFAR showed less stable performance, with substantial fluctuations in accuracy and ranking across datasets.

Overall, F-BP-BRB outperforms both K-BP-BRB and other existing approaches in terms of accuracy, stability, and adaptability, confirming that integrating FCM with BP-BRB offers greater advantages in handling fuzzy classification problems. To further validate the effectiveness of the proposed model, we compare the F-BP-BRB-a BP-BRB model based on Fuzzy C-Means clustering-with several classical machine learning classifiers, including Artificial Neural Network (ANN) and k-Nearest Neighbors (KNN) [49], Naive Bayes (NB) [56], Decision Tree (DT) [57], and Support Vector Machine (SVM) [58].

Table 6 summarizes the classification accuracy and ranking results across 23 publicly available datasets, where the numbers in parentheses indicate the model's ranking for each dataset. Overall, F-BP-BRB ranked first on 13 out of 23 datasets, achieving an average rank of 1.5, which is markedly superior to traditional methods. In most datasets, F-BP-BRB achieved high classification accuracy, with particularly strong performance on medium- to large-scale datasets such as Phoneme, Magic, Satimage, and Penbased. For example, on the Penbased dataset, F-BP-BRB achieved an accuracy of 99.04%, significantly outperforming ANN (94.39%).

Even on relatively challenging small-sample datasets such as Glass and Cleveland, F-BP-BRB maintained high accuracy. Notably, although neural networks performed close to optimally on certain datasets (e.g., Iris, Wine), their training process was highly sensitive to data size and parameter settings. In contrast, BP-BRB achieved consistently high performance across multiple datasets without extensive parameter tuning, demonstrating strong generalization capability and robustness.

Table 6. Comparison of CBRBS with classical ML-related classifiers.

Dataset	KNN	NB	DT	SVM	ANN	K-BP-BRB	F-BP-BRB
Heart	74.81(5)	83.70(1)	77.41(4)	55.56(6)	82.22(3)	83.33(2)	83.70(1)
Diabetes	74.09(4)	76.30(2)	73.82(5)	65.10(7)	75.39(3)	73.57(6)	76.43(1)
Wdbc	95.96(4)	92.97(6)	93.32(5)	62.74(7)	96.31(2)	96.30(3)	97.19(1)
Transfusion	76.20(5)	75.40(6)	78.34(3)	75.27(7)	76.34(4)	78.87(2)	79.41(1)
Pima	68.83(7)	77.27(1)	71.82(6)	75.45(3)	73.77(5)	74.22(4)	77.21(2)
Mammographic	79.04(7)	82.41(2)	83.98(1)	79.88(5)	80.60(4)	79.76(6)	81.69(3)
Phoneme	79.44(6)	76.05(7)	86.42(2)	84.49(4)	80.98(5)	89.01(1)	84.51(3)
Twonorm	94.86(6)	97.83(1)	85.12(7)	97.69(3)	96.96(5)	97.58(4)	97.78(2)
Magic	74.79(5)	72.69(6)	81.17(4)	65.88(7)	83.73(3)	86.29(2)	86.32(1)
Iris	94.67(3)	94.00(4)	94.67(3)	97.33(1)	97.33(1)	94.67(3)	96.00(2)
Wine	97.19(2)	96.63(4)	92.13(5)	44.38(6)	97.17(3)	97.19(2)	97.75(1)
Cleveland	55.56(4)	54.88(5)	56.57(3)	53.87(6)	52.53(7)	57.30(1)	56.62(2)
Glass	66.36(6)	48.60(7)	66.82(5)	68.69(3)	67.76(4)	85.50(1)	83.59(2)
Satimage	86.67(3)	75.39(7)	82.34(6)	82.71(5)	85.61(4)	90.92(2)	91.14(1)
Penbased	55.90(6)	85.68(4)	74.35(5)	13.71(7)	94.39(3)	98.03(2)	99.04(1)
Vowel	15.25(7)	67.07(5)	46.97(6)	88.48(3)	84.14(4)	96.97(1)	93.13(2)
Tae	–	34.44(4)	49.01(3)	34.37(5)	–	55.54(2)	56.92(1)
Wisconsin	–	65.01(4)	91.51(3)	65.01(4)	–	97.07(1)	96.48(2)
Bupa	–	57.97(3)	57.39(4)	57.97(3)	–	67.24(2)	67.76(1)
Page-Blocks	–	89.78(4)	89.44(5)	89.82(3)	–	96.95(1)	96.47(2)
Balance	–	45.76(5)	59.52(3)	53.60(4)	–	86.72(2)	90.57(1)
Contraceptive	–	42.70(4)	43.65(3)	42.70(4)	–	50.71(2)	54.04(1)
Yeast	–	31.22(4)	56.81(3)	31.20(5)	–	58.22(2)	59.02(1)
Num No.1	0	3	1	1	1	6	13
Average rank	5	4.2	4.1	4.7	3.8	2.3	1.5

K-BP-BRB also showed competitive performance, ranking among the top methods on several datasets with an average rank of 2.6. This further illustrates that rule-based models can achieve a favorable balance between interpretability and predictive performance, while the introduction of fuzzy membership in F-BP-BRB enhances modeling capability.

In summary, F-BP-BRB not only outperforms traditional methods and neural networks in terms of accuracy but also exhibits superior stability and adaptability, indicating that constructing the rule reasoning mechanism as a trainable, parameterized structure is an effective modeling strategy.

5 Conclusion and Future Work

This study proposes a Backpropagation-based Belief Rule Base (BRB) model with parameterized rule centers. By integrating parameter optimization with rule-based reasoning, the model combines interpretability and learning capability, successfully embedding backpropagation into the inherently interpretable BRB structure. The proposed approach can automatically optimize parameters, offering stronger generalization ability, clearer interpretability, high training efficiency, and a smooth and stable inference process-realizing a new paradigm of neuro-symbolic integration.

Experimental results demonstrate that clustering-based BRB can more accurately combine rules, thereby improving inference performance. Parameterizing the original rule logic into rule centers represents a key step in transforming BRB from an expert system into a data-driven, and scalable machine learning model. Future work will focus on extending the method to more complex tasks, incorporating transfer learning to enhance performance in data-scarce scenarios. In addition, the model's structure allows for the easy introduction of domain constraints or controllable rules-for example, fixing certain rule weights or setting upper and lower bounds-facilitating the development of controllable AI systems.

References

1. Hassija, V., et al.: Interpreting black-box models: a review on explainable artificial intelligence. Cogn. Comput. **16**(1), 45–74 (2024)
2. Cong-Lem, N., Soyoof, A., Tsering, D.: A systematic review of the limitations and associated opportunities of ChatGPT. Int. J. Hum.-Comput. Interact. **41**(7), 3851–3866 (2025)
3. Sun, R.: Robust reasoning: integrating rule-based and similarity-based reasoning. Artif. Intell. **75**(2), 241–295 (1995)
4. Yang, J.-B., et al.: Belief rule-base inference methodology using the evidential reasoning approach-RIMER. IEEE Trans. Syst. Man Cybern.-Part A Syst. Hum. **36**(2), 266–285 (2006)
5. Dempster, A.P.: A generalization of Bayesian inference. J. Roy. Stat. Soc.: Ser. B (Methodol.) **30**(2), 205–232 (1968)
6. Shafer, G.: A Mathematical Theory of Evidence, vol. 42. Princeton University Press (1976)
7. Hwang, C.-L., et al.: Methods for multiple attribute decision making. Multiple attribute decision making: methods and applications a state-of-the-art survey, pp. 58–191 (1981)
8. Zadeh, L.A.: Fuzzy sets. Inf. Control **8**(3), 338–353 (1965)
9. Liu, J., et al.: Self-tuning of fuzzy belief rule bases for engineering system safety analysis. Ann. Oper. Res. **163**, 143–168 (2008)
10. Yang, J.-B., et al.: Optimization models for training belief-rule-based systems. IEEE Trans. Syst. Man Cybern.-Part A Syst. Hum. **37**(4), 569–585 (2007)
11. Zhou, Z.-J., et al.: Online updating belief rule based system for pipeline leak detection under expert intervention. Expert Syst. Appl. **36**(4), 7700–7709 (2009)
12. Wu, J., et al.: AutoBRB: an automated belief rule base model for pathologic complete response prediction in gastric cancer. Comput. Biol. Med. **140**, 105104 (2022)

13. Xu, X., et al.: A belief rule-based expert system for fault diagnosis of marine diesel engines. IEEE Trans. Syst. Man Cybern. Syst. **50**(2), 656–672 (2017)
14. Zhou, Z.-J., et al.: A survey of belief rule-base expert system. IEEE Trans. Syst. Man Cybern. Syst. **51**(8), 4944–4958 (2019)
15. Leyva, H., Bojórquez, J., Bojórquez, E., Reyes-Salazar, A., Carrillo, J., López-Almansa, F.: Multi-objective seismic design of BRBs-reinforced concrete buildings using genetic algorithms. Struct. Multidiscip. Optim. **64**(4), 2097–2112 (2021). https://doi.org/10.1007/s00158-021-02965-5
16. Qian, B., et al.: An effective soft computing technology based on belief-rule-base and particle swarm optimization for tipping paper permeability measurement. J. Ambient. Intell. Humaniz. Comput. **10**, 841–850 (2019)
17. Werbos. Backpropagation: Past and future. IEEE 1988 International Conference on Neural Networks. IEEE (1988)
18. LeCun, Y., et al.: Handwritten digit recognition with a back-propagation network. In: Advances in Neural Information Processing Systems, vol. 2 (1989)
19. Liu, J., et al.: A novel belief rule base representation, generation and its inference methodology. Knowl.-Based Syst. **53**, 129–141 (2013)
20. Yang, J.-B., Dong-Ling, X.: On the evidential reasoning algorithm for multiple attribute decision analysis under uncertainty. IEEE Trans. Syst. Man Cybern.-Part A Syst. Hum. **32**(3), 289–304 (2002)
21. Hu, G., et al.: Hierarchical belief rule-based model for imbalanced multi-classification. Expert Syst. Appl. **216**, 119451 (2023)
22. Chang, L., et al.: Structure learning for belief rule base expert system: a comparative study. Knowl.-Based Syst. **39**, 159–172 (2013)
23. You, Y., et al.: Ensemble belief rule-based model for complex system classification and prediction. Expert Syst. Appl. **164**, 113952 (2021)
24. Yang, L.-H., et al.: Belief rule-base expert system with multilayer tree structure for complex problems modeling. Expert Syst. Appl. **217**, 119567 (2023)
25. Zhenjie, Z., et al.: A novel nonlinear causal inference approach using vector-based belief rule base. Int. J. Intell. Syst. **36**(9), 5005–5027 (2021)
26. Badhon, B., et al.: IRAF-BRB: an explainable AI framework for enhanced interpretability in project risk assessment. Expert Syst. Appl. 127979 (2025)
27. Gutiérrez-Urzúa, F., Freddi, F., Tubaldi, E.: Seismic risk and failure modes assessment of steel BRB frames under earthquake sequences. Struct. Saf. **115**, 102598 (2025)
28. Zhang, J., et al.: Effects of BRB on cyclic behavior of resilient RAC composite frames with ultra-high strength steel bars. Eng. Struct. **332**, 120059 (2025)
29. Calzada, A., et al.: A new dynamic rule activation method for extended belief rule-based systems. IEEE Trans. Knowl. Data Eng. **27**(4), 880–894 (2014)
30. Yang, L.-H., et al.: A data envelopment analysis (DEA)-based method for rule reduction in extended belief-rule-based systems. Knowl.-Based Syst. **123**, 174–187 (2017)
31. Fu, Y.-G., et al.: A novel extended rule-based system based on K-Nearest Neighbor graph. Inf. Sci. **662**, 120158 (2024)
32. Zhang, A., et al.: A new rule reduction and training method for extended belief rule base based on DBSCAN algorithm. Int. J. Approximate Reasoning **119**, 20–39 (2020)
33. Ren, T.-Y., et al.: Dynamic rule activation method based on activation factor for extended belief rule-based systems. In: 2021 16th International Conference on Intelligent Systems and Knowledge Engineering (ISKE). IEEE (2021)

34. Gao, F., et al.: A greedy belief rule base generation and learning method for classification problem. Appl. Soft Comput. **98**, 106856 (2021)
35. Bi, W., et al.: A framework for extended belief rule base reduction and training with the greedy strategy and parameter learning. Multimedia Tools Appl. **81**(8), 11127–11143 (2022)
36. Yang, L.-H., et al.: Highly explainable cumulative belief rule-based system with effective rule-base modeling and inference scheme. Knowl.-Based Syst. **240**, 107805 (2022)
37. Zhou, Z., et al.: A novel belief rule-based fault diagnosis method with interpretability. Comput. Model. Eng. Sci. (CMES) **136**(2) (2023)
38. Zisad, S.N., et al.: An integrated deep learning and belief rule-based expert system for visual sentiment analysis under uncertainty. Algorithms **14**(7), 213 (2021)
39. Kabir, S., et al.: An integrated approach of belief rule base and deep learning to predict air pollution. Sensors **20**(7), 1956 (2020)
40. You, Y., et al.: Interpretability and accuracy trade-off in the modeling of belief rule-based systems. Knowl.-Based Syst. **236**, 107491 (2022)
41. Zhuang, J., et al.: Extended belief rule-base optimization base on clustering tree and parameter optimization. IEEE Access **9**, 12533–12544 (2021)
42. Fu, Y.-G., et al.: Construction of EBRB classifier for imbalanced data based on Fuzzy C-Means clustering. Knowl.-Based Syst. **234**, 107590 (2021)
43. Yang, L.-H., et al.: A novel modeling approach for cumulative belief rule-base with joint optimization and rule synthesis. IEEE Trans. Syst. Man Cybern. Syst. (2025)
44. Yang, L.-H., et al.: Multi-attribute search framework for optimizing extended belief rule-based systems. Inf. Sci. **370**, 159–183 (2016)
45. Yang, L.-H., Wang, Y.-M., Yang-Geng, F.: A consistency analysis-based rule activation method for extended belief-rule-based systems. Inf. Sci. **445**, 50–65 (2018)
46. Zhang, A., et al.: A new rule reduction and training method for extended belief rule base based on DBSCAN algorithm. Int. J. Approximate Reasoning **119**, 20–39 (2020)
47. Zhu, H., et al.: A structure optimization method for extended belief-rule-based classification system. Knowl.-Based Syst. **203**, 106096 (2020)
48. Fu, Y.-G., et al.: A framework for optimizing extended belief rule base systems with improved Ball trees. Knowl.-Based Syst. **210**, 106484 (2020)
49. Aghaeipoor, F., Javidi, M.M., Fernández, A.: IFC-BD: an interpretable fuzzy classifier for boosting explainable artificial intelligence in big data. IEEE Trans. Fuzzy Syst. **30**(3), 830–840 (2021)
50. Yang, L.-H., et al.: A micro-extended belief rule-based system for big data multiclass classification problems. IEEE Trans. Syst. Man Cybern. Syst. **51**(1), 420–440 (2018)
51. Yang, L.-H., Chen, J.-H., Ye, F.-F.: Extended belief rule base inference model based on rule clustering and parameter learning. Control Decis. **39**(8), 2685–2693 (2024)
52. Gonzblez, A., Pérez, R.: SLAVE: a genetic learning system based on an iterative approach. IEEE Trans. Fuzzy Syst. **7**(2), 176–191 (2002)
53. Ishibuchi, H., Yamamoto, T., Nakashima, T.: Hybridization of fuzzy GBML approaches for pattern classification problems. IEEE Trans. Syst. Man Cybern. Part B (Cybern.) **35**(2), 359–365 (2005)
54. Mansoori, E.G., Zolghadri, M.J., Katebi, S.D.: SGERD: a steady-state genetic algorithm for extracting fuzzy classification rules from data. IEEE Trans. Fuzzy Syst. **16**(4), 1061–1071 (2008)

55. Chen, Z., Chen, G.: Building an associative classifier based on fuzzy association rules. Int. J. Comput. Intell. Syst. **1**(3), 262–273 (2008)
56. John, G.H., Langley, P.: Estimating continuous distributions in Bayesian classifiers. arXiv preprint arXiv:1302.4964 (2013)
57. Quinlan, J.R.: C4.5: programs for machine learning. Elsevier (2014)
58. Chang, C.-C., Lin, C.-J.: LIBSVM: a library for support vector machines. ACM Trans. Intell. Syst. Technol. (TIST) **2**(3), 1–27 (2011)

Rule Extraction Methods Based on Triadic Fuzzy Linguistic Formal Context

Mingyang Shao, Huaizhe Zhao, Yi Liu, Qing Guo(✉), and Li Zou

School of Computer and Artificial Intelligence, Shandong Jianzhu University, Jinan 250101, Shandong, China
2024110104@stu.sdjzu.edu.cn

Abstract. The processing of complex data in uncertain environments has long been a research hotspot for numerous scholars. Aiming at the problems of multidimensional data processing in fuzzy linguistic environments and mining implication rules among attributes described by linguistic values, this paper proposes a rule extraction method based on the triadic fuzzy linguistic formal context. First, a triadic fuzzy linguistic formal context is constructed by integrating linguistic term sets with triadic concept analysis. Subsequently, triadic fuzzy linguistic concepts are defined based on derivation operators, and a knowledge discovery algorithm is designed via an incremental construction method to extract conceptual knowledge with semantic information from fuzzy triadic relations. On this basis, a decision rule extraction method based on the triadic fuzzy linguistic decision context is proposed, which constructs a decision rule base by analyzing the semantic implication relationships between conditional and decision attributes. This method provides a novel approach with both semantic interpretability and credibility for the decision analysis of multidimensional data in fuzzy linguistic environments, effectively making up for the deficiencies of traditional methods in handling linguistic value preference information and triadic multidimensional correlations, and offering new ideas for knowledge discovery and rule mining in uncertain environments.

Keywords: Linguistic term set · Triadic concept analysis · Triadic fuzzy linguistic formal context · Incremental construction · Rule extraction

1 Introduction

With the rapid development of social information technology, various applications generate massive amounts of data, making knowledge extraction from complex data a focal point of scholarly research. Formal Concept Analysis (FCA), also known as concept lattice theory, was introduced by Wille [1] in 1982 as a significant mathematical theory. As a pivotal visualization-oriented approach based on formal contexts, FCA derives potential conceptual knowledge through unique relationships between objects and attributes, hierarchically representing

J. Lu et al. (Eds.): FLINS-ISKE 2026, LNCS 16752, pp. 509–523, 2027.
https://doi.org/10.1007/978-981-92-2369-5_31

specialization and generalization of concepts within concept lattices. Due to its concise yet powerful nature, FCA has found extensive applications in fields such as information retrieval, data mining, and knowledge management [2]. To address increasingly complex data, Wille et al. [3] combined Peirce's triadic categories to propose Triadic Concept Analysis (TCA), offering a new perspective for information processing and three-dimensional data analysis. Subsequently, numerous scholars have conducted research on TCA with remarkable achievements [4]. For instance, Wang et al. [5,6] introduced Boolean matrix representation and incremental construction methods for triadic concepts; Liu et al. [7] applied heuristic construction methods of triadic concepts to social recommendation systems; Li et al. [8,9] simplified triadic concepts through triadic factor analysis and analyzed their relationship with formal concepts; Tang et al. [10] established a cognitive system model under triadic formal concept analysis, proposing effective methods to transform any information granule into sufficient, necessary, and sufficient-necessary information granules; Qi et al. [11] proposed an information simplification method for concept trilattice; Li et al. [12] introduced text classification methods based on triadic concept analysis.

In some cases, people express uncertainty information in their lives through linguistic values. For instance, individuals might describe the taste of food as "slightly sweet" or "very bitter". The processing of linguistic values has emerged as a prominent research area in intelligent information processing, leading to significant advancements. Zadeh [13] introduced the theory of linguistic variables and computing with words in 1996, offering an effective method for uncertainty reasoning. Zhang et al. [14] applied linguistic fuzzy sets to handle multi-attribute decision-making problems with linguistic preferences, demonstrating its utility in risk assessment for sustainable supply chains. Pang et al. [15] proposed an association rule mining with fuzzy linguistic information based on attribute partial ordered structure. Zou et al. [16] proposed a kind of linguistic truth-valued intuitionistic fuzzy lattice based on the point view of intuitionistic fuzzy set and linguistic truth-valued lattice implication algebra. In order to solve MADM problem, Zhou et al. [17] integrated linguistic concepts into traditional intuitionistic fuzzy theory, proposing a TOPSIS method based on intuitionistic fuzzy linguistic concepts to address multi-attribute decision-making problems.

Rule extraction, as a core component of knowledge discovery, aims to transform massive heterogeneous information into interpretable decision support. Dwivedi et al. [18] systematically classified existing explainable artificial intelligence methods, providing a theoretical basis for researchers to select appropriate interpretable modeling schemes. Abbas et al. [19] conducted research on explainable artificial intelligence for clinical decision support systems in healthcare. Bourne [20] proposed an inference model for conceptual rule learning based on analyzing human cognitive processes in conceptual problem-solving. To address the fuzziness and uncertainty inherent in real-world decision-making scenarios, this paper proposes a decision rule extraction method based on triadic fuzzy linguistic decision contexts. By analyzing the semantic implicative relationships between conditional and decision attributes, the method constructs an interpretable rule base, providing reliable support for complex decision-making. This approach form a unified rule extraction framework for handling uncertain linguistic information, offering an effective tool for knowledge discovery in complex data environments.

The main work of this paper is as follows:

(1) A triadic fuzzy linguistic formal context is proposed by integrating fuzzy linguistic concepts with triadic formal contexts, where triadic fuzzy linguistic concepts are defined through derivation operators to effectively handle complex preference information in multidimensional data under fuzzy linguistic environments.

(2) A knowledge discovery algorithm based on the triadic fuzzy linguistic formal context is designed, which incorporates incremental construction ideas to extract conceptual knowledge with fuzzy triadic relations and semantic information.

(3) A decision rule extraction method based on triadic fuzzy linguistic decision contexts is proposed, which analyzes semantic implications between conditional and decision attributes to construct an interpretable decision rule base, providing support for complex decision-making environments.

The rest of this paper is organized as follows. Section 2 provides a brief survey of linguistic term set and triadic concept analysis. Section 3 proposes a knowledge discovery algorithm integrating incremental construction idea. Section 4 introduces a decision rule extraction method based on triadic fuzzy linguistic decision contexts. Finally, Sect. 5 concludes the paper and outlines future research directions.

2 Preliminaries

This section briefly recalls the basic knowledge of linguistic term set and triadic concept analysis.

Definition 1. *[21] Let $S = \{s_t | t = -\tau, \dots, -1, 0, 1, \dots, \tau\}$ be a finite linguistic term set (LTS) with an odd number of linguistic terms, where τ is a positive integer. The LTS S satisfies the following characteristics.*

(1) Ordering: $s_i \geq s_j \iff i \geq j$;

(2) Negation Operator: $Neg(s_{-i}) = s_i$, particularly, $Neg(s_0) = s_0$;

(3) Maximum Operator: $\max\{s_i, s_j\} = s_i \iff s_i \geq s_j$;

(4) Minimum Operator: $\min\{s_i, s_j\} = s_i \iff s_i \leq s_j$.

Definition 2. *[22] A formal context is a triple (U, A, I), where $U = \{x_1, x_2, \dots, x_n\}$ is a set of objects, $A = \{a_1, a_2, \dots, a_m\}$ is a set of attributes, and $I \subseteq U \times A$ is a binary relation between U and A. For any $x \in U$ and $a \in A$, $(x, a) \in I$ means that x possesses attribute a, denoted by xIa.*

Definition 3. *[3] A triadic formal context is a quadruple (G, M, B, Y), where G is a set of objects, M is a set of attributes, B is a set of conditions, and $Y \subseteq G \times M \times B$ is a ternary relation among G, M and B. For any $g \in G$, $m \in M$, and $b \in B$, $(g, m, b) \in Y$ denotes that object g possesses attribute m under condition b.*

Definition 4. *[3] Let* $K = (K_1, K_2, K_3, Y)$ *be a triadic formal context.*

For $\{i, j, k = 1, 2, 3\}$ *with* $j < k$, $X \subseteq K_i$ *and* $Z \subseteq K_j \times K_k$, *the* (i)*-derivation operators are defined by*

$$X^{(i)} = \{(a_j, a_k) \in K_j \times K_k \mid a_i, a_j, a_k \text{ are related by } Y \text{ for all } a_i \in K_i\}, \quad (1)$$

$$Z^{(i)} = \{a_i \in K_i \mid a_i, a_j, a_k \text{ are related by } Y \text{ for all } (a_j, a_k) \in Z\}. \quad (2)$$

For $\{i, j, k = 1, 2, 3\}$, $X_i \subseteq K_i$, $X_j \subseteq K_j$ *and* $A_k \subseteq K_k$, *the* (i, j, A_k)*-derivation operators are defined by*

$$X_i^{(i,j,A_k)} = \{a_j \in K_j \mid a_i, a_j, a_k \text{ are related by } Y \text{ for all } (a_i, a_k) \in X_i \times A_k\}, \quad (3)$$

$$X_j^{(i,j,A_k)} = \{a_i \in K_i \mid a_i, a_j, a_k \text{ are related by } Y \text{ for all } (a_j, a_k) \in X_j \times A_k\}. \quad (4)$$

Definition 5. *[3] Let* $K = (K_1, K_2, K_3, Y)$ *be a triadic formal context,* $A_i \subseteq K_i$, $i = 1, 2, 3$. *If* $A_i = (A_j \times A_k)^{(i)}$, *for* $\{i, j, k = 1, 2, 3\}$ *and* $j < k$, *then* (A_1, A_2, A_3) *is called a triadic concept, where* A_1, A_2 *and* A_3 *are called the extent, intent, and modus of* (A_1, A_2, A_3), *respectively. The family of all triadic concepts of the triadic formal context* K *is denoted by* $\mathfrak{T}(K)$.

Definition 6. *[3] Let there be arbitrary elements a, b, and c in a set that satisfy the following properties:*

(1) Reflexivity: $a \prec a$;

(2) Transitivity: If $a \prec b$ *and* $b \prec c$, *then* $a \prec c$.

Then, the relation $\prec$ *is called a preorder, and the set with this preorder is referred to as a preordered set. If a preorder relation further satisfies symmetry (i.e., if* $a \prec b$, *then* $b \prec a$*), then it is an equivalence relation.*

3 Knowledge Discovery Based on Triadic Fuzzy Linguistic Formal Context

In the era of big data, Internet data is complex and multi-dimensional. Due to the fuzziness of human thinking, people tend to use linguistic values to express preference information. In order to deal with this situation, this section proposes the definition and properties of triadic fuzzy linguistic formal context, gives the knowledge discovery algorithm based on the idea of incremental construction, and describes the relationship between concepts by triadic fuzzy linguistic diagram.

3.1 Triadic Fuzzy Linguistic Formal Context

Definition 7. *A triadic fuzzy linguistic formal context is a quadruple* $(G, L_{S_a}, T, \tilde{I})$, *where* $G = \{x_1, x_2, \ldots, x_n\}$ *is a non-empty finite set of objects,* $L_{S_a} = \{l^j_{S_k} \mid j \in \{1, 2, \ldots, m\}, k \in \{-\tau, \ldots, -1, 0, 1, \ldots, \tau\}\}$ *is a non-empty finite set of linguistic concept set,* $T = \{y_1, y_2, \ldots, y_n\}$ *is a non-empty finite set of conditions,* $\tilde{I}$ *is a fuzzy ternary relation among* G, L_{S_a} *and* T, *where* $\tilde{I} = \{\langle (x, l_{S_k}, y), \mu_{\tilde{I}}(x, l_{S_k}, y)\rangle \mid (x, l_{S_k}, y) \in G \times L_{S_a} \times T\}$, $\mu_{\tilde{I}} : G \times L_{S_a} \times T \to [0, 1]$.

Definition 8. *Let $(G, L_{S_a}, T, \tilde{I})$ be a triadic fuzzy linguistic formal context, $\lambda \in [0,1]$ is the credibility level of object and linguistic concept under condition, $F^{L_{S_a} \times T}$ denotes the set of all fuzzy sets of L_∞ defined under the condition T. For $X \subseteq G$ and $Z \subseteq L_{S_a} \times T$, $\tilde{Z} \in F^{L_{S_a} \times T}$ the $[i]$-derivation operators are defined by*

$$X^{[i]} = \left\{ (\langle l_{S_k}, \mu_X(l_{S_k}) \rangle, y) \mid \forall x \in G, \tilde{I}(x, l_{S_k}, y) \geq \lambda \right\} \tag{5}$$

$$\tilde{Z}^{[i]} = \left\{ x \in G \mid \forall (l_{S_k}, y) \in Z, \tilde{I}(x, l_{S_k}, y) \geq \tilde{Z}(l_{S_k}, y) \wedge \tilde{I}(x, l_{S_k}, y) \geq \lambda \right\} \tag{6}$$

where $X^{[i]}$ denotes that the set of linguistic concepts that all the objects of X satisfy the credibility level λ under the condition y, $\mu_X(l_{S_k}) = \min_{x \in X} \tilde{I}(x, l_{S_k}, y)$ and $\tilde{Z}^{[i]}$ denotes that the set of objects that can be described by all the linguistic concepts under the condition y that satisfy the credibility level λ.

Definition 9. *Let $(G, L_{S_a}, T, \tilde{I}_\lambda)$ be a triadic fuzzy linguistic formal context under the credibility level $\lambda \in [0,1]$. For any $X \subseteq G$, $C \subseteq T$, $B_{S_a} = \langle l_{S_k}, \mu_X(l_{S_k}) \rangle$, where $l_{S_k} \subseteq L_{S_a}$, $\mu_X(l_{S_k}) = \min_{x \in X} \tilde{I}(x, l_{S_k}, y)$. If $X^{[i]} = Z$, $\tilde{Z}^{[i]} = X$, then (X, B_{S_a}, C) is called a triadic fuzzy linguistic concept, where X, B_{S_a} and C are called the extent, intent and modus of the triadic fuzzy linguistic concept, respectively. The set of all triadic fuzzy linguistic concepts of $(G, L_{S_a}, T, \tilde{I}_\lambda)$ under the credibility level λ is denoted by $TFLC(G, L_{S_a}, T, \tilde{I}_\lambda)$.*

For any (A_1, A_2, A_3), $(B_1, B_2, B_3) \in TFLC(G, L_{S_a}, T, \tilde{I}_\lambda)$, the quasiorder relation and equivalence relation between (A_1, A_2, A_3) and (B_1, B_2, B_3) are defined as follow.

$$(A_1, A_2, A_3) \prec_i (B_1, B_2, B_3) :\Leftrightarrow A_i \subseteq B_i \tag{7}$$

$$(A_1, A_2, A_3) \sim_i (B_1, B_2, B_3) :\Leftrightarrow A_i = B_i \tag{8}$$

where $(i = 1, 2, 3)$.

Property 1. *Let $(G, L_{S_a}, T, \tilde{I})$ be a triadic fuzzy linguistic formal context. For any (A_1, A_2, A_3), $(B_1, B_2, B_3) \in TFLC(G, L_{S_a}, T, \tilde{I}_\lambda)$, if $(A_1, A_2, A_3) \prec_i (B_1, B_2, B_3)$ and $(A_1, A_2, A_3) \prec_j (B_1, B_2, B_3)$, then $(A_1, A_2, A_3) \succ_k (B_1, B_2, B_3)$, where $\{i, j, k\} = \{1, 2, 3\}$.*

Proof. According to Definition 5 and Definition 9, Property 1 can be obtained.

Example 1. Table 1 presents an example of a triadic fuzzy linguistic formal context $(G, L_{S_a}, T, \tilde{I})$, where $G = \{A, B, C\}$ represents three objects, $L_{S_a} = \{l^1_{S_{-1}}, l^1_{S_0}, l^1_{S_1}, l^2_{S_{-1}}, l^2_{S_0}, l^2_{S_1}\}$ denotes the linguistic concept set that describes two attributes using the linguistic term set $S = \{S_{-1} = \text{low}, S_0 = \text{moderate}, S_1 = \text{high}\}$, and $T = \{T_1, T_2, T_3\}$ stands for three conditions.

Table 1. Triadic fuzzy linguistic formal context $(G, L_{S_a}, T, \tilde{I})$

	T_1						T_2						T_3					
	$l^1_{S_{-1}}$	$l^1_{S_0}$	$l^1_{S_1}$	$l^2_{S_{-1}}$	$l^2_{S_0}$	$l^2_{S_1}$	$l^1_{S_{-1}}$	$l^1_{S_0}$	$l^1_{S_1}$	$l^2_{S_{-1}}$	$l^2_{S_0}$	$l^2_{S_1}$	$l^1_{S_{-1}}$	$l^1_{S_0}$	$l^1_{S_1}$	$l^2_{S_{-1}}$	$l^2_{S_0}$	$l^2_{S_1}$
A	0.1	0.2	0.7	0.2	0.2	0.6	0	0.1	0.9	0.1	0.2	0.7	0.3	0.7	0	0.2	0.4	0.4
B	0.7	0.2	0.1	0	0.2	0.8	0.1	0.3	0.6	0.2	0.2	0.6	0.2	0.3	0.5	0	0.1	0.9
C	0.1	0.4	0.5	0.4	0.5	0.1	0.1	0.4	0.5	0	0.8	0.2	0	0.6	0.4	0.2	0.5	0.3

3.2 Incremental Construction Algorithm

The effective construction of concepts plays a crucial role in knowledge discovery, recommendation research, and other related fields. Currently, the acquisition of triadic concepts mainly relies on methods such as progressive construction, heuristic construction, and incremental construction. In uncertain environments, people tend to express preference information using linguistic values. To acquire conceptual knowledge expressed through linguistic values, which results from the fuzziness and uncertainty of human thinking, a knowledge discovery algorithm integrating the idea of incremental construction is proposed.

Definition 10. *Let $(G, L_{S_a}, T, \tilde{I})$ be a triadic fuzzy linguistic formal context. For any $X_i \subseteq G$ and $C_k \subseteq T$, define: $a_{ik}(X_i, C_k) = (X, B_{S_a}, C)$, where $X = (B_{S_a} \times C_k)^{[i]}$, $B_{S_a} = (X_i \times C_k)^{[j]}$, $C = (X \times B_{S_a})^{[k]}$, where B_{S_a} is represented in the following form: $\langle l_{S_k}, \mu_X(l_{S_k})\rangle$ and $\mu_X(l_{S_k}) = \min_{x \in X} \tilde{I}(x, l_{S_k}, y)$.*

Definition 11. *Let $(G, L_{S_a}, T, \tilde{I})$ be a triadic fuzzy linguistic formal context, if $C_k \neq \emptyset$, then $(G, L_{S_a}, C_k, \tilde{I}^{C_k})$ is called a sub context of the triadic fuzzy linguistic formal context under condition C_k, where $\tilde{I}^{C_k} = \tilde{I} \cap (G \times L_{S_a} \times C_k)$. For $(G, L_{S_a}, C_k, \tilde{I}^{C_k})$, where $X \subseteq G$, $B_{S_a} = \langle l_{S_k}, \mu_X(l_{S_k})\rangle$, $l_{S_k} \subseteq L_{S_a}$, $\mu_X(l_{S_k}) = \min_{x \in X} \tilde{I}(x, l_{S_k}, y)$ and $C^\circ \subseteq C_k$, derivation operators are provided as follows:*

$$X^{(1)\tilde{I}^{C_k}} = \left\{(\langle l_{S_k}, \mu_X(l_{S_k})\rangle, y) \in L_{S_a} \times C_k \mid \forall x \in G, \tilde{I}^{C_k}(x, l_{S_k}, y) \geq \lambda\right\} \quad (9)$$

$$B_{S_a}^{(2)\tilde{I}^{C_k}} = \left\{(x, y) \in G \times C_k \mid \forall l_{S_k} \in B_{S_a}, \tilde{I}^{C_k}(x, l_{S_k}, y) \geq \lambda\right\} \quad (10)$$

$$C^{\circ(3)\tilde{I}^{C_k}} = C^{\circ(3)} \quad (11)$$

$$(B_{S_a} \times C^\circ)^{(1)\tilde{I}^{C_k}} = (B_{S_a} \times C^\circ)^{(1)} \quad (12)$$

$$(X \times C^\circ)^{(2)\tilde{I}^{C_k}} = (X \times C^\circ)^{(2)} \quad (13)$$

$$(X \times B_{S_a})^{(3)\tilde{I}^{C_k}} = \left\{y \in C_k \mid \forall x \in G, \forall l_{S_k} \in L_{S_a}, \tilde{I}^{C_k}(x, l_{S_k}, y) \geq \lambda\right\} \quad (14)$$

This algorithm considers the impact and generation of concepts as the number of conditions continues to increase. First, consider the concepts generated by the fuzzy linguistic context under a single condition. As conditions are added one by one, analyze the changes in the original triadic fuzzy linguistic concepts after adding a new condition and the impact of the newly added condition on the concepts generated by the fuzzy linguistic context under that condition. The specific algorithmic process is outlined in Algorithm 1:

Algorithm 1. Knowledge discovery algorithm based on triadic fuzzy linguistic concept

Input: $(G, L_{S_a}, T, \tilde{I})$ and credibility level λ
Output: concepts knowledge $TFLC(G, L_{S_a}, T, \tilde{I}_\lambda)$
Initialization $\mathfrak{I}(K) = \emptyset$, $\rho(K) = \emptyset$, $n =$ number of conditions, $i = 2$;
$\mathfrak{I}(K) \leftarrow \{(X, B_{S_a}, C) \mid (X, B_{S_a}, C) \in \mathfrak{I}(X, B_{S_a}, y_1, \tilde{I}_\lambda^{y_1})\}$;
while$(i \leq n)$
  for (X, B_{S_a}, C) in $\mathfrak{I}(K)$ **do**
    if $l_{S_k} \cap (X \times \{y_i\})^{(2)} = l_{S_k}$ **then**
      $\mathfrak{I}(K) \leftarrow \mathfrak{I}(K) - (X, B_{S_a}, C)$;
      $\mathfrak{I}(K) \leftarrow \mathfrak{I}(K) \cup (X, B_{S_a}, C \cup \{y_i\})$;
    end if
    if $l_{S_k} \cap (X \times \{y_i\})^{(2)} \neq l_{S_k}$ and $l_{S_k} \cap (X \times \{y_i\})^{(2)} \neq \emptyset$ **then**
      $\mathfrak{I}(K) \leftarrow \mathfrak{I}(K) \cup a_{23}(l_{S_k} \cap (X \times \{y_i\})^{(2)}, C \cup \{y_i\})$;
    end if
  end for
  $\rho(K) \leftarrow \{(x, b_{S_a}, c) \mid (x, b_{S_a}, c) \in \mathfrak{I}(X, B_{S_a}, y_i, \tilde{I}_\lambda^{y_i})\}$;
  for (x, b_{S_a}, c) in $\rho(K)$ **do**
    if $x \cap X \neq x$ or $b_{S_a} \cap B_{S_a} \neq b_{S_a}$ **then**
      $\mathfrak{I}(K) \leftarrow \mathfrak{I}(K) \cup (x, b_{S_a}, c)$;
    end if
  end for
  $\rho(K) = \emptyset$, $i = i + 1$;
end
$TFLC(G, L_{S_a}, T, \tilde{I}_\lambda) \leftarrow \mathfrak{I}(K) \cup (G, L_{S_a}, \emptyset) \cup (G, \emptyset, T) \cup (\emptyset, L_{S_a}, T)$;
return $TFLC(G, L_{S_a}, T, \tilde{I}_\lambda)$;

The algorithm takes the triadic fuzzy linguistic formal context and the credibility level as inputs, and adopts the idea of incremental construction to realize the dynamic acquisition of conceptual knowledge. Conditions are independent and order-independent. First, the concept set and temporary set are initialized, and the initial concepts are generated from the sub-context under a single condition. Then, new conditions are added one by one, the existing concepts are traversed to judge the attribute intersection relationship, and the concepts satisfying the conditions are updated or merged. Meanwhile, the new concepts generated under the newly added conditions are added to the concept set to remove redundant results. Finally, boundary concepts such as the full object set, full attribute set, and empty set are supplemented to obtain the complete triadic fuzzy linguistic concept knowledge set. Let the number of objects be $|G|$, the number of linguistic attributes be $|L_{S_a}|$, and the number of conditions be $|T|$. The outer loop of the algorithm traverses all conditions, while the inner loop performs linguistic concept verification and concept update operations on each concept. The computational overhead of a single condition iteration is $O(|G| \cdot |L_{S_a}|)$, and a total of $|T|$ iterations are executed. Therefore, the time complexity of the algorithm is $O(|T| \cdot |G| \cdot |L_{S_a}|)$. Since the number of conditions $|T|$ is usually much smaller than the number of objects $|G|$, the algorithm has high execution efficiency.

Applying Algorithm 1 to process data in Example 1 of the triadic fuzzy linguistic formal context, provide the steps for obtaining the following concepts:
(1) Initialization $\mathfrak{I}(K) = \emptyset$, $\rho(K) = \emptyset$, $n = 3$, $i = 2$, credibility level $\lambda = 0.6$;

Under a single condition in the formal context $(X, B_{S_a}, y_1, \tilde{I}_\lambda^{y_1})$, can obtain the following concepts:

$$\begin{aligned}\mathfrak{I}(K) = \{&\left(A, \langle (l_{S_1}^1, 0.7), (l_{S_1}^2, 0.6)\rangle, T_1\right), \\ &\left(B, \langle (l_{S_{-1}}^1, 0.7), (l_{S_1}^2, 0.8)\rangle, T_1\right), \left(AB, \langle (l_{S_1}^2, 0.6)\rangle, T_1\right)\}.\end{aligned}$$

(2) After adding the condition T_2, the concept $\left(A, \langle (l_{S_1}^1, 0.7), (l_{S_1}^2, 0.6)\rangle, T_1\right)$ changes as follows:

$$\{l_{S_1}^1, l_{S_1}^2\} \cap (A \times T_2)^{(2)} = \{l_{S_1}^1, l_{S_1}^2\},$$

therefore, the concept is updated to $\left(A, \langle (l_{S_1}^1, 0.7), (l_{S_1}^2, 0.6)\rangle, T_1T_2\right)$.

For the concept under condition T_2,

$$\begin{aligned}\rho(K) = \{&\left(C, \langle (l_{S_0}^2, 0.8)\rangle, T_2\right), \\ &\left(AB, \langle (l_{S_1}^1, 0.6), (l_{S_1}^2, 0.6)\rangle, T_2\right)\}.\end{aligned}$$

For any $(X, B_{S_a}, C) \in \mathfrak{I}(K)$, with $C \cap X \neq C$ or $l_{S_0}^2 \cap B_{S_a} \neq l_{S_0}^2$ and $AB \cap X \neq AB$ or $\{l_{S_1}^1, l_{S_1}^2\} \cap B_{S_a} \neq \{l_{S_1}^1, l_{S_1}^2\}$ established, the concepts

$$\left(C, \langle (l_{S_0}^2, 0.8)\rangle, T_2\right), \left(AB, \langle (l_{S_1}^1, 0.6), (l_{S_1}^2, 0.6)\rangle, T_2\right)$$

are all added concepts.

Then,

$$\begin{aligned}\mathfrak{I}(K) = \{&\left(A, \langle (l_{S_1}^1, 0.7), (l_{S_1}^2, 0.6)\rangle, T_1T_2\right), \\ &\left(B, \langle (l_{S_{-1}}^1, 0.7), (l_{S_1}^2, 0.8)\rangle, T_1\right), \\ &\left(AB, \langle (l_{S_1}^2, 0.6)\rangle, T_1T_2\right), \left(C, \langle (l_{S_0}^2, 0.8)\rangle, T_2\right), \\ &\left(AB, \langle (l_{S_1}^1, 0.6), (l_{S_1}^2, 0.6)\rangle, T_2\right)\}.\end{aligned}$$

(3) After adding the condition T_3, the concept $\left(B, \langle (l_{S_{-1}}^1, 0.7), (l_{S_1}^2, 0.8)\rangle, T_1\right)$ changes as follows:

$$\{l_{S_{-1}}^1, l_{S_1}^2\} \cap (B \times T_3)^{(2)} = \{l_{S_1}^2\} \neq \{l_{S_{-1}}^1, l_{S_1}^2\},$$

$$a_{23}\left(l_{S_k} \cap (B \times T_3)^{(2)}, C \cup T_3\right) = a_{23}(l_{S_1}^2, T_1T_3),$$

Because $(l_{S_1}^2 \times T_1T_3)^{(1)} = \{B\}$, $l_{S_1}^2 \subset (B \times T_2)^{(2)} = \{l_{S_1}^1, l_{S_1}^2\}$, the concept remains unchanged, and add concept $\big(B, \langle(l_{S_1}^2, 0.6)\rangle, T_1T_2T_3\big)$.

For the concept under condition T_3,

$$\rho(K) = \left\{\left(B, \langle(l_{S_1}^2, 0.9)\rangle, T_3\right), \left(AC, \langle(l_{S_0}^1, 0.6)\rangle, T_3\right)\right\},$$

Concept $\big(AC, \langle(l_{S_0}^1, 0.6)\rangle, T_3\big)$ is the newly added concept.

Based on the above ideas, all triadic fuzzy linguistic conceptual knowledge can be obtained as shown in Table 2.

For example, concept 1# $\big(A, \langle(l_{S_1}^1, 0.7), (l_{S_1}^2, 0.6)\rangle, T_1T_2\big)$ indicates that the object A can be described by $l_{S_1}^1$ (with a credibility of 0.7) and $l_{S_1}^2$ (with a credibility of 0.6) under the conditions of T_1 and T_2.

Table 2. Triadic fuzzy linguistic conceptual knowledge

1#	$(A, \langle(l_{S_1}^1, 0.7), (l_{S_1}^2, 0.6)\rangle, T_1T_2)$
2#	$(B, \langle(l_{S_{-1}}^1, 0.7), (l_{S_1}^2, 0.8)\rangle, T_1)$
3#	$(B, \langle(l_{S_1}^2, 0.6)\rangle, T_1T_2T_3)$
4#	$(C, \langle(l_{S_0}^2, 0.8)\rangle, T_2)$
5#	$(AB, \langle(l_{S_1}^2, 0.6)\rangle, T_1T_2)$
6#	$(AB, \langle(l_{S_1}^1, 0.6), (l_{S_1}^2, 0.6)\rangle, T_2)$
7#	$(AC, \langle(l_{S_0}^1, 0.6)\rangle, T_3)$
8#	$(\emptyset, \langle(l_{S_{-1}}^1, 1), (l_{S_0}^1, 1), (l_{S_1}^1, 1), (l_{S_{-1}}^2, 1), (l_{S_0}^2, 1), (l_{S_1}^2, 1)\rangle, T_1T_2T_3)$
9#	$(ABC, \emptyset, T_1T_2T_3)$
10#	$(ABC, \langle(l_{S_{-1}}^1, 1), (l_{S_0}^1, 1), (l_{S_1}^1, 1), (l_{S_{-1}}^2, 1), (l_{S_0}^2, 1), (l_{S_1}^2, 1)\rangle, \emptyset)$

The proposed triadic fuzzy linguistic concepts knowledge discovery algorithm effectively extracts conceptual knowledge from fuzzy preference information described with linguistic values. By analyzing triadic fuzzy linguistic concepts, valuable conceptual knowledge can be obtained under various influencing conditions.

4 Decision Rule Extraction Method Based on Triadic Fuzzy Linguistic Decision Context

One of the core goals of knowledge discovery is to transform massive and scattered information into interpretable decision-making bases, and rule extraction is a key technology for processing information and decisions. Its value lies not only in removing redundant information from complex data and obtaining core laws, but also in providing a quantifiable analysis tool for analyzing the cognitive logic of human decision-making. In practical decision-making scenarios, object features and evaluation dimensions often exhibit significant fuzziness and uncertainty. This section proposes a decision rule extraction method based on the triadic fuzzy linguistic decision context. By studying the implication relationships between triadic fuzzy linguistic conceptual knowledge with conditional attributes and decision attributes under the decision context, decision rules are mined and a decision rule base is constructed, thereby providing a new method for making high-credibility decisions.

Definition 12. *Let* $(G, L_{S_a}, T, \tilde{I})$ *and* $(G, D_{S_a}, K, \tilde{J})$ *be two triadic fuzzy linguistic formal contexts, with their respective sets of triadic fuzzy linguistic concept knowledge denoted as* $TFLC(G, L_{S_a}, T, \tilde{I})$, $TFLC(G, D_{S_a}, K, \tilde{J})$. *If* $\forall (Y, M_{S_a}, N) \in TFLC(G, D_{S_a}, K, \tilde{J})$, $\exists (X, B_{S_a}, C) \in TFLC(G, L_{S_a}, T, \tilde{I})$ *and* $X \subseteq Y$, $X \neq \emptyset$, *then is said* $TFLC(G, D_{S_a}, K, \tilde{J})$ *to be finer than* $TFLC(G, L_{S_a}, T, \tilde{I})$, *denoted as* $TFLC(G, D_{S_a}, K, \tilde{J}) \leq TFLC(G, L_{S_a}, T, \tilde{I})$.

Definition 13. *Let* $(G, L_{S_a}, T, \tilde{I})$ *and* $(G, D_{S_a}, K, \tilde{J})$ *be two triadic fuzzy linguistic formal contexts sharing the same object domain. Let* L_{S_a} *be the set of conditional linguistic concepts describing conditional attributes, and* D_{S_a} *be the set of decision linguistic concepts describing decision attributes. Given that* $L_{S_a} \cap D_{S_a} = \varnothing$ *and* $\tilde{I}$ *are fuzzy triadic relations among* G, L_{S_a} *and* T, *and* $\tilde{J}$ *are fuzzy triadic relations among* G, D_{S_a} *and* K, *the septuple* $(G, L_{S_a}, T, \tilde{I}, D_{S_a}, K, \tilde{J})$ *is referred to as a triadic fuzzy linguistic decision context.*

Definition 14. *Let* $(G, L_{S_a}, T, \tilde{I}, D_{S_a}, K, \tilde{J})$ *be a triadic fuzzy linguistic decision context, and* $TFLC(G, D_{S_a}, K, \tilde{J}) \leq TFLC(G, L_{S_a}, T, \tilde{I})$. *For* $Y \neq \varnothing$ *and* $Y \neq G$, *if* $(Y, M_{S_a}, N) \in TFLC(G, D_{S_a}, K, \tilde{J})$, $(X, B_{S_a}, C) \in TFLC(G, L_{S_a}, T, \tilde{I})$ *and* $X \subseteq Y$, *then* $(B_{S_a}, C) \to (M_{S_a}, N)$ *is referred to as a triadic fuzzy linguistic decision rule, denoted as* $B_{S_a} \times C \to M_{S_a} \times N$. *That is, "Under the conditions in set* C, *if the conditional attributes can be described by the conditional linguistic concept* B_{S_a}, *then under the conditions in set* N, *the decision attributes can be described by the decision linguistic concept* M_{S_a}.

Definition 15. *Let* $(G, L_{S_a}, T, \tilde{I}, D_{S_a}, K, \tilde{J})$ *be a triadic fuzzy linguistic decision context,* $C \subseteq T$ *and* $N \subseteq K$, *let* $R(L_{S_a}, T, D_{S_a}, K) = \{B_{S_a} \times C \to M_{S_a} \times N \mid (X, B_{S_a}, C) \in TFLC(G, L_{S_a}, T, \tilde{I}), (Y, M_{S_a}, N) \in TFLC(G, D_{S_a}, K, \tilde{J})\}$ *be the set of triadic fuzzy linguistic decision rules. Then* $R(L_{S_a}, T, D_{S_a}, K)$ *is referred to as a triadic fuzzy linguistic decision rule base.*

Definition 16. *Let* $(G, L_{S_a}, T, \tilde{I}, D_{S_a}, K, \tilde{J})$ *be a triadic fuzzy linguistic decision context, and* $C \subseteq T$, $N \subseteq K$, $(X, B_{S_a}, C) \in TFLC(G, L_{S_a}, T, \tilde{I})$, $(Y, M_{S_a}, N) \in TFLC(G, D_{S_a}, K, \tilde{J})$. *If* $r : B_{S_a} \times C \to M_{S_a} \times N$ *is a triadic fuzzy linguistic decision rule, then the support and confidence of this rule are defined as:*

$$\mu_{Support}(r) = \frac{|X \cap Y|}{|G|} \tag{15}$$

$$\rho_{Confidence}(r) = \frac{|X \cap Y|}{|X|} \tag{16}$$

Theorem 1. *In the triadic fuzzy linguistic decision context* $(G, L_{S_a}, T, \tilde{I}, D_{S_a}, K, \tilde{J})$, $C \subseteq T$, $N \subseteq K$. *If* $r : B_{S_a} \times C \rightarrow M_{S_a} \times N$ *is a triadic fuzzy linguistic decision rule, then* $\mu_{Support}(r) \in [0, 1]$ *and* $\rho_{Confidence}(r) \in [0, 1]$.

Proof. From Definition 15, we know that $\mu_{\text{Support}}(r) = \frac{|X \cap Y|}{|G|}$. Since X is the extent of the conditional concept and Y is the extent of the decision concept, both are subsets of the object universe G. Thus, $\emptyset \subseteq X \cap Y \subseteq G$, which implies $0 \leq |X \cap Y| \leq |G|$. Moreover, since the object universe G is a non-empty finite set, we have $|G| \geq 1$. Therefore, $0 \leq \frac{|X \cap Y|}{|G|} \leq 1$, that is, $\mu_{\text{Support}}(r) \in [0, 1]$.

Similarly, since $X \cap Y \subseteq X$, we have $0 \leq |X \cap Y| \leq |X|$. As X is the extent of the conditional concept corresponding to the decision rule, and the rule is a valid decision rule, it follows that $X \neq \emptyset$, i.e., $|X| \geq 1$. Hence, $0 \leq \frac{|X \cap Y|}{|X|} \leq 1$, that is, $\rho_{\text{Confidence}}(r) \in [0, 1]$.

Definition 17. *Let* $(G, L_{S_a}, T, \tilde{I}, D_{S_a}, K, \tilde{J})$ *be a triadic fuzzy linguistic decision context. For the triadic fuzzy linguistic decision rules* $r_1 : B^1_{S_a} \times C^1 \rightarrow M^1_{S_a} \times N^1$ *and* $r_2 : B^2_{S_a} \times C^2 \rightarrow M^2_{S_a} \times N^2$, *if one of the following conditions is satisfied:*

1. $B^1_{S_a} \subseteq B^2_{S_a}$ *and* $M^1_{S_a} = M^2_{S_a}$;
2. $B^1_{S_a} = B^2_{S_a}$ *and* $M^2_{S_a} \subseteq M^1_{S_a}$.

Then the triadic fuzzy linguistic decision rule r_1 *is said to imply* r_2, *denoted as* $r_1 \rightarrow r_2$, *and* r_2 *is referred to as a redundant rule in the rule base.*

To efficiently obtain credible and non-redundant decision rules from the triadic fuzzy linguistic decision context, a decision rule extraction algorithm based on the triadic fuzzy linguistic decision context is designed according to the implication relationship between the aforementioned conditional concepts and decision concepts, as shown in Algorithm 2. The algorithm achieves the acquisition of high-confidence decision rules by generating conditional concepts and decision concepts, judging the implication relationship, calculating support and confidence, and removing redundant rules.

Algorithm 2. Decision Rule Extraction Algorithm Based on Triadic Fuzzy Linguistic Decision Context

Input: $(G, L_{S_a}, T, \tilde{I}, D_{S_a}, K, \tilde{J})$, credibility level λ, support ς, confidence ν
Output: triadic fuzzy linguistic decision rule base $R(G)$
Initialization $R(G) = \emptyset$, ς, ν, λ;
for $(G, L_{S_a}, T, \tilde{I})$ **do**
    obtain TFLC$(G, L_{S_a}, T, \tilde{I})$ from Algorithm 1;
end for
for $(G, D_{S_a}, K, \tilde{J})$ **do**
    obtain TFLC$(G, D_{S_a}, K, \tilde{J})$ from Algorithm 1;
end for
for $(X, B_{S_a}, C) \in$ TFLC$(G, L_{S_a}, T, \tilde{I})$ **do**
    for $(Y, M_{S_a}, N) \in$ TFLC$(G, D_{S_a}, K, \tilde{J})$ **do**
        if $X \subseteq Y$ **then**

$R(G) = R(G) \cup (B_{S_a} \times C \rightarrow M_{S_a} \times N)$;
end if
end for
end for
for r in $R(G)$ **do**
calculate $\mu_{Support}(r)$, $\rho_{Confidence}(r)$ according to Definition 4.5;
if $\mu_{Support}(r) < \varsigma$ and $\rho_{Confidence}(r) < \nu$ **then**
$R(G) = R(G) - r$;
if $B^i_{S_a} \subseteq B^{r-i}_{S_a}$ and $M^i_{S_a} = M^{r-i}_{S_a}$ **or** $B^i_{S_a} = B^{r-i}_{S_a}$ and $M^{r-i}_{S_a} \subseteq M^i_{S_a}$ **then**
$R(G) = R(G) - (r - i)$;
end if
end if
end for
return $R(G)$;

Time Complexity Analysis of the Algorithm: Let the number of objects in the triadic fuzzy linguistic decision context be $|G|$, the number of conditional concept sets be $|K_1|$, the number of decision concept sets be $|K_2|$, and the number of linguistic concepts be $|L_{S_a}|$. The algorithm needs to traverse all conditional concepts and decision concepts to generate candidate rules, with a traversal overhead of $O(|K_1| \cdot |K_2|)$. The complexity of judging the inclusion relationship between concept extents is $O(|G|)$. The calculation of rule support, confidence, and the removal of redundant rules are all linear operations. Therefore, the overall time complexity of the algorithm is: $O(|K_1| \cdot |K_2| \cdot |G|)$.

Example 2. Based on Table 1, decision attributes are introduced into the triadic fuzzy linguistic formal context of Example 3.1 to form a triadic fuzzy linguistic decision context $(G, L_{S_a}, T, \tilde{I}, D_{S_a}, K, \tilde{J})$, as shown in Table 3. Here, $D_{S_a} = \{d^1_{S_{-1}}, d^1_{S_0}, d^1_{S_1}, d^2_{S_{-1}}, d^2_{S_0}, d^2_{S_1}\}$ represents the set of linguistic concepts describing the two decision attributes using the linguistic term set $S = \{S_{-1} =$ Lower, $S_0 =$ Medium, $S_1 =$ Higher$\}$, and $K = \{K_1\}$ represents the conditions.

Table 3. Triadic fuzzy linguistic decision context $(G, L_{S_a}, T, \tilde{I}, D_{S_a}, K, \tilde{J})$

	T_1						T_2						T_3						K_1					
	$l^1_{S_{-1}}$	$l^1_{S_0}$	$l^1_{S_1}$	$l^2_{S_{-1}}$	$l^2_{S_0}$	$l^2_{S_1}$	$l^1_{S_{-1}}$	$l^1_{S_0}$	$l^1_{S_1}$	$l^2_{S_{-1}}$	$l^2_{S_0}$	$l^2_{S_1}$	$l^1_{S_{-1}}$	$l^1_{S_0}$	$l^1_{S_1}$	$l^2_{S_{-1}}$	$l^2_{S_0}$	$l^2_{S_1}$	$d^1_{S_{-1}}$	$d^1_{S_0}$	$d^1_{S_1}$	$d^2_{S_{-1}}$	$d^2_{S_0}$	$d^2_{S_1}$
A	0.1	0.2	0.7	0.2	0.2	0.6	0	0.1	0.9	0.1	0.2	0.7	0.3	0.7	0	0.2	0.4	0.4	0.2	0.2	0.6	0.2	0.3	0.5
B	0.7	0.2	0.1	0	0.2	0.8	0.1	0.3	0.6	0.2	0.2	0.6	0.2	0.3	0.5	0	0.1	0.9	0.6	0.3	0.1	0.1	0.2	0.7
C	0.1	0.4	0.5	0.4	0.5	0.1	0.1	0.4	0.5	0	0.8	0.2	0	0.6	0.4	0.2	0.5	0.3	0.2	0.4	0.4	0.2	0.6	0.2

The triadic fuzzy linguistic concepts $TFLC(G, L_{S_a}, T, \tilde{I})$ computed under the triadic fuzzy linguistic formal context $(G, L_{S_a}, T, \tilde{I})$ are shown in Table 2, while the triadic fuzzy linguistic concepts $TFLC(G, D_{S_a}, K, \tilde{J})$ under the triadic fuzzy linguistic formal context $(G, D_{S_a}, K, \tilde{J})$ are presented in Table 4.

Table 4. Triadic fuzzy linguistic conceptual knowledge $TFLC(G, D_{S_a}, K, \tilde{J})$

1#	$(A, <(d^1_{S_0}, 0.6)>, K,)$
2#	$(B, <(d^1_{S_0}, 0.6), (d^2_{S_1}, 0.7)>, K,)$
3#	$(C, <(d^2_{S_0}, 0.6)>, K,)$
4#	$(ABC, \emptyset, K,)$
5#	$(\emptyset, <(d^1_{S_{-1}}, 1), (d^1_{S_0}, 1), (d^1_{S_1}, 1), (d^2_{S_{-1}}, 1), (d^2_{S_0}, 1), (d^2_{S_1}, 1)>, K,)$
6#	$(ABC, <(d^1_{S_{-1}}, 1), (d^1_{S_0}, 1), (d^1_{S_1}, 1), (d^2_{S_{-1}}, 1), (d^2_{S_0}, 1), (d^2_{S_1}, 1)>, \emptyset)$

To ensure that the extracted rules cover sufficient objects and avoid poor representativeness caused by excessively low support, the minimum support is set as $\varsigma = 0.3$. To guarantee the complete certainty of the generated rules and meet the requirements of high-reliability decision-making scenarios, rules with a minimum confidence $\nu = 1.0$ are selected. The triadic fuzzy linguistic decision rule base for the triadic fuzzy linguistic decision context $(G, L_{S_a}, T, \tilde{I}, D_{S_a}, K, \tilde{J})$, obtained by Algorithm 2, is shown in Table 5

Table 5. Triadic fuzzy linguistic decision rule base $R(L_{S_a}, T, D_{S_a}, K)$

Decision Rule	Support	Confidence
$r_1 : \langle (l^1_{S_1}, 0.7), (l^2_{S_1}, 0.6)\rangle \times T_1T_2 \rightarrow \langle (d^1_{S_1}, 0.6)\rangle \times K_1$	0.33	1.0
$r_2 : \langle (l^1_{S_{-1}}, 0.7), (l^2_{S_1}, 0.8)\rangle \times T_1 \rightarrow \langle (d^1_{S_{-1}}, 0.6), (d^2_{S_1}, 0.7)\rangle \times K_1$	0.33	1.0
$r_3 : \langle (l^2_{S_1}, 0.6)\rangle \times T_1T_2T_3 \rightarrow \langle (d^1_{S_{-1}}, 0.6), (d^2_{S_1}, 0.7)\rangle \times K_1$	0.33	1.0
$r_4 : \langle (l^2_{S_0}, 0.8)\rangle \times T_2 \rightarrow \langle (d^2_{S_0}, 0.6)\rangle \times K_1$	0.33	1.0

The rule $r_1 : \langle (l^1_{S_1}, 0.7), (l^2_{S_1}, 0.6)\rangle \times T_1T_2 \rightarrow \langle (d^1_{S_1}, 0.6)\rangle \times K_1$ indicates that under conditions T_1 and T_2, if attribute can be described by $l^1_{S_1}$ (with a credibility of 0.7) and $l^2_{S_1}$ (with a credibility of 0.6), then under condition K_1, attribute can be described by $d^1_{S_1}$ (with a credibility of 0.6).

Due to the high complexity of human cognitive processes, decision-making problems often exist in uncertain environments. The decision rule extraction method based on triadic fuzzy linguistic decision contexts proposed in this section not only effectively reduces information loss but also extracts non-redundant rule sets with varying levels of trust and confidence from uncertain linguistic information, providing decision-makers with credible actionable insights. By deeply analyzing the implicative relationships between conditional and decision attributes, it uncovers hidden correlations and characteristics within the data. Furthermore, the method employs triadic fuzzy linguistic representations that align closely with human natural cognition and expression habits, enhancing the interpretability and practicality of the model. Consequently, this approach holds broad application prospects in fields such as linguistic knowledge reasoning and intelligent recommendation systems.

5 Conclusions

This paper first introduces a triadic fuzzy linguistic formal context by integrating linguistic term sets with triadic concept analysis, which can handle preference information described by linguistic terms in multidimensional data. Secondly, based on the derivation operators, triadic fuzzy linguistic concepts are defined and their related properties are discussed. Furthermore, by incorporating the idea of incremental construction, a knowledge discovery algorithm for triadic fuzzy linguistic concepts is proposed to mine conceptual knowledge with fuzzy triadic relations and semantic information.

Further, the paper extends the framework by proposing a decision rule extraction method based on triadic fuzzy linguistic decision contexts. This method investigates the implicative relationships between triadic fuzzy linguistic conceptual knowledge involving conditional and decision attributes under a decision background, enabling the mining of decision rules and the construction of a decision rule base. This provides a novel approach for high-confidence decision-making.

For future research, dynamic updates of conceptual knowledge in response to changes in objects or attributes will be explored. The methods proposed in this paper will be applied to construct rule bases for tasks such as recommendation systems and knowledge reasoning.

Acknowledgements. This work is partially supported by the National Natural Science Foundation of China (62176142).

References

1. Ferré, S., Rudolph, S. (eds.): ICFCA 2009. LNCS (LNAI), vol. 5548. Springer, Heidelberg (2009). https://doi.org/10.1007/978-3-642-01815-2
2. Guo, X., Zhao, S., Zhao, J., Liu, J.: Visualization of multi-valued attribute association rules based on concept lattice. J. Comput. Appl. **33**(8), 2198–2203 (2013)
3. Lehmann, F., Wille, R.: A triadic approach to formal concept analysis. In: Ellis, G., Levinson, R., Rich, W., Sowa, J.F. (eds.) ICCS-ConceptStruct 1995. LNCS, vol. 954, pp. 32–43. Springer, Heidelberg (1995). https://doi.org/10.1007/3-540-60161-9_27
4. Wei, L., Wan, Q., Qian, T., Qi, J.: An overview of triadic concept analysis. J. Northwest Univ. (Nat. Sci. Ed.) **44**(5), 689–699 (2014)
5. Wang, X., Li, J., Wu, W.: Boolean matrix representation of triadic concepts. Comput. Sci. **50**(6), 109–115 (2023)
6. Wang, X., Quan, Y., Li, J., Wu, W.: Incremental construction method of triadic concepts. J. Nanjing Univ. (Nat. Sci.) **58**(1), 19–28 (2022)
7. Liu, Z., Zhao, Q., Zou, L., Min, F.: Heuristic construction of triadic concept and its application in social recommendation. Comput. Sci. **48**(6), 234–240 (2021)
8. Li, J., Li, X., Wang, X., Wu, W.: Reduction of triadic concepts based on triadic factor analysis. J. Nanjing Univ. (Nat. Sci.) **56**(4), 480–493 (2020)
9. Li, J., Zhu, R., Wang, X., Wu, W.: The relationship between triadic concepts and formal concepts. J. Nanjing Univ. (Nat. Sci.) **54**(4), 786–793 (2018)

10. Tang, Y., Fan, M., Li, J.: Cognitive system model and approach to transformation of information granules under triadic formal concept analysis. J. Shandong Univ. (Nat. Sci.) **59**(8), 102–106 (2014)
11. Qi, J., Wei, L.: Simplification of triadic contexts and concept trilattices. Comput. Sci. **44**(9), 53–57 (2017)
12. Li, Z., Zhang, Z., Wang, L.: Research on text classification algorithm based on triadic concept analysis. Comput. Sci. **44**(8), 207–215 (2017)
13. Zadeh, L.: Fuzzy logic = computing with words. IEEE Trans. Fuzzy Syst. **4**(10), 103–111 (1996)
14. Zhang, X., Sun, B., Chen, X., Chu, X., Yang, J.: An approach to evaluating sustainable supply chain risk management based on BWM and linguistic value soft set theory. J. Intell. Fuzzy Syst. **39**(3), 4369–4382 (2020)
15. Pang, K., Li, S., Lu, Y., Kang, N., Zou, L., Lu, M.: Association rule mining with fuzzy linguistic information based on attribute partial ordered structure. Soft. Comput. **27**(23), 17447–17472 (2023)
16. Zou, L., Shi, P., Pei, Z., Xu, Y.: On an algebra of linguistic truth-valued intuitionistic lattice-valued logic. J. Intell. Fuzzy Syst. **24**(3), 447–456 (2013)
17. Zhou, Y., Zou, L.: TOPSIS method based on intuitionistic fuzzy linguistic concept. In: International Conference on AI Logic and Applications. Springer, Singapore (2022)
18. Dwivedi, R., Dave, D., Naik, H., et al.: Explainable AI (XAI): core ideas, techniques, and solutions. ACM Comput. Surv. **55**(9), 1–33 (2023)
19. Abbas, Q., Jeong, W., Lee, S.W.: Explainable AI in clinical decision support systems: a meta-analysis of methods, applications, and usability challenges. Healthcare **13**(17), 2154 (2025)
20. Bourne, L.E.: An inference model for conceptual rule learning. In: Theories in Cognitive Psychology, pp. 231–256. Routledge (2024)
21. Herrera, F., Herrera, V., Verdegay, J.: A model of consensus in group decision making under linguistic assessments. Fuzzy Sets Syst. **78**, 73–87 (1996)
22. Cui, H., Deng, A., Yue, G., et al.: The linguistic concept's reduction methods under symmetric linguistic-evaluation information. Symmetry **15**(4), 813 (2023)

Evidence-Driven and Game-Theoretic Consensus Modeling for Group Decision Making: A Case Study of Retail Marketing Strategy Selection

Qihao Zhang(✉) and Wen He

Hunan Institute of Science and Technology, Yueyang 414006, Hunan, China
822511120452@vip.hnist.edu.cn

Abstract. This paper proposes an Evidence-Driven Game-Theoretic Decision-Making Consensus Model (EDCB) to address the practical challenges in group decision-making where significant differences in experts' subjective preferences hinder stable consensus formation, and traditional consensus methods lack objective guidance mechanisms, often triggering opinion conflicts and complicating the implementation of decisions. First, the Analytic Hierarchy Process (AHP) is employed to map the structure of expert subjective preferences. Then, it constructs an adaptive step function based on subjective-objective bias to precisely correct deviations, driving iterative convergence of expert preferences toward the data-based benchmark through multiple rounds of iteration. Subsequently, a behavioral consensus criterion rooted in game steady states is proposed, incorporating individual utility functions and adopting Nash equilibrium as the stopping rule for iterations. Finally, the TOPSIS method is applied to select the optimal alternative from the consensus weights after meeting standards. Experimental results validate the effectiveness of the EDCB and demonstrate its effectiveness through a case study on evaluating marketing strategies for retail enterprises.

Keywords: Evidence-driven consensus model · Analytic hierarchy process · Random forest regression · Nash equilibrium · TOPSIS

1 Introduction

In complex and dynamic decision-making environments, group decision making (GDM) aims to find collective solutions to decision problems based on the preferences (or opinions) expressed by a group of decision-makers (or individuals) [1]. The requirements for its processes are becoming increasingly complex [2]. In real organizational contexts, decision-makers often belong to different interest units, and their subjective preferences are easily influenced by multiple factors such as departmental affiliations, functional roles, perspective biases, and cognitive distortions. This makes it difficult to achieve effective and stable consensus

J. Lu et al. (Eds.): FLINS-ISKE 2026, LNCS 16752, pp. 524–546, 2027.
https://doi.org/10.1007/978-981-92-2369-5_32

during the critical stage of allocating weight to decision criteria. The consensus-reaching process (CRP) is designed to assist decision-makers with social relationships in reaching consensus [3]. The traditional group decision-making consensus process relies heavily on repeated interactions and adjustments of opinions among experts [1]. At its core, this approach involves the negotiation, compromise, and integration of subjective opinions. In the context of big data, such a model not only tends to produce one-sided judgments with cognitive biases but also yields consensus outcomes that lack stability. Furthermore, existing group decision-making methods primarily rely on aggregation techniques like weighted averaging and optimization models to achieve opinion convergence. While these can formally reduce group disagreements, they generally overlook the behavioral mechanisms and psychological costs inherent in the consensus-forming process. In real-world decision-making scenarios, when individuals perceive their opinions as being systematically diminished, they are prone to exhibit non-cooperative behaviors and cognitive resistance, thereby undermining the stability of group consensus [4]. Therefore, how to establish a methodologically rigorous, stable, and practically implementable consensus-building mechanism among diverse stakeholders has become a critical issue in group decision-making research.

In recent years, significant progress has been made in researching the consensus-reaching process within group decision-making. Overall, existing literature primarily focuses on two directions: improving consensus measurement methods and optimizing feedback regulation mechanisms. Regarding consensus measurement, Guo et al. [5] have constructed a two-layer consensus model that quantifies decision-maker divergence by measuring both the similarity of evaluation scales and the similarity of ranking outcomes, thereby making the consensus measure more comprehensive. Hou et al. [6] have constructed consensus evaluation metrics using multiple distance measures to enhance the stability of measuring group opinion divergence. While such research significantly improves the accuracy of divergence identification at the technical level, it still lacks in-depth analysis of extreme preference persistence and dynamic evolutionary stability at the mechanism level. As research deepens, scholars have gradually recognized that traditional linear feedback mechanisms implicitly assume experts passively accept adjustment suggestions, overlooking their strategic behavior. Therefore, game theory methods [7] have been introduced into consensus-reaching research. Meng et al. [8] have modeled the consensus process as a prisoner's dilemma game, depicting the mechanism by which individuals balance group interests against personal costs. In addressing minority challenges, Shen et al. [9] have integrated social network structures with regret theory, dynamically adjusting opinion weights through regret feedback to enhance the expression and integration of minority opinions in consensus processes. Yuan et al. [10] have proposed a minimal conflict consensus framework based on social network analysis, while Yang et al. [11] have introduced prospect theory in large-scale group settings to construct psychological value functions characterizing non-cooperative behavior. These studies have made significant progress in identifying minority opinions and managing non-cooperative behavior. However, the following shortcomings remain: Most models analyze only behavioral outcomes, lack mechanisms for linking to external objective data. They also lack systematic strategy guidance design, causing minority factions to often passively adjust their opinions under group pressure. Their corrective actions lack logical support for utility improvement. Additionally, some models exhibit high complexity, increasing computational costs and resulting in low efficiency for equilibrium solutions. At the same time, existing research generally employs a linear mechanism with a uniform adjustment step size [12], overlooking the significant heterogeneity among individual experts in terms of knowledge background, risk preferences, and other factors [11]. This feedback mechanism can easily provoke resistance among some experts, causing them to passively conform to group consensus or even gradually abandon their own valuable professional judgments during iterative adjustments, resulting in a consensus that is difficult to implement in practice. In existing research, when determining whether consensus has been reached, Li et al. [13] have proposed a Consensus Reaching Process (CRP) method based on finite confidence and social networks to manage expert opinions, while Xu et al. [14] have proposed a dynamic method for automatically adjusting decision-makers' (DM) weights in Group Decision Making (GDM) problems. However, current models typically use the group consensus level reaching a preset threshold as the algorithm's termination condition, without considering the optimization of individual utility. Consequently, once the decision enters the implementation phase, individual preferences may shift or decision-makers may change their minds, causing the consensus that has been reached to gradually unravel. Therefore, how to construct a regulatory mechanism with behavioral adaptability and clearly define steady-state criteria for consensus formation has become a key issue to be addressed in current group decision-making research.

In the objective evidence extraction phase, this paper selects Random Forest [15] as the core algorithm for feature importance analysis. Compared to traditional linear regression and single decision trees, Random Forest demonstrates greater adaptability when handling complex decision-making data. Linear regression models [16] rely on the assumption of linear relationships among variables, making it difficult to capture the nonlinear characteristics and variable interaction effects commonly found in multi-attribute decision-making, and they are relatively sensitive to outliers; while single decision trees [17] can fit nonlinear relationships, their model structure is susceptible to data perturbations, leading to overfitting and resulting in insufficient generalization ability. Furthermore, while Support Vector Machines (SVMs) [18] perform well in classification tasks involving small sample sizes and high-dimensional data, they are highly sensitive to kernel function types and parameter settings, resulting in high tuning costs in practical applications and poor model interpretability. In contrast, Random Forests, by constructing multiple decision trees and incorporating a random sampling strategy, can effectively control model variance. This approach enhances the model's overall robustness and predictive stability while retaining its ability to fit nonlinear relationships. Additionally, this algorithm directly outputs feature importance metrics, facilitating the analysis of the relative contributions of each feature to the decision-making process. Based on these advantages, this paper employs Random Forests to extract objective weight vectors from large-scale historical decision data, providing data-driven support for the subsequent expert consensus-building process.

To address the limitations of the aforementioned studies, this paper proposes an evidence-driven consensus enhancement model (EDCB). This model resolves the dual challenges of deviating from objective principles and conflicting with minority perceptions inherent in traditional consensus models through the deep integration of subjective and objective evidence and the application of Nash equilibrium game theory logic. The primary research content and innovations of this paper are as follows:

(1) A dual-driven evidence consensus logic framework integrating subjective and objective elements is established. In this framework, the importance of criteria features extracted via machine learning serves as objective factual evidence, providing a data-driven basis for rational persuasion and consensus building; building;
(2) A dynamic threshold-based asymmetric guidance mechanism was designed, constructing a dynamic threshold function to identify minority experts whose judgments deviate from objective facts. Through adaptive coefficient adjustment, this mechanism enables the gradual iterative alignment of individual preferences with objective patterns while fully preserving the diversity of expert expertise;
(3) A behavioral consensus criterion based on game steady states is proposed, introducing Nash equilibrium theory to define the endpoint conditions for consensus attainment. This reveals the intrinsic mechanism of behavioral evolution enabling rational compromise by minority opinion parties, thereby providing stability assurance for achieving strong consensus.

The remainder of this paper is structured as follows. Section 2 reviews the fundamental concepts. The data-evidence-driven approach and the group game model are presented in Sect. 3. Section 4 offers a case study that illustrates the method's effectiveness and practicality through comparative analysis. Finally, Sect. 5 summarizes the conclusions and outlines directions for future research.

2 Fundamental Knowledge

2.1 Building a Group Consensus Framework

For a group consensus problem [19], we first define the set of experts participating in decision-making as: $E = \{e_1, e_2, \ldots, e_k\}$, $k \geq 2$, the set of evaluation attributes as: $C = \{c_1, c_2, \ldots, c_n\}$, $n \geq 2$, the set of alternatives is defined as: $X = \{x_1, x_2, \ldots, x_m\}$, $m \geq 2$. The evaluation given by expert e_k for proposal x_i under attribute c_j is denoted as: $x_{ij}^{(k)}$. e_k performs pairwise comparisons of attributes using the Analytic Hierarchy Process (AHP) [20], constructs a judgment matrix, and calculates the eigenvector corresponding to the maximum eigenvalue. This yields the expert's initial subjective weight vector: $W_{sub}^{(k)} = (w_{k1}, w_{k2}, \ldots, w_{kn})$, where w_{kj} denotes the importance rating of the jth attribute by the kth expert, satisfying the normalization constraint $\sum_{j=1}^{n} w_{kj} = 1$. Simultaneously, machine learning methods [21] are introduced as core metrics for objective facts. The Random Forest (RF) algorithm is selected, leveraging its feature importance evaluation principle. Training is conducted using large-scale historical data to construct an objective weight vector: $W_{obj} = (w_{obj,1}, w_{obj,2}, \ldots, w_{obj,n})$. The 4 steps for constructing the group consensus framework in this paper are as follows:

Step 1 Quantify expert subjective preferences using the AHP method, combine with Random forest to extract objective attribute weights, and establish a factual evidence baseline.
Step 2 Establish a similarity-based consensus evaluation system, leveraging dynamically contracting divergence thresholds to accurately identify minority experts deviating from objective facts.
Step 3 Introduce a game utility function, infuse objective evidence, guide outlier experts to correct biases, and drive group opinion toward strong consensus.
Step 4 The system performs macro-level assessment based on group consensus levels, conducts micro-level stability testing using Nash equilibrium, aggregates weights upon achieving consensus and steady-state utility, and outputs the optimal alternative.

2.2 Definitions of Relevant Consensus Measurement Theories

In multi-criteria decision-making [22], attribute weight vectors reflect the decision-maker's position within a multidimensional preference space.

Definition 1. *[23] Euclidean distance serves as a metric tool for measuring discrepancies among expert opinions and the degree of deviation between experts' subjective preferences and objective factual evidence. Let the weight vectors of experts e_k and e_l be denoted as W_k and W_l, respectively. The Euclidean distance d_{kl} between the opinions of experts e_k and e_l is defined as:*

$$d(W_k, W_l) = \sqrt{\sum_{j=1}^{n} (w_{kj} - w_{lj})^2} \tag{1}$$

The smaller the distance, the closer the decision preferences of the two experts are. This also defines $d(W_k, W_{obj})$, which calculates the deviation between expert e_k and the objective evidence benchmark W_{obj}, thereby assessing whether the expert's preferences deviate from objective principles.

Definition 2. *Let the decision-making group consist of m experts, where the weight vector of expert e_k is $\mathbf{W}_k = (w_{k1}, w_{k2}, \ldots, w_{kn})$. Then the group's average opinion vector is:*

$$W_{avg} = \frac{1}{m} \sum_{k=1}^{m} W_k \tag{2}$$

Definition 3. *[24] The individual consensus level CL_k of expert e_k characterizes the similarity between their opinion and the group's mainstream opinion (i.e., the group's average opinion vector), defined as:*

$$CL_k = 1 - \frac{d(W_k, W_{avg})}{\sqrt{2}} \tag{3}$$

where $d(W_k, W_{avg})$ denotes the distance metric function between the expert's opinion vector and the group's average opinion vector. A larger value of CL_k indicates that the expert's opinion is closer to the group's average opinion, while a smaller value indicates a greater deviation.

Definition 4. *[25] Group Consensus Level (GCL) is a macro-level indicator measuring the overall state of agreement within a decision-making group. A higher value indicates greater convergence of group opinions. It is obtained by weighting and aggregating the individual consensus levels of all members:*

$$GCL = \sum_{k=1}^{m} \lambda_k \cdot CL_k \tag{4}$$

Here, λ_k denotes the weight assigned to expert e_k, with $\lambda_k \geq 0$ and $\sum_{k=1}^{m} \lambda_k = 1$.

In this work, we choose uniform weights, i.e., $\lambda_k = 1/m$ for all k.

2.3 Theory of Consensus in Game Steady States

During the evolution of consensus, the adjustment of expert opinions is characterized through utility functions. The expert utility function comprises two components: compliance utility and the cost of factual deviation.

Definition 5. *During the consensus evolution process, the Agreement Utility (AU) of expert e_k at time t quantifies the positive benefit derived from the degree of alignment between the expert's opinion and the prevailing opinion within the group. Its formula is as follows:*

$$AU_k(t) = f(-d(W_k(t), W_{avg}(t))) \tag{5}$$

where $d(\cdot,\cdot)$ denotes the Euclidean distance; $W_k(t)$ represents the weight vector of expert e_k at time t; $f : \mathbb{R} \to \mathbb{R}$ is a monotonically increasing function. When the expert opinion is closer to the group average opinion (i.e., $d(W_k(t), W_{\text{avg}}(t))$ is smaller), the compliance utility $AU_k(t)$ increases.

Definition 6. *The deviation cost $DC_k(t)$ of expert e_k measures the risk cost arising from the deviation of their subjective preferences from the objective evidence benchmark W_{obj}. It is defined as:*

$$DC_k(t) = g(d(W_k, W_{obj})) \tag{6}$$

where $g : \mathbb{R} \rightarrow \mathbb{R}$ is a monotonically increasing function. As the deviation increases, i.e., $d(W_k(t), W_{\text{obj}})$ grows larger, the factual deviation cost $DC_k(t)$ also increases.

Definition 7. *During the consensus evolution process, the total utility function U_k of expert e_k comprehensively measures the benefits and costs derived from its decision-making behavior. It consists of two components: compliance utility and deviation cost:*

$$U_k(W_k, W_{-k}) = \alpha \cdot AU_k(t) - (1 - \alpha) \cdot DC_k(t) \tag{7}$$

where $\alpha \in [0, 1]$ is the preference balance coefficient, adjusting the weighting between consistency gains and deviation costs, and W_{-k} denotes the set of strategies adopted by experts other than e_k.

1. In the computational experiments of this paper, the preference balancing coefficient is set to $\alpha = 0.6$. In the expert utility evaluation, the weight of the group consensus benefit AU_k accounts for 60%, while the deviation cost DC_k of maintaining individual initial opinions accounts for 40%. From the perspective of consensus games, $\alpha = 0.6$ effectively balances the need to incorporate experts' heterogeneous views while preventing consensus deadlocks caused by excessive stubbornness among experts.
2. The utility function U_k provides guidance for the design of subsequent weight update rules. If experts could optimize autonomously, they would adjust weights along the gradient of the utility function, moving toward positions of higher utility. The consensus iteration algorithm in Sect. 3 of this paper is, in essence, a proxy implementation of this gradient ascent process, where in the system performs the utility-maximizing search on behalf of the experts.
3. The utility function U_k is a theoretical characterization of the experts' decision-making preferences, rather than an objective function directly optimized by the algorithm. In the specific algorithm presented in Sect. 3, the system executes each weight update based on the degree of deviation, eliminating the need for experts to explicitly compute or solve for the optimal response.

A Nash equilibrium [26] refers to a state of strategy combinations formed in a game process [27], where no participant can unilaterally adjust their strategy to increase their utility level while assuming other participants maintain their strategies unchanged.

Definition 8. *If there exists an optimal policy combination* $W^* = (W_1^*, W_2^*, \ldots, W_m^*)$ *such that for any expert* $e_k \in E$*, the following inequality holds within its policy space:*

$$U_k(W_k^*, W_{-k}^*) \geq U_k(W_k, W_{-k}^*), \quad \forall W_k \in \Omega_k \tag{8}$$

where Ω_k *denotes the policy space corresponding to the k expert, the feasible domain of their weight vector.* W_{-k}^* *represents the policy combination of all other members except expert* e_k*. At this point,* W^* *constitutes the Nash equilibrium of the game, and the system thereby reaches a stable state.*

3 Data-Evidence-Driven Approach and Group Game Model Construction Methodology

To resolve the dilemma of mismatch between experts' subjective preferences and objective patterns in complex decision-making, this section constructs the algorithmic core of the EDCB model through dual dimensions of dynamic guidance logic and convergence termination criteria. On one hand, it precisely captures cognitive biases via dynamic thresholds while guiding minority opinions toward rational convergence through adaptive step sizes. On the other hand, it proposes behavioral consensus determination criteria, identifying steady-state endpoints for consensus iteration from a behavioral game theory perspective based on Nash equilibrium theory. The organic coupling of these two approaches achieves the unification of subjective experience and objective principles.

This paper proposes an evidence-driven framework for reaching group consensus, defining "evidence" as a multi-level dynamic system that permeates the entire consensus process, rather than a simple combination of subjective and objective weights. This system is composed of historical evidence, behavioral evidence, and equilibrium evidence, the specific components of which are described as follows:

- Historical Evidence: Historical evidence is derived from large-scale historical operational data collected in real-world scenarios. Using the supervised learning model known as Random Forests, we extract importance scores for each metric to quantify their actual driving effects on decision outcomes. Within this framework, historical evidence serves as an objective reference point, providing a factual benchmark for calibrating and iterating expert subjective opinions, thereby preventing subjective preferences from deviating excessively from actual decision-making patterns.
- Behavioral evidence: Behavioral evidence is a dynamically generated control signal during the consensus iteration process, addressing the limitations of traditional models that use fixed thresholds and fixed adjustment steps. This paper introduces two types of behavioral evidence: first, a dynamic threshold function $\tau(t)$, which adapts based on the iteration round and real-time consensus level; second, an adaptive adjustment step size $\beta(t)$, which calculates the magnitude of opinion adjustment differentially based on the degree of

deviation between individual opinions and historical evidence to ensure the rationality of the regulation.

- Equilibrium evidence: Equilibrium evidence is used to determine whether the system has reached a stable convergence state, addressing the shortcoming of traditional models that rely solely on the level of group consensus as a termination condition. A single consensus level can only ensure that expert opinions appear to converge on the surface; it cannot guarantee that individual experts have further motivation to adjust their views. To address this, this paper constructs a dual convergence criterion: at the macro level, it requires that the divergence of group opinions be below a set threshold; at the micro level, it requires that the change in an expert's utility $\Delta U_k(t)$ approach 0.

3.1 Adaptive Regulation Mechanism Based on Subjective and Objective Bias Identification

This section guides a minority of experts whose views deviate from objective facts, facilitating the formation of consensus in group decision-making. Based on this, it proposes a comprehensive process for achieving consensus driven by data evidence, establishing a pathway for calibrating subjective decision preferences and reaching collective agreement.

To scientifically delineate the scope of divergence between experts' subjective preferences and objective facts, this paper introduces a dynamic threshold function $\tau(t)$ that evolves with iteration round t. Its formula is defined as:

$$\tau(t) = \tau_{max} \cdot e^{-\lambda t} \tag{9}$$

Here, τ_{max} represents the system's maximum tolerance for disagreement during the initial decision-making phase, while λ serves as the convergence control coefficient. A larger λ accelerates the decline rate of the threshold τ_{max}, indicating higher demands for the timeliness of expert opinions aligning with objective facts and stricter time constraints for correcting expert biases.

This model first requires identifying the minority of experts deviating from objective facts. Let $d(W_k, W_{obj})$ denote the Euclidean distance between the subjective weight W_k of expert e_k and the objective evidence W_{obj} during the t iteration. The identification logic for minority experts is as follows:

- When $d(W_k, W_{obj}) > \tau(t)$, the expert is deemed to be deviating and must enter the adjustment phase.
- When $d(W_k, W_{obj}) \leq \tau(t)$, the expert is deemed to be a retainer, and the system acknowledges the validity of their experience.

For the identified deviation experts, this paper introduces a Sigmoid-type adaptive step size function to further optimize the adjustment effect and avoid decision oscillations or insufficient fine-tuning caused by fixed adjustment steps. This function satisfies:

$$\beta(t) = \frac{\beta_{base}}{1 + e^{\gamma \cdot (d(t) - \tau)}} \tag{10}$$

Among these, β_{base} represents the upper limit of adjustment intensity. γ controls the degree of step size variation in response to deviation magnitude. When the expert's subjective preference deviates significantly from the target ($d(t) > \tau$), the function's denominator increases, causing $\beta(t)$ to decrease rapidly; As the expert's preference gradually approaches the target range, $\beta(t)$ converges toward the baseline step size β_{base}. This dynamic adjustment effectively prevents decision oscillation caused by excessive step sizes when the expert deviates too far from the target, while maintaining stable fine-tuning as the target is approached, ensuring smooth weight updates.

Based on this, the subjective weight update for outlier experts employs the following adaptive iterative algorithm to progressively correct and optimize preferences:

$$W_i(t+1) = (1 - \beta(t))W_i(t) + \beta(t)[\delta W_{obj} + (1 - \delta)W_{avg}(t)] \tag{11}$$

Here, the confidence coefficient $\delta \in [0, 1]$ is used to adjust the relative weights of objective evidence and group experience in the update direction: when $\delta \to 1$, the update direction is primarily driven by objective data; when $\delta \to 0$, the update converges more toward the group's average opinion.

Remark 1. Equation (11) is mathematically equivalent to a gradient ascent along the gradient direction of the utility function with a step size of $\beta(t)$. Although the algorithm is executed and adjusted via feedback by the system at the operational level, the direction and magnitude of each update correspond exactly to the behavior an expert would exhibit when autonomously performing a utility-maximizing search.

This mechanism satisfies the following three major characteristics:

- Asymmetry: The system computes a nonzero $\beta(t)$ only for outlier experts satisfying $D(t) > \tau(t)$, while the weights of other experts remain unchanged.
- Evidence Injection: Ensures the update direction consistently converges toward objective evidence W_{obj} through the confidence coefficient δ.
- Adaptive Evolution: As the iterative process progresses, expert preferences gradually converge toward objective reality, with $\beta(t)$ dynamically fine-tuned until the system reaches a game equilibrium.

3.2 Determination of Behavioral Convergence Criteria Based on Game Steady State

Within the framework of group decision-making, the formation of consensus is manifested not only as numerical convergence but also as a behavioral steady state achieved by decision-makers under the combined influence of subjective preferences, group constraints, and objective evidence. To accurately characterize this steady state, the model incorporates Nash equilibrium theory, formalizing the consensus iteration process as a non-cooperative game. By adopting the theoretical framework of non-cooperative games to model the process of reaching group consensus, each expert is treated as an independent decision-maker. The role of game theory here is to provide a normative benchmark—the Nash equilibrium defines which consensus states can be considered stable.

During dynamic iteration, individual behavior is driven by the utility function U_k. The trade-off between compliance gains and deviation costs determines the direction of weight adjustment, while the step size function $\beta(t)$ controls the magnitude of adjustment. When the utility loss from deviating from the objective benchmark exceeds the gains from maintaining the original position, adjustment becomes a rational choice for individuals. Due to the model's asymmetric adjustment mechanism, expert weights within the rational range remain stable, while deviating agents gradually converge toward the objective benchmark and group average during iteration. As iterations progress, the utilities of the various agents tend to stabilize, and the system gradually approaches the Nash equilibrium—a steady state under the specified conditions.

To precisely identify the steady state of the game at the computational level, this paper establishes a dual convergence criterion:

- Macro-level judgment criteria: Group consensus level meets requirements:

$$GCL \geq \theta \tag{12}$$

  Among these, the predefined consensus threshold $\theta \in [0, 1]$ (set to 0.98 in this paper)
- Micro-steady-state criterion: Calculate the change in expert utility between two consecutive iterations:

$$\Delta U_k(t) = |U_k(t+1) - U_k(t)| \tag{13}$$

  If for all k it holds that:

$$\Delta U_k(t) \leq \varepsilon \tag{14}$$

Given the current population state, no expert can achieve a utility improvement exceeding ε by unilaterally adjusting its own weight. The system is therefore deemed to have converged to a steady state. Here, $\varepsilon > 0$ represents a given tolerance error (set to 10^{-3} in this paper).

Remark 2. The Nash equilibrium provides a normative definition of the convergence result, while the algorithm achieves dynamic progression through iterative feedback. The former provides theoretical support for the algorithm's termination conditions, while the latter serves as the specific computational path to reach a stable state.

3.3 Solution Ranking

After completing the game convergence of expert weights, the final optimal ranking must be performed on the set of evaluated alternatives $X = \{x_1, x_2, \ldots, x_m\}$, $m \geq 2$. To ensure the structural rationality and stability of the decision outcome within the multi-criteria space, this paper employs the TOPSIS [28] method for comprehensive evaluation. This approach precisely reflects the relative merits of each alternative by characterizing the geometric distance between them and the ideal solution, while fully accounting for the influence of indicator weights. The specific steps are as follows:

Step 1 Compile the comprehensive evaluation results of the expert panel for each proposal under each evaluation criterion to construct a comprehensive decision matrix $Z = [z_{ij}]_{m\times n}$, where z_{ij} represents the comprehensive score of proposal x_i under criterion j. Simultaneously, perform normalization and standardization on the matrix as follows.

$$r_{ij} = \frac{z_{ij}}{\sqrt{\sum_{i=1}^{m} z_{ij}^2}}, \quad R = [r_{ij}]_{m\times n} \tag{15}$$

Step 2 After completing the expert weighting game and obtaining the group consensus weight vector $W^* = (w_1^*, w_2^*, \ldots, w_n^*)$ applied to the normalization matrix R, construct a weighted normalized decision matrix $V = [v_{ij}]_{m\times n}$ where $v_{ij} = w_j^* \cdot r_{ij}$. Subsequently, determine the ideal reference points within the multi-criteria decision space to characterize the relative strengths and weaknesses of each alternative. For benefit-oriented criteria, define positive ideal solutions and negative ideal solutions as follows:

- The optimal solution is defined as the vector composed of the optimal values under each criterion:

$$V^+ = (v_1^+, v_2^+, \ldots, v_n^+), \quad v_j^+ = \max_i\{v_{ij}\} \tag{16}$$

- The negative ideal solution is defined as the vector formed by the worst possible values under each criterion:

$$V^- = (v_1^-, v_2^-, \ldots, v_n^-), \quad v_j^- = \min_i\{v_{ij}\} \tag{17}$$

Among these, the positive ideal solution represents a hypothetical scenario achieving optimal performance across all evaluation criteria, while the negative ideal solution denotes a reference scenario exhibiting the most unfavorable conditions across all criteria.

Step 3 Calculate the Euclidean distances D_i^+ and D_i^- between each alternative and the positive ideal solution and negative ideal solution, respectively, according to Eq. (1). Then compute the composite score C_i using the following formula:

$$C_i = \frac{D_i^-}{D_i^+ + D_i^-} \tag{18}$$

The closer the value of C_i is to 1, the closer the plan is to the ideal decision state and the farther it is from the risk state.

Step 4 Sort in descending order based on the values of C_i. The alternative corresponding to the largest C_i represents the optimal choice under the current group consensus.

In the consensus-building process, the feedback mechanism is the concrete process that drives the formation of consensus, while the game-theoretic equilibrium serves as the theoretical goal of this process. The EDCB model guides the decision-making system to gradually evolve toward a Nash equilibrium state by introducing feedback rules. Game theory provides a clear normative benchmark for consensus outcomes, and the Nash equilibrium clearly defines the state conditions that a stable consensus must satisfy. The dynamic threshold, adaptive step size, and weight update rules designed in this section serve as the computational means to achieve this benchmark. Game theory specifies the conditions that the final state must satisfy, while the feedback mechanism drives the system to gradually converge to this state. To facilitate system implementation, we have encapsulated the aforementioned algorithmic logic into the following pseudocode:

Algorithm 1: Data Evidence-Driven Consensus Reaching Algorithm

---

Input: Expert matrix set $E = \{e_1, e_2, \ldots, e_k\}$, $k \geq 2$, Alternative set $X = \{x_1, x_2, \ldots, x_m\}$, $m \geq 2$, Historical dataset D, Expert evaluation matrix $x_{ij}^{(k)}$, Objective benchmark weight vector W_{obj}, Consensus threshold θ, Convergence tolerance ε, Confidence coefficient δ, Step size baseline β_{base}, Dynamic threshold parameters $\tau_{\max}, \lambda, \gamma$

Output: Final consensus weight vector W^*

Phase 1: Objective Evidence Extraction

Train the Random Forest model on dataset D to calculate feature importance Normalize the importance vector to obtain the objective evidence W_{obj}

Phase 2: Iterative Game and Guidance

Initialize iteration round $t = 0$ Calculate the initial group average weight $W_{\text{avg}}^{(0)}$ using Eq. (2) Measure the initial consensus level $\text{GCL}^{(0)}$

while $GCL^{(t)} < \theta$ ***and*** $\max(\Delta U^{(t)}) > \varepsilon$ **do**

  Update the dynamic threshold $\tau(t)$ using Eq. (9)

  foreach *expert* $e_k \in \{e_1, \ldots, e_m\}$ **do**

    Calculate the distance $d_k^{(t)}$ between expert k and objective evidence using Eq. (1)

    if $d_k^{(t)} > \tau(t)$ **then**

      `// Execute asymmetric adjustment logic`

      Calculate the adaptive step size $\beta_k^{(t)}$ using Eq. (10)

      Set the guidance target: $V_{\text{target}}^{(t)} = \delta W_{\text{obj}} + (1 - \delta) W_{\text{avg}}^{(t)}$

      Update the expert weight $W_k^{(t+1)}$ using Eq. (11)

      `// Update individual utility and increment`

      Calculate $U_k^{(t+1)}$ and $\Delta U_k(t)$ using Eq. (7) and Eq.(13)

    else

      `// Reserve reasonable opinions`

      Set individual utility change: $\Delta U_k(t) = 0$

  Update iteration round: $t = t + 1$

  Update group average weight: $W_{\text{avg}}^{(t)} = \frac{1}{m} \sum_{k=1}^{m} W_k^{(t)}$

  Re-measure the group consensus level $\text{GCL}^{(t)}$

Calculate the final consensus weight vector: $W^* = W_{\text{avg}}^{(t)}$

Phase 3: Alternative Ranking

Calculate the comprehensive score of each alternative based on W^*

Output the optimal ranking result of alternatives

To provide a more intuitive illustration of the implementation process of the group consensus method described in this paper, Fig. 1 shows the detailed workflow of the model.

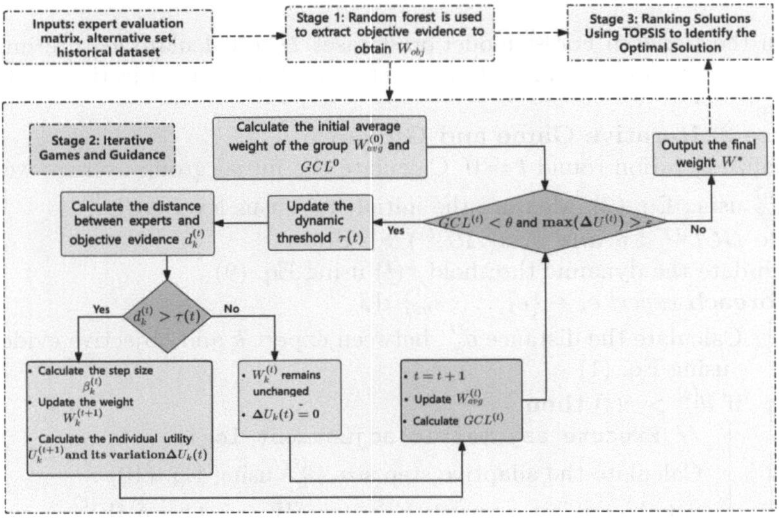

Fig. 1. EDCB Algorithm Flowchart.

4 Case Study Analysis of Marketing Strategy Scenarios for Retail Enterprises

This section applies the dynamic consensus decision-making model constructed earlier to a real-world retail marketing strategy selection scenario. Through empirical analysis of the Kaggle retail dataset, it validates the model's effectiveness and robustness in extracting objective patterns as data evidence within complex environments involving multi-departmental conflicts of interest. This enables precise identification of outlier minority factions and facilitates efficient achievement of group consensus. This study established a computational environment based on Python 3.11, running on Windows 11 Professional with an Intel Core i5-13450HX processor and 32 GB of memory.

4.1 Example Description

This study draws on the publicly accessible "Retail Sales Trends and Marketing Effectiveness Dataset" available on Kaggle. After performing data sampling and preprocessing, marketing strategy records from continuous time periods for the enterprise were chosen for experimental analysis.

This study utilizes the "Retail Sales Trends and Marketing Effectiveness" dataset, publicly available on the Kaggle platform, as its source of objective data. The raw data includes weekly sales figures, promotional discount rates, the Consumer Price Index, regional unemployment rates, holiday indicators, as well as environmental and cost factors such as temperature and fuel prices. To align with the evaluation framework of this study, targeted feature engineering was performed based on the original fields. Three composite features—advertising intensity, discount intensity, and seasonal alignment—were constructed and normalized to provide reliable data support for subsequent objective evidence extraction. Let $E = \{e_1, e_2, \ldots, e_{20}\}$ be a set of 20 experts involved in cross-departmental decision-making, comprising 7 experts from the Marketing Department ($e_1 - e_7$) who focus on brand promotion and market share, 6 experts from the Financial Control Department ($e_8 - e_{13}$) who focus on cost control and return on investment, and 7 experts from the Operations Management Department ($e_{14} - e_{20}$) who emphasize operational efficiency and seasonal management. $C = \{c_1, c_2, c_3\}$ is the set of evaluation attributes, where c_1 represents advertising intensity, c_2 represents discount intensity, and c_3 represents seasonal alignment. $X = \{x_1, x_2, x_3\}$ is the set of alternative strategies, where x_1 represents a strategy combining moderate advertising investment, high discount intensity, and high seasonal alignment; x_2 represents a strategy combining moderate advertising investment, moderate discount intensity, and moderate seasonal alignment; and x_3 represents a strategy combining high advertising investment, low discount intensity, and low seasonal alignment. To quantify the initial weight preferences among experts, this paper employs the AHP judgment matrix method to obtain expert subjective weights. Each expert is required to conduct pairwise comparisons and assign scores on a 19 scale for each attribute across three dimensions: marketing effectiveness, cost-effectiveness, and implementation feasibility. The key parameters set in this paper are as follows: Let the consensus threshold be $\theta = 0.98$, $\varepsilon = 10^{-3}$, the initial value of the dynamic threshold be $\tau_{max} = 0.45$, $\beta_{\text{base}} = 0.35$, the dynamic threshold decay coefficient be $\lambda = 0.12$, and the confidence coefficient be $\delta = 0.55$. The specific steps of the EDCB consensus model process are as follows:

Step 1. Use the Random Forest algorithm to train the model and obtain the objective weight benchmark. Use the three standardized decision feature variables $[c_1, c_2, c_3]$ as input variables, and sales revenue as the target variable. The objective weight benchmark obtained is $W_{obj} = [0.294, 0.568, 0.138]$.

Step 2. The expert weight matrix is normalized according to the standardized method for benefit-oriented indicators.

Step 3. The system first identifies experts based on the dynamic threshold $\tau(t)$, accurately pinpointing those deviating from objective facts. Subsequently, it calculates the adaptive adjustment step size $\beta_k^{(t)}$ according to Equation (10) to drive deviations toward objective facts. Each iteration computes GCL and ΔU to achieve the final steady-state consensus. The iterative results for each round are shown in Table 1:

Table 1. Results of Consensus Iteration Rounds

Iteration Round	$GCL^{(t)}$	$\max(\Delta U^{(t)})$	Group Average Weight
1st Round	0.8234	0.0847	$[0.315, 0.425, 0.260]$
2nd Round	0.8924	0.0423	$[0.308, 0.432, 0.260]$
3rd Round	0.9456	0.0187	$[0.299, 0.425, 0.276]$
4th Round	0.9852	0.0089	$[0.293, 0.420, 0.287]$
5th Round	0.9876	0.0023	$[0.292, 0.419, 0.289]$
6th Round	0.9889	0.0008	$[0.291, 0.419, 0.290]$

Step 4. After six rounds of iteration, the final composite weight vector is $W^* = [0.298, 0.508, 0.194]$. The deviation from the objective benchmark vector is 0.061, which falls within the acceptable range. Simultaneously, the group consensus level GCL $= 0.9889 > 0.98$ and the maximum utility function $\max(\Delta U(t)) = 0.0008 < 0.001$ are satisfied, achieving the consensus level. The weights and consensus levels for each department at this point are shown in Table 2.

Table 2. Departmental Weight Allocation and Average Consensus Level

Department	Adv. Wt.	Disc. Wt.	Seas. Wt.	Avg. Cons.
Marketing Department	0.292	0.418	0.290	0.986
Financial Control Department	0.288	0.422	0.290	0.989
Operations Management Department	0.293	0.417	0.290	0.985

Step 5. Select the optimal solution based on the TOPSIS method.
First, calculate the standardized matrix R according to Eq. (15) and construct the weighted matrix V by combining it with W^*. Subsequently, determine the positive ideal solution $A^+ = \{0.217, 0.407, 0.150\}$ composed of the maximum values of each attribute and the negative ideal solution $A^- = \{0.145, 0.136, 0.062\}$. Next, the Euclidean distances from each alternative to the positive and negative ideal solutions are computed. Finally, C_i is calculated using formula (18), as shown in Table 3, confirming that $x_1 \succ x_2 \succ x_3$ represents the optimal alternative ranking.

Table 3. TOPSIS Method Decision Ranking Results

Rank	Alternative	C_i	Alternative Description
1	x_1	0.708	Medium advertising input, high discount intensity, and high season matching combination strategy
2	x_2	0.473	Medium advertising input, medium discount intensity, and medium season matching combination strategy
3	x_3	0.202	High advertising input, low discount intensity, and low season matching combination strategy

4.2 Sensitivity Analysis

To evaluate the robustness and adaptability of the EDCB model under different parameter settings and to observe its impact on the final ranking of proposals, this section conducts a systematic sensitivity analysis from two key dimensions: sensitivity to the dynamic identification threshold (by adjusting τ_{max} or λ) and the preference balancing coefficient α. This paper first investigates the impact of the dynamic consensus threshold τ_{max} and the decay coefficient λ on the model's consensus performance. A sensitivity analysis was conducted using the proportion of deviating parties as an evaluation metric, with τ_{max} taking values in the range $[0.25, 0.65]$, and λ is set to $[0.08, 0.16]$. As shown in Fig. 2, the results indicate that as τ_{max} increases from 0.25 to 0.65, the proportion of deviating parties exhibits a significant nonlinear trend of first decreasing and then rebounding. It reaches a global minimum within the interval $\tau_{max} \in [0.45, 0.55]$, it reaches a global minimum; simultaneously, there is a significant interaction effect between λ and τ_{max}. When τ_{max} is at a moderate level, a moderate value of $\lambda \in [0.10, 0.13]$ is more conducive to reducing the proportion of deviants, effectively promoting the convergence of group consensus.

To further verify the robustness of the proposed method, a sensitivity analysis was conducted with respect to the preference balancing coefficient α. The analysis results are shown in Fig. 3. As the preference balancing coefficient α gradually increases within the interval $[0.1, 0.9]$, the ranking of the alternatives remains consistently $x_1 \succ x_2 \succ x_3$ without any changes. This indicates that the method exhibits high stability in decision outcomes when this parameter is adjusted, remaining unaffected by fluctuations in the preference balancing coefficient, thereby further validating the model's reliability.

4.3 Comparative Analysis

To validate the effectiveness of the proposed model, we conducted a comparative analysis with existing group consensus methods. Jing et al. [29] have proposed a consensus-reaching strategy based on a game-theoretic equilibrium mechanism, quantifying group consensus through a consensus level metric; however,

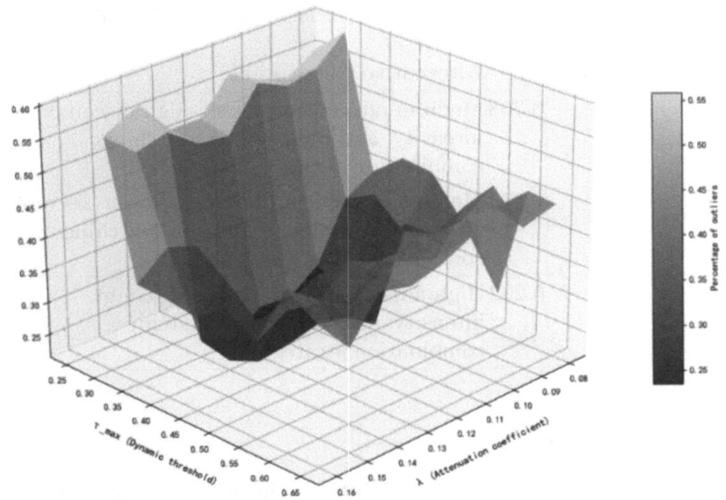

Fig. 2. Sensitivity Analysis of Consensus Effect to Dynamic Threshold and Attenuation Coefficient.

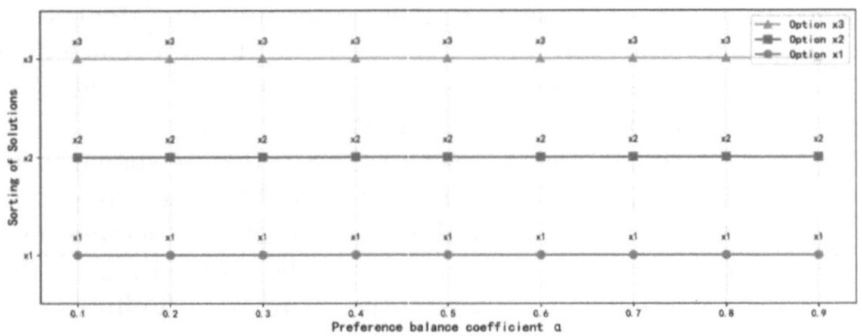

Fig. 3. Analysis of the Impact of the Preference Balancing Coefficient α on Alternative Ranking.

this method did not establish an explicit mechanism for integrating subjective and objective factors, and its weight allocation lacked objective justification and interpretability. Liao et al. [30] have introduced subgroup cohesion and subgroup size to achieve structural fusion of weights, and designed an adaptive feedback strategy based on different consensus levels, demonstrating a certain degree of dynamic adjustment capability. However, the convergence criterion of this method relies solely on a single group consensus level metric, making it prone to situations where the overall target is met but local imbalances persist. Tang et al. [31] have employs differentiated adjustment strategies based on different consensus levels, achieving adaptive optimization of the consensus feedback process. However, the weighting structure of this method does not stem from a clear evidence-driven fusion mechanism, making it difficult to adapt to complex and uncertain decision-making scenarios. Zhang et al. [32] have addressed the challenge of group consensus under uncertain costs by optimizing the consensus process through robust optimization modeling, data-driven weighting, social network trust propagation, and fair utility constraints. However, this method does not incorporate game theory as a theoretical foundation, lacking the theoretical basis to ensure the stability of the consensus equilibrium, making it difficult to achieve long-term stability in consensus outcomes. Overall, most existing models focus on improving consensus quality through structural or behavioral optimization, yet rarely consider both the integration of external objective evidence and Nash equilibrium conditions simultaneously, resulting in significant improvements in both consensus quality and stability.

To validate the superiority of the proposed EDCB consensus model, this section designs four sets of comparative methods for quantitative comparison experiments. The EDCB model constructed in this paper is designated as Method 1; Method 2 is a subjective-weight-only model, which obtains expert subjective weights solely through AHP, does not incorporate data-driven objective information, and relies exclusively on traditional consensus mechanisms to iterate opinions; Method 3 is an objective-weight-only model, which relies solely on objective weights obtained through Random Forest learning for decision-making, without integrating expert subjective preferences; Method 4 is a traditional consensus model, employing a basic opinion adjustment strategy that does not fuse subjective and objective weights and is not guided by game-theoretic equilibrium. By comparing the performance of these four models across metrics such as the final consensus level, the number of convergence iterations, the consistency of decision rankings, and algorithm execution time, we can systematically validate the advantages of the EDCB model in terms of integrating subjective and objective information, accelerating consensus convergence, and enhancing decision stability. The experimental results are shown in Table 4:

Table 4. Comparative Analysis Results

Method	Final Consensus Level	Number of Convergence Iterations	Algorithm Execution Time	Solution Ranking
Method 1	0.9889	6	13.39	$x_1 \succ x_2 \succ x_3$
Method 2	0.9651	22	39.05	$x_1 \succ x_2 \succ x_3$
Method 3	0.9612	22	37.97	$x_1 \succ x_2 \succ x_3$
Method 4	0.9767	6	22.40	$x_1 \succ x_2 \succ x_3$

As shown in the comparison results in Table 4, Method 1 proposed in this paper demonstrates significant advantages in both consensus performance and computational efficiency. In terms of consensus achievement, Method 1 reaches a final consensus level of 0.9889, representing improvements of 2.47%, 2.88%, and 1.25% over Methods 2, 3, and 4, respectively, effectively enhancing the consistency of group consensus; in terms of convergence efficiency, Method 1 achieves convergence in only 6 iterations, reducing the number of iterations by 72.7% compared to Methods 2 and 3, and decreasing the algorithm execution time by 65.7%, 64.7%, and 40.2% respectively. Furthermore, the ranking results of the solutions obtained by the four methods were completely consistent, further validating the reliability of the method.

5 Conclusions and Future Prospects

This paper has tackled several challenges in group decision-making, including the divergence between experts' subjective preferences and objective realities, the lack of objective guidance in traditional consensus approaches for experts who deviate from factual benchmarks, and instability in the implementation of existing methods. To address these issues, it has introduced an Evidence-Driven Game-Theoretic Decision-Making Consensus Model (EDCB) and has demonstrated its effectiveness through a case study on evaluating marketing strategies for retail enterprises. The primary contributions have been summarized as follows:

(1) An evidence-based consensus guidance mechanism has been developed: By employing the random forest algorithm to derive objective weighting benchmarks from historical data, the consensus-reaching process has been firmly anchored in reliable empirical evidence.
(2) An adaptive adjustment mechanism has been constructed: Outlier experts are accurately detected using dynamic thresholds, while directional guidance has been provided via adaptive step sizes grounded in the Sigmoid function. This design effectively maintains the valuable expertise of rational experts and avoids excessive homogenization of opinions in the consensus.
(3) A steady-state criterion rooted in Nash equilibrium has been introduced: The feedback process has been modeled as a non-cooperative game among experts, incorporating individual utility functions and adopting Nash equilibrium as the stopping rule for iterations. This guarantees that the consensus results not only meet overall consistency requirements at the group level but also exhibit behavioral stability for each individual expert.

Despite the progress made in terms of consensus formation and stability, several aspects warrant further improvement:

(1) In the current model, the expert confidence coefficient δ has been treated as a fixed constant. Future studies could investigate dynamic confidence updating, allowing experts to adjust their confidence during the game process, thereby forming a dynamic confidence adjustment mechanism that more closely reflects real-world decision-making behavior.
(2) For large-scale expert groups, the computational burden associated with calculating game equilibria becomes substantial. Subsequent work may incorporate clustering algorithms and distributed computing frameworks to enhance the model's computational efficiency.

(3) Concerning evidence sources, the present framework relies exclusively on historical data as objective evidence. Future research could consider fusing multi-source and uncertain information to construct a more comprehensive and robust objective benchmark.

Acknowledgment. This research was funded by the Excellent Youth Project of the Hunan Provincial Department of Education (Grant No. 24B0608) and by the Basic Research Program of the Yueyang Municipal Bureau of Science and Technology (Grant No. 2024091000008).

References

1. Zhang, H., Zhao, S., Kou, G., Li, C.-C., Dong, Y., Herrera, F.: An overview on feedback mechanisms with minimum adjustment or cost in consensus reaching in group decision making: research paradigms and challenges. Inform. Fusion **60**, 65–79 (2020)
2. Tran, T.N.T., Felfernig, A., Le, V.M.: An overview of consensus models for group decision-making and group recommender systems. User Model. User-Adap. Inter. **34**(3), 489–547 (2024)
3. Dong, Y., et al.: Consensus reaching in social network group decision making: research paradigms and challenges. Knowl.-Based Syst. **162**, 3–13 (2018). Special Issue on Intelligent Decision-making and consensus under uncertainty
4. Chen, Y., Zhou, M., Liu, F.: A consensus model with non-cooperative behavior adaptive management based on cognitive psychological state computation in large-scale group decision. Cogn. Comput. **16**(6), 3344–3363 (2024)
5. Guo, J., Wang, Z., Zhang, Z.: A dual-level consensus model for large-scale group decision-making driven by trust relationships in social networks. Eng. Appl. Artif. Intell. **136**, 109033 (2024)
6. Hou, X., Xu, T., Zhang, C.: Building consensus with enhanced k-means++ clustering: a group consensus method based on minority opinion handling and decision indicator set-guided opinion divergence degrees. Electronics **14**(8), 1638 (2025)
7. Meng, F., Tang, J., Li, X.: Consensus group decision making based on coalitional nash-bargaining game. Expert Syst. Appl. **255**, 124756 (2024)
8. Meng, F.-Y., Gong, Z.-W., Pedrycz, W., Chu, J.-F.: Selfish-dilemma consensus analysis for group decision making in the perspective of cooperative game theory. Eur. J. Oper. Res. **308**(1), 290–305 (2023)
9. Shen, Y., Ma, X., Zhang, H., Zhan, J.: Fusion social network and regret theory for a consensus model with minority opinions in large-scale group decision making. Inform. Fusion **112**, 102548 (2024)
10. Yuan, Y., Wang, C., Cheng, D., Zhang, F., Zhou, Z., Cheng, F.: Minimum conflict consensus models for group decision-making based on social network analysis considering non-cooperative behaviors. Inform. Fusion **99**, 101855 (2023)
11. Yang, W., Liang, C.: A large-scale consensus decision-making model for non-cooperative behavior based on incomplete probabilistic hesitant fuzzy information in social trust networks. Inf. Sci. **714**, 122196 (2025)

12. Liang, H., Dong, Y., Xu, Y.: Minimum cost consensus model with loss aversion for group decision making. J. Oper. Res. Soc. **74**(9), 2182–2196 (2023)
13. Li, Y., Kou, G., Li, G., Peng, Y.: Consensus reaching process in large-scale group decision making based on bounded confidence and social network. Eur. J. Oper. Res. **303**(2), 790–802 (2022)
14. Xu, Y., Dou, R., Wang, H.: A dynamically weight adjustment in the consensus reaching process for group decision-making with hesitant fuzzy preference relations. Int. J. Syst. Sci. **48**(6), 1311–1321 (2017)
15. Salman, H., Kalakech, A., Steiti, A.: Random forest algorithm overview. Babylon. J. Mach. Learn. **69–79**, 2024 (2024)
16. Huang, X.: Predictive models: regression, decision trees, and clustering. Appl. Comput. Eng. **79**, 124–133 (2024)
17. Blockeel, H., Devos, L., Frénay, B., Nanfack, G., Nijssen, S.: Decision trees: from efficient prediction to responsible AI. Front. Artif. Intell. **6** (2023)
18. Nagar, K., Chawla, M.P.S.: A survey on various approaches for support vector machine based engineering applications. Int. J. Emerg. Sci. Eng. **11**(11), 6–11 (2023)
19. Wan, S., Zou, W., Dong, J., et al.: Dual strategies consensus reaching process for ranking consensus based probabilistic linguistic multi-criteria group decision-making method. Expert Syst. Appl. **262**, 125342 (2025)
20. Lee, Y.-J.: Unveiling key criteria for effective leadership: a multi-criteria decision-making framework using topic modeling and analytic hierarchy process (AHP). Int. J. Anal. Hierarchy Process **17**(2) (2025)
21. Trigka, M., Dritsas, E.: A comprehensive survey of machine learning techniques and models for object detection. Sensors **25**(1), 214 (2025)
22. Sahoo, S.K., Goswami, S.S.: A comprehensive review of multiple criteria decision-making (MCDM) methods: advancements, applications, and future directions. Decis. Mak. Adv. **1**(1), 25–48 (2023)
23. Mussabayev, R.: Optimizing euclidean distance computation. Mathematics **12**(23), 3787 (2024)
24. Zhou, Y., Zhou, L.: Dynamic adaptive consensus reaching process with risk attitude for multi-criteria group decision making. Inform. Fusion **127**, 103760 (2026)
25. Zhang, H., Dong, Y., Chiclana, F., Yu, S.: Consensus efficiency in group decision making: a comprehensive comparative study and its optimal design. Eur. J. Oper. Res. **275**(2), 580–598 (2019)
26. Niazi, M.U.B., Özgüler, A.B., Yıldız, A.: Consensus as a nash equilibrium of a dynamic game, arXiv preprint (2017)
27. Yang, H.-X.: A consensus opinion model based on the evolutionary game, arXiv preprint (2018)
28. Parida, P.K., Baral, S.P., Sahoo, S.K.: Topsis method for multi-criteria decision making in fuzzy environment. Int. J. Electr. Eng. Technol. **12**(11), 122–130 (2021)
29. Jing, F., Chao, X.: Fairness concern: an equilibrium mechanism for consensus-reaching game in group decision-making. Inform. Fusion **72**, 147–160 (2021)
30. Tang, M., Liao, H., Xu, J., Streimikiene, D., Zheng, X.: Adaptive consensus reaching process with hybrid strategies for large-scale group decision making. Eur. J. Oper. Res. **282**(3), 957–971 (2020)
31. Tang, J., Meng, F.: An adaptive core-nash bargaining game consensus mechanism for group decision making. Group Decis. Negot. **33**, 1–33 (2024)
32. Zhang, X., Liang, H., Qu, S.: Robust consensus modeling: concerning consensus fairness and efficiency with uncertain costs. Mathematics **12**(8), 1266 (2024)

From Gene–Disease Discovery to Personalised Recommendations: AI for Translational Genomics

Kairui Guo[1], Mengjia Wu[1], Yi Zhang[1], Guangquan Zhang[1], Hua Lin[2], Mark Grosser[2], and Jie Lu[1](✉)

[1] Australian Artificial Intelligence Institute, University of Technology Sydney, Sydney, NSW, Australia
{Kairui.Guo,Mengjia.Wu,Yi.Zhang,Guangquan.Zhang,Jie.Lu}@uts.edu.au
[2] 23Strands, Sydney, Australia
{Hua.Lin,Mark.Grosser}@23strands.com

Abstract. Genomic and gene expression analytics pose a complex translational challenge. Biological mechanisms are only partly understood. Evidence evolves rapidly. Real-world datasets are high-dimensional, heterogeneous, and often sparsely labelled. This paper presents AI methodological advances in translational genomics, focusing on four components: evidence-traceable knowledge mining, gene–disease association modelling, transfer learning across cohort and population shifts, and personalised recommendations. Gene–disease association is the core inference task. Risk prediction is the main deployable output. Recommendation extends prediction into decision support. We review methods that structure large-scale biomedical evidence using bibliometrics and natural language processing, model genomic associations using graph-based and deep learning approaches, improve robustness through transfer learning across cohorts and populations, and translate risk estimates into evidence-grounded outputs using recommender systems. These components define a workflow for genomic analysis that emphasises traceable evidence, biologically informed modelling, and reliable performance across cohorts.

Keywords: Artificial Intelligence · Genetics and genomics · Transfer learning · Recommender systems · Disease risk prediction

1 Introduction

AI for biomedical genomics has advanced quickly; however, translating it into real clinical settings remains difficult [4,9]. In practice, implementing AI genomic models is complicated by several challenges. Genomic datasets are high-dimensional, labelled outcomes are often limited, and predictive models do not always transfer well across cohorts. A recent review of AI-driven biomedical genomics summarised recently developed models, the main computational tasks

J. Lu et al. (Eds.): FLINS-ISKE 2026, LNCS 16752, pp. 547–558, 2027.
https://doi.org/10.1007/978-981-92-2369-5_33

in the field, and proposed barriers to deployment, including weak external validation, inconsistent endpoint definitions, and limited solutions to uncertainty [6]. These challenges highlight the need for methods that generalise across cohorts and produce interpretable outputs with clear links to supporting evidence.

Recent advances in AI have broadened the scope of genomic modelling. Foundation-scale regulatory models, such as AlphaGenome, demonstrate the ability to predict variant-level functional effects from genomic sequence context at scale [2]. However, variant-effect prediction alone is not sufficient for translation into real-world genomic and health applications. Practical deployment also requires methods that can organise fast-changing biomedical evidence, model gene–disease relationships across multiple data types, transfer across populations and cohorts, and support follow-up decisions.

This paper presents a set of AI methodological advances through a translation-centred workflow, as shown in Fig. 1. Gene–disease association serves as the central inference task, risk prediction is the principal deployable output, and recommendation provides the decision-support layer that connects modelling to downstream use. The remainder of the paper is organised as follows. Section 2 reviews the main methodological components of the proposed workflow, including evidence-traceable knowledge mining, disease-adapted modelling, transfer learning, and personalised recommendation. Section 3 considers their broader implications for translational genomics, as well as current limitations and future directions. Section 4 concludes the paper.

2 Methodological Advances in Translational Genomics

This section presents the main methodological components of the proposed workflow, as shown in Fig. 1. We begin with evidence-traceable knowledge mining, which converts biomedical literature into structured evidence for genomic analysis. Then, we consider gene–disease association modelling, where model design is matched to data structure and disease context. We next address transfer learning under population and cohort shift, with emphasis on robustness across heterogeneous settings. The final component is personalised recommendation, where predictive outputs are translated into actionable guidance.

2.1 Evidence-Traceable Disease Genetic Knowledge Mining

Rapid growth in genomic evidence is accompanied by substantial complexity: findings are high-dimensional (variants, genes, pathways, phenotypes), span multiple biological scales, and may be incomplete or even contradictory as new studies revise earlier conclusions [3]. Importantly, most such genomic evidence is documented in research literature, where clinical associations between genes, diseases, phenotypes, and chemicals are reported in a fragmented and unstructured manner. These pieces of evidence are often distributed across individual studies with varying contexts, methodologies, and populations, making it difficult to systematically aggregate, compare, and validate them. This motivates the

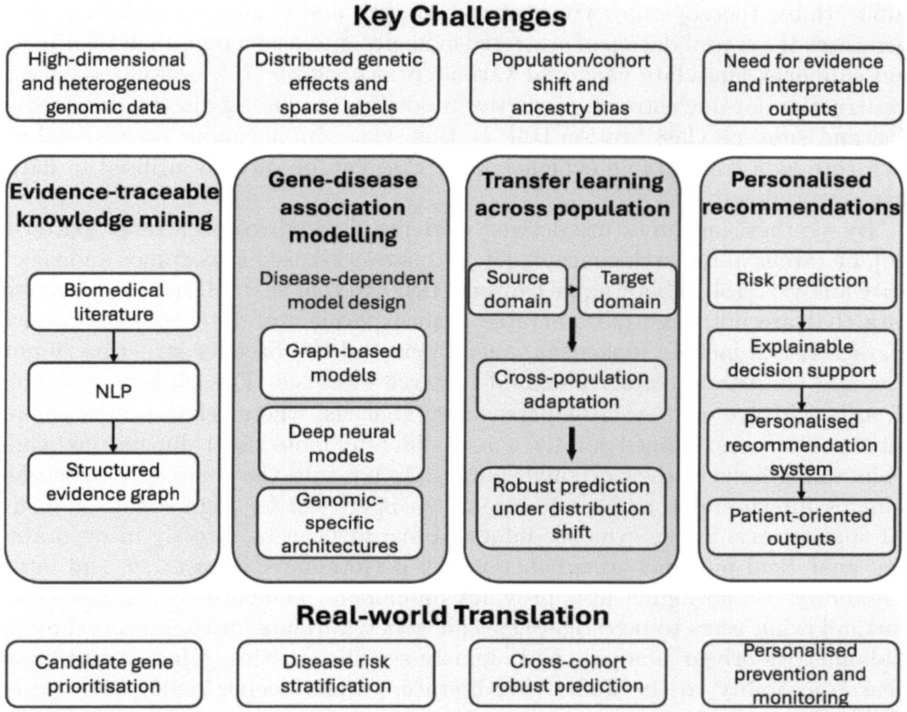

Fig. 1. Overview of the translational genomics workflow. Key challenges motivate four methodological components: evidence-traceable knowledge mining, gene–disease association modelling, transfer learning across populations, and personalised recommendation. These components support real-world translation in candidate gene prioritisation, disease risk stratification, cross-cohort robust prediction, and personalised prevention and monitoring.

need for an explicit, evidence-traceable layer that not only organises and contextualises heterogeneous findings, but also preserves their provenance and enables transparent navigation from derived knowledge back to original sources. Such a layer allows the literature to be treated as an active and structured evidence base rather than a static background.

BiblioEngine utilises intelligent bibliometrics and NLP-driven knowledge mining to support disease genetic knowledge discovery and structured analysis [19]. It provides a highly efficient and effective framework to extract biomedical entities and their relationships from large-scale literature, and to organise them into structured, machine-readable representations across multiple perspectives, including genes, diseases, chemicals, and mutations. Through the integration of entity extraction, concept mapping, and literature metadata, the platform enables each piece of derived knowledge to remain traceable to its supporting publications, thereby supporting reproducibility and evidence validation. This facilitates the consolidation of scattered evidence into a coherent analytical space and supports candidate gene and variant prioritisation [15], as well as feature construction for downstream predictive modelling, including disease risk prediction and subtype classification [16]. In this sense, BiblioEngine operationalises literature as a computable evidence layer that can be directly utilised in data-driven workflows.

By synthesising literature-derived evidence through co-occurrence patterns [20], hierarchical research concepts [6], and network-based importance and specificity analysis [18], BiblioEngine captures rich external characteristics of genomic data that are not directly observable from experimental datasets alone. These characteristics include biological associations and the broader structure, intensity, and consistency of accumulated research evidence. The explicit modelling of such evidence enables cross-perspective analysis, where relationships among entities can be examined jointly across different biomedical dimensions while maintaining links to their original sources. As a result, the generated knowledge remains interpretable and verifiable, which is critical for high-stakes biomedical applications. These evidence-informed features can be directly incorporated into analytical pipelines to enhance model performance, robustness, and interpretability. BiblioEngine also provides on-demand evidence for decision support, allowing users to investigate specific genes, variants, or conditions through bibliometrics-driven analysis and domain-specific queries, while maintaining clear traceability to the underlying literature when deeper review is required (Table 1).

2.2 Gene–Disease Association Modelling

Gene–disease association is a central task in biomedical genomics as it links genomic variation to disease mechanisms, candidate genes, and downstream risk prediction. This task becomes more complex as diseases become increasingly multifactorial and the genetic contribution can range from direct high-penetrance effects to subtle polygenic influences. Disease settings also differ in data modality, cohort size, signal-to-noise characteristics, and the extent to which prior biological structure is available. These differences determine what kind of representation is needed, which form of inductive bias is helpful, and how readily a model can be interpreted or applied in practice. Model choice should therefore follow the statistical and biological properties of the task.

Graph-based modelling is particularly suitable when the underlying problem is relational. This approach treats genetic variants, genes, phenotypes, and supporting evidence as components of a structured multi-layer system. Geno-GCN utilises a genome-specific graph convolutional network for diabetes prediction, demonstrating that genome-informed graph construction can improve representation learning beyond existing clinical guidelines and conventional machine learning models [5]. Another example is heterogeneous graph learning for candidate gene prediction in endometriosis, where multi-relational disease–gene evidence is encoded explicitly in a graph framework to prioritise biologically plau-

Table 1. Representative studies in the translational genomics workflow.

Study	Application	Method	Role
BiblioEngine [19]	Disease knowledge mining	Bibliometrics + NLP	Structures biomedical literature into machine-readable evidence for disease genetic knowledge discovery and downstream analysis.
HGNN candidate gene prediction [15]	Endometriosis	Heterogeneous graph neural network	Prioritises candidate genes through heterogeneous disease–gene evidence.
Geno-GCN [5]	Diabetes	Genome-specific graph convolutional network	Uses genome-informed graph structure for disease risk prediction beyond conventional tabular models.
Deep polygenic prediction [17]	Cardiovascular disease	Deep neural network	Models nonlinear polygenic effects for cardiovascular risk stratification.
Multi-variant DNN [16]	Endometriosis	Deep multi-variant neural network	Captures complex multi-variant genomic effects for endometriosis prediction.
Deep cross networks [21]	Prostate cancer	Deep cross network	Learns higher-order genomic interactions for prostate cancer risk stratification.
Multi-model transfer learning [22]	Subtype classification	Transfer learning + genotypic analysis	Transfers complementary representations for subtype prediction under limited target data.
Deep transfer learning [14]	Diabetes complications	Deep transfer learning	Adapts external genomic representations to improve prediction under limited labelled data.
MetaGeno [23]	Ischaemic stroke risk	Chromosome-wise multi-task framework	Embeds chromosome-level biological knowledge into a multi-task risk model.
Cancer recommender [12]	Personalised recommendation	Genomic risk prediction + LLM-driven recommender	Extends genomic risk prediction into explainable recommendation and decision support.

sible candidates [15]. In both cases, the graph is constructed so that learning follows known relationships among variants, genes, and phenotypes that are difficult to express in tabular form. Thus, the candidate gene prioritisation and disease risk modelling preserve biologically meaningful structure.

Deep neural models are advantageous when the dominant challenge is high-dimensional multivariate effects and nonlinear interactions. Predictive performance depends on the capacity to learn complex combinations of genomic features without overly restrictive assumptions. This is illustrated by deep neural modelling for cardiovascular disease [17], extensive multi-variant modelling for endometriosis [16], and deep cross networks for prostate cancer risk prediction [21]. These studies show that the main value of deep architectures lies in their ability to represent distributed, nonlinear disease signals that are difficult to recover with conventional approaches. These methods also show that the choice of architecture can be tailored to the translational goal. For cardiovascular disease, the emphasis is on polygenic risk stratification. For endometriosis, the focus is on improved prediction in a complex and under-characterised condition. For prostate cancer, learning higher-order interactions that support more refined cancer risk modelling becomes the key task.

A further methodological step is to incorporate biological prior knowledge directly into architectural design. One approach is to build models that reflect statistical regularities in the data and the biological organisation of genomic information, so that the architecture itself becomes part of the modelling hypothesis. MetaGeno provides a representative example of a chromosome-wise multi-task genomic framework for ischaemic stroke risk prediction, modelling stroke jointly with correlated, modifiable risk factors [23]. By embedding chromosome-level biological knowledge directly into the architecture, the model is better aligned with the organisation of genetic information and more capable of representing shared and task-specific genomic effects across related phenotypes. These studies show that gene–disease modelling works best when the architecture matches the structure of the task. Graph-based methods are useful when relations among entities are central, deep multi-variant models are useful when nonlinear genomic effects dominate, and biologically informed architectures are useful when prior genomic structure should directly shape learning.

2.3 Transfer Learning Under Population and Cohort Shift

Portability across cohorts remains a major challenge in biomedical genomics. Models developed in one dataset often lose reliability in another because genomic prediction is sensitive to ancestry composition, recruitment strategy, phenotype definition, outcome prevalence, and data quality [8]. These differences determine whether a model can be used beyond the setting in which it was first developed. Transfer learning is therefore important for maintaining prediction quality when the training and target cohorts differ.

This issue is especially important in genomics because datasets are rarely balanced across populations. Large training resources are often dominated by particular ancestry groups or research environments, while target cohorts may be smaller, clinically heterogeneous, or sparsely labelled. Training only on the target cohort is often unstable, whereas direct reuse of a source model may fail because the underlying data distributions are different. Transfer learning provides a middle path. It enables the reuse of informative representations from related tasks or cohorts while allowing adaptation to the target setting. This is valuable because underrepresented populations often have the least data on which standard transfer assumptions can fail.

This problem is explored in two representative settings. First, multi-model transfer learning is applied to disease subtype classification, where related source models are integrated to improve prediction with limited target data [22]. In the seizure subtype classification study, transfer is treated as model integration rather than as the reuse of a single pretrained representation. By combining complementary representations across models, the framework is better able to capture subtype variation in highly imbalanced settings, including categories with fewer than 100 samples alongside major subtypes with more than 10,000 cases. This is particularly relevant for clinically heterogeneous cohorts, where labelled data are sparse for rare subtypes and direct application of a source model may not be sufficiently reliable.

Second, deep transfer learning is used to predict diabetes complications from genomic data under limited labelled target data [14]. The study transfers knowledge from Type 2 Diabetes to Type 1, where complication prediction is constrained by a much smaller cohort. Using UK Biobank data, a deep neural network is first trained on Type 2 genomic data to preserve representations relevant to diabetes and its associated complications. The pretrained network is then adapted to Type 1 through transfer learning with width and depth modifications [14]. The significance of this design lies in its practical relevance: it improves complication prediction in a data-limited target setting without assuming that Type 1 and Type 2 Diabetes share identical pathophysiology. This makes transfer learning a useful strategy for improving robustness and reducing reliance on large labelled target cohorts in clinically selective settings.

These models show that cross-cohort prediction is an important translational problem in genomic medicine. Transfer learning improves performance in small target cohorts by adapting knowledge learned from related source cohorts to data-limited target settings, whereas biologically informed architectures strengthen generalisation by embedding genomic structure directly into model design. This makes it possible to reduce dependence on large labelled target datasets, in turn, improve portability across heterogeneous populations. Transfer learning supports more reliable prediction where clinical and demographic variation cannot be ignored.

2.4 From Genomic Risk Prediction to Personalised Recommendation

Risk prediction is a necessary capability in genomic medicine, but it is not a complete translational output. A model score identifies elevated susceptibility, yet clinical utility depends on a controlled score-to-action mapping. A given risk level should be followed by targeted screening, confirmatory testing, counselling, lifestyle interventions, or referral to a medical specialist. From a recommender-systems viewpoint, this is a constrained decision-support problem in which a predictive model provides risk estimates, and a downstream policy layer selects and communicates guideline-compatible actions under safety and governance constraints.

Genomics-enhanced cancer risk prediction for personalised recommendation provides a representative instance of this design pattern [12]. In this work, risk inference is coupled to recommendation generation through an explicit interpretability bridge. A mixture-of-experts risk model produces patient-specific cancer risk estimates, SHAP analysis identifies the SNPs/genes and non-genetic covariates most responsible for the score, and these high-impact factors are used to parameterise the recommendation request. Recommendation is therefore conditioned on the drivers of the prediction, rather than being a generic text template appended to a classifier. The recommender system also grounds prompts in curated medical evidence, which makes recommendations easier to review and check clinically."

By applying the proposed workflow to recommender systems, recommendation generation can be treated as grounded generation. Retrieval-augmented generation supports this formulation by conditioning outputs on an explicit evidence set, improving factual consistency and enabling provenance-aware review [10]. Clinical deployment requires evaluation beyond fluency, with attention to factual consistency, bias, and likelihood of harm, consistent with multi-axis human evaluation frameworks for medical LLM outputs [13]. Foundation-scale regulatory models such as AlphaGenome provide richer variant-level priors [2] that can strengthen both risk modelling and evidence retrieval. Thus, the translational challenge is to produce recommendations that remain evidence-grounded, auditable, and compatible with real care pathways.

3 Discussion and Future Directions

The studies reviewed in this paper motivate a workflow-oriented view of translational genomics in which evidence-traceable knowledge mining, gene–disease association modelling, transfer learning, and recommendation are integrated. Under this workflow, AI for biomedical genomics centres on four linked tasks: constructing structured evidence representations, designing models that reflect biological and statistical structure, adapting representations across cohorts and populations, and transforming predictive outputs into decision-support signals. The core design principle is the alignment between the choice of method and the translational objective.

The results indicate that no single model class is sufficient across translational settings. Instead, performance and utility depend on matching the modelling strategy to the structure of the task. Graph-based methods are appropriate when inference depends on relationships among genes, variants, phenotypes, and supporting evidence. Deep multi-variant architectures are better suited to distributed, nonlinear disease signals in high-dimensional settings. Transfer learning is important when cohorts differ in size, composition, or endpoint definition. Recommendation methods are necessary when predictive outputs need to be converted into actionable guidance. The value of this workflow is that it makes model choice explicit from evidence construction through to recommendation.

A critical challenge is deployment at scale. Genomic datasets remain unevenly distributed across populations, which limits portability and increases the risk of ancestry-related bias. Outcome definitions are also inconsistent across studies, making evaluation and cross-study comparison difficult. Literature-derived evidence improves coverage and traceability, but it also inherits publication bias, incomplete reporting, and context-dependent claims. These constraints make rigorous evaluation essential before deployment. Models intended for deployment should therefore be assessed for robustness across cohorts, calibration under domain shift, and the quality of downstream outputs, particularly when recommendations are part of the pipeline.

Deployment also depends on factors outside the model architecture. In healthcare environments, implementation requires trust, validation, and user competence. Public concern about automated systems has increased in the era of generative models, raising the bar for transparency and responsible communication [1]. For academic and industry leadership, this implies that methodological advances should be paired with clearer evaluation standards, explicit statements of model scope and limitations, and training for clinicians and other end users who interpret system outputs. From a systems perspective, the goal is to support expert decision-making by developing models that identify relevant evidence and assist analysis when expert capacity is limited or when task complexity exceeds the reliability of manual workflows.

These considerations make women's health a strong target for future work. Many conditions across the female life course remain under-characterised, clinically heterogeneous, and poorly supported by current data resources [11]. This indicates that fragmented evidence, sparse labels, cohort shift, and limited decision support often co-occur. The workflow developed in this paper is well-suited to these conditions and applications. Evidence-traceable knowledge mining can support structured synthesis of rapidly evolving literature in areas such as endometriosis, polycystic ovary syndrome, preeclampsia, and gestational diabetes. Disease-adapted modelling is able to capture heterogeneous genomic mechanisms in fertility-related conditions and variability in assisted reproduction outcomes [7]. Transfer learning can improve robustness in data-limited cohorts, particularly for rare pregnancy complications and under-studied gynaecological disorders. Recommendation connects prediction with prevention, monitoring, treatment selection, and longitudinal care planning across these settings. Future work should prioritise women's health where both technical impact and translational value are likely to be high.

4 Conclusion

This paper presents a translation-centred perspective on AI in biomedical genomics organised around four linked methodological components: evidence-traceable knowledge mining, gene–disease association modelling, transfer learning across populations and cohorts, and personalised recommendation. Translational genomics advances through methods designed for the structure and constraints of real clinical problems. Computational models contribute by capturing biological structure, adapting across heterogeneous data settings, and generating outputs that support downstream interpretation and action. This workflow integrates evidence synthesis, modelling, adaptation, decision support, and recommendations.

Across the studies reviewed here, these components address genomic tasks that require traceable evidence, problem-matched architectures, robustness under cohort variation, and decision-relevant outputs. Knowledge mining converts fragmented literature into structured evidence resources. Disease-adapted modelling captures the statistical and biological characteristics of specific genomic tasks. Transfer learning improves robustness in data-limited and heterogeneous cohorts. Recommendation translates prediction into communication, prevention, monitoring, and care planning. This perspective is especially relevant to women's health. In future works, we plan to adapt the proposed workflow in under-characterised conditions across the women's life course, including pregnancy, mental health, menopause, and ageing. This would present clear opportunities for evidence-grounded, robust, and clinically relevant methodological development.

Acknowledgments. This work was supported by the Australian Research Council Linkage Project (LP210100414).

Disclosure of Interests. The authors declare no competing interests.

References

1. Afroogh, S., Akbari, A., Malone, E., Kargar, M., Alambeigi, H.: Trust in AI: progress, challenges, and future directions. Hum. Soc. Sci. Commun. **11**(1), 1568 (2024). https://doi.org/10.1057/s41599-024-04044-8
2. Avsec, Ž, Latysheva, N., Cheng, J., et al.: Advancing regulatory variant effect prediction with alphagenome. Nature **649**(8099), 1206–1218 (2026). https://doi.org/10.1038/s41586-025-10014-0
3. Dattani, S., Howard, D.M., Lewis, C.M., Sham, P.C.: Clarifying the causes of consistent and inconsistent findings in genetics. Genet. Epidemiol. **46**(7), 372–389 (2022)
4. Duong, D., Solomon, B.D.: Artificial intelligence in clinical genetics. Eur. J. Hum. Genet. **33**(3), 281–288 (2025). https://doi.org/10.1038/s41431-024-01782-w
5. Guo, K., Lin, H., Grosser, M., Zhang, G., Lu, J.: Geno-GCN: a genome-specific graph convolutional network for diabetes prediction. In: Proceedings of the 46th

Annual International Conference of the IEEE Engineering in Medicine and Biology Society (EMBC), pp. 1–4. IEEE (2024). https://doi.org/10.1109/EMBC53108.2024.10782451
6. Guo, K., et al.: Artificial intelligence-driven biomedical genomics. Knowl.-Based Syst. **279**, 110937 (2023). https://doi.org/10.1016/j.knosys.2023.110937
7. Hamel, M., Dufort, I., Robert, C., Léveillé, M.C., Leader, A., Sirard, M.A.: Genomic assessment of follicular marker genes as pregnancy predictors for human IVF. MHR: Bas. Sci. Reprod. Med. **16**(2), 87–96 (2009)
8. Kachuri, L., et al.: Polygenic risk methods in diverse populations (PRIMED) consortium methods working group, Witte, J.S., Ge, T.: Principles and methods for transferring polygenic risk scores across global populations. Nat. Rev. Genetics **25**(1), 8–25 (2024). https://doi.org/10.1038/s41576-023-00637-2
9. Kelly, C.J., Karthikesalingam, A., Suleyman, M., Corrado, G., King, D.: Key challenges for delivering clinical impact with artificial intelligence. BMC Med. **17**(1), 195 (2019). https://doi.org/10.1186/s12916-019-1426-2
10. Lewis, P., et al.: Retrieval-augmented generation for knowledge-intensive NLP tasks. Adv. Neural Inf. Process. Syst. (NeurIPS) (2020). https://arxiv.org/abs/2005.11401
11. Lu, J., Dawson, A., Guo, K.: AI for women's health: framework and focus areas. In: Companion Proceedings of the ACM Web Conference 2025, pp. 845–848. ACM (2025). https://doi.org/10.1145/3701716.3718981
12. Lu, K., et al.: Genomics-enhanced cancer risk prediction for personalized llms-driven healthcare recommender systems. ACM Trans. Inf. Syst. **43**(6) (2025). https://doi.org/10.1145/3745022
13. Singhal, K., et al.: Large language models encode clinical knowledge. Nature **620**, 172–180 (2023). https://doi.org/10.1038/s41586-023-06291-2
14. Soo, Z., et al.: A deep transfer learning approach for predicting diabetes complications using genomic data. In: Health Information Science, pp. 140–151. Springer (2024). https://doi.org/10.1007/978-981-96-5597-7_13
15. Soo, Z., et al.: A heterogeneous graph neural network for candidate gene prediction in endometriosis. J. Biomed. Inf. 105028 (2026). https://doi.org/10.1016/j.jbi.2026.105028
16. Soo, Z., et al.: An extensive multi-variant deep neural network approach to enhance genomic prediction of endometriosis. Neurocomputing **656**, 131496 (2025). https://doi.org/10.1016/j.neucom.2025.131496
17. Soo, Z., Lin, H., Yang, Y., Grosser, M., Zhang, Y., Lu, J.: Deep neural network-empowered polygenic disease prediction on cardiovascular diseases. In: Proceedings of the 18th International Conference on Intelligent Systems and Knowledge Engineering (ISKE), pp. 309–315. IEEE (2023). https://doi.org/10.1109/ISKE60036.2023.10481444
18. Wu, M., et al.: Profiling covid-19 genetic research: a data-driven study utilizing intelligent bibliometrics. Front. Res. Metrics Anal. **6**, 683212 (2021)
19. Wu, M., Zhang, Y., Lin, H., Grosser, M., Zhang, G., Lu, J.: Biblioengine: an AI-empowered platform for disease genetic knowledge mining. In: Health Information Science, pp. 187–198. Springer (2023). https://doi.org/10.1007/978-981-99-7108-4_16
20. Wu, M., Zhang, Y., Zhang, G., Lu, J.: Exploring the genetic basis of diseases through a heterogeneous bibliometric network: a methodology and case study. Technol. Forecast. Soc. Chang. **164**, 120513 (2021)

21. Xu, H., Zhang, G., Lin, H., Grosser, M., Lu, J.: Deep cross networks for intelligent prostate cancer risk prediction. In: Web Information Systems Engineering - WISE 2025, Lecture Notes in Computer Science, vol. 16367. Springer, Singapore (2026). conference chapter; add page range/DOI if available
22. Yang, Y., Guo, K., Fang, Z., Lin, H., Grosser, M., Lu, J.: Multi-model transfer learning and genotypic analysis for seizure type classification. In: Health Information Science, pp. 223–234. Springer (2023). https://doi.org/10.1007/978-981-99-7108-4_19
23. Yang, Y., et al.: Metageno: a chromosome-wise multi-task genomic framework for ischaemic stroke risk prediction. Briefings Bioinf. **26**(4), bbaf348 (2025). https://doi.org/10.1093/bib/bbaf348

Properties of Inclusion Relationship Between Clause Sets in Propositional Logic

Xiaomei Zhong[1,2(✉)], Hanhui Zhan[1], and Yang Xu[1,2]

[1] School of Mathematics, Southwest Jiaotong University, Chengdu 610031, China
[2] National-Local Joint Engineering Laboratory of System Credibility Automatic Verification, Southwest Jiaotong University, Chengdu, Sichuan, China
zhongxm2013@home.swjtu.edu.cn

Abstract. Theorem machine proving holds a fundamental, historical, and interdisciplinary position in the field of artificial intelligence. It is not only one of the core directions of early AI research but also serves as a crucial technical support for modern intelligent systems to achieve logical reasoning and symbolic processing capabilities. The automatic reasoning method based on resolution is a crucial approach for theorem machine proving, and it has been widely applied in multiple domains. To further enhance the efficiency of resolution reasoning, this paper focuses on the research of inclusion relationships between clause sets in propositional logic, derives some properties of inclusion relationships, and illustrates these properties through examples. These conclusions provide certain theoretical support for further establishing corresponding resolution reasoning methods based on inclusion relationships.

Keywords: Inclusion relationship · Clause set · Propositional logic

1 Introduction

Artificial intelligence is a branch of computer science, and it is dedicated to building intelligent systems capable of performing tasks that typically require human intelligence. With the development of computing power, algorithms, and technology, artificial intelligence has been widely applied in various fields and has given rise to a rich variety of research directions. Theoretical machine proof holds a fundamental, historical, and interdisciplinary position in the field of artificial intelligence. It is not only one of the core directions of early AI research but also serves as a crucial technical support for modern intelligent systems to achieve logical reasoning and symbolic processing capabilities.

In the field of automated theorem proving, since Robinson proposed the resolution principle in 1965 [13], deductive reasoning methods based on the resolution principle have undergone development by numerous researchers and have yielded abundant achievements. For instance, from 1967 to 1992, three classic resolution reasoning methods emerged: semantic resolution, linear resolution, and lock resolution [1,2,6,14]. The core idea of these three types of resolution

J. Lu et al. (Eds.): FLINS-ISKE 2026, LNCS 16752, pp. 559–569, 2027.
https://doi.org/10.1007/978-981-92-2369-5_34

inference methods is to select the literals or clauses involved in the resolution through certain strategies, so as to minimize the generation of redundant literals and clauses during the resolution process, thereby achieving the goal of improving the efficiency of resolution inference. In addition, further research has been conducted on the compatibility among these three types of resolution inference methods, yielding some relevant conclusions [7–10]. Although these three types of resolution inference methods can improve the efficiency of resolution inference to some extent, efficiency remains the core issue in the process of resolution inference, severely restricting their practical application. Therefore, since the 1990s, researchers have been conducting studies on this issue and have achieved a series of results. For example, in 1994, L. Backmair and H. Ganzinger proposed the superposition rule based on mediation reasoning [3]. The core of this rule lies in reducing redundant reasoning steps through the combination of ordered resolution and equality elimination. It can effectively accommodate various redundancy elimination rules. In 1998, Chen Bin introduced the support set strategy into generalized resolution and proposed the generalized support set resolution method [4]. In 2006, Xia Shifen [15] proposed a method for matrixizing clause sets in propositional logic systems, and subsequently introduced four rules and elementary transformation rules for clause set matrices. These rules enable the simplification of clause set matrices, thereby reducing the difficulty of problem solving. In 2018, Xu Yang et al. proposed the contradiction separation rule based on multi-clause collaborative participation in deduction [16], establishing a deduction theoretical system based on contradiction separation. Originally, only two clauses could be resolved, but under this system, multiple clauses can be dynamically and collaboratively resolved. In 2019, Liu Qinghua proposed SOL resolution [12], which effectively addressed the bottlenecks in efficiency and scalability of traditional resolution methods by introducing dynamic symbol ordering, conflict learning, and path constraint mechanisms. This resolution method is an improvement from ordered linear resolution, further enhancing the efficiency of linear resolution. In 2021, Zhong Jian proposed a dynamic multi-clause collaborative deductive algorithm architecture based on the theory of contradiction separation deductive reasoning [19]. In 2026, Xu Yang et al. extended the theory of contradiction-separation-based deduction by systematically constructing and applying standard contradictions for multi-clause deduction and automated theorem generation. The methods for building the maximum triangular and triangular-type standard contradictions are proposed [17]. Most of the above resolution inference methods are universally applicable. Their processing starts from the problem to be solved itself, and conclusions are drawn through multiple resolution inferences. The resolution process does not utilize relevant information from other known conclusions in the same field. To further enhance the efficiency of resolution reasoning, it is imperative to investigate how to utilize relevant information from known conclusions to ascertain the correctness of unknown problems. To achieve this objective, in 2024, Ji Ruixin [5] introduced the definition of inclusion relationship between clause sets and separately explored the connections between the inclusion relationship and the satisfiability of clause sets, as well as the connections between the inclusion relationship and the unsatisfiability of clause sets in propositional logic. Building upon this foundation, this paper focuses on exploring the properties of the inclusion relationship between clause sets in propositional logic.

2 Preliminary

This section enumerates some fundamental knowledge in propositional logic covered in this paper, as well as the definition of the matrix corresponding to a clause set. For more detailed information, readers are referred to reference [11,15].

A proposition is a declarative statement that can be judged as true or false. The truth value of a proposition can only take two values: true (or represented by 1), indicating a true proposition; false (or represented by 0), indicating a false proposition.

Definition 1. *In propositional logic,* $x_1, x_2, \cdots$ *are generally used to represent propositional variables, and propositional variables are referred to as atoms. The formula is a symbol string recursively defined as follows:*

(1) Atoms are formulas;

(2) If G *and* H *are formulas, then* $\neg G, G \vee H, G \wedge H, G \rightarrow H and G \leftrightarrow H$ *are formulas;*

(3) All formulas are symbol strings obtained by finite use of (1) and (2).

Definition 2. *Atoms or non-atoms are called literals.*

Definition 3. *The disjunction of a finite number of literals is called a clause, the conjunction of a finite number of literals is called a phrase, and the clause set* $S = \{C_1, C_2, \cdots, C_m\}$ *is the logical formula* $C_1 \wedge C_2 \wedge \cdots \wedge C_m$*, where* $C_1, C_2, \cdots, C_m$ *are clauses in propositional logic.*

Definition 4. *The disjunction of a finite number of phrases is called a disjunctive normal form, denoted as*

$$C_1 \vee C_2 \vee \cdots \vee C_m,$$

where C_i *is a phrase,* $i = 1, \cdots, m$.

Definition 5. *The conjunction of a finite number of clauses is called a conjunctive normal form, denoted as*

$$C_1 \wedge C_2 \wedge \cdots \wedge C_m,$$

where C_i *is a clause,* $i = 1, \cdots, m$.

Definition 6. *If a propositional logic formula is true under any assignment, it is called a tautology (tautological formula).*

Definition 7. P *is an atom. Literal* P *and its negation* $\neg P$ *are called complementary pairs.*

Definition 8. *A and B are two propositional logical formulas. If the truth values of A and B are equal under any assignment, it is said that A and B are equivalent, denoted as* $A = B$.

Definition 9. *S is a clause set in propositional logic. If there exists an assignment v such that the truth value of S under v is true, then v is called satisfiable; otherwise, it is called unsatisfiable.*

Definition 10. $S = \{C_1, C_2, \cdots, C_m\}$ *is a clause set in propositional logic. If S contains n propositional variables, then S corresponds to an* $m \times n$ *matrix with elements -1, 0 and 1, where each row of the matrix represents a clause and each column represents a propositional variable. If a clause contains the propositional variable, the corresponding matrix element is 1; If the clause does not contain the propositional variable, the corresponding matrix element is 0; If the clause contains a negation of the propositional variable, the corresponding matrix element is -1.*

For example, $S = \{x_1 \vee x_2 \vee x_3, x_2 \vee \neg x_3, x_1 \vee \neg x_3\}$ is a clause set in proposition logic, where x_1, x_2, x_3 are propositional variable. Then S corresponds to the matrix S' and the elements of S' are as follows.

$$S' = \begin{bmatrix} 1 & 1 & 1 \\ 0 & 1 & -1 \\ 1 & 0 & -1 \end{bmatrix}.$$

3 Properties of Inclusion Relationship

This section discusses the inclusion relationship between clause sets using their corresponding matrices, derives some properties of the inclusion relationship, and illustrates these properties through examples. The following Definitions 11, 12 and 13, Theorem 1 and Lemma 1 are all existing conclusions. For more detailed information, readers are referred to reference [5,11].

Definition 11. *Let* S_1 *and* S_2 *be two clause sets in propositional logic.* M_1 *and* M_2 *are the matrix representations corresponding to* S_1 *and* S_2*, respectively.* M_1' *is the matrix obtained by swapping some rows and columns of* M_1*. If* M_2 *is a sub block of a block matrix of* M_1'*, and* M_2 *satisfies the following form:*

$$M_1' = \begin{bmatrix} M_2 & 0 \\ A & B \end{bmatrix},$$

it is called that clause set S_1 *truly contains clause set* S_2*, where A is an arbitrary matrix, and B is a non-zero matrix.*

Theorem 1. *Let* S_1 *and* S_2 *be two clause sets in propositional logic.* M_1 *and* M_2 *are the matrix representations corresponding to* S_1 *and* S_2*, respectively.* M_1' *is the matrix obtained by swapping some rows and columns of* M_1*. If* $M_2 = M_1'$*, then the clause sets A and B are equal, that is, A=B.*

Definition 12. *If $x_1, x_2, \cdots, x_n$ are different proposition variables in the proposition logic, and y_i is a proposition variable different from x_i , where i =1, 2,..., n, then the finite set*

$$\{y_1/x_1, y_2/x_2, \cdots, y_n/x_n\}$$

is called a renaming replacement.

Theorem 2. *Let S_1 and S_2 be two clause sets in propositional logic. Then clause set S_1 truly contains clause set S_2 if and only if there exists a renaming replacement θ such that every clause in S_2 is a clause in ${S_1}^\theta$*

Proof. Let M_1 and M_2 be the matrix representations corresponding to S_1 and S_2, respectively.

($\Rightarrow$) If clause set S_1 truly contains clause set S_2, then according to Definition 11, the matrix ${M_1}'$ obtained by swapping some rows and columns of M_1 satisfies the following form:

$${M_1}' = \begin{bmatrix} M_2 & 0 \\ A & B \end{bmatrix}.$$

The clause set corresponding to ${M_1}'$ is denoted as ${S_1}'$. Therefore, there exists the renaming replacement θ to make ${S_1}^\theta = {S_1}'$. According to matrix ${M_1}'$, each clause in S_2 is a clause in ${S_1}^\theta$.

($\Leftarrow$) Assuming there is a renaming replacement θ such that each clause in clause set S_2 is a clause of ${S_1}^\theta$. The clause set corresponding to ${S_1}^\theta$ is denoted as W. Then W=M_1 and the matrix ${M_1}'$ obtained by swapping some rows of W satisfies the following form:

$${M_1}' = \begin{bmatrix} M_2 & 0 \\ A & B \end{bmatrix}.$$

According to Definition 11, clause set S_1 truly contains clause set S_2. This concludes the proof.

Corollary 1. *Let S_1 and S_2 be two clause sets in propositional logic. If clause set S_1 truly contains clause set S_2, then there exists clause set ${S_1}'$ that is equivalent to S_1 and each clause in S_2 is a clause of ${S_1}'$.*

Proof. Because clause set S_1 truly contains clause set S_2, it can be concluded that there exists a renaming replacement θ such that every clause in S_2 is a clause in ${S_1}^\theta$ by Theorem 2. Since $S_1 = {S_1}^\theta$, the conclusion holds.

Example 1. Let $S_1 = C_1, C_2\}$ and $S_2 = \{D_1, D_2, D_3, D_4\}$ be two clause sets in propositional logic, where

$$\begin{aligned} C_1 &= y_1 \vee y_2, \\ C_2 &= y_1 \vee \neg y_2, \\ D_1 &= x_1 \vee x_3, \\ D_2 &= x_1 \vee \neg x_3, \\ D_3 &= x_1 \vee \neg x_2 \vee x_3 \vee \neg x_4, \\ D_4 &= \neg x_1 \vee x_2 \vee x_4, \end{aligned}$$

and $y_1, y_2, x_1, x_2, x_3, x_4$ are proposition variables. M_1 andM_2 are the matrix representations corresponding toS_1 and S_2, respectively.

Therefore, $M_1 = \begin{bmatrix} 1 & 1 \\ 1 & -1 \end{bmatrix}$, $M_2 = \begin{bmatrix} 1 & 0 & 1 & 0 \\ 1 & 0 & -1 & 0 \\ 1 & -1 & 1 & -1 \\ -1 & 1 & 0 & 1 \end{bmatrix}$

After swapping the second and third columns of M_2, a new matrix M_2' is obtained. The elements of M_2' are as follows.

$$M_2' = \begin{bmatrix} 1 & 1 & 0 & 0 \\ 1 & -1 & 0 & 0 \\ 1 & 1 & -1 & -1 \\ -1 & 0 & 1 & 1 \end{bmatrix}.$$

According to M_2', M_1 is composed of the elements at the intersection of the first 2 rows and first 2 columns of M_2'. Therefore, clause set S_2 contains clause set S_1 by Definition 11.

Let θ be the renaming replacement $\{y_1/x_1, y_2/x_3\}$. In fact, after replacing the proposition variables in clause set S_2 with θ , the clause set

$$S_2{}' = D_1{}', D_2{}', D_3{}', D_4{}'\}$$

is obtained, where

$$\begin{aligned} D_1{}' &= y_1 \vee y_2, \\ D_2{}' &= y_1 \vee \neg y_2, \\ D_3{}' &= y_1 \vee x_2 \vee y_2 \vee \neg x_4, \\ D_4{}' &= \neg y_1 \vee x_3 \vee x_4. \end{aligned}$$

Since clauses C_1, C_2 are equal to clauses $D_1{}', D_2{}'$ respectively, clause set S_2 contains clause set S_1 by Theorem 2.

Definition 13. *Let S_1 and S_2 be two clause sets in propositional logic. M_1 and M_2 are the matrix representations corresponding to S_1 and S_2, respectively. M_1' is the matrix obtained by swapping some rows and columns of M_1. If M_2 is a sub block of a block matrix of M_1', and M_2 satisfies the following form:*

$$M_1' = \begin{bmatrix} M_2 & C \\ A & B \end{bmatrix},$$

it is called that clause set S_1 contains clause set S_2, where A and B are matrices that are not both zero matrices at the same time, and C is a non-zero matrix.

Theorem 3. *Let S_1 and S_2 be two clause sets in propositional logic. Then clause set S_1 contains clause set S_2 if and only if there exists a renaming replacement θ such that each clause of S_2 is a partial clause of a clause in ${S_1}^{\theta}$.*

Proof. Let M_1 and M_2 be the matrix representations corresponding to S_1 and S_2, respectively.

($\Rightarrow$) If clause set S_1 contains clause set S_2, then according to Definition 13, the matrix $M_1{}'$ obtained by swapping some rows and columns of M_1 satisfies the following form:

$$M_1{}' = \begin{bmatrix} M_2 & C \\ A & B \end{bmatrix},$$

where C is a non-zero matrix. The clause set corresponding to $M_1{}'$ is denoted as $S_1{}'$. Therefore, there exists the renaming replacement θ to make $S_1{}^{\theta}{=}S_1{}'$. According to matrix $M_1{}'$, each clause in S_2 is a partial clause of a clause in $S_1{}^{\theta}$.

($\Leftarrow$) Assuming there is a renaming replacement θ such that each clause in clause set S_2 is a partial clause of a clause in $S_1{}^{\theta}$. The clause set corresponding to $S_1{}^{\theta}$ is denoted as W. Then $W = M_1$ and the matrix $M_1{}'$ obtained by swapping some rows of W satisfies the following form:

$$M_1{}' = \begin{bmatrix} M_2 & C \\ A & B \end{bmatrix},$$

where C is a non-zero matrix. According to Definition 13, clause set S_1 contains clause set S_2.

This concludes the proof.

Corollary 2. *Let S_1 and S_2 be two clause sets in propositional logic. If clause set S_1 contains clause set S_2, then there exists clause set $S_1{}'$ that is equivalent to S_1 and each clause in S_2 is a partial clause of a clause in $S_1{}'$.*

Proof. Because clause set S_1 contains clause set S_2, it can be concluded that there exists a renaming replacement θ such that every clause in S_2 is a partial clause of a clause in $S_1{}^{\theta}$ by Theorem 3. Since $S_1 = S_1{}^{\theta}$, the conclusion holds.

Example 2. Let $S_1 = C_1, C_2, C_3, C_4, C_5\}$ and $S_2 = \{D_1, D_2, D_3, D_4, D_5, D_6\}$ be two clause sets in propositional logic, where

$$\begin{aligned} C_1 &= x_1 \vee x_2, \\ C_2 &= x_1 \vee \neg x_2, \\ C_3 &= \neg x_1 \vee x_2, \\ C_4 &= \neg x_1 \vee \neg x_2, \\ C_5 &= x_1 \vee x_3, \end{aligned}$$

$$\begin{aligned} D_1 &= y_1 \vee y_2 \vee y_3, \\ D_2 &= y_1 \vee \neg y_2 \vee \neg y_3, \\ D_3 &= \neg y_1 \vee y_2 \vee y_3 \vee y_5, \\ D_4 &= \neg y_1 \vee \neg y_2 \vee \neg y_3, \\ D_5 &= y_1 \vee \neg y_2 \vee y_3 \vee \neg y_4, \\ D_6 &= y_1 \vee y_4 \vee \neg y_5, \end{aligned}$$

and $x_1, x_2, x_3, y_1, y_2, y_3, y_4, y_5$ are proposition variables. M_1 and M_2 are the matrix representations corresponding to S_1 and S_2, respectively.

Therefore, $M_1 = \begin{bmatrix} 1 & 1 & 0 \\ 1 & -1 & 0 \\ -1 & 1 & 0 \\ -1 & -1 & 0 \\ 1 & 0 & 1 \end{bmatrix}$, $M_2 = \begin{bmatrix} 1 & 1 & 1 & 0 & 0 \\ 1 & -1 & -1 & 0 & 0 \\ -1 & 1 & 1 & 0 & 1 \\ -1 & -1 & -1 & 0 & 0 \\ 1 & -1 & 1 & -1 & 0 \\ 1 & 0 & 0 & 1 & -1 \end{bmatrix}$.

After swapping the second and third columns of M_2, a new matrix M_2* is obtained. The elements of M_2* are as follows.

$$M_2* = \begin{bmatrix} 1 & 1 & 1 & 0 & 0 \\ 1 & -1 & -1 & 0 & 0 \\ -1 & 1 & 1 & 0 & 1 \\ -1 & -1 & -1 & 0 & 0 \\ 1 & 1 & -1 & -1 & 0 \\ 1 & 0 & 0 & 1 & -1 \end{bmatrix}.$$

Then further swap the third and fourth columns of M_2* to obtain matrix M_2** and the elements of M_2** are as follows.

$$M_2** = \begin{bmatrix} 1 & 1 & 0 & 1 & 0 \\ 1 & -1 & 0 & -1 & 0 \\ -1 & 1 & 0 & 1 & 1 \\ -1 & -1 & 0 & -1 & 0 \\ 1 & 1 & -1 & -1 & 0 \\ 1 & 0 & 1 & 0 & -1 \end{bmatrix}.$$

Finally, swapping the 4th and 5th rows of M_2** yields matrix$M_2{}'$ and the elements of $M_2{}'$ are as follows.

$$M_2{}' = \begin{bmatrix} 1 & 1 & 0 & 1 & 0 \\ 1 & -1 & 0 & -1 & 0 \\ -1 & 1 & 0 & 1 & 1 \\ -1 & -1 & 0 & -1 & 0 \\ 1 & 0 & 1 & 0 & -1 \\ 1 & 1 & -1 & -1 & 0 \end{bmatrix}.$$

According to $M_2{}'$, M_1 is composed of the elements at the intersection of the first 5 rows and first 3 columns of $M_2{}'$. Therefore, clause set S_2 contains clause set S_1 by Definition 13.

Let θ be the renaming replacement$\{x_1/y_1, x_2/y_2, x_3/y_3\}$. In fact, after replacing the proposition variables in clause set S_2 with θ, the clause set $S_2{}^{\theta} = D_1{}', D_2{}', D_3{}', D_4{}', D_5{}', D_6{}'\}$ is obtained, where

$$\begin{aligned} D_1{}' &= x_1 \vee x_2 \vee x_3, \\ D_2{}' &= x_1 \vee \neg x_2 \vee \neg x_3, \\ D_3{}' &= \neg x_1 \vee x_2 \vee x_3 \vee y_5, \\ D_4{}' &= \neg x_1 \vee \neg x_2 \vee \neg x_3, \\ D_5{}' &= x_1 \vee \neg x_2 \vee x_3 \vee \neg y_4, \\ D_6{}' &= x_1 \vee y_4 \vee \neg y_5. \end{aligned}$$

Since each clause of S_1 is a partial clause of a clause in $S_1{}^{\theta}$, clause set S_2 contains clause set S_1 by Theorem 3.

Lemma 1. *Let S_1 and S_2 be two clause sets in propositional logic and S_1 truly contains S_2. If S_1 is satisfiable, then S_2 is satisfiable.*

Theorem 4. *Let S_1 be a clause set in propositional logic. The matrix corresponding to S_1 is M_1. If M_1' is the matrix obtained by swapping some rows and columns of M_1, and M_1' satisfies the following form:*

$$M_1{}' = \begin{bmatrix} M_2 & O \\ O & M_3 \end{bmatrix},$$

then S_1 is satisfiable if and only if both S_2 and S_3 are satisfiable, where O is a zero matrix, S_2 and S_3 are the clause sets corresponding to M_2 and M_3, respectively.

Proof. ($\Rightarrow$) According to Definition 11, clause set S_1 truly contains clause sets S_2 and S_3 . If S_1 is satisfiable, then clause sets S_2 and S_3 are satisfiable by Lemma 1.

($\Leftarrow$) If clause sets S_2 and S_3 are satisfiable, then there exist assignments v_1 and v_2 such that $v_1(S_2) = 1, v_2(S_3) = 1$. According to the structure of M_1', S_2 and S_3 do not have the same proposition variable. Therefore, it can be assumed that the proposition variables of S_2 is $x_1, \cdots, x_m$, and the proposition variables of S_3 is $x_{m+1}, \cdots, x_n$. So the proposition variables of S_1 are $x_1, \cdots, x_n$.

Furthermore, there exists an assignment v of S_1: $v(x_1) = v_1(x_2), \cdots, v(x_m) = v_1(x_m)$,$v(x_{m+1}) = v_2(x_{m+1}), \cdots, v(x_n) = v_2(x_n)$, such that $v(S_1) = v_1(S_2) \wedge v_2(S_3) = 1$. Therefore, S_1 is satisfiable.

This concludes the proof.

Example 3. Let $S_1 = \{C_1, C_2, C_3, C_4\}$ be a clause set in propositional logic, where

$$C_1 = x_1 \vee \neg x_3,$$
$$C_2 = \neg x_2 \vee x_4,$$
$$C_3 = \neg x_1 \vee x_3,$$
$$C_4 = x_2 \vee \neg x_4,$$

and x_1, x_2, x_3, x_4 are proposition variables. M_1 is the matrix corresponding to S_1.

Therefore,

$$M_1 = \begin{bmatrix} 1 & 0 & -1 & 0 \\ 0 & -1 & 0 & 1 \\ -1 & 0 & 1 & 0 \\ 0 & 1 & 0 & -1 \end{bmatrix}.$$

After swapping the second and third rows of M_1, a new matrix M_1* is obtained. The elements of M_1* are as follows.

$$M_1* = \begin{bmatrix} 1 & 0 & -1 & 0 \\ -1 & 0 & 1 & 0 \\ 0 & -1 & 0 & 1 \\ 0 & 1 & 0 & -1 \end{bmatrix}.$$

Then further swap the second and third columns of M_1* to obtain matrix M_1' and The elements of M_1' are as follows.

$$M_1' = \begin{bmatrix} 1 & -1 & 0 & 0 \\ -1 & 1 & 0 & 0 \\ 0 & 0 & -1 & 1 \\ 0 & 0 & 1 & -1 \end{bmatrix}.$$

Furthermore, M_1' can be written in the following form

$$M_1' = \begin{bmatrix} M_2 & O \\ O & M_3 \end{bmatrix}.$$

Let S_2 and S_3 be the clause sets corresponding to M_2 and M_3 respectively, x_1, x_2 be the proposition variables of S_2, and x_3, x_4 be the proposition variables of S_3. According to M_2 and M_3, it can be concluded that $S_2 = \{x_1 \vee \neg x_2, \neg x_1 \vee x_2\}$, $S_3 = \{\neg x_3 \vee x_4, x_3 \vee \neg x_4\}$, $S_1{}' = \{x_1 \vee \neg x_2, \neg x_1 \vee x_2, \neg x_3 \vee x_4, x_3 \vee \neg x_4\}$ and $S_1 = S_1{}'$.

Meanwhile, there exist assignments v_1 and v_2 for S_2 and S_3 respectively: $v_1(x_1) = v_1(x_2) = 1$,$v_2(x_3) = v_2(x_4) = 1$, such that $v_1(S_2) = v_2(S_3) = 1$. So S_2 and S_3 are satisfiable. Therefore, S_1 is satisfiable by Theorem 4.

In fact, there exists an assignment v of $S_1{}' : v(x_1) = v(x_2) = v(x_3) = v(x_4) = 1$, such that $v(S_1{}') = 1$. So $S_1{}'$ is satisfiable, which means S_1 is also satisfiable.

Acknowledgments. This work is partially supported by the Youth Fund Project for Humanities and Social Sciences Research of Ministry of Education of the People's Republic of China (Grant No. 20XJCZH016) and National Natural Science Foundation of China (Grant No. 12271319).

References

1. Anderson, R., Bledsoe, W.W.: A linear format for resolution with merging and a new technique for establishing completeness. J. ACM **17**(3), 525–534 (1970)
2. Boyer R.S.: Locking: a restriction of resolution (dissertation). University of Texas at Austin (1971)
3. Bachmair, L., Ganzinger, H.: Rewrite-based equational theorem proving with selection and simplification. J. Log. Comput. **4**(3), 217–247 (1994)
4. Bin, C.: Principle of Generalized Support Set Reduction. J. Northern Jiaotong Univ. **22**(2), 70–72 (1998) (in Chinese)
5. Ruixin, J.: Inclusion Relation between Clause Sets in Propositional Logic. Southwest Jiaotong University, Bachelor's Degree Thesis (2024) (in Chinese)
6. Kowaski, R., Kuehner, D.: Linear resolution with selection function. Artif. Intell. **2**, 227–260 (1971)
7. Xuhua, L.: Semantic resolution of locks using lemmas - LI resolution. J. Jilin Univ. **4**, 129–136 (1979). (in Chinese)
8. Xuhua, L.: Input semilock resolution on horn sets. Chin. Sci. Bull. **16**, 1201–1202 (1985). (in Chinese)
9. Xuhua, L.: A new principle of semantic resolution. J. Jilin Univ. **2**, 112–117 (1987). (in Chinese)
10. Xuhua, L.: Compatibility issues among three resolution principles. J. Softw. **2**, 60–64 (1992). (in Chinese)
11. Xuhua, L.: Automated Reasoning Based on Resolution Method. Science Press (1994) (in Chinese)
12. Qinghua, L., Yang, X.: An Improved OL Resolution - SOL Resolution. Comput. Eng. Sci. **41**(01), 102–107 (2019) (in Chinese)

13. Robinson, J.A.: A machine-oriented logic based on the resolution principle. J. ACM **12**(1), 23–41 (1965)
14. Slagle, J.R.: Automatic theorem proving with renamable and semantic resolution. J. ACM **14**(4), 687–697 (1967)
15. Shifen, X.: Research on Automated Reasoning Based on Petri Net Models. Southwest Jiaotong University, Doctoral Dissertation (2006) (in Chinese)
16. Xu, Y., Liu, J., Chen, S., Zhong, X., He, X.: Contradiction separation based dynamic multi-clause synergized automated deduction. Inf. Sci. **462**, 93–113 (2018)
17. Yang, X., Shuwei, C., Xiaomei, Z., Jun, L., Xingxing, H.: Towards multi-clause automated deduction and theorem generation: Constructing and applying standard contradictions. Knowledge-Based Systems,**334**, ID:114985 (2026)
18. Jian, Z.: Satisfiability Decision of Logical Formulas: Methods, Tools, and Applications. Science Press (2000) (in Chinese)
19. Jian, Z.: Research on an Automated Theorem Proving System Based on Dynamic Multi-agent Collaborative Reasoning. Southwest Jiaotong University, Doctoral Dissertation (2021) (in Chinese)

Quasi-Projective Synchronization of Quaternion-Valued Non-identical Cohen-Grossberg Neural Networks with Interaction Terms and Mismatched Parameters

Zili Jiang, Mingbing Luo, and Manchun Tan(✉)

Department of Mathematics, Jinan University, 510632 Guangzhou, Guangdong, China
tanmc@jnu.edu.cn

Abstract. The exponential quasi-projective synchronization (QPJS) is investigated about quaternion-valued non-identical Cohen-Grossberg neural networks (QVNNs) featuring interaction terms and mismatched parameters. By constructing appropriate Lyapunov functions and designing an effective nonlinear controller, sufficient conditions for achieving QPJS of the studied QVNNs have been rigorously established. The non-decomposition technique is employed, since separating QVNNs demands sophisticated calculations and derivation steps. Furthermore, the error bound of the considered system is estimated. Finally, numerical examples are simulated to verify the effectiveness and feasibility of the proposed results.

Keywords: Quasi-projective synchronization · Neural networks · Interaction terms · Mismatched parameters

1 Introduction

In recent years, significant progress has been made in the study of synchronization in neural networks. Synchronization is a fundamental dynamical behavior in complex networks, referring to the process by which a drive network and a response network achieve state consistency through interaction. Faced with increasing demands for storage capacity and the ability to handle complex scenarios, traditional real-valued neural networks and complex-valued neural networks often struggle due to inherent limitations in their representational capacity. Against this backdrop, quaternion-valued neural networks (QVNNs) have emerged. The synchronization phenomena in QVNNs have attracted considerable attention from scholars [1–6].

Synchronization encompasses various types, such as finite-time synchronization, fixed-time synchronization, and projective synchronization. Among these,

J. Lu et al. (Eds.): FLINS-ISKE 2026, LNCS 16752, pp. 570–585, 2027.
https://doi.org/10.1007/978-981-92-2369-5_35

the quasi-projective synchronization (QPJS), as a specialized variant of projective synchronization, has garnered widespread research interest [7–9]. QPJS relaxes the strict error constraints of classical projective synchronization by allowing bounded synchronization errors between the drive and response networks. In neural network synchronization, interaction terms reflect the dynamic connectivity characteristics between neurons, describing their mutual influences through adjustable coupling weights and time-varying signal intensities. The connection patterns directly determine the network's synchronization capability, while multi-layer interaction structures explain the formation mechanisms of collective behaviors in complex systems.

As a generalized model, the Cohen–Grossberg neural networks model was first proposed by Cohen and Grossberg [10], which includes cellular neural network and Hopfield neural network as special cases of it. These networks have been widely applied in various fields, including image processing, medical diagnosis, and transportation systems. Researchers are paying increasing attention to this and have achieved some important results [11–13].

Furthermore, in practical applications, due to various factors such as external disturbances or measurement errors, the connection weights or parameters between two systems often fail to achieve perfect matching, which represents the most common scenario in the real world. However, to the best of our knowledge, the QPJS problem for quaternion-valued non-identical Cohen-Grossberg neural networks (QVNICGNNs) with interaction terms and mismatched parameters has not been explored in the existing literature. Therefore, this paper investigates the QPJS of such networks with time delays using the direct method. The main contributions to be presented in this article can be described as follows:

(1) For the first time, this paper investigates the QPJS of QVNICGNNs, which extends the understanding of dynamic behaviors of Cohen-Grossberg neural networks with interaction terms and mismatched parameters, a topic that has remained untouched in earlier works.
(2) In contrast to references [14–16], instead of decomposing the quaternion-valued system into four real-valued systems, we treat it as an integrated entity and employ a non-decomposition approach by utilizing the properties and theoretical framework of quaternions.
(3) Unlike previous quasi-synchronization studies [17–19], this work tackles QPJS for QVNNs with interaction terms and parameter mismatches, presenting a more complex universal case with practical relevance.

Notations: $\mathbb{R}$ and $\mathbb{R}_+$ denote the sets of all real and nonnegative numbers, respectively. $\mathbb{M}$ denotes the set of all quaternions. The set of all N-dimensional quaternion is denoted by $\mathbb{M}^N$. $\mathbb{M}^{N\times N}$ represents all $N \times N$ quaternion matrices' set. $C([y-c, y], \mathbb{M}^N)$ denotes the set of all continuous function mapping from the interval $[y-c, y]$ to a range of $\mathbb{M}^N$. For $z = z^{\mathcal{R}} + \mathbf{i}z^{\mathcal{I}} + \mathbf{j}z^{\mathcal{J}} + \mathbf{k}z^{\mathcal{K}} \in \mathbb{M}$, where $z^S \in \mathbb{R}$, $S \in \Pi = \{\mathcal{R}, \mathcal{I}, \mathcal{J}, \mathcal{K}\}$, the conjugate of z is denoted by $z^* = z^{\mathcal{R}} - \mathbf{i}z^{\mathcal{I}} - \mathbf{j}z^{\mathcal{J}} - \mathbf{k}z^{\mathcal{K}}$, and the module of z is written as $|z| = \sqrt{z^*z} = \sqrt{zz^*} = \sqrt{(z^{\mathcal{R}})^2 + (z^{\mathcal{I}})^2 + (z^{\mathcal{J}})^2 + (z^{\mathcal{K}})^2}$. Here $\mathbf{i}, \mathbf{j}$ and $\mathbf{k}$ are standard imaginary units in $\mathbb{M}$ which obey the Hamilton rules: $\mathbf{ij} = -\mathbf{ji} = \mathbf{k}$, $\mathbf{jk} = -\mathbf{kj} = \mathbf{i}$, $\mathbf{ki} = -\mathbf{ik} = \mathbf{j}$ and $\mathbf{i}^2 = \mathbf{j}^2 = \mathbf{k}^2 = -1$. For quaternion vector $z = (z_1, z_2, \cdots, z_N) \in \mathbb{M}^N$, the norm of which are defined as $\|z\| = \sqrt{\sum_{i=1}^N |z_i|^2}$.

2 Preliminaries

Consider the QVNICGNNs with interaction terms, mismatched parameters as the drive system

$$\begin{aligned}\dot{\varphi}_i(t) =& - a_i(\varphi_i(t))\Big(b_i(\varphi_i(t)) - \sum_{\vartheta=1}^N c_{i\vartheta} f_\vartheta(\varphi_\vartheta(t)) \\ & - \sum_{\vartheta=1}^N d_{i\vartheta} g_\vartheta(\varphi_\vartheta(t-\iota(t))) - \gamma \sum_{\vartheta=1}^N k_{i\vartheta} h_\vartheta(\psi_\vartheta(t)) - I_i(t)\Big), i \in \Gamma, \end{aligned} \tag{1}$$

and the corresponding response system

$$\begin{aligned}\dot{\psi}_i(t) =& - \tilde{a}_i(\psi_i(t))\Big(\tilde{b}_i(\psi_i(t)) - \sum_{\vartheta=1}^N \tilde{c}_{i\vartheta} f_\vartheta(\psi_\vartheta(t)) \\ & - \sum_{\vartheta=1}^N \tilde{d}_{i\vartheta} g_\vartheta(\psi_\vartheta(t-\iota(t))) - \gamma \sum_{\vartheta=1}^N \tilde{k}_{i\vartheta} h_\vartheta(\varphi_\vartheta(t)) - \tilde{I}_i(t)\Big) + u_i(t), i \in \Gamma, \end{aligned} \tag{2}$$

where $\varphi_i(t) \in \mathbb{M}, \psi_i(t) \in \mathbb{M}$ mean the state of the ith neuron, $c_{i\vartheta}, \tilde{c}_{i\vartheta}, d_{i\vartheta}, \tilde{d}_{i\vartheta}$ represent the connection weights. $k_{i\vartheta}, \tilde{k}_{i\vartheta}$ denote the interaction weight, and γ represents the interaction strength. $a_i(\cdot), \tilde{a}_i(\cdot)$ and $b_i(\cdot), \tilde{b}_i(\cdot)$ denote the amplification function and behaved function, respectively. $f_\vartheta(\cdot), g_\vartheta(\cdot)$ denote the activation function. $h_\vartheta(\cdot)$ denotes the interaction function. $\iota(t)$ represents time delays. $I_i(t), \tilde{I}_i(t)$ are the external control input. $u_i(t)$ is the control to be designed. Γ represents the set $\{1, ..., N\}$.

Denote $\mathfrak{e}_i(t) = \psi_i(t) - \lambda\varphi_i(t)$, where λ represents the projective coefficient. Then, through systems (1) and (2), the error system can be calculated as:

$$\begin{aligned}
\dot{\mathfrak{e}}_i(t) =& - \tilde{a}_i(\psi_i(t))\Big(\tilde{b}_i(\psi_i(t)) - \sum_{\vartheta=1}^{N} \tilde{c}_{i\vartheta} f_\vartheta(\psi_\vartheta(t)) \\
& - \sum_{\vartheta=1}^{N} \tilde{d}_{i\vartheta} g_\vartheta(\psi_\vartheta(t-\iota(t))) - \gamma \sum_{\vartheta=1}^{N} \tilde{k}_{i\vartheta} h_\vartheta(\varphi_\vartheta(t)) - \tilde{I}_i(t)\Big) + u_i(t) \\
& - \lambda\Big[- a_i(\varphi_i(t))\Big(b_i(\varphi_i(t)) - \sum_{\vartheta=1}^{N} c_{i\vartheta} f_\vartheta(\varphi_\vartheta(t)) \\
& - \sum_{\vartheta=1}^{N} d_{i\vartheta} g_\vartheta(\varphi_\vartheta(t-\iota(t))) - \gamma \sum_{\vartheta=1}^{N} k_{i\vartheta} h_\vartheta(\psi_\vartheta(t)) - I_i(t)\Big)\Big] \\
=& - \tilde{a}_i(\psi_i(t))\tilde{b}_i(\psi_i(t)) + \lambda a_i(\varphi_i(t)) b_i(\varphi_i(t)) \\
& + \sum_{\vartheta=1}^{N} \tilde{a}_i(\psi_i(t))\tilde{c}_{i\vartheta} f_\vartheta(\psi_\vartheta(t)) - \lambda \sum_{\vartheta=1}^{N} a_i(\varphi_i(t)) c_{i\vartheta} f_\vartheta(\varphi_\vartheta(t)) \\
& + \sum_{\vartheta=1}^{N} \tilde{a}_i(\psi_i(t))\tilde{d}_{i\vartheta} g_\vartheta(\psi_\vartheta(t-\iota(t))) - \lambda \sum_{\vartheta=1}^{N} a_i(\varphi_i(t)) d_{i\vartheta} g_\vartheta(\varphi_\vartheta(t-\iota(t))) \\
& + \gamma \sum_{\vartheta=1}^{N} \tilde{a}_i(\psi_i(t))\tilde{k}_{i\vartheta} h_\vartheta(\varphi_\vartheta(t)) - \lambda\gamma \sum_{\vartheta=1}^{N} a_i(\varphi_i(t)) k_{i\vartheta} h_\vartheta(\psi_\vartheta(t)) \\
& + \tilde{a}_i(\psi_i(t))\tilde{I}_i(t) - \lambda a_i(\varphi_i(t)) I_i(t) + u_i(t) \\
=& - \tilde{a}_i(\psi_i(t))\tilde{b}_i(\psi_i(t)) + \lambda a_i(\varphi_i(t)) b_i(\varphi_i(t)) \\
& + \sum_{\vartheta=1}^{N} \tilde{a}_i(\psi_i(t))\tilde{c}_{i\vartheta} f_\vartheta(\mathfrak{e}_\vartheta(t)) + \sum_{\vartheta=1}^{N} \tilde{a}_i(\psi_i(t))\tilde{d}_{i\vartheta} g_\vartheta(\mathfrak{e}_\vartheta(t-\iota(t))) \\
& - \lambda\gamma \sum_{\vartheta=1}^{N} a_i(\varphi_i(t)) k_{i\vartheta} h_\vartheta(\mathfrak{e}_\vartheta(t)) \\
& + \sum_{\vartheta=1}^{N} \tilde{a}_i(\psi_i(t))\tilde{c}_{i\vartheta} f_\vartheta(\lambda\varphi_\vartheta(t)) + \sum_{\vartheta=1}^{N} \tilde{a}_i(\psi_i(t))\tilde{d}_{i\vartheta} g_\vartheta(\lambda\varphi_\vartheta(t-\iota(t))) \\
& - \lambda\gamma \sum_{\vartheta=1}^{N} a_i(\varphi_i(t)) k_{i\vartheta} h_\vartheta(\lambda\varphi_\vartheta(t)) - \lambda \sum_{\vartheta=1}^{N} a_i(\varphi_i(t)) c_{i\vartheta} f_\vartheta(\varphi_\vartheta(t)) \\
& - \lambda \sum_{\vartheta=1}^{N} a_i(\varphi_i(t)) d_{i\vartheta} g_\vartheta(\varphi_\vartheta(t-\iota(t))) + \gamma \sum_{\vartheta=1}^{N} \tilde{a}_i(\psi_i(t))\tilde{k}_{i\vartheta} h_\vartheta(\varphi_\vartheta(t)) \\
& + \tilde{a}_i(\psi_i(t))\tilde{I}_i(t) - \lambda a_i(\varphi_i(t)) I_i(t) + u_i(t), i \in \Gamma,
\end{aligned} \tag{3}$$

where $f_\vartheta(\mathfrak{e}_\vartheta(t)) = f_\vartheta(\psi_i(t)) - f_\vartheta(\lambda\varphi_i(t)), g_\vartheta(\mathfrak{e}_\vartheta(t)) = g_\vartheta(\psi_i(t)) - g_\vartheta(\lambda\varphi_i(t))$, $h_\vartheta(\mathfrak{e}_\vartheta(t)) = h_\vartheta(\psi_i(t)) - h_\vartheta(\lambda\varphi_i(t))$.

Next, assumption and several lemmas are needed to get the QPJS criteria of QVNICGNNs.

Assumption 1. *For any $\vartheta \in \Gamma$, the activation functions $f_\vartheta(\cdot)$, $g_\vartheta(\cdot)$, and $h_\vartheta(\cdot)$ admit the quaternion representation:*

$$X_\vartheta(\varphi_\vartheta) = X_\vartheta(\varphi_\vartheta)^{\mathcal{R}} + \mathbf{i}X_\vartheta(\varphi_\vartheta)^{\mathcal{I}} + \mathbf{j}X_\vartheta(\varphi_\vartheta)^{\mathcal{J}} + \mathbf{k}X_\vartheta(\varphi_\vartheta)^{\mathcal{K}},$$

where X can represent f, g, or h.

There exist positive constants $l_\vartheta^X \in \mathbb{R}_+$ such that for any ψ_ϑ, φ_ϑ:

$$\big(X_\vartheta(\psi_\vartheta) - X_\vartheta(\varphi_\vartheta)\big)^* \big(X_\vartheta(\psi_\vartheta) - X_\vartheta(\varphi_\vartheta)\big) \le l_\vartheta^X (\psi_\vartheta - \varphi_\vartheta)^*(\psi_\vartheta - \varphi_\vartheta),$$

where $l_\vartheta^f = l_\vartheta$, $l_\vartheta^g = \sigma_\vartheta$, *and* $l_\vartheta^h = \pi_\vartheta$ *correspond to the original constants for* f, g, *and* h, *respectively.*

Assumption 2. *There exists* $\mathcal{F}_\vartheta, \mathcal{G}_\vartheta, \mathcal{H}_\vartheta \in \mathbb{R}_+$ *such that:*

$$f_\vartheta^*(\varphi_\vartheta)f_\vartheta(\varphi_\vartheta) \leq \mathcal{F}_\vartheta, g_\vartheta^*(\varphi_\vartheta)g_\vartheta(\varphi_\vartheta) \leq \mathcal{G}_\vartheta, h_\vartheta^*(\varphi_\vartheta)h_\vartheta(\varphi_\vartheta) \leq \mathcal{H}_\vartheta.$$

Assumption 3. *The amplification function* $a_i(\cdot), \tilde{a_i}(\cdot)$ *satisfies:*

$$a_i^*(\cdot)a_i(\cdot) \leq \alpha_i, \tilde{a_i}^*(\cdot)\tilde{a_i}(\cdot) \leq \tilde{\alpha_i},$$

where $\alpha_i, \tilde{\alpha_i} \in \mathbb{R}_+$.

Lemma 1. ([20]) *If* $p_1, p_2 \in \mathbb{M}$, *then for any* $\epsilon > 0$, *the following inequality holds:*

$$p_1^*p_2 + p_2^*p_1 \leq \epsilon p_1^*p_1 + \frac{1}{\epsilon}p_2^*p_2.$$

Lemma 2. ([21]) *Suppose that* $\zeta(t)$ *is non-negative and satisfies:*

$$D^+\zeta(t) \leq -a\zeta(t) + b\zeta(t - \iota(t)),$$

where $t > 0$, *and* $a, b \in \mathbb{R}_+$ *with* $a > b$, *then*

$$\zeta(t) \leq \sup_{-\iota \leq \varrho \leq 0} \zeta(\varrho)e^{-\omega t},$$

ω *is the positive smallest solution of the equation:* $\omega = a - be^{-\omega\iota}$.

Definition 1. ([22]) *Systems* (1) *and* (2) *will attain global QPJS if there exists* P, ω *and* $\Omega \in \mathbb{R}_+$ *such that:*

$$\|\mathfrak{e}(t)\| \leq P \sup_{-\iota \leq \varrho \leq 0} \|\mathfrak{e}(\varrho)\|e^{-\omega t} + \Omega$$

holds for $t > 0$.

3 Main Results

Theorem 1. *Suppose that Assumptions 1-3 hold. If there exist positive constants*

$$\tilde{F}_1 = \min_i \Big[\big(2\varpi_i - (2(\gamma+1)\lambda\alpha_i + (4+\gamma)\tilde{\alpha}_i)\big) - \sum_{\vartheta=1}^{N}(\tilde{c}_{\vartheta i}^*\tilde{c}_{\vartheta i}l_i + \lambda\gamma k_{\vartheta i}^*k_{\vartheta i}\pi_i)\Big] > 0,$$

$$\tilde{F}_2 = \max_i \sum_{\vartheta=1}^{N} \tilde{d}_{\vartheta i}^*\tilde{d}_{\vartheta i}\sigma_i > 0,$$

$$\begin{aligned}\tilde{F}_3 = \sum_{i=1}^{N}\sum_{\vartheta=1}^{N} \big(&\tilde{c}_{i\vartheta}^*\tilde{c}_{i\vartheta}\mathcal{F}_\vartheta + \tilde{d}_{i\vartheta}^*\tilde{d}_{i\vartheta}\mathcal{G}_\vartheta + \lambda\gamma k_{i\vartheta}^*k_{i\vartheta}\mathcal{H}_\vartheta + \lambda c_{i\vartheta}^*c_{i\vartheta}\mathcal{F}_\vartheta \\ &+ \lambda d_{i\vartheta}^*d_{i\vartheta}\mathcal{G}_\vartheta + \gamma\tilde{k}_{i\vartheta}^*\tilde{k}_{i\vartheta}\mathcal{H}_\vartheta\big) > 0\end{aligned}$$

such that $\tilde{F}_1 > \tilde{F}_2$ *holds, then systems* (1) *and* (2) *achieve global QPJS under controller:*

$$\begin{aligned}u_i(t) = &-\varpi_i\mathfrak{e}_i(t) - \tilde{a}_i(\psi_i(t))(\tilde{I}_i(t) - \tilde{b}_i(\psi_i(t)) \\ &-\lambda a_i(\varphi_i(t))(b_i(\varphi_i(t)) - I_i(t)).\end{aligned} \tag{4}$$

Proof. Choose the Lyapunov function:

$$\mathcal{V}(t) = \sum_{i=1}^{N} \mathfrak{e}_i{}^*(t)\mathfrak{e}_i(t). \tag{5}$$

Calculating the upper right Dini derivative of $\mathcal{V}(t)$, one obtains:

$$\begin{aligned}
D^+\mathcal{V}(t) =& \sum_{i=1}^{N}(D^+\mathfrak{e}_i{}^*(t)\mathfrak{e}_i(t) + \mathfrak{e}_i{}^*(t)D^+\mathfrak{e}_i(t))\\
=& \sum_{i=1}^{N}\Bigg(\sum_{\vartheta=1}^{N} \tilde{a}_i(\psi_i(t))\tilde{c}_{i\vartheta} f_\vartheta(\mathfrak{e}_\vartheta(t)) + \sum_{\vartheta=1}^{N} \tilde{a}_i(\psi_i(t))\tilde{d}_{i\vartheta} g_\vartheta(\mathfrak{e}_\vartheta(t-\iota(t)))\\
&- \lambda\gamma \sum_{\vartheta=1}^{N} a_i(\varphi_i(t))k_{i\vartheta}h_\vartheta(\mathfrak{e}_\vartheta(t))\\
&+ \sum_{\vartheta=1}^{N} \tilde{a}_i(\psi_i(t))\tilde{c}_{i\vartheta} f_\vartheta(\lambda\varphi_\vartheta(t)) + \sum_{\vartheta=1}^{N} \tilde{a}_i(\psi_i(t))\tilde{d}_{i\vartheta} g_\vartheta(\lambda\varphi_\vartheta(t-\iota(t)))\\
&- \lambda\gamma \sum_{\vartheta=1}^{N} a_i(\varphi_i(t))k_{i\vartheta}h_\vartheta(\lambda\varphi_\vartheta(t)) - \lambda \sum_{\vartheta=1}^{N} a_i(\varphi_i(t))c_{i\vartheta} f_\vartheta(\varphi_\vartheta(t))\\
&-\lambda \sum_{\vartheta=1}^{N} a_i(\varphi_i(t))d_{i\vartheta} g_\vartheta(\varphi_\vartheta(t-\iota(t))) + \gamma \sum_{\vartheta=1}^{N} \tilde{a}_i(\psi_i(t))\tilde{k}_{i\vartheta}h_\vartheta(\varphi_\vartheta(t))\Bigg)^* \mathfrak{e}_i(t)\\
&+ \mathfrak{e}_i^*(t)\sum_{i=1}^{N}\Bigg(\sum_{\vartheta=1}^{N} \tilde{a}_i(\psi_i(t))\tilde{c}_{i\vartheta} f_\vartheta(\mathfrak{e}_\vartheta(t)) + \sum_{\vartheta=1}^{N} \tilde{a}_i(\psi_i(t))\tilde{d}_{i\vartheta} g_\vartheta(\mathfrak{e}_\vartheta(t-\iota(t)))\\
&- \lambda\gamma \sum_{\vartheta=1}^{N} a_i(\varphi_i(t))k_{i\vartheta}h_\vartheta(\mathfrak{e}_\vartheta(t))\\
&+ \sum_{\vartheta=1}^{N} \tilde{a}_i(\psi_i(t))\tilde{c}_{i\vartheta} f_\vartheta(\lambda\varphi_\vartheta(t)) + \sum_{\vartheta=1}^{N} \tilde{a}_i(\psi_i(t))\tilde{d}_{i\vartheta} g_\vartheta(\lambda\varphi_\vartheta(t-\iota(t)))\\
&- \lambda\gamma \sum_{\vartheta=1}^{N} a_i(\varphi_i(t))k_{i\vartheta}h_\vartheta(\lambda\varphi_\vartheta(t)) - \lambda \sum_{\vartheta=1}^{N} a_i(\varphi_i(t))c_{i\vartheta} f_\vartheta(\varphi_\vartheta(t))\\
&-\lambda \sum_{\vartheta=1}^{N} a_i(\varphi_i(t))d_{i\vartheta} g_\vartheta(\varphi_\vartheta(t-\iota(t))) + \gamma \sum_{\vartheta=1}^{N} \tilde{a}_i(\psi_i(t))\tilde{k}_{i\vartheta}h_\vartheta(\varphi_\vartheta(t))\Bigg)\\
=& \sum_{i=1}^{N}\sum_{\vartheta=1}^{N}\Big(f_\vartheta^*(\mathfrak{e}_\vartheta(t))\tilde{c}_{i\vartheta}^*\tilde{a}_i^*(\psi_i(t))\mathfrak{e}_i(t) + \mathfrak{e}_i^*(t)\tilde{a}_i(\psi_i(t))\tilde{c}_{i\vartheta} f_\vartheta(\mathfrak{e}_\vartheta(t))\Big)\\
&+ \sum_{i=1}^{N}\sum_{\vartheta=1}^{N}\Big(g_\vartheta^*(\mathfrak{e}_\vartheta(t-\iota(t)))\tilde{d}_{i\vartheta}^*\tilde{a}_i^*(\psi_i(t))\mathfrak{e}_i(t) + \mathfrak{e}_i^*(t)\tilde{a}_i(\psi_i(t))\tilde{d}_{i\vartheta} g_\vartheta(\mathfrak{e}_\vartheta(t-\iota(t)))\Big)\\
&+ \sum_{i=1}^{N}\sum_{\vartheta=1}^{N}\Big(-\lambda\gamma h_\vartheta^*(\mathfrak{e}_\vartheta(t))k_{i\vartheta}^* a_i^*(\varphi_i(t))\mathfrak{e}_i(t) - \lambda\gamma a_i(\varphi_i(t))k_{i\vartheta}h_\vartheta(\mathfrak{e}_\vartheta(t))\mathfrak{e}_i(t)\Big)
\end{aligned}$$

$$
\begin{aligned}
&+\sum_{i=1}^{N}\sum_{\vartheta=1}^{N}\Big(f_{\vartheta}^{*}(\lambda\varphi_{\vartheta}(t))\tilde{c}_{i\vartheta}^{*}\tilde{a}_{i}^{*}(\psi_{i}(t))\mathfrak{e}_{i}(t)+\mathfrak{e}_{i}^{*}(t)\tilde{a}_{i}(\psi_{i}(t))\tilde{c}_{i\vartheta}f_{\vartheta}(\lambda\varphi_{\vartheta}(t))\Big)\\
&+\sum_{i=1}^{N}\sum_{\vartheta=1}^{N}\Big(g_{\vartheta}^{*}(\lambda\varphi_{\vartheta}(t-\iota(t)))\tilde{d}_{i\vartheta}^{*}\tilde{a}_{i}^{*}(\psi_{i}(t))\mathfrak{e}_{i}(t)+\mathfrak{e}_{i}^{*}(t)\tilde{a}_{i}(\psi_{i}(t))\tilde{d}_{i\vartheta}g_{\vartheta}(\lambda\varphi_{\vartheta}(t-\iota(t)))\Big)\\
&+\sum_{i=1}^{N}\sum_{\vartheta=1}^{N}\Big(-\lambda\gamma h_{\vartheta}^{*}(\lambda\varphi_{\vartheta}(t))k_{i\vartheta}^{*}a_{i}^{*}(\varphi_{i}(t))\mathfrak{e}_{i}(t)-\lambda\gamma\mathfrak{e}_{i}^{*}(t)a_{i}(\varphi_{i}(t))k_{i\vartheta}h_{\vartheta}(\lambda\varphi_{\vartheta}(t))\Big)\\
&+\sum_{i=1}^{N}\sum_{\vartheta=1}^{N}\Big(-\lambda f_{\vartheta}^{*}(\varphi_{\vartheta}(t))c_{i\vartheta}^{*}a_{i}^{*}(\varphi_{i}(t))\mathfrak{e}_{i}(t)-\lambda\mathfrak{e}_{i}^{*}(t)a_{i}(\varphi_{i}(t))c_{i\vartheta}f_{\vartheta}(\varphi_{\vartheta}(t))\Big)\\
&+\sum_{i=1}^{N}\sum_{\vartheta=1}^{N}\Big(-\lambda g_{\vartheta}^{*}(\varphi_{\vartheta}(t-\iota(t)))d_{i\vartheta}^{*}a_{i}^{*}(\varphi_{i}(t))\mathfrak{e}_{i}(t)-\lambda\mathfrak{e}_{i}^{*}(t)a_{i}(\varphi_{i}(t))d_{i\vartheta}g_{\vartheta}(\varphi_{\vartheta}(t-\iota(t)))\Big)\\
&+\sum_{i=1}^{N}\sum_{\vartheta=1}^{N}\Big(\gamma h_{\vartheta}^{*}(\varphi_{\vartheta}(t))\tilde{k}_{i\vartheta}^{*}\tilde{a}_{i}^{*}(\psi_{i}(t))\mathfrak{e}_{i}(t)+\gamma\mathfrak{e}_{i}^{*}(t)\tilde{a}_{i}(\psi_{i}(t))\tilde{k}_{i\vartheta}h_{\vartheta}(\varphi_{\vartheta}(t))\Big)\\
&-2\varpi_{i}\sum_{i=1}^{N}\mathfrak{e}_{i}^{*}(t)\mathfrak{e}_{i}(t)
\end{aligned}\tag{6}
$$

According to Lemma 1 and Assumptions 1-3, one obtains:

$$
\begin{aligned}
D^{+}\mathcal{V}(t)\leq&\sum_{i=1}^{N}\Big[\big(-2\varpi_{i}+2(\gamma+1)\lambda\alpha_{i}+(4+\gamma)\tilde{\alpha}_{i}\big)+\sum_{\vartheta=1}^{N}(\tilde{c}_{\vartheta i}^{*}\tilde{c}_{\vartheta i}l_{i}+\lambda\gamma k_{\vartheta i}^{*}k_{\vartheta i}\pi_{i})\Big]\mathfrak{e}_{i}^{*}(t)\mathfrak{e}_{i}(t)\\
&+\sum_{i=1}^{N}\sum_{\vartheta=1}^{N}\tilde{d}_{\vartheta i}^{*}\tilde{d}_{\vartheta i}\sigma_{i}\mathfrak{e}_{i}^{*}(t-\iota(t))\mathfrak{e}_{i}(t-\iota(t))+\sum_{i=1}^{N}\sum_{\vartheta=1}^{N}\big(\tilde{c}_{i\vartheta}^{*}\tilde{c}_{i\vartheta}\mathcal{F}_{\vartheta}+\tilde{d}_{i\vartheta}^{*}\tilde{d}_{i\vartheta}\mathcal{G}_{\vartheta}\\
&+\lambda\gamma k_{i\vartheta}^{*}k_{i\vartheta}\mathcal{H}_{\vartheta}+\lambda c_{i\vartheta}^{*}c_{i\vartheta}\mathcal{F}_{\vartheta}+\lambda d_{i\vartheta}^{*}d_{i\vartheta}\mathcal{G}_{\vartheta}+\gamma\tilde{k}_{i\vartheta}^{*}\tilde{k}_{i\vartheta}\mathcal{H}_{\vartheta}\big)\\
\leq&-\tilde{F}_{1}\mathcal{V}(t)+\tilde{F}_{2}\mathcal{V}(t-\iota(t))+\tilde{F}_{3},
\end{aligned}
$$

i.e.,

$$
D^{+}\Big(\mathcal{V}(t)-\frac{\tilde{F}_{3}}{\tilde{F}_{1}-\tilde{F}_{2}}\Big)\leq-\tilde{F}_{1}\Big(\mathcal{V}(t)-\frac{\tilde{F}_{3}}{\tilde{F}_{1}-\tilde{F}_{2}}\Big)+\tilde{F}_{2}\Big(\mathcal{V}(t-\iota(t))-\frac{\tilde{F}_{3}}{\tilde{F}_{1}-\tilde{F}_{2}}\Big).
$$

Consider the variable substitution:

$$
H(t)=\mathcal{V}(t)-\frac{\tilde{F}_{3}}{\tilde{F}_{1}-\tilde{F}_{2}},
$$

then one obtains:

$$
D^{+}H(t)\leq-\tilde{F}_{1}H(t)+\tilde{F}_{2}H(t-\iota(t)).
$$

From Lemma 2 one obtains:

$$
H(t)\leq\sup_{-\iota\leq\varrho\leq0}H(\varrho)e^{-\mu t},
$$

where μ is the positive smallest solution of the equation: $\mu=\tilde{F}_{1}-\tilde{F}_{2}e^{-\mu\iota}$.

Then we have:

$$\mathcal{V}(t) \leq \sup_{-\iota \leq \varrho \leq 0} \mathcal{V}(\varrho) e^{-\mu t} + \frac{\tilde{F}_3}{\tilde{F}_1 - \tilde{F}_2},$$

i.e.,

$$\|\mathfrak{e}(t)\| \leq \sup_{-\iota \leq \varrho \leq 0} \sum_{i=1}^{N} \|\mathfrak{e}(\varrho)\| e^{-\frac{\mu t}{2}} + \sqrt{\frac{\tilde{F}_3}{\tilde{F}_1 - \tilde{F}_2}}.$$

Thus, based on Definition 1 one can conclude that system (1) is QPJS with system (2).

In fact, if we assume that the external input $I_i(t), \tilde{I}_i(t)$ are bounded, that is, $I_i^*(t) I_i(t) \leq \mathcal{I}_i, \tilde{I}_i^*(t) \tilde{I}_i(t) \leq \tilde{\mathcal{I}}_i$, where $\mathcal{I}, \tilde{\mathcal{I}} \in \mathbb{R}_+$, then one can relax the controller as:

$$\tilde{u}_i(t) = -\tilde{\varpi}_i \mathfrak{e}_i(t) + \tilde{a}_i(\psi_i(t)) \tilde{b}_i(\psi_i(t)) - \lambda a_i(\varphi_i(t)) b_i(\varphi_i(t)). \tag{7}$$

Theorem 2. *Suppose that Assumptions 1-3 hold. If there exist positive constants*

$$\tilde{\eta}_1 = \min_i \Big[\big(2\tilde{\varpi}_i - (2(\gamma + 3/2)\lambda\alpha_i + (5+\gamma)\tilde{\alpha}_i)\big) - \sum_{\vartheta=1}^{N} (\tilde{c}_{\vartheta i}^* \tilde{c}_{\vartheta i} l_i + \lambda\gamma k_{\vartheta i}^* k_{\vartheta i} \pi_i) \Big] > 0,$$

$$\tilde{\eta}_2 = \max_i \sum_{\vartheta=1}^{N} \tilde{d}_{\vartheta i}^* \tilde{d}_{\vartheta i} \sigma_i > 0,$$

$$\begin{aligned}\tilde{\eta}_3 = \sum_{i=1}^{N} \Big[\tilde{\mathcal{I}}_i + \lambda \mathcal{I}_i + \sum_{\vartheta=1}^{N} \big(& \tilde{c}_{i\vartheta}^* \tilde{c}_{i\vartheta} \mathcal{F}_\vartheta + \tilde{d}_{i\vartheta}^* \tilde{d}_{i\vartheta} \mathcal{G}_\vartheta + \lambda\gamma k_{i\vartheta}^* k_{i\vartheta} \mathcal{H}_\vartheta + \lambda c_{i\vartheta}^* c_{i\vartheta} \mathcal{F}_\vartheta \\ & + \lambda d_{i\vartheta}^* d_{i\vartheta} \mathcal{G}_\vartheta + \gamma \tilde{k}_{i\vartheta}^* \tilde{k}_{i\vartheta} \mathcal{H}_\vartheta \big) \Big] > 0\end{aligned}$$

such that $\tilde{\eta}_1 > \tilde{\eta}_2$ *holds, then systems* (1) *and* (2) *can achieve global QPJS.*

Proof. Choose the Lyapunov function:

$$\mathcal{V}(t) = \sum_{i=1}^{N} \mathfrak{e}_i{}^*(t) \mathfrak{e}_i(t).$$

Similar to the proof the Theorem 1, one obtains:

$$\begin{aligned}D^{+}\mathcal{V}(t) \leq &\sum_{i=1}^{N}\Big[\big(-2\tilde{\varpi}_i+2(\gamma+3/2)\lambda\alpha_i+(5+\gamma)\tilde{\alpha}_i\big)\\&+\sum_{\vartheta=1}^{N}(\tilde{c}^{*}_{\vartheta i}\tilde{c}_{\vartheta i}l_i+\lambda\gamma k^{*}_{\vartheta i}k_{\vartheta i}\pi_i)\Big]\mathfrak{e}^{*}_i(t)\mathfrak{e}_i(t)\\&+\sum_{i=1}^{N}\sum_{\vartheta=1}^{N}\tilde{d}^{*}_{\vartheta i}\tilde{d}_{\vartheta i}\sigma_i\mathfrak{e}^{*}_i(t-\iota(t))\mathfrak{e}_i(t-\iota(t))\\&+\sum_{i=1}^{N}\Big[\tilde{\mathcal{I}}_i+\lambda\mathcal{I}_i+\sum_{\vartheta=1}^{N}\big(\tilde{c}^{*}_{i\vartheta}\tilde{c}_{i\vartheta}\mathcal{F}_\vartheta+\tilde{d}^{*}_{i\vartheta}\tilde{d}_{i\vartheta}\mathcal{G}_\vartheta\\&+\lambda\gamma k^{*}_{i\vartheta}k_{i\vartheta}\mathcal{H}_\vartheta+\lambda c^{*}_{i\vartheta}c_{i\vartheta}\mathcal{F}_\vartheta+\lambda d^{*}_{i\vartheta}d_{i\vartheta}\mathcal{G}_\vartheta+\gamma\tilde{k}^{*}_{i\vartheta}\tilde{k}_{i\vartheta}\mathcal{H}_\vartheta\big)\Big]\\ \leq &-\tilde{\eta}_1\mathcal{V}(t)+\tilde{\eta}_2\mathcal{V}(t-\iota(t))+\tilde{\eta}_3.\end{aligned}$$

Repeat the process in Theorem 1, one obtains:

$$\|\mathfrak{e}(t)\| \leq \sup_{-\iota\leq N\leq 0}\sum_{i=1}^{N}\|\mathfrak{e}(N)\|e^{-\frac{\delta t}{2}}+\sqrt{\frac{\tilde{\eta}_3}{\tilde{\eta}_1-\tilde{\eta}_2}},$$

where δ is the positive smallest solution of the equation: $\delta=\tilde{\eta}_1-\tilde{\eta}_2e^{-\delta\iota}$.

Then, based on Definition 1 one can demonstrate that QVNICGNN (1) is QPJS with QVNICGNN (2).

4 Illustrative Examples

Two examples are provided to demonstrate the validity of the conclusions in this paper: Example 1 validates the accuracy of Theorem 1, and Example 2 confirms the correctness of Theorem 2.

Example 1: Considering the following models:

$$\begin{aligned}\dot{\varphi}_i(t)=&-a_i(\varphi_i(t))\Bigg(b_i(\varphi_i(t))-\sum_{\vartheta=1}^{2}c_{i\vartheta}f_\vartheta(\varphi_\vartheta(t))\\&-\sum_{\vartheta=1}^{2}d_{i\vartheta}g_\vartheta(\varphi_\vartheta(t-\iota(t)))-\gamma\sum_{\vartheta=1}^{2}k_{i\vartheta}h_\vartheta(\psi_\vartheta(t))-I_i(t)\Bigg),\end{aligned}$$

$$\begin{aligned}\dot{\psi}_i(t)=&-\tilde{a}_i(\psi_i(t))\Bigg(\tilde{b}_i(\psi_i(t))-\sum_{\vartheta=1}^{2}\tilde{c}_{i\vartheta}f_\vartheta(\psi_\vartheta(t))\\&-\sum_{\vartheta=1}^{2}\tilde{d}_{i\vartheta}g_\vartheta(\psi_\vartheta(t-\iota(t)))-\gamma\sum_{\vartheta=1}^{2}\tilde{k}_{i\vartheta}h_\vartheta(\varphi_\vartheta(t))-\tilde{I}_i(t)\Bigg)+u_i(t),\end{aligned}$$

where $i = 1, 2$.

$$C = (c_{i\vartheta})_{2\times 2} = \begin{pmatrix} 0.3 - 0.2\mathbf{i} + 0.5\mathbf{j} + 0.7\mathbf{k} & -0.8 + 0.5\mathbf{i} - 0.1\mathbf{j} + 0.9\mathbf{k} \\ 0.7 - 0.9\mathbf{i} - 0.4\mathbf{j} - 0.3\mathbf{k} & -0.6 - 0.8\mathbf{i} + 0.2\mathbf{j} - 0.7\mathbf{k} \end{pmatrix},$$

$$D = (d_{i\vartheta})_{2\times 2} = \begin{pmatrix} -0.9 + 0.3\mathbf{i} + 0.6\mathbf{j} + 0.8\mathbf{k} & 0.4 - 0.5\mathbf{i} - 0.3\mathbf{j} + 0.8\mathbf{k} \\ 0.2 - 0.7\mathbf{i} + 0.1\mathbf{j} - 0.2\mathbf{k} & -0.4 + 0.3\mathbf{i} + 0.2\mathbf{j} + 0.5\mathbf{k} \end{pmatrix},$$

$$K = (k_{i\vartheta})_{2\times 2} = \begin{pmatrix} 0.1 - 0.4\mathbf{i} + 0.7\mathbf{j} - 0.9\mathbf{k} & -0.1 + 0.8\mathbf{i} - 0.3\mathbf{j} + 0.1\mathbf{k} \\ 0.5 + 0.3\mathbf{i} - 0.6\mathbf{j} + 0.4\mathbf{k} & -0.2 - 0.9\mathbf{i} + 0.7\mathbf{j} - 0.2\mathbf{k} \end{pmatrix},$$

$$\tilde{C} = (\tilde{c}_{i\vartheta})_{2\times 2} = \begin{pmatrix} 0.6 - 0.8\mathbf{i} + 0.5\mathbf{j} + 0.9\mathbf{k} & -0.3 + 0.7\mathbf{i} + 0.4\mathbf{j} - 0.2\mathbf{k} \\ -0.5 + 0.1\mathbf{i} + 0.8\mathbf{j} + 0.2\mathbf{k} & 0.7 - 0.2\mathbf{i} + 0.9\mathbf{j} - 0.6\mathbf{k} \end{pmatrix},$$

$$\tilde{D} = (\tilde{d}_{i\vartheta})_{2\times 2} = \begin{pmatrix} -0.1 + 0.5\mathbf{i} - 0.1\mathbf{j} + 0.3\mathbf{k} & 0.9 - 0.6\mathbf{i} + 0.8\mathbf{j} + 0.7\mathbf{k} \\ 0.8 + 0.9\mathbf{i} + 0.3\mathbf{j} - 0.4\mathbf{k} & -0.7 - 0.3\mathbf{i} - 0.2\mathbf{j} + 0.1\mathbf{k} \end{pmatrix},$$

$$\tilde{K} = (\tilde{k}_{i\vartheta})_{2\times 2} = \begin{pmatrix} 0.2 - 0.5\mathbf{i} - 0.7\mathbf{j} + 0.8\mathbf{k} & -0.3 + 0.3\mathbf{i} + 0.1\mathbf{j} - 0.9\mathbf{k} \\ 0.4 - 0.5\mathbf{i} + 0.3\mathbf{j} + 0.6\mathbf{k} & 0.9 + 0.1\mathbf{i} - 0.5\mathbf{j} + 0.8\mathbf{k} \end{pmatrix}.$$

Choose

$$\begin{aligned}
&a_1(\varphi_1(t)) = 0.1\sin(\varphi_1^{\mathcal{R}}(t)) + 0.1\cos(\varphi_1^{\mathcal{I}}(t)) + 0.1\sin(\varphi_1^{\mathcal{J}}(t)) + 0.1\cos(\varphi_1^{\mathcal{K}}(t)),\\
&a_2(\varphi_2(t)) = 0.1\sin(\varphi_2^{\mathcal{R}}(t)) + 0.1\cos(\varphi_1^{\mathcal{I}}(t)) + 0.1\sin(\varphi_1^{\mathcal{J}}(t)) + 0.1\cos(\varphi_1^{\mathcal{K}}(t)),\\
&\tilde{a}_1(\psi_1(t)) = 0.1\cos(\psi_1^{\mathcal{R}}(t)) + 0.1\sin(\psi_1^{\mathcal{I}}(t)) + 0.1\cos(\psi_1^{\mathcal{J}}(t)) + 0.1\sin(\psi_1^{\mathcal{K}}(t)),\\
&\tilde{a}_2(\psi_2(t)) = 0.1\cos(\psi_2^{\mathcal{R}}(t)) + 0.1\sin(\psi_2^{\mathcal{I}}(t)) + 0.1\cos(\psi_2^{\mathcal{J}}(t)) + 0.1\sin(\psi_2^{\mathcal{K}}(t)),\\
&b_1(\varphi_1(t)) = 2\varphi_1(t), b_2(\varphi_2(t)) = 2\varphi_2(t), \tilde{b}_1(\psi_1(t)) = 2\psi_1(t), \tilde{b}_2(\psi_2(t)) = 2\psi_2(t),\\
&I_1(t) = \sin(t) + \cos(t)\mathbf{i} + \cos(t)\mathbf{j} + \sin(t)\mathbf{k}, I_2(t) = \sin(t) + \cos(t)\mathbf{i} + \cos(t)\mathbf{j} + \sin(t)\mathbf{k},\\
&\tilde{I}_1(t) = \cos(t) + \sin(t)\mathbf{i} + \sin(t)\mathbf{j} + \cos(t)\mathbf{k}, \tilde{I}_2(t) = \cos(t) + \sin(t)\mathbf{i} + \sin(t)\mathbf{j} + \cos(t)\mathbf{k}.
\end{aligned}$$

Let $\iota(t) = 0.1 + 0.1\sin(t)$, $l_\vartheta = \sigma_\vartheta = \pi_\vartheta = 2$, $\mathcal{F}_\vartheta = \mathcal{G}_\vartheta = \mathcal{H}_\vartheta = 2$, $\alpha_i = \tilde{\alpha}_i = 0.1$, where $i, \vartheta = 1, 2$. The initial condition is of drive system is $(\varphi_1(s), \varphi_2(s)) = (0.1 + 1.3\mathbf{i} - 0.5\mathbf{j} + 0.5\mathbf{k},\ 0.6 - 1.5\mathbf{i} + 0.5\mathbf{j} + 0.4\mathbf{k})^T$, and the initial condition of the response system is $(\psi_1(s), \psi_2(s)) = (0.4 + 0.2\mathbf{i} + 0.1\mathbf{j} + 0.3\mathbf{k}, 0.3 - 0.3\mathbf{i} + 0.1\mathbf{j} - 0.4\mathbf{k})^T$.

Set $\gamma = 0.1, \lambda = 0.8, \varpi_1 = \varpi_2 = 20$, then one can calculate that:

$$\begin{aligned}
\tilde{F}_1 &= \min_i \Big[\big(2\varpi_i - (2(\gamma+1)\lambda\alpha_i + (4+\gamma)\tilde{\alpha}_i)\big) \\
&\quad - \sum_{\vartheta=1}^{2} \big(\tilde{c}^*_{\vartheta i}\tilde{c}_{\vartheta i} l_i + \lambda\gamma k^*_{\vartheta i} k_{\vartheta i}\pi_i\big)\Big] = 33.0412 > 0, \\
\tilde{F}_2 &= \max_i \sum_{\vartheta=1}^{2} \tilde{d}^*_{\vartheta i}\tilde{d}_{\vartheta i}\sigma_i = 5.8600 > 0, \\
\tilde{F}_3 &= \sum_{i=1}^{2}\sum_{\vartheta=1}^{2} \Big(\tilde{c}^*_{i\vartheta}\tilde{c}_{i\vartheta}\mathcal{F}_\vartheta + \tilde{d}^*_{i\vartheta}\tilde{d}_{i\vartheta}\mathcal{G}_\vartheta + \lambda\gamma k^*_{i\vartheta}k_{i\vartheta}\mathcal{H}_\vartheta + \lambda c^*_{i\vartheta}c_{i\vartheta}\mathcal{F}_\vartheta \\
&\quad + \lambda d^*_{i\vartheta}d_{i\vartheta}\mathcal{G}_\vartheta + \gamma\tilde{k}^*_{i\vartheta}\tilde{k}_{i\vartheta}\mathcal{H}_\vartheta\Big) = 38.3636 > 0,
\end{aligned}$$

then the conditions in Theorem 1 are satisfied.

Figure 1 depicts the phase trajectories of $\mathfrak{e}^{\mathcal{R}}, \mathfrak{e}^{\mathcal{I}}, \mathfrak{e}^{\mathcal{J}}, \mathfrak{e}^{\mathcal{K}}$ in system (3) without controller and with controller (4), respectively. The phase trajectories of $\varphi^{\mathcal{R}}, \varphi^{\mathcal{I}}, \varphi^{\mathcal{J}}, \varphi^{\mathcal{K}}$ and $\psi^{\mathcal{R}}, \psi^{\mathcal{I}}, \psi^{\mathcal{J}}, \psi^{\mathcal{K}}$ with controller (4) are illustrated in Fig. 2. Theoretical and numerical analyses demonstrate that the controller (4) performs effectively.

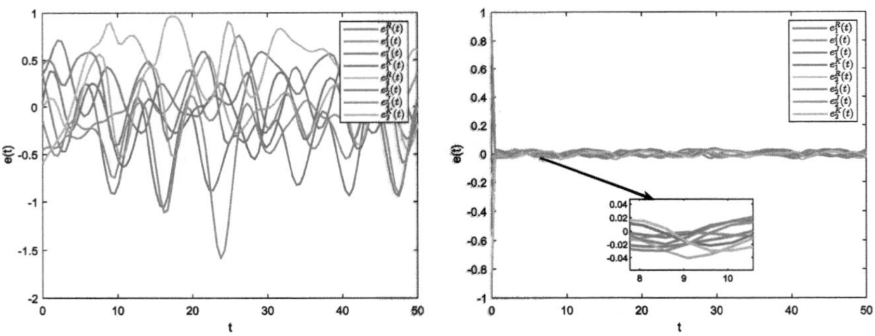

(a) (b)

Fig. 1. Phase trajectories of $\mathfrak{e}^{\mathcal{R}}, \mathfrak{e}^{\mathcal{I}}, \mathfrak{e}^{\mathcal{J}}, \mathfrak{e}^{\mathcal{K}}$ (a) without controller and (b) with controller (4).

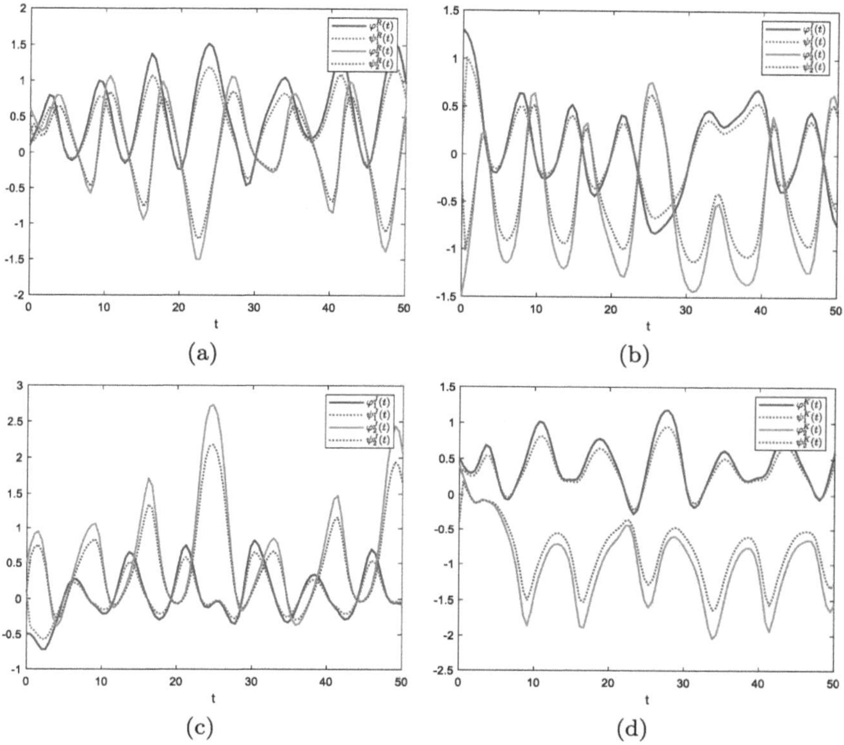

Fig. 2. Phase trajectories of $\varphi^{\mathcal{R}}, \varphi^{\mathcal{I}}, \varphi^{\mathcal{J}}, \varphi^{\mathcal{K}}$ and $\psi^{\mathcal{R}}, \psi^{\mathcal{I}}, \psi^{\mathcal{J}}, \psi^{\mathcal{K}}$ with controller (4).

Example 2: The connection weight matrix are selected as:

$$C = (c_{i\vartheta})_{2\times 2} = \begin{pmatrix} 0.3 + 1.2\mathbf{i} + 0.5\mathbf{j} + 0.7\mathbf{k} & -0.8 + 1.5\mathbf{i} - 1.1\mathbf{j} + 1.9\mathbf{k} \\ 0.7 - 0.9\mathbf{i} - 1.4\mathbf{j} - 1.3\mathbf{k} & -0.6 - 1.5\mathbf{i} + 0.2\mathbf{j} - 1.7\mathbf{k} \end{pmatrix},$$

$$D = (d_{i\vartheta})_{2\times 2} = \begin{pmatrix} -0.9 + 1.3\mathbf{i} + 0.6\mathbf{j} + 0.8\mathbf{k} & 0.4 - 0.5\mathbf{i} - 0.3\mathbf{j} + 1.8\mathbf{k} \\ 1.2 - 0.7\mathbf{i} + 0.1\mathbf{j} - 0.2\mathbf{k} & -0.4 + 1.1\mathbf{i} + 1.2\mathbf{j} + 1.5\mathbf{k} \end{pmatrix},$$

$$K = (k_{i\vartheta})_{2\times 2} = \begin{pmatrix} 1.1 + 1.4\mathbf{i} + 1.7\mathbf{j} - 1.9\mathbf{k} & -0.1 + 0.8\mathbf{i} - 0.3\mathbf{j} + 1.1\mathbf{k} \\ 1.5 + 1.3\mathbf{i} - 1.6\mathbf{j} + 0.4\mathbf{k} & -0.2 - 0.9\mathbf{i} + 0.7\mathbf{j} - 1.2\mathbf{k} \end{pmatrix},$$

$$\tilde{C} = (\tilde{c}_{i\vartheta})_{2\times 2} = \begin{pmatrix} 1.6 + 0.8\mathbf{i} + 1.5\mathbf{j} + 2.9\mathbf{k} & 0.3 + 0.7\mathbf{i} + 1.4\mathbf{j} - 0.2\mathbf{k} \\ -0.5 + 1.1\mathbf{i} + 0.8\mathbf{j} + 1.2\mathbf{k} & 0.7 - 0.2\mathbf{i} + 0.9\mathbf{j} - 0.6\mathbf{k} \end{pmatrix},$$

$$\tilde{D} = (\tilde{d}_{i\vartheta})_{2\times 2} = \begin{pmatrix} -0.1 + 0.5\mathbf{i} + 1.1\mathbf{j} + 0.3\mathbf{k} & 1.9 - 1.6\mathbf{i} + 1.8\mathbf{j} + 0.7\mathbf{k} \\ 0.8 + 1.9\mathbf{i} + 0.3\mathbf{j} - 0.4\mathbf{k} & 0.7 - 1.3\mathbf{i} + 0.2\mathbf{j} + 1.1\mathbf{k} \end{pmatrix},$$

$$\tilde{K} = (\tilde{k}_{i\vartheta})_{2\times 2} = \begin{pmatrix} 0.2 + 0.5\mathbf{i} - 0.7\mathbf{j} + 0.8\mathbf{k} & 0.3 + 0.3\mathbf{i} + 1.1\mathbf{j} - 0.9\mathbf{k} \\ 0.4 - 0.5\mathbf{i} + 0.3\mathbf{j} + 0.6\mathbf{k} & 0.9 + 1.1\mathbf{i} - 0.5\mathbf{j} + 1.8\mathbf{k} \end{pmatrix}.$$

Choose

$$\begin{aligned}
a_1(\varphi_1(t)) &= 0.2\cos(\varphi_1^{\mathcal{R}}(t)) + 0.2\cos(\varphi_1^{\mathcal{I}}(t)) + 0.2\sin(\varphi_1^{\mathcal{J}}(t)) + 0.2\sin(\varphi_1^{\mathcal{K}}(t)),\\
a_2(\varphi_2(t)) &= 0.2\cos(\varphi_2^{\mathcal{R}}(t)) + 0.2\cos(\varphi_1^{\mathcal{I}}(t)) + 0.2\sin(\varphi_1^{\mathcal{J}}(t)) + 0.2\sin(\varphi_1^{\mathcal{K}}(t)),\\
\tilde{a}_1(\psi_1(t)) &= 0.2\sin(\psi_1^{\mathcal{R}}(t)) + 0.2\sin(\psi_1^{\mathcal{I}}(t)) + 0.2\cos(\psi_1^{\mathcal{J}}(t)) + 0.2\cos(\psi_1^{\mathcal{K}}(t)),\\
\tilde{a}_2(\psi_2(t)) &= 0.2\sin(\psi_2^{\mathcal{R}}(t)) + 0.2\sin(\psi_2^{\mathcal{I}}(t)) + 0.2\cos(\psi_2^{\mathcal{J}}(t)) + 0.2\cos(\psi_2^{\mathcal{K}}(t)),\\
b_1(\varphi_1(t)) &= 3\varphi_1(t), b_2(\varphi_2(t)) = 3\varphi_2(t), \tilde{b}_1(\psi_1(t)) = 3\psi_1(t), \tilde{b}_2(\psi_2(t)) = 3\psi_2(t),\\
I_1(t) &= I_2(t) = 0.1\sin(t) + 0.1\cos(t)\mathbf{i} + 0.1\cos(t)\mathbf{j} + 0.1\sin(t)\mathbf{k},\\
\tilde{I}_1(t) &= \tilde{I}_2(t) = 0.1\cos(t) + 0.1\sin(t)\mathbf{i} + 0.1\sin(t)\mathbf{j} + 0.1\cos(t)\mathbf{k}.
\end{aligned}$$

Let $f(\cdot) = g(\cdot) = h(\cdot) = 1.5\tanh(\cdot)$, $\iota(t) = 0.25 - 0.05\cos(t)$, $l_\vartheta = \sigma_\vartheta = \pi_\vartheta = 1.5$, $\mathcal{F}_\vartheta = \mathcal{G}_\vartheta = \mathcal{H}_\vartheta = 1.5$, $\alpha_i = \tilde{\alpha}_i = 0.2$, $\mathcal{I}_1 = \mathcal{I}_2 = 0.1$, $\tilde{\mathcal{I}}_1 = \tilde{\mathcal{I}}_2 = 0.1$, where $i, \vartheta = 1, 2$. The initial condition is of drive system is $(\varphi_1(s), \varphi_2(s)) = (0.8+0.4\mathbf{i}+0.8\mathbf{j}+0.6\mathbf{k}, -0.4-0.2\mathbf{i}+0.2\mathbf{j}+0.4\mathbf{k})^T$, and the initial condition of the response system is $(\psi_1(s), \psi_2(s)) = (-0.5+0.1\mathbf{i}-0.4\mathbf{j}-0.3\mathbf{k}, 0.9+0.2\mathbf{i}+0.2\mathbf{j}-0.6\mathbf{k})^T$.

Set $\gamma = 0.2$, $\lambda = -1$, $\tilde{\varpi}_1 = 24$, $\tilde{\varpi}_2 = 26$, then one can calculate that:

$$\begin{aligned}
\tilde{\eta}_1 &= \min_i\Big[\big(2\tilde{\varpi}_i - (2(\gamma + 3/2)\lambda\alpha_i + (5+\gamma)\tilde{\alpha}_i)\big)\\
&\quad - \sum_{\vartheta=1}^{2}\big(\tilde{c}_{\vartheta i}^*\tilde{c}_{\vartheta i}l_i + \lambda\gamma k_{\vartheta i}^* k_{\vartheta i}\pi_i\big)\Big] = 26.4390 > 0,\\
\tilde{\eta}_2 &= \max_i \sum_{\vartheta=1}^{2} \tilde{d}_{\vartheta i}^*\tilde{d}_{\vartheta i}\sigma_i = 19.9950 > 0,\\
\tilde{\eta}_3 &= \sum_{i=1}^{N}\Big[\tilde{\mathcal{I}}_i + \lambda\mathcal{I}_i + \sum_{\vartheta=1}^{2}\big(\tilde{c}_{i\vartheta}^*\tilde{c}_{i\vartheta}\mathcal{F}_\vartheta + \tilde{d}_{i\vartheta}^*\tilde{d}_{i\vartheta}\mathcal{G}_\vartheta + \lambda\gamma k_{i\vartheta}^* k_{i\vartheta}\mathcal{H}_\vartheta + \lambda c_{i\vartheta}^* c_{i\vartheta}\mathcal{F}_\vartheta\\
&\quad + \lambda d_{i\vartheta}^* d_{i\vartheta}\mathcal{G}_\vartheta + \gamma\tilde{k}_{i\vartheta}^*\tilde{k}_{i\vartheta}\mathcal{H}_\vartheta\big)\Big] = 6.1590 > 0,
\end{aligned}$$

then the conditions in Theorem 2 are satisfied.

The phase trajectories of $\mathfrak{e}^{\mathcal{R}}, \mathfrak{e}^{\mathcal{I}}, \mathfrak{e}^{\mathcal{J}}, \mathfrak{e}^{\mathcal{K}}$ in uncontrolled and controlled system (3) are shown in Fig. 3. When applying controller (7), the phase trajectories of $\varphi^{\mathcal{R}}, \varphi^{\mathcal{I}}, \varphi^{\mathcal{J}}, \varphi^{\mathcal{K}}$ and $\psi^{\mathcal{R}}, \psi^{\mathcal{I}}, \psi^{\mathcal{J}}, \psi^{\mathcal{K}}$ are displayed in Fig. 4. Theoretical and numerical analyses demonstrate that the controller (7) performs effectively.

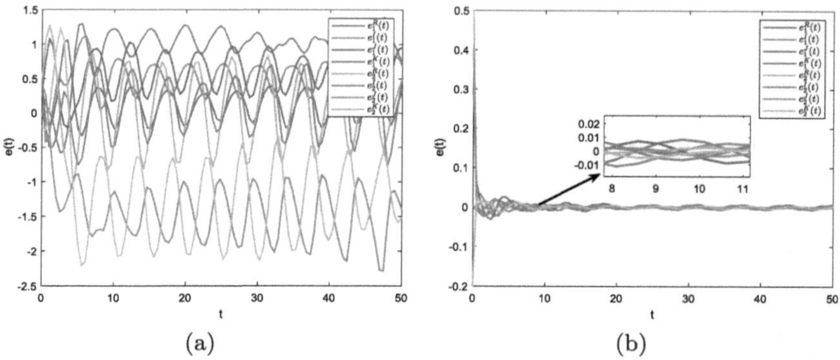

Fig. 3. Phase trajectories of $\mathfrak{e}^{\mathcal{R}}, \mathfrak{e}^{\mathcal{I}}, \mathfrak{e}^{\mathcal{J}}, \mathfrak{e}^{\mathcal{K}}$ (a) without controller and (b) with controller (7).

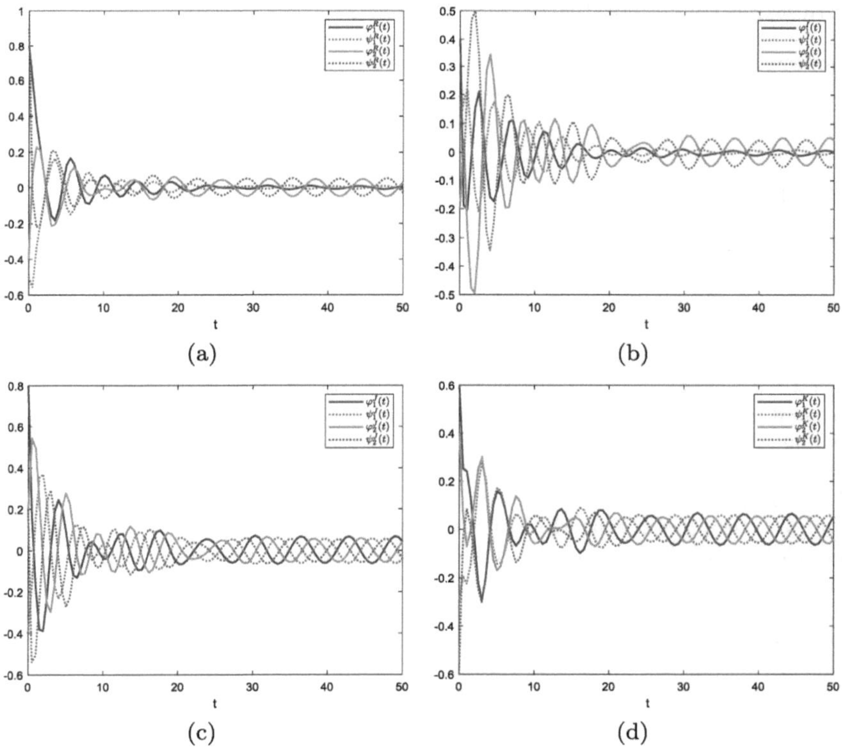

Fig. 4. Phase trajectories of $\varphi^{\mathcal{R}}, \varphi^{\mathcal{I}}, \varphi^{\mathcal{J}}, \varphi^{\mathcal{K}}$ and $\psi^{\mathcal{R}}, \psi^{\mathcal{I}}, \psi^{\mathcal{J}}, \psi^{\mathcal{K}}$ with controller (7).

5 Conclusion

This paper addresses the QPJS problem in QVNNs with interaction terms and mismatched parameters. A structurally simple yet highly efficient controller is developed. By constructing appropriate Lyapunov functions and exploiting the inherent properties of quaternion elements within their theoretical framework, we establish sufficient conditions for achieving QPJS and rigorously estimate the upper bounds of the error. However, there remains room for exploring new Lyapunov functions and inequality techniques. Consequently, future research will prioritize examining the fixed-time synchronization of QVNNs subject to time delays and impulsive effects, as well as exploring novel Lyapunov functions and inequality techniques.

Acknowledgment. The research is supported by grants from the National Natural Science Foundation of China (No. 62172188) and the Guangdong Basic and Applied Basic Research Foundation (No. 2024A1515010272).

Disclosure of Interests. The authors declare that they have no known competing financial interests or personal relationships that could have appeared to influence the work reported in this paper.

References

1. Zhao, N., Qiao, Y., Xu, C.: New results on synchronization control of memristor-based quaternion-valued fuzzy neural networks with delayed impulses. Fuzzy Sets Syst. **484**, 108940 (2024)
2. Jiang, Z., Luo, M., Tan, M.: Design and analysis of quaternion-valued artificial neural networks for image encryption/decryption. Neural Process. Lett. **58**, 6 (2026)
3. Qu, L., Jiang, Z., Wu, L.: Color image storage and retrieval via sliding mode control of quaternion-valued neural networks. Axioms **15**, 72 (2026)
4. Xiang, J., Tan, M.: Fixed-time synchronization for delayed quaternion-valued stochastic fuzzy neural network with reaction–diffusion terms. Neural Process. Lett. **54**(6), 5483–5523 (2022)
5. Du, W., Xiang, J., Tan, M.: Multistability of quaternion-valued recurrent neural networks with discontinuous nonmonotonic piecewise nonlinear activation functions. Neural Process. Lett. **55**(5), 5855–5884 (2023)
6. Tan, M., Liu, Y., Xu, D.: Multistability analysis of delayed quaternion-valued neural networks with nonmonotonic piecewise nonlinear activation functions. Appl. Math. Comput. **341**, 229–255 (2019)
7. Meng, X., Fei, Y., Li, Z.: Quasi-projective synchronization control of delayed stochastic quaternion-valued fuzzy cellular neural networks with mismatched parameters. Cogn. Comput. **16**(5), 2206–2221 (2024)
8. Kumar, A., Singh, S., Das, S., Cao, Y.: Projective quasi-synchronization of complex-valued recurrent neural networks with proportional delay and mismatched parameters via matrix measure approach. Eng. Appl. Artif. Intell. **126**, 106800 (2023)
9. Baluni, S., Yadav, V.K., Das, S.: Quasi-projective synchronization of non-identical time-varying delayed quaternion-valued neural networks with interaction terms: direct method. Eng. Appl. Artif. Intell. **136**, 108866 (2024)

10. Cohen, M., Grossberg, S.: Absolute stability of global pattern formation and parallel memory storage by competitive neural networks. IEEE Transactions On Systems, Man, and Cybernetics **SMC-13**(5), 815-826 (1983)
11. Chen, Y., Shi, Y., Guo, J., Cai, J.: Finite-time synchronisation of quaternion-valued memristor-based Cohen-Grossberg neural networks via event-triggered control by direct method. Int. J. Control **99**(2), 476–493 (2026)
12. Gan, B., Chen, H., Xu, B., Kang, W.: A norm stability condition of neutral-type Cohen-Grossberg neural networks with multiple time delays. Chaos, Solitons Fractals **175**, 113958 (2023)
13. Cheng, Y., Shi, Y., Guo, J.: Exponential synchronization of quaternion-valued memristor-based Cohen-Grossberg neural networks with time-varying delays: norm method. Cogn. Neurodyn. **18**, 1943–1953 (2024)
14. Chen, Y., Shi, Y., Guo, J., He, X.: Synchronisation of quaternion-valued fuzzy neural networks with time-varying delays via novel event-triggered mechanism. Int. J. Dyn. Control **14**, 31 (2026)
15. Deng, H., Bao, H.: Fixed-time synchronization of quaternion-valued neural networks. XXPhys. A **527**, 121351 (2019)
16. Song, X., Man, J., Song, S., Ahn, C.K.: Finite/fixed-time anti-synchronization of inconsistent Markovian quaternion-valued memristive neural networks with reaction-diffusion terms. IEEE Trans. Circuits Syst. I Regul. Pap. **68**(1), 363–375 (2020)
17. Li, R., Cao, J., Xue, C., Manivannan, R.: Quasi-stability and quasi-synchronization control of quaternion-valued fractional-order discrete-time memristive neural networks. Appl. Math. Comput. **395**, 125851 (2021)
18. Zhao, M., Li, H., Zhang, L., Hu, C., Jiang, H.: Quasi-synchronization of discrete time fractional-order quaternion-valued memristive neural networks with time delays and uncertain parameters. Appl. Math. Comput. **453**, 128095 (2023)
19. Liu, Y., Zhang, Q., Li, R.: Exponentially quasi-synchronization control of quaternion-valued memristive neural networks: matrix measure strategies and Frobenius norm methods. J. Supercomput. **81**, 204 (2025)
20. Li, H., Jiang, H., Cao, J.: Global synchronization of fractional-order quaternion-valued neural networks with leakage and discrete delays. Neurocomputing **385**, 211–219 (2020)
21. Yang, X., Cao, J., Yu, W.: Exponential synchronization of memristive Cohen-Grossberg neural networks with mixed delays. Cogn. Neurodyn. **8**, 239–249 (2014)
22. Li, R., Cao, J., Li, N.: Quasi-synchronization control of quaternion-valued fuzzy memristive neural networks. Fuzzy Sets Syst. **472**, 108701 (2023)

Online Fatigue-Aware Power Dispatch for Wind Farms with Hard Feasibility Guarantees

Jiangtao Zhao[1], Lin Zhang[1], Tiantian He[1], Tianyi Zhang[1], Ximeng Wang[2], Bin Wang[3](✉), and Feng Hong[3]

[1] Sanya Oceanographic Institution, Ocean University of China, Qingdao, China
{zhaojiangtao,zhanglin6624,hetiantian,zhangtianyi9288}@stu.ouc.edu.cn
[2] Cyber Finance Department, Postal Savings Bank of China, Beijing, China
[3] College of Computer Science and Technology, Ocean University of China, Qingdao, China
{wangbin9545,hongfeng}@ouc.edu.cn

Abstract. The rapid expansion of utility-scale wind farms, coupled with high-frequency grid dispatch updates (e.g., 1-second cadence), poses a critical challenge: flexible, large-capacity turbines are increasingly susceptible to amplified cyclic loading and accelerated fatigue accumulation. Traditional uniform power allocation strategies often exacerbate cross-turbine load imbalances, while high-fidelity fatigue assessment methods like Rainflow Counting (RFC) remain computationally prohibitive for real-time control loops. To address this fidelity-latency gap, this study proposes a novel online fatigue-aware dispatch framework that integrates a low-complexity fatigue surrogate with a hard-feasible projection-based scheduler. Unlike existing soft-penalty approaches, our method strictly enforces grid-mandated constraints—including power balance, turbine bounds, and ramp rates—at every time step, guaranteeing operational safety. Crucially, we validate that the proposed lightweight surrogate maintains a high ranking consistency (Spearman correlation > 0.97) with offline RFC benchmarks, enabling effective turbine-level risk prioritization without compromising real-time performance. Experiments on two wind farms demonstrate that the proposed strategy significantly mitigates fatigue damage concentration on critical hotspot turbines (reducing top-K damage variance by up to 48%) while maintaining near-neutral farm-level total damage and strict feasibility. These results establish the framework as a practical and deployable solution for real-time fatigue-aware wind farm control under strict operational constraints.

Keywords: Wind Farm Dispatch · Fatigue-Aware Control · Hard Feasibility · Real-Time Optimization · Load Balancing

J. Zhao and L. Zhang—Co-first authors with equal contributions.

J. Lu et al. (Eds.): FLINS-ISKE 2026, LNCS 16752, pp. 586–603, 2027.
https://doi.org/10.1007/978-981-92-2369-5_36

1 Introduction

Wind power has emerged as a cornerstone of global decarbonization efforts [1], driven by the deployment of increasingly larger and more flexible utility-scale turbines [2]. In modern Automatic Generation Control (AGC) systems, this trend necessitates second-level dispatch updates (typically at a 1-second cadence) that must simultaneously satisfy stringent farm-level power tracking requirements and individual turbine operational limits. While larger rotors enhance energy capture, their inherent flexibility amplifies cyclic structural loading, making fatigue-aware dispatch under hard feasibility constraints a pivotal issue for operational reliability and Levelized Cost of Energy (LCOE) reduction [3].

A fundamental bottleneck in current practice is the mismatch between evaluation fidelity and control latency. Rainflow Counting (RFC), the gold standard for fatigue assessment [4–6], relies on full-cycle identification from historical data, rendering it computationally intractable for online, second-level decision-making [7]. Consequently, real-time dispatch systems often operate without a validated, fatigue-related risk indicator, relying instead on simplified proxies or offline diagnostics that fail to capture dynamic load variations.

At the farm level, prevailing dispatch strategies remain predominantly power-centric [8–10]. Under heterogeneous inflow conditions and complex wake effects, uniform or naive power allocation can induce persistent cross-turbine load imbalances, accelerating fatigue accumulation on specific "hotspot" units [11,12]. While some recent studies have attempted to incorporate fatigue objectives, they largely rely on soft penalty functions within optimization frameworks. Such approaches lack hard feasibility guarantees, risking constraint violations under aggressive wind fluctuations—a scenario unacceptable for grid compliance and equipment safety. Thus, the practical challenge remains: How to enforce strict farm-level balance and turbine-level box/ramp constraints while actively suppressing fatigue concentration in real-time?

To bridge these gaps, we propose an online fatigue-aware dispatch framework tailored for second-level decision-making under hard constraints. Unlike existing fatigue-aware dispatch approaches that rely on soft penalties or offline evaluation, our approach integrates a validated online fatigue ranking signal into a strictly feasible real-time dispatch loop. Our approach combines a computationally efficient fatigue surrogate with a projection-based scheduling algorithm. The technical contributions of this work are threefold:

- We develop a lightweight fatigue surrogate that provides a validated online ranking signal consistent with offline RFC (Spearman > 0.97), enabling real-time turbine-level risk prioritization.
- We design a projection-based scheduler that guarantees strict per-step feasibility (power balance, turbine bounds, and ramp rates), distinguishing the proposed approach from soft-penalty-based methods.
- We integrate online ranking, hard-feasible scheduling, and unified offline RFC evaluation into a coherent real-time dispatch framework, enabling effective fatigue load redistribution under strict operational constraints.

Together, these contributions provide a unified solution for safe, real-time, fatigue-aware dispatch, transforming fatigue management from an offline diagnostic tool into an active, online control layer.

2 Related Work

Power-centric dispatch and control methods substantially improve tracking and grid-level coordination [8,10,12–17], but fatigue is commonly handled as a secondary penalty or an offline diagnostic. As a result, fatigue concentration can persist even when power-level objectives are well satisfied.

System-level reviews summarize broader farm-control advances, including flow control, reactive-power coordination, and offshore scenarios [18–20]. These studies clarify capability landscapes, but they do not directly resolve second-level fatigue-aware dispatch with strict per-step feasibility.

Fatigue modeling studies provide physically grounded cycle-level analysis and life-oriented interpretation [2,21–24]. Their evaluation fidelity is high, yet the resulting pipelines are typically not constructed for low-latency dispatch loops with hard equality/box/ramp constraints.

Data-driven wind-power models improve short-term state observability and can support dispatch decisions [9,25]. However, most are optimized for prediction accuracy rather than for fatigue-aware online redistribution under a strictly feasible dispatch set.

Taken together, the literature still leaves three practical gaps for real-time deployment. First, online fatigue signals are rarely validated for ranking consistency against offline reference damage. Second, fatigue guidance is often detached from a dispatch-feasible second-level control loop. Third, strategy comparisons are frequently reported without a unified post-processing protocol. In particular, existing studies rarely provide an online fatigue signal validated against RFC-based ranking, nor do they integrate such signals into a strictly feasible per-step dispatch framework. The present work addresses these gaps through integrated online ranking, hard-feasible scheduling, and unified offline RFC evaluation.

3 Preliminaries

3.1 Model Assumptions

We use the following assumptions to keep second-level dispatch updates and unified post-processing computationally tractable:

- *Homogeneous turbine type.* Turbines are modeled with identical rated parameters; inter-turbine differences are represented by inflow and command trajectories.
- *Independent per-turbine accounting.* Main-shaft and tower-channel damage indices are accumulated per turbine without explicit cross-turbine structural coupling.

- *Accurate tracking at dispatch timescale.* At the 1 s resolution, issued power commands are treated as accurately tracked; delay-aware closed-loop dynamics are left for future work.
- *Zero-initialized comparative index.* The normalized online index starts from zero for all turbines as a computational baseline for relative comparison.

These assumptions define the modeling scope for second-level dispatch and controlled cross-strategy comparison; absolute structural-life reconstruction remains outside the current scope.

4 Proposed Methodology

4.1 Estimation of Fatigue Damage

This subsection defines the control-oriented online fatigue proxy used in dispatch updates. The proxy is used as a comparative ranking signal across turbines, while final strategy comparison is performed by offline surrogate-based RFC post-processing. We use two load channels throughout: main-shaft torque proxy T_i (shaft channel) and tower-channel load proxy H_i (tower channel). Representative load trajectories for both channels are shown in Fig. 1.

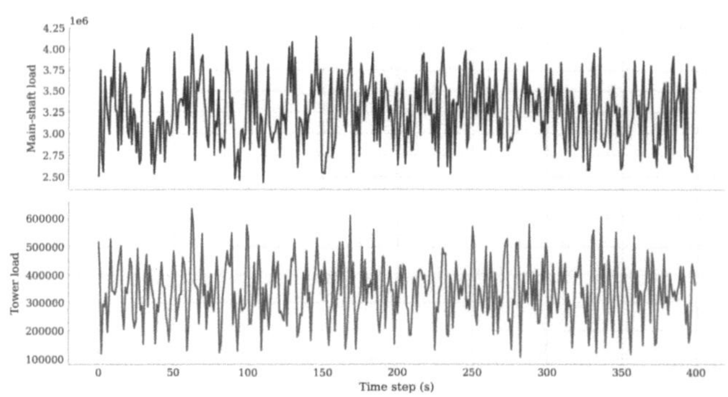

Fig. 1. Representative main-shaft and tower-channel load trajectories for Farm B.

According to the *Palmgren-Miner* linear cumulative damage theory [26], the fatigue damage index D_F can be expressed as:

$$D_F = \frac{n_F}{N_F} \tag{1}$$

where n_F is cycle count, and N_F is fatigue life at that stress level. In online dispatch, L_t denotes the per-channel load proxy at second t, with fixed mapping $L_t^{\text{shaft}} = T_i(t)$ and $L_t^{\text{tower}} = H_i(t)$.

To obtain a low-latency control signal, we treat each second as one proxy increment window and set $n_F = 1$ in Eq. (1). The resulting per-step increment and cumulative proxy are

$$D_{F,t} = \frac{1}{N_{F,t}} \tag{2}$$

$$D_F(T) = \sum_{t=1}^{T} D_{F,t}. \tag{3}$$

The corresponding per-step amplitude and mean stress for channel L_t are

$$S_{a,t} = |L_t - L_{t-1}| \tag{4}$$

$$S_{m,t} = \frac{L_t + L_{t-1}}{2}. \tag{5}$$

Because operating points have nonzero mean stress, we apply Goodman correction [27] to obtain equivalent stress amplitude:

$$S_{\text{eq},i} = \frac{S_{a,i}}{1 - \frac{S_{m,i}}{\sigma_b}} \tag{6}$$

where σ_b is the tensile strength. Using the equivalent stress amplitude $S_{\text{eq},i}$, the fatigue life is estimated via the Basquin S–N relation [28–30]:

$$N_F = \frac{C}{S_{\text{eq},i}^k} \tag{7}$$

with material constants C and k. Accordingly, $N_{F,t}$ follows from Eqs. (6) and (7), and $D_{F,t}$ follows from Eq. (2).

The online proxy is used for comparative ranking in control, and final strategy comparison uses strict offline RFC on reconstructed channels $\hat{T}_i$ and $\hat{H}_i$ under a unified protocol.

Define channel-wise online cumulative proxy as $D^{\text{cum}}_{\text{shaft},i}(t)$ and $D^{\text{cum}}_{\text{tower},i}(t)$, per-step increments as $\Delta D_{\text{shaft},i}(t)$ and $\Delta D_{\text{tower},i}(t)$, and dispatch proxy score as

$$d_i(t) = D^{\text{cum}}_{\text{shaft},i}(t) + D^{\text{cum}}_{\text{tower},i}(t) + \Delta D_{\text{shaft},i}(t) + \Delta D_{\text{tower},i}(t). \tag{8}$$

Here $d_i(t)$ is a dimensionless control index rather than a calibrated physical-life measure: cumulative terms encode long-horizon exposure, while increment terms increase sensitivity to current trend changes. This additive form is a heuristic ranking design for online relative prioritization. Online scheduling uses T_i, H_i, and $d_i(t)$ for relative risk guidance. Offline comparison uses reconstructed $\hat{T}_i$ and $\hat{H}_i$ in strict RFC under the same shaft/tower channel definitions.

4.2 Model of Power Dispatch

The traditional stress-unaware dispatch scheme uses uniform allocation, where the per-turbine power P_{avg} is

$$P_{avg} = \frac{P}{N} \tag{9}$$

where P denotes the total power dispatch instruction issued by the grid to the wind farm at a time step, and N is the total number of turbines in the farm.

However, uniform allocation does not take into account the relationship between turbine operational states and cumulative fatigue damage, resulting in some turbines being subjected to high fatigue loads for extended periods. This could compromise the operational safety of their mechanical components.

To formulate fatigue-aware dispatch as a control problem, command updates must be connected to fatigue-relevant load responses. We therefore introduce control-oriented load proxies before presenting the dispatch objective and constraints.

Control-Oriented Load Proxies. The scheduler acts on power commands, whereas fatigue balancing is guided by predicted load channels. In implementation, both channels are given by fitted surrogate models:

$$T_i(t) = \mathcal{M}_T(\phi_i(t)), \qquad H_i(t) = \mathcal{M}_H(\phi_i(t)), \tag{10}$$

where $\phi_i(t)$ contains $v_i(t), p_i(t), p_i(t-1)$, and optionally includes state-variation features such as pitch, ω_r, ω_f, Δv, and Δp when available. This matches the piecewise-linear load estimator used in the codebase and keeps online dispatch and offline evaluation directionally consistent.

Guiding Dispatch Objective. The dispatch objective is treated as a design principle that balances concentration suppression and command smoothness. At each step, the online fatigue proxy provides balancing guidance, and load-fluctuation regularization controls command variability:

$$J_t \propto \frac{1}{N}\sum_{i=1}^{N}(\Delta D_{\text{shaft},i}(t) + \Delta D_{\text{tower},i}(t)) + \frac{1}{N}\sum_{i=1}^{N}|p_i(t) - p_i(t-1)|. \tag{11}$$

Eq. (11) specifies the intended trade-off between concentration suppression and command smoothness. It is not solved as an independent optimizer; instead, it serves as a control-oriented design principle that guides the construction of the heuristic dispatch updates. Section 4.3 implements this principle via proxy-guided anchoring, smoothness blending, and hard-feasible projection. In implementation, w_f is retained as a fairness-monitoring term in the reported objective proxy, while per-step updates follow the same proxy-guided anchoring, smoothness blending, and hard-feasible projection procedure.

Hard-Feasible Constraint Set. Online dispatch is solved under explicit per-step constraints:

- Farm-level balance:

$$\sum_{i=1}^{N} p_i(t) = P_t \tag{12}$$

- Turbine power bounds:

$$0 \le p_i(t) \le P_{\text{rated}} \quad \forall i \tag{13}$$

- Deviation from farm mean is limited:

$$\left| p_i(t) - \frac{P_t}{N} \right| \le \delta_{\max}, \quad \forall i, \tag{14}$$

- Inter-step ramp is limited:

$$|p_i(t) - p_i(t-1)| \le r_{\max}, \quad \forall i. \tag{15}$$

In the reported setting, $p_{\min} = 0$ MW, $p_{\max} = P_{\text{rated}} = 5$ MW, $\delta_{\max} = 1$ MW, and $r_{\max} = 0.5$ MW/s. For efficient implementation, the three bound types are merged into dynamic per-turbine limits:

$$l_i(t) = \max\left(p_{\min}, \frac{P_t}{N} - \delta_{\max}, p_i(t-1) - r_{\max} \right), \tag{16}$$

$$u_i(t) = \min\left(p_{\max}, \frac{P_t}{N} + \delta_{\max}, p_i(t-1) + r_{\max} \right). \tag{17}$$

If $\sum_i l_i(t) \le P_t \le \sum_i u_i(t)$, the bounded-simplex projection step admits a feasible point and returns a strictly feasible dispatch command.

4.3 Heuristic Dispatch Solver

To realize Eq. (11) under hard constraints, dispatch is computed by a projection-based scheduler with explicit per-step feasibility enforcement.

At each step, let $d_i(t)$ denote the current proxy score defined in Sec. 4.1, combining cumulative terms and the latest increments over shaft and tower channels. The solver first constructs a proxy-guided anchor:

$$\bar{p}_i(t) = \frac{P_t}{N} - \gamma \frac{d_i(t) - \bar{d}(t)}{\sigma_d(t) + \varepsilon}, \tag{18}$$

where $d_i(t)$ is the online proxy score, $\bar{d}(t)$ and $\sigma_d(t)$ are its cross-turbine mean and standard deviation, and $\gamma > 0$ controls balancing strength. To reduce command chattering, the anchor is blended with the previous command:

$$\tilde{p}_i(t) = (1 - \eta)\bar{p}_i(t) + \eta p_i(t-1), \quad 0 \le \eta < 1. \tag{19}$$

This yields a smooth candidate command with built-in fatigue balancing.

Power-balance dual updates are then performed under box/deviation/ramp bounds:

$$p_i^{(k)}(t) = \mathrm{clip}\Big(\tilde{p}_i(t) - \lambda^{(k)},\ l_i(t),\ u_i(t)\Big), \tag{20}$$

$$\lambda^{(k+1)} = \lambda^{(k)} + \alpha\left(\sum_{i=1}^{N} p_i^{(k)}(t) - P_t\right), \tag{21}$$

where α is the dual step size. After iterations, a final hard projection onto the bounded simplex is applied:

$$\Pi_{\mathcal{F}_t}(\mathbf{x}) = \arg\min_{\mathbf{p}\in\mathcal{F}_t} \|\mathbf{p} - \mathbf{x}\|_2^2, \quad \mathcal{F}_t = \left\{\mathbf{p}\ \middle|\ \sum_i p_i = P_t,\ l_i \le p_i \le u_i\right\}, \tag{22}$$

which guarantees strict feasibility at each step under the feasibility condition above.

The dual loop serves as a fast pre-balance stage that reduces equality residual and correction magnitude. The final bounded-simplex projection enforces strict hard feasibility.

In implementation, η is the smoothness coefficient from w_s, damage_gain is a base scaling factor, and $\gamma = w_d \times$ damage_gain is the single effective damage-balancing coefficient used in Eq. (18). Thus, γ is derived from configured weights rather than tuned independently.

4.4 Formulation–Implementation Consistency

Equations (12)–(15) define the implemented feasible set. The dispatch solver is a projection-based heuristic with an explicit fatigue-balancing mechanism, and online dispatch and offline assessment are separated.

For offline assessment, the calibrated surrogate reconstructs loads from wind speed and dispatch commands:

$$\begin{aligned} \hat{T}_i(t) &= \mathcal{M}_T(v_i(t), p_i(t)), \\ \hat{H}_i(t) &= \mathcal{M}_H(v_i(t), p_i(t)). \end{aligned} \tag{23}$$

In practice, both MT and MH are implemented as lightweight fitted surrogate models calibrated on historical data, ensuring low computational complexity and real-time applicability.

Offline damage is then computed by strict rainflow counting with Goodman correction and Miner accumulation. For strategy s, turbine i, and exponent m_{off}, the damage index is

$$D_{i,s}^{\text{off}}(m_{\text{off}}) = \sum_{c\in\mathcal{R}_{i,s}^{\text{shaft}}} \left(A_{i,s,c}^{\text{shaft}}\right)^{m_{\text{off}}} + \sum_{c\in\mathcal{R}_{i,s}^{\text{tower}}} \left(A_{i,s,c}^{\text{tower}}\right)^{m_{\text{off}}}, \tag{24}$$

with farm-level total $D_s^{\text{off}} = \sum_{i=1}^{N} D_{i,s}^{\text{off}}$. Improvement against baseline b is

$$I_{s\leftarrow b}^{\text{off}}(m_{\text{off}}) = 100 \cdot \frac{D_b^{\text{off}}(m_{\text{off}}) - D_s^{\text{off}}(m_{\text{off}})}{D_b^{\text{off}}(m_{\text{off}})}. \tag{25}$$

All strategy comparisons are performed under this unified protocol. The offline RFC channels are fixed to reconstructed shaft and tower signals $\hat{T}_i$ and $\hat{H}_i$, corresponding one-to-one to the online shaft and tower channels.

5 Experiments

5.1 Data Source

The dataset used in this study contains two real wind-farm sequences, Farm A and Farm B, each with 100 turbines over 2000 s. Per turbine, we use wind speed v, planned or issued power command P_{ref}, generated power P_{out}, main-shaft torque channel T_{shaft}, tower-channel load F_t, and auxiliary states including pitch angle and shaft-speed signals. We evaluate four scenarios: A temporal split, B temporal split, A→B transfer, and B→A transfer. The cross-farm transfer settings (A→B and B→A) are designed to evaluate the generalization ability of the surrogate ranking and dispatch behavior across different wind farm conditions. For each temporal split, the first 1000 steps are used for surrogate fitting and the last 1000 steps are used for online dispatch and offline post-processing. For each transfer setting, fitting is performed on one farm and evaluation uses the first 1000 steps of the other farm. Using the first 1000 steps in transfer avoids mixing source-farm fitting and target-farm temporal-holdout segments.

Table 1 lists the measurement fields, their units, and how each field is used in the dispatch/evaluation pipeline.

Table 1. Data description, units, and channel mapping.

Field	Meaning	Unit	Usage
P_{ref}	planned/issued turbine power command	MW	dispatch input
v	wind speed	m/s	dispatch and surrogate input
P_{out}	measured generated power	MW	tracking/diagnostic check
T_{shaft}	shaft-torque channel	Nm	shaft channel reference
F_t	tower/thrust channel	N	tower channel reference
pitch angle	blade pitch state	deg	auxiliary state
ω_r, ω_f	shaft-speed states	rad/s	torque proxy/auxiliary

All fields are used under a fixed normalization/comparison protocol; no per-scenario preprocessing retuning is applied.

5.2 Experimental Setup

We use a 1-second sampling interval and evaluate all strategies on the same 1000-step test horizon. The compared strategies are:

- *Original farm allocation as the operational reference*: $p_i \propto v_i^3$, followed by hard projection.
- *Uniform comparison baseline*: equal turbine allocation followed by hard projection.
- *Proposed*: damage-guided anchor, dual balance update, and final bounded-simplex projection.

All strategies share the same hard-feasible set defined by Eqs. (12)–(15), and all dispatch/proxy hyperparameters are fixed across scenarios. Notably, these parameters are shared across both temporal and transfer scenarios without per-case retuning, suggesting a degree of cross-scenario robustness.

For evaluation, offline post-processing applies strict rainflow counting with Goodman correction and Miner accumulation under $m_{\text{off}} = 3$. Experiments are run on an Intel i5-9300H CPU with 4 cores, 8 threads, and 16 GB RAM.

Table 2 lists the key symbols, fixed values, and functional roles used in online scheduling and offline post-processing.

Table 2. Key symbols and parameter values used in all scenarios.

Category	Symbol	Value	Role
Power bound	$p_{\min}, p_{\max}$	0, 5 MW	constraint
Deviation/ramp	$\delta_{\max}, r_{\max}$	1 MW, 0.5 MW/s	constraint
Dual update	α, K	0.01, 15	solver hyperparameter
Objective weights	w_d, w_f, w_s	0.6, 0.3, 0.1	solver hyperparameter
Damage scaling	damage_gain	0.25	base scaling factor
Effective coefficient	$\gamma = w_d \times$ damage_gain	0.15	used in Eq. (18)
Numerical floor	ϵ	10^{-12}	stabilization
S–N scale/exponent	C, k	10^{10}, 3	normalized setting
Online proxy	window, m_{on}	10, 3	online setting
Goodman parameter	σ_b	10^9	normalized constant
Offline RFC	mode, m_{off}	strict, 3	unified protocol

To improve reproducibility, all methods use one protocol: identical scenario splits, feasible-set projection, post-processing pipeline, and fixed hyperparameters with no per-scenario retuning. The uniform strategy is the quantitative baseline for Top-K reduction, while Orig.-prop. is reported as the operational reference.

Runtime and feasibility are reported in Table 3, where the columns separate proposed-only latency statistics from all-strategy feasibility checks. Time is per-step wall-clock latency for the online pipeline, including proxy update, dispatch solve, and final projection, while excluding offline fitting and offline RFC. All methods remain strictly feasible at every step, and the proposed method stays well within the 1-s budget.

Table 3. Runtime and feasibility summary across four scenarios with 1000 steps each.

Metric	Proposed	All strategies
Solve time mean (ms)	8.70–9.28	$\leq$ 9.28
Solve time p99 (ms)	10.60–10.81	$\leq$ 20.43
Solve time max (ms)	13.47–19.37	$\leq$ 27.91
Feasible steps	4000/4000	12000/12000
Balance MAE	$\approx 10^{-14}$, numerical precision level	

5.3 Experimental Results

Results are organized to answer three questions: whether the surrogate is reliable for evaluation, whether the online proxy preserves turbine ranking, and where dispatch gains occur.

We first verify surrogate reliability, because all offline RFC comparisons depend on reconstructed channels.

Table 4. Surrogate fitting quality for two-farm calibration splits.

Fit farm	Split	RMSE_T	R^2_T	RMSE_F	R^2_F
A	Train	85540.52	0.9569	69641.91	0.5937
A	Holdout	87892.00	0.9546	71125.27	0.5878
B	Train	85452.38	0.9566	69382.63	0.5916
B	Holdout	84955.79	0.9572	68603.41	0.6006

In Table 4, the fit quality columns RMSE_T, R^2_T, RMSE_F, and R^2_F indicate $R^2_T \approx 0.95$ for both farms with small train–holdout gaps, suggesting limited overfitting. Thrust-channel fitting is lower but stable across splits.

We then test whether the online proxy preserves turbine-level ranking against the offline RFC reference.

Table 5. Online fatigue proxy vs offline RFC damage index consistency for turbine-level ranking under the proposed strategy.

Scenario	Spearman	Kendall	Top-10 overlap
A temporal	0.977	0.884	0.80
A→B	0.971	0.861	0.80
B temporal	0.980	0.885	0.90
B→A	0.953	0.844	0.70

For Table 5, the three ranking columns Spearman, Kendall, and Top-10 overlap jointly indicate high agreement in both temporal and transfer settings; Notably, this consistency remains strong under cross-farm transfer settings, indicating that the online proxy maintains robust ranking performance under dataset shifts. Figure 2 provides the corresponding scatter view.

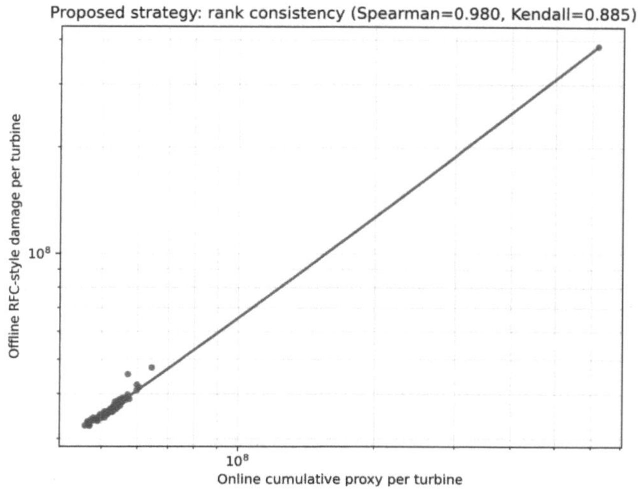

Fig. 2. Turbine-level consistency between online fatigue proxy and offline RFC damage index under the proposed strategy.

We finally evaluate strategy effects under the shared hard-feasible setting.

Across Table 6, the primary improvement is in distributional metrics $D_{\max}$ and $\text{Var}(D_i)$. In the transfer scenarios A→B and B→A, proposed D_{farm} is slightly higher than uniform (4.013 vs 3.989, and 4.114 vs 4.103), while $D_{\max}$ and $\text{Var}(D_i)$ are both reduced. We therefore interpret the method as a redistribution strategy that prioritizes hotspot suppression under hard feasibility, not as a strict minimizer of farm-level total damage.

To interpret where the gains appear in the distribution, Fig. 3 visualizes the B-temporal case. The proposed strategy produces a more compact turbine-level damage distribution and a lighter upper tail than uniform dispatch.

To separate rank-tail behavior from overall distribution shape, Fig. 4 focuses on sorted turbine damage in B-temporal. Gains concentrate in the high-damage tail. Figure 5 then checks whether this tail pattern is stable across scenarios for $K = 1, 3, 5, 10, 20$.

The cross-scenario Top-K curves show monotonic reduction across rank sets, indicating that the effect is not limited to a single extreme turbine.

5.4 Robustness and Mechanism Checks

This subsection reports compact checks on ranking consistency, cross-scenario stability, and bound activity.

Table 7 compares strategy-level rank consistency statistics and shows that the proposed strategy keeps strong proxy–RFC alignment, close to uniform and above the original allocation reference.

Table 6. Scenario-wise comparison with $m_{\text{off}} = 3$ and an evaluation horizon of 1000 steps. Orig.-prop. denotes the original farm allocation reference. Underlined values denote the best result within each scenario group. Units are 10^9 for D_{farm}, 10^8 for $D_{\max}$, and 10^{15} for $\text{Var}(D_i)$.

Scenario	Strategy	D_{farm}	$D_{\max}$	$\text{Var}(D_i)$
A-temp	Orig.-prop.	174.804	26.343	29.647
	Uniform	5.087	5.255	6.878
	Proposed	4.935	4.717	5.254
B-temp	Orig.-prop.	172.789	25.538	14.208
	Uniform	4.039	5.193	2.320
	Proposed	4.000	3.845	1.205
A→B	Orig.-prop.	173.462	26.023	13.850
	Uniform	3.989	5.163	2.296
	Proposed	4.013	4.365	1.593
B→A	Orig.-prop.	173.509	26.659	16.955
	Uniform	4.103	5.200	2.320
	Proposed	4.114	4.329	1.555

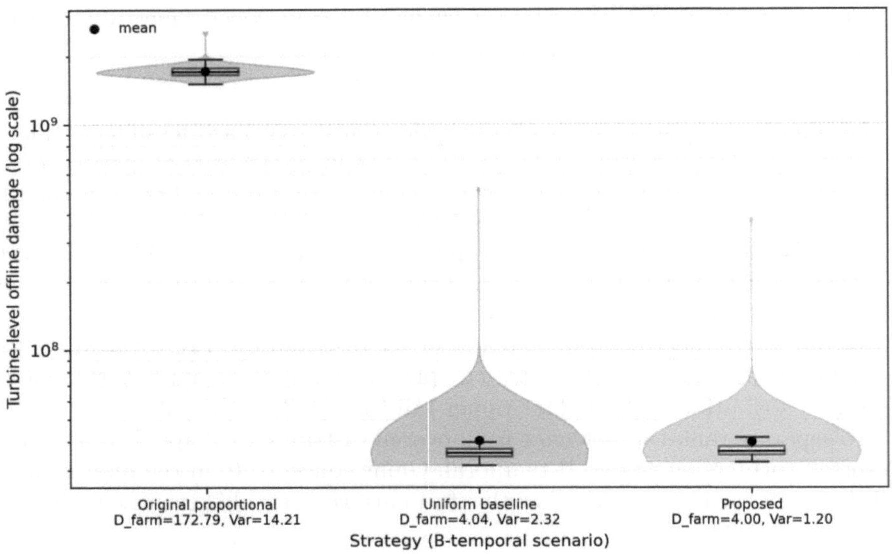

Fig. 3. Distribution of turbine-level offline damage on a log scale in the B-temporal scenario. Violin and box overlays compare the original farm allocation reference, uniform baseline, and proposed strategy.

Table 7. Proxy-vs-RFC ranking consistency across all strategies across four scenarios. Values are reported as mean with minimum-to-maximum range.

Strategy	Spearman	Kendall	Top-10 overlap
Original alloc. (ref.)	0.814 (0.706–0.852)	0.630 (0.522–0.676)	0.58 (0.50–0.70)
Uniform baseline	0.981 (0.978–0.985)	0.887 (0.877–0.905)	0.83 (0.70–0.90)
Proposed	0.970 (0.953–0.980)	0.869 (0.844–0.885)	0.80 (0.70–0.90)

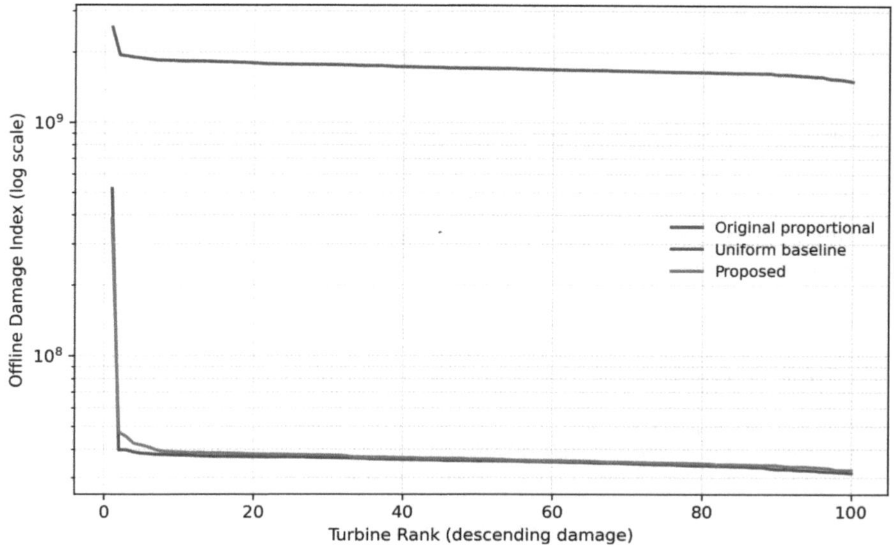

Fig. 4. Sorted turbine offline damage curves in descending order with a log-scale y-axis under three strategies in the B-temporal scenario.

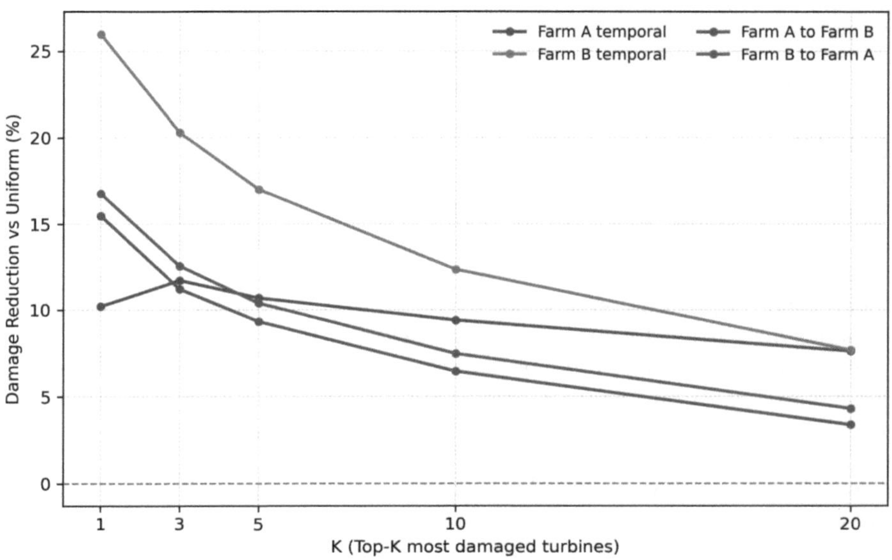

Fig. 5. Cross-scenario Top-K hotspot reduction of proposed strategy vs uniform baseline.

Table 8. Cross-scenario stability summary.

Scenario	Top-1 red. (%)	Top-10 red. (%)	Disp.-ref.	Disp.-uniform	Disp.-proposed
A temporal	10.22	9.41	0.452	≈ 0	0.127
B temporal	25.96	12.35	0.454	≈ 0	0.142
A→B	15.46	6.45	0.457	≈ 0	0.135
B→A	16.74	7.48	0.454	≈ 0	0.127

From Table 8, the two reduction columns Top-1 and Top-10 remain positive in all scenarios, while the dispersion columns show moderate proposed command spread.

Table 9. Constraint activity ratios under shared hard-feasible settings.

Scenario	Orig.-prop.		Uniform		Proposed	
	Dev. hit	Ramp hit	Dev. hit	Ramp hit	Dev. hit	Ramp hit
A temporal	0.0195	0.5804	0.0000	0.0000	0.0152	0.0001
B temporal	0.0194	0.5788	0.0000	0.0000	0.0018	0.0000
A→B	0.0203	0.5783	0.0000	0.0000	0.0034	0.0000
B→A	0.0200	0.5818	0.0000	0.0000	0.0049	0.0000

In Table 9, the bound-hit columns deviation hit and ramp hit indicate that hotspot suppression is not driven by clipping, with near-zero ramp hits and low deviation hits under proposed dispatch.

6 Discussion and Limitations

The evidence supports a specific claim: under shared hard-feasible constraints, online fatigue-guided redistribution can reduce high-damage concentration without materially changing farm-level total normalized damage. This positions the method as an operational balancing layer for second-level dispatch rather than as a replacement for high-fidelity structural-life analysis.

Three aspects remain outside the current scope. *Execution dynamics*: issued commands are assumed to be accurately tracked at the 1-s timescale, and delay-aware actuator dynamics are not modeled. *Damage calibration*: offline RFC assessment depends on surrogate reconstruction quality, and reported values are normalized comparative indices rather than calibrated Miner-life consumption. *External validity*: current evidence is limited to two farms and four scenarios with a restricted comparator set. While transfer experiments provide initial evidence of generalization, broader validation across more diverse wind farm datasets remains future work.

Future work will incorporate delay-aware closed-loop dynamics, improve load reconstruction for tighter offline calibration, and extend validation to multi-site datasets with uncertainty-aware effect-size analysis.

7 Conclusion

This study presents a practical second-level fatigue-aware dispatch framework that combines a lightweight online ranking proxy with hard-feasible projection scheduling. Under a unified offline RFC protocol, the method consistently suppresses high-damage concentration while preserving strict per-step feasibility and near-neutral farm-level total normalized damage. The framework is best interpreted as a comparative control layer for turbine-level fatigue balancing under operational constraints.

Acknowledgments. This work was supported by the National Natural Science Foundation of China under Grant No. 62402463 and Grant No. 41976185.

References

1. Veers, P., et al.: Grand challenges: wind energy research needs for a global energy transition. Wind Energy Sci. **7**(6), 2491–2496 (2022). https://doi.org/10.5194/wes-7-2491-2022
2. Njiri, J.G., Beganovic, N., Do, M.H., Söffker, D.: Consideration of lifetime and fatigue load in wind turbine control. Renew. Energy **131**, 818–828 (2019). https://doi.org/10.1016/j.renene.2018.07.109
3. Loew, S., Bottasso, C.L.: Lidar-assisted model predictive control of wind turbine fatigue via online rainflow counting considering stress history. Wind Energy Sci. **7**(4), 1605–1625 (2022). https://doi.org/10.5194/wes-7-1605-2022
4. Downing, S., Socie, D.: Simple rainflow counting algorithms. Int. J. Fatigue **4**(1), 31–40 (1982). https://doi.org/10.1016/0142-1123(82)90018-4
5. Amzallag, C., Gerey, J., Robert, J., Bahuaud, J.: Standardization of the rainflow counting method for fatigue analysis. Int. J. Fatigue **16**(4), 287–293 (1994). https://doi.org/10.1016/0142-1123(94)90343-3
6. Musallam, M., Johnson, C.M.: An efficient implementation of the rainflow counting algorithm for life consumption estimation. IEEE Trans. Reliab. **61**(4), 978–986 (2012). https://doi.org/10.1109/TR.2012.2221040
7. Barradas-Berglind, J.J., Wisniewski, R.: Representation of fatigue for wind turbine control. Wind Energy **19**(12), 2189–2203 (2016). https://doi.org/10.1002/we.1975
8. Zhang, Z., Kusiak, A., Song, Z.: Scheduling electric power production at a wind farm. Eur. J. Oper. Res. **224**(1), 227–238 (2013). https://doi.org/10.1016/j.ejor.2012.07.043
9. Xu, Z., Chu, B., Geng, H., Nian, X., Zhang, C.: Model-guided learning for wind farm power optimization. IEEE Trans. Control Syst. Technol. **32**(2), 428–439 (2024). https://doi.org/10.1109/TCST.2023.3315547
10. Hong, P., Qin, Z.: Distributed active power optimal dispatching of wind farm cluster considering wind power uncertainty. Energies **15**(7) (2022). https://doi.org/10.3390/en15072706
11. Zhao, R., et al.: An improved power control approach for wind turbine fatigue balancing in an offshore wind farm. Energies **13**(7) (2020). https://doi.org/10.3390/en13071549

12. Cai, W., Hu, Y., Fang, F., Yao, L., Liu, J.: Wind farm power production and fatigue load optimization based on dynamic partitioning and wake redirection of wind turbines. Appl. Energy **339**, 121000 (2023). https://doi.org/10.1016/j.apenergy.2023.121000
13. Zong, H., Porté-Agel, F.: Experimental investigation and analytical modelling of active yaw control for wind farm power optimization. Renew. Energy **170**, 1228–1244 (2021). https://doi.org/10.1016/j.renene.2021.02.059
14. Song, D., et al.: Application of surrogate-assisted global optimization algorithm with dimension-reduction in power optimization of floating offshore wind farm. Appl. Energy **351**, 121891 (2023). https://doi.org/10.1016/j.apenergy.2023.121891
15. Zhang, G., et al.: Coordinated active and reactive power dynamic dispatch strategy for wind farms to minimize levelized production cost considering system uncertainty: A soft actor-critic approach. Renew. Energy **218**, 119335 (2023). https://doi.org/10.1016/j.renene.2023.119335
16. Zhou, Y., Li, Z., Wang, G.: Study on leveraging wind farms' robust reactive power range for uncertain power system reactive power optimization. Appl. Energy **298**, 117130 (2021). https://doi.org/10.1016/j.apenergy.2021.117130
17. Yuan, J., He, S., Cheng, J., Wang, W.: Application of multi-objective control in power and load optimization of wind turbines. IEEE Access **13**, 25783–25793 (2025). https://doi.org/10.1109/ACCESS.2025.3539575
18. Meyers, J., et al.: Wind farm flow control: prospects and challenges. Wind Energy Sci. **7**(6), 2271–2306 (2022). https://doi.org/10.5194/wes-7-2271-2022
19. Adegoke, S.A., Sun, Y., Wang, Z., Stephen, O.: A mini review on optimal reactive power dispatch incorporating renewable energy sources and flexible alternating current transmission system. Electr. Eng. **106**(4), 3961–3982 (2024). https://doi.org/10.1007/s00202-023-02199-2
20. Didier, F., Liu, Y.C., Laghrouche, S., Depernet, D.: A comprehensive review on advanced control methods for floating offshore wind turbine systems above the rated wind speed. Energies **17**(10) (2024). https://doi.org/10.3390/en17102257
21. Haselibozchaloee, D., Correia, J., Mendes, P., de Jesus, A., Berto, F.: A review of fatigue damage assessment in offshore wind turbine support structure. Int. J. Fatigue **164**, 107145 (2022). https://doi.org/10.1016/j.ijfatigue.2022.107145
22. Farid, M.: Data-driven method for real-time prediction and uncertainty quantification of fatigue failure under stochastic loading using artificial neural networks and gaussian process regression. Int. J. Fatigue **155**, 106415 (2022). https://doi.org/10.1016/j.ijfatigue.2021.106415
23. Horn, J.T., Leira, B.J.: Fatigue reliability assessment of offshore wind turbines with stochastic availability. Reliab. Eng. Syst. Safety **191**, 106550 (2019). https://doi.org/10.1016/j.ress.2019.106550
24. Yang, Y., Liang, F., Zhu, Q., Zhang, H.: An overview on structural health monitoring and fault diagnosis of offshore wind turbine support structures. J. Marine Sci. Eng. **12**(3) (2024). https://doi.org/10.3390/jmse12030377
25. Cakiroglu, C., Demir, S., Hakan Ozdemir, M., Latif Aylak, B., Sariisik, G., Abualigah, L.: Data-driven interpretable ensemble learning methods for the prediction of wind turbine power incorporating shap analysis. Expert Syst. Appl. **237**, 121464 (2024). https://doi.org/10.1016/j.eswa.2023.121464
26. Miner, M.A.: Cumulative damage in fatigue (1945)
27. Goodman, J.: Mechanics Applied to Engineering, vol. 1. Longmans, Green (1930)
28. Basquin, O.H.: The exponential law of endurance tests. Proc. Am. Society Test. Mater. **10**, 625–630 (1910)

29. Murakami, Y., Takagi, T., Wada, K., Matsunaga, H.: Essential structure of s-n curve: Prediction of fatigue life and fatigue limit of defective materials and nature of scatter. Int. J. Fatigue **146**, 106138 (2021). https://doi.org/10.1016/j.ijfatigue.2020.106138
30. Kwofie, S., Chandler, H.: Fatigue life prediction under conditions where cyclic creep–fatigue interaction occurs. Int. J. Fatigue **29**(12), 2117–2124 (2007). https://doi.org/10.1016/j.ijfatigue.2007.01.022

Author Index

J. Lu et al. (Eds.): FLINS-ISKE 2026, LNCS 16752, pp. 605–606, 2027.
https://doi.org/10.1007/978-981-92-2369-5

MIX
Papier aus verantwortungsvollen Quellen
Paper from responsible sources
FSC® C105338

If you have any concerns about our products,
you can contact us on
ProductSafety@springernature.com

In case Publisher is established outside the EU,
the EU authorized representative is:
Springer Nature Customer Service Center GmbH
Europaplatz 3, 69115 Heidelberg, Germany

Printed by Libri Plureos GmbH
in Hamburg, Germany